DUST MITES

DUST
MITES

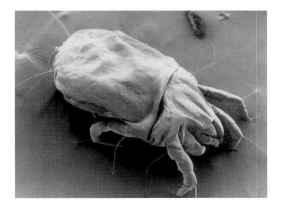

MATTHEW J. COLLOFF

CSIRO
PUBLISHING

 Springer

Co-published by Springer Science + Business Media B.V., Dordrecht, The Netherlands and **CSIRO** PUBLISHING, Collingwood, Australia

Sold and distributed:
In the Americas, Europe and Rest of the World by Springer Science + Business Media B.V., with ISBN 978-90-481-2223-3
springer.com

In Australia and New Zealand by **CSIRO** PUBLISHING, with ISBN 978-0-6430-6589-5
www.publish.csiro.au

Front cover: photograph by Matthew J. Colloff, background woodcut by August Hauptmann
Back cover: illustration by Antony van Leeuwenhoek

Set in 10/13 Minion
Cover and text design by James Kelly
Edited by Anne Findlay
Typeset by Planman Technologies
Index by Russell Brooks
Printed in Australia by Ligare

Every effort has been made to contact copyright holders for their permission to reprint / reproduce material in this book. The publishers would be grateful to hear from any copyright holder who is not acknowledged here and will undertake to rectify any errors or omissions in future editions of this book.

Contents

Acknowledgements

I owe a great debt of gratitude to those research scientists who have worked on dust mites and allergy. Their discoveries shaped my thoughts, and form the subject matter of this book. I thank the following people for their interest, wit, insight and for sharing their knowledge of dust mites, acarology, allergy and related issues over the last 20 years or so: John Andrews, Larry Arlian, Rob de Boer, Harry Morrow Brown, Martin Chapman, Julian Crane, Roy Crowson, Alex Fain, Enrique Fernández-Caldas, Malcolm Cunningham (who also gave permission to use his figures of thermohygrographic recordings), Peter Friedman, Barbara Hart, David Hay, Mike Hill, Stephen Holgate, Willi Knülle, Jens Korsgaard, Peter McGregor, Charlie McSharry, Terry Merrett, Bruce Mitchell, Roy Norton, Tom Platts-Mills, Heather Proctor, Rob Siebers, Frits Spieksma, Richard Sporik, Geoff Stewart, Wayne Thomas, Walter Trudeau, David Walter, Manfred Walzl (who gave permission to use his figures of the gut and reproductive organs of dust mites), and Ann Woolcock (who persuaded me to compile global datasets on distribution and abundance of dust mites and their allergens).

I owe a particular debt of gratitude to two people. Euan Tovey (Woolcock Institute for Medical Research, Sydney) contributed to this book in many ways. He has provided continued support and enthusiasm, sent me copies of numerous papers and manuscripts, tolerated dozens of queries over the years, and allowed me to reproduce several of his figures. Christina Luczynska (King's College, London) maintained a regular correspondence with me on many aspects of dust mites, allergy and epidemiology of asthma from the early 1990s until her death in October, 2005. I thank her for her friendship, critical insight and honesty. During her short life she inspired and motivated many of her scientific colleagues and friends. I am privileged to have been one of them.

Kevin Jeans, latterly commissioning editor of CSIRO Publishing, was a source of inspiration and a pleasure to work with. His successor, Anne Crabb, wisely left me alone to get on with it. John Manger and Briana Elwood did likewise, and saw the book through to completion with humour, tolerance and goodwill. Tracey Millen, Anne Findlay and James Kelly provided efficient, sensitive editorial and design support and encouragement. I thank them all.

For providing me with climate data used in Chapter 4, I thank Peter Jones (Centro Internacional de Agricultura Tropical, Cali, Colombia), Amos Porat (Climatology Branch, Israel Meterorology Service, Bet Dagan, Israel), and William Brown (Climate Services Division, National Climatic Data Centre, Asheville, NC, USA). I am grateful to Richard Brenner, Martin Chapman and Kosta Mumcuoglu for permission to reproduce illustrations. Photographs from Papua New Guinea were taken by Yvon Perouse, and supplied courtesy of Geoff Clarke (CSIRO). Of my other CSIRO colleagues, I am very grateful to Bob Sutherst, Anne Bourne and Ric Bottomly, for their assistance in databasing the global distribution of dust mite species. Andrew Whiting, at short notice and with a high degree of professional skill, translated these records into elegant distribution maps. Saul Cunningham patiently helped with queries on statistics and data analysis, Kim Pullen let me run ideas and numbers past him and gently corrected me when they were wrong, and Anne Hastings provided Figures 3.4 and 4.2. Mike Lacey gave me unstinting assistance with the chemistry of pheromones, lipids and cuticular hydrocarbons. His cheerful encouragement helped me think deeper about how the chemical properties of these compounds influence the biology of the mites. I thank the staff of CSIRO Library Services for their ever-prompt and efficient assistance in obtaining many obscure, hard-to-find publications.

I am grateful to Roy Norton and Tomoyo Sakata for editing the pheromone section and Frank Radovsky generously answered my queries about mites associated with ancient human remains. Sam Killen helped me with maths problems when I got stuck, and Alison Killen provided support, encouragement and generally put up with me.

My research on dust mites would not have been possible without the facilities, support and funding provided by The University of Glasgow, The Medical

Research Council, The Wellcome Trust, The Royal Society, The Stobhill Hospital Trust, Glasgow, and CSIRO Entomology, Canberra. I am especially grateful to Ron Dobson for his many kindnesses and years of wise counsel while I was at the Department of Zoology, University of Glasgow.

I thank those volunteers who have allowed me into their homes in search of mites. It is to them and the many thousands like them that this book is dedicated. Lastly, Huw Smith and Tony Girdwood of the Scottish Parasite Diagnostic Laboratory, Stobhill Hospital Trust, Glasgow, I thank for their friendship over the years. They provided support, inspiration and instilled in me the unabashed joy of doing science.

Preface

Research on mites and allergies has grown enormously since 1964 when dust mites were confirmed as the source of allergens capable of inducing allergic reactions. Studies have become multidisciplinary, drawing on the skills of molecular biologists, clinicians, immunologists, acarologists, architects and engineers, epidemiologists, hygienists and pest controllers. It has become rather difficult for practitioners of one speciality to become familiar with the literature generated by another.

In 1987 a group of scientists and clinicians met in Bad Kreuznach, Germany, to discuss the state of the house dust mite allergy problem. They made recommendations about research collaboration, standardisation of methods, and set guidelines on the level of allergen exposure that was perceived to represent a risk for the development of asthma. It was evident from those discussions that knowledge gaps existed between the clinical and allergological researchers and those working on dust mite biology and ecology. That gap still exists today, though people are more aware of it and doing more to bridge it. The purpose of this book is to provide a reference work for all those with an involvement or interest in house dust mite research, incorporating in a single volume the topics of systematics, physiology, ecology, epidemiology, allergen biochemistry and mite control and allergen avoidance. This task has been a little overwhelming at times, especially since the book was written in my spare time. I make no apologies for a rather basic treatment of some of the clinical and immunological aspects. A detailed review is beyond my scope. I hope I have demonstrated that research on the biology and ecology of house dust mites is most useful when integrated within the broader context of epidemiology and management of disease, rather than as an end in itself, and that the control of dust mites is subject to the same ecological principles as any other problem in pest management.

One reason for writing this book was to tackle some of the myths and misconceptions about house dust mites that have appeared in the literature and on the Internet, some of which have generated misunderstanding of what these animals do and how they live. Most are harmless generalisations, but inaccuracies tend to be cumulative and lead to bias. The control of dust mites is a significant area where the need for high-quality objective data has been downplayed, partly due to interests related to commercial anti-mite products, but also due to a lack of appreciation about the manner in which dust mite populations behave in response to environmental variables.

I have attempted to make this book as comprehensive as possible. The intention is, first and foremost, that it is a work both of reference and synthesis. I have tried to explain basic biological and ecological phenomena for the benefit of medical researchers who may not be familiar with them. More experienced biologists can skip these sections. Putting dust mite research into an historical context is important to me because the first point-of-contact for the advance of knowledge is what has already been written. The sections on the history of research show what has been done, how the subject has progressed and therefore what is likely to be productive for future investigators.

Matthew J. Colloff
Canberra, December 2008

Introduction

What are dust mites and why are they important?

House dust mites are arachnids, not insects, and are related to ticks, spiders and harvestmen. They are found in almost every home, where they live in dust which accumulates in carpets, bedding, fabrics and furniture. As well as providing a habitat for the mites, house dust also contains their food source: shed human skin scales which become colonised by moulds, yeasts and bacteria. The principal dust mite species belong to the family Pyroglyphidae, with *Dermatophagoides pteronyssinus*, *D. farinae* and *Euroglyphus maynei* being the top three pyroglyphid species in terms of global frequency and abundance. *D. farinae*, though common in continental Europe and North America, is rare in the UK and Australia. *Blomia tropicalis* (family Echimyopodidae) has emerged as a particularly important species in the tropics and subtropics. In rural homes in temperate latitudes, species of *Glycyphagus* and *Lepidoglyphus* (family Glycyphagidae) may be very abundant. Traditionally, the common name 'house dust mite' has been used to include those members of the family Pyroglyphidae that live permanently in house dust. Terms such as 'domestic mites' have been used to include pyroglyphid mites as well as stored products species such as *Lepidoglyphus destructor*.

Allergens from dust mites and other indoor allergens – those from domestic pets and cockroaches are the most common – are ubiquitous allergens to which people are exposed and become sensitised. They have been found at an Antarctic research station (Siebers *et al.*, 1999) and on the Mir Space Station (Ott *et al.*, 2004). An association between mites and asthma has long been suspected and because of this, dust mites have been the subject of intense study for more than three decades. A considerable body of data on dust mite ecology, physiology, allergy, allergen chemistry and molecular biology has now been collected, and a more complete understanding of the principal dust mite species and their allergens has emerged.

As a result of dusting, vacuuming, bed-making, or any other activity that causes settled dust to become airborne, the faecal pellets and smaller allergen-bearing particles become temporarily suspended in the air – the faecal pellets are too large to stay there for very long – and may become inhaled. Those people who are atopic (i.e. are genetically predisposed to develop allergic reactions to common allergens like those derived from pollens, dust mite and animal skin scales) respond to this exposure either by making IgE antibodies, which then bind with immunologically active cells to cause the release of mediators such as histamine, and the development of localised inflammation. The allergic reactions are manifest as symptomatic asthma, eczema, rhinitis and conjunctivitis. Although the estimate is by no means reliable, and probably conservative, roughly 1–2% of the world population (65–130 million people) suffer from allergy to house dust mites.

In this book I attempt to cover some major issues of house dust mite biology that have relevance to allergy and asthma *per se*. Specifically, I address the theme of the biological properties of dust mites that make them such important agents of human disease. This approach is somewhat different from that of other reviews of dust mite biology (van Bronswijk and Sinha, 1971; Wharton, 1976; Arlian, 1989; Spieksma, 1991; Hart, 1995), which have presented the basic facts of dust mite biology. We can only go some way toward answering this question by looking at the physiology, reproduction, ecology and evolution of other, related, mite taxa. Dust mites should not be studied in isolation or their study viewed as a discrete discipline. The major biological attributes that have contributed to the success of dust mites are their body water balance, digestive physiology and population dynamics.

Life cycle

Mites are poikilothermic (they cannot control their body temperature) so the length of their life cycle varies with the temperature of their habitat. The stages in the life cycle are the egg, a six-legged larva, two eight-legged nymphal stages and adult males and females. In the laboratory, at optimum conditions (75–80% RH at 25–30°C), egg-to-adult development of *D. pteronyssinus* takes 3–4 weeks. The adults live for about 4–6 weeks, during which time the females each produce 40–80 eggs.

Ecology

Nobody has estimated accurately the total numbers of mites in mattresses or carpets. To do this, the item would have to be cut up, washed thoroughly and each mite removed and counted – an almost impossible task. Instead, estimates of mite population size are made by sampling small areas with a vacuum cleaner or sticky trap. Numbers of mites fluctuate according to season. In northern Europe, populations are generally largest in late summer and autumn and smallest in winter. The autumn increase correlates with greater production of allergens and some indication in some studies of a worsening of allergic symptoms. Larger mite populations tend to be found at places with damper climates than dry ones, thus allergy to mites tends to be rarer among people living in continental interiors or mountainous regions than among people living at low-altitude maritime localities, although there are many exceptions.

Water balance – the key to survival

Mite body water loss constrains colonisation and population growth. It is the ability of house dust mites to survive at humidities well below saturation that accounts for their successful colonisation of human dwellings worldwide. Dust mites live in conditions where temperature and humidity is far from constant. Fluctuations occur in beds due to body heat and sweating by the occupant, and when the bed is vacated, temperature and humidity fall until they match those of the ambient air. Dust mites survive these large fluctuations in microclimate by burrowing down into areas of the mattress where moisture may be retained, or they can cluster together and remain still to minimise body water loss. Additionally, they possess a simple mechanism that extracts water from unsaturated air. At the base of the first pair of legs are glands full of a solution of sodium and potassium chloride. This fluid absorbs water from the air and, as humidity falls, water evaporates from the glands and the salts crystallise, blocking the entrance of the gland and reducing further water loss. As humidity increases again, the salts re-dissolve and water is absorbed by the hygroscopic salts to replenish that lost during the dry period.

Allergens

During digestion, cells bud off from the wall of the midgut, engulf food particles and travel along the gut lumen breaking down the food as they go. The products of digestion are absorbed throughout the gut epithelium into the haemolymph. By the time they reach the hindgut, the cells start to dehydrate and die, packaging themselves into faecal pellets surrounded by a peritrophic membrane that protects the delicate hindgut from damage by abrasion. This mode of digestion results in relatively large quantities of enzymes accumulating in the faecal pellets. The pellets, some 20–50 μm in diameter, are egested and accumulate in the textiles which the mites inhabit. The enzymes, being proteins, are immunogenic – capable of eliciting an immune response when humans are exposed to them. The first mite allergen that was identified and purified is called Der p 1, and is found mainly in the faeces. Many more are now known, from many more species.

Why do house dust mites make allergens? Clearly allergens are biologically functional proteins within the mites and the allergenic activity is incidental; an unfortunate consequence of their ubiquity and abundance in human dwellings. The association of Der p 1 with the gut and faecal pellets strongly indicates a digestive function, as does the sequence of their amino acids. Several other allergens of mites are also functional enzymes, including amylase (group 4 allergens). These allergenic enzymes have been found in extracts enriched with mite faecal pellets, suggesting they are associated with digestion. Group 2 allergens are not found in large concentrations in faecal pellets and are probably derived from a source other than the gut. Other allergens have no known functional role and database searches for comparisons of their amino acid sequences yield few clues. Tovey *et al.* (1981) estimated that *D. pteronyssinus* in laboratory cultures produced about 20 faecal pellets per mite per day, each containing an average of 100 picograms of Der p 1. Faecal pellets and Der p 1 are relatively stable at room temperature and therefore accumulate in house dust. Group 1 allergens are highly water-soluble and become denatured at temperatures above 75°C, whereas group 2 allergens are heat-resistant.

Epidemiology

Dust mite allergy has been shown to be an independent risk factor for the development of asthma (reviewed by Platts-Mills *et al.*, 1987; International Workshop Report, 1988; Platts-Mills *et al.*, 1989). Allergy to house dust mites and other indoor allergens is a major cause of ill health worldwide. The prevalence of asthma in Australia is among the highest in the world. In 1993, approximately 23% of children in the 7–11-year-old age group had asthma, compared with 17% from New Zealand and about 15% from the UK. A significant proportion of cases, perhaps between a third and a half, can be attributed to allergens of dust mites. Globally, the prevalence of asthma has been increasing markedly since the 1960s, and had risen about 1.5–2 times in Australia by 2000. Are more people being exposed to dust mite allergens than previously and are greater concentrations present within their homes? What else might be going on that could explain this phenomenon?

The distribution and abundance of dust mites is not uniform – houses next door to each other and of the same design can have vastly different mite population densities and species-composition. Thus patterns of exposure to the allergens will vary also. These differences, extended regionally and globally, translate into epidemiological variables such as the proportion of people who develop mite-mediated allergies, the age at which symptoms are manifest, the severity of symptoms and their morbidity, and the risk of development of allergic diseases in newborn children. Furthermore, there is evidence to suggest that the symptoms and pathology of allergic disease can influence the nutrition, reproductive physiology and population dynamics of the mites. For example, people with atopic dermatitis tend to have very dense dust mite populations in their beds compared with those of healthy non-atopics. They also have lower levels of certain lipids in their skin scales, which probably more closely match lipid dietary optima for dust mites than fresh scales from non-atopics. They shed more scales and lose more body water at night through sweating and transcutaneous transpiration. All these factors result in microhabitat changes that are potentially advantageous to dust mites.

In recent years it has been shown that reduction in allergen exposure can result in improvement of clinical symptoms of allergy. As this implies that the condition is avoidable, it would seem reasonable to use allergen avoidance measures in clinical management, although such intervention is by no means reliable or reproducible. Furthermore, acquisition of sensitivity to allergens of mites and pets during infancy may increase the risk of developing asthma, and it has been suggested that allergen eradication be directed toward infants at high risk to attempt to prevent sensitisation and symptoms. However, recommendation of allergen avoidance has been constrained by conflicting results of published clinical trials, a bewildering profusion of different methods and products, with little clear information about where and how often to use them or which patients are likely to benefit. Additionally, there is no universal agreement on how to monitor allergen exposure that may be relevant both to primary sensitisation and to triggering of symptoms.

The association between dust mites and humans has, I suspect, been a very long one, probably commencing with human settlement and the development of agricultural systems and food storage. But there is no way of knowing whether early human communities harboured dust mites in their homes and suffered mite-induced asthma and allergies. Stored products mites have been found in Neolithic remains from archaeological sites in Europe and even in the gut contents of mummified human remains. Dust mites have been found in low densities in dwellings of isolated tribal societies in Amazonia and Papua New Guinea, though in the latter case the mite populations only really took off after the tribespeople started using blankets and Western-style clothing. Why should the early association between dust mites and humans be of any consequence? Apart from the fact that historical problems have a curious attractiveness to many biologists (myself included) that vastly outweighs the likelihood of their solubility, it would make a tremendous difference to our understanding of the biology of dust mites to know if they evolved in tandem with Neolithic societies or whether mite allergy is a 20th-century phenomenon brought about by favourable (for the mites) changes in housing design and construction. Both hypotheses have been made, and both are somewhat difficult to test. There is little doubt that in many parts of the world houses are warmer, moister and less well-ventilated than they used to be, partly due to double-glazing, central heating and insulation.

Mite control and allergen avoidance

A number of products aimed at reducing exposure to allergens of mites and pets are currently available for sale direct to the public, without medical supervision of their use, and, in several instances, without independent evaluation of their efficacy or safety. This is a matter of concern. It is also worrying that these products can be purchased and used by people who may have symptoms that are not attributable to mite and pet allergens. Reduction in exposure to allergens can improve symptoms of asthma and reduce the need for drugs. Although well-designed trials have demonstrated clinical benefit, and several control treatments are available commercially, relatively few physicians give patients advice on mite and allergen control.

Allergen exposure in bedrooms can be reduced by using a mattress cover, replacing old pillows and, if possible, removing the carpet. Mites can be killed in all manner of ways, but standardised, routine methods for reproducibly and reliably controlling mites and their allergens and consistently alleviating allergic asthma have yet to be designed. This objective requires better knowledge of the biology and ecology of these extraordinary creatures than we have at present.

Perceptions of dust mites and allergic diseases

In the 1970s nobody had heard about dust mites apart from a few scientists and doctors and a handful of asthma patients. As an undergraduate reading zoology in the late 1970s and early 1980s, I cannot recall myself or my fellow students having any awareness through our course work or reading of these minute, blind arachnids that shared our lodgings. Dust mites may have merited half a page or so in the medical entomology textbooks, whereas ticks and chigger mites had entire chapters devoted to them. And the term allergy evoked no association with asthma, but with 'Total Allergy Syndrome' and a generally held view that this disorder, as with other allergies, was partially or wholly psychosomatic. At school, there were a few asthmatic children but rarely more than one in a class of 30 pupils.

Since the early 1990s, public perceptions about asthma, allergy and dust mites have changed completely. Articles on these topics in the media have been largely responsible for educating people that asthma is a major public health issue, that it can be fatal, and that its prevalence has increased considerably. Schoolteachers are versed in first aid provision for sufferers and are aware of symptoms and medication use. Their classrooms may each now contain four or more asthmatics. It is accepted that a sizeable proportion of asthma cases have an allergic basis and that allergic reactions are not 'all in the mind'. Many people have heard of dust mites and know they live in their beds and carpets and produce allergens in their faecal pellets. Publicity campaigns by medical charities, fundraising events to support research and the publication of new research findings have formed the basis for the rise in media interest, together with the unending public fascination with human disease and the life that cohabits their homes.

Attitudes within the medical profession have changed too. In the preface to their book explaining the discovery of dust mites and their role in asthma and allergy, *House Dust Atopy and the House Dust Mite*, Voorhorst and colleagues (1969) stated starkly that they had been unable to persuade their professional colleagues of the connection between mites and disease, because many of them were not acquainted with the frame of reference within which the discovery of dust mites in homes had taken place. At the time, the notion that dust mites cause asthma was perceived as not biologically plausible and regarded with suspicion or derision (Spieksma, 1992; Spieksma and Dieges, 2004). This attitude persisted throughout much of the 1970s and 1980s. Nowadays there can be few medical practitioners who do not take seriously the role of mites in allergic disease. However, it would be wrong to assume that dust mites were *the* most important source of allergens in relation to diseases with an atopic basis, or that the relationship between allergen exposure, development of allergy and appearance of disease is anything other than complex and multi-factorial.

Allergy has received recognition as a medical discipline in its own right rather than being regarded as a branch of clinical immunology, and postgraduate specialist training courses exist. Allergy and asthma clinics have become more common and widespread, and there are doctors and nurses in general practice with specialist knowledge and training. Professional and learned societies such as the British Society of Allergy and Clinical Immunology and the American Academy of Allergy and Immunology have campaigned hard and successfully to rid the discipline of its former public image of pseudoscience and overtones of alternative medicine.

1. Identification and taxonomy, classification and phylogeny

The main service which the present day world expects of its systematists remains ... the speedy and reliable identification of organisms.

R.A. Crowson, 1970

Discrimination and identification have value beyond the obvious separation of edible from poisonous, valuable from worthless, or safe from dangerous. This is a means to gain an appreciation of the richness of the environment and our human place within it ... We start to understand our history by seeking to collect and classify.

Richard Fortey, 1997

1.1 What is the use of taxonomy?

The subclass Acari – the mites – contains about 45 000 species that have been formally named and described. This is a small percentage of the total global diversity of mites, estimated to be between 540 000 and 1 132 000 species (Walter and Proctor, 1999), making it the most diverse group of arthropods after the insects. The science dealing with the study of mites is called Acarology. To make sense of the enormous diversity of living organisms a system of description and ordering is required.

Taxonomy (literally, the naming of taxa, or groups of phylogenetically related organisms – subspecies, species, genera, families and so forth: see Table 1.1) is the science that deals with the recognition, description and defining of organisms. It involves providing taxa with an 'identity' that allows them to be recognised, hopefully in a reliable and repeatable manner. For practical purposes, the identity of a species is defined by comparing it with related species and by characters that are unique to that taxon. In the majority of animal taxa, and especially arthropods, such characters have been mostly morphological ones because traditionally the vast bulk of taxonomic work was done using dead specimens from museum collections. However, characters based on behaviour, ecology, biochemistry, gene sequences, protein characteristics and biogeography are also used by taxonomists to great effect. Nevertheless, most newly described species are defined by morphological differences between themselves and previously described species, and are referred to as morphospecies. The morphospecies represent the taxonomist's 'first cut' in terms of accuracy of definition. A single morphospecies may, on closer investigation through the comparison of different populations of that morphospecies, turn out to contain several biological species, not separable by morphological differences but with unique characters of behaviour and biology and, if sexually reproducing rather than parthenogenetic, only capable of producing fertile offspring by mating with other members of the same biological species. So, the definition of species at a higher resolution than morphospecies requires the taxonomist to make detailed observations on the life history and biology of live populations. An example of such a study on dust mites is that showing a lack of interbreeding of populations of *Dermatophagoides farinae* and *D. microceras* by Griffiths and Cunnington (1971).

Definitions of taxonomy are numerous and some include taxonomy and systematics as separate but overlapping activities, others do not. Systematics involves the study of the diversity of organisms and their phylogenetic relationships: how they are related through evolutionary history. Taxonomy supplies the data for studies in systematics and phylogeny. I will try to explain how taxonomy works in practice, as well as to attempt a definition. It is important to state at the outset that taxonomy provides the basis for the identity of species. Its practitioners seek to separate and characterise species, even if they are morphologically very similar. Thus, when operating effectively, taxonomic procedure provides scientists in other disciplines with as much assurance as possible that they are studying a single entity and not a complex of species. Why is this important? Imagine studying the allergens of what had been thought of as a single species of dust mite, but which turned out to be two following a taxonomic investigation. Suppose they have specific allergens and their distribution and biology are different? The result would be that one would draw inaccurate conclusions about the clinical importance of each species; how many people are exposed to it and in which centres of human population, with all the ensuing consequences for the management of allergic reactions caused by those species. This situation has happened, to a limited extent, with

Table 1.1 Classification of the grain mite *Acarus siro* Linnaeus, showing major categories of the taxonomic hierarchy. (Note that not all categories, or taxa, have common names. The ordinal-subordinal classification of the mites is currently unstable: the Astigmata has been proposed to have been derived from within the oribatid sub-order Desmonomata and some of its characters are shared with this group of oribatids (Norton, 1998).)

Category of classification (taxon)	Scientific and common name (in brackets) of taxon, and important defining characters
Kingdom	Animalia (animals, i.e. those multicellular, heterotrophic organisms that develop from a ball of cells – the blastula).
Phylum	Arthropoda (arthropods, i.e. those animals with external skeletons and jointed limbs).
Sub-phylum	Chelicerata (i.e. those arthropods with chelicerate mouthparts and no antennae).
Class	Arachnida (arachnids, i.e. those chelicerates with eight legs and a body divided into two distinct regions).
Sub-class	Acari (mites, i.e. those arachnids with chelicerate mouthparts plus a subcapitulum, with reduced segmentation of the posterior body region, and with a six-legged larva).
Order	Acariformes (i.e. those mites with leg coxae fused to the body, anisotropic setae, a dorso-sejugal furrow and anamorphic postembryonic development).
Infra-order	Sarcoptiformes (i.e. those Acariformes with a toothed rutellum, prodorsal differentiation and no solenidia on tarsus IV).
Sub-order	Astigmata (i.e. those Sarcoptiforms with lateral glands and reduced setation of the opisthosoma).
Superfamily	Acaroidea (i.e. those Astigmata with a clear propodosomal and hysterosomal division).
Family	Acaridae (i.e. those Acaroidea with solenidion ω_1 at the base of the tarsus and usually with a rectangular prodorsal shield).
Genus	*Acarus* (i.e. those Acaridae with 12 pairs of notogastral setae and solenidion σ_1 on Genu I more than 3 x longer than σ_1).
Species	*Acarus siro* (i.e. that species of *Acarus* with dorsal setae d_1 not more than 2 x length of h_1 and with setae d_1 or e_1 no longer than the distance between its base and the base of the seta immediately posterior to it).

at least one pair of dust mite species (*Dermatophagoides farinae* and *D. microceras*), as we will see later, and has caused some confusion.

E.O. Wilson in his autobiography, *Naturalist* (1994), makes clear the importance of identification and taxonomic skills in the armoury of the evolutionary biologist:

> *If they are also naturalists – and a great many of the best evolutionary biologists are naturalists – they go into the field with open eyes and minds, complete opportunists looking in all directions for the big questions, for the main chance. To go this far the naturalist must know one or two groups of plants or animals well enough to identify specimens to genus or species. These favoured organisms are actors in the theater of his vision. The naturalist lacking such information will find himself lost in a green fog, unable to tell one organism from another, handicapped by his inability to distinguish new phenomena from those already well known. But if well-equipped, he can gather information swiftly while continuously thinking, every working hour, 'What patterns do the data form? What is the meaning of the patterns? What is the question they answer? What is the story I can tell?'*

The message of this chapter is that taxonomy is of relevance equally to ecologists, epidemiologists and biochemists, indeed all life scientists, because they need to know the identity of the animals they are working with as accurately as possible if they are to make any progress with their research. Knowing what something is called unlocks the door to the library of research that has been done on that organism and its relatives. Biologists ignore the taxonomy of the organisms they study at their peril.

1.2 How taxonomy works

1.2.1 Perceptions of taxonomy

Providing the best, most accurate information on the identity of organisms carries with it a big responsibility, especially so if the taxonomist is working with a group that is of economic or medical importance. Taxonomy has gained a reputation as an arcane science; practised in cluttered, dusty rooms in museums by elderly people with no interest beyond the group on which they work. They are uncommunicative (except to other taxonomists), and unresponsive to the needs of other researchers. They cause confusion by incessant changing of names of organisms, and take perverse delight in so doing, with little thought to the effect their deliberations have on other researchers. They know little or nothing of the biology of the organisms they study, or biology in general, because they only work with dead specimens (Mound, 1983). Some would question whether taxonomy even merits the status of a science, since some of its practitioners treat it more like a craft. This is exemplified by the arbitrary and polarised approach they take to their methodology of defining taxa, classifying themselves as 'lumpers' who tend to group variable taxa together, or 'splitters' who break existing taxa up into new ones on the basis of the slightest differences (often perceptible only to themselves), according to their propensity to view morphological variation between individuals, populations and species as an asset or a menace.

I would argue that these accusations are largely based on ignorance, outdated notions or are simply untrue. Taxonomy *is* a science in its own right, based on the phylogenetic species concept and the testing of hypotheses of character distributions, transformations and evolutionary similarity of taxa (Wheeler, 2007). But taxonomists have not been very good at promoting a positive public image. The value of taxonomic research to other scientists is absolute: without it there can be no progress. Yet taxonomists have been slow to capitalise on this fact as well as to recognise the true value of their knowledge and expertise. As an illustration both of the utility of taxonomy and the way in which taxonomists work, let us examine one example: Fain's (1966a) study of the taxonomy of *Dermatophagoides pteronyssinus*.

When Voorhorst *et al.* (1964) first reported their hypothesis that mites were the cause of allergy to house dust, Spieksma and Spieksma-Boezeman had isolated mites from dust samples taken from houses in Leiden and sent them to Alex Fain in Antwerp for identification. Fain reported back that they included a species belonging to the genus *Dermatophagoides*. As stated here, this sounds like an almost pedestrian event; the day-to-day stuff of research – one scientist seeking advice and information from another. But it conceals a phenomenal amount of detective work on the part of Fain. By the time his provisional identification was reported to Spieksma, he had embarked on a detailed investigation to discover the identity of the mites. First, he had to determine how many species were present in the sample. In fact there were two:

Euroglyphus maynei, described by Cooreman in 1950 from samples of cottonseed collected in Belgium, and another which he identified tentatively as *Dermatophagoides pteronyssinus*, first described by Trouessart in 1897 under the name *Paralges pteronyssoides*. He had seen the species before, having collected it from animal skins in France. But to be sure of his identification Fain had to track down and compare the type specimens (the ones that Trouessart used for his original description) with the Leiden specimens. There is no central register of type specimens, and Fain tried two different museums before he found them in the Berlese Collection in Florence. In order to examine them he had to travel to Italy (the Berlese Collection is too important and valuable to allow for the loan of specimens). Using an unfamiliar microscope and without the convenience of working in his own laboratory, Fain identified which stages in the life cycle were present (there was a larva, 17 nymphs, six males and seven females in the type series). Since Trouessart had described the species before the rules of nomenclature became formalised (see section 1.2.2), no formal type had been designated. Fain chose one of the specimens, an adult female, to serve as the lectotype; the type designated as part of a revisionary work, and he redescribed the species, making a series of some 28 drawings of its external anatomy, including minute details of the positions of the setae on each of the legs and the variation in the shape of the propodosomal shield. Before he visited Florence, Fain had already embarked on a comprehensive examination of the literature on taxonomic acarology. The purpose of this was partly to discover more information about the species but also to find out whether anyone else, not knowing of Trouessart's work, had described *Dermatophagoides pteronyssinus* under a different name. His search uncovered two such examples, *Mealia toxopei*, described by Oudemans in 1928 and *Visceroptes satoi*, described by Sasa in 1950 and a further two, *Dermatophagoides scheremetewskyi* Bogdanov, 1864 and *Pachylichus crassus* Canestrini, 1894, which may well have been *D. pteronyssinus* but which could not be confirmed as such. This investigation involved further examination of the type specimens of Oudemans, as well as the descriptions and figures by the other authors (types were unavailable for study for various reasons). Finally, Fain searched for previously unidentified specimens in his and other mite collections and made an inventory of them. This provided not only data on the habitats in which the species was found but also on its geographical distribution.

This sort of investigation is standard work for taxonomists. So what makes it special and important? Apart from the unique blend of scholarship, history, iconography, detective work, linguistics, comparative morphology and morphometrics, the identity of *Dermatophagoides pteronyssinus* was determined and defined. This provided a benchmark for other taxonomists; a basis for comparison with other species of *Dermatophagoides* as they were discovered. Since 1966, nine new species have been described, including several that are of considerable importance in dust mite allergy. Clear definitions of species allowed for the production of identification keys, allowing non-taxonomists to identify specimens and opening up the field of dust mite research to ecologists and physiologists. Differences were found between the distributions of *Dermatophagoides pteronyssinus* and *D. farinae*, which are now known to relate to their different temperature and humidity tolerance and water-balance capabilities. As allergens came to be isolated from *Dermatophagoides* spp., the identity of species in culture could be checked for contamination with other species, and there is now a vast knowledge of the allergen repertoires of different species, and a recognition that each allergen is capable of eliciting a highly specific immune response, which is the basis of studies on dust mite T-cell immunity and immunotherapy, as well as monoclonal antibody production. Allergens from four *Dermatophagoides* species and several other astigmatid mite species have been isolated and purified (discussed further in Chapter 7). The allergen genes have been sequenced, cloned and expressed to produce recombinant allergen products. Patterns of exposure to allergens from different species by different patient populations are being investigated, providing a basis for extending the epidemiology of dust mite allergy. Simply put, without the basic taxonomic research that Fain conducted in the 1960s, none of these other studies would have been possible because confusion would have reigned. As new allergens are sought in previously uninvestigated species, the necessity for sound taxonomic identification remains.

1.2.2 Names and nomenclature

The naming and renaming of species and other taxa is governed by a complex set of rules called the International Code of Zoological Nomenclature, published in French and English in the same volume. The fourth edition (1999) is the valid edition for current use, and came into effect on 1st January 2000. The Code is administered by a group of commissioners (known as the International Commission for Zoological Nomenclature), who meet to hear appeals and submissions on nomenclatorial matters and publish a journal of their deliberations (*The Bulletin of Zoological Nomenclature*). Similar codes exist for viruses, bacteria, fungi and plants.

To some, nomenclature appears to be an arcane, jurisprudential discipline, having more in common with law than with biological sciences. However, the aim of nomenclature is to achieve stability of scientific names and to minimise taxonomic confusion. Most taxonomists will have listened *ad nauseam* to tea-time discussions in which their non-taxonomist colleagues bemoan some name change to their favourite study group of organisms foisted upon them by the International Commission and the 'confusion' that this is going to cause. What is often not apparent to non-taxonomists is that names and species are, in the eyes of taxonomists, two separate and independent entities which are linked together by the concept of the type specimen (explained more fully in 1.2.2.c) and it is vitally important to be clear about which correct, valid name is attached to which organism. For a practical guide to biological nomenclature, see Jeffrey (1989).

a Binomial nomenclature

Every one of the 45 000 described species of mites has a published description and a scientific name, consisting of a genus name, such as *Dermatophagoides*, and a species name, such as *microceras*. This compound name is called a binomial and the idea of binomial names was introduced by Linnaeus (the year of publication of the 10th edition of his *Systema Naturae*, 1758, is the official starting point of animal nomenclature) partly to prevent the confusion of earlier taxonomists, who used multiple, descriptive names. Berenbaum (1995) cites an example of a butterfly known as *Papilio media alis pronis praefertim interioribus maculis oblongis argenteis perbelle depictis*. The binomial has the advantage that it can be descriptive without being too cumbersome: it can be remembered relatively easily. *Dermatophagoides* is derived from the Greek words *dermis* meaning 'skin', *phagos* referring to feeding and the suffix *-oides* meaning 'to look like' and roughly translates as 'thing that looks like those that eat skin'. The reason for this name is that Bogdanoff

(1864), the describer of the genus *Dermatophagoides*, thought the mites resembled (though were distinct from) those in the genus *Dermatophagus*, described by Fürstenburg (1861) in his treatise on mites associated with scabietic-type skin diseases. The name *microceras* is derived from the Latin *micro* meaning 'small' and *ceres* meaning 'wax'; *cera* meaning wax image or figure, thus 'small waxy figure'.

Making the name easy to remember and pronounce is part of the job of taxonomists and there are specific instructions in the International Code of Zoological Nomenclature about the formulation of names. Furthermore, the binomial is rooted into a classification: each species is contained within a genus, each genus within a family, each family within an order and so on (refer to Table 1.1). Most importantly, each binomial name is unique. The binomial is always written in italics, but the family or other group names are not. After the name one often sees a surname and a date (never written in italics). This refers to the person or persons who first described the species and when they did so, such as *Dermatophagoides microceras* Griffiths and Cunnington, 1971. The reason for including this information is so that the description can be looked up in the literature and also to clarify the exact identity of the species being referred to. It cannot be confused with another species, in the same genus, that has been given the same name by somebody else who was unaware that the name had already been used. This occasionally happens, and is called homonymy. It is often brought about by taxonomists using common, simple descriptive names like *spinatus*, *ovatus* or *magna* when spinyness, oval body shape or large size may be common characters within the genus. Thankfully, there are no examples of this as yet in dust mite taxonomy. Where the name of the describer and the date are enclosed in brackets, it means that the identity of the species had been examined by someone other than the person who originally described it and that they decided, for any number of possible good reasons, that it belonged in a different genus. For example *Euroglyphus maynei* (Cooreman, 1950) was originally placed in the genus *Mealia* by Cooreman, but was moved to the genus *Euroglyphus* by Fain in 1965. Although the details of the 'mover' are absent in zoological nomenclature (though included in abbreviated form in botanical nomenclature), there is a formal method of citation which gives exact details of who has done what, taxonomically, to the species, when and where. It is called a list of combinations and

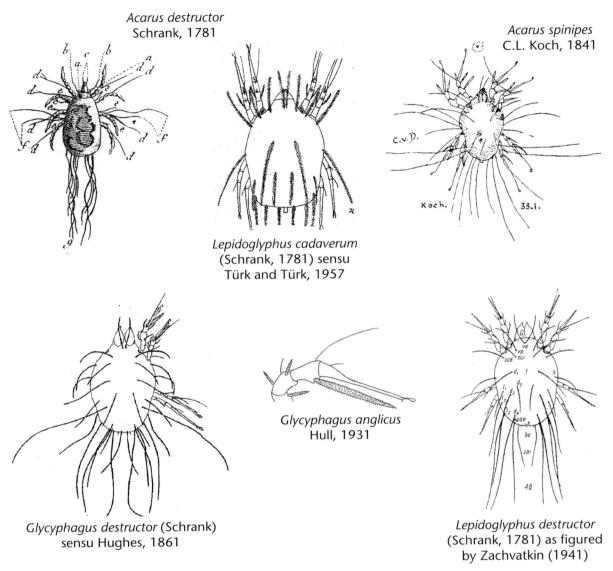

Acarus destructor
Schrank, 1781

Acarus spinipes
C.L. Koch, 1841

Lepidoglyphus cadaverum
(Schrank, 1781) sensu
Türk and Türk, 1957

Glycyphagus anglicus
Hull, 1931

Glycyphagus destructor (Schrank)
sensu Hughes, 1861

Lepidoglyphus destructor
(Schrank, 1781) as figured
by Zachvatkin (1941)

Figure 1.1 Figures and descriptions of *Lepidoglyphus destructor* by various authors (see list of combinations and synonymies in Table 1.2). The differences between them illustrate one of the major problems for taxonomists in determining the identity of classical species. Only the leg of *Glycyphagus anglicus* was ever illustrated.

synonymies (see Table 1.2 below). Some of the illustrations of these various species are provided (Figure 1.1) to show how drawings of the same species can look very different.

b Synonymy and the oldest name

When there is more than one name for a taxon, they are called synonyms. The oldest available name has priority, as stated by the International Code of Zoological Nomenclature. Subsequent names are called junior synonyms. There are exceptions to the priority rule, such as when a name is in widespread use and its replacement by an earlier name, though valid and available, would cause more confusion than it would solve. For example, *Dermatophagoides pteronyssinus*: *pteronyssinus* is not the oldest available name for the species. According to Gaud (1968) and Domrow (1992) it is *pteronyssoides* (see Appendix 1). Furthermore, Oshima (1968) considered the name *Dermatophagoides* as invalid because when the genus was redefined by Fain (1967b) the type species *D. scheremetewskyi* was not redescribed, as required by the International Code, and therefore the next available name, *Mealia*, has priority. *Mealia pteronyssoides* may well be

Table 1.2 A list of combinations and synonymies for the genus *Lepidoglyphus* and the species *Lepidoglyphus destructor* (a frequent inhabitant of damp houses) and how to make sense of it.

Lepidoglyphus Zachvatkin, 1936

Lepidoglyphus Zachvatkin, 1936 ← The name, author and year of description of the genus.

Type species: *Acarus cadaverum* Schrank, 1781, 512. ← The number after the date is the page number where the
↑   species name was first used.

The type species of a genus is that species which helps define the genus and is usually the first species to be described.

Glycyphagus (*Lepidoglyphus*): Zachvatkin, 1941 ← This is a recombination, or change of status. In 1941 Zachvatkin decided *Lepidoglyphus* was only a sub-genus of *Glycyphagus* and did not merit a genus of its own.

Glycyphagus Hering, 1838 sensu Hughes, 1961 (in part) ← This means part of the genus *Glycyphagus*, originally described by Hering, used in a restricted sense, as defined by Hughes, is synonymous with *Lepidoglyphus*.

Lepidoglyphus destructor (Schrank, 1781)

Acarus destructor Schrank, 1781 ← This is the name under which Schrank published the original description of the species, in the genus *Acarus* (in 1781 there was only one genus of mites described).

Lepidoglyphus cadaverum (Schrank, 1781) sensu Türk and Türk, 1957 ← Schrank's species was redescribed by Türk and Türk. The species they redescribed is the same as *Lepidoglyphus destructor*. This does not mean that *L. cadaverum*, as originally described, is a junior synonym of *L. destructor*, only the species used 'in the sense of Türk and Türk'.

Acarus spinipes C.L. Koch, 1841 ← A junior synonym.

Glycyphagus anglicus Hull, 1931 ← Another junior synonym.

Glycyphagus destructor (Schrank) sensu Hughes, 1961 ← A recombination. Hughes, in her restricted definition of *Glycyphagus*, recombined *L. destructor* to the genus *Glycyphagus*.

Lepidoglyphus destructor (Schrank) sensu Hughes, 1976 ← In 1976 Hughes changed her mind about *Lepidoglyphus* and regarded it as a valid genus, as originally conceived by Zachvatkin in 1936. She then re-recombined *Glycyphagus destructor* back into *Lepidoglyphus*.

Glycyphagus cadaverum (Schrank, 1781): Domrow, 1992 ← A recombination, but different from the previous one. Domrow regarded *cadaverum* as a valid species and not a junior synonym of *destructor*. Further, by placing it in *Glycyphagus* he demonstrated that he did not recognise the genus *Lepidoglyphus* as valid either (like Hughes, 1961). Listing Domrow's combination here, indicates I do not consider *G. cadaverum* a valid species.

---

more nomenclatorially correct than *Dermatophagoides pteronyssinus*, but nobody except a few taxonomists would know what it was. One of the consequences would be that all the allergens of *Dermatophagoides pteronyssinus* would all have to be re-named 'Mea p 1, Mea p 2 ...' and so on, according to the rules of allergen nomenclature (see section 7.4.1).

In his essay 'Bully for Brontosaurus' Gould (1991) points out some of the consequences of the legalistic side of taxonomic practice, specifically concerning changes of names of taxa and the Laws of Priority of the International Code of Zoological Nomenclature. He makes the point that taxonomy defines its major activity by the work of the least skilled, and of the Law of Priority he says:

When new species are introduced by respected scientists, in widely read publications, people take notice and the names pass into general use. But when Ignaz Doofus publishes a new name with a crummy drawing and a few lines of telegraphic and muddled description in the Proceedings of the Philomathematical Society of Pfennighalbpfennig *(circulation 533), it passes into well-deserved oblivion. Unfortunately under the Strickland code of strict priority, Herr Doofus's name, if published first, becomes the official moniker of the species – so long as Doofus didn't break any rules in writing his report. The competence and usefulness of his work have no bearing on the decision.*

A fair amount of the taxonomist's time is spent sorting out the ambiguities and confusions created by what Gould refers to as 'the veritable army of Doofuses', and requires the sort of bibliographic archaeology illustrated in Table 1.2 below. However, what many critics forget is that taxonomic practice is a consequence of its times. For taxonomists in the 19th century the poor optical quality of microscopes, compared with those of the present day, was a considerable hindrance, especially to acarologists dealing with such small and morphologically complex organisms. The fewer species known at that time and the consequent greater taxonomic 'distance' between them meant that a paragraph of Latin description, with no figures, was all that was necessary for an adequate description of a new species. Thankfully this is no longer so, but taxonomists in 100 years' time will probably be cursing those of us working today, saying, 'they had the technology in the 1990s to be able to produce complete DNA sequences, so why did they stick to those awful, detailed morphological descriptions, with page after page of diagrams and scanning electron micrographs?'

c Type specimens

To taxonomists, a name and a species are separate entities. The means of associating a name with a species is to designate type specimens. These are representative individuals of a species that demonstrate the key character states by which that species is defined. They are usually chosen by the taxonomist during his or her description of the species (though taxonomists are able to designate particular kinds of types, under strict guidelines, during the process of revisionary work). Usually they are selected from within the group of specimens that will form the basis of the description of the new species.

1.3 Classification and taxonomy of domestic mites

Classification is not quite the same thing as taxonomy. It represents the next step after species and genera have been described, named and defined. A classification of a group of organisms represents a conceptualisation of the hierarchy of the component taxa. It is formed by identifying particular shared characters and grouping organisms in hierarchies according to whether they possess those characters. If that classification is based on a phylogenetic analysis (see section 1.4 below), then classification and phylogeny are congruent, at least in theory. In practice, the classification of many groups of organisms may often be artificial and have very little to do with phylogeny, representing little more than a 'pigeon-holing' system based on relatively few characters. Classifications are intended to help make sense of the diversity of living organisms, and serve as working hypotheses of their relatedness. Those classifications that are not based on phylogenetic analyses have very limited predictive value, and their major utility is for identification purposes, reflecting the compulsive human desire to place things into categories, meaningful or otherwise (Crowson, 1970).

The classification of mites found in house dust is a tale of three superfamilies: the Glycyphagoidea, the Acaroidea and the Analgoidea (which contains the family Pyroglyphidae). It is within these three taxa that the vast majority of allergenically important species are found. Furthermore, each of these superfamilies is associated with other animals. Relatively few astigmatid species are not associated with other animals for part or all of their life cycles (see Chapter 5). The Glycyphagoidea are predominantly associated with mammals; the Acaroidea with insects, birds and mammals and the Analgoidea almost entirely with birds. These associations have independently brought members of each superfamily into contact with humans and their dwellings through the activities of their anthropophilic hosts, and from whence habitat shifts have occurred to house dust (see section 1.4.4 below).

1.3.1 Classification of the Astigmata

Figure 1.2a shows the 10 superfamily-group divisions within the Astigmata, based on the phylogenetic analysis by OConnor (1981) (cf. also Norton et al., 1993, their Figure 1.6). This classification was based on an extensive phylogenetic analysis of the non-psoroptidid Astigmata. The Psoroptidia include the feather mite superfamilies Pterolichoidea and Analgoidea (containing the dust mite family Pyroglyphidae) and the superfamily of skin parasites, the Sarcoptoidea. Gaud and Atyeo (1996) include a third superfamily of feather mites, the Freyanoidea. OConnor (1981) gives a comprehensive account of the complex history of the classification of the Astigmata.

1.3.2 Classification of the Glycyphagoidea and Acaroidea

Mites belonging to the superfamiles Acaroidea and Glycyphagoidea have been referred to traditionally as 'storage mites' or 'stored products mites'. More

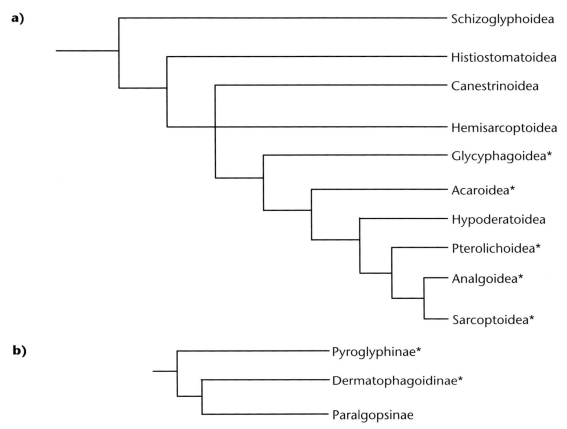

Figure 1.2 **a)** Phylogenetic relationships of the superfamilies of the Astigmata, modified after OConnor (1981) and Norton *et al.* (1993). Asterisks indicate those superfamilies containing genera and species found in house dust and/or known to produce allergens; **b)** possible phylogenetic relationships within the analgoid family Pyroglyphidae. Subfamily classification based on Gaud and Atyeo (1996).

confusingly, they are also referred to collectively along with all other true house dust-dwelling mites as 'domestic mites', reflecting the fact that they are often found associated with stored products and dust within homes as well as within commercial storage premises such as granaries, warehouses and barns.

a   The Glycyphagoidea

The Glycyphagoidea contains seven monophyletic families according to the phylogenetic classification of OConnor (1981; 1982a; see Figure 1.3 below). Four families and six genera are found in house dust: Chortoglyphidae (containing the genus *Chortoglyphus*), Echimyopodidae (genus *Blomia*), Glycyphagidae (genera *Gohieria*, *Glycyphagus* and *Lepidoglyphus*) and Aeroglyphidae (genus *Glycycometus* [what used to be called *Austroglycyphagus*]).

The biology and ecology of glycyphagoid mites is the least well known of all the major taxa of domestic mites, yet in recent years they have become recognised

as second only to the Pyroglyphidae as major sources of allergens (see Chapter 8). About 10 species are known to produce allergens of clinical significance.

The genus *Chortoglyphus* has deutonymphs that are endofollicular parasites of rodents in North and Central America, while the adults are presumably nest-dwellers (OConnor, 1981). One species, *C. arcuatus*, is cosmopolitan and associated with stored food and houses (Hughes, 1976) and is of allergenic importance (Puerta *et al.*, 1993). It has been redescribed by Moreira (1978).

The genus *Blomia*, like *Chortoglyphus*, consists predominantly of associates of New World mammals. Again the deutonymphs are endofollicular parasites, while the adults are presumed to be nest-dwellers. Several species have been described from house dust, mostly in the tropics and subtropics. An appreciation of the allergenic importance of *Blomia* has increased greatly in the last 10 years (Arruda and Chapman, 1992; Fernández-Caldas and Lockey, 1995), but unfortunately the taxonomy of the genus is currently badly muddled.

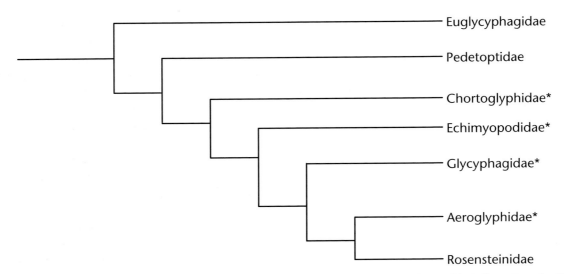

Figure 1.3 Phylogenetic relationships of families of Glycyphagoidea (after OConnor, 1981). Asterisks indicate those families containing genera and species found in house dust and/or known to produce allergens.

Allergens have been isolated from *B. tropicalis* and *B. kulagini*, which are probably the same species (refer to section 1.3.3a below). It would considerably aid an understanding of the epidemiology of *Blomia*-induced allergy if the taxonomic identity and global distribution of its constituent species were delineated clearly. A catalogue of *Blomia* species is given in Appendix 2.

Gohieria species are found in nests of Holarctic rodents and insectivores (OConnor, 1981), and one species, *G. fusca*, is of cosmopolitan distribution, associated with stored products and house dust (Zachvatkin, 1941; Hughes, 1976).

Lepidoglyphus is regarded as a junior synonym of *Glycyphagus* by some authors – for example OConnor (1981) who states that to recognise *Lepidoglyphus* as valid, together with a number of other genera that are cladistically part of *Glycyphagus*, would render *Glycyphagus* paraphyletic. In his opinion, a large number of the undescribed *Glycyphagus* he has examined bridge many of the 'gaps' between these other genera. In other words, OConnor tends toward a 'lumper' approach (see the earlier section 1.2.1) to *Glycyphagus*. However, until the genus is fully revised and the new species formally described, there are some practical reasons of identification for retaining *Lepidoglyphus* and *Glycyphagus* as separate genera: *Lepidoglyphus* lacks a *crista metopica* on the prodorsal region and has a large, feathery subtarsal-scale, whereas *Glycyphagus* has the crista but lacks the scale (Hughes, 1976; see Key 1.12 at the end of this chapter). The two genera contain several species of economic importance as pests of stored products (Zachvatkin, 1941; Hughes, 1976). The most widespread species found in house dust are *Lepidoglyphus destructor* and *Glycyphagus domesticus*. Both are cosmopolitan, but found predominantly in temperate latitudes in houses, barns and grain storage facilities where they are a major cause of allergies among farmers (Cuthbert *et al.*, 1984; Cuthbert, 1990; van Hage-Hamsten *et al.*, 1985).

The genus *Glycycometus* (= *Austroglycyphagus*) contains species mainly associated with nests of birds and with bat guano (Fain 1976; 1978), but three species have been found in house dust in the tropics: *G. lukoschusi* from Surinam (Fain, 1976), *G. malaysiensis* and *G. kualalumpurensis* from Malaysia (Fain and Nadchatram, 1980). *G. geniculatus* is associated with stored grain and the nests of bees and birds and may be cosmopolitan in distribution (Hughes, 1976). The genus has been pretty much neglected from an allergological standpoint, and although there are sporadic records in the literature on dust mite surveys of the tropics (usually recorded as 'Austroglycyphagus sp.') it has probably been confused with the better-known genus *Blomia*, and may well be more common and widespread than indicated previously.

b The Acaroidea

The Acaroidea contains four families according to OConnor (1982a, b). Three families and five genera are found in house dust and are known to be allergenic: Lardoglyphidae (containing the genus *Lardoglyphus*), Suidasiidae (genus *Suidasia*) and Acaridae (genera *Acarus*, *Tyrophagus*, *Aleuroglyphus*).

The most important genera of the Acaroidea found in house dust and known to produce allergens are *Acarus* and *Tyrophagus*. They are taxonomically complex, containing species that are morphologically very similar to each other and that are notoriously difficult to identify with any degree of certainty. In this book, I have not attempted to provide keys to *Acarus* and *Tyrophagus* species. In dust mite surveys the most commonly cited species of Acaridae are *Acarus siro* and *Tyrophagus putrescentiae*. Most of these identifications are unreliable for the simple reason that the experience required to reliably and repeatedly differentiate these has always been rare among taxonomic acarologists.

The genus *Acarus* was revised by Griffiths (1964; 1970), the latter paper containing a key to the 10 species then recognised. Hughes (1976) gives a key to eight species. The taxonomy of *Tyrophagus* has been reviewed by Zachvatkin (1941); Türk and Türk (1957); Robertson (1959, 1961); Samsinák (1962); Johnston and Bruce (1965); Griffiths (1984); Lynch (1989) and Fan and Zhang (2007). Griffiths (1984) recognised some 21 species of which 15 were regarded as rare or very rare, restricted to one or two records and limited geographical ranges. The remainder are common, cosmopolitan, economically important pests, some of them found in a very wide variety of habitats including soil, stored products, live plants and dwellings. The most up-to-date keys are those by Fan and Zhang (2007) and Lynch (1987); the latter contains modifications of keys by Johnston and Bruce (1965) and Hughes (1976).

1.3.3 Classification of the Pyroglyphidae

The possible phylogenetic relationships of the three subfamilies of the family Pyroglyphidae recognised by Gaud and Atyeo (1996) are shown in Figure 1.2b (see above). The reason for my having catalogued all the known members of the family Pyroglyphidae (Appendix 1), even though many of them are not inhabitants of domestic dust, may at first seem obscure. The intent is to provide complete documentation of the taxonomic literature of the family, to give an overview of its current classification, and to have a basis from which to discern patterns of evolutionary affinities within the family and with other astigmatid taxa. I am not suggesting that a catalogue of species can alone provide a framework for theories of the evolutionary origins of dust mites, but it is the starting point of the process.

Historically, the classification of the members of the family Pyroglyphidae as it is currently conceived, as with so many other mite taxa, is a litany of uncertainty.

Its species and genera have been shuffled and re-shuffled with each review. The definition and concept of what constitutes Pyroglyphidae have changed considerably since Cunliffe (1958) proposed the family to contain a single species of one genus, *Pyroglyphus*. Members of the genus *Dermatophagoides* were placed in the family Epidermoptidae by Dubinin (1953) and by subsequent authors until Fain's (1965) revision. He recognised the affinities between *Dermatophagoides* and *Pyroglyphus* and transferred *Dermatophagoides* to the Pyroglyphidae. Fain also included in the Pyroglyphidae the genera *Pyroglyphus* (and *Hughesiella* as a subgenus within *Pyroglyphus*), *Bontiella* and *Euroglyphus* (and *Gymnoglyphus* as a subgenus within *Euroglyphus*). Fain (1967b), in a second revision, divided the Pyroglyphidae into two subfamilies, the Dermatophagoidinae (containing *Pyroglyphus*, *Bontiella* and *Euroglyphus*) which he proposed in 1963, previously within the family Psoroptidae, and the Pyroglyphinae (containing *Dermatophagoides* and *Sturnophagoides*). This formed the basis of the classification of the Pyroglyphidae until Fain's (1988b) third revision, where he created three new subfamilies, the Paralgopsinae (containing *Paralgopsis*), the Onychalginae (containing *Kivuicola*, *Onychalges* and *Paramealia*) and the Guatemalichinae (containing *Fainoglyphus*, *Guatemalichus* and *Pottocola*).

OConnor (1982b) elevated the Pyroglyphidae to superfamily status as a sister group of the Analgoidea, and included within it two other families – the Turbinoptidae which are parasites of the nasal cavities of birds and the feather mite family Ptyssalgidae.

Gaud and Atyeo (1996) used different characters from those of Fain (1988b) to define subfamilies, and do not recognise Fain's subfamilies Onychalginae and Guatemalichinae as valid, synonymising them with Dermatophagoidinae. Neither do they recognise OConnor's (1982b) superfamily Pyroglyphoidea because of the six characters used by OConnor to define the Pyroglyphoidea, only one – the dorsoterminal position of placement of solenidion ω_1 on tarsus I (see Figure 1.8) – is not found elsewhere in the Analgoidea. They consider that the Pyroglyphidae has greater affinities with the Analgoidea, and place the Pyroglyphidae within it. Gaud and Atyeo (1996) state:

The Pyroglyphidae occupies a special place amongst the feather mites, but the group is still relatively unknown. This family not only contains a few taxa that are true feather mites, *but a large number of nidicoles and detritivores that are occasionally found in the plumage … The Pyroglyphidae, with few specialisations, are probably similar to the common ancestors of the Analgoidea and Psoroptoidea. The small number of derived character states makes it difficult to diagnose subfamilies and genera … The placement of solenidia ω_1 on tarsi I defines, without ambiguity, the Pyroglyphidae among other families of the Analgoidea. This character is not limited to the Pyroglyphidae, it occurs in various Psoroptoidea associated with mammals (Psoroptidae, Cebalgidae, Marsupalidae).*

1.3.4 Examples of taxonomic problems with certain dust mites, and some solutions

I make the distinction here between problems of identification and of taxonomy. Examples of identification problems are common with dust mites. One of the most significant involves the commonly experienced difficulty in telling apart the sibling species *Dermatophagoides farinae* and *D. microceras* (see Cunnington *et al.*, 1987). This is despite both species having been clearly and thoroughly described and shown to be reproductively isolated. By contrast, problems of taxonomy arise when there is some doubt over the identity of the species, often due to inadequate descriptions, two or more people having described the same species independently, or the same person having described the same species more than once.

a The identity of Blomia tropicalis, B. kulagini and B. gracilipes

Our current understanding of the taxonomy of *Blomia* is based largely on the work of van Bronswijk *et al.* (1973a, b). In the first of these papers they described *Blomia tropicalis* from house dust in the tropics and subtropics. In the second, they compared the species of *Blomia*, redescribing *B. kulagini* and presenting a key to species. What was intended as a work of clarification has had the opposite effect as a result of not following the standard taxonomic practice of redescribing *B. kulagini* based either on the type material or on specimens collected from the same locality as the type (so-called topotypic material). Fain *et al.* (1977) succinctly summed up the taxonomic confusion within *Blomia*. They found the types of *Chortoglyphus gracilipes* Banks 1917 belonged to the genus

Blomia and recombined the species accordingly. They state:

B. gracilipes *belongs to the group which possess a long copulatory tube and has solenidia ω_1 and ω_2 of tarsus I situated at the same distance from the base of the tarsus. These characters are shared by* B. kulagini *Zachvatkin, 1936 and* B. tropicalis *van Bronswijk [et al.], 1973.*

B. gracilipes *lacks the pair of cuticular projections ('wrats' of van Bronswijk) on the posterior region of the opisthosoma but this character might not be visible owing to the poor condition of the specimens. With this exception it appears to be extremely close to* B. kulagini. *Unfortunately the type of* B. kulagini, *along with the others of Zachvatkin, has been lost so it is impossible to decide if it should fall into synonymy with* B. gracilipes.

Van Bronswijk [et al.] have chosen what they believe to be specimens representative of B. kulagini *from Japan but they have noted several differences between the original description [by Zachvatkin] and their material. We think therefore that the identity of the true kulagini could be ascertained only after examination of new specimens collected from the typical locality (wheat stored in Moscow granaries).*

OConnor (1981) points out that the various species of *Blomia* associated with stored products and house dust are extremely similar morphologically and suspects that they constitute only a single, synanthropic species. I agree with him. All the *Blomia* specimens I have in my collections (from Burma, Colombia, Brazil, Australia and the Philippines) are very similar and can be referred to as *Blomia tropicalis* van Bronswijk *et al.*, 1973a. If *B. kulagini* were to be redescribed from either topotypic material from Moscow granaries, or from rediscovery of the type material, and found to be synonymous with *B. tropicalis*, then *B. kulagini* would be the valid name because of priority (refer to section 1.2.2b).

The problem of the identity of *Blomia* species highlights a major recurrent problem for taxonomists working on species whose type series were deposited in European institutions before 1939. Many taxonomic collections in major cities were destroyed or lost in World War II, and it seems highly unlikely that the types of *B. kulagini* will be rediscovered.

b The correct name of Dermatophagoides pteronyssinus

Baker *et al.* (1956) say it was *Dermatophagoides scheremetewskyi*. Oshima (1968) says *Mealia pteronyssina*. Fain (1966a) says *Dermatophagoides pteronyssinus* and Domrow (1992) says *Dermatophagoides pteronyssoides*. All have a case, but who is right? Gaud (1968) and Domrow (1992) have concluded that *Paralges pteronyssoides* Trouessart, 1886 is the senior synonym of *Dermatophagoides pteronyssinus* (Trouessart, 1897), on the basis of priority (see section 1.2.2b). Gaud's reasoning is based on the list of specimens of *D. pteronyssinus* and their geographical distribution examined by Fain (1966a). Item 7 in the list is of several male and female specimens of *D. pteronyssinus* on a slide from the Trouessart collection at the Muséum national d'Histoire naturelle de Paris, together with a specimen of *Microlichus avus charadricola* Fain and other specimens of a *Thyreophagus* sp. from a species of shrike (*Gallinago nigripennis*) from the Cape of Good Hope, South Africa. This is the host and locality data given by Trouessart (1886) for *Paralges pteronyssoides*. Fain *et al.* (1974) state that since the name *pteronyssoides* was not used for 50 years after its description it should, under Article 23b of the International Code of Zoological Nomenclature, be regarded as a *nomen oblitum* or forgotten name. Article 23b has since been revoked. However, the argument against the use of *pteronyssoides* Trouessart, 1886 as the oldest (and therefore valid) name for the species is the same as that preventing the use of *scheremetewskyi* Bogdanoff, 1864: it would cause too much confusion.

Gaud (1968) goes on to say that *Paralgoides anoplopus* Gaud and Mouchet, 1959 should probably be considered synonymous with *Paralges pteronyssoides* Trouessart, 1887, and there are no differences between the type material and that collected by Trouessart. This is an important point that tells us something about the biology of *Dermatophagoides pteronyssinus*. If, as the evidence suggests, *Paralgoides anoplopus* is synonymous with *Paralges pteronyssoides*, which is in turn synonymous with *D. pteronyssinus*, then this extends the range of avian hosts of *D. pteronyssinus*. *P. anoplopus* has been found on *Cinnyris chloropygius* by Gaud and Mouchet (1959), as well as on *Anthus pratensis*, *Balearica pavonina*, *Eutoxeres aquila*, *Pica pica*, *Sylvia communis* and *Upupa epops* by Gaud (1968). Also, *P. pteronyssoides* has been found on *Aulacorhynchus coeruleicinctus* from South America (Trouessart, 1886), and on *Gallinago nigripennis* from the Cape of Good Hope. Gaud (1968) reckons that this long list of species, combined with very small numbers of mite specimens found on each bird, indicates that *D. pteronyssinus* is 'accidental' in plumage and not a true feather 'parasite'. This accidental occurrence on feathers may suggest *D. pteronyssinus* is a true nest commensal.

c Races, varieties or sibling species within Dermatophagoides pteronyssinus, D. farinae and D. microceras

Evidence of morphological variation in populations of *D. farinae* and *D. microceras* as identified by Fain (1990), and molecular polymorphisms of *D. pteronyssinus* and *D. farinae* populations recorded by Thomas *et al.* (1992), have led to speculation that each of these three species actually represent a far more complex series of separate, closely related, morphologically very similar, so-called sibling species. For example, I have heard anecdotal evidence of populations of *D. pteronyssinus* from North America and Europe having locality-specific numbers of lobes in the receptaculum seminis of the females (Figure 2.12a). I examined this character in a laboratory population of dust mites from Glasgow. In the first 10 mites the number of lobes varied from eight to 11. In species with a broad global distribution there is a tendency to undergo adaptive radiations and for geographically isolated populations to speciate. However, with synanthropic species there is less likelihood of geographic isolation. More outbreeding could be anticipated because humans are a highly mobile species, moving to new areas and bringing their mites with them.

1.4 Biodiversity, phylogeny and evolution

What are dust mites and where have they come from? Central to the answer is the consideration of evolutionary origins. To get some kind of basic picture we need to construct the phylogenetic relationships of the family Pyroglyphidae and, based on this phylogeny plus inferences from the biology and ecology of dust mites and their relatives, an hypothesis of how dust mites came to live in human dwellings.

1.4.1 Phylogeny and the value of predictive classifications

Phylogeny means, loosely, 'the history of the tribe'. Reconstructing a phylogeny is similar to compiling a genealogical tree in that both indicate, first and foremost, the degree of relatedness between the members of the 'tribe'. However, a family tree is based on known fact, documentary evidence, whereas a phylogeny is only ever, at best, an hypothesis of the most likely evolutionary history of the taxon, based on fossil evidence (where available), comparative morphology, genomic sequences or other data such as reproductive behaviour.

The value of classifications that are based on phylogeny is that they offer a simple, concise information retrieval system that is predictive in nature. For example, if *Dermatophagoides microceras* is more closely related to *D. farinae* than to *D. pteronyssinus*, one may be able to predict that the allergens of *D. microceras* will have higher sequence homology with those of *D. farinae* than with those of *D. pteronyssinus*, even though there is currently no data available on the sequences of allergens of *D. microceras*. Why is this information of use?

1.4.2 Background – phylogenetic relationships of the sarcoptiform mites

Mites are an order within the Class Arachnida, along with the spiders, scorpions, harvestmen, ricinuleids, whip-scorpions, schizomids, sun-spiders, palpigrades, false-scorpions and amblypygids. The arachnids are an ancient group, and fossil scorpions of the Silurian were probably among the first terrestrial arthropods. Mites too are an ancient group: the earliest fossils are from Lower Devonian deposits, ca. 400 million years before present (mybp) (reviewed by Bernini, 1991).

The majority of species that have evolved in association with humans and their dwellings belong to the suborder Astigmata, first recorded as fossils in amber some 28 million years ago, though probably much older. The Astigmata are thought to have evolved from within an ancient, soil-dwelling group of mites, the Oribatida, fossils of which have been recovered from 376–379 mybp Devonian mudstones from Gilboa, New York State (Norton *et al.*, 1988). The evidence for the origin of the Astigmata from within the Oribatida is based on an extensive series of shared characters (Norton, 1998).

1.4.3 Biodiversity and phylogenetic relationships within the Pyroglyphidae – out of Africa in the nests of birds?

There has been no phylogenetic analysis of the family Pyroglyphidae and I am not about to attempt one here. However, we can summarise some basic trends in morphology and biology within the

different subfamilies that may give some clues to their evolution, as well as draw on biogeographical and habitat data from the catalogue in Appendix 1. From Appendix 1 and Table 1.3, only members of the subfamilies Pyroglyphinae (three out of eight genera: *Euroglyphus*, *Hughesiella* and *Gymnoglyphus*) and Dermatophagoidinae (all four genera: *Dermatophagoides*, *Hirstia*, *Malayoglyphus* and *Sturnophagoides*) are found in house dust, with 13 species in all. A number of other points emerge:

- the vast bulk of pyroglyphid diversity has been found exclusively on birds or in their nests in the tropics (28 spp.), especially west and central Africa (16 spp.), Latin America and Cuba (9 spp.);
- parrots, swallows and martins, sparrows, waxbills, weavers, woodpeckers and starlings are among the most frequent hosts of pyroglyphid taxa;
- several species, now known to be important inhabitants of dust, were first recorded in atypical habitats (e.g. *Euroglyphus maynei* in mouldy cottonseed cake);
- virtually all bird-associated taxa have been recorded only from the type locality or have restricted distributions;
- only dust-associated species are known to be geographically widespread;
- a large proportion of all the known pyroglyphid species have been described by just two people: Alex Fain and Jean Gaud, accounting for 23 species between them. The systematics of the family, especially subfamilial and generic concepts, is predominantly due to the work of Fain.

Both Fain and Gaud had strong historical research links with Africa and Gaud worked exclusively on feather mites. Is the high diversity of African avian associates a real phenomenon or is it an artefact of the research interests of these two authors? Evidence from research on feather mite taxonomy shows that these mites are often highly host-specific, and that any bird species (of some 8000) may have between two and eight different feather mites associated with it. By way of contrast, house dust is a relatively homogenous microhabitat, so one would expect a higher diversity of bird-associated pyroglyphids than dust-associated pyroglyphids. Furthermore, birds have a far older evolutionary history than humans. The 'host' relationships of pyroglyphid species in Table 1.3 are summarised in Figure 1.4 which shows the phylogenetic

Table 1.3 Pyroglyphid mite species known from birds and their geographical distribution.

Birds	Pyroglyphids	Distribution
Psittaciformes		
Psittacide (true parrots)		
Agapornis pullaria	Dermatophagoidinae	
	Dermatophagoides anisopoda	Central Africa
Ara macao	Paralgopsinae	
	Paralgopsis ctenodontus	South America
Pyrrhura leucotis	Paralgopsinae	
	Paralgopsis paradoxus	South America
Apodiformes		
Apodidae (swifts)		
Tachornis phoenicobia iradii	Dermatophagoidinae	
	Guatemalichus tachornis	Central America
Piciformes		
Picidae (woodpeckers)		
Meiglyptes tristis	Pyroglyphinae	
	Asiopyroglyphus thailandicus	South-East Asia
Phloeoceastes rubricollis	Pyroglyphinae	
	Campephilocoptes atyeoi	South America
P. leucepogon	*C. paraguayensis*	South America
Capitonidae (barbets)		
Pogonoiulus scolopaceus	Dermatophagoidinae	
	Pottocola (Capitonocoptes) longipilis	West Africa
	Onychalges spinitarsis	Central Africa
Lybius dubius	Dermatophagoidinae	
	Pottocola (Capitonocoptes) lybius	Central Africa
L. vielloti	*P. (C.) lybius*	Central Africa
L. torquatus	*P. (C.) lybius*	Central Africa
L. rubrifacies	*P. (C.) lybius*	Central Africa
L. bidentatus	*P. (C.) ventriscutata*	Central Africa
Passeriformes		
Estrildidae (finches and waxbills)		
Lonchura cucullatus	Pyroglyphinae	
	Bontiella bouilloni	Central Africa
Clytospiza monteiri	Dermatophagoidinae	
	Onychalges asaphospathus	West Africa
Euschistospiza dybovskyi	*O. asaphospathus*	West Africa
Spermestes bicolor	*O. asaphospathus*	West Africa
Spermophaga haematina	Dermatophagoidinae	
	Onychalges odonturus	West Africa
Estrilda melpoda	Dermatophagoidinae	
	Onychalges pachyspathus	West Africa
E. atricapilla	*O. pachyspathus*	West Africa
E. nonnula	*O. pachyspathus*	West Africa

Table 1.3 continued

Table 1.3 Continued

Birds	Pyroglyphids	Distribution
E. astrild	O. pachyspathus	South Africa
Lagonostica rubricata	O. schizurus	West Africa, South Africa
Hirundinidae (swallows and martins)		
Hirundo neoxena	Pyroglyphinae	
	Weelawadjia australis	Australia
Delichon urbica	Dermatophagoidinae	
	Hirstia chelidonis	Europe
Petrochelidon fulva	Dermatophagoidinae	
	Sturnophagoides petrochelidonis	Central America
Ploceidae (sparrows and weavers)		
Passer griseus	Dermatophagoidinae	
	Dermatophagoides aureliani	Central Africa
Passer domesticus	Dermatophagoidinae	
	Dermatophagoides simplex	South America
	Onychalges nidicola	South America
Ploecus nigricollis brachypterus	Dermatophagoidinae	
	Paramealia ovata	West Africa
Sturnidae (starlings and mynahs)		
Buphagus africanus	Dermatophagoidinae	
	Dermatophagoides rwandae	Central Africa
Buphagus erythrorynchus	Dermatophagoidinae	
	Dermatophagoides sclerovestibulatus	South Africa
Sturnus vulgaris	Dermatophagoidinae	
	Sturnophagoides bakeri Fain, 1967	North America
Furnariidae (woodcreepers)		
Certhiaxis erythrops	Dermatophagoidinae	
	Fainoglyphus magnasternus	South America

relationships of the birds. There is no obvious pattern of phylogenetic relatedness between mites and avian 'hosts' as there is with ectoparasitic arthropods such as lice, but then the biological nature of the association of pyroglyphids and birds is not sufficiently defined to be sure of how taxon-specific it really is. The data in Table 1.3 indicate that the two subfamilies of pyroglyphids which contain species found in house dust, the Pyroglyphinae and Dermatophagoidinae, are the most widespread geographically, the most species-rich and are associated with a higher diversity of avian taxa than the subfamily that does not contain species that are found in house dust. This tends to suggest that the Pyroglyphinae and Dermatophagoidinae may represent the more ancestral taxa within the family (see Figure 1.2b).

1.4.4 Evolutionary inferences from the biology and ecology of pyroglyphids and other Astigmata

Probably about 300 species of mites are associated with humans and their homes as parasites, commensals or pests of stored food and other products. The advent of permanent human settlement, agriculture and food storage systems occurred about 10 000 years ago, so the association between astigmatid mites and human habitation has endured for about 0.044% of the duration of the fossil record of the Astigmata; equivalent to 38 seconds in 24 hours. The opportunities afforded to mites by the development of permanent settlement, agriculture and the domestic technologies of food storage and weaving are explored in more detail in Chapter 4.

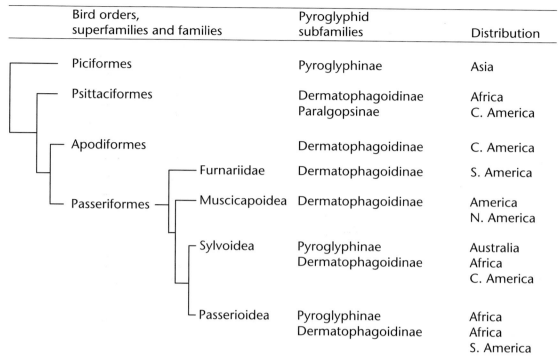

Bird orders, superfamilies and families	Pyroglyphid subfamilies	Distribution
Piciformes	Pyroglyphinae	Asia
Psittaciformes	Dermatophagoidinae Paralgopsinae	Africa C. America
Apodiformes	Dermatophagoidinae	C. America
Furnariidae	Dermatophagoidinae	S. America
Muscicapoidea	Dermatophagoidinae	America N. America
Sylvoidea	Pyroglyphinae Dermatophagoidinae	Australia Africa C. America
Passerioidea	Pyroglyphinae Dermatophagoidinae	Africa Africa S. America

Figure 1.4 Summary of phylogenetic relationships of orders of birds and the suborders of pyroglyphid mites found on them. The phylogenetic relationships of the birds follow Sibley and Ahlquist (1990).

a Free-living soil dwellers and the effects of the development of agriculture

The Astigmata consists of species mostly involved in commensal, symbiotic or ectoparasitic relationships with other animals for at least part of their life cycle. Relatively few species are completely free-living, although the Astigmata were probably derived from free-living fungus feeders in rotting logs and vegetation. The most important, the Glycyphagoidea and Acaroidea are free-living as adults and found in diverse habitats including nests of birds and mammals. They differ from the Psoroptidia in that the deutonymph is heteromorphic, non-feeding, resistant to desiccation, and is thus well adapted for dispersal to new habitats via attachment to other animals – a practice known as phoresy. Phoresy is characteristic of species associated with temporary or restricted habitats: to exploit patchy resources successfully the mites need efficient dispersal. Species that are commensal, parasitic or exploit resources in widespread, contiguous habitats have lost the ability to form heteromorphic deutonymphs. This includes all of the Psoroptidia, which have no deutonymphal stage at all, some *Tyrophagus* spp. and all *Aleuroglyphus* spp.

The relatively short time since the inception of agriculture and permanent settlement is inadequate for the evolution of such a diverse array of morphologies exhibited by the mite taxa that are found in association with human dwellings and stored products. Only rarely are more than one or two species in each genus associated with humans. The rest may be free-living in natural habitats. This suggests that many taxa, from different lineages, have colonised human habitats at different times, rather than a single lineage, on one occasion, followed by an adaptive radiation. This being so, those taxa that have successfully colonised human dwellings and stored products must have been pre-adapted to these habitats, i.e. they lived in ones that provided very similar resources before they made the shift to human dwellings. OConnor (1979) found that of those genera of Astigmata associated with stored products and house dust, six contained species widespread in field habitats, 12 contained mammal nest inhabitants, five were bird-nest dwellers and 12 were found in rare or ephemeral habitats. The trophic specialisations of dust mites and stored products mites in their non-human habitats are mirrored by niches that they have occupied within human dwellings. Thus the pyroglyphids, originally bird-nest dwellers, have encountered similar skin scales in the human nest, the bed, as they have in the feral, avian habitat, together with warmth and humidity generated by the occupant. Similarly, the glycyphagoid and acaroid inhabitants of human dwellings have encountered seeds, cereals and other plant materials that would also have been stored by rodents and other mammals in their nests. Given the similarity of trophic niches of human dwellings to those available naturally, it is not surprising that a number of mite species have become associated with human habitations. Most importantly, many species appear to have retained the ancestral ability to feed on fungi (OConnor, 1982a). In fact the Glycyphagidae are regarded as primarily fungivorous and not graminivorous, *Acarus siro* is a fungivore that also feeds on grain germ, and pyroglyphids are skin scale feeders that also feed on fungi and yeasts, whereas *Tyrophagus* spp. are mainly field species attacking stems, roots and fruits of plants, but which also feed on fungi.

b   Habitat shifts to commensalism and parasitism

What is the evidence for the habitat shift to human dwellings? Were dust mites really colonisers of the domestic environment of early civilisations or are they recent invaders resulting from changes in housing design during the 20th century? The evidence for an early association with humans is as follows:

- Those species of Astigmata associated with dust in dwellings today are the Pyroglyphidae, Glycyphagidae, Echymyopodidae and Acaridae. All these families contain species associated with the nests of small mammals (Glycyphagidae, Echimyopodidae), birds (Pyroglyphidae) or are abundant in arable soils (Acaridae). The predominant birds were Passeriformes, including a number of anthropophilic species that build their nests near, on or in human habitation.
- Evidence, both morphological (Fain, 1979) and allergenic (Stewart and Fisher, 1986; Arlian *et al.*, 1988), indicates that families containing species which are dermal parasites of vertebrates (Psoroptidae and Sarcoptidae) are closely related to the Pyroglyphidae.
- Members of the families Acaridae and Glycyphagidae were the first mites to be documented as present in homes by pioneer microscopists such as Hooke (1665) and van Leeuwenhoek (see Heniger, 1976; other early literature reviewed by Oudemans, 1926). They have also been found in archaeological remains (Radovsky, 1970; Kliks,

1988; Baker, 1990; Mehl, 1998) and in homes of members of rural, tribal communities in Papua New Guinea (Anderson and Cunnington, 1974) and Colombia (Sánchez-Medina *et al.*, 1993; see Chapter 4).
- Several lineages of Astigmata have evolved to live in human dwellings (OConnor, 1979; 1994), and species-diversity of the domestic acarofauna can be quite high (Bronswijk, 1981). The development of a relatively species-rich acarine community within homes points toward a longer time course for their association with humans, with multiple colonisation events.

Countering these arguments, and in support of the fact that dust mites are a recent phenomenon associated with changes in dwellings such as wall-to-wall carpets, double-glazing and central heating are:

- the paucity of evidence of pyroglyphid mites in association with humans prior to 1865, and in dwellings prior to 1964;
- the rapid rise in prevalence and incidence of asthma since the 1960s, some of which may be attributable to dust mites;
- the apparent increase in dust mite allergen concentrations in homes since the 1970s.

Woolcock *et al.* (1995) have presented a case for the latter view, but there is little direct evidence for it (see Chapter 8). On closer examination, the 'ancient' and 'recent' theories are not mutually exclusive at all but are parts of the same continual process, the evolution of the ecological relationship between dust mites and humans.

c   The evolution of resistance to desiccation

There are differences in the water balance capabilities of the different lineages of mites that have colonised human habitation (Chapter 3). Although the critical equilibrium humidity (CEH; the humidity at which body water loss balances water gain) of pyroglyphid dust mites is not considerably lower than that of stored products species, population growth of pyroglyphids is optimal at drier conditions than for acaroids and glycyphagoids. The acaroids and glycyphagoids are, predominantly, still tied nutritionally to fungi (OConnor, 1982a). Thus they require a high degree of atmospheric water not only in which to survive, but without which their major food source, the hyphae and spores of moulds, would fail to thrive. By making a dietary shift to keratinaceous material, its lipids and associated xerophilic microorganisms, the Pyroglyphidae are liberated from the humidity constraints of fungal feeding. If body fat content is related to body water content, as is the case with insects, mites with high lipid diets, such as pyroglyphids, will have lower body water content than mites feeding on low lipid diets such as acaroids. What are the consequences for body water loss? Does a high fat diet predispose to lower CEHs? If so, then the shift to skin-feeding would represent an important precursor of the high colonising potential that is derived from the ability of pyroglyphids to withstand desiccation.

How have dust mites evolved a tolerance to lower humidities than other mites that have colonised human dwellings? The shift of pyroglyphids to feeding on skin could have adapted them to greater tolerance of low humidity because, with an increased dietary lipid component, they were able to gain metabolic water from the hydrolysis of fat (oxidation of 1 g of fat yields 1.1 g of water). Lipids constitute a major dietary component that pyroglyphid dust mites share with feather mites, which appear to feed on oils and waxes secreted onto feathers. Perhaps it is significant that the lowest CEH yet recorded for a mite is for the feather mite *Proctophyllodes troncatus* (Gaede and Knülle, 1987). Other differences between pyroglyphid dust mites and stored products species that relate to water balance, and which dust mites have in common with feather mites are:

- relatively large eggs, laid in small numbers, as opposed to medium-sized eggs laid in large numbers;
- striated cuticle, consisting of ridges and troughs, which allows for flexibility and hydrostatic-related properties as well as acting as a means of holding in the troughs a film of moist air close to the body surface;
- dorso-ventral flattening;
- relatively long-lived, with long population doubling time;
- absence of Malphigian tubules (present in the Acaridae and where ionic regulation and resorption of salts takes place). These structures would appear to be unnecessary in the Pyroglyphidae because resorption takes place via the mouth;
- water gain from saturated air may have evolved from an alternative osmoregulatory mechanism.

The nests of birds and human dwellings share certain properties relating to moisture. Both are essentially arid, though with moist spots distributed spatially and temporally which can be exploited by mite populations. Additionally, food resources in both habitats are not evenly distributed.

d Differences in fecundity and population dynamics between pyroglyphids, acaroids and glycyphagoids

As with water balance capabilities, there are some important differences between the reproductive capabilities and population dynamics of pyroglyphid mites, acaroid and glycyphagoid mites (Chapter 5). Fashing (1994) and OConnor (1994) identified life-history and demographic modifications of various groups of Astigmata that relate to their capacity to exploit temporary habitats and resources. Most of the Acaroidea and some Glycyphagoidea fall into a category with short generation time, oviposition period and egg development time, and with high fecundity − what have been called 'r-selected' taxa (Pianka, 1970) − exhibit extremely rapid population increase in response to resources becoming available. By way of contrast, the Pyroglyphidae are predominantly what has been called 'K-selected', i.e. with longer generation times, oviposition periods and egg development, and with low or moderate fecundity (though see Stearns (1992) for a discussion of the deficiencies of r and K-selection theory). This comparison holds only within the Astigmata. Compared with the long-lived, low-fecundity soil-dwelling oribatid mites, the Pyroglyphidae appear capable of decidedly rapid population increase.

Acaroid and glycyphagoid mites have a deutonymphal instar, highly morphologically modified for dispersal and survival of adverse conditions. Several species have both a phoretic, dispersive form and a quiescent, non-dispersive form (Knülle, 1995). Both forms are non-feeding. This stage is absent from the life cycle of pyroglyphid mites, though the quiescent protonymphal stage of *D. farinae* appears to have a similar survival function (though it is not dispersive) in that it is more resistant to desiccation than other stages in the life cycle (Ellingsen, 1975). Quiescent protonymphs of *D. pteronyssinus* appear not to share this capability. It is noteworthy that *D. farinae* has a slightly lower CEH than *D. pteronyssinus* and occurs far more frequently than *D. pteronyssinus* in the dry climate of inland California (Lang and Mulla, 1977a).

The loss of the dispersive/survivor deutonymphal instar in the Pyroglyphidae, combined with their lower reproductive rates, indicates that house dust represents a fairly stable habitat in terms of availability of resources compared with those habitats occupied by acaroids and glycyphagoids. This begs the question of why houses are not full of acaroid and glycyphagoid mites. In homes with a high moisture content, populations of acaroids and glycyphagoids can reach enormous numbers of active instars very quickly, though populations can decline equally as fast if the humidity falls, with the formation of phoretic or quiescent deutonymphs. Pyroglyphid mites simply appear to have a different strategy for survival of dry conditions: the active, reproducing stages of the life cycle can cope better with substantial water loss so there is less need for a dispersive or a quiescent, drought-resistant stage. One important consequence of this strategy is that even in relatively dry conditions the pyroglyphid population comprises fully functional instars, capable of feeding and egesting and producing allergens, whereas in acaroid and glycyphagoid mites much of the population will be represented by non-feeding, non-egesting, and hence non-allergen producing stages.

1.5 Characters used in taxonomy and identification – external anatomy

Accurate identification of most organisms is dependent upon the ability to discriminate morphological characters of taxonomic importance. With mites, as for many other arthropods, the majority of characters relate to the exoskeleton and structures growing from it or through it. So most characters of taxonomic importance relate to external anatomy. For the Pyroglyphidae these include hairs, or setae on the legs and body, plus other sensory structures, the microsculpture of the cuticle, the structure of the genitalia, and the shape and dimensions of the legs and their segments.

1.5.1 Divisions of the body

In acarological terminology, the body of a mite is called the idiosoma. It is divided into two major parts (see Figure 1.5), the propodosoma which is the anterior part, bearing the mouthparts or gnathosoma, and the first two pairs of legs, and the hysterosoma, bearing the third and fourth pairs of legs, plus the region behind the fourth pair of legs, known as the opisthosoma. The metapodosoma, a term rarely used

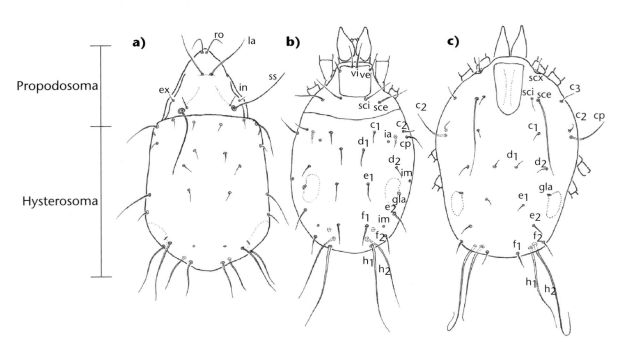

Figure 1.5 Dorsal views of some acariform mites with setal nomenclature. **a)** an oribatid; **b)** an acarid; **c)** a pyroglyphid. Hysterosomal setal signatures are according to their hysterosomal segments C, D, E, F, H and PS. Thus p_1 and p_2 are the setae of the pseudanal segment (cf. Figure 1.6). gla, lateral opisthosomal gland; ia, im, ih, ip, hysterosomal cupules; in, interlamellar setae; le, lamellar setae; ro, rostral setae; sce, internal scapular setae; sci, external scapular setae; ss, sensillus; scx, supracoxal seta; ve, vertical external setae; vi, vertical internal setae.

in keys to mites, is the region bearing the third and fourth pairs of legs. The podosoma consists of the propodosoma plus the metapodosoma – the region bearing all the legs. Compared with other groups of arachnids, mites have undergone reduction in body segmentation; the signs of which are indicated only by transverse whorls of setae. The boundary of the propodosoma and hysterosoma is not always clearly demarcated. In some taxa, there is a clear division, the sejugal furrow, separating the two.

1.5.2 Idiosomal chaetotaxy – the names of the hairs of the body

Every seta and other hair-like sensory structure on the body of the vast majority of mites has a name, usually abbreviated to a so-called signature or notation. These are abbreviations of names of setae, such as *vi* for internal vertical setae. In some of the acarological literature the names of setae are written in Latin, hence *setae verticales internae*. Translation is usually as obvious as in this example. Some mites have too many setae to name, but most, including all the Astigmata have relatively few, which are fairly easy, with practice, to recognise and count. The arrangement of these structures, their nomenclature, their development and appearance in different stages in the life cycle and the study of their homologies between different taxa is called chaetotaxy. For those

researchers involved in a lot of identification work, the nomenclature of setae is well worth learning. When one is examining specimens down a microscope, setae provide a very convenient series of signposts to assist in navigation over the body of a mite, as well as being important taxonomic characters in their own right, with many different shapes, lengths and patterns of ornamentation.

There are many different systems of setal notation, mostly as a consequence of taxonomists developing, in isolation, their own system for the particular group of mites that they are interested in. This can be confusing for the beginner when looking at the older literature. However, in recent years, a lot of effort has gone into working out homologies of setae within the Acariformes, the group that contains the Prostigmata, and Sarcoptiformes (Oribatida + Astigmata). Setal nomenclature for the Mesostigmata is completely different from that of the acariform mites. The system used here for Sarcoptiformes follows Griffiths *et al.* (1990) and is preferred because it uses the same notation for both the Astigmata and Oribatida for almost all regions of the body. It represents an attempt at recognition of setal homologies, in accordance with the common evolutionary origins of the Astigmata and Oribatida (see Figures 1.5 and 1.6).

The chaetotaxy of different stages in the life cycle is not the same. Larvae have the fewest setae of all instars,

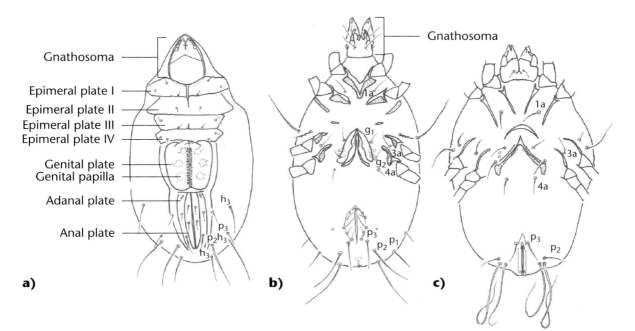

Figure 1.6 Ventral views of some acariform mites. **a)** an oribatid; **b)** a female acarid; **c)** a female pyroglyphid. 1a, 1b, 1c, setae of epimeron 1 (epimeral plate 1); 2a, setae of epimeron 2; ad_{1-3}, adanal setae; g_1, g_2, genital setae 1 and 2; p_{1-3}, pseudanal setae.

then nymphs and then adults. Thus various setae appear following moults. Tracing these appearances is called developmental chaetotaxy. The full complement of setae for any developmental stage is called the chaetome. The sections on external anatomy of body divisions below refer to the adult chaetome because it is the best known. Many species of mites have been described only from adult specimens and the immatures are undescribed.

1.5.3   The dorsal surface of the idiosoma

The dorsal surface of the propodosoma is often referred to as the prodorsum. Dorsally, the most anterior structures visible on many Astigmata are the dorsal surfaces of the paired chelicerae (see Figure 1.5), although in some species (such as in subfamily Pyroglyphinae) the chelicerae may be partially covered by an anterior projection of the propodosoma, the tegmen. In many (though not all) Astigmata there is a prodorsal shield, which is more heavily tanned, or sclerotised, than the surrounding cuticle area, often ovoid or sub-rectangular with a fine, pitted microsculpture. The prodorsal setae are usually distributed around the edges of the prodorsal shield and only rarely do they arise from it. On the prodorsum there are four pairs of setae in the Glycyphagoidea and Acaroidea. The most anterior pair are the internal verticals (*vi*) situated medially either on the anterior or the centre of the prodorsum. Lateral to these are the external verticals (*ve*). Behind the verticals are the scapular setae, the internal scapular setae (*sci*) positioned medially and the external scapular setae (*sce*) located laterally from them. In the Pyroglyphidae *sce* is often very long, especially in the Dermatophagoidinae, whereas *sci* is a short, thin seta, except in the Paralgopsinae where it may be long and well developed. The Pyroglyphidae have only two or three pairs of prodorsal setae. Setae *ve* and *vi* have been lost, except in the Paralgopsinae where *vi* is present; a character that is used to define this subfamily.

It is worth mentioning here that the supracoxal seta (*scx*) emerges from the point where the first leg meets the propodosoma. It is not a prodorsal seta but can give the impression in some microscope preparations of specimens that it emerges from the lateral margin of the prodorsum. It can usually be recognised in many species of Acaroidea because it is elaborately branched or spinose.

Posterior to the setae of the *sc* series is the sejugal furrow, if present, and the setae that comprise the chaetome of the hysterosoma; the c, d, f, h, p (or ps) and ad series. The latter two are ventral in position and will be described in the next section. In the Pyroglyphidae there are four pairs of c setae, c_1 the median pair, c_2 the pair anteriolateral to these, c_3, positioned even further laterally, around on the ventral surface just anterior to the bases of legs III, with seta cp posterior to c_3. The d series consists of two pairs of setae, d_1 more or less in the middle of the hysterosoma and d_2 lateral to these, more or less perpendicular from c_2. In many Pyroglyphidae there is an hysterosomal shield, especially in males, similar in microsculpture to the prodorsal shield, extending from around the region of setae d_2 to the caudal margin of the opisthosoma. In the Dermatophagoidinae the hysterosomal shield is largest in the males. In females it is often absent or reduced. Setae of the e and f series consist of two and one pair of setae respectively, e_2 and f_2, positioned posteriolaterally, behind the opening of the lateral opisthosomal gland (*gla*) with e_1 medially, posterior of d_1 medially, posterior of d_1. Some or all of the h series (three pairs) are very long in the Pyroglyphidae and Acaridae and are positioned along the caudal margin of the opisthosoma, often originating ventrally, posterior to the anus. In the Pyroglyphidae seta h_2 and h_3 are developed in this manner. Seta h_1 is invariably small, setiform and the most medial of the h series, positioned behind and slightly lateral of the anus. Just anterior to the h series on the ventral surface are the pseudanal (p or ps) setae. In the Pyroglyphidae there are two pairs of p setae: an anterior pair, p_3, lateral of the anterior margin of the anus and p_2, more posteriolaterally positioned.

1.5.4   The ventral surface of the idiosoma

The gnathosoma is divided into two parts, the chelicerae which are positioned dorsally, and the subcapitulum which consists of the rest of the mouthparts. Laterally are the palps which are two-segmented in nearly all the Astigmata and bear a series of setae and other sensory structures. The palps emerge from the mentum, which forms the base of the subcapitulum and bears a pair of setae. Anterior of the mentum and fused to it, medial of the palps, is the paired rutellum.

The podosoma bears the legs. The coxae of acariform mites have become fused with the ventral surface and are known as epimera or epimeres. Epimera I and II are adjacent to each other and there is a gap between them and epimera III and IV, also adjacent. The epimera may have areas of porose

Key 1   Stages in the life cycle of astigmatid mites found in house dust

In this key the stages of the life cycle of domestic mites are dealt with and it applies to the Astigmata found in houses. It will also work for the immature stages of most of the other major groups of mites. The hypopial stage, unique to the Astigmata, is absent from the life cycle of members of some families, e.g. the Pyroglyphidae.

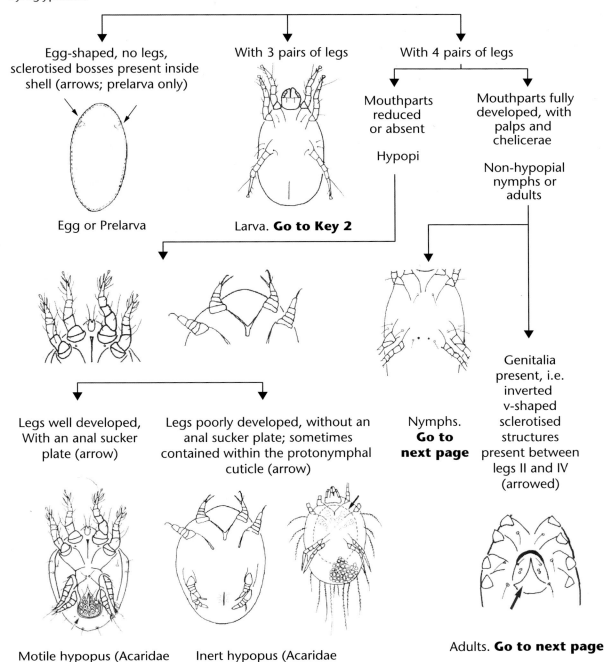

Egg-shaped, no legs, sclerotised bosses present inside shell (arrows; prelarva only)

With 3 pairs of legs

With 4 pairs of legs

Mouthparts reduced or absent

Hypopi

Mouthparts fully developed, with palps and chelicerae

Non-hypopial nymphs or adults

Egg or Prelarva

Larva. **Go to Key 2**

Legs well developed, With an anal sucker plate (arrow)

Legs poorly developed, without an anal sucker plate; sometimes contained within the protonymphal cuticle (arrow)

Nymphs. **Go to next page**

Genitalia present, i.e. inverted v-shaped sclerotised structures present between legs II and IV (arrowed)

Motile hypopus (Acaridae and Glycyphagoidea)

Inert hypopus (Acaridae and Glycyphagoidea)

Adults. **Go to next page**

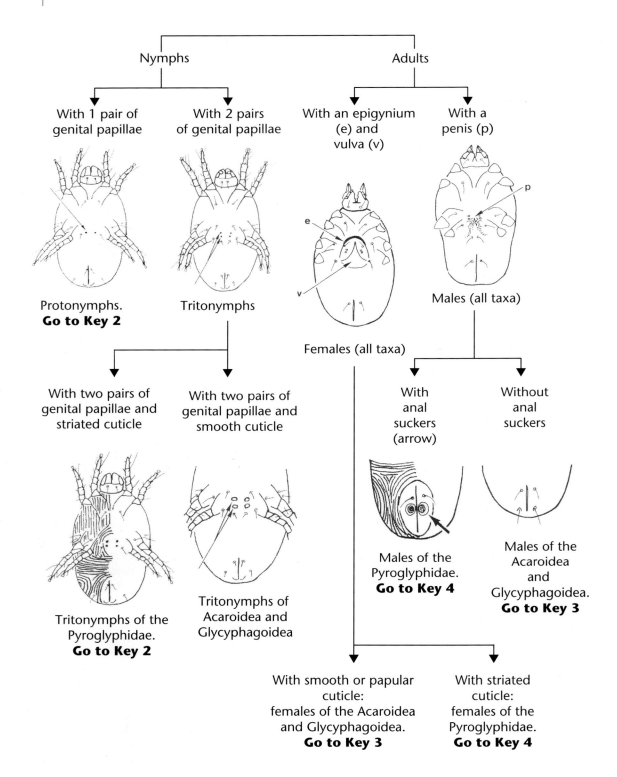

Nymphs

Adults

With 1 pair of genital papillae

With 2 pairs of genital papillae

With an epigynium (e) and vulva (v)

With a penis (p)

Protonymphs.
Go to Key 2

Tritonymphs

Males (all taxa)

Females (all taxa)

With two pairs of genital papillae and striated cuticle

With two pairs of genital papillae and smooth cuticle

With anal suckers (arrow)

Without anal suckers

Tritonymphs of the Pyroglyphidae.
Go to Key 2

Tritonymphs of Acaroidea and Glycyphagoidea

Males of the Pyroglyphidae.
Go to Key 4

Males of the Acaroidea and Glycyphagoidea.
Go to Key 3

With smooth or papular cuticle: females of the Acaroidea and Glycyphagoidea.
Go to Key 3

With striated cuticle: females of the Pyroglyphidae.
Go to Key 4

Key 2 Larvae and nymphs of pyroglyphid mites found in house dust (modified after Mumcuoglu, 1976)

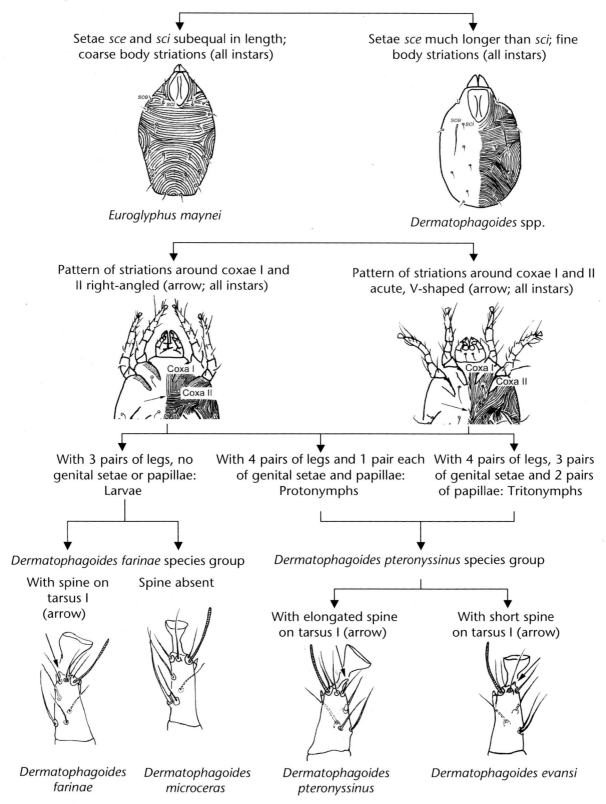

Setae *sce* and *sci* subequal in length; coarse body striations (all instars)

Euroglyphus maynei

Setae *sce* much longer than *sci*; fine body striations (all instars)

Dermatophagoides spp.

Pattern of striations around coxae I and II right-angled (arrow; all instars)

Pattern of striations around coxae I and II acute, V-shaped (arrow; all instars)

With 3 pairs of legs, no genital setae or papillae: Larvae

With 4 pairs of legs and 1 pair each of genital setae and papillae: Protonymphs

With 4 pairs of legs, 3 pairs of genital setae and 2 pairs of papillae: Tritonymphs

Dermatophagoides farinae species group

Dermatophagoides pteronyssinus species group

With spine on tarsus I (arrow)

Spine absent

With elongated spine on tarsus I (arrow)

With short spine on tarsus I (arrow)

Dermatophagoides farinae

Dermatophagoides microceras

Dermatophagoides pteronyssinus

Dermatophagoides evansi

Key 3 The adults of major groups of mites most frequently found in house dust (modified after Colloff and Spieksma, 1992)

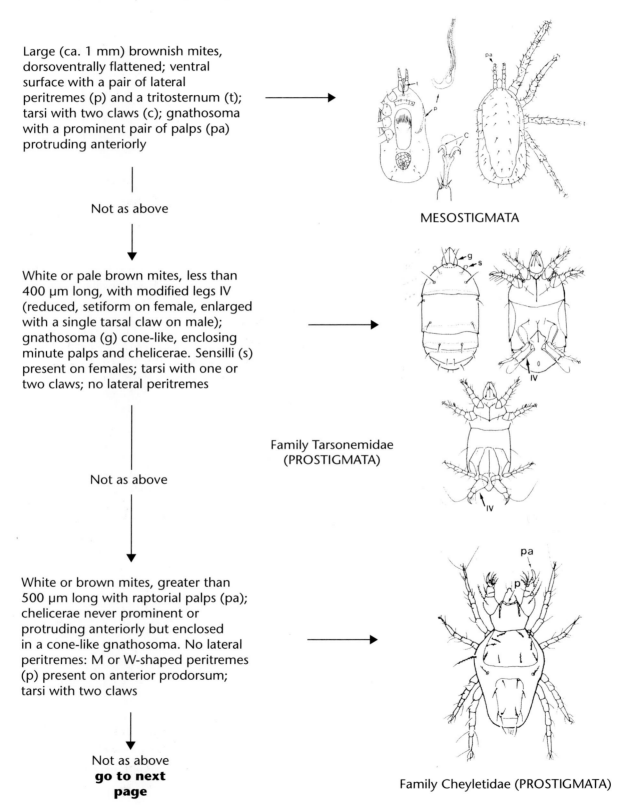

Large (ca. 1 mm) brownish mites, dorsoventrally flattened; ventral surface with a pair of lateral peritremes (p) and a tritosternum (t); tarsi with two claws (c); gnathosoma with a prominent pair of palps (pa) protruding anteriorly

Not as above

MESOSTIGMATA

White or pale brown mites, less than 400 μm long, with modified legs IV (reduced, setiform on female, enlarged with a single tarsal claw on male); gnathosoma (g) cone-like, enclosing minute palps and chelicerae. Sensilli (s) present on females; tarsi with one or two claws; no lateral peritremes

Family Tarsonemidae (PROSTIGMATA)

Not as above

White or brown mites, greater than 500 μm long with raptorial palps (pa); chelicerae never prominent or protruding anteriorly but enclosed in a cone-like gnathosoma. No lateral peritremes: M or W-shaped peritremes (p) present on anterior prodorsum; tarsi with two claws

Not as above
go to next page

Family Cheyletidae (PROSTIGMATA)

White, worm-shaped mites with
short, stumpy legs and an
elongated, annulate opisthosoma (o)

Not as above

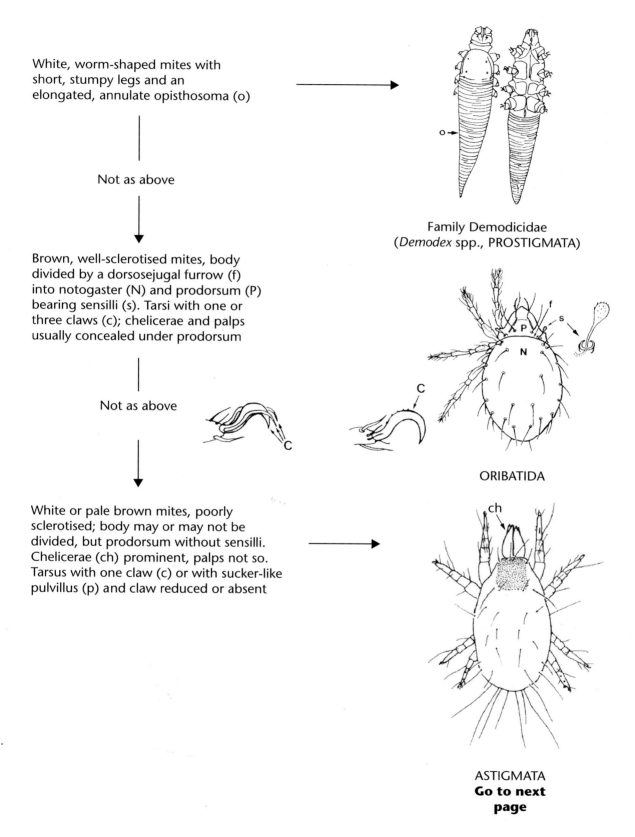

Family Demodicidae
(*Demodex* spp., PROSTIGMATA)

Brown, well-sclerotised mites, body
divided by a dorsosejugal furrow (f)
into notogaster (N) and prodorsum (P)
bearing sensilli (s). Tarsi with one or
three claws (c); chelicerae and palps
usually concealed under prodorsum

Not as above

ORIBATIDA

White or pale brown mites, poorly
sclerotised; body may or may not be
divided, but prodorsum without sensilli.
Chelicerae (ch) prominent, palps not so.
Tarsus with one claw (c) or with sucker-like
pulvillus (p) and claw reduced or absent

ASTIGMATA
**Go to next
page**

With 1–2 pairs of vertical setae (vi); one solenidion (s) on tarsus I at distal end of segment, the other at the proximal end; cuticle reticulate or with minute papillae, or smooth

Vertical setae absent (except in the Paralgopsinae); both solenidia on tarsus I at distal end of segment (arrowed); cuticle with fine striations

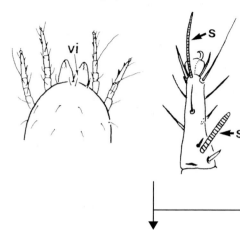

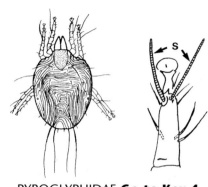

PYROGLYPHIDAE **Go to Key 4**

Prodorsal shield (s) and dorsal transverse groove present (f) (hard to see in some specimens); cuticle smooth; claw attached to tarsus by pair of sclerites (sl)

Prodorsal shield absent; dorsal transverse groove absent; cuticle papillate or reticulate (though smooth in *Chortoglyphus*); claw attached to tarsus by thin tendon (t)

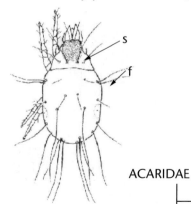

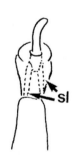

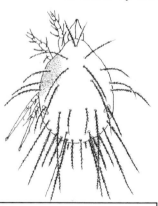

GLYCYPHAGOIDEA
Go to Key 12

ACARIDAE

External vertical setae (ve) considerably shorter than internal vertical setae (vi)

External vertical setae (ve) subequal in length to internal vertical setae (vi)

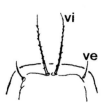

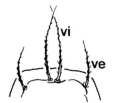

Acarus

Tyrophagus

Key 4 Adults of the family Pyroglyphidae found in house dust

Cuticle with coarse, irregular striations and/or punctations. Setae *sce* and *sci* short. Tegmen present (t)

Cuticle with fine lines and ridges or smooth. Tegmen absent. Setae *sci* not long and well developed; *sci* and *sce* either subequal in length or *sce* much longer.

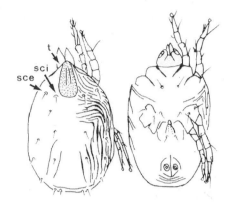

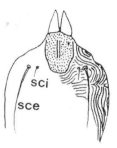

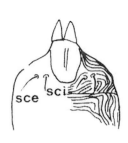

PYROGLYPHINAE

DERMATOPHAGOIDINAE
Females: **go to key 5**
Males: **go to key 9**

Tegmen (t) entire. Prodorsal shield (p) longer than broad. Vulva (v) not incised; lateral margin curved (arrow); Setae *g* and *4a* present, *3a* absent

Tegmen (t) notched. Prodorsal shield as broad as long. Vulva (v) incised; lateral margin straight (arrowed); ventral setae *3a* and *4a* present, *g* absent

Tegmen sharp, bifid. apex. Vulva not incised apically; lateral margin curving posteriomedially; ventral setae *3a*, *g* and *4a* present

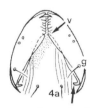

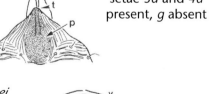

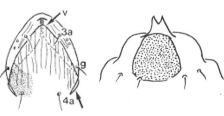

Euroglyphus maynei

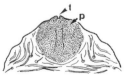

Gymnoglyphus longior

Hughesiella africana

Key 5   Females of the subfamily Dermatophagoidinae (after Fain, 1990, with modifications)

Cuticle appearing almost smooth, but with very fine, faint striations, ca. 1 μm or less apart in anteriodorsal region (arrow)

Cuticle with well-defined striations, ca. 1.5–2.5 μm apart in anteriodorsal region (arrow)

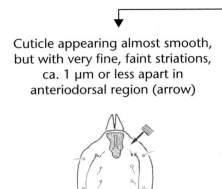

Hirstia

Dorsal striations not beaded; median shield absent; vulva lacking median incision

Dorsal striations of beaded appearance. Median shield (s) present; vulva (v) incised medially

Ca. 400–430 μm long; posterior opisthosoma not punctate or sclerotised

Ca. 300–310 μm long; posterior opisthosoma punctate and sclerotised

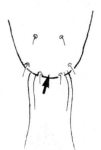

Hirstia chelidonis

Hirstia domicola

Go to next page

Sturnophagoides 1 species in house dust: *S. brasiliensis*

Setae *sci* and *sce* short, subequal in length; epigynium (e) short, transverse, poorly sclerotised

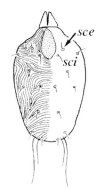

Malayoglyphus

Setae *sci* very short, *sce* long and well developed; epigynium ∩-shaped and well sclerotised

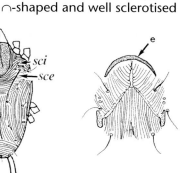

Dermatophagoides

Ca. 220–240 µm long; setae *sci* and *sce* subequal in length

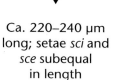

Malayoglyphus intermedius

Ca. 320–350 µm long; setae *sci* about half the length of *sce*

Malayoglyphus carmelitus

Cuticle anterior of epigynium (e) with transverse striae strongly arched anteriorly or Λ shaped. Dorsal cuticle with transverse striae between setae d_1 and e_1 (arrowed); striae Λ–shaped posterior of e_1

Cuticle anterior of epigynium (e) striated longitudinally. Dorsal cuticle with longitudinal striations between setae d_1 and e_1 (arrowed)

Cuticle anterior of epigynium (e) striated transversely; striae weakly arched anteriorly. Dorsal cuticle with transverse striations between setae d_1 and e_1 (arrowed); striae ∩-shaped posterior of e_1

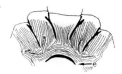

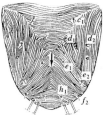

Aureliani species group
(*D. aureliani, D. neotropicalis, D. rwandae, D. sclerovestibulans, D. simplex*)
Go to key 8

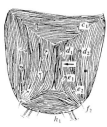

Pteronyssinus species group
(*D. pteronyssinus, D. evansi*)
Go to key 7

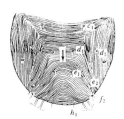

Farinae species group (*D. farinae, D. siboney, D. microceras*)
Go to key 6

Key 6 Females of the Farinae species group

Ca. 250–310 μm long. Prodorsal shield (p) ca. 2 times as long as wide. Tarsi I and II with large apical spine (s). Bursa copulatrix (large arrow) broad; strongly sclerotised next to external opening (small arrow); sclerotised section rounded anteriorly. Ratio of distance between posterior end of genital apodeme and base of seta *4a* (a), to distance between bases of setae *4a* (g) 1:1.5 to 1:2

Ca. 400–420 μm long. Prodorsal shield (p) ca. 1.5 times as long as wide. Tarsus I with short, blunt spine; tarsus II lacking spine. Bursa copulatrix (large arrow) weakly sclerotised in region next to external opening (small arrow). Ratio of distance between posterior end of genital apodeme and base of seta *4a* (a), to distance between bases of setae *4a* (g) 1:3 to 1:4

Ca. 390–440 μm long. Prodorsal shield (p) ca. 1.5 times as long as wide. Tarsi I and II with large apical spine (s). Bursa copulatrix (large arrow) broad and strongly sclerotised in region next to external opening (small arrow); sclerotised section pointed anteriorly. Ratio of distance between posterior end of genital apodeme and base of seta *4a* (a), to distance between bases of setae *4a* (g) 1:2 to 1:2.5

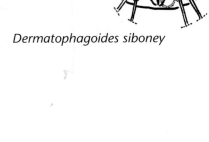

Dermatophagoides siboney

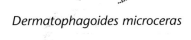

Dermatophagoides microceras

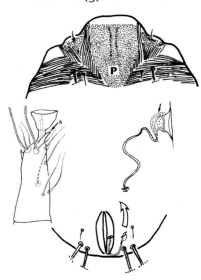

Dermatophagoides farinae

Key 7   Females of the Pteronyssinus species group

Sclerotised base of receptaculum seminis ∪-shaped in cross section (r), broader apically than basally; circular with ca. 10–13 lobes (l) when viewed from above; ductus bursae (d) of uniform thickness

Dermatophagoides pteronyssinus

Sclerotised base of receptaculum seminis ∪-shaped (r), broader at base than at apex; without lobes; ductus bursae (d) about twice as thick in posterior half as in anterior half

Dermatophagoides evansi

Key 8 Females of the Aureliani species group

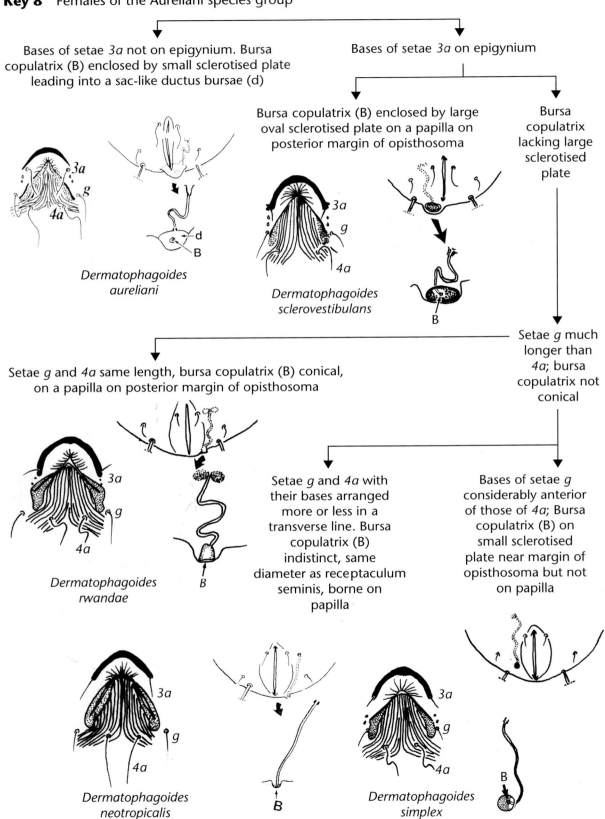

Bases of setae *3a* not on epigynium. Bursa copulatrix (B) enclosed by small sclerotised plate leading into a sac-like ductus bursae (d)

Dermatophagoides aureliani

Bases of setae *3a* on epigynium

Bursa copulatrix (B) enclosed by large oval sclerotised plate on a papilla on posterior margin of opisthosoma

Dermatophagoides sclerovestibulans

Bursa copulatrix lacking large sclerotised plate

Setae *g* much longer than *4a*; bursa copulatrix not conical

Setae *g* and *4a* same length, bursa copulatrix (B) conical, on a papilla on posterior margin of opisthosoma

Dermatophagoides rwandae

Setae *g* and *4a* with their bases arranged more or less in a transverse line. Bursa copulatrix (B) indistinct, same diameter as receptaculum seminis, borne on papilla

Bases of setae *g* considerably anterior of those of *4a*; Bursa copulatrix (B) on small sclerotised plate near margin of opisthosoma but not on papilla

Dermatophagoides neotropicalis

Dermatophagoides simplex

Key 9   Males of the subfamily Dermatophagoidinae

Notes: i) The male of *Dermatophagoides rwandae* has not been described. ii) The following species are known to have heteromorphic males (see section 1.5.7b): *Dermatophagoides anisopoda, D. farinae, D. neotropicalis, D. sclerovestibulatus, D. simplex, Sturnophagoides bakeri, S. brasiliensis* and *S. petrochelidonis.*

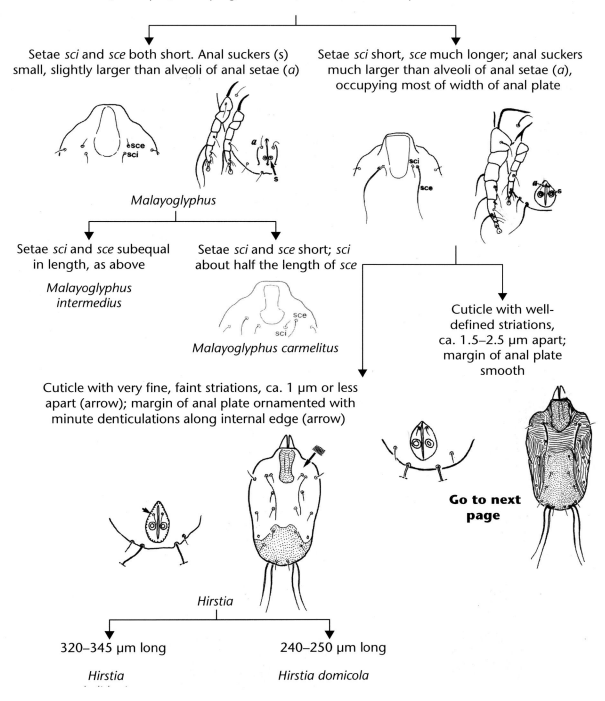

Setae *sci* and *sce* both short. Anal suckers (*s*) small, slightly larger than alveoli of anal setae (*a*)

Malayoglyphus

Setae *sci* short, *sce* much longer; anal suckers much larger than alveoli of anal setae (*a*), occupying most of width of anal plate

Setae *sci* and *sce* subequal in length, as above

Malayoglyphus intermedius

Setae *sci* and *sce* short; *sci* about half the length of *sce*

Malayoglyphus carmelitus

Cuticle with well-defined striations, ca. 1.5–2.5 μm apart; margin of anal plate smooth

Cuticle with very fine, faint striations, ca. 1 μm or less apart (arrow); margin of anal plate ornamented with minute denticulations along internal edge (arrow)

Go to next page

Hirstia

320–345 μm long

Hirstia

240–250 μm long

Hirstia domicola

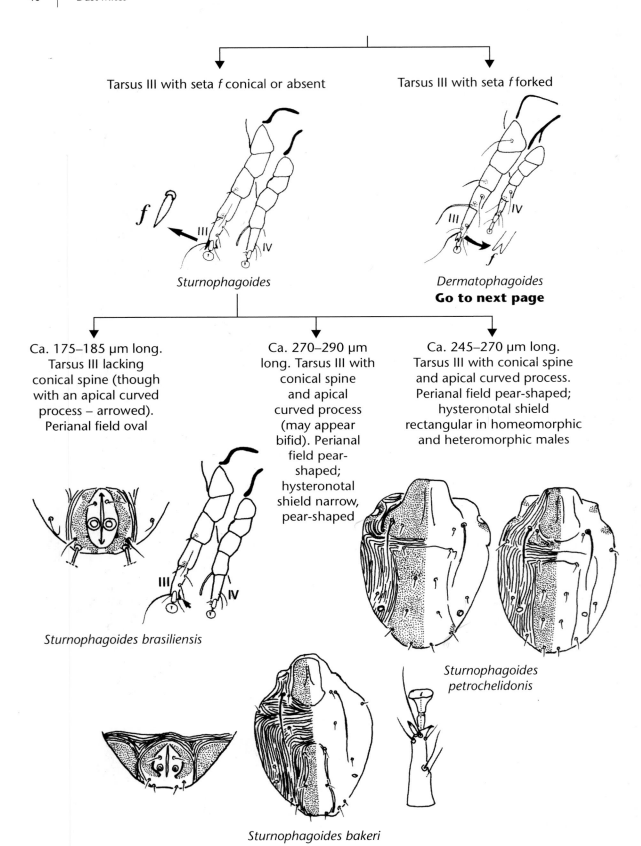

Tarsus III with seta *f* conical or absent

Tarsus III with seta *f* forked

f

III

IV

Sturnophagoides

III

IV

f

Dermatophagoides
Go to next page

Ca. 175–185 μm long. Tarsus III lacking conical spine (though with an apical curved process – arrowed). Perianal field oval

Ca. 270–290 μm long. Tarsus III with conical spine and apical curved process (may appear bifid). Perianal field pear-shaped; hysteronotal shield narrow, pear-shaped

Ca. 245–270 μm long. Tarsus III with conical spine and apical curved process. Perianal field pear-shaped; hysteronotal shield rectangular in homeomorphic and heteromorphic males

Sturnophagoides brasiliensis

III

IV

Sturnophagoides petrochelidonis

Sturnophagoides bakeri

Species groups of *Dermatophagoides* (Males)

Hysteronotal shield (H) longer than broad, extending to point equidistant between setae c_1 and d_1; setae *4a* vestigial, only their alveoli present; apodemes II and III fused to form oblique Y-shaped structure (arrow); apodemes I not fused. Setae ps_2 longer than anal plate (a), positioned far behind anal suckers (s)

Hysteronotal shield (H) as long as broad, extending to point between setae d_1 and e_1 or slightly anterior of d_1; setae *4a* short, not vestigial; apodemes II and III not fused; apodemes I not fused in homeomorphic males; heteromorphic males with apodemes I fused to form Y-shaped median structure (arrow); setae ps_2 shorter than anal plate (a), positioned lateral to anal plate, more or less at level of anal suckers (s)

Hysteronotal shield (H) longer than broad, extending to point between d_1 and e_1; or slightly anterior of d_1; setae *4a* short (vestigial in *D. aureliani*); heteromorphic males (and homeomorphic males of some spp.) with apodemes I fused to form X-shaped median structure (arrow). Setae ps_2 longer than anal plate, positioned considerably posterior of anal suckers

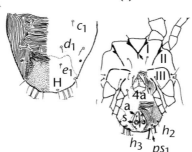

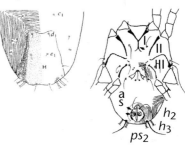

Pteronyssinus species group
(*D. pteronyssinus, D. evansi*)

Farinae species group
(*D. farinae, D. microceras, D. siboney*)

Go to key 10

Aureliani species group
(*D. aureliani,
D. neotropicalis,
D. rwandae,
D. sclerovestibulans,
D. simplex*)

Go to key 11

Legs III 1.5 times longer than legs IV and femur (F) 1.3 times broader. Narrow sclerotised perianal region (P) encircling anal plate (a); striations surrounding perianal region extending posteriorly as far as alveoli of setae ps_2

Legs III 1.6 times longer than legs IV and femur (F) 1.8 times broader. Broad sclerotised perianal region (P) covering most of posterior opisthosoma; striations surrounding perianal region extending posteriorly only as far as anal suckers (s)

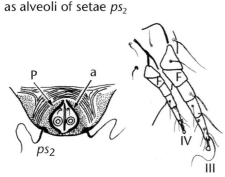

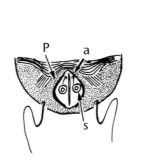

Dermatophagoides pteronyssinus

Dermatophagoides evansi

Key 10 Males of the Farinae species group

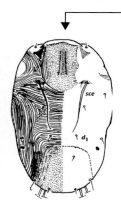

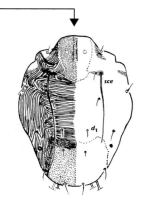

Ca. 200–250 μm long. Setae *sce* shorter extending posteriorly not as far as setae d_1. Males homeomorphic

Ca. 280–350 μm long. Setae *sce* extending posteriorly well beyond setae d_1. Males homeomorphic or heteromorphic

Dermatophagoides siboney

Tarsus I with small apical protuberance (process *S*) and curved apical spine (f). Tarsus II with process *S* and no spine

Tarsus without small apical protruberance (process *S*) but with curved spine (f). Tarsus II without process *S* or spine

Dermatophagoides farinae

Dermatophagoides microceras

Key 11 Males of the Aureliani species group

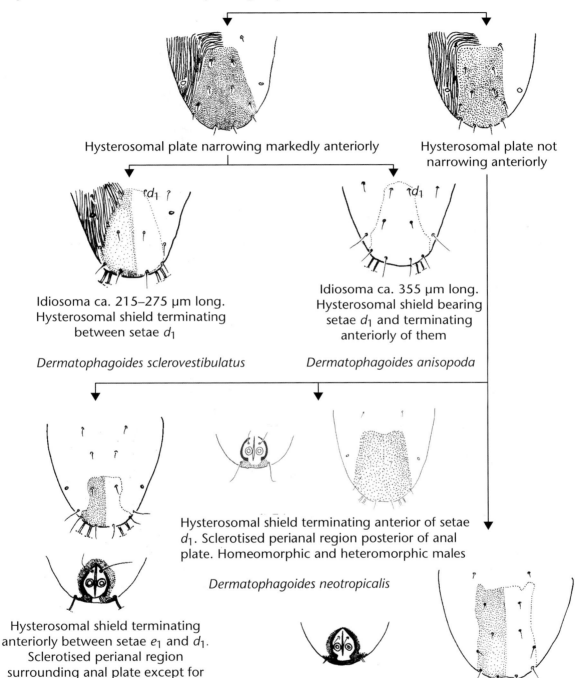

Hysterosomal plate narrowing markedly anteriorly

Hysterosomal plate not narrowing anteriorly

Idiosoma ca. 215–275 μm long. Hysterosomal shield terminating between setae d_1

Dermatophagoides sclerovestibulatus

Idiosoma ca. 355 μm long. Hysterosomal shield bearing setae d_1 and terminating anteriorly of them

Dermatophagoides anisopoda

Hysterosomal shield terminating anterior of setae d_1. Sclerotised perianal region posterior of anal plate. Homeomorphic and heteromorphic males

Dermatophagoides neotropicalis

Hysterosomal shield terminating anteriorly between setae e_1 and d_1. Sclerotised perianal region surrounding anal plate except for narrow posterior region. Homeomorphic males

Dermatophagoides aureliani

Hysterosomal shield terminating anteriorly midway between setae d_1 and c_1. Sclerotised perianal region completely surrounding anal plate. Heteromorphic males

Dermatophagoides simplex

Key 12 Key to the Glycyphagoidea most frequently found in house dust

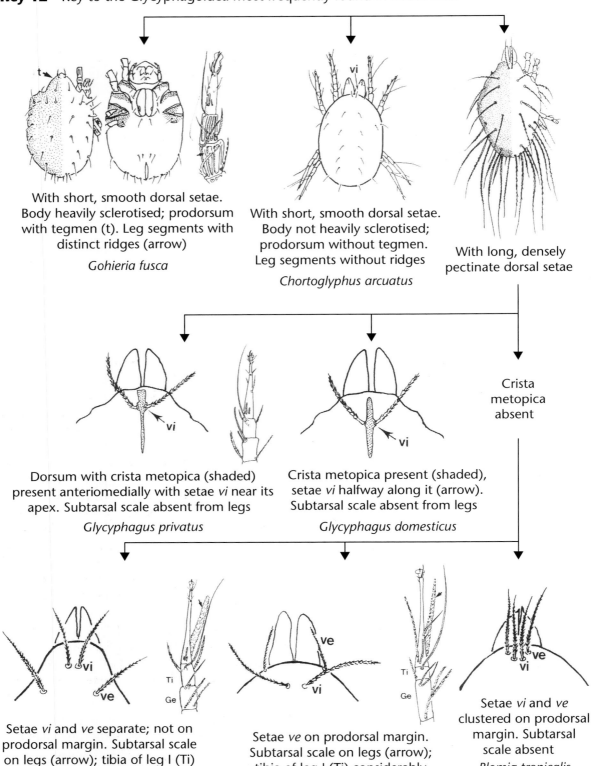

With short, smooth dorsal setae. Body heavily sclerotised; prodorsum with tegmen (t). Leg segments with distinct ridges (arrow)

Gohieria fusca

With short, smooth dorsal setae. Body not heavily sclerotised; prodorsum without tegmen. Leg segments without ridges

Chortoglyphus arcuatus

With long, densely pectinate dorsal setae

Crista metopica absent

Dorsum with crista metopica (shaded) present anteriomedially with setae *vi* near its apex. Subtarsal scale absent from legs

Glycyphagus privatus

Crista metopica present (shaded), setae *vi* halfway along it (arrow). Subtarsal scale absent from legs

Glycyphagus domesticus

Setae *vi* and *ve* separate; not on prodorsal margin. Subtarsal scale on legs (arrow); tibia of leg I (Ti) ca. same length as genu (Ge)

Lepidoglyphus destructor

Setae *ve* on prodorsal margin. Subtarsal scale on legs (arrow); tibia of leg I (Ti) considerably shorter than genu (Ge)

Austroglycyphagus

Setae *vi* and *ve* clustered on prodorsal margin. Subtarsal scale absent

Blomia tropicalis

2. Physiology and internal anatomy

What is characteristic of life is not the individual details of structure or behaviour, but the way in which they cohere to form a self-regulating and self-preserving whole.

J.B.S. Haldane, 1927

2.1 Physiology – the functional biology of organisms

Physiology explores the functions of organisms, their organ systems, and the way they work. It concerns how they contend with their environment, but is far more than just an account of the physics, chemistry and mechanics of the functional processes of life. That the external environment has a crucial influence on the lives and workings of organisms is a very old idea in the history of human culture (Mayr, 1982). The concepts of the environment as inherently hostile, and that the degree of adaptation of a species to its environment is a measure of its success, are integral to the central tenet of Darwinian evolution.

Schmidt-Nielsen (1975) defined physiology as dealing with an exploration of how animals live under circumstances where the environment seems to place insurmountable problems in their way. But he warned that the study of physiological specialisation, arrived at as solutions to particular problems imposed by environmental constraints, can lead to the descriptive study of peculiarities among a few rare and unusual animals if these specialisations are not related to general principles of how all animals function. One way of avoiding such a bias is through using the comparative approach. An inherent strength of comparative physiology is that it shows very clearly that there is more than one way to solve a problem imposed by the environment. It also shows that not all solutions are as elegant nor are they as efficient as has been suggested by biological teaching in the past. The idea that organisms are perfectly adapted to their environments, that the mechanisms that bring about survival and reproduction are paragons of adaptive virtue – the faultless culmination of millions of years of evolution – is a very common one, frequently portrayed by the media in contemporary popular natural history, and a view held by a surprising number of biologists. Furthermore, the arrival at a particular solution, for example the use of gills (as opposed to lungs) as organs for obtaining oxygen from the environment by diffusion, imposes additional challenges and constraints on an organism, with implications for other organ systems.

In 1939, in the preface to the first edition of *The Principles of Insect Physiology*, Wigglesworth pointed out that anatomical and ecological specialities in insects have their physiological counterparts, and this link means the disciplines of anatomy, physiology and ecology are inseparable. He noted that, from an applied perspective, a knowledge of the ecology of insects and, indeed, all pest organisms, is required for their effective control (Wigglesworth, 1972). Their ecology can only be properly understood when their physiology is known. Wigglesworth used the appropriate integumentary metaphor of finding weak spots in the physiological armour of insects. With mites in houses this concept applies most clearly in relation to water balance (see Chapter 3). An understanding of how mites prevent themselves from dying of desiccation has stimulated the search for methods of control.

The physiology of dust mites provides pointers to possible control measures, as well as clues to why these mites are widespread and abundant, and why they produce allergens to which people become exposed and sensitised. For example, in dust mites, digestion takes place not in a specialised stomach but throughout a large proportion of the length of the gut, within the lumen, caecae and free-floating digestive cells (see section 2.2.2). This process appears to be relatively energy-inefficient compared with other digestive systems (Penry and Jumars, 1987). Having done their job, the digestive cells die and their remains are egested, together with large amounts of enzymes they contain, all wrapped in a peritrophic membrane that has to be synthesised *de novo*. There is little evidence that mites reclaim faecal enzymes, and the process appears messy and wasteful. But, were it not for this combination – intracellular digestion, a small faecal pellet (composed of a series of smaller pellets) capable of becoming airborne, together with inefficient resorption of digestive enzymes – mite faecal pellets may well be less allergenically important than they are.

In this chapter, the anatomy of dust mites is explained insofar as it is necessary to understand the associated physiological principles and mechanisms.

2.2 Nutrition, feeding and the digestive system

Why is an organism's mode of feeding important? To quote Slansky and Scriber (1985), the consumption and utilisation of food constitute a *sine qua non* for growth, development and reproduction. The evolution of different lifestyles is moulded by constraints placed on the consumption and utilisation of food. One of the most common forms of interaction between organisms is that they eat each other. The most basic effect is that the size of the population of the consumer increases at the expense of that of the consumed. In a community of organisms there is a complex network of feeding interactions called a food web (see 4.3.5). The food web represents a model of who eats whom and how energy and nutrients are transferred within the community. So, the type of feeding has a major impact on the population dynamics and energy flow of an ecological community.

Food quality and type, presence of metabolic inhibitors and rate of consumption have direct effects on mite growth, rates of development, survival and thus population growth. Food quality and utilisation have knock-on effects for various organ systems. For example, the water and nitrogen content of the food are directly relevant to the performance required of the osmoregulatory and excretory system (see section 2.3).

2.2.1 Mouthparts and feeding

a Anatomy of the mouthparts

The mouthparts of all mites consist of morphological variations of a number of basic structures and appendages. The mouthparts are borne on a segment of the body, unique to the mites, called the gnathosoma. It is not equivalent to the insect head: mites have no 'head' in the generally accepted sense of the word. Rather, the gnathosoma represents the fusion of a number of body segments situated in front of the first pair of legs. It may or may not articulate freely with the posterior body segment, the opisthosoma, to which it is joined.

On the gnathosoma (see Figure 2.1), positioned dorsally, is a pair of chelicerae (singular: chela). Chelicerae are functionally equivalent to the mandibles of insects, though they are not structurally or developmentally analogous. They are the organs that carry out the grasping, cutting or piercing of the food, depending on the exact mode of feeding of the mites in question.

Among dust-dwelling mites there are varying degrees of feeding (or trophic) specialisation. The tarsonemid and cheyletid mites have specialised stylelike chelicerae designed for piercing (see Figure 2.1a). The tarsonemids pierce the hyphae of fungi in order to extract liquid contents, while the cheyletids ambush and suck the fluid contents out of other mites. Liquid mycophagy and predation are highly specialised methods of gaining nutrition and it follows that the laboratory culture of tarsonemids and cheyletids also requires the careful maintenance of the living organisms that form the basis of their diet. Glycyphagoid mites are also fungal feeders, but they feed on solid chunks of fungal mycelium and on spores, a less-specialised habit than feeding on liquid contents of fungal hyphae, and one considered ancestral in the Astigmata (OConnor, 1979). Furthermore, their mouthparts are not strongly modified or adapted, but

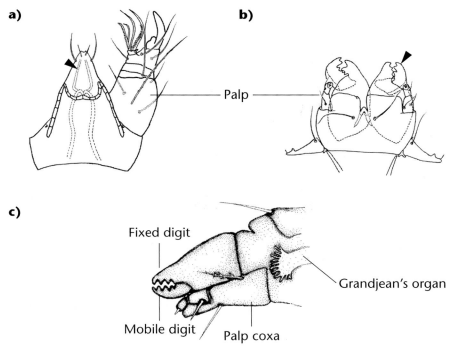

Figure 2.1   a) Dorsal view of chelicerae of *Cheyletus eruditus* showing piercing chelicerae (arrow); **b)** dorsal view of *Dermatophagoides pteronyssinus* showing chelate-dentate chelicerae (arrow); **c)** lateral view of *Acarus siro* showing chelate-dentate chelicerae.

are like shears – so-called 'chelate-dentate' chelicerae – as are those of most of the Oribatida and the Acaroidea. The pyroglyphid mites also have chelate-dentate chelicerae (see Figure 2.1b) and appear to feed on a much wider range of solid food than other domestic mite taxa (refer to section 2.2.4).

In many mites, including the Pyroglyphidae, the chela consists of two components: an elongated cylindrical or barrel-shaped element or shaft (the 'basal segment' of Brody *et al.* (1972)) on the anterior end of which is a dorsally positioned immovable arm or blade called the *digitus fixtus*, or fixed digit (see Figure 2.1c). Lying ventral of the fixed digit is a similarly shaped one that hinges at the junction between the fixed digit and the cylindrical part of the chela. This structure is called the *digitus mobilis* or mobile digit (Figure 2.1c shows this structure). The two digits are usually roughly serrated or spurred along their interior edges, and these serrations interlock, like those on a pair of dressmaker's crimping shears. Their function is to grasp and cut the food into portions of manageable size for ingestion. They do this slightly differently from a pair of scissors, with which they are often compared, because only the lower blade is mobile, just like the claw of a crab. The

mobile digit moves in the vertical plane only, operated by levator and depressor muscles. It is not capable of lateral or circular movement like the chewing action of human jaws.

Each chela articulates with the idiosoma via the cheliceral sheath, a piece of arthrodial membrane attached to the cheliceral shaft, which allows for greater mobility of the chelicerae because they are not joined together at the base as in some Prostigmata. Each chela is capable of being retracted into the body and extended to grasp food. The chelicerae can move independently of each other (Brody *et al.*, 1972) though they may also move alternately with one extended forward while the other is retracted. Movement is achieved by the powerful retractor muscles, attached at their anterior end to the ventral surface of the cheliceral shaft and at their posterior end to the inside of the dorsal idiosomal integument. The extension of the chelicerae is thought to be a passive mechanism that occurs automatically as a result of hydrostatic pressure manifested through body turgor. When dust mites lose body water in dry conditions, they cease to feed. This has been regarded as a mechanism for reducing further body water loss by lowering the amount of activity. However, it may

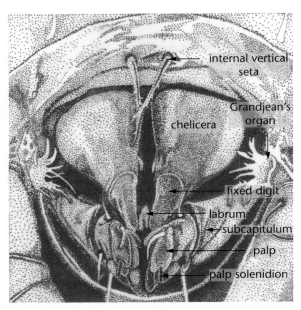

internal vertical seta

Grandjean's organ

chelicera

fixed digit

labrum

subcapitulum

palp

palp solenidion

Figure 2.2 Front view of gnathosoma and mouthparts of *Acarus siro*. Re-drawn from a scanning electron micrograph by D.A. Griffiths in Evans (1992; Figure 5.1B).

be that they are actually incapable of feeding in this state because their body turgor pressure is too low to allow them to extend their chelicerae.

Lateral of the chelicerae are the segmented palps (shown in Figures 2.1c, 2.2). They probably assist in directing food towards the mouth (T.E. Hughes, 1950) but their major function is sensory. They have a battery of chemoreceptors and mechanoreceptors on them and are responsible for tasting the food. Most of the chemoreceptors are concentrated on the apical segment, the palp-tarsus (see section 2.7.1; Figure 2.2). The ventral surfaces of the chelicerae rest in gutter-like grooves called cheliceral furrows that form the internal surface of the infracapitulum or hypostome. The prebuccal cavity is delineated dorsally by the ventral surfaces of the cheliceral shafts. Situated medially between the cheliceral grooves is the labrum, a long thin structure extending posteriorly from the wall of the pharynx, and that covers the dorso-median part of the prebuccal cavity.

b The ingestion of food

The food material is released from the fully retracted chela into the prebuccal cavity. How the food gets into the mouth is unclear, though it is logical that it is guided there by the constraining morphology of the cheliceral grooves, and is squeezed in by the pressure of subsequent mouthfuls, or rather 'chela-fuls', of food dragged onto the cheliceral grooves behind it.

When the prebuccal cavity is full of food it is sealed anteriorly and dorsally by the labrum, dorsally by the chelicerae and laterally by cuticular ridges on the lateral edges of the chelicerae where they meet the lateral edge of the hypostome (see Figure 2.2). The prebuccal cavity needs to be sealed in order that the pharyngeal pump can suck the food in. It is highly likely that the food needs to be at least semi-fluid prior to ingestion, in order to provide the lubrication required for it to pass down the narrow pharynx and oesophagus. Thus, saliva-associated carbohydrases, such as amylase, become mixed with the food in the prebuccal cavity, before the food is ingested (Brody *et al.*, 1972). The saliva plays the dual roles of predigestion and lubrication. The salivary glands, which appear to open into the prebuccal space, consist of structures shaped like bunches of grapes positioned anteriodorsally and anterioventrally within the idiosoma (Brody *et al.*, 1972; Figure 2.3).

2.2.2 Digestion, egestion and the functional anatomy of the gut

The gut together with the cuticle are major boundaries between the mite and its environment. Hence, the gut represents the route of entry not only for food, but for toxins and pathogens also. How the gut functions should be important to an understanding of methods of dust mite control that operate via the digestive

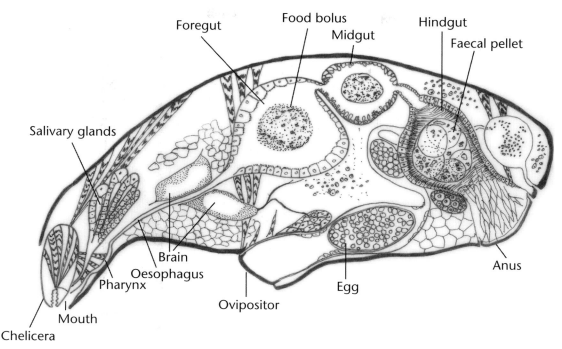

Figure 2.3 Saggital section through an astigmatid mite showing the major components of the digestive system, based mostly on figures by Brody *et al.* (1972, Figure 1), Michael (1901, Plate C, Figure 1), and Woodring and Cook (1962, Figure 1).

system, such as the toxins benzyl benzoate and disodium octaborate. In fact, almost nothing is known about the effect of these acaricides on the dust mite gut (refer to Chapter 9).

The functional anatomy, morphology and histology of the digestive system of *Dermatophagoides farinae* have been studied in detail by Brody *et al.* (1972), and that of *Acarus siro* by T.E. Hughes (1950) and Šobotník *et al.* (2008). The following account is largely based on their work. Other accounts I have found of particular use are those of Michael (1901); Woodring and Cook (1962); Rhode and Oemick (1967); Kuo and Nesbitt (1970); Baker (1975) and Tongu *et al.* (1986). Šobotník *et al.* (2008) provide details of other references on acarine gut anatomy. There is some difference in terminology of the various regions of the digestive system (see Table 2.1) but the basic anatomy is more or less the same throughout the acariform mites so far examined (as seen in Figure 2.3). The key characteristics of the major regions of the gut are summarised in Table 2.2.

a The pharynx and oesophagus

Food is pumped into the cuticle-lined pharynx from the prebuccal cavity. Not all the material entering the pharynx is semi-liquid. The pharynx can be distended from ca. 2–3 μm up to ca. 20 μm in order to

accommodate solid structures, such as fungal spores and pieces of skin and mite cuticle. In transverse section the pharynx is flattened, with the lateral edges curved dorsally (see Figure 2.4a). It provides the mechanical force to allow the food to pass into the midgut. A series of dilator muscles are attached to the dorsal surface of the pharynx and it can be opened by their action to allow food in (T.E. Hughes, 1950). Food is rendered into a sludge-like consistency by alternating contraction of the dilator muscles and the action of salivary enzymes and saliva, as well as the regurgitation of midgut secretions (Šobotník *et al.*, 2008). The sludge is then squeezed along the pharynx. At the junction between the pharynx and the cuticle-lined oesophagus, the gut is star-shaped in outline, indicating the walls are highly elastic (Figures 2.4b, c). Bands of constrictor and dilator muscles are present, which suggests that food in the oesophagus is moved by peristalsis. Wharton and Brody (1972) observed the most frequent contractions of the gut of live specimens of *Dermatophagoides farinae* occurred in the pharyngeal region with 20 contractions per min. (range 7–27), compared with the anterior midgut with 8 per min. (range 3–18) and 3 per min. for the posterior midgut and 2 per min. for the hindgut. Kuo and Nesbitt (1970) observed an oesophageal valve in the acarid mite *Caloglyphus mycophagus*, as did Šobotník *et al.*

Table 2.1 An attempt to homologise differences in the nomenclature of the regions of the gut of acariform mites, according to various authors. Modified after Brody *et al.* (1972).

Foregut		Midgut		Hindgut			Reference
Pharynx	Oesophagus	Ventriculus	Colon	Rectum		Anus	Michael (1901)
Pharynx	Oesophagus	Midgut	Small intestine	Colon	Rectum	Anus	Vitzthum (1940)
Pharynx	Oesophagus	Stomach	Colon	Post-colon	Rectum	Anus	T.E. Hughes (1950)
Pharynx	Oesophagus	Stomach	Colon	Rectum		Anus	Kuo and Nesbitt (1970)
Pharynx	Oesophagus	Anterior midgut	Posterior midgut	Anterior hindgut	Posterior hindgut	Anus	Brody *et al.* (1972)
Pharynx	Oesophagus	Ventriculus	Colon	Post-colon	Rectum	Anus	Akimov (1973)
Pharynx	Oesophagus	Stomach	Colon	Post-colon	Rectum	Anus	R.A. Baker (1975)
Pharynx	Oesophagus	Ventriculus	Colon	Rectum		Anus	G.T. Baker & Krantz, (1985)

(2008) in *Acarus siro*. The valve prevents re-entry of food into the pharynx. The oesophageal-pharyngeal region is surrounded by the brain, and separated from it by a basement membrane and the external invaginations of the oesophagus, which form part of the haemocoel and are filled with haemolymph.

b The midgut and midgut caeca

Food from the pharynx moves into the oesophagus and thence into the anterior midgut. The majority of digestion and absorption takes place in the midgut and its lateral, paired, blind-ended caeca (singular, caecum). Intracellular digestion takes place in the anterior midgut (see Figures 2.5a–c), whereas extracellular digestion appears to take place in the midgut caeca (Brody *et al.*, 1972; Baker, 1975; Figure 2.4d). In arthropods that feed on keratin-rich diets, the midgut caeca probably function as maceration and digestion chambers to reduce the disulphide bonds in the keratin prior to further digestion. The caeca become so full in engorged specimens of *Dermatophagoides* spp. that they become almost indistinguishable from

Table 2.2 The regions of the gut of acariform mites: a summary of their characteristics and functions. ND = no data available.

Region	Histology and cell characteristics	Function	pH	Enzyme production
Pharynx	Cuticle-lined	Pumping food to midgut	ND	No
Oesophagus	Cuticle-lined and/or with thin squamous epithelium	Elastic link between pharynx and midgut – passage of large food particles	ND	No
Anterior midgut	Cells phagocytic: highly vacuolated; with microvilli	Intracellular ingestion	5.0–6.0*	Yes
Midgut caecae	Cells without vacuoles or microvilli but packed with electron-dense structures: budding off of anucleate spheres	Extracellular digestion	5.0–6.0*	Yes
Posterior midgut	Cells with very long microvilli	Peritrophic encapsulation; extracellular digestion and resorbtion	7.0–7.5* 6.8–7.4**	Yes
Anterior hindgut	Cuticle-lined	Water resorbtion, compaction of faecal pellet	7.0–8.0* 6.2–8.5**	No
Posterior hindgut	Cuticle-lined	Passage of faecal pellet to anus	ND	No
Anus	Cuticle-lined	Defaecation	ND	No

*Based on T.E. Hughes (1950) for *Acarus siro* and **van Bronswijk and Berrens (1979) for *Dermatophagoides* spp.

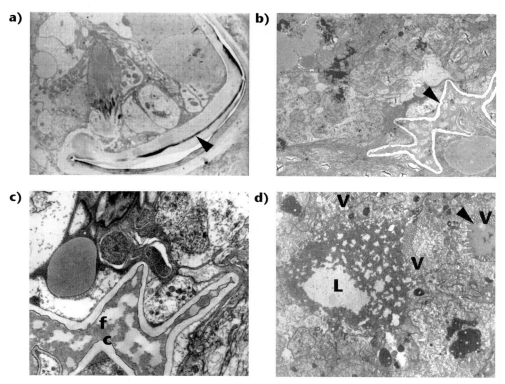

Figure 2.4 Transmission electron micrographs of the digestive system of *Dermatophagoides pteronyssinus*. **a)** Cuticle-lined pharynx (arrow); **b)** oesophagus, star-shaped in cross-section (arrow); **c)** detail of cuticle-lined (c) oesophagus showing contents of macerated food (f); **d)** midgut showing food-filled lumen (L) and opening of midgut caecae (arrow). V = villi. Original micrographs by the author.

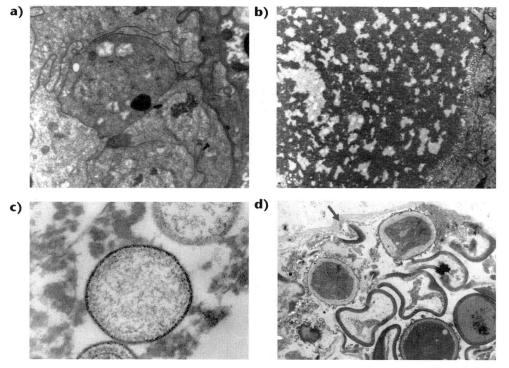

Figure 2.5 Transmission electron micrographs of the digestive system of *Dermatophagoides pteronyssinus*. **a)** Midgut cells and gut wall; **b)** midgut lumen; **c)** guanine spherules showing concretions; **d)** faecal pellet filled with whole and partially digested yeast cells (y). Arrow = mucus coat. Original micrographs by the author.

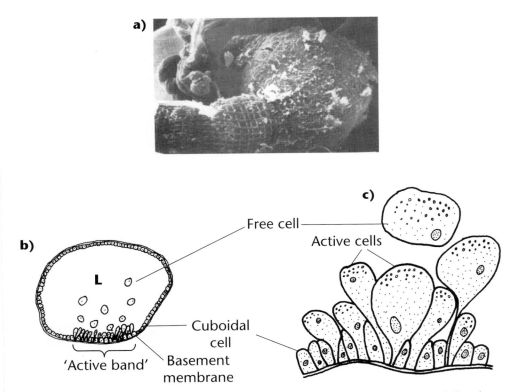

Figure 2.6 Midgut musculature and digestive cells. **a)** Scanning electron micrograph of the midgut of *Glycyphagus domesticus* after ultrasonic dissection, showing network of longitudinal and circular muscle fibres, from Walzl (2001, Figure 3); **b)** transverse section through anterior midgut, drawn from a micrograph by Brody *et al.* (1972, Figure 21). L = midgut lumen; **c)** detail of active cells and undifferentiated cuboidal cells.

the rest of the midgut (Tongu *et al.*, 1986; Tovey and Baldo, 1990).

The epithelial tissues of the anterior midgut of dust mites, as with most mites studied to date, consist of a layer of cuboidal cells attached to a basement membrane (Šobotník *et al.*, 2008) underlain by a regular network of longitudinal and transverse muscle, elegantly visualised by Walzl (2001) in fixed specimens of *Lepidoglyphus destructor* that had been dissected by ultrasonication (see Figures 2.5a, 2.6a). There also exist areas of what Brody *et al.* (1972) refer to as 'active cells' (i.e. cells containing active cytoplasm – digestive cells) arranged in rows along the ventral surface of the anterior midgut (Figure 2.6b). The cells are mononucleate, with many microvilli, rough endoplasmic reticulum and are highly vacuolated. Some of the digestive cells are expanded and balloon-shaped, projecting some distance into the lumen, attached only by a narrow stalk to the basement membrane. The stalked, balloon-shaped cells bud off into the lumen of the midgut (shown in Figure 2.6c). Numerous degenerating cells have been observed floating free in the midgut lumen (Brody *et al.*, 1972). By way of contrast with the

ventral epithelium, the dorsal and lateral epithelium of the midgut consists of mononucleate cuboidal cells (Figure 2.6b) with microvilli, lysosomes, mitochondria and extensive rough endoplasmic reticulum. The cells are separated from the gut contents by a layer of undefined material and their role is not clear (Brody *et al.*, 1972). Baker (1975) also found what appeared to be two cell types, cuboidal and globular, in the anterior midgut of *Histiogaster carpio*, as did Tongu *et al.* (1986) with *Dermatophagoides farinae* and *D. pteronyssinus*. Prassé (1967), working with *Caloglyphus* spp., referred to them as 'digestive cells' (*Verdaumgszell*) and 'gland cells' (*Drüsenzell*). Kuo and Nesbitt (1970) described two types of cell from the anterior midgut: flattened, basophilic, squamous cells with a granular cytoplasm and cuboidal cells full of small vacuoles. The alternative view (T.E. Hughes, 1950; Perron, 1954) is that there is only one cell type in the midgut epithelium and that the different cell morphologies are simply different stages of development and differentiation. Thus, the cuboidal epithelial cells become differentiated into the larger, vacuolated cells that then bud off from the epithelium into the midgut lumen.

The cells of the midgut caeca are different from those found in the anterior midgut. They have many electron-dense bodies, which are probably lysosomes, and dense arrays of microvilli. The cells do not contain food vacuoles, indicating that they are not involved in phagocytosis. Brody *et al.* (1972) suggest that the midgut caeca are the primary site of extracellular digestion. Baker (1975) reported cells in the midgut caeca of *Histogaster carpio* that differ from those just described. They were large, club-shaped cells that produced an apical swelling from which spheres are budded off. These anucleate spheres were found to almost fill the caecal lumen. These observations suggest that the cells may be budding off lysosome-containing portions of themselves that then burst, releasing their enzymes and facilitating extracellular digestion. Kuo and Nesbitt (1970) reported similar cells in *Caloglyphus mycophagus* that budded off a large, single vacuole into the caecal lumen.

c The posterior midgut and the peritrophic envelope

The posterior midgut is separated from the anterior midgut by a valve or sphincter and is, like the midgut, composed of epithelial tissue. The valve is capable of considerable distension when the gut is full (Tongu *et al.*, 1986). Cells in the posterior midgut have very long microvilli (up to 3.5 μm) and very extensive rough endoplasmic reticulum. Contraction of the anterior midgut effects the movement of gut contents through the valve. The gut contents consist of dying digestive cells plus bits of food debris and undigested material. Prassé (1967), Akimov (1973) and Brody *et al.* (1972) have suggested that some extracellular digestion may take place in the posterior midgut, but its most important function is the absorption of the products of digestion.

As the food moves through the valve into the posterior midgut it enters the peritrophic envelope ('retaining membrane' of Wharton and Brody, 1972), which expands posteriorly as more material passes through. The gut contents in the posterior chamber form a spherical mass with increasing bulk. Eventually the surrounding peritrophic envelope breaks away from the valve, and the food ball floats free. After some time, the ball is pushed through into the hindgut (as shown in Figure 2.3).

The term *peritrophic envelope* is used here rather than peritrophic membrane, because the structure is morphologically and functionally different from a membrane. Wharton and Brody (1972) describe it as 'a more or less consolidated network or mat of fibres containing chitin that is embedded in a matrix'. The envelope surrounding the food balls in the posterior midgut is 1–2 μm thick and separates them from the gut epithelium. The outer wall of the envelope consists of extremely fine, felt-like arrangement of fibrils (3.5 nm in diameter, 300 nm long). This permeable, fibrous structure allows the passage of smaller molecules. It represents the final, solidified structure of the envelope from protein and chitin components secreted in liquid form from midgut secretory cells (Richards and Richards, 1977).

In dust mites the solid peritrophic envelope has only been located in the posterior midgut, though T.E. Hughes (1950) identified a mucus-like secretion from cells in the anterior midgut of *Acarus siro*, which may represent a fluid precursor of the envelope. In many insects, the envelope lines the midgut throughout its entire length, thus dividing the midgut lumen into two compartments, the ectoperitrophic space between the midgut epithelial cells and peritrophic envelope and the endoperitrophic space between the peritrophic envelope and the midgut lumen proper (Terra, 1990). This compartmentalisation of the midgut allows for an anteriorly directed flow of water, enzymes and the products of digestion small enough to get through the pores of the peritrophic membrane. These molecules flow in the ectoperitrophic space towards the midgut caecae where intermediary and final, extracellular, digestion occurs. It is not certain whether dust mites have a midgut that is compartmentalised in this manner: two possible models of midgut digestion are discussed below (see section 2.2.6). Šobotník *et al.* (2008) found the peritrophic envelope cells in *Acarus siro* were localised laterally in the ventriculus, close to the caecal openings.

The functions of the peritrophic envelope, a structure unique to the arthropods, have been the subject of debate (Richards and Richards, 1977; Peters, 1992; Terra, 2001) and include provision of a barrier against microbial infection of the midgut, protection of the posterior midgut epithelium from mechanical damage by undigested pieces of solid food (Brody *et al.*, 1972), provision of lubrication of the food bolus, and the binding and exclusion of toxins. Terra and Ferreira (1994) drew analogies in function between the peritrophic envelope and the gastrointestinal mucus of mammals. An anti-oxidant role for the peritrophic envelope of the moth *Helicoverpa zea* has been

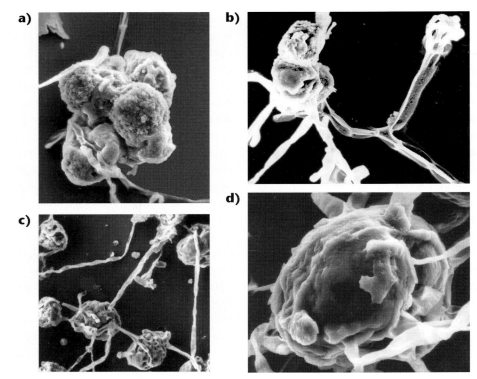

Figure 2.7 Scanning electron micrographs of faecal pellets of *Dermatophagoides pteronyssinus* and associated fungi. **a)** Group of four to five pellets with fungal hyphae; **b)** pellets with hyphae and conidia; **c)** hyphae connected to a series of pellets; **d)** pellet showing emergent hyphae. Original micrographs by the author.

demonstrated (Summers and Felton, 1996), whereby the gut is protected from oxidative damage by dietary pro-oxidants due to the effective scavenging of hydroxyl radicals by the peritrophic envelope. Tellam *et al.* (1999) provide a detailed treatment of the functional molecular characteristics of the peritrophic envelope.

d The hindgut and the anus

In the cuticle-lined anterior hindgut the food ball becomes reduced from about 30 μm to 15 μm in diameter, an eight-fold reduction in volume, probably due to resorption of water. The ball changes colour, from light to dark brown, and is now referred to as a faecal particle (see Figure 2.7). The faecal particle is compacted with three to five other particles to form a faecal pellet, with a diameter of ca. 30–50 μm. The individually enveloped faecal particles are held together with what appears to be mucus, not encapsulated within a single peritrophic envelope as has been assumed by some authors (see Figure 2.5d). This fact has important consequences for particularisation of the pellet post-egestion in relation to the size range of allergenic particles and allergen exposure (refer to Chapter 8). The faecal pellet is then egested through the anus. The mucus

that binds the pellets together may be secreted from the cells underlying the hindgut cuticle, entering the hindgut through numerous cuticular pores.

e Gut pH

The pH of the various parts of the gut can give some important clues about the mode of digestion and the enzymes involved, since all enzymes have a pH optimum. In an extreme example, the wool-feeding larva of the webbing clothes moth, *Tineola bisselliella*, has a highly alkaline midgut pH and its major digestive enzyme is a keratinolytic metalloproteinase with a pH optimum of about 9.4 (Ward, 1975). The measurement of gut pH in something as small as a mite is not straightforward and few attempts have been made. An estimate of the gut pH of *Acarus siro* was made by T.E. Hughes (1950) by culturing the mite in food material stained with various indicators including litmus, phenol red, neutral red and universal indicator and observing colour changes through the cuticle of the live mites. The results are summarised in Table 2.2 and show pH varies from slightly acid in the midgut to neutral or slightly alkaline in the hindgut. The gut of dust mites, which are also keratin-feeders (or, at least, keratin ingestors), also becomes more alkaline in the hindgut: van Bronswijk and Berrens (1979) reported the pH of the midgut of *Dermatophagoides* spp. as 6.8–7.4 and the hindgut as 6.2–8.5. This indicates that if dust mites are able to digest keratin, they do not use enzymes that operate at high pH like those of clothes moths (see section 2.2.4b).

The pH for optimal activity of various enzymes from *Dermatophagoides* spp. (pH 6–8.5; Table 2.3 below) fits with the neutral to slightly acidic conditions of the midgut. Barabanova and Zheltikova (1985) provide graphs of the activity of amylase, invertase, cellulase and chitinase in homogenates of *D. farinae* and *D. pteronyssinus* and found optimal activity of pH 5.0–8.5. The pH optima of carbohydrases tend to be slightly more acidic than that of the proteases, which may reflect differences in sites of activity in the gut. The carbohydrases would seem better suited to the slightly more acidic conditions in the anterior midgut than the neutral pH of the posterior midgut.

f Redox potential

Redox potential (or oxidation-reduction potential, ORP) is a measure of the electrical potential of the gut to produce oxidation and reduction reactions. The measurement is a relative one, against a standard hydrogen electrode potential (E_0), measured in millivolts (mV), where $E_0 = 0$ mV. A negative redox potential indicates a reducing environment; a positive potential indicates an oxidising environment. Insects that feed on keratin have highly negative redox potentials: the keratinophagous moths are between –100 to 400 mV, and the carpet beetles –170 to –230. Redox potential, like pH on which it depends, increases from foregut to posterior midgut (Christeller, 1996).

Redox potential in the dust mite gut has not been measured, but the behaviour of digestive enzymes *in vitro* indicates that the gut is a strongly reducing environment. Dust mite lysozyme shows considerably enhanced activity in the presence of the reducing agents EDTA and dithiothreitol (DTT) (G.A. Stewart,

Table 2.3 Biochemical and physicochemical properties of digestive enzymes of *Dermatophagoides pteronyssinus*.

	Allergen Group	Mol. wt. (kDa)	Distribution in mites**		pH optimum	Reference
			Faeces	**Bodies**		
Amylase	4	60	++	+	6.4	Lake *et al.*, 1991; Stewart *et al.*, 1991; Stewart, Bird, Krska *et al.*, 1992
Glucoamylase		52	+	+	6–7	Stewart *et al.*, 1991; Stewart, Bird, Krska *et al.*, 1992
Lysozyme		10	+	+	=6	Stewart *et al.*, 1991; Stewart, Bird, Krska *et al.*, 1992
Trypsin	3	28–30	+++	–	7–8.5	Stewart *et al.*, 1991, 1994; Stewart, Bird, Krska *et al.*, 1992
Chymotrypsin	6	25	+	+	ND	Stewart *et al.*, 1991, 1994; Stewart, Bird, Krska *et al.*, 1992
Serine	2	14	+	++		Stewart *et al.*, 1994
Protease and Collagenase*	9	28			ND	King *et al.*, 1996
Cysteine protease	1	25.134	+	+	7	Chua *et al.*, 1988; Stewart *et al.*, 1991; Stewart, Bird, Krska *et al.*, 1992
Lipase		52			ND	Stewart *et al.*, 1991

*This enzyme was referred to as elastase by Stewart *et al.* (1994; see text).

**The number of crosses gives a semi-quantitative indication of the relative proportion of enzyme, judged by relative activity against specific substrates, in faecal extracts (spent growth medium) and in extracts made from mite bodies. ND = not determined.

pers. comm.). Also, Der p 1 requires activation with reducing agents such as DTT or cysteine before enzymatic activity is detectable (Stewart *et al.*, 1989; 1991; see section 2.2.3e). These findings resonate with that of Christeller (1996), that DTT greatly increased the rate of solubilisation of wool *in vitro* by midgut extracts of the brown house moth (*Hoffmanophila pseudospretella*), and also maintained a highly negative redox potential. While DTT is not present *in vivo*, physiological reducing agents that are capable of maintaining negative redox potential include the sulphydryl-containing amino acid cysteine, which is a hydrolysis/reduction product of keratin (see section 2.2.4b).

2.2.3 Digestive enzymes of domestic mites – the repertoire

Dust mites possess a broad spectrum of digestive enzymes, several with overlapping substrate specificities (see Table 2.4). Several of these enzymes are also allergens (see Chapter 7). In general, it has been regarded that the diversity of digestive enzymes detected in the digestive system of an animal depends on its trophic specialisation and capacities: the enzymes reflect the diet (e.g. Bowman, 1981; Wigglesworth, 1972). This adaptive view may be an oversimplification. The absence of an enzyme from Table 2.4 probably indicates only a failure to detect it, not that it does not occur in the mite. I suspect that the enzyme repertoires of most domestic mites are broadly similar, but there are species-specific, or population-specific differences in the relative amounts of enzymes expressed in the gut and salivary glands. Trophic specialisation may reflect quantitative differences in enzyme expression and activity, not qualitative ones. In their review of insect digestive enzymes, Terra and Ferreira (1994) have argued that digestive physiology should not be studied only in relation to types of diet. If one seeks phylogenetic explanations for enzyme repertoires and compartmentalisation, the pattern seems to be that adaptive features in related arthropods with different feeding habits have evolved from the same basic digestive anatomy and physiology.

One example where trophic specialisation is possibly reflected by a qualitative difference in the enzyme repertoire is in *Acarus siro*, which feeds on the highly proteinaceous germ of damaged grain, in preference to the endosperm (Solomon, 1962). This is a specialised diet and one that most likely emerged after the development of agricultural production and grain storage (see section 4.1.5). *A. siro* appears to have no cysteine proteinase activity comparable with that of dust mites (see Table 2.4). Barber *et al.* (1996) found that the major cysteine proteinase of *Dermatophagoides pteronyssinus*, Der p 1, was bound and inactivated by water-insoluble seed storage proteins, prolamins, which are present in wheat, barley and rye flour. Furthermore, the proteolytic activity of Der p 1 is inhibited by water-soluble components from wheat. In order to digest wheat proteins, it is possible that *Acarus siro* has evolved the use of alternative proteinases that are not inactivated or inhibited by compounds in grain. Cysteine proteinase inhibitors are also present in human epidermis and stratum corneum (Järvinen, 1978) though it is not known whether ingestion of skin scales by dust mites has any inhibitory effect on the hydrolytic properties of *Dermatophagoides* group 1 allergens.

In addition to those digestive enzymes known to be allergenic, dust mites have a number of polysaccharidases capable of digesting fungal cell wall components such as chitin and mannans.

a Localisation of digestive enzyme-allergens

A detailed review of the various attempts to localise enzyme-allergens within dust mites is given in Chapter 7. The regions of the body that stain positive for allergens are the buccal-pharyngeal region, the anterior and posterior midgut and the cuticular and subcuticular tissues. The enzymatic activity of the cuticle is related to moulting (see section 2.8.3).

The buccal-pharyngeal region is where the salivary glands are found. The most frequent salivary gland-associated enzymes of arthropods are the carbohydrases amylase and invertase (Wigglesworth, 1972). The buccal-pharyngeal region stained strongly positive for the Group 2 allergen of *Lepidoglyphus destructor* (van Hage-Hamsten *et al.*, 1995), formerly known as Lep d 1. Ichikawa *et al.* (1998) found the structure of Der f 2 is most similar to the two regulatory domains of the aminoacyltransferase enzyme, transglutaminase (EC 2.3.2.13), and suggested that Group 2 allergens were involved in mite immune responses to bacteria (see section 2.5.2). However, the endogenous function of group 2 allergens is still unclear (Stewart and McWilliam, 2001; section 7.4.3).

The midgut, its caecae, and the faecal pellets stain positive for group 1 allergens. These cysteine

Table 2.4 Digestive enzymes isolated from dust mites and stored products mites.

	EC No.	Pyroglyphidae			Acaroidea				Glycyphagoidea					
		D.pt	D.f	E.m	A.s	T.p	T.l	A.o	G.d	L.d	G.f	B.k	B.t	C.a
Esterases	3.1													
Carboxylic ester hydrolases	3.1.1													
Triacylglycerol lipase	3.1.1.3	+	+		−	−	−		−	−,+[6]	+	+		−
Phospholipase A$_2$	3.1.1.4	+												
Phospholipase A$_1$	3.1.1.32	+												
Esterase/lipase activity		+	+		+	+	+		+	+	+	+	+	+
Phosphoric monoester hydrolases	3.1.3													
Alkaline phosphatase	3.1.3.1	+	+		−,+[4]	−	+		+	−,+[4]	−	−,+	−	+
Acid phosphatase[1]	3.1.3.2	+	+		−,+[4]	+	+		+	+	+	−,+	−	+
Phosphoric diester hydrolases	3.1.4													
Phospholipase C	3.1.4.3	+												
Glycosidases	3.2													
α-Amylase[2]	3.2.1.1	+	+	+	+		+	+	+	+				
Glucoamylase	3.2.1.3	+	+	+					+					
Cellulase	3.2.1.4	−,+[3]	+		−		+	+	−	−				
Chitinase[2]	3.2.1.14	−,+[3]	+		+		+	+	+	+				
Lysozyme	3.2.1.17	+	+	+	+	+	+	+	+	+				+
α-D-Glucosidase[7]	3.2.1.20	+	+		+		+		−	+		+	+	
β-D-Glucosidase[7]	3.2.1.21	+	+		+		+		+	+				
α-D-Galactosidase[7]	3.2.1.22	+	+		+		+		−	+		+	+	
β-D-Galactosidase[7]	3.2.1.23	+	+		+		+		+	+		+	+	
α-D-Mannosidase	3.2.1.24	+	+		+		+		+	+		+	+	
β-*N*-Acetyl-D-glucosaminidase	3.2.1.30	+	+									+	+	
β-Glucuronidase	3.2.1.31	+	+		+	−	+		+	+	−	+	+	−
Invertase	3.2.1.48	+	+											
α-L-Fucosidase	3.2.1.51	+			+		+		+	+		+	+	
β-*N*-Acetyl hexosaminidase	3.2.1.52	+			+		+		+	+				
Peptidases	3.4													
Aminopeptidases	3.4.11													
Leucine aminopeptidase	3.4.11.1	+	+		+	+	+		+	+		+	+	
Cystinyl aminopeptidase	3.4.11.3	+			+		+		+	+		+	+	
Valine aminopeptidase		+	+		+		+		+	+		+	+	
Metallocarboxy-peptidases	3.4.17													
Carboxypeptidase A	3.4.17.1	+	+	−		+								
Carboxypeptidase B	3.4.17.2	+	+	−		+								

continued

Table 2.4 Continued

	EC No.	Pyroglyphidae			Acaroidea				Glycyphagoidea					
		D.pt	D.f	E.m	A.s	T.p	T.l	A.o	G.d	L.d	G.f	B.k	B.t	C.a
Serine endopeptidases	3.4.21													
Chymotrypsin	3.4.21.1	+	+	+	−,+[5]	−,+[8]	+		−	−,+[5]	−	+	+	−
Trypsin[2]	3.4.21.4	+	−,+[5]	+	+	+	+	+	+	+	−	+	+	−
Cysteine endopepti-dases/proteinases[2]	3.4.22	+	+	+				+		+				
Cathepsin B	3.4.22.1					+								
Cathepsin D	3.4.23.5					+								
Metalloendopepti-dases	3.4.24													
Collagenase activity		+	+							+				
Non-specific peptidase activity														
'Hide proteinase' activity					+		+		+	+				
'Keratinase' activity		±	±											
Enzymes acting on phosphorous-nitro-gen bonds	3.9													
Phosphoamidase	3.9.1.1	+	+										−	−

D.pt = *Dermatophagoides pteronyssinus*; D.f = *D. farinae*; E.m = *Euroglyphus maynei*; A.s = *Acarus siro*; T.p = *Tyrophagus putrescentiae*; T.l = *Tyrophagus longior*; A.o = *Aleuroglyphus ovatus*; G.d = *Glycyphagus domesticus*; L.d = *Lepidoglyphus destructor*; G.f = *Gohieria fusca*; B.k = *Blomia kulagini*; B.t = *Blomia tropicalis*; C.a = *Chortoglyphus arcuatus*. Data from Barabanova and Zheltikova (1985); Bowman (1981, 1984); Bowman and Childs (1982); Bowman and Lessiter (1985); Cardona *et al.* (2006); Edwards *et al.* (1992); King *et al.* (1996); Lake *et al.* (1991); Martínez *et al.* (1999); McCall *et al.* (2001); Mori and Sajiki (1997); Ortego *et al.* (2000); Rigamonti *et al.* (1993); Stewart *et al.* (1986, 1991, 1992, 1994, 1998); Wolden *et al.* (1982). The data of Bousquet *et al.* (1980) were not included because assays were done on commercially available allergen preparations, rather than on extracts of live mites, and thus lacked standardisation of extraction procedures.

[1]Non-allergenic; [2]known allergenicity; suspected allergenicity; [3]chitinase and cellulase activity detected by Barabanova and Zheltikova (1985) but not by Stewart *et al.* (1991); [4]acid phosphatase activity detected by Bowman (1984) but not Rigamonti *et al.* (1993); [5]enzyme activity not detected by Rigamonti *et al.* (1993); [6]enzyme activity detected by Rigamonti *et al.* (1993); [7]data of Rigamonti *et al.* (1993) not included: no differentiation between α- and β-forms; [8]enzyme activity detected by Ortego *et al.* (2000).

proteinases are probably localised within midgut digestive cells. Other important midgut enzymes of arthropods include lysozyme; the serine proteinases trypsin (Group 3 allergens in dust mites) and chymotrypsin (Group 6 allergens in dust mites); pepsin; elastase; aminopeptidases; carboxypeptidases and glucosidases (Terra and Ferreira, 1994) as well as chitinase (McCall *et al.*, 2001). The cellular sites of synthesis of dust mite cysteine proteinases are not known, but the majority of enzymes in Table 2.4 are known in other organisms to be associated with lysosomes (Kirschke and Barrett, 1987). The exceptions are amylase, which is usually salivary-gland derived, and the exopeptidases and alkaline phosphatase, which are associated with plasma membranes.

In the sections below details are given of those digestive enzymes isolated from dust mites, with emphasis on their biology and enzymatic properties. Many general biochemical and enzymatic properties are taken from the publications by Dixon and Webb (1979) and the International Union of Biochemistry and Molecular Biology Nomenclature Committee (1992). A summary of their biochemical and physico-chemical properties is given in Table 2.3. I have attempted to minimise the degree of overlap with Chapter 7 on individual allergens by stressing allergological properties therein and minimising enzymatic properties.

b Glycosidases

α-*Amylase (EC 3.2.1.1)* hydrolyses 1,4-α-glucan chains of polysaccharides, such as starch and glycogen, to glucose. The *Dermatophagoides* group 4 allergens are amylase-like enzymes (Lake *et al.*, 1991). *Dermatophagoides* spp. amylase has been isolated from faecal

linked with impaired epidermal barrier function in people with atopic dermatitis due to disrupted biosynthesis of ceramides, so that lipid lamellae in the stratum corneum do not form properly (Melnick *et al.*, 1990). Further, Jin *et al.* (1994) found lower ceramide levels in the stratum corneum of non-atopic people, which they ascribed to greater ceramidase activity and enhanced ceramide degradation. The phenomenon was associated with dry skin and was worse in older people. So, skin scales from people with atopic dermatitis, or older people with dry skin may be of greater nutritional value to dust mites than those of healthy, non-atopics or younger people without dry skin because:

- the stratum corneum lipids are only partially assembled and easier to digest than fully assembled lipid lamellae;
- the total lipid content may match more closely the dietary optima for the mites;
- the skin scales contain lower concentrations of those free fatty acids that inhibit mite population growth.

c Bacteriophagy – feeding on bacteria

Any organism that feeds on skin scales is bound to ingest, as a matter of course, the so-called normal human skin flora (Marples, 1965), consisting mostly of Gram-positive *Staphylococcus*, *Corynebacterium*, *Sarcina* and *Micrococcus* spp. There is some evidence that dust mites may harbour endosymbiotic bacteria in the midgut (Douglas and Smith, 1989; Douglas and Hart, 1989). The bacteria were found to be closely associated with the microvilli of the epithelial cells and could not be cultured on nutrient agar. In order to survive in the midgut caecae, they would probably have to be anaerobes because of the low redox potential.

Endosymbiotic bacteria have been postulated to be the source of cellulolytic and ligninolytic enzymes of soil-dwelling Astigmata that feed saprophytically on tissues of higher plants (Stefaniak and Seniczak, 1976). Barabanova and Zheltikova (1986) suggested that the cellulase activity they detected in *D. pteronyssinus* and *D. farinae* was due to the presence of gut symbiotic bacteria. Several bacteria are keratinolytic including various *Bacillus* and *Streptomyces* spp., as well as a number of Gram-negative bacteria. Lucas *et al.* (2003) demonstrated the capacity of several bacterial isolates from soils to degrade feather keratin. Given the diversity, abundance and ubiquity of such bacteria, it is biologically plausible that they may be partly responsible for keratin digestion in dust mites.

Most of the information on bacteria found in dust mites comes from literature on the aetiology of Kawasaki disease (see section 8.2.2b). Oh *et al.* (1986) isolated and identified *Bacillus* spp., *Staphylococcus* spp., as well as Gram-negative non-fermenting rods and Gram-positive rods from homogenates of *D. pteronyssinus* and *D. farinae*, and Ushijima *et al.* (1983) isolated at least 12 different bacteria from dust mites, and Kato *et al.* (1983) identified 20 but cited only *Proprionibacterium acnes*, *Staphylococcus epidermidis*, *Bacillus subtilis* and *Pseudomonas* spp. Valerio *et al.* (2005) isolated bacterial 16S ribosomal RNA genes from dust mites belonging to various *Bartonella* spp. and alpha-proteobacteria. *Dermatophagoides farinae* had 11–24-fold more copies of 16S genes than *D. farinae* though it is not known why.

The presence of lysozyme activity in many domestic mites (Table 2.4) suggests they can and do feed on bacteria. Erban and Hubert (2008) found 14 species of acaroids, glycyphagoid and pyroglyphid mites contained high lysozyme activity (in whole mite extracts and spent growth medium). The authors were able to demonstrate *in vivo* digestion of fluorescein-labelled cells of *Micrococcus lysodeikticus* by the mites, as well as enhanced population growth of mites fed on bacteria compared with controls.

d Mycophagy – feeding on fungi

Moulds

Van Bronswijk and Sinha (1973) suggested that certain moulds in house dust, in particular *Aspergillus amstelodami*, achieved partial digestion of a laboratory diet of defatted human skin scales and yeast, thus rendering these 'predigested' scales more palatable and nutritious for dust mites due to decreased lipid content. This hypothesis was not tested experimentally. They found that when mites were fed on fungal cultures alone, none of the 45 species tested were adequate for the reproduction of *D. farinae*. This does not prove that dust mites do not feed on fungi in the wild. Hay *et al.* (1992) demonstrated that fungi in laboratory cultures are at very high densities and that at such densities they are likely to have an antagonistic effect on dust mites due to the production of mycotoxins.

Enhanced population growth of *D. pteronyssinus* after *Aspergillus penicilloides* or *Aspergillus repens* were added to human skin scale and house dust cultures was reported by van der Lustgraaf (1978a). However, Douglas and Hart (1989) considered that enhanced mite population growth is not a consequence of fungal predigestion of skin scales but because *A. penicilloides* is a food source for the mites. They make the point that although van Bronswijk and Sinha (1973) demonstrated the improved performance of dust mites on a diet that had been pre-incubated with *Aspergillus amstelodami* than in the absence of this fungus, they failed to perform the critical experiment to determine the effect of the fungus, namely assessment of mite performance on diet supplemented with *A. amstelodami* but not pre-incubated with it. Hay *et al.* (1993) point out further basic flaws in the experimental design of van Bronswijk and Sinha (1973), including that small starting populations of mites would limit the number of mating encounters, the lack of fungus-free control mites and the fact that the skin scales, on which the fungi were supposed to predigest the lipids, had already been defatted by washing in ether prior to use.

Lustgraaf (1978a) hypothesised a loose mutualistic association in which the mites benefit from feeding on skin scales colonised by fungi and the fungi benefit because the condidospores are concentrated within the faecal pellet and dispersed, when voided, together with a package of faecal nutrients. Whether dispersal of fungi is enhanced by the spores becoming concentrated in the faecal pellet is open to question. I have found spores survive passage through the gut and are able to germinate within the pellet, eventually the hyphae penetrating the peritrophic membrane (Colloff, unpublished data; Figure 2.7). Faecal pellets deposited onto glass coverslips remained intact following germination of spores and were able to provide sufficient nutrients to the fungi to allow development of hyphae and formation of conidiophores in the absence of any other food source (Figures 2.7a–d).

Saint Georges-Gridelet (1984) stated that *Aspergillus penicilloides* was 'essential' in the diet of *Dermatophagoides pteronyssinus*, and in a later paper (Saint Georges-Gridelet, 1987a) that, 'A strong relationship between *D. pteronyssinus* and xerophilic fungi in culture and in mattress dust has been established.' One might assume these papers provided evidence to support the hypothesis of van Bronswijk and Sinha (1973) on the importance of fungi for dust mites. In fact, neither contains any experimental data on fungi. A subsequent paper (Saint Georges-Gridelet, 1987b) showed that *D. pteronyssinus* seemed to prefer loose,

aerated substrates to dense powdery substrates that may become clogged with fungal mycelia, referred to as 'the antagonistic effects of the endemic mould, *Aspergillus penicilloides*, with which the mite establishes a symbiotic relationship'. But it provided no additional evidence to substantiate the claim, repeated again therein that: 'Xerophilic fungi play an important role by predigesting the substrate.' The reason I dwell on this issue is because it is a classic example of how a hypothesis in one research paper can be treated in subsequent publications as if it were proven scientific fact. The promulgation of this misinformation has had far-reaching consequences. It led to the development, patenting, commercialisation, sale and usage of mite control measures, based on the principle that dust mites could be controlled by using fungicides to kill the moulds on which, allegedly, the mites were so closely dependent. This approach has largely been discredited by clinical trials (see Chapter 9).

Hay *et al.* (1993) examined the performance of *D. pteronyssinus* over two generations in the presence of *Aspergillus penicilloides* and in fungus-free control mites that were bred in sterilised yeast and liver powder. This is the only study on dust mite – fungal relationships that has used fungus-free control mites. The fungus reduced survival, development rate, adult duration and fecundity of the mites, and the detrimental effects were greater with increased density of fungi, probably due to clogging of the substrate by mycelia. Despite these effects, a fungal requirement, probably for nutrients, was indicated by the poor performance of fungus-free mites in the second generation. Hay *et al.* (1993) concluded that the fungi exert both detrimental and beneficial effects on the mites and therefore the application of selective fungicides for mite control could result in either increased or decreased mite populations.

So far, this account has dealt exclusively with fungi and pyroglyphid mites. But acaroid and glycyphagoid mites also feed on fungi. Research on fungal feeding by stored products mites has concentrated on determining which fungi are preferred by which mites (Sinha, 1964; Armitage and George, 1986; Parkinson *et al.*, 1991). Some of these studies have been done using pure fungal cultures grown on agar slopes. This approach introduces several confounding factors, highlighted by Hay *et al.* (1992), most notably that the densities of fungi in pure culture can be considerably higher than the mites may encounter in the wild. The fungi at these densities would have an adverse effect

Table 2.7 Biochemical composition of fungi and skin scales. DW = dry weight; ND = no data.

	% DW in fungi[1]	% DW in spores of *Aspergillus oryzae*[2]	% DW in mycelium of *Penicillium chrysogenum*[3]	% DW in skin scales
Carbohydrates	16–85	9.0	60	
Lipids	0.2–87	2.2	1.4	5[4]
Proteins	14–44	22.8	16.4	2.6[5]
RNA	1–10	ND	ND	
DNA	0.15–0.3	ND	ND	
Ash	1–29	5.3	ND	

[1]Griffin (1994); [2,3]Cochrane (1958); [4]Elias (1981); [5]Berrens (1972).

on the mites, either through production of mycotoxins, resource competition, or adverse modification of the physical nature of the substratum. At lower fungal densities, the mites may respond differently.

One study attempted to avoid this problem by feeding mites with fragments of mycelia and spores (Parkinson *et al.,* 1991). The authors assessed egg production of pairs of *Acarus siro, Lepidoglyphus destructor* and *Tyrophagus longior* that were fed on a control diet of yeast and wheat germ compared with mycelial pellets of *Cladosporium cladosporoides, Aspergillus ruber, A. repens* and *Penicillium cyclopium*, as well as spores of the latter two fungi. The mites produced fewer eggs on fungal diets, with *T. longior* doing best and *L. destructor* worst. *Aspergillus ruber* was the most suitable fungus for the mites, while *C. cladosporoides* was least suitable. Mycelial pellets provided a better diet than spores.

OConnor (1984), commenting on the study by Sinha (1964), noted that two glycyphagoid species, *Lepidoglyphus destructor* and *Aeroglyphus robustus* fed and reproduced on relatively few of the fungi tested. This indication of restricted fungal preference was supported by gut contents analysis by OConnor of some naturally occurring glycyphagoids, which revealed that each mite species tended to contain spores and mycelia from only one fungal species. By way of contrast, *Tyrophagus putrescentiae*, which is a common contaminant of laboratory cultures of fungi (as well as various cell and tissue lines, nematodes, mites and insects), fed on 22 of 24 species of fungi tested by Sinha (1964), and *Acarus siro* fed on 20. The capacity for a more generalised feeding habit by these two acarids is reflected by the wide variety of natural and synanthropic habitats in which these mites occur (A.M. Hughes, 1976).

Armitage and George (1986) summed up the situation in the relationship between stored products mites and fungi, stating that although stored products mites may suffer toxic effects from fungi and may not reproduce very well on fungi as their sole food source, they can be attracted to them, graze on them and digest what they ingest. Grazing by stored products mites may limit growth of certain fungi, either by feeding or by production of anti-fungal compounds (see section 2.9.1g).

Fungi in the diet appear to be advantageous to dust mites, providing minerals, vitamins and other nutrients that they would be hard pressed to obtain from other naturally occurring dietary sources, as well as a source of water (see section 3.4.3c). Griffin (1994) pieced together biochemical analyses of fungi from many reports (Table 2.7). Of note is the wide variation in percentage composition of most of the major components and the very low DNA concentrations (about a tenth of what is found in other eukaryotes and bacteria). Many of the proteins in fungi are covalently bound to carbohydrates to form glycoproteins and peptidoglycans, which are common cell wall components. Carbohydrates in fungi are primarily polysaccharides (monosaccharides are present in very low concentrations), and many of them are cell wall constituents. Glycogen is the principal cellular storage polysaccharide.

There is still no published data that shows fungi contribute to improved performance of dust mites by predigesting their dietary components. What becomes clear even from a cursory examination of the humidity optima of mites and moulds is that for mites there is a fine line between fungi as a food resource and fungi as a toxic environmental hazard, or as a physical barrier to their mobility. *Tyrophagus putrescentiae* and *Caloglyphus rodriguezi* larvae were killed by mycotoxins when challenged with a variety of moulds in axenic culture (Rodriguez *et al.*, 1984). Lustgraaf (1978a) showed that at 75% and 80% RH on wheat germ, *Aspergillus penicilloides* grew abundantly and suppressed population growth of *D. pteronyssinus*, whereas at 71% it grew less well and was not antagonistic. On skin scales, the mould grew moderately at 71% and 75% and stimulated growth of *D. pteronyssinus* populations. There is no evidence that this enhanced growth was due to modification by fungi of the diet of the mites, but clear evidence exists that they fed on the fungi because their guts contained conidia of *A. penicilloides*.

Yeasts

Yeasts are unicellular fungal morphs that reproduce by cell division involving either budding or fission. Yeast morphs occur in many different groups of fungi, although the term usually refers to the yeast-like Hemiascomycete fungi. Like certain bacteria, yeasts are part of the normal human skin flora. Common species on human skin include *Pityrosporum* spp. and *Candida* spp. Some 60–90% of adults carry yeasts and populations on the scalp and other sebaceous skin areas can be in the order of 0.5 million cells cm^{-2} (Andrews, 1976). Yeasts are very abundant on shed skin scales, in house dust (Andersen, 1985) and in dwellings (Hunter *et al.*, 1988). Hallas and Korsgaard (1983) found corresponding autumn population maxima of yeasts and *Dermatophagoides farinae* in mattress dust. Yeasts tend to prefer much higher humidities than dust mites, and the population increase of both yeasts and mites could be due to increased atmospheric humidity at the onset of Autumn, or it could be that the increase in mites mirrored the increase in yeasts because they were using the yeasts as food, as concluded by Andersen (1985).

Dust mites perform reasonably well in cultures of baker's yeast (*Saccharomyces cervisiae*). I grew continuous cultures of *D. pteronyssinus*, *Blomia tropicalis* and *T. putrescentiae* in only granulated baker's yeast for several years. The nutritional value to *D. pteronyssinus* of *Saccaromyces cervisiae* was assessed by Andersen (1991), who found extracts of culture medium (yeast and skin scales; 1:1) contained 15% extractable protein, of which 83% was from yeast and 17% from the sparingly extractable skin scales (much of the protein in skin scales is structural, i.e. keratin). During initial mite population growth there is a corresponding decrease in the amount of extractable protein in cultures, indicating that the mites first consume easily available, digestible protein. As the population growth rate flattens out, the protein content reflects the density of the mite population.

e Coprophagy – feeding on faeces

The assumption that dust mites are routinely autocoprophagous – they eat their own faeces – has found its way into the literature. The primary evidence for coprophagy is based on videotape taken by Dr John Rees, exhibited to delegates at the Second International Workshop on Dust Mite Allergens and Asthma, Minster Lovell, UK, 1990 (see van Moerbeke, 1991), showing a dust mite from a laboratory culture consuming what appears to be a faecal pellet. Coprophagy has also been observed in old, crowded laboratory cultures of *D. pteronyssinus* that had not been subcultured, nor had new culture medium added for some weeks (Colloff, unpublished data). Populations were beginning to decline sharply and it is possible that under these circumstances coprophagy represented a response to lack of food (see also McGregor and Peterson, 2000). There is no evidence that dust mites are routinely coprophagic in their natural habitat. No mites in the Acaroidea or Glycyphagoidea have been reported to be coprophagic. Further research is required to clarify this issue because there are some implications for digestive physiology and allergen production.

The reason that animals are coprophagous is that certain components in their diets are not easily digested and require a second passage through the gut in order to be broken down. Rabbits, with their high intake of cellulose from plant material, represent the classic mammalian example of a coprophage. The potential adaptive benefits for dust mites in routine coprophagy, would include the recycling of nutrients, the facilitation of digestion from re-ingesting high levels of digestive enzymes present in the faecal pellets, and the potential for transfer, incubation and growth of microorganisms which may also facilitate digestion. The process of digestion in dust mites gives the appearance of being rather inefficient (see section 2.2.2). Perhaps this is so only if viewed as an event involving only a single passage of the food through the gut.

f Cannibalism

Pieces of dust mite cuticle have been found in faecal pellets (Halmai, 1994). This is not direct evidence of cannibalism, since many terrestrial arthropods ingest pieces of their cuticle after they have moulted. Brody *et al.* (1972) and Wharton (1976) state that cannibalism is not unusual and particles the size of entire leg segments may be passed into the gut, though the authors do not state whether they have *seen* entire leg segments in the gut. There are no reported observations of dust mites feeding on other dust mites, live or dead. By way of contrast, *Cheyletus* spp. will readily prey on their own offspring if no other prey is available (Barker, 1991).

g   Detritivory – feeding on organic debris

This section deals with organic remains in house dust. Skin scales and dust mite faeces are technically detritus, but have been dealt with above (see sections 2.2.4b and e). Organic debris in house dust varies in diversity and origin according to whether the dust is from carpets (high diversity) or beds (low diversity). The organic components found in carpet dust (see section 4.6.1) include potential food items for dust mites other than skin scales. Mostly these are bits of human food debris that find their way from the plate to the floor. Bearing in mind that dust mites can be cultured *in vitro* on a wide range of foods (see section 6.6), it is probable that they can feed on some of our spillages, though there is no quantitative data on this.

Another human spillage that mites feed on is human semen in beds (Colloff 1988). Nearly 80% of dust samples from beds of volunteers in Glasgow contained either prostatic acid phosphatase activity (an indicator of semen) or sperm heads (detected by microscopy). When female mites were fed on house dust only, the mean numbers of eggs they produced declined from 1.4 to 1.3 over 12 days, compared with an increase to 2.5 eggs per day by mites fed a 2:1 mix of dust and dried semen. This suggested that under comparable microclimatic conditions, mattresses that receive regular input of semen are likely to have greater mite population densities than ones that do not.

h   Summary

We still do not know what a dust mite feeds on in its natural habitat, what its feeding preferences are, and which nutrients it gains from its various dietary components. We know nothing about the rates and efficiencies of food consumption and assimilation (Slansky and Rodriguez, 1987). We assume that skin scales and microorganisms are the major source of nutrients, making dust mites both detritivores and consumers of saprophages. The lines of argument for and against dust mites gaining their nutrients predominantly from digestion of skin scales are rather short on hard evidence, but are as set out below.

For

- Fungal predigestion of skin scale lipids would facilitate the process of digestion.
- Physiological capacity – the enzyme repertoire and redox potential of the gut are adequate to allow digestion of α-keratin.
- There is a super-abundance of skin scales in the microhabitat of dust mites: they are the 'obvious' food source.
- Dust mites will feed on a diet of skin scales when grown in culture.

Against

- Keratinolysis by dust mites has not been demonstrated physiologically: intact pieces of skin scale can be observed in faecal pellets.
- The catholic feeding habits of dust mites observed *in vitro* include feeding on hyphal fungi and yeasts, fishmeal, dog food, dried *Daphnia* sp., wheat germ, various artificial diets, and so forth. Why should such omnivory not be countenanced in their natural habitat?
- The keratin in skin scales may not be the main source of nutrients, but the lipids, or microorganisms growing on the lipids are, and the dust mites ingest skin scales but do not digest them.

2.2.5   Dietary requirements

a   Vitamins and minerals

One of the major functions of vitamins and minerals in the metabolism of animals is to act as cofactors of enzymes. B vitamins (niacin, thiamine, riboflavin) and D vitamins (ergosterol) were added to mattress dust and floor dust and found to significantly increase fecundity and population growth of *D. pteronyssinus* in floor dust, but less growth occurred in mattress dust (Saint Georges-Gridelet, 1987a). The reason given for the difference was that the mattress dust contained higher population densities of *Aspergillus penicilloides*, from which the mites could obtain the vitamins. Folic acid (vitamin B_c), riboflavin (vitamin B_2), thiamine (vitamin B_1), niacin (vitamin B_3), pyridoxine (vitamin B_6) and biotin (vitamin B_{12}) were found to be essential for growth and reproduction of *Acarus siro* (Levinson *et al.*, 1992).

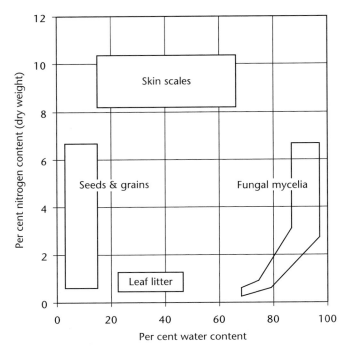

Figure 2.8 Relationship between water and nitrogen content of common food sources of stored products mites (farines – grains and flours; fungi), dust mites (fungi, skin scales) and soil-dwelling astigmatid and oribatid mites (fungi, leaf litter). Re-drawn and modified after Slansky and Scriber (1985; Figure 1).

Mineral salts containing barium, cadmium, cobalt, copper, iodine, iron, lithium, manganese, molybdenum, silver and zinc were found to suppress oviposition of *T. putrescentiae* when added to a wheat germ diet at concentrations of 0.25–6.0% by weight (Boczek *et al.*, 1984).

b Proteins and amino acids

Amino acids are a major source of dietary nitrogen. The nitrogen content of common foodstuffs of astigmatid and oribatid mites is shown in Figure 2.8. Essential amino acids for *T. putrescentiae* are arginine, isoleucine, histidine, leucine, lysine, methionine, phenylanaline, tyrosine, threonine, tryptophan, valine, the classical 11 amino acids essential for the nutrition of the rat (Rodriguez and Lasheen, 1971; Table 2.5).

c Lipids

Matsumoto (1975) found that population growth of *Dermatophagoides farinae* started to show suppression when fed on diets containing 8% lipid or more. Growth was most rapid at 4–6% (see Figure 2.9a). The reproduction of *D. farinae* is suppressed by castor oil (containing 85% ricinoleic acid), coconut oil (50% lauric acid) and lavender oil, the triglycerides triacetin and tributyrin, and the monoglycerides monolaurin

and monostearin at concentrations of 2–10% of diet. Reproduction was stimulated by the plant oils that were rich in oleic and linoleic acid, such as linseed (20% each of oleic and linoleic acid), soybean, cottonseed, sesame seed, peanut (60% oleic acid) and olive (80% oleic acid), and the animal fats cod liver oil, butter (25% each of palmitic and oleic acid), lanolin and lard (25% palmitic and 50% oleic acid). Oleic, linoleic and palmitic acids administered as individual dietary additives also increased reproduction, but the best lipid source was found to be the phosphoglyceride lecithin, presumably because of the phosphorus the mites get from it. Saint Georges-Gridelet (1984) obtained similarly enhanced population growth of *D. pteronyssinus*, as well as improved survival at sub-optimum humidities, using lard added to diets of either human skin scales and yeast or dried *Daphnia* sp. Populations grew highest at 8.5% lipid content in partly defatted skin scales/yeast and at 14% lipid in partly defatted *Daphnia* sp. The difference in optimal lipid level in the diets was explained by the higher carbohydrate content of dried *Daphnia* sp.

Table 2.7 shows that short chain fatty acids are more effective inhibitors of development when fed to *Tyrophagus putrescentiae* than longer chain fatty acids (Rodriguez, 1972). The inhibition of development

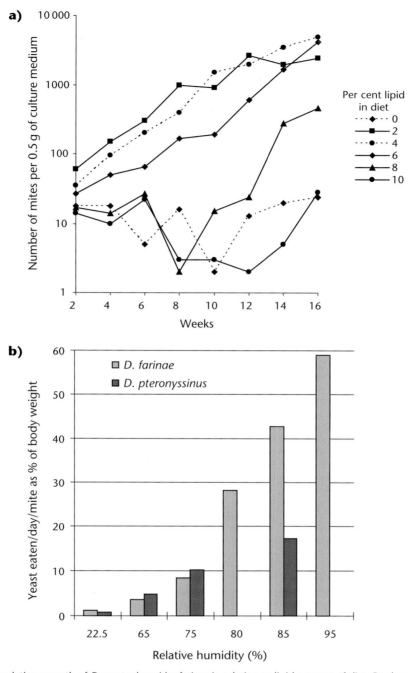

Figure 2.9 **a)** Population growth of *Dermatophagoides farinae* in relation to lipid content of diet. Re-drawn from Matsumoto (1975, Figure 1); **b)** feeding rates of *D. farinae* and *D. pteronyssinus* on yeast at different humidities (data from Arlian, 1977).

was measured by reduction in body weight. But when added to a casein diet, the methyl esters of the longer chain fatty acids laurate, myristate and oleate were more inhibitory than the methyl esters of the short chain fatty acids caproate, caprylate and caprate. *Calglyphus rodriguezi* is more tolerant than *T. putrescentiae* to proprionic acid and methyl esters but less tolerant to caprylic acid (Rodriguez, 1972).

One of the biochemical characteristics of insects and arachnids is their dependence on dietary sources of sterols (reviewed by Svoboda and Thompson, 1985). With a few exceptions, arthropods appear to lack the capacity to synthesise squalene, the 30-carbon lipid precursor of triterpenes, and steroids like the moulting hormones ecdysone and 20-hydroxyecdysone. Other organisms, such as the symbiotic gut bacteria of certain insects, are able to synthesise squalene from the reductive tail-to-tail coupling of two 15-carbon farnesyl pyrophosphate molecules. Farnesyl pyrophosphate is the precursor of all sesquiterpenes, and is synthesised from the 5-carbon isopentyl pyrophosphate and geranyl pyrophosphate – the 10-carbon precursor of all monoterpenes. The monoterpenes are the major pheromones of astigmatid mites (see section 2.9). This biosynthetic pathway leads back to acetyl coenzyme A and the citric acid cycle (shown in Figure 2.10), and demonstrates the interconnected nature of lipid metabolism and the multiple functions of lipids in the lives of arthropods. Sterols were found to be essential dietary lipids for three acarid species (Boczek, 1964) and van Bronswijk and Sinha (1973) suggested sterols may be released, along with free fatty acids, as a consequence of the digestion of complex lipids of fresh skin scales by moulds such as *Aspergillus amstelodami*. Since ergosterol is the principle fungal steroid (Griffin, 1994), mites could obtain it from consuming fungi direct, and probably do.

Sato *et al.* (1993) investigated the attractant properties of palmitic, heptadecanoic and stearic acids, and their methyl esters, to eight species of domestic mites. *Dermatophagoides pteronyssinus*, *D. farinae*, *T. putrescentiae* and *Lardoglyphus konoi* responded to most or all of these, whereas *Carpoglyphus lactis*, *Aleuroglyphus ovatus*, *Glycyphagus domesticus* and *Acarus immobilis* showed no response to any of the substances tested. They also tested the effects of carbon chain length of saturated fatty acids (C_7-C_{20}) on attractiveness to *D. farinae*, *L. konoi* and *T. putrescentiae* and found sporadic attractiveness, with C_7, C_{11} and C_{20} inactive for all species and C_{14}-C_{16} and C_{18} active for all species. It is hard to know what to make of this data, as the authors give only brief, partial explanations of the biological significance of what is an exacting, and I suspect important, series of behavioural observations on the chemical ecophysiology of domestic mites. Apart from the obvious fact that any attractant has potential applications in baits or traps for mite control purposes, the authors seem to indicate the following:

- triacylglycerols (triglycerides) are important dietary and storage lipids for domestic mites;
- the fatty acids or their methyl esters that were tested are hydrolysis products of triacylglycerols;
- the alkanes and alkenes from extracts of the lateral opisthosomal glands and cuticle of various mite species (Kuwahara, Leal *et al.*, 1992; section 2.9)

include major components that could be derived from decarboxylation of fatty acids following hydrolysis of dietary triacylglycerols;
- the biological implication is that the capacity of some domestic mites (*D. pteronyssinus*, *D. farinae*, *T. putrescentiae* and *L. konoi*) to be able to detect various fatty acids and their methyl esters in potential foodstuffs allows them to exploit foods that are rich in fats and oils, whereas other species (*Carpoglyphus lactis*, *Aleuroglyphus ovatus*, *Glycyphagus domesticus* and *Acarus immobilis*) lack this ability;
- the mites that feed on foods that are rich in fats and oils produce in their bodies fatty acids and their methyl esters as hydrolysis products and secrete some of these compounds onto the surface of the cuticle for use either as pheromones (Ninomiya and Kawasaki, 1988) or as cuticular waterproofing agents (see section 2.8.2).

Does this accord with the preferred diets of the species concerned? Certainly for the fatty acid responders it would appear to. *Dermatophagoides* spp. do well on 4–6% lipid (see above), *L. konoi* is a pest of dried fish with a reputation for preferring foodstuffs with a high fat content (Matsumoto, 1978) and *T. putrescentiae* likes cheese and other stored products high in oils like linseed, copra and peanuts (A.M. Hughes, 1976). With the nonresponders the situation is less clear. *Carpoglyphus lactis* prefers diets high in sugar, such as dried fruit; *Glycyphagus domesticus* is probably a fungal feeder, but like *Acarus immobilis* is found on the same stored products as *T. putrescentiae*, whereas *Aleuroglyphus ovatus* is more often found in bran, wheat and flour (A.M. Hughes, 1976).

Body lipid content of *T. putrescentiae*, *C. lactis*, *A. ovatus*, *D. farinae* and *L. konoi* was reported by Kobayashi *et al.* (1979; see Table 2.8). *Dermatophagoides farinae* had far lower triglyceride content compared with the other mites, but much higher levels of free fatty acids and cholesterol. The high level of free fatty acids in *D. farinae* may be an artefact caused by the hydrolysis of triglyceride, or there may be fundamental differences in the dynamic equilibrium between fatty acids and storage fats in pyroglyphid mites compared with acaroids. The levels of phospholipids encountered by Kobayashi *et al.* (1979) are typically higher than those found in insects, whereas the levels of triglyceride are lower. The

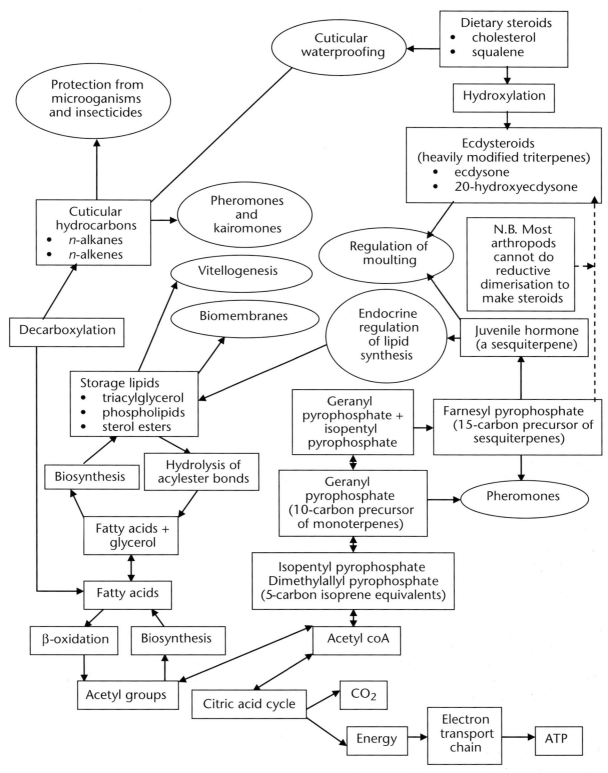

Figure 2.10 Diagram of processes in lipid metabolism of arthropods.

Table 2.8 Types of lipids as a percentage of total body lipid content of some adult astigmatid mites (modified after Kobayashi *et al.*, 1979). ND = not done.

	Total body lipid	Cholesteryl esters	Triglyceride	Fatty acids	Cholesterol	Mono-glyceride	Phospholipids
Dermatophagoides farinae	1.6	8.0	5.0	29.3	6.7	0.2	51.1
Lardoglyphus konoi	2.9	1.2	22.8	2.4	2.6	0.5	70.9
Tyrophagus putrescentiae	2.1	1.2	42.0	3.4	1.3	1.4	50.7
Carpoglyphus lactis	2.6	1.5	31.5	2.9	3.4	ND	60.7
Aleuroglyphus ovatus	2.4	1.8	41.6	1.8	1.8	ND	53.1

phospholipid fraction, which may well be the most important storage lipid in mites, consists of phosphatidylserine, phosphatidylcholine, phosphatidylinositol, phosphatidylethanolamine, cardiolipin and phosphatidic acid. Mori and Sajiki (1997) identified phospholipases in *D. pteronyssinus* that catalyse phosphatidylcholine hydrolysis to free fatty acids and acylglycerol.

Astigmatid mites have a fat body, although it differs in structure and arrangement from that of insects. Lipid globules can be seen floating around in the haemocoel and lodged in intercellular spaces in the parenchymatous tissue under the light microscope, appearing as large, lightly osmiophilic inclusions under the transmission electron microscope. In insects, shuttling of lipids from the site of storage to the site of use is conducted by a unique group of lipoproteins, the lipophorins, which are different from mammalian plasma lipoprotein (Chino, 1985). Group 14 allergens may be lipophorin-like proteins (Chapter 7).

d Carbohydrates

Although the effect of sugars and starch on the growth of several stored products mite species was studied by Matsumoto (1964, 1965a, b; 1966), no research has been done on carbohydrate nutrition of dust mites although their digestive enzyme repertoire contains several carbohydrases (see Table 2.4).

2.2.6 Feeding rates

There is relatively little data on the feeding rates of dust mites. One of the most comprehensive studies is that by Arlian (1977), who investigated how much water could be gained by *D. farinae* and *D. pteronyssinus* from feeding on yeast at 25°C and humidities from 22.5–95% RH (see section 3.4.3c; Figure 2.9b).

2.2.7 Conclusions: two models of dust mite digestion

From the above account it is apparent that our understanding of the digestive and nutritional physiology and biochemistry of dust mites is still far from complete. Little is known of precisely where and how the processes of digestion take place. The functions of different cells and tissues in the gut are still unclear, as are the dietary requirements of dust mites. However, two basic models of digestion in dust mites emerge. The first is a relatively unsophisticated, inefficient and potentially wasteful process; the second is more complex, efficient and similar to that in insects.

a Uncompartmentalised model

In this model (Figure 2.11a) it is assumed that most processing of undigested food takes place intracellularly in the midgut and midgut caecae (though this does not rule out some extracellular digestion occurring in the caecae) and that the digestive cells, together with the enzymes they contain, die after digestion and are compressed into faecal pellets and voided. Hence, there are high levels of digestive enzymes in the faeces. It is therefore expensive of energy and relatively inefficient. Products of digestion are absorbed into the haemocoel through the anterior and posterior midgut but not the midgut caecae. The peritrophic envelope commences posterior of the openings of the midgut caecae, so that the only way that food can enter the caecae is anteriorly via the oesophagus. Any partially digested food that has not been phagocytosed in the caecae may be squeezed out into the midgut where it may be further processed.

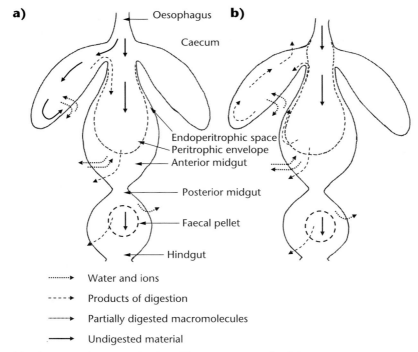

a)

Oesophagus

Caecum

b)

Endoperitrophic space
Peritrophic envelope
Anterior midgut

Posterior midgut

Faecal pellet

Hindgut

········▶ Water and ions

- - - -▶ Products of digestion

··········▶ Partially digested macromolecules

——▶ Undigested material

Figure 2.11 Models of **a)** uncompartmentalised, and **b)** compartmentalised digestive systems.

b Compartmentalised model

Intracellular and extracellular digestion takes place, with the peritrophic membrane performing more effective compartmentalisation of the gut (see Figure 2.11b). Extracellular digestion allows for partially broken down macromolecules that are small enough to pass through the peritrophic membrane to enter the endoperitrophic space. Re-circulation occurs within the endoperitrophic space, with partially digested macromolecules being moved anteriorly from the midgut through the endoperitrophic space into the caecae where final digestion takes place. Products of digestion are absorbed into the haemocoel through the anterior and posterior midgut and the midgut caecae. This model assumes the peritrophic envelope commences anterior of the opening of the midgut caecae and there is no direct access to the caecae from the oesophagus for unprocessed food. Enzymes are re-circulated and re-used or digested. The process is more energy efficient and involves a much larger area of the surface of the gut in sorption of digestive products into the haemocoel.

2.3 Excretion and osmoregulation

The functions of excretion include the maintenance of ionic concentrations at correct levels, maintenance of water content, and, hence, osmotic concentration

of body fluids; removal of waste products of metabolism and foreign substances, toxins or their metabolites. Excretory systems are thus all about maintaining constant concentrations of substances within the body and tissue fluids by removing excess amounts.

The coxal glands are the primary organs of excretion and osmoregulation in arthropods, and for several major groups they also function as a means of extracting water from unsaturated air. This role is of great relevance to water balance, dealt with in detail in Chapter 3. Since there is considerable overlap between physiological considerations of osmoregulation and water balance, these topics are dealt with together in Chapter 3, and the following section focuses on excretion of nitrogenous waste.

2.3.1 Excretion

a The removal of excess nitrogen

The major nitrogenous excretory product of arachnids is guanine. In insects it is generally uric acid or urea. A few terrestrial arthropods such as woodlice (Isopoda) and cockroaches excrete their excess nitrogen in the form of gaseous ammonia. The reason that animals need to excrete nitrogen is because they obtain in their diet, especially in the form of amino acids, far more than they can readily use. The excretory system is also intimately involved in water balance. This function of the excretory system is covered in more detail in Chapter 3. Edney (1977) and Hadley (1994) give good, readable reviews of excretion by terrestrial arthropods.

b Guanine

Guanine (2-amino-6-hydroxypurine) is one of the least water-soluble forms in which to excrete nitrogenous waste. It is about five times less soluble than uric acid, itself barely water-soluble. In *Acarus siro* it was found to act as an aggregation pheromone (Levinson *et al.*, 1991; see section 2.9.1) as well as an excretory product.

In the Astigmata, guanine crystals have been observed in faecal pellets in the posterior midgut (see Figure 2.5c), and crystals accumulate in cells associated with the body wall in the mid-dorsal region of *Dermatophagoides farinae*, as well as floating free in the haemolymph (Wharton, 1976). This suggests storage as well as removal. Mullins and Cochrane (1974) found cockroaches stored uric acid in the fat body and hypothesised that it acted, together with sodium, potassium and ammonium, as an ion sink, whereby ions could be stored or released as needed, thus aiding in osmotic regulation of the haemolymph. Perhaps stored guanine spherules in mites can play a similar function.

In *A. siro* guanine was found in the same tissue – parenchymatous cells – in individuals fed on a high-protein diet of wheat germ. T.E. Hughes (1950) gives details of the chemical determination of guanine in faecal pellets of *A. siro*. However, the mechanisms of guanine formation and excretion remain unclear. This contrasts with the considerable research effort that has gone into the quantification of guanine in house dust samples as a surrogate measure of allergen exposure.

Kort *et al.* (1997) found that on artificial test substrates seeded with house dust and *Dermatophagoides pteronyssinus* or *Glycyphagus domesticus* there was a significant correlation between *D. pteronyssinus* population density and guanine concentration after 8 weeks, but the same relationship was not found with *G. domesticus*. The authors suggested this was due to *G. domesticus* feeding on carbohydrate-rich fungi that colonised the artificial substrate, whereas *D. pteronyssinus* was feeding on protein-rich house dust. However, even if this were so, the available protein content of house dust is not necessarily higher than that of fungi: it depends on species and whether mites are feeding on mycelia or spores (see Table 2.7). Some fungi have up to 44%

protein dry weight (Griffin, 1994), and Andersen (1991) found the vast bulk of extractable protein in a yeast/skin scale culture came from the yeast. Since house dust is predominantly skin scale, it cannot be assumed that the lower guanine production by *G. domesticus* was due to lower dietary protein intake. It would be expected that mites feeding on a high protein diet would produce more nitrogenous waste than ones feeding on a carbohydrate diet, but this would best be determined using chemically defined diets.

Considering that guanine has been used widely as a surrogate measurement of mite allergen exposure (see section 8.4.3), one would perhaps expect that as a spin-off, some research would have been devoted to guanine production in dust mites. We are almost as much in the dark about this now as when Edney (1977) commented that little was known about the formation of guanine in arachnids.

c Organs involved in nitrogenous excretion

The primary role of the coxal glands as excretory and osmoregulatory organs in arthropods has already been mentioned. In dust mites, the supracoxal glands, assumed to be either modified coxal glands or other organs that have adopted the function of the coxal glands, are associated mainly with osmoregulation and water vapour uptake from ambient air and are dealt with in detail in Chapter 3 (see section 3.4.2a). An analysis of extracts of the plugs of solidified secretions in the openings of the supracoxal glands of *Tyrophagus putrescentiae* indicated that while they were mostly composed of potassium chloride, traces of ammonia were also present (Wharton and Furumizo, 1977), hinting at an ancestral function of the glands in removal of nitrogenous waste.

The posterior midgut, the hindgut and the Malphigian tubules of *Acarus siro* appear to form a functional unit in which the faeces solidify as water is resorbed and guanine becomes present in the gut contents (T.E. Hughes, 1950). The Malphigian tubules of *A. siro* contain no guanine (T.E. Hughes, 1950), neither do those of *Caloglyphus mycophagus* (Kuo and Nesbitt, 1970), and appear not to play any role in guanine formation and nitrogenous excretion. The excretory function appears to have been taken over by the wall of the posterior midgut. According to T.E. Hughes (1959), excretion takes place in *Glycyphagus* by the diffusion of wastes from the haemocoel into the posterior midgut. However, Baker (1975) recorded the presence of crystalline contents in the Malphigian

Table 2.9 Distribution of Malphigian tubules in the Astigmata.

	Tubules present or absent	Tubules with crystal-line inclusions?	Reference
Pyroglyphidae			
Dermatophagoides farinae	–		Brody *et al.*, 1972
D. pteronyssinus	–		Tongu *et al.*, 1986
Glycyphagidae			
Talpacarus platygaster	–		Michael, 1901
Glycyphagus domesticus	–		A.M. & T.E. Hughes, 1939
Acaridae			
Acarus siro	+ but reduced	–	Berlese, 1897a; T.E. Hughes, 1950
Tyrophagus longior	+ but reduced		Nalepa, 1884
Carpoglyphus lactis	+ but reduced		Nalepa, 1884
Rhizoglyphus robini	+	not mentioned	Michael, 1901; G.T. Baker & Krantz, 1985
R. echinopus	+ but reduced	not mentioned	Akimov, 1973
Caloglyphus mycophagus	+	±	Rohde & Oemick, 1967; Kuo & Nesbitt, 1970
Histiogaster carpio	+	+	R.A. Baker, 1975
Aleuroglyphus ovatus	–		Vijayambika and John, 1977

tubules of *Histiogaster carpio* and suggested that the tubules did indeed function as excretory organs. Rohde and Oemick (1967) mention 'particles' in the tubules of *Caloglyphus mycophagus*, but state that they were not tested for guanine.

Malphigian tubules occur sporadically within the Astigmata (refer to Table 2.9). Where present, they connect with the gut at the junction between posterior midgut and hindgut and are a useful point of demarcation between the epithelium-lined midgut and the cuticle-lined hindgut. Malphigian tubules were not found in *Dermatophagoides farinae* by Brody *et al.* (1972) nor in *Glycyphagus domesticus* by A.M. and T.E. Hughes (1939). They occur in the Acaridae (Baker and Krantz, 1985; T.E. Hughes, 1950), but are reduced in some species. Michael (1901) said of them:

> In the Tyroglyphidae [= Acaridae] they appear to
> be far smaller [than in gamasid mites], more
> delicate, less evidentially functional, and
> altogether strike the observer as more rudimen-
> tary, or, perhaps, nascent would be a more
> appropriate word.

If they do not function as organs of excretion in the Acaridae, then what do they do?

The best description of the histology of the Malphigian tubules of an astigmatid mite is for *Caloglyphus mycophagus* by Kuo and Nesbitt (1970), who illustrate a thin epithelium with strongly basophilic and acidophilic cells containing indistinctly stained nuclei. A series of very fine microvilli project from the cells and fill the lumen of the tubule.

2.4 Respiration and gas exchange

Animals gain their energy from the oxidation of their food, which yields carbon dioxide and water. Thus animals need to take up oxygen and get rid of carbon dioxide, a process called gas exchange, often referred to as respiration. Strictly speaking, respiration is the cellular metabolic activity that results in the production of energy, though it is used so commonly to refer to gas exchange that it is used in that sense here.

In insects and many mites, the respiratory organs consist of branching tubes called tracheae that link the surface of the cuticle with the internal organs. In the Astigmata the tracheae are absent and no other obvious respiratory organs have been identified, so how do they achieve gas exchange? The assumption has been that it occurs directly through the cuticle. Herein lies a dilemma. The need for gas exchange and the need to gain water and retain it are potentially in conflict. The cuticle is water-proofed against water loss, so can it also be sufficiently permeable to allow for efficient gas exchange? The reason that the Astigmata lack tracheae may be because the cost of tracheal water loss would outweigh the benefit of efficient gas exchange. Mites that possess tracheae and that live in dry conditions, such as ticks and some Mesostigmata, have evolved various mechanisms for reducing the rate of tracheal water loss while optimising gas exchange through a specialised series of grooves, channels, spiracles and humidity-regulating mechanisms including diffusion barriers and complex sub-spiracular chambers (reviewed by Evans, 1992). But do the Astigmata really have no specialised structures capable of gas exchange? It is possible that the supracoxal glands have the dual function of water uptake and gas exchange. This is discussed in Chapter 3.

2.5   Blood and circulation

Mite haemolyph has been poorly studied. What little is known indicates a fascinating, multi-functional system, a more detailed study of which would be likely to yield findings with important implications for mite control. The blood and circulatory system of mites consists of haemocytes suspended in haemolymph, or attached to tissues. The haemolymph is the fluid in which the tissues are bathed and is contained within the haemocoel. The circulatory system is a 'lacunar' (lake-like) one, in that it consists of a pool of fluid that is moved by the body musculature and internal organs, rather than a system of vessels through which blood is pumped to the tissues. The arrangement of the brain around the oesophagus is relevant here. Wharton (1976) observed that movement of food material through the oesophagus would be accompanied by a significant flow of blood through the brain from the anterior ventral region to the posterior dorsal region, thus providing for circulation from the neurohaemal junction to the rest of the body.

The haemolymph functions in the transport of nutrients from the gut as well as in the dissemination of hormones. It is the site for the accumulation of waste products, and it provides an important mechanical role via its hydrostatic properties in supporting the tissues and exoskeleton. Furthermore, it is the cell culture medium for haemocytes.

2.5.1   Haemocytes

Haemocytes of arthropods are known to play a role in protection against infection, in immune responses, clotting and wound repair, synthesis of respiratory pigments such as haemoglobin and haemocyanin, and in moulting. Kanungo and Naegle (1964) recognised some eight different cell types in the haemolymph of *Caloglyphus berlesei*, similar in general to those found in insects. By far the most abundant were amoebocytes (probably analogous to type I granulocytes of trombiculid mites observed by Shatrov, 1997) and these were mostly attached to muscles and other tissues. These cells play an important role in moulting (see section 2.8.3).

Kanungo and Naegle (1964) also investigated the role of haemocytes in detoxification of pesticides following from observations on the role of insects' haemocytes in insecticide resistance. They found that the haemocytes were more reactive to the acaricide 2,2-dichlorovinyl methyl phosphate (DDVP) than were other cells and tissues. There were gross pathological changes in the haemocytes of mites dipped in a 0.02% DDVP solution well before onset of any symptoms of toxicity in the mites, including distortion and eventual rupture of the nuclear membrane, vesiculation, vacuolation and breakdown of the cell membrane. They pointed out that haemocytes can therefore be used as physiological indicators to ascertain the degree of toxicity of a given chemical. This observation perhaps has potential for development of a bioassay for toxicity of house dust mite acaricides.

2.5.2   Immune responses of dust mites

Compared with insects, there is precious little information on the immune system of dust mites, or mites in general. Ichikawa *et al.* (1998) suggested that Der f 2 is part of an innate anti-bacterial defence mechanism.

2.6   Reproduction

With a few exceptions, the Astigmata are obligatory sexual reproducers, and copulation is the norm. The desmonomate oribatid mites, from which the Astigmata may have evolved, contain many thelytokous species. Thelytokes reproduce asexually (parthenogenetically) most of the time, but occasionally mate and produce exclusively male offspring. In sexually reproducing oribatids, sperm transfer occurs via the deposition by the males of stalked packets of sperm (called spermatophores), which the female then locates and inserts into her reproductive tract. In at least one species of galumnid oribatid, sperm is transferred by the male passing the spermatophore direct to the female, but there are no documented cases of intromittent sperm transfer within the Oribatida. In some acaroid species, intromittent sperm transfer occurs, but a spermatophore is involved (Griffiths and Boczek, 1977). In the Pyroglyphidae, intromittent sperm transfer occurs without a spermatophore.

The Astigmata have a posterior opening to the female reproductive system into which sperm is ejaculated or spermatophores deposited, and an anterior-ventral opening through which eggs are laid. The Oribatida has only an anterio-ventral opening, enclosed by a pair of sclerotised plates, through which spermatophores are taken up and eggs are laid. In oribatid males, the penis (also called 'the male genital organ', 'the ejaculatory complex' and 'the aedeagus') is also enclosed by genital plates, whereas in the Astigmata the penis is at least partly externalised and genital plates are absent in both males and females. There is partial evidence that the Astigmata may have evolved from within the Oribatida (Norton, 1998; see section 1.4.2). The anatomy of the genitalia of these taxa may appear superficially similar, and one can envisage how the reduction of the astigmatid ovipositor and partial externalisation

of the male genitalia could have evolved. But it is a mystery how the double-orificed astigmatid female genital tract evolved from a single-orificed oribatid ancestor.

The anatomy and ultrastructure of the reproductive systems of pyroglyphid mites are known in exquisite detail from the meticulous work of Walzl (1978; 1991a; 1992) on *Dermatophagoides farinae* and *D. pteronyssinus*. The reproductive systems of acaroid and glycyphagoid mites are essentially similar (see section 2.6.3).

2.6.1 The female reproductive system of the Pyroglyphidae

The posterior part of the female reproductive system is concerned with receipt, storage and internal movement of sperm. It comprises the bursa copulatrix, the ductus bursae, the receptaculum seminis and the ducti receptaculi. The anterior part is concerned with yolk synthesis, the production of the eggshell, and the development and deposition of the eggs. It consists of the ovaries, oviducts, accessory gland, ovipositor and oviporus (see Figure 2.12a). There is some difference in terminology between that used by Walzl (1992), which is followed here, and that used by Fain (1990). For bursa copulatrix, Fain uses 'vestibule'; for ductus bursae, he uses 'bursa' and for receptaculum seminis he uses 'spermatheca'.

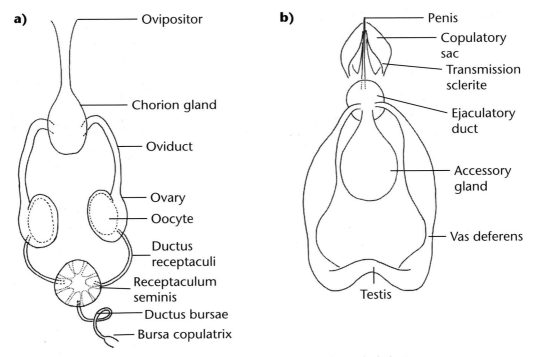

Figure 2.12 Diagram of **a)** the female and **b)** male reproductive systems of pyroglyphid mites.

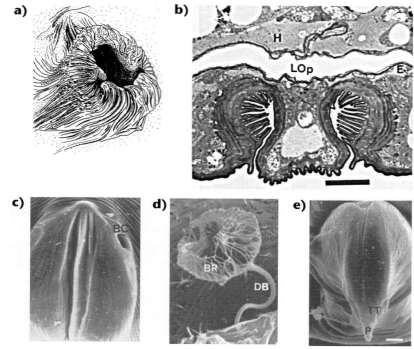

Figure 2.13 Reproductive organs of *Dermatophagoides* spp. **a)** Ovipositor of *D. farinae*, re-drawn from Walzl (1991, Figure 30.1b); **b)** cross-section of ovipositor of *D. pteronyssinus*. H = haemocoel; LOp = lumen, E = epidermis. Scale bar = 5 μm. From Walzl (1992; Figure 24); **c)** opening of bursa copulatrix (BC) of *D. pteronyssinus*. Scale bar = 2 μm. From Walzl (1992; Figure 14a); **d)** receptaculum seminis (BR) and ductus bursae (DB) of *D. pteronyssinus*. Scale bar = 10 μm. From Walzl (1992; Figure 15a); **e)** penis (P) of *D. pteronyssinus*. TT = transmission sclerites. Scale bar = 2 μm. From Walzl (1992; Figure 9b).

a The bursa copulatrix

This is the posterior, copulatory opening of the female reproductive system (Figure 2.12a, 2.13c). Some authors have used the term to include both the opening, the ductus bursae, and even the receptaculum seminis. In *Dermatophagoides* spp. the bursa copulatrix is positioned on the caudal region of the opisthosoma just to one side of the anus. In some species such as *Dermatophagoides neotropicalis* and *D. aureliani* it may be positioned on a raised papilla, but more usually the external opening is flush with the cuticle or slightly recessed into it, and about 5 μm in diameter. Walzl (1992) considers the external opening to have species-specific characteristics.

b The ductus bursae

This structure is the sclerotised tube, 1–2 μm in diameter and about 50 μm long in most *Dermatophagoides* spp., that leads from the bursa copulatrix and connects with the base of the receptaculum seminis (see Figure 2.12a, 2.13d). In *Dermatophagoides aureliani* it takes the form of a sausage-shaped unsclerotised sac immediately internal of the bursa copulatrix, with a narrow sclerotised tube emerging from it anteriorly. In *D. evansi* the

ductus bursae is broadened in the half most proximal to the bursa copulatrix, but is not sac-like. In many *Dermatophagoides* spp. the region immediately anterior of the bursa copulatrix is cone-shaped or expanded to accommodate the penis and differs in its degree of sclerotisation between species. In *D. farinae* it is heavily sclerotised, in *D. microceras* only lightly so.

c The receptaculum seminis

The tubular ductus bursae joins onto a large membranous, elastic sac, the receptaculum seminis, ca. 70 μm long and 30 μm broad in *Dermatophagoides* spp. (shown in Figure 2.13d). This sac lies free in the haemocoel and is bathed in haemolymph, so its position, though usually median, above the rectum, may be displaced laterally. The sac consists of epithelial tissue bound by a basal lamina. Where it joins the ductus bursae there is a sclerotised region of distinctive morphology. In *D. pteronyssinus*, floral comparisons are often used: this region is crescent-shaped or tulip-shaped in cross-section and daisy-shaped when viewed from above, comprising some 10–13 lobes, each corresponding to the base of a cell, connected by thickened regions of the basal lamina.

d The ductus receptaculi

Two ducts extend from the base of the receptaculum seminis towards the ovary, commencing anteriolateral of the junction with the ductus bursae. It is the ductus receptaculi that join the posterior, sperm-handling region of the reproductive system with the anterior, egg-making part and are attached to the posterior ends of the ovaries. In transverse section, the wall of the ductus is bounded by three cells.

e The ovaries

The two ovaries are situated each on either side of the ventral region of the opisthosoma lateral to the rectum. Each ovary may contain up to 12 oocyctes, connected to the central syncytial mass by what Walzl (1992) refers to as a nutritive cord.

f The oviducts

Vitellogenesis takes place within the oviducts and the oocyte volume increases during passage down the oviduct. The ova grow alternately in each oviduct so that one from the right and then one from the left oviduct move down into the accessory or chorion gland.

g The chorion gland

The chorion (or accessory) gland is positioned dorsal of the point of termination of the two oviducts. Eggs can be observed to remain in this gland for up to two days, and it is here that the chorion or 'shell' of the egg is secreted from vesicles within the gland cells. Muscles then expel the chorionated egg towards the ovipositor.

h The ovipositor

The ovipositor is an expandable tube through which the egg is laid (see Figures 2.12a, b). During oviposition the edges of the V-shaped vulva are opened in a posterior direction, and the short, tubular ovipositor is extruded (refer to Figures 2.12a, 2.13a, b). The egg is pushed out through the oviporus. The scanning electron micrographs of Walzl (1991a) clearly show the edges of the V-shaped vulva are actually part of the posterior-lateral wall of the extruded ovipositor.

2.6.2 The male reproductive system of the Pyroglyphidae

The male reproductive system consists of a single testis, two vasa differentia, an accessory gland, ejaculatory duct and the penis (shown in Figures 2.12b, 2.13e).

a The testis

The single testis is probably derived from a pair that has fused. Other Astigmata possess paired testes. It hangs suspended in the haemocoel close to the rectum in the region of the anal suckers, and contains a syncytial mass in which spermatogenesis takes place. At the posterior end spermatogonia are visible, while anteriorly, mature, closely packed sperm are present. The sperm are irregular, ellipsoid, aflagellate, and measure ca. 4 µm in length by 2 µm in breadth (Walzl, 1991a; Alberti, 1995).

b The vasa deferentia

The paired vasa deferentia originate at the anterior end of the testis. The lumen of the posterior part of the vasa deferentia may be filled with a plug of mature sperm. The walls are surrounded by minute circular muscle fibres that contract and pump the sperm towards the penis. The vasa deferentia are fused with the ejaculatory duct at the level of Legs IV, ventral of the accessory gland.

c The accessory gland

The accessory gland appears to function in the secretion of seminal fluid: its cells are packed with Golgi complexes and rough endoplasmic reticulum. Walzl (1992) found the seminal secretion to be contained in vesicles ca. 0.2 µm in diameter, which fused in the apical part of the gland. The pH in this region was found to be acidic, whereas in the basal part of the gland it was alkaline. Walzl surmised that the fusion of vesicles and the change in pH represented final stages in the maturation of the seminal fluid, prior to the blending with sperm as the final ejaculate.

d The ejaculatory duct and penis

The ejaculatory duct is located directly after the junction of the vasa deferentia with the accessory gland. Distally the duct merges into the penis, becomes more thick walled, with a lumen diameter near the apex of the penis of ca. 1 µm. I use the term penis here, rather than aedeagus, as used by several authors, not least because it is easier to remember how to spell. When not in use the bulk of the penis is contained within the body. During copulation, the penis is extruded (Walzl, 1991a), and it can be rendered so artificially for study by scanning electron microscopy by using microwave treatment (Walzl, 1991b). The penis bears laterally a pair of transmission sclerites – bladelike structures which may be involved in the location of the female bursa copulatrix immediately prior to copulation (see Figures 2.12b, 2.13e). The morphology of the penis is species-specific, for example in *D. pteronyssinus* the tips of the penis and the transmission sclerites are blunted whereas in *D. farinae* they are tapered to a point.

2.6.3 The reproductive systems of some acaroid and glycyphagoid mites

In some of the glycyphagoid mites there are some intriguing variations in morphology of the reproductive system. Instead of a bursa copulatrix, females of the genus *Blomia* have an elongated copulatory tube that extends some 30–50 μm from the posterior margin of the opisthosoma. The shape of the copulatory tube is of taxonomic significance according to van Bronswijk *et al.* (1973a, b). In *Chortoglyphus arcuatus* the penis is located between the first and second pair of legs and is disproportionately long for the size of the males.

2.6.4 Copulatory behaviour in the Pyroglyphidae

The following account is based on a combination of personal observations, using a stereobinocular microscope, of copulation events among male and female *Dermatophagoides pteronyssinus* kept in laboratory cultures; on observations recorded in the literature and on interpretations based on morphology of the male and female genitalia as described above.

Male *D. pteronyssinus* will mate either with active females or may take up station close to a quiescent tritonymph and wait for it to moult. The males are able to determine which of the quiescent tritonymphs will moult into females with a very high degree of accuracy. As part of the series of observations described above, I took eight couples in copula, killed them, macerated, cleared and mounted them in gum chloral. The series consisted of eight males, seven females (one gravid) and one tritonymph. An examination of the ventral surface of the tritonymph revealed the adult cuticle visible beneath the tritonymphal cuticle: details of the ovipositor could be seen clearly, as well as the rudiments of the ductus bursae, although not the bursa copulatorix. Male pyroglyphids mount tritonymphs that will later moult into females. This behaviour may be adaptive to maximise mating success. It seems unlikely that the males are able to select their future mates on the basis of morphological characteristics, although Oshima

and Sugita (1965) noticed the potential female tritonymphs had larger coxae than the potential males. Rather, it seems likely that the major sensory cue is chemical and that males are able to tell the tritonymphs apart on the basis of sex pheromones secreted from the lateral opisthosomal glands (see section 2.9).

During copulation the male and female are positioned facing opposite directions, as is typical of sexually reproducing astigmatid mites, as observed by van Leeuwenhoek (see Figure 4.4). This position is referred to as retroconjugate. The penis is everted from between the transmission sclerites, bent backward between legs IV of the male, and inserted into the bursa copulatrix. The male is held in position by its anal suckers attached to the posterior opisthosoma of the female and also, possibly, by the suckers on the tarsi of legs IV. The female continues to move about, even feeding, while the male is dragged behind her. The pair may remain in copula for up to 48 hours. This very long duration is because the diameter of the oval, immotile sperm is about the same as the tip of the penis (ca. 1.5–2 μm) and sperm are probably ejaculated in single file. The sperm are pumped by muscles at the base of the penis at the rate of about 150–230 contractions per minute.

Males readily detach from the females following even slight mechanical disturbance. I have observed instances of what appeared to be an act of stridulation: the production of a mechanical stimulus (invariably sound in insects – as demonstrated by crickets and grasshoppers) by the rubbing of one part of the body against another. The male used the forked seta on tarsus III to repeatedly rub backwards and forwards across the lateral region of the female's opisthosoma; the direction of movement of the tarsus running more or less transverse to the direction of the cuticular striae. The ridges and troughs of the striae would likely result in a resonant amplifying effect similar to that of running a stick across a piece of corrugated iron. The exact function of this behaviour is not known. Clearly it serves as a signal to the female, but of what? That the male is still hanging on? That copulation still has some way to go before completion? Or does it serve to stimulate the production of pheromones from the lateral opisthosomal glands (see section 2.9), and, if so, why? To ward off rival males? To trigger some physiological process in the female? These questions serve to demonstrate what little we know even about such basic events of dust mite biology as copulation, though they also serve to indicate the significant potential for discovery using simple methods and at little cost.

Females and males of *D. pteronyssinus* copulate more than once. Furumizo (1975a) observed that female *D. farinae* copulated about four times and males fertilised as many as five females. Pyroglyphid sperm is not stored internally within spermatophores as in the Acaridae (Griffiths and Boczek, 1977), so it must remain loose in the receptaculum seminis until required. How the immotile sperm make it through the long, thin ductus bursae is not known. Perhaps the long copulation time is required not only to transfer large numbers of sperm in single file but also to push the immotile sperm by application of hydrostatic pressure up the ductus bursae and into the receptaculum seminis.

2.7 Nervous system and sensory organs

2.7.1 The brain and central nervous system

The brain of dust mites is among the smallest in the animal kingdom. In adult females of *Dermatophagoides pteronyssinus* it is a tightly consolidated blob of tissue, about 30–40 μm in diameter. In other astigmatid mites the brain is described variously as wedge-shaped or sub-octagonal (Michael, 1895, 1901; Wooley, 1988). It is located in the anterior region of the body with the posterior part of the pharynx and oesophagus passing through it. Due to their small size, the structure of the brains of mites is not well known compared with that of insects. In beautifully prepared sections

for light microscopy (some of which still exist in the acarology collection at the Natural History Museum, London), Michael (1895) identified the four pairs of what he called the 'great nerves' leading to the legs (Figures 2.14a, b) and a pair of genital or splanchnic nerves. In most arthropods the sub-oesophageal and supraoesophageal ganglia are joined by circumoesophageal commissures, but in mites these structures are fused into a dense mass of tissue, penetrated by the oesophagus. Michael (1985) also described an external layer of small, round, deeply staining cortical cells and an inner fibrous mass that stains less intensely. He also gave details of relative brain volume of different species in relation to body volume. Mesostigmata had brains occupying 1–1.6% of body volume, Prostigmata 0.2–1.5%, Oribatida 0.32–0.35% and the only astigmatan species, the glycyphagid *Talpacarus platygaster*, 0.2%. The brain of *Dermatophagoides pteronyssinus* occupies about 1.5–1.6% of body volume.

The consolidation of the brain and central nervous system of mites parallels the loss of body segmentation (T.E. Hughes, 1959). In insects by comparison, in addition to the brain, each body segment has a pair of ganglia connected transversely and joined to the ganglia of adjacent body segments. This is similar to the arrangement of the central nervous system of annelid worms. The anatomy and histology of the acarine brain is reviewed by Wooley (1988). The histology and endocrine function of acarine brains has only been studied in any detail in ticks.

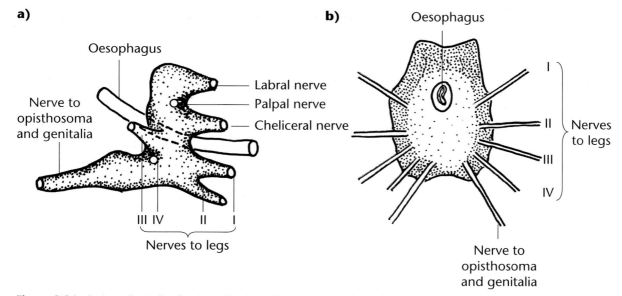

Figure 2.14 Brain and ganglia of astigmatid mites. **a)** *Acarus siro*, re-drawn from T.E. Hughes (1959); **b)** horizontal section through brain of *Hericia hericia* Robin (= *H. robini* Canestrini), re-drawn from Michael (1895).

2.7.2 Sensory organs

As mentioned in Chapter 1, every hair or seta on the body of most mites has a name. Traditionally they have been grouped according to their optical properties under polarised light. What Evans (1992) calls 'true' setae (ordinary setae, eupathidia and famuli – see below) are birefringent, containing a core of anisotropic chitin, whereas solenidia are not birefringent.

a Mechanoreceptors

Most of the setae on the body function as mechanoreceptors, detecting vibration of the air or substrate, aiding in locomotion and navigation through the detection of other objects and gravity. The very long setae of certain glycyphagoids (*Glycyphagus* spp., *Blomia* spp., *Lepidoglyphus* spp.) may offer some protection against predatory cheyletid mites. A similar arrangement of dorsal setae is found in some oribatids, such as *Neotrichozetes spinulosa*, and some heterochthoniid mites (e.g. *Heterochthonius gibbus*) where the long dorsal setae are erectile.

b Photoreceptors

Mites are blind in the sense that they lack organs that are capable of forming an image. However, they have photoreceptors, called ocelli, which have been referred to as eyes by some authors. Very few Astigmata possess ocelli. Wooley (1988) lists only the family Saproglyphidae as having them. Dust mites exposed to an incandescent light source may show what appear to be photonegative responses, but it is difficult to differentiate the effects of heat and light. Furumizo (1975a) conducted an experiment on the response of *D. farinae* to different wavelengths of 350–800 nm. There was no response at 350–475 nm and 700–800 nm. The response was positive at 500–575 nm (green part of the spectrum) and negative at 600–675 nm (orange/red part of the spectrum). These data indicate that dust mites possess photoreceptors. Where they are is not known.

c Chemoreceptors

External scapular setae

The external scapular setae of astigmatid mites (known by their notation *sce*, see section 1.5.3, Figure 1.5c) were confirmed as receptors for the sex pheromone β-acaridial in males of *Caloglyphus polyphyllae* by the microsurgical removal of one or a pair of setae (Leal *et al.*, 1989; see section 2.9.1). Mites with fully excised setae did not respond to concentrations of the pheromone even at 1000 ppm, whereas those with partially excised setae responded at concentrations of about 100 ppm, compared with ca. 1 ppm for mites with intact setae; indicating a dose-response proportional to the degree of damage to the seta.

Leal and Mochizuki (1990) extended these observations using the alarm pheromone of *Rhizoglyphus robini*, neryl formate, to elicit an escape response in non-excised individuals, but no response in those with the external scapular setae removed. Further, they discovered that these setae in *R. robini* and *Tyrophagus putrescentiae* were multiporose (unlike mechanoreceptors, which have no pores) and emerged from a membranous socket, like the sensillus of oribatid mites or the trichobothria of other acariform mites.

The external scapular setae have been regarded by Leal and Mochizuki (1990) as homologous with the sensilla (singular: sensillus; also known as 'pseudostigmatic organs', trichobothria, bothridial setae) of oribatid mites (Figure 1.5a). Yet there is not much evidence that the sensilla are chemosensory. They have been assumed to be anemometric – they detect and measure air currents – on account of the fact that they waggle about when blown upon (R.A. Norton, *pers. comm.*, 1997). Arboreal oribatids, living in habitats where air currents are likely to be strong and frequent, have regressed sensilla, whereas soil oribatids, where air currents are slight and rare have long, often elaborate sensilla (Aoki, 1973; Colloff and Niedbala, 1996). In fact, their function is not known (gravity perception has been suggested by Alberti, 1998), and the waggling behaviour could be explained as olfactory air sampling behaviour for detection of semiochemicals (see p. 91 for a definition).

So are the external scapular setae of astigmatids homologous with the sensilla of oribatids? According to Norton (1998), following OConnor (1981), the prodorsal setae of the Astigmata, the vertical internals (*vi*), vertical externals (*ve*), internal scapular setae (*sci*) and external scapular setae (*sce*) are homologous with the rostrals (*ro*), lamellars (*le*), interlamellars (*in*) and the anterior exobothridials (*exa*) of oribatids respectively, and he stated that 'these homologies are almost certainly correct'. In other words, the Astigmata are missing two pairs of setae, the sensilla (or bothridial setae, *bo*) and the posterior exobothridial setae (*exp*) from the ancestral, oribatid complement (see Figure 1.5). The evidence for sensillar loss in the Astigmata is that in the desmonomate oribatids, from within which the Astigmata are postulated to have evolved, the sensillus is often regressed. So did the hollow, multiporose, chemosensory, external scapular setae and associated alveoli of the Astigmata evolve from the solid, non-porose, mechanoreceptive, non-alveolar anterior exobothridial setae of the Oribatida? That *sce* are modified sensilla, the alternative to the *sce = exa* hypothesis preferred by OConnor (1981) and Norton (1998), seems to me to be more likely because *sce* are multiporose, have alveoli and have been demonstrated experimentally to be chemosensory.

The external scapular setae of *Dermatophagoides* spp. and other members of the subfamily Dermatophagoidinae are long and very well developed, much longer than the adjacent internal scapular setae, whereas those of the *Euroglyphus* spp. and other Pyroglyphinae are short, about the same length as the internal scapulars. Is the function of the external scapulars of these two subfamilies of pyroglyphids different? Have the external scapular setae of the Pyroglyphinae become regressed and lost their olfactory function?

Leg solenidia

The first and second pairs of legs of astigmatid and oribatid mites bear a series of hair-like structures different in appearance from the other setae (see section 1.5.6) – the solenidia. Under the light microscope these structures appear to have lateral striae arranged in a tight spiral. Under the scanning electron microscope, the tarsal solenidia, ω_1 and ω_2, of *Rhizoglyphus robini* were found to have numerous pores, indicating an olfactory function. They were similar in morphology and ultrastructure to sensilla of the chemosensory Haller's organ found on Tarsus I of ticks (Sonenshine *et al.*, 1986). But the long whip-like solenidion ϕ had no pores at all (Leal and Mochizuki, 1990).

Eupathidia

Eupathidia are terminal-pore sensilli found on the palp-tarsus. They appear to be hollow and filled with protoplasm and have been postulated to have a gustatory and/or mechanosensory function (Evans, 1992).

Famuli

The famuli are tiny sense organs, stubby or spinose, found singly on the tarsus, usually associated with the solenidia. Like eupathidia, they are hollow. Again, their function is not clear, but an olfactory role has been suggested. Alberti (1998) remarked that because of their small size they were unlikely to function as contact chemoreceptors.

d Hygroreceptors

Dust mites respond readily to changes in humidity, but organs that detect these changes have not been identified. The best educated guess is the supracoxal setae, based on their proximity to the opening of the supracoxal glands that are known to function in water balance (see section 3.5).

2.8   The cuticle and moulting

2.8.1   The cuticle

In the preface to his classic monograph 'Biology of the Arthropod Cuticle', Neville (1975) laconically observed: 'Mention the words "arthropod cuticle" to most biologists and they usually provoke a glazed expression. This is because the cuticle is commonly regarded as an inert substance.' It is also because the arthropod cuticle, together with the underlying epidermis that secretes it, is a bit like mammalian skin – it tends to get taken for granted, and is thought of as a rather dull organ system. Neville admirably dispelled these fallacies with a detailed treatment of the many structures and functions that characterise the cuticle. In Acarology texts, coverage of the cuticle is patchy. It rates not much of a mention from Evans *et al.* (1961) and Walter and Proctor (1999), though T.E. Hughes (1959), Wooley (1988) and Evans (1992) provide good coverage.

The cuticle, or integument, is the protective exoskeleton that characterises all arthropods, allowing sites for the attachment of muscles. It lines the tracheal and genital systems, as well as certain glands and parts of the gut. The cuticle has to be sufficiently permeable to allow gas exchange but impermeable enough to prevent water loss. Conservation of body water is absolutely critical to the biological success of terrestrial arthropods. The cuticle varies in thickness, hardness, composition and elasticity on different parts of the body, and in different species, in order to provide both mechanical protection and sufficient flexibility for locomotion. The disadvantage of having an exoskeleton is that it has to be shed in order for individuals to grow. This introduces a period of some considerable vulnerability into the life cycle of arthropods. During moulting and immediately afterwards the animal is immotile and thus susceptible to predation. Immediately after moulting the cuticle is soft and permeable and liable to water loss until the cuticle hardens.

The different parts of the cuticle are, like just about every other arthropod organ system, the subject of different terminology by different authors

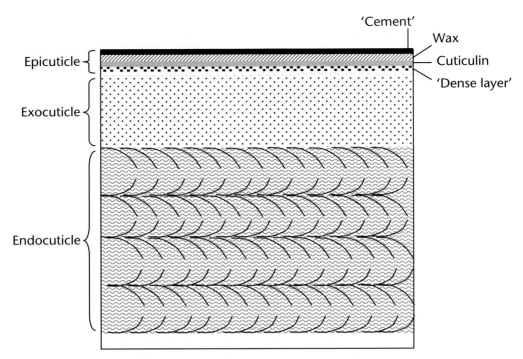

Figure 2.15 Diagram of the various layers of the cuticle of *Acarus siro*. Re-drawn after Klag (1971). Not to scale.

(see Figure 2.15; Table 2 of Alberti *et al.*, 1981). Since the ultrastructure of the cuticle is generally the same in mites as in other arthropods (Alberti *et al.*, 1981), the terminology of Neville (1975) is followed herein. In mites the wax layer is often referred to as the cerotegument.

Wharton (1976) in his review of dust mites gives a general account of acarine cuticle development, stating that the cuticulin is the first layer deposited by the epidermal cells during moulting. Beneath this the procuticle develops and differentiates into the exocuticle and endocuticle. There is continuous secretion of the extracuticulin layer of the epicuticle (which consists of the cuticulin and the wax layer above it). The wax layer is secreted onto the surface through the pore canals that traverse the cuticulin and terminate at the inner surface. Klag (1971) found the epicuticle of *Acarus siro* consisted of four layers (shown in Figure 2.15).

T.E. Hughes (1959) made a clear distinction between dark, tanned cuticle where rigidity is conferred by orthoquinone tanning and smooth, whitish flexible cuticle of most astigmatid mites, which is 'sulphur tanned'. He determined that the cuticle of *Acarus siro* contains 3–5% sulphur, which is not present as cysteine, as it is in keratin.

There are characteristic morphological differences in the external appearance of the cuticle of astigmatid mites. Acaroids such as *Acarus siro* and *Tyrophagus putrescentiae* have a smooth cuticle. Many glycyphagoids such as *Glycyphagus domesticus* have a cuticle that is covered in minute papillae (see Alberti *et al.*, 1981, their Figure 13), though in *Gohieria fusca* it is pitted and well tanned, as it is in the dispersive hypopi of both acaroids and glycyphagoids. In pyroglyphid mites the cuticle is invariably covered in fine striations consisting of minute ridges and troughs. This characteristic is also found in cheyletid mites. The surface microsculpture extends no deeper than the upper lamellae of the endocuticle (see Figure 2.16b), so in the case of mites with striated cuticle, the function of the striae would appear at first glance to be unrelated to increased elasticity. There is no apparent change in width of striae according to the hydrostatic pressure of the body, but this has not been studied in any detail. With the exception of certain specialised areas, such as the genitalia (see section 2.6), the cuticle of the Pyroglyphidae appears to be elastic only in the sense that reduction in hydrostatic pressure of the body in dehydrated individuals is accompanied by appearance of large furrows that develop laterally along the opisthosoma and transversely behind the prodorsal shield (Figure 2.16a). The same furrowing was observed by T.E. Hughes (1959) in the cuticle of *Acarus siro* which was deemed stiff and inelastic. These

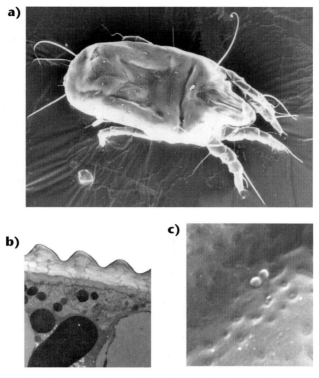

Figure 2.16 Cuticle of *Dermatophagoides pteronyssinus*. **a)** Partially dehydrated female showing cuticular folds; **b)** transmission electron micrograph of section of cuticle; **c)** scanning electron micrograph of surface of egg showing microsculpture. Original micrographs by the author.

furrows disappear when mites become rehydrated. The cuticle in the region of the female pyroglyphid ovipositor is, by contrast, highly convoluted and elastic (see Figures 2.13a, b; Walzl, 1991a). It has to be in order for the egg to get through it. The extreme convolutions of the cuticle of the ovipositor, like the rest of the striated body cuticle, do not extend further than the upper endocuticle. With the exception of the ovipositor, the function of striated body cuticle of pyroglyphids (mirrored by glycyphagoids with papillate cuticle) is, I suspect, more to do with maintaining a barrier layer of moist air close to the body surface than with flexibility.

In pyroglyphid mites, at least four different kinds of cuticle are present. There is the cuticle of the eggs (Figure 2.16c). There is the arthrodial membrane, which is flexible and elastic, and links the various leg segments with each other and the legs with the body, and at the bases of certain setae. Then there is non-striated cuticle found on the prodorsal shield and on the opisthosomal shields of males, parts of the genital and anal regions, the mouthparts, subcapitulum and legs, the lining of the pharynx and hindgut and the apodemes bordering the coxal fields. The rest of the body is covered in striated cuticle. I can find no

published images of sections through arthrodial membrane of dust mites, although Neville (1975) provides illustrations of the insect equivalent.

There are striking ultrastructural differences between striated and non-striated cuticle of *Dermatophagoides* spp. The latter, as illustrated by Tongu *et al.* (1986), is ca. 0.8 μm thick, the endocuticle has clearly defined parallel lamellae, and is penetrated by broad pore canals spaced at regular intervals. The pore canals extend from the epithelium to part of the way through the exocuticle but do not appear to open through the epicuticle onto the surface. Wharton (1976, his Figure 5B) showed that the pore canals do penetrate the exocuticle and cerotegument and open to the surface of the cuticle. The puncta in which the very narrow openings of the pore canals are situated can be visualised following electron bombardment of the cuticle of the prodorsal shield. In contrast, striated cuticle has fewer well-defined lamellae in the endocuticle, and has much narrower helicoidal pore canals spaced at irregular intervals. The reason for the difference in the structure of the pore canals would appear to be because in striated cuticle the chitin microfibrils in the lamellae are stacked on top of one another in a helicoidal arrangement (see Neville, 1975, his Figure 5.22), whereas in non-striated cuticle they are unidirectional (see Evans, 1992).

2.8.2 Cuticular hydrocarbons

Alberti *et al.* (1981) state that of the 71 species of mites they examined in their monograph of the ultrastructure of the acarine cuticle, only the tick genus *Ixodes* has wax canals in the epicuticle. I suspect this depends on terminological distinctions and what kind of cuticle one is looking at. Certainly, as Wharton (1976) has shown, the pore canals of non-striated cuticle of *Dermatophagoides farinae* penetrate the epicuticle, but do they secrete waxes? Astigmatid mites most certainly have epicuticular waxes. Leal *et al.* (1990) reported a benzoate ester, hexyl-2-formyl-3-hydroxybenzoate (common name: hexyl rhizoglyphinate) from *Rhizoglyphus robini*, a compound with antifungal properties that was recovered from cuticular extracts. Subsequently Leal and Kuwahara (1991) reported on a variety of wax-like compounds, inferring these were secreted from the lateral opisthosomal glands (see Table 2.10). What differentiated these compounds from the semiochemicals also secreted by the glands (see section 2.9.1) was their longer retention times (>10 min. compared with <8 min. for semiochemicals) when hexane-chloroform extracts of mites were analysed by gas liquid chromatography.

Leal and Kuwahara (1991) divided the wax layer compounds into four groups with their constituent species as follows:

1. Ester type compounds (Neryl myristate, hexyl linolate, hexyl rhizoglyphinate)
   Aleuroglyphus ovatus, Tyrophagus spp., *Rhizoglyphus robini.*
2. Insect hydrocarbon type (C_{27}-C_{33})
   Carpoglyphus lactis, Histiostoma laboratorium, Glycyphagus domesticus, Acarus immobilis.
3. Medium sized hydrocarbons (C_{17}-C_{19})
   Suidasia medanensis, Caloglyphus spp.
4. Compounds derived from food (cholesterol, squalene)
   Dermatophagoides farinae, Lardoglyphus konoi.

In certain respects this classification tells us more about what we do not know than what we do. The inventory of cuticular hydrocarbons listed for each species (Leal and Kuwahara, 1991, their Figures 1–3; see Table 2.10) is almost certainly incomplete, making it more difficult to discern patterns that may relate to common aspects of biology or phylogeny. Also, the function of individual cuticular hydrocarbons is unclear. Waterproofing, pheromonal and antifungal functions have been identified for some, but we are short on detail.

Where melting points are available, they are given in Table 2.10. In isolation, some compounds would be liquid at room temperature, others solid. But when mixed together, the lower molecular weight compounds act as solvents for the higher molecular weight compounds. This simple capacity to alter viscosity is likely to be related to the capacity to distribute the hydrocarbons over the entire body surface and to regulate wax consistency with changes in ambient temperature. Otherwise solid hydrocarbons will become greasy when part of a complex mixture. It is tempting to suggest that the striations of the cuticle may aid the dispersal of cuticular hydrocarbons through capillarity, although the mechanical feasibility of this has not been worked out.

So is there any relationship between the species contained in the four groups and the capacity of those species to detect fatty acids and their methyl esters (Sato *et al.*, 1993; see section 2.2.5c)? The fatty acid responders (*Dermatophagoides farinae, D. pteronyssinus, Lardoglyphus konoi* and *Tyrophagus putrescentiae*) have C_{13}, C_{15} and/or C_{17} as the longest chain hydrocarbons. The non-responders (*Carpoglyphus lactis, Aleuroglyphus ovatus, Glycyphagus domesticus* and *Acarus immobilis*) have longer chain alkanes or alkenes (C_{25}-C_{33}). The biological significance of this is unclear. It is not the whole story, and more information on the complete repertoires of cuticular hydrocarbons will be required in order to unravel another tangle in the chemical ecophysiology of astigmatid mites.

2.8.3 Ecdysis

The process of moulting in arthropods is called ecdysis and is controlled by the interaction of three major groups of hormones: neuropeptides, ecdysteroids and juvenile hormone. Ecdysone is released, presumably from the prothoracic or moulting gland. The moulting process has not been well studied in mites though in general terms it appears to resemble that of insects. The first report of the presence of a moulting hormone in Acari other than ticks was 2-deoxyecdysone, the precursor of ecdysone, purified from *Tyrophagus putrescentiae* by Sakagami *et al.* (1992).

The following stages have been identified in ticks and the account below follows that given by Evans

(1992, p. 15). Commencement of moulting is indicated by epidermal mitosis, which spreads from the anterior to posterior end of the body, concomitant with separation of the cuticle from the epidermis – the process known as apolysis. This separation is caused by the hydrolysis of the procuticular lamellae nearest the epidermis. In other words, the cuticle is broken down, which indicates the activity of a chitinase. As a result of the breakdown of the cuticular lamellae, a series of exuvial cavities are formed. Enzyme-containing moulting fluid is produced by epidermal cells and fills the exuvial cavities. Lysed cuticular components are then taken up by other epidermal cells and a new layer of epicuticle is formed, secreted from other epidermal cells.

Abnormal, post-imaginal (adult) moulting has been observed in females of *Dermatophagoides farinae* by Furumizo and Wharton (1975) and in *D. pteronyssinus* by Halmai (1989).

2.8.4   The role of haemocytes in ecdysis

The exuvial space contains cells that dissolve certain parts of the cuticle as a result of their secretory action, as mentioned above. According to Jones (1950; 1954), the cells are derived from the epidermis which splits into two layers. The outer layer lies in the exuvial spaces and those of the inner layer form the epidermis of the next instar. However, Kanungo (1969) described the activity and behaviour of cells in the moulting cycle of *Caloglyphus berlesei* and found that the exuvial space was not filled with epidermally derived cells but with haemocytes that had migrated through the epidermis. In freshly moulted tritonymphs, the haemocytes (which are basophilic at this stage) are attached to various organs but start to float free in the haemocoel after about 10 hours and aggregate below the epidermis around the anal area. Next they start to put out pseudopodia (amoebomorphosis) and begin to migrate through the epidermis (diapedesis). They are now acidophilic, and Kanungo and Naegle (1964) observed these cells in the exuvial space and described them as amoebocytes. After 16 hours there has been pronounced diapedesis of amoebocytes through the ventral epidermis, which then retracts from the cuticle. Diapedetic movement continues laterally around the body and dorsally. By 30 hours the tritonymph has become quiescent. Feeding ceases and histolysis commences. The quiescent period lasts about 15 hours. The dorsal cuticle retracts from the epidermis and the movement and aggregation of amoebocytes can be observed, by phase-contrast microscopy, within the exuvial space directly below the cuticle. By this stage histolysis of muscles has commenced, accompanied by aggregations of amoebocytes containing granular cytoplasmic components. These cells, which are larger than their counterparts in the ecdysial spaces, start to degenerate and die towards the end of the quiescent period.

The amoebocytes probably secrete enzymes that dissolve the cuticle (Jones, 1954) and also play a role in the phagocytosis of histolysing muscle tissue (Kanungo and Naegle, 1964). It is not known what factors initiate the migration of haemocytes to the exuvial space or trigger amoebomorphosis, but it is likely that the process is under hormonal control. Shatrov (1997) stated that the high level of activity of the haemocytes of trombiculid and acarid mites during moulting could be explained by the absence of moulting glands. Interestingly, in spider mites, which possess moulting glands, the haemocytes do not migrate into the exuvial space and moulting occurs very rapidly (Mothes-Wagner, 1984), indicating that the moulting glands of spider mites produce chitinolytic enzymes.

2.9   Chemical communication – pheromones and behaviour

Research on semiochemicals has grown enormously in the last 30 years and is now widely recognised as part of a discipline known as chemical ecology, with its own international association and journal. It has become a research field with important applications in agriculture and medicine, such as the management and manipulation of chemical signals for the control of populations of injurious arthropods.

The sense of smell – olfaction – is particularly important to mites and insects in communication. Chemicals used for communication are called semiochemicals (Greek: *semeion* – sign; *sema* – mark) and fall into two major categories, those used to communicate between individuals of the same species, termed pheromones (Greek: *phero* – convey), and those used to communicate between different species in which case they are called allelochemicals (Greek: *allel* – one another). This latter group includes kairomones, substances which benefit the receiver but disadvantage the producer (e.g. plants damaged by grazing insects produce terpenes that attract more insects); allomones, which benefit the producer but with neutral effects on the receiver (e.g. defensive or repugnatorial

Table 2.10 Lipids and other hydrocarbons which may function as components of the epicuticular wax layer, or as solvents for pheromones, isolated from the cuticle and lateral opisthosomal glands of various astigmatid mites (based on Leal and Kuwahara, 1991).

	Melting point (°C)	Dermatophagoides farinae	D. pteronyssinus	Carpoglyphus lactis	Histiostoma laboratorium	Suidasia medanensis	Lardoglyphus konoi	Glycyphagus domesticus	Aleuroglyphus ovatus	Tyrophagus putrescentiae	T. neiswanderi	T. perniciosus	T. similis	Acarus siro	A. immobilis	Rhizoglyphus robini	Caloglyphus polyphyllae	C. moniezi	C. rodriguezi	Caloglyphus sp.	Schwiebea sp.
Steroids and steroid precursors																					
Cholesterol	148.5	×				×	×														
Squalene	−20.0	×	×				×		×	×	×	×	×				×	×		×	
Terpene and fatty acid esters																					
Neryl myristate	ND								×												
Hexyl linolate	ND									×	×	×	×								
Alkanes																					
Tridecane C_{13}	−6.0			×				×	×	×	×	×	×	×		×	×		×		×
Tetradecane C_{14}	5.5			×					×	×	×	×	×	×	×	×					
Pentadecane C_{15}	10.0	×	×	×						×	×	×	×		×		×				
Heptadecane C_{17}	22.0														×						
Pentacosane C_{25}	53.3				×			×							×						
Hexacosane C_{26}	56.4														×						
Heptacosane C_{27}	59.5			×	×			×							×						
Octacosane C_{28}	61.4														×						
Nonacosane C_{29}	63.7			×	×			×							×						
Hentriacontane C_{31}	66.0			×																	
Tritriacontane C_{33}	72.0			×																	

	Melting point (°C)	Dermatophagoides farinae	D. pteronyssinus	Carpoglyphus lactis	Histiostoma laboratorium	Suidasia medanensis	Lardoglyphus konoi	Glycyphagus domesticus	Aleuroglyphus ovatus	Tyrophagus putrescentiae	T. neiswanderi	T. perniciosus	T. similis	Acarus siro	A. immobilis	Rhizoglyphus robini	Caloglyphus polyphyllae	C. moniezi	C. rodriguezi	Caloglyphus sp.	Schwiebea sp.
Alkenes																					
(Z)-5-tridecene C_{13}	−22.2			×						×	×										
(Z)-7-tetradecene C_{14}	−12.0									×	×										
Pentadecene C_{15} (isomer not known)	−2.8				×																
(Z,Z)-6,9-pentadecene C_{15}	ND		×	×						×	×										
Heptadecene C_{17} (isomer not known)	11.2		×		×																
Heptadecadiene C_{17} (isomer not known)	ND				×																
(Z)-8-heptadecene C_{17}	ND														×		×	×		×	
(Z,Z)-6,9-heptadecadiene C_{17}	ND														×		×	×		×	
(Z)-8-heptadecenyl formate C_{17}	ND																			×	
(Z,Z)-8,11-heptadecenyl formate C_{17}	ND																			×	
(Z)-9-nonadecene C_{19}	32.0					×															
(Z,Z)-6,9-nonadecadiene C_{19}	ND					×															
(?)-9-heneicosene C_{21}	ND					×															
(?,?)-6,9-heneicosadiene C_{21}	3.0					×															
Aldehydes																					
(Z,Z)-farnesal	ND					×															
(Z,E)-farnesal	ND					×															

chemicals that advertise the distastefulness of an organism to potential predators); and synomones, which benefit both producer and receiver (e.g. plants damaged by grazing insects produce terpenes that attract parasitoid wasps that kill the grazing insects). Incidentally, the term kairomone is widely used in a more general sense to mean an inter-specific communication substance, with no connotation of effect on producer or receiver.

Pheromones are produced by exocrine glands, are volatile, and evoke a highly specific response in the individual that detects their presence. They can be grouped by function into sex pheromones, involved in attracting a mate from some distance away or in courtship at close quarters; aggregation pheromones which evoke crowding behaviour in response to changes in habitat conditions such as sudden availability of food, or reduction in humidity or temperature; and spacing or dispersion pheromones used in keeping individuals at a distance from one another, for example during early phases of colonisation. Additionally, there are trail-marking pheromones (typically used by social insects such as ants) and alarm pheromones used, for example, as anti-predator devices, evoking rapid dispersal, and commonly found in social insects. Reviews of most of these different functional groups of pheromones in insects can be found in Kerkut and Gilbert (1985) Volume 9 (Behaviour).

Arthropod pheromones are typically small, simple, organic molecules, synthesised from more complex lipids and lipid-like compounds. Pheromones released by arthropods often consist of blends of several compounds. This is partly because some of the biosynthetic pathways result in multiple products (such as those for acetate ester sex pheromones of moths reviewed by Jurenka and Roelofs, 1993). Such blends can produce chemical messages that are more complex and subtle than those communicated by single compounds. The temptation here is to draw an analogy between arthropod pheromones and human language: the greater the vocabulary, the more scope for variety and complexity in communication. There is no need, and it is not accurate anyway: insects communicate by sound and auditory reception as well as by pheromones. We are highly pheromonal species too, and chemical communication and smell are of vital importance in human biology. We are just not very aware of it because, for most of us, communication by language is so dominant in everyday modern life.

Many molecules that mites use as pheromones are widespread in nature, occurring in plants as well as other arthropods. The monoterpenoid citral, which exists as two isomers, geranial and neral, is found in flowering plants, bees, wasps, butterflies and mites. Several of these compounds have common chemical names that are clearly derived from those of the plants from which they were first isolated. This should not be taken to imply that the mites derive these substances from feeding on plants (relatively few Astigmata feed directly on flowering plants), but rather that the capacity for their synthesis has evolved independently in several unrelated organisms.

2.9.1 Pheromones of astigmatid mites and behavioural responses

Investigations of the pheromones of astigmatid mites have been dominated by the research team of Yasumasa Kuwahara and co-workers. This pioneering work has persisted for over 25 years and represents a remarkable achievement in furthering our understanding of the chemical ecology of arthropods and the basis of mite behaviour. During this period, Kuwahara and colleagues have investigated over 60 astigmatid mite species belonging to 10 families, have identified over 90 compounds and characterised about 10 previously unknown ones, as well as a series of cuticular hydrocarbons (see section 2.8.2). What is summarised here on the chemical and biological properties of astigmatid pheromones is derived almost entirely from their research publications. These are summarised in reviews by Kuwahara (1991, 1999, 2004).

a The chemical nature of mite pheromones

Table 2.11 details the distribution of pheromones, and other compounds of unknown function, found in the lateral opisthosomal glands of astigmatid mites. Figure 2.17 gives the molecular structures of these pheromones.

Monoterpenes: the majority of the compounds in Figure 2.17 are monoterpenes (C_{10} compounds synthesised from two isoprene units, C_5H_8), comprising open-chain (acyclic) molecules such as the acaridials and single closed-ring (monocyclic) compounds like robinal and isopiperitenone. Terpenes are a group of lipids, first derived by steam distillation of plants, also known as essential oils. They are often highly fragrant, and cultures of mites containing these substances may also have a very characteristic floral, or vegetal odour. For example, *Tyrophagus putrescentiae* infestations in stored grain smell like mint.

Table 2.11 Lipids and other organic compounds with known or putative pheromone activity, isolated from the cuticle and lateral opisthosomal glands of various astigmatid mites. Edited version of full dataset in Kuwahara (2004).

Compound	Kuwahara (2004) no.	Dermatophagoides farinae	D. pteronyssinus	Carpoglyphus lactis	Histiostoma laboratorium	Suidasia medanensis	Lardoglyphus konoi	Glycyphagus domesticus	Blomia tropicalis	Aleuroglyphus ovatus	Tyrophagus putrescentiae	T. perniciosus	T. similis	T. longior	Acarus siro	A. immobilis
Acyclic monoterpenes																
Neryl formate	1	1	1								1					
Neral*	2	1	1	1		1	1	1	1	1	1	1	1			
Geranyl formate	5		1													
Geranial*	6	1	1	1	1	1	1	1	1	1		1	1			
2(S),3(S)-epoxyneral	9									1						
α-acaridial*	11										1			1		
β-acaridial*	13									1	1	1		1		
Monocyclic monoterpenes																
(S),(+)-isopiperitenone	16												1			
Furanoterpenes																
Rosefuran	24										1					1
Alcohols and phenols																
(2,6-HMBD)	31	1								1	1	1			1	1
(3-H-1,2-D)	32	1		1		1		1		1			1			1
Rhizoglyphinyl formate	35		1			1										

continued

Table 2.11 Continued

Compound	Kuwahara (2004) no.	Dermatophagoides farinae	D. pteronyssinus	Carpoglyphus lactis	Histiostoma laboratorium	Suidasia medanensis	Lardoglyphus konoi	Glycyphagus domesticus	Blomia tropicalis	Aleuroglyphus ovatus	Tyrophagus putrescentiae	T. pernicosus	T. similis	T. longior	Acarus siro	A. immobilis
Lardolure	44						1									
Hexyl linolate	49										1	1	1	1		
Alkanes and alkenes																
Dodecane	64														1	
Tridecane	65	1		1			1	1	1		1	1	1	1	1	1
Tetradecane	68	1									1	1	1	1		1
Pentadecane	71	1	1	1							1		1	1		1
Z-6-pentadecene	72	1	1							1	1					
Z-7-pentadecene	73			1							1			1		
Heptadecane	76		1													1
(Z)-8-heptadecene	77		1													1
(Z,Z)-6,9-heptadecadiene	78		1	1												1
Pentacosane	86			1				1								1
Heptacosane	87			1				1								1
Nonacosane	88			1				1								1

Male sex pheromone
Aggregation pheromone
Alarm pheromone

* Compounds known to have anti-fungal activity.

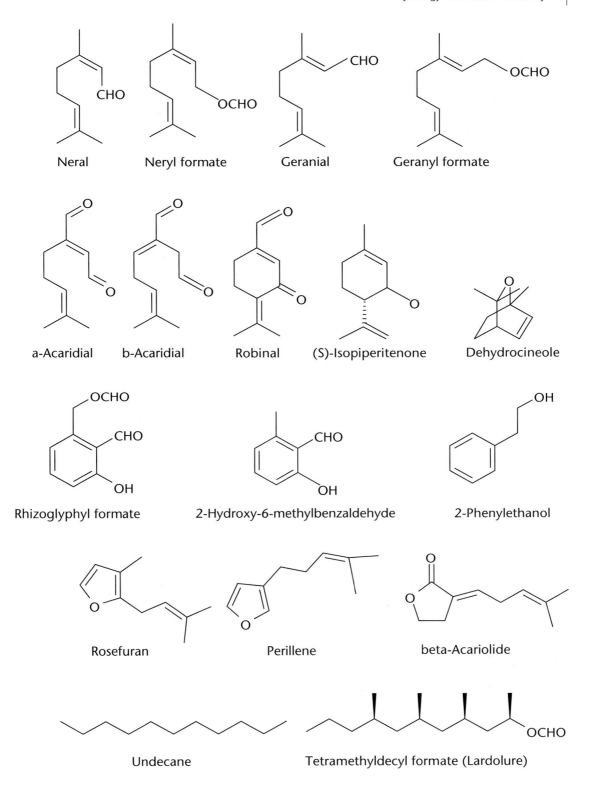

Figure 2.17 Pheromones of astigmatid mites.

Phenols: the phenols are compounds with an hydroxyl group (OH) bonded to an aromatic ring, such as rhizoglyphyl formate and the rhizoglyphinates.

Furanoterpenes: these are monoterpenes with five-membered heterocyclic ring compounds that contain oxygen such as rosefuran and perillene.

Alkanes and alkenes: this group consists of the saturated straight chain alkanes (C_{11}-C_{17}), such as undecane. The unsaturated branched chain analogues of alkanes are the alkenes. Only one is currently known as a mite pheromone – lardolure.

b   Alarm pheromones

When mites are disturbed they emit alarm pheromones that disperse any aggregations of mites of the same species. The first alarm pheromone to be isolated from astigmatid mites was neryl formate, (Z)-3,7-dimethylocta-2,6-dien-1-ol formate, from *Tyrophagus putrescentiae* by Kuwahara *et al.* (1975). A piece of filter paper was soaked in dilutions of a pentane solution of extracts of mite bodies and of isolated neryl formate and placed in the centre of aggregations of mites. The pheromone was active at concentrations of about 10 parts per million (ppm) (Kuwahara *et al.*, 1975; their Table III).

Kuwahara *et al.* (1980) extended these observations to other astigmatid species and found that *Lardoglyphus konoi* and *Carpoglyphus lactis* were highly sensitive to neryl formate and citral (a mixture of neral and geranial); the (Z) and (E) isomers of citral, both at 1 ppm and 100 ppm respectively, whereas *Aleuroglyphus ovatus* and *Dermatophagoides farinae* reacted only at concentrations of 1000 and 10 000 ppm respectively for neryl formate and 1000 ppm for citral. The presence of citral in all species, but the low reactivity of these latter two, indicates that it serves a function other than as an alarm pheromone, probably as a mould inhibitor (Cole *et al.*, 1975; see section 2.9.1g). Alarm pheromone activity of crude hexane extracts of each of the abovementioned species revealed a degree of species-specificity that was absent when pre-prepared neryl formate and citral were used in the filter paper bioassay. Therefore the signal that triggers dispersal behaviour is a complex one, reflecting either a species-specific concentration range of citral and/or neryl formate, possibly in different ratios, or the presence of other, unidentified components. The composition of discharged volatiles in disturbed mites was 10–50 times higher than that of undisturbed controls, and had a higher concentration of citral

containing a much higher ratio of neral to geranial (Kuwahara *et al.*, 1980), indicating that the mites control the relative concentrations of the components in order to generate chemical signals of enhanced specificity.

Neral and neryl formate are the most common alarm pheromones of astigmatid mites (Kuwahara, 2004). Geranial is much rarer – only *Histiostoma laboratorium* shows an alarm response to it. Other rarer alarm pheromones include β-acaridial in *Tyrophagus longior* (which functions as the female sex pheromone in *Caloglyphus polyphyllae*), isopiperitenone in *Tyrophagus similis*, 2,6-HMBD in *T. perniciosus*, a mix of C_{13}, C_{14} and C_{15} monoenes in *T. neiswanderi* and heptadecadiene in *Tortonia* sp.

c   Aggregation pheromones

Astigmatid mites grown in culture will aggregate into piles, often in response to lowered humidity as a water conserving behavioural mechanism (see section 3.4.4). Only three aggregation pheromones have been isolated from astigmatid mites: 1,3,5,7-tetramethyldecyl formate (common name, lardolure) from *Lardoglyphus konoi* (Kuwahara *et al.*, 1982; Mori and Kuwahara, 1986a, b; My-Yen *et al.*, 1980b) which also acts as an aggregation kairomone for *Carpoglyphus lactis*, *Aleuroglyphus ovatus* and *Tyrophagus putrescentiae* (My-Yen *et al.*, 1980a), 2-phenylethanol from *Caloglyphus* sp. (Mori and Kuwahara, 2000) and guanine, which has a dual function in *Acarus siro* as the major excretory product, as in other arachnids, and as an aggregation pheromone (Levinson *et al.*, 1991). 2-phenylethanol has also been identified as an aggregation pheromone of the bark beetle, *Ips paraconfusus*, by Renwick *et al.* (1976) and as a sex pheromone in the cabbage looper moth, *Trichoplusia ni* by Jacobson *et al.* (1976).

d   Sex pheromones

Female sex pheromones are produced by females and stimulate males. Male sex pheromones are produced by males and stimulate females. 2(E)-(4-methyl-3-pentenylidene)-butanedial (common name, β-acaridial), a novel acyclic monoterpene, was the first sex pheromone identified from an astigmatid mite *Caloglyphus polyphyllae* (Leal *et al.*, 1989; Kuwahara, 1995). When exposed to the pheromone, males stopped feeding and sought females to copulate with. Both virgin and mated females were found to possess β-acaridial, and males responded to concentrations of ca. 1 part per million.

2-Hydroxy-6-methyl-benzaldehyde (2,6-HMBD) was identified as the female sex pheromone of *Aleuroglyphus ovatus* and *Acarus immobilis* by Kuwahara *et al.* (1992) and Sato *et al.* (1993) respectively, whereas it functions as an alarm pheromone of *Tyrophagus perniciosus* (Leal *et al.*, 1988).

One of the more remarkable outcomes of the research on astigmatid semiochemicals has been the finding of a simple hydrocarbon molecule, undecane ($C_{11}H_{24}$), a straight-chain alkane, which is highly biologically active even though the molecule is saturated (Mori *et al.*, 1995). It is the female sex pheromone of *Caloglyphus rodriguezi*. The obvious questions that arise, but remain unanswered, are what is the active site of the molecule and how does it interact with the receptor? Undecane is also a sex pheromone and alarm pheromone in ants (Walter *et al.*, 1993).

Other female sex pheromones include (R)-epoxyneral and rosefuran in *Caloglyphus* spp. and β-acaridial in *Caloglyphus polyphyllae*. The only male sex hormones isolated so far has been an alkane-alkene mix (C_{13}, C_{15}, $C_{17}d$, $C_{17}e$) from *Acarus immobilis* (Sato *et al.*, 1993).

e Pheromone mixtures, solvents and signal specificity

The terpenes and phenols represent only minor components of the pheromone mixtures by volume. The major components of the lateral opisthosomal glands of *Tyrophagus putrescentiae* are *n*-alkanes (C_{13}-C_{15}) and the corresponding monoenes and a pentadecadiene (Kuwahara *et al.*, 1979). Howard *et al.* (1988) found complex mixtures of alkanes, alkenes and alkadienes in six species of acarid mites, and consider that the contents of the opisthosomal glands are best characterised as dilute solutions of oxygenated terpenes in hydrocarbon solvents. The role of the hydrocarbons, other than as solvents, is unclear, although undecane and Z,Z-6,9-heptadecadiene function as sex pheromones and alarm pheromones respectively. Howard *et al.* (1988) consider a defensive, anti-predator role is possible, drawing upon functional analogies from similar glandular hydrocarbon mixtures found in other arthropods.

The greater the complexity of the chemical message, the greater the prospects are for species-specificity. Single-compound messages would appear to be rare. In fact, a glance at Table 2.11 indicates that many compounds could not operate effectively as pheromones on their own without also acting as kairomones. The methods of increasing the complexity and species-specificity of the signal are to use several different compounds, or different isomers and to vary the ratios of the ingredients (Mori and Kuwahara, 2000).

One of the reasons that a relatively small repertoire of compounds can be used to send specific messages is because species are often ecologically partitioned, and do not come into contact with pheromones of another species that would elicit a kairomonal response in themselves. For example, 2,6-HMBD is released as a female sex pheromone by *Aleuroglyphus ovatus* and *Acarus immobilis*. But *A. ovatus* is a true stored products species, found in bran, flour, wheat germ and dried fish, whereas *A. immobilis* is found outdoors in birds' nests, vegetation and farm storage premises.

f Chemical, stereochemical and isomeric requirements for biological activity

The alarm pheromone of *Tyrophagus putrescentiae*, neryl formate, is an acyclic monoterpene (see Figure 2.17). In order to investigate the relationship between chemical structure of monoterpenes and biological activity, Kuwahara and Sakuma (1982) synthesised 15 compounds, based on modifications to the structure of neryl formate, particularly the main carbon chain and substitutions at C_3, and tested them for their alarm pheromone activity. They found there was no need to preserve the monoterpene structure and the prerequisite for biological activity detectable at ≥1000 ppm was the presence of the (Z)-allylic primary alcohol formate on an octane or nonane carbon chain (12 and 13 carbon chains were inactive). Substitution of the methyl residue at C_3 increased sensitivity to 10–100 ppm and methyl substitution at C_7 on a (Z)-2-n-octenol carbon chain increased sensitivity to 100–1000 ppm. The native pheromone, neryl formate, possesses all these attributes.

There appear to be highly specific stereochemical requirements for some pheromones if they are to show any biological activity. *Carpoglyphus lactis* and *Lardoglyphus konoi* showed far less sensitivity to a synthetic mixture of synthesised stereoisomers of lardolure than they did to the natural aggregation pheromone (Kuwahara *et al.*, 1982). (1R,3R,5R,7R)-lardolure was found to be the active natural pheromone, whereas the isomer (1S,3S,5S,7S)-lardolure was found to be completely inactive (Mori and Kuwahara, 1986).

g Anti-fungal activity of pheromones

Table 2.11 shows the compounds isolated from the lateral opisthosomal glands of various astigmatid mites. Those marked with an asterisk are known to have anti-fungal activity. Of the various groups of compounds, the acyclic monoterpenes have a high proportion of anti-fungal agents, including neral and geranial (Cole *et al.*, 1975; Okamoto *et al.*, 1978) and acaridial, which are common among pyroglyphid, acaroid and glycyphagoid species alike. The cuticular terpene esters methyl rhizoglyphinate and hexyl rhizoglyphinate from *Caloglyphus* sp. and *Rhizoglyphus setosus* respectively (Table 2.10) also have potent anti-fungal activity (Leal *et al.*, 1990).

2.9.2 Pheromone production and release – the lateral opisthosomal glands

These paired, fluid-filled glands are, as the name suggests, situated one on either side of the opisthosoma. They have also been called oil glands, expulsory vesicles, dermal glands, latero-abdominal glands and opisthonotal glands. They occur generally in adult and immature Astigmata and Oribatida (except Enarthronota and Palaeosomata). Michael (1901) noted that the glands were present in all life history stages subsequent to the egg, including the hypopus. In lightly sclerotised mites, the glands often show up as brownish or yellowish hemispherical or pouch-like structures. The gland consists of a single cell with a cuticle-lined lumen, linked by a duct to a crescent-shaped external opening with a cuticular flap which opens and closes the gland (Brody and Wharton, 1970). The shape of the opening of the gland and its flap varies between species (Howard *et al.*, 1988). The entire structure is moulted and new glands are formed from an undifferentiated hypodermal cell. The hypodermal cytoplasm surrounding the lumen of the gland contains large lipid droplets, glycogen granules and rough endoplasmic reticulum, indicating a secretory function. The contents of the gland contain both volatile and viscous components that appear to be expelled intermittently by contraction of the muscles surrounding the gland.

The lateral opisthosomal glands were first described by 19th century acarologists (see Michael, 1901) but their function was not known until Kuwahara *et al.* (1979) noticed that the colourless glands of live *Tyrophagus putrescentiae* darkened after the mites were mounted on microscope slides in de Fauré's medium (see section 6.6.7). The researchers had identified neryl formate as the alarm pheromone (see section 2.9.1b), and found a much lower frequency of darkening among those mites that had been disturbed and had discharged the contents of their glands than among undisturbed mites. They reasoned that the darkening was due to the reaction of neryl formate with de Fauré's medium, which they confirmed experimentally, and that the alarm pheromone was emitted from the glands. My-Yen *et al.* (1980a) found the lateral opisthosomal glands of *T. putrescentiae* contained citral as a minor component and stained the glands using Purpald (4-amino-3-hydrazino-5-mercapto-1,2,4-triazole), an aldehyde-detecting reagent, which gives a red-purple colour reaction in the presence of citral at alkaline conditions. Kuwahara *et al.* (1980) extended these observations to *Lardoglyphus konoi, Carpoglyphus lactis, Aleuroglyphus ovatus* and *Dermatophagoides farinae* all of which were found to contain citral in the lateral opisthosomal glands.

2.10 Consequences of acarine physiology for ecological interactions

An abiding theme in the physiology of pyroglyphid, acaroid and glycyphagoid mites is the multiple functions played by many components, especially of the digestive, excretory and communication systems. This versatility is not necessarily indicative of some elegant economy of nature; rather, it gives an indication of how some physiological systems could have evolved. For example, terpenoid, alkane and alkene pheromones, which are the mainstay of acarine communication, represent the end-products of an intricate series of lipid biochemical pathways (see Figure 2.10). Some of them are also potent anti-fungal agents (see section 2.9.1g). Did mite pheromones evolve as a side-product of the metabolism of dietary lipid in an environment where it is important to keep fungi at bay? Are pheromones a means of excretion of excess lipid moieties? In insects, excess fatty acids from triglyceride hydrolysis may be egested in the faeces, possibly due to inefficient lipid absorptive capacity of the gut (Turunen, 1979). We know that mites have the chemosensory capacity to detect fatty acids in potential food sources and that some fatty acids have pheromonal properties both in insects and mites (see Sato *et al.*, 1993). In any event, terpenoid pheromones determine behaviour and interactions of domestic mites and thus set the baseline for some fundamentals of their ecology.

Lipids have such a vital role in the lives of terrestrial arthropods, but are often neglected or underestimated by researchers. They function as internal communicators – hormones – signalling and mediating metabolic processes such as moulting, reproduction and immune responses; they are the signal molecules for communication with other individuals of the same and different species – pheromones and kairomones; they make the cuticle waterproof, act as antimicrobial agents, affect the rate of uptake of pesticides and function as energy storage compounds.

Another example of multiple roles of components in acarine physiology is the dual function of guanine in *Acarus siro*, as an excretory product and as a pheromone. Levinson *et al.* (1991) postulate a mutualistic relationship between *A. siro* and the moulds and yeasts that colonise the endosperm of wheat, a preferred food for this mite species. *Acarus siro* tends not to

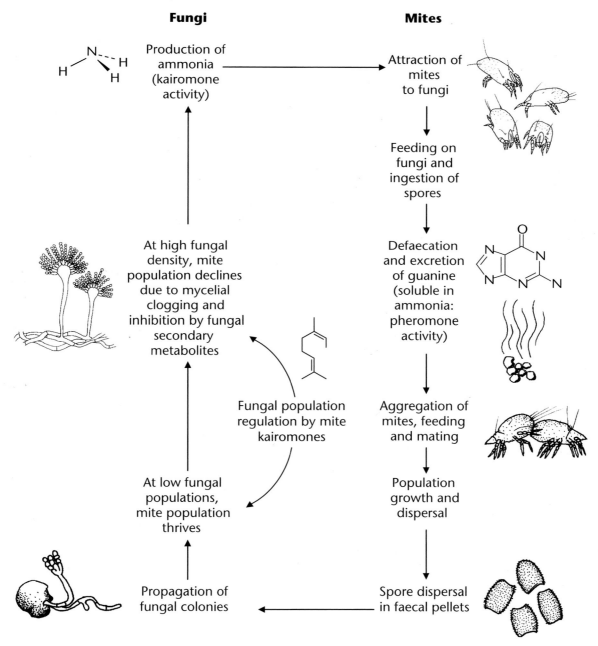

Fungi **Mites**

Production of ammonia (kairomone activity) → Attraction of mites to fungi

Feeding on fungi and ingestion of spores

At high fungal density, mite population declines due to mycelial clogging and inhibition by fungal secondary metabolites

Defaecation and excretion of guanine (soluble in ammonia: pheromone activity)

Fungal population regulation by mite kairomones

Aggregation of mites, feeding and mating

At low fungal populations, mite population thrives

Population growth and dispersal

Propagation of fungal colonies ← Spore dispersal in faecal pellets

Figure 2.18 Conceptual model of the relationship between domestic mites and fungi.

consume the endosperm unless it is mouldy (Solomon, 1962). The fungi produce ammonia as a terminal metabolite of the amino acids and purines from broken and sprouted wheat. Ammonia acts as a highly attractive kairomone for colonising mites, and provides a means by which the otherwise insoluble, non-volatile, guanine can be dissolved and dispersed. Fungi also produce volatile octadienols that are strongly attractive to some acarid mites (Vanhalen et al., 1980). The colonising mites feed on fungi and defecate. Other mites aggregate at the site in response to the detection of guanine, which indicates that other mites are present and there is the opportunity to mate. They mate and disperse, having fed on mouldy endosperm, reproduce, egest and disseminate the fungal spores in their guts. Thus new fungal cultures are propagated (see Figure 2.18).

This association brings to mind the relationship between dust mites and fungi, hypothesised by van Bronswijk and Sinha (1971) which Saint Georges-Gridelet (1984, 1987) asserts is a 'true symbiotic relationship', whereby shed human skin scales are colonised by moulds such as members of the Aspergillus glaucus group, which break down the lipids in the scales to form free fatty acids. The partially digested skin scales are then supposed to be more palatable to the mites. Bearing in mind the reservations of Douglas and Hart (1989) that the proof for this dietary modification theory is a little thin, and considering the mites can probably do just as well from feeding on the fungi direct, it is tempting to suggest that the dust mite–fungus relationship may revolve less around predigestion of skin scales as a major advantage for dust mites and more around the capacity for the guanine in its faeces to be solubilised by ammonia produced by fungi, in order to attract partners for mating. The scabies mite Sarcoptes scabiei was found to be attracted to a variety of nitrogenous compounds including guanine, purine, adenine, allantoin, xanthine, uric acid and various ammonium salts (Arlian and Vyszenski-Moher, 1996) and there is evidence for an aggregation pheromone in the faeces of Dermatophagoides farinae, although it is not yet known whether it is guanine (Reka et al., 1992). One can envisage intense selection pressure on a minute blind arthropod that lives in house dust, and is obligately sexually reproducing, to evolve a fail-safe mechanism for finding a mate. In contrast, the high abundance of potential food items in house dust would suggest less intense selection pressure to evolve mechanism to find food.

3. Water balance

Life was born in water and is carrying on in water. Water is life's mater *and* matrix, *mother and medium. There is no life without water.*

Albert Szent-Györgyi, 1971

3.1 Introduction – water and survival

What ultimately determines where dust mites can live? Access to adequate food and a favourable habitat are inconsequential unless the mites can fend off dehydration, reach maturity and reproduce. The ability to control body water under fluctuating conditions of temperature and humidity is a key constraint that determines where in homes dust mites can survive; in which homes, and in which parts of the world. It is commonly assumed that humidity is *the* major factor controlling distribution and abundance of dust mites. This is only partly true. Water balance depends on atmospheric humidity, but also on temperature. In fact, it is impossible to consider the effects of humidity without also taking temperature into account.

The literature on water balance in terrestrial arthropods is extensive, with the monograph by Edney (1977) representing an essential source, as well as the synthesis of more recent research by Hadley (1994). Most research on water balance of mites has been done by Wharton, Arlian *et al.* in the USA and Knülle *et al.* in Germany. Both groups have published comparative reviews of water balance in mites and insects, differing in emphasis. Wharton and Arlian (1972a) focus on critical equilibrium activity and methods of determining water fluxes. Wharton and Richards (1978) concentrate on the kinetics of transpiration and sorption. Arlian and Veselica (1979) deal with the physical and chemical

aspects of diffusion. Wharton (1978) and Wharton *et al.* (1979) cover measurement of water loss and uptake using tritiated water. Knülle (1984) reviews evolutionary and ecophysiological implications of water vapour uptake in mites and insects. Wharton (1985) takes a comparative approach to water balance in insects and mites, with emphasis on active water uptake. Arlian (1992) gives a concise account of all the key aspects of water balance in mites.

3.2 Water and water vapour

Water balance is a crucial part of the lives of dust mites. But it is one of the most difficult topics for non-specialists to understand because of its basis in physical chemistry. I have tried to use simple terms and formulae where possible, and have not dealt in detail with the physical and biological properties of water and water vapour. This can be found in the monograph by Edney (1977) and a detailed treatment of atmospheric humidity is given by Rosenberg (1974). Nor have I covered in any depth the mathematics used to derive water exchange kinetics using tritiated water. These are given by Arlian and Wharton (1974) and by Arlian and Veselica (1979).

3.2.1 Biological measures – water activity and osmolarity

The colligative properties of water and aqueous solutions relate to the concentration of molecules or ions in that solution. They include water activity, osmotic pressure, freezing point depression and lowering of vapour pressure. The reason they are important in biology is because they determine in what direction water will move and the energy required for water to be moved against a concentration gradient. If body water intake is uncontrolled, the haemolymph and cellular fluid will become dilute. Cells will swell and eventually burst, and the dilution of body fluids will disrupt chemical reaction rates and metabolic processes. If body fluids become too concentrated as a result of uncontrolled body water loss, then those metabolites that are only sparingly soluble will precipitate out causing cell damage and eventual death. The water content of the haemolymph and tissues has to be maintained within a range that allows metabolic processes to function. But at the same time, arthropods have to undergo substantial seasonal and diurnal changes in ambient temperature and humidity that can dramatically affect rates of body water uptake and loss, and dust mites are no exception. The capacity for the haemolymph and tissues of dust mites to survive and regulate changes in concentration is the reason that an understanding of the mechanisms of water uptake and loss is so important. The water volume of haemolymph and tissues is regulated by haemolymph osmoregulation, by which the solutes responsible for regulation are moved around within the body; and excretory osmoregulation, by which solutes are expelled.

The concentration of water within organisms is expressed as water activity (a_w) which is the ratio of water molecules in the body to all other molecules, also called the mole fraction. The activity of pure water is 1, and where water is absent it is 0. The water activity of most terrestrial arthropods is about 0.99, that is 99% of all molecules in their bodies are water (Edney, 1977). Air has a relative humidity significantly lower than 99% so terrestrial arthropods lose water to the air. The a_w of the haemolymph of *Dermatophagoides farinae* is 0.987 (Davidson, unpublished, quoted by Arlian and Wharton, 1974), yet the a_w at which water loss equals water gain in these mites is 0.7. In order to maintain body water at a_w 0.987, water has to be actively pumped into the body from the air because the tendency is for water to move from high activity to low activity. Water activity (that is the mole fraction of water in the liquid phase), water vapour activity (a_v; that is

the mole fraction of water in the vapour phase) and humidity are related as follows:

$$a_w = \frac{RH}{100} = a_v$$

In other words, a_v is the same as relative humidity (RH) but is expressed as a ratio rather than as a percentage.

The relationship between water activity and osmolarity is particularly important (see Wharton, 1985). Arlian and Veselica (1979) give a conversion table of water activity to osmolarity. The number of moles of water in 1000 g is 56 (i.e. 1000 divided by 18, the molecular weight of water), so the formula for conversion is as follows:

$$a_w = \frac{55.508}{55.508 + osmoles}$$

Osmotic pressure is the force required to prevent a solution from mixing with the pure solvent (measured in osmoles per kg, bars, atmospheres or millimetres of mercury – mm Hg). In haemolymph with an a_w of 0.987, the osmotic pressure is 0.731 osmoles kg^{-1}. In haemolymph with an a_w of 0.99 it is 0.561 osmoles kg^{-1}.

3.3 Water uptake versus loss, and the critical equilibrium activity

Wharton and Arlian (1972) made the crucial point that water is unique among major 'nutrients' because it is significantly volatile in the biological temperature range, existing as liquid and vapour, and thus terrestrial arthropods exchange water with the air. The higher water activity within arthropods than in surrounding air not only means that water is lost to the atmosphere, but that any water gain from the atmosphere has to take place against a steep concentration gradient, as explained above.

Populations of *Acarus siro* were found to lose weight at low humidities and gain weight when moved from low to high humidity (Solomon, 1962). Knülle (1962) demonstrated the same phenomenon, and found gain of water vapour by *A. siro* took place at 75% RH but not at 70% RH. He fixed the point at which water loss balanced water gain at 71%. This is the so-called critical equilibrium humidity (CEH) (Wharton, 1963; Knülle and Wharton, 1964), also known as the critical equilibrium activity (CEA)

a)

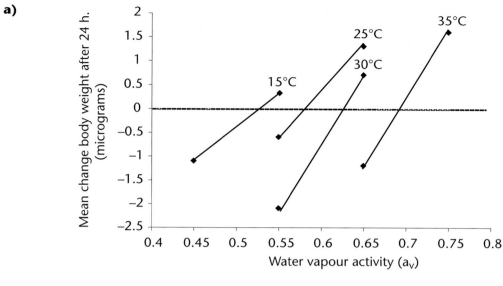

b)

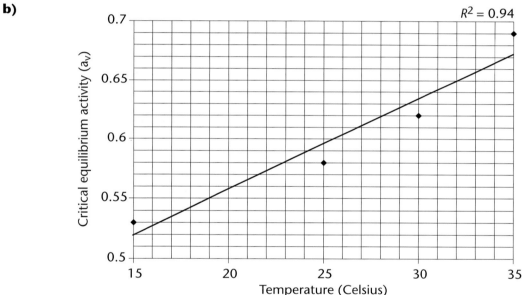

Figure 3.1 a) The effect of temperature and water vapour activity on change in weight of partially dehydrated adult female *Dermatophagoides farinae*. Intercepts of slopes with dashed horizontal line passing through zero correspond with the critical equilibrium activity at each temperature. Re-drawn from Arlian and Veselica (1981a); **b)** critical equilibrium activities of *D. farinae* females at different temperatures, based on data in Figure 3.1a.

(see Figures 3.1a, b). At humidities below the CEA, mites lose water, dehydrate and die. They can only take up water above their critical equilibrium activity.

A variety of mites and insects have the capacity to take up water from unsaturated air (Edney, 1977; Hadley, 1994). They remove and concentrate water from the air and imbibe it as an alternative to drinking liquid water. Many of the arthropods that are capable of doing this have a tendency to live in habitats in which liquid water is rare or absent.

Others are parasitic (ticks, some Mesostigmata, fleas) and are able to gain water from their blood meal when they are able to find a host, but may otherwise live in relatively dry conditions. But other arthropods capable of water vapour uptake may live in humid environments, such as certain beetles and larval Lepidoptera. As Knülle (1984) remarked, shortage of free water in the environment has not necessarily been a prerequisite for the acquisition of the faculty of water vapour uptake.

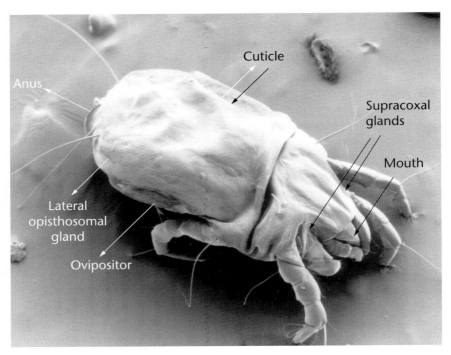

Figure 3.2 Routes of water uptake (black) and loss (white) from an adult female dust mite.

The water balance equation for dust mites, that is rate of total water uptake versus rate of total water loss, can be represented simply as follows (see Figure 3.2):

$$\text{Ratio}_{\text{uptake: loss}} = (A + B + C) / (W + X + Y + Z)$$

Where A = loss through oviposition (females), B = loss through cuticular transpiration; C = loss through egestion; W = active uptake through the supracoxal glands; X = passive uptake by transcuticular sorption; Y = uptake through eating and drinking; Z = uptake through generation of metabolic water. Each of the components of this equation is examined in more detail below.

3.3.1 Determination of critical equilibrium activity and water flows

CEA and water gain and loss have been determined in various ways. Knülle (1962) used a sort of choice chamber method, allowing mites to move to different humidities. Wharton, Arlian and co-workers have used gravimetric techniques using a microbalance. This method suffers from the disadvantage that it is relatively non-specific. Weight losses and gains can be due to water movement via different, independent routes. More importantly, it cannot differentiate between the net effect of a small water gain and a small water loss versus a large water gain and a large water loss (Arlian

and Veselica, 1979). In order to measure actual rates of water flux, tritiated water techniques have been used, predominantly to measure transpiration rates (see Arlian and Wharton, 1974; and reviews by Wharton and Arlian, 1972; Arlian and Veselica, 1979; Wharton *et al.*, 1979). If mites are exposed to tritiated water vapour (i.e. water containing the radioactive hydrogen isotope tritium, ^{3}H) they take up tritium until the concentration of tritium in the body comes into equilibrium with that in the air. If the mite is removed to tritium-free air, it transpires the tritiated vapour and the radioactivity can then be measured to give an estimation of the rate of transpiration (see Arlian and Veselica, 1979, for more details of the technique).

3.3.2 Effects of temperature on critical equilibrium activity

Critical equilibrium humidities for various astigmatid mites are given in Table 3.1. In some arthropods CEA does not vary with temperature, whereas in others it is strongly temperature-dependent (Edney, 1977). The CEA of *Dermatophagoides farinae* was found to be proportional to temperature (Arlian and Veselica, 1981a; 1981b [The data in these two papers is basically the same. The first is a brief publication; the second is a longer version]). Until this investigation, all CEA determinations of mites had been done at a single temperature (see Table 3.1). The CEA of partially dehydrated adult female *D. farinae* (loss of 30–50% of body water) was determined gravimetrically at combinations of a_v from 0.45–0.95 and temperatures of 15–40°C. Dehydrated mites had regained most water at 0.55 a_v at 15°C, 0.65 a_v at 25–30°C, and 0.75 a_v at 35°C (refer to Figure 3.1a). As temperature increased, so did the a_v at which mites could achieve net water gain. In other words, the CEA lowered with decrease in temperature from an a_v of 0.69 at 35°C to 0.52 at 15°C (shown in Figure 3.1b). The actual CEA at any given temperature is the mean between the lowest a_v at which water uptake occurred and the highest a_v at which water was lost, represented in Figure 3.1a by the intercept at 0.

The discrepancy between the CEA of 0.58 a_v at 25°C and the CEA of 0.7 a_v at the same temperature calculated by Larson (1969) may have been due to differences in hydration. Arlian and Veselica (1981a) used mites that had lost up to 50% of their body water

after deliberate dehydration for 24 hours at 0 a_v prior to determination of the CEA. The relatively long period of standardisation in a desiccating environment allowed for measurable net water gain. Larson (1969) used mites that were dehydrated for only 6 hours, resulting in far less water loss, so it was more difficult to detect water gain when mites were returned to a hydrating a_v near their CEA. Considering that by 1981 the CEA of *Dermatophagoides pteronyssinus* (0.73 a_v) had only been assessed at one temperature (25°C; Arlian, 1975a), the question arises as to whether *D. pteronyssinus* has a similar pattern of variation in CEA with temperature as *D. farinae*. Prior to determination of CEA, adult female *D. pteronyssinus* were standardised at 75% RH and 25°C for 24 hours, conditions at which they would have lost relatively little water. By invoking the same argument that Arlian and Veselica (1981a) used to explain the discrepancy between their results and Larson's (1969) results, it follows that a gravimetric determination of CEA using

Table 3.1 Water relations of some stored products mites (Acaridae and Glycyphagidae), some dust mites (Pyroglyphidae) and a feather mite (Proctophyllodidae). ND = no data provided.

	CEA a_v	T°	Reference	Optima for population increase		Reference
				T°C	RH%	
Acaridae						
Acarus siro	0.75 0.71	20 22	Solomon, 1962, 1966; Knülle, 1962, 1965	20–25	90	Cunnington, 1985
Tyrophagus putrescentiae	0.75–0.84	25	Cutcher, 1973	32	90	Cunnington, 1969; Barker, 1967
Glycyphagidae						
Glycyphagus domesticus	0.7–0.76	15	Seethaler *et al.*, 1979	25	90	Hora, 1934
Lepidoglyphus destructor	0.7–0.75	20	Eckert, in Gaede & Knülle, 1987			
Pyroglyphidae						
Dermatophagoides farinae	0.7 0.69 0.63 0.58 0.52	25 35 30 25 15	Larson, 1969 Arlian & Veselica, 1981a	27	75	Furumizo, 1975a
D. pteronyssinus	0.73 0.57	25 16	Arlian, 1975a; de Boer *et al.*, 1998	23–25	75–80	Spieksma, 1967; Arlian *et al.*, 1990
Proctophyllodidae						
Proctophyllodes troncatus	0.57	25	Gaede & Knülle, 1987	ND	ND	

hydrated specimens of *D. pteronyssinus* would be less accurate than if partially dehydrated mites were used. Critical equilibrium activity of *D. pteronyssinus* was determined as 0.57 a_v at 16°C by de Boer *et al.* (1998), suggesting that its CEA varies with temperature in a similar way to that of *D. farinae*.

Change in temperature is likely to affect rates of transpiration and sorption in the following ways:

- alteration of cuticular permeability due to changes in viscosity of cuticular hydrocarbons;
- changes in kinetic energy and chemical potential of water that drive rates of transpiration and sorption;
- variation in efficiency of production and cycling of salt solution to the supracoxal glands.

The ecological implications of the discovery of the increase in CEA with temperature are considerable. As Arlian and Veselica (1981a) observed, populations of *D. farinae* had often been reported in homes where the ambient humidity was rarely above the estimated CEA of *D. farinae* of 0.7 a_v. This disparity was explained by assuming humidity increased periodically above 70% within the microhabitat of the mites. The re-evaluation of the CEA in relation to temperature indicates that *D. farinae* can maintain water and survive at 0.45–0.55 a_v at 15°C, 0.55–0.65 a_v at 25–30°C and 0.65–0.75 a_v at 35°C, allowing populations to thrive in much drier, cooler conditions than had been thought possible. These observations have been extended to the survival of mites in fluctuating humidities that mimic the conditions in natural habitats. They indicate that mites can survive desiccating conditions for weeks if exposed to hydrating conditions for periods of an hour or two a day.

3.3.3 Body water compartmentalisation – how much and where is it?

The mass of exchangeable water can be estimated in arthropods by comparing wet weights and dry weights. Tables of water content of arthropods are given by Edney (1977) and Hadley (1994) for the same six species of mites. Wharton (1985) noted the water content of these mites is about 66%, compared with about 70% in the insects. He suggested the difference in water content between mites and insects was because mites, with their high surface area to volume ratio, have a larger proportion of cuticle with a water content of less than 50%. However, the average water content

Table 3.2 Relationship between surface area, body mass and water content for some species of mites found in houses. ND = no data provided.

	Life cycle stage	°C	RH%	Mean body mass fresh weight (µg)	Mean body water mass (µg)	Mean body water mass % of body weight	Reference
Dermatophagoides farinae	F	25	75	12.7 ± 1.9	10.4	81.9	Arlian, 1972
	F	25	75	12.99 ± 0.5	9.71	74.7	Arlian & Wharton, 1974
	M			4.1 ± 1.0	3.1	75	Arlian & Wharton, 1974
D. pteronyssinus	F	25	75	5.83 ± 0.16	4.3	73.8	Arlian, 1972, 1975a
	M	25	75	3.48 ± 0.18	2.52	72.4	Arlian, 1972, 1975a
D. microceras	F			6.9 ± 0.9	5.1	73.9	Yoshikawa, 1979
Euroglyphus maynei	F			1.9 ± 0.4	1.4	76	Colloff, unpublished data
	M			0.6 ± 0.2	0.4	73	Colloff, unpublished data
Acarus siro	ND	20	95.5	ND	ND	70.5	Solomon, 1966
	ND	20	75	ND	ND	66.5	Solomon, 1966

of the scorpions listed by Hadley (1994) is 67%, and most scorpions are larger than most insects (though they tend to have much thicker cuticle). The mean water content of all the species listed by Hadley (1994) is 68.8 ± 8.8%, with a range from 40.8% in a Namib Desert beetle to 85.2% in a house spider. There is no apparent relationship between the body water content of the arthropod species and the relative dryness of their habitat. Water content is highly variable within and between species, and is related to physiological condition, developmental stage, fat content (the more body fat, the less body water), humidity, temperature and diet. The water content of dust mites is given in Table 3.2, with a mean of 73.7%.

In terms of rates of water loss and gain, mites have two major compartments of water. At humidities above the CEA, *D. farinae* lost water from a large, slow compartment that also gained water via an active pump. The water in this pump represents a second, smaller water compartment (Arlian and Wharton, 1974). This pump is what we now know to be the supracoxal glands and water conducting mechanism (see section 3.5.2a). The large compartment is the rest of the body water. Below the CEA the small compartment loses water rapidly because it is transpired through the openings of the supracoxal glands, until the glands become blocked with salt and the large compartment loses water slowly because it is lost through the less permeable cuticle. There is about 3 μg of water in the small, fast compartment (Arlian and Wharton, 1974). Of the body water, there are two compartments: the haemolymph and tissues. Haemolymph is the reservoir from which water can be exchanged with the tissues. In insects this movement is accomplished by moving solutes around between the haemolymph and the fat body.

There appears to be a discrepancy between the body water content by weight and the body water activity of 0.99, but this can be explained by the fact that water molecules are relatively light compared with other molecules in the body.

3.3.4 Water balance of different stages of the life cycle

Most research on water relations has been done using adults. Males of *D. pteronyssinus* transpire water faster than females (Arlian, 1975a) above and below the CEA, and males are more susceptible to dehydration, possibly due to their smaller size

or their cuticle being more porous to water. Males have more non-striated cuticle than females, due to the presence of the opisthosomal shield, lacking in females.

Given the differences in transpiration between males and females, it is likely that water relations of eggs, larvae and nymphs differ also. Cunnington (1984) found the eggs of *Acarus siro* were much more resistant to desiccation than other stages in the life cycle. Growth and survival of a dust mite population in relation to humidity is determined by water relations of the most susceptible and the most resistant stages: it is the extremes that matter. As will be seen, adults do not represent the extremes when it comes to water balance. This might explain some of the discrepancies between CEAs of adults and optimal humidity for population increase in Table 3.1.

Ellingsen (1975), using tritiated water, found the quiescent protonymphs of *D. farinae* exchanged water about 10 times slower than active protonymphs for the first 16 days. Water exchange then virtually ceased. In other words, quiescent protonymphs (which have a very low metabolic rate) are almost watertight. The half-life of body water exchange of quiescent protonymphs at 75% RH, 25°C is about 160 days, compared with 20 hours for active protonymphs and 28 hours for adult females (Arlian and Wharton, 1974; Ellingsen, 1975). The CEA of active protonymphs at 25°C was estimated to be 0.75 a_v, compared with 0.7 a_v for adult females.

Several domestic mite species have a facultative desiccation-resistant stage, either a homeomorphic protonymph as in *D. farinae*, but not *D. pteronyssinus* or *Euroglyphus maynei*; or a heteromorphic deutonymph (hypopus) as in *Glycyphagus domesticus* (Hora, 1934; Wallace, 1960; Knülle, 1995), *Lepidoglyphus destructor* (Wallace, 1960; Knülle, 1987, 1991, 1995) and *Acarus siro* (Knülle, 1995), but not *Tyrophagus putrescentiae*. The role of the desiccation-resistant stage was summarised by Knülle (1991):

> [The hypopus] provides a 'fail safe device' both for survival as well as development in irregularly fluctuating environments, and facilitates the adaptation of populations to local conditions.

More detail of the biology of these 'survivor nymphs' is given in Chapter 5.

3.4 Water loss

The major routes of water loss are evaporation from the supracoxal glands and water conducting channels, transpiration across permeable cuticle; in faeces, eggs and body fluids such as semen, saliva and gastrointestinal secretions. Dust mites differ from many terrestrial arthropods in that they lack spiracles and tracheae for the exchange of respiratory gases, so water is not lost via tracheal gas exchange.

3.4.1 Rates of water loss

The mean water mass of female *D. farinae* at equilibrium, that is above the CEA, is in the range of 6–10.7 µg (Arlian and Wharton, 1974; Arlian and Veselica, 1982; Arlian, 1992). The mean water mass, measured just prior to death, after 14 hours in a desiccating atmosphere, was 4.68 µg (48% of the equilibrium water weight; Arlian and Wharton, 1974). Mites were still alive after losing 52% of body water, and the minimum viable water content was estimated to be about 5.73 µg or 59% of body weight. *D. farinae* lost water by transpiration at a constant rate below the CEA (Arlian and Wharton, 1974; see Figures 3.11, 3.12) but *D. pteronyssinus* lost more water during the initial period of exposure (Arlian, 1975a).

3.4.2 Water loss and body size

Small size imposes considerable constraints on water balance because of the relationship between surface area and body volume. Surface area is directly proportional to the rate of evaporation. Small mites, with a high surface-to-volume ratio, are prone to losing water at a rate which is more difficult to compensate than for a mite of larger mass with greater body water reserves. For similar shaped organisms, there is a proportional increase in surface area as body size decreases, as described by the following equation:

$$\text{Surface area} = \text{body volume}^{2/3}$$

The relationship between surface area, as a *square* function and volume, as a *cube* function and, hence, surface being equivalent to two-thirds of volume, holds true even for organisms of complex shape. Although dust mites may be of different size, they are basically geometrically similar. Large adult female *D. farinae* (with ca. 8.5 µg of body water) and small individuals (with ca. 6.1 µg of body water) transpired water at about the same rate, but the survival times of large mites under dehydrating conditions was longer. Mean viable water mass of large mites was 56%, while for small mites it was 30% (Arlian and Veselica, 1982).

3.4.3 Routes of water loss

a Oviposition

The females of *Euroglyphus maynei* are considerably smaller than those of *D. pteronyssinus*, but their eggs appear to be about the same dimensions (see Figures 3.3, 3.4). This indicates that *E. maynei* loses a

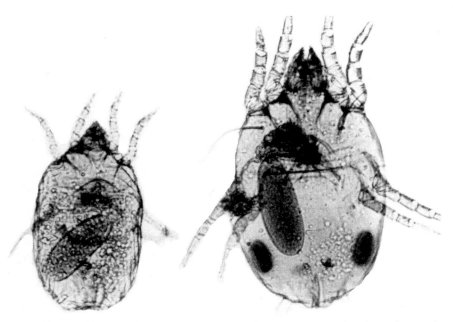

Figure 3.3 *Euroglyphus maynei* (left) and *Dermatophagoides pteronyssinus* (right) adult females with eggs showing relative dimensions.

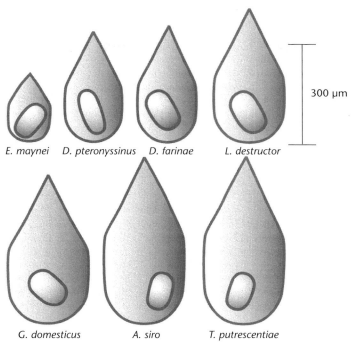

Figure 3.4 Diagram showing relative dimensions of adult females and their eggs of seven species of domestic mite. Lengths (L) and breadths (B) are in microns. Sources of dimensions: ♀♀ *Dermatophagoides pteronyssinus* and *Euroglyphus maynei*: Colloff (1991b); *Glycyphagus domesticus* and *Lepidoglyphus destructor*: Michael (1901); *Acarus siro*: Colloff (personal observations); eggs of *D. farinae*: Furumizo (1975a&b); *D. pteronyssinus* and *E. maynei*: Colloff (1991b); *G. domesticus*, *A. siro*, *Tyrophagus putrescentiae* and *L. destructor*: Colloff (personal observations); eggs/day: *E. maynei*: mean of values given by Taylor (1975) and Hart and Fain (1988); *D. pteronyssinus*: Arlian *et al.* (1990); *D. farinae*: Furumizo (1975a); *G. domesticus* and *L. destructor*: Chmielewski (1987, 1988); *A. siro*: mean of values given by Cunnington (1985); *T. putrescentiae*: Rivard (1961a, b).

significantly greater percentage of acquired body water per oviposition event than does *D. pteronyssinus*. In order to make eggs, the female has to be able to take in enough water for her own physiological requirements as well as those of the eggs. These are likely to be considerably higher than when she is not reproducing, placing additional pressure on the capacity of mites to maintain their body water balance. Assuming the water activity of eggs is the same as that of the adult females, 0.99, it is possible to calculate the rate of water loss during oviposition from an estimation of egg and adult volume and from fecundity (see Table 3.3). The association between water loss, the size of females and their eggs and fecundity provides some useful insights into the connection between water relations of domestic mites and their reproduction, population growth rates and life history strategies. These will be explored in more detail in Chapter 5. The following section deals only with the implications for water balance.

Table 3.3 gives details of several variables likely to have an impact upon comparative amounts of water lost by females of different domestic mite species

during oviposition. The relationship between female volume and egg volume for the seven species is linear (that is small mites lay small eggs; see Figure 3.5), although the ratio of female volume to egg volume varies from 5.4 for *E. maynei* to 38.7 for *A. siro* (that is small mites lose more of their body water volume through oviposition). Large mites can afford to lose a larger percentage of their body water via oviposition, and the relationship between female volume and total body water loss via oviposition is exponential (see Figure 3.6).

In relation to water loss from oviposition, there are four main categories of mites that emerge (shown in Figure 3.7).

1. Small mites (ca. 230 μm long) that lay few, relatively large eggs (*Euroglyphus maynei*): the ratio of the volume of females to eggs is ca. 5:1, with an average of 1.3 eggs oviposited per day for a relatively short oviposition period (ca. 11–15 days), representing an average loss of 24% of female body water mass.

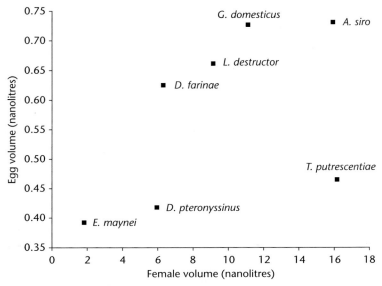

Figure 3.5 Relationship between volume of eggs and volume of females.

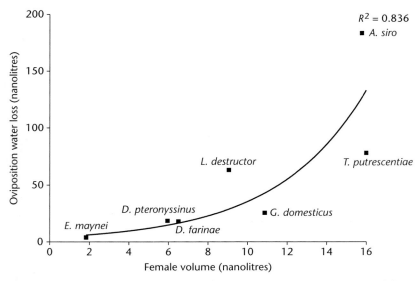

Figure 3.6 Relationship between volumes of adult females of seven species of domestic mites and the amount of water contained within eggs that is lost from females during the entire oviposition period (based on data in Table 3.3).

2. Medium-sized mites (ca. 385 μm long) that lay few, medium-sized eggs relative to the size of the female (*Dermatophagoides farinae* and *D. pteronyssinus*): the average ratio of the volume of females to eggs is 14:1, with ca. 2 eggs oviposited per day for a relatively long oviposition period of 27 days, representing an average loss of 15% of female body water mass.

3. Large mites (ca. 435 μm long) that lay a moderate number of medium-sized eggs relative to the size of the female (*Lepidoglyphus destructor* and

Glycyphagus domesticus): the average ratio of the volume of females to eggs is 16:1, with ca. 4 eggs oviposited per day for an average oviposition period of ca. 24 days, representing an average loss of 27% of female body water mass.

4. Very large mites (ca. 515 μm long) that lay large numbers of small eggs relative to the size of the female (*Acarus siro* and *Tyrophagus putrescentiae*): the average ratio of the volume of females to eggs is 31:1, with ca. 18 eggs oviposited per day, for an average oviposition period of ca. 20

Table 3.3 Surface area and volume ratios of adult female domestic mite species and their eggs in relation to water loss via oviposition

	Body volume (nL)	Female body surface area (mm²)	Units area per unit volume	Volume water per female (nL)	Egg volume (nL)	Egg surface area (mm²)	Units area per unit volume	Volume water per egg (nL)	Ratio female volume : egg volume	Mean fecundity (eggs/day)	Mean oviposition period (days)	% daily water loss via oviposition	Total fecundity (no. eggs)	Water content of all eggs laid (nL)
Euroglyphus maynei	1.84	0.09	20.44	1.356	0.342	0.025	13.68	0.252	5.38	1.3	11	24.2	15	3.8
Dermatophagoides farinae	6.51	0.24	27.13	4.798	0.575	0.036	15.97	0.424	11.32	1.6	28	14.1	42	17.8
D. pteronyssinus	5.96	0.22	27.09	4.393	0.368	0.029	12.69	0.271	16.20	2.5	26	15.4	68	18.4
Acarus siro	15.78	0.41	38.49	11.630	0.680	0.038	17.90	0.501	23.21	24.4	17	105.1	365	182.9
Tyrophagus putrescentiae	16.00	0.43	37.21	11.792	0.414	0.029	14.28	0.305	38.65	11.0	23	28.5	255	77.8
Glycyphagus domesticus	10.87	0.33	32.94	8.011	0.676	0.040	16.90	0.498	16.08	3.3	18	20.5	51	25.4
Lepidoglyphus destructor	9.06	0.28	32.36	6.677	0.611	0.037	16.51	0.450	14.83	4.9	29	33.0	140	63.0

Data sources: volume of water of females and eggs: 73.7% (mean of values in column 6 of Table 3.2) of female or egg volume. Mean lengths and breadths of mites used to calculate areas and volumes is given in Figure 3.4.

Calculation of volumes and surface areas: abbreviations: r = radius, h = height, d = diameter. Legs were not included. Volume of legs represents <0.5% of total body volume. Units are nanolitres, nL (1 nL = 1 million cubic micrometres).

Female body volume = volume of a cone + volume of a cylinder + volume of a hemisphere = $\frac{1}{3}\pi r^2 h + \pi r^2 h + \frac{\pi d^3/6}{2}$

Female body area = curved area of a cone + volume of a cylinder + curved surface area of a cylinder + curved area of a hemisphere = $\pi r h + 2\pi r h + 2\pi r^2$

Egg volume = volume of a sphere + volume of a cylinder = $(\pi d^3/6) + \pi r^2 h$

Egg surface area = curved area of a sphere + curved area of a cylinder = $4\pi r^2 + 2\pi r h$

A similar, alternative set of formulae for calculating body volume is given by Nelson in Michael (1895). Using Nelson's formula gives a volume of 6.73 nL for the adult female of *D. pteronyssinus*.

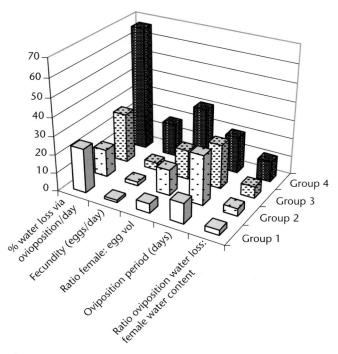

Figure 3.7 Characteristics of categories of mites based on oviposition water loss.

days, representing an average loss of 67% of female body water mass.

Water within the developing egg is compartmentalised from the mother's body water by the eggshell and is effectively not available to her. Strictly speaking it is not the *loss* of water from oviposition that imposes the constraint, but the need to replenish sufficient body water for subsequent eggs to develop. The constraints imposed by oviposition water loss mean that small mites with relatively large eggs cannot afford to lay very many eggs very often, whereas large mites can lay large numbers of eggs over longer periods of time. This pattern seems to hold true. Total fecundity of

E. maynei is only 15 eggs over an 11-day oviposition period, compared with 365 eggs over a 17-day period for *Acarus siro*. Total oviposition water loss can be mitigated by regulating fecundity, the oviposition period and the size of the egg, and the relationship between these variables is a complex one. The degree of flexibility that females have to alter these parameters has not been investigated experimentally, but is likely to depend on their size, CEA and microclimatic optima for population increase.

There is an obvious association between these categories and phylogeny (see Figure 3.8). Groups 1 and 2 are members of the family Pyroglyphidae: group 1 represents the subfamily Pyroglyphinae, group 2 the

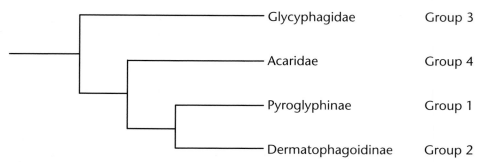

Figure 3.8 Phylogenetic relatedness of categories of mites based on oviposition water loss.

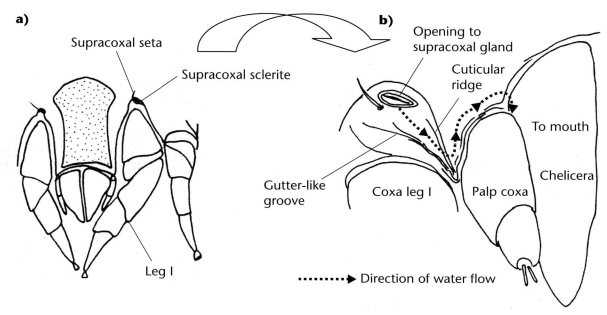

Figure 3.9   **a)** Position of supracoxal seta, sclerite and opening of supracoxal gland; **b)** water-conducting channels.

Dermatophagoidinae. Group 3 are members of the Glycyphagidae, and group 4 the Acaridae.

b   Cuticular transpiration

The water vapour activity of the air affects the permeability of the cuticle. The amount of water lost is the product of the force (or vapour pressure) that causes its loss, multiplied by the surface area to volume ratio that the force acts upon, multiplied by the permeability of the cuticle (Arlian and Wharton, 1974). The diffusion rate of water or gases across the cuticle is roughly inversely proportional to the cube root of the molecular weight of molecules that are diffusing. If the permeability of the cuticle decreases, then the diffusion rates will also decrease. Transpiration is also affected by temperature. The transition temperature of insect cuticle represents that below which permeability is relatively low and above which it increases rapidly (Edney, 1977). It is thought to be due to changes in the orientation of cuticular lipids, from an ordered polar monolayer to a less ordered and more permeable state. Arlian and Veselica (1982) determined cuticular permeability of *D. farinae* at 0.75 and 0.95 a_v and 15–52°C and found transpiration increased exponentially with increase in temperature, indicating the cuticle of *D. farinae* females does not have a transition temperature (see Figures 3.11 and 3.4). In mites exposed to 0.75 and 0.95 a_v at 15, 25, 30 and 35°C the water loss is equalled by water uptake and the body water pool behaves as a single compartment. Tempera-

tures of 42 and 47°C were dehydrating at both a_v values and although the water pool exhibits two compartments, the small, fast compartment associated with the supracoxal glands is lost so rapidly that the transpiration measured is predominantly from the large, slow compartment of body water.

Arlian and Veselica (1981b) concluded that the lack of a transition temperature, combined with a very efficient cuticular waterproofing mechanism and lack of a spiracular respiration system (through which water can be lost) were the major factors accounting for the high degree of resistance to desiccation of *D. farinae*.

c   Egestion, excretion and osmoregulation

Dermatophagoides pteronyssinus has been estimated to egest ca. 20 faecal pellets per day at conditions above the CEA (Tovey *et al.*, 1981). The water content of the pellets is not known, but it is presumed to be less than the mean body water content of 73.5% because water is resorbed from faeces in the hindgut. Also, the amount of water must be sufficient to allow for germination of spores and development of fungal hyphae, as illustrated in Figure 2.7, although it is possible that germination occurred as a result of the pellets passively sorbing water post-defecation. Crude estimates of the range of water loss from egestion can be made from calculating the volume of faecal pellets and fitting different percentages for water content. If water content is 10%, and each faecal pellet consists

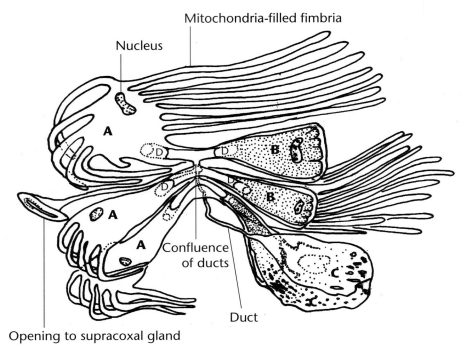

Figure 3.10 Morphology of the supracoxal gland. Re-drawn based on Brody *et al.* (1976). A = A cells, B = B cells, C = C cells, D = ducts of A–B cell units.

of ca. three to five spherical faecal particles of ca. 15 μm diameter (see section 2.2.2d), then the amount of water in each pellet represents only 0.24% of mean body water mass for female *D. pteronyssinus*, or almost 5% body water loss, based on an egestion rate of 20 pellets per day. If water content of pellets is 40%, then almost 20% of body water is lost through daily egestion.

Arlian (1977, 1992) found that *D. farinae* consumed 10 times the amount of food at 85% RH than at 75% RH and that defecation rate was proportional to feeding. At humidities below the CEA, mites fed and defecated relatively little. This observation suggests that under these conditions, the rate of water loss associated with feeding and egestion is significant, and is not offset by the production of metabolic water.

Water loss through excretion is limited by using guanine as the major excretory product. If the Malphigian tubules of the Acaridae do not function in nitrogenous excretion (see section 2.3.1c), perhaps they play a role in osmoregulation. The presence of dense microvilli would suggest a possible role in the resorption of ions and water.

Fashing (1989) found that the Claparède organs and genital papillae of the aquatic astigmatid mite

Naidacarus arboricola were characterised by cells containing plicate cell membrane in close contact with densely packed mitochondria, typical of cells involved in active transport, and similar to the osmoregulatory chloride cells of aquatic insects. Claparède organs and genital papillae in freshwater mites probably also function in osmoregulation, but their function in terrestrial mites is less clear. Alberti and Coons (1999) suggested they may function in the uptake of water from liquid films, but such films are absent in house dust, and it may be significant that the genital papillae of pyroglyphid mites are considerably smaller than those of acarid mites and may not therefore play a significant role in water uptake.

d   Lateral opisthosomal glands

The contents of the lateral opisthosomal glands consist of a series of terpenoid pheromones contained within a complex mix of alkanes, alkenes and other hydrocarbons (see section 2.9). The water content of these glands is likely to be non-existent since many of these compounds are not miscible with water, and the volatility of the pheromones would be reduced substantially by the presence of water. The lateral opisthosomal glands cannot be considered as a route for body water loss.

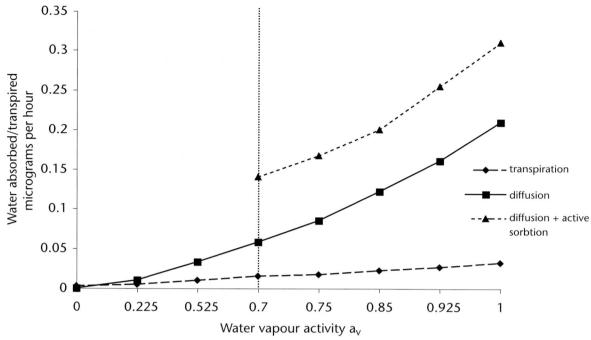

Figure 3.11 Relationship between water absorption/transpiration rates and water vapour activity for female *Dermatophagoides farinae*. Vertical dotted line = critical equilibrium humidity. Re-drawn after Arlian and Wharton, 1974.

3.4.4 Behavioural limitation of water loss

Possibly the first experimental work on acarine physiology was carried out by Robert Boyle (1627–1691) and Robert Hooke (1635–1703) with cheese mites (probably *Acarus* or *Tyrophagus* spp.) and recorded by Boyle in the *Transactions of the Royal Society* in 1670. Boyle had secured the services of Hooke (see section 4.2.1) for the design and construction of an air pump for the investigation of animal respiration. The pump had a series of glass receptacles into which various animals could be inserted (including a duck, a viper and a 'large lusty frog') and subjected to a partial vacuum. I came across the account of these experiments in an 1828 compendium of scientific writings, entitled *Readings in Natural Philosophy; or, a Popular Display of the Wonders of Nature; Exclusively Selected from the Transactions of the Royal Society of London, From its Foundation to the present Time (Chiefly Intended for the Use of Schools)* edited by the Rev. C.C. Clarke. As well as a 'Description of a monstrous Animal cast out of the Stomach by Vomit' by Dr Lister of York (1682), and several contributions by van Leeuwenhoek on microscopy, the book contains the following passage (p. 27) about Boyle and Hooke's observations on the behaviour of animals *in vacuo*:

We conveyed a number of mites, together with the mouldy cheese they were bred in to nourish them, into three or four portable receivers (which were all of them very small) not much differing in size. As soon as ever one of the receivers was removed from the engine, I looked with great attention on it; and though just before the withdrawing of the air the mites were seen to move up and down in it, yet within a few minutes after the receiver was applied to the engine, I could discern in them no life at all, nor was any perceived by some younger eyes than mine, whereunto I exposed them. Nay, by the help of a double convex glass (that was so set in a frame as to serve me as a microscope on such occasions) I was not able to see any of them stir up and down.

Boyle's discernment of 'no life at all' requires interpretation. The formation of a vacuum involves the withdrawal of air from a container to create a low pressure space. The only other work I know of concerning the effects of a vacuum on mites was conducted in the early 1980s as part of an honours degree project at the Department of Zoology, University of Glasgow, supervised by Dr R.M. Dobson. The aim was to examine the potential of low pressure as a means of controlling dust mites in mattresses and bedding, with the intention, if the experiments were successful, of modifying an industrial-sized hospital autoclave for the purpose. This alternative to autoclaving proper was designed to avoid the damage caused to the bedding fabrics, especially foam rubber, by the effects of superheated steam. Specimens of *Dermatophagoides pteronyssinus* were subjected to vacuums at pressures obtainable by a standard electrical laboratory vacuum pump for periods of up to 24 hours. Despite repeated attempts to kill them, the mites refused to succumb.

The removal of air, and the water vapour it contains, means that Boyle's receivers would have undergone a drastic reduction in humidity. Astigmatid mites have a number of behavioural mechanisms for responding to reductions in atmospheric humidity that seem to be linked to body water conservation. Arlian (1977) showed that below their CEA, mites virtually ceased eating and defecating. Van Bronswijk *et al.* (1971) described how *D. pteronyssinus*, *Acarus siro* and *Glycyphagus domesticus* responded to dry conditions by escaping to hide beneath culture medium or in cracks. Blythe (1976) noted that *D. pteronyssinus* sought cracks or side walls of the culture vessel against which to press themselves, presumably to reduce water loss by transpiration. The mites ceased movement, withdrew their legs close to the body and applied the flattened under-surface of the body to the substrate. He also noticed a marked crowding, or aggregation response to desiccation, with the mites pressing their bodies together in piles two or three deep (as illustrated by van Bronswijk, 1981). Kuwahara *et al.* (1975) showed that these responses are initiated by aggregation pheromones (see section 2.9.1c). The humidity in the spaces between the mites is higher than ambient and this is thought to help reduce water loss (Glass *et al.*, 1998).

In conditions where mites have undergone body water loss to the point where the cuticle has folded considerably (Figure 2.16a), the mites seem unwilling or unable to move when prodded with a fine needle. The flexibility of the cuticle in response to changes in humidity means that the mites have a partially hydrostatic exoskeleton. It may be that movement is impaired because the reduced hydrostatic pressure of integument impairs the working of the muscles that are attached to it.

3.5 Water uptake

Body water can be gained by drinking liquid water, consumption of hydrated food, production of metabolic water and uptake of water from the air either passively across the cuticle or actively via the supracoxal glands.

3.5.1 Availability of water to dust mites

In house dust there is rarely, if ever, any liquid water present, so gaining water via drinking is not an option available to dust mites. Dust mites feed on solid food, though nutritionally derived water appears to contribute little to their overall water requirements (Arlian, 1977). However, it is interesting that their diet contains a relatively high volume of lipids. Oxidation of a gram of lipid can derive 1.1 g of water, compared with 0.4–0.5 g of water from a gram of protein and 0.6 g from a gram of glucose (Schmidt-Nielsen, 1975). There have been observations that dust mites are coprophagic (J. Rees, *pers. comm.*, 1992), though this has only ever been observed with *in vitro* cultures, and the behaviour may be either a response to a shortage of suitable food or an attempt to derive any residual moisture present in the faecal pellets. The main source of water for dust mites is in vapour form in the air, and the mites have evolved an ingenious mechanism for its extraction and uptake.

3.5.2 Mechanisms of water gain

The osmotic pressures of body fluids of terrestrial arthropods are in the order of 0.2–1.0 osmoles kg^{-1} (Arlian and Veselica, 1979), which translates into a chemical potential, or mole fraction, or water activity, of 0.95–0.99, that is 95–99% of all molecules comprising the body fluid are water (Wharton and Richards, 1978). The water content of air (water activity, a_w) in the home is almost always considerably lower than this. This means the mechanism of water uptake has to be an active, energy-expensive one in order to move water into the mite against this concentration gradient.

a Active uptake – the supracoxal glands

Coxal glands of arthropods (also referred to in various groups of mites as oral glands, coxal organs, mouth glands, tubular glands, podocephalic glands and nephridia), play a role in water balance, osmoregulation and excretion, and their biology and morphology was reviewed by Wooley (1988), Evans (1992) and Alberti and Coons (1999). Coxal glands probably originated from the nephridial glands of the annelid ancestors of arthropods (Alberti and Coons, 1999). The supracoxal glands of astigmatid mites are probably not 'true' coxal glands because they are not of the typical tubular type. They may be derived from other glands in the prosomal region that have adopted some of the functions of coxal glands (Alberti and Coons, 1999).

Supracoxal glands of astigmatid mites are paired structures opening onto the supracoxal sclerite, located dorsally at the apex of the groove between the gnathosoma and the coxa of Leg I (see Figure 3.9a; Grandjean, 1937). The variable anatomy of the supracoxal glands in astigmatid mites (see Alberti and Coons, 1999 for many references), plus confusion about their origin and homology with coxal glands of other arthropods, led early workers to ascribe respiratory, accessory and salivary functions for them. Brody *et al.* (1976) made the first detailed ultrastructural study of the supracoxal glands of an astigmatid mite, *Dermatophagoides farinae*. The motivation for this research was to discover whether the supracoxal glands were a source of allergens, as it had been for an earlier ultrastructural study of the lateral opisthosomal glands (Brody and Wharton, 1970).

Wharton *et al.* (1979), reviewing the work of their research group, stated that by 1974 they knew that dust mites had an active water pump to extract water from unsaturated air, but that it was localised somewhere and not distributed over the entire surface of the body. When mites were below their critical equilibrium activity, they lost water from a small, fast compartment of water and larger slow one, and the so-called fast compartment was postulated to be due to transpiration from the active water pump (Arlian and Wharton, 1974). Of possible candidate sites, the supracoxal glands seemed to be the most likely because the gland opens onto the podocephalic canal that leads to the mouth (see Figure 3.11). Rudolph and Knülle (1974) found that ticks secrete hygroscopic saliva and imbibe it after it has become diluted by absorption of water from the air. The saliva contained high concentrations of potassium and sodium ions. Similarly, mites that had died of desiccation were found to have the supracoxal glands blocked by a plug containing a high proportion of potassium and chloride, determined by electron probe microscopy (Wharton and Furumizo, 1977). This clue suggested that dust mites, like ticks, secrete hygroscopic salty saliva from the supracoxal glands.

Secretions from the supracoxal glands flow from the opening of the gland into the podocephalic canal, which runs medially along the edge of the supracoxal sclerite, over the dorsal surface of the palp, and into the mouth (see Figure 3.9b). When the humidity falls below the CEA, the solutes precipitate as water is lost from the gland, and a plug of salt is formed. The supracoxal glands of *Lepidoglyphus destructor* are not closed by a salt plug. The opening to the gland is recessed and may be closed by the trochanter of Leg I (Wharton, 1985). The salt plugs dissolve when the mites are exposed to high humidities. The plugs were observed by Wharton and Furumizo (1977) in *D. farinae, Acarus siro* and *Tyrophagus putrescentiae*. Different ratios of sodium chloride and potassium chloride have equilibrium water activities that match the critical equilibrium activities of different acarid and pyroglyphid mites. The equilibrium value of saturated potassium chloride, which was the major salt isolated from supracoxal plugs of *T. putrescentiae*, is 0.85 a_w, corresponding with the CEA of 0.75–0.84 for this species (Cutcher, 1973), and a 1:1 mixture of saturated NaCl and KCl has an equilibrium activity of 0.7 a_w, which matches exactly the CEA of *D. farinae* (Arlian and Wharton, 1974).

Wharton and Furumizo (1977) point out that the hygroscopic secretion from the supracoxal glands would have to contain a wetting agent in order to aid the flow along the podocephalic canal, which is only 1 μm wide in parts. They suggest, based on Arlian and Wharton's (1974) estimates, that *D. farinae* can imbibe about 0.55 μg of water per hour from supracoxal gland secretions. This amount consists of about 0.45 μg of the secretion from the gland plus an extra 0.1 μg gained from the air while the secretion is in the podocephalic canal. The pumping of this water down the oesophagus would provide a means of circulating the haemolymph between the midgut and the supracoxal glands. Water would be absorbed through the midgut and/or the midgut caecae into the haemolymph, pumped forward through the neurohaemal canal – the haemolymph-filled space between the oesophagus and the brain – and recycled into the supracoxal glands.

In *D. farinae* each gland is composed of eight cells, four types of cell divided into four two-celled units (see Figure 3.10; Brody *et al.*, 1976). Three of the units are identical whereas the fourth has distinctly different morphology and has been interpreted as an acinous salivary gland by Alberti and Coons (1999). Macromolecules of >30 kDa found in the extracts of supracoxal pore plugs (Wharton and Furumizo, 1977) would be consistent with the presence of some salivary enzymes. However, Brody *et al.* (1972) have illustrated that while the salivary glands are adjacent to the supracoxal glands, they are far more extensive than a two-cell unit.

Wharton *et al.* (1979) stated that the three identical units are salt secreting, while the other secretes other materials, such as a surfactant or wetting agent to reduce surface tension. Of the three identical units, each unit consists of a single so-called A-cell. These cells are elongated, fimbriate and filled with strings of mitochondria. Each of these cells partially surrounds a B-cell which is smaller and filled with microtubules that flow into a proximal Y-shaped duct. The ducts of the three sets of units fuse into a single duct that leads to the opening of the gland. The fourth unit, the acinous salivary gland, comprises what Brody *et al.* (1976) refer to as a C-cell (containing vesicles, lysosomes and large amounts of rough endoplasmic reticulum), which also vents into the common duct, and a smaller D-cell. Ventral of the opening is the supracoxal seta, which in *Dermatophagoides* spp. are curved and setiform (Oshima, 1968), whereas in acarid and glycyphagid mites they are expanded at the bases and covered with ornate pectinations (Hughes, 1976). The function of these setae has not been worked out, though their often ornate structure and cryptic placement, often within an invagination of the supracoxal plate, would tend to suggest they are not mechanoreceptors. Their proximity to the openings to the supracoxal glands begs the question of a hygroreceptive function.

b Supracoxal glands, salt concentrations and flow rates

Wharton *et al.* (1979) define the primary function of the supracoxal glands as the excretion of excess salt. It is now known that water vapour can be extracted from unsaturated air by mites belonging to the Pyroglyphidae (*Dermatophagoides farinae* and *D. pteronyssinus*), Acaridae (*Acarus siro* and *Tyrophagus putrescentiae*) and Glycyphagidae (*Glycyphagus domesticus*). The involvement of hygroscopic salt solutions has been demonstrated for pyroglyphids and acarids (Wharton and Furumizo, 1977), but was not investigated in glycyphagids (Seethaler *et al.*, 1979), though it is the most likely mechanism.

Adult female *D. farinae* examined by Wharton *et al.* (1979) contained 9.589 μg of water, of which it was estimated that 0.2 μg was a 1:1 mixture of NaCl and KCl in solution, and available to operate the water pump. Salt concentrations were estimated using laser-generated X-ray microradiography (Epstein *et al.*, 1979).

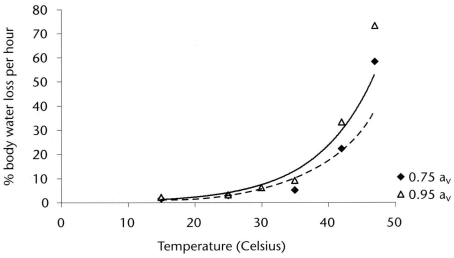

Figure 3.12 Relationship between transpiration rate of *Dermatophagoides farinae* and temperature at 0.75 a$_v$ and 0.95 a$_v$. Re-drawn from Arlian and Veselica (1981b).

When dehydrated mites are returned to humidities above the CEA, the precipitated salt deliquesces and the water pump is reactivated after about 3 hours. The flow rate of the pump is at least 0.1 µg per hour at an atmospheric a_v of 1.00 and 0.07 µg per hour at the CEA (Arlian and Wharton, 1974). In order for the pump to work efficiently the flow rate has to be faster than the rate of water loss by diffusion.

c Passive uptake – transcuticular sorption

Unlike active uptake, passive sorption does not cease below the CEA. It may proceed at a rate lower than when it is lost by cuticular transpiration, but diffusion of water into the mite across the cuticle still takes place even at very low humidities (see Figure 3.12).

d Eating and drinking

The only sources of free water in house dust are atmospheric water vapour and dust-dwelling organisms such as other mites and hyphal fungi, both of which have tissue water activity of 0.99. Dust mites are not predators so cannot gain water through the consumption of mites, though predatory taxa such as *Cheyletus* spp. can. Cheyletid mites capture other mites in their raptorial palps, pierce the cuticle with their needle-like chelicerae and suck out the body fluids. Similarly, the minute, fungal-feeding tarsonemid mites are able to pierce the walls of fungal hyphae and imbibe the liquid contents.

Dust mites ingest some water with their food, though not very much compared with predators and fungal fluid feeders. In soil-dwelling fungal feeding mites with chelate chelicerae, the gut contents often can be observed to contain quantities of fungal hyphae snipped off into neat lengths by the chelicerae. The guts of dust mites tend not to contain such characteristic masses of fungal hyphae and it seems therefore unlikely that they gain significant dietary water from fungal feeding, although they almost certainly feed on a certain amount of fungi (see section 2.2.4d). The water content of skin scales (as stratum corneum) increases exponentially with increase in relative humidity as water is sorbed (Spencer *et al.*, 1975; Grice, 1980). The rate of increase in sorption is relatively slow up to about 75% RH and then increases rapidly. At 0.75 a_v and 30°C the water content is about 15%; at 0.85 a_v it is about 35%, and at 0.95 a_v it is 55% (see Figure 3.13a). The water content of dried yeast granules is lower: about 15% by weight

at 25°C and 0.75 a_v and 20% at 0.95 a_v (shown in Figure 3.13b; Arlian, 1977).

Most of the early work on water balance of dust mites has been done on mites that were fasting. In order to rectify this, Arlian (1977) studied the significance of a_v on feeding rate and the quantity of water gained from food between 0.225 and 0.95 a_v at 25°C. He made the point that if food is available and its water content is sufficiently high, then the extra water the mite gains through feeding could possibly decrease the critical equilibrium activity. This too was investigated. Arlian (1977) found that above their CEA *D. farinae* and *D. pteronyssinus* consumed 8.4–58.9% and 10.3–50.9% of their body weight in yeast granules daily (see Figure 3.14a). When a_v was below the CEA, both species consumed less than 5% of their body weight per day (equivalent to 2.2–6.3% of maximum food intake for *D. farinae* and 1.8–9.4% for *D. pteronyssinus*). The amount of water gained from dietary intake for *D. farinae* was 0.01–1.55 µg/mite/day and was 0.003–0.51 µg/mite/day for *D. pteronyssinus* (shown in Figure 3.14b). *D. farinae* gained almost twice as much dietary water as *D. pteronyssinus*: about 5% of total water uptake at 0.225 a_v and 20% at 0.95 a_v. Below their CEA, both species were not strongly attracted to their food, were seldom observed feeding and produced fewer faecal pellets. Above the CEA, mites were strongly attracted to food, fed almost continuously and faecal material in the culture vessels was abundant.

Arlian (1977) concluded that below the CEA the amount of water gained from food by both species replaced less than 1% of water lost through transpiration and, for *D. farinae*, only 4–5% of the water gained by sorption (compared with 4–19% above the CEA). Therefore water gained from food below the CEA is not sufficient to replace water lost by transpiration. Also, dietary water could not have the effect of lowering the CEA under the experimental conditions used by Arlian (1977) and the CEA of fasting mites and feeding mites is the same. Above the CEA dietary water is, of course, surplus to requirements, since water balance can be maintained by sorption alone. In terms of water budgets, it is far more cost-effective for mites to cease feeding below their CEA and start to cluster in order to conserve water (see section 3.4.3) than it is to go on actively feeding and transpiring.

The fact that mites fed below their CEA, albeit at a considerably reduced rate, has some anatomical and

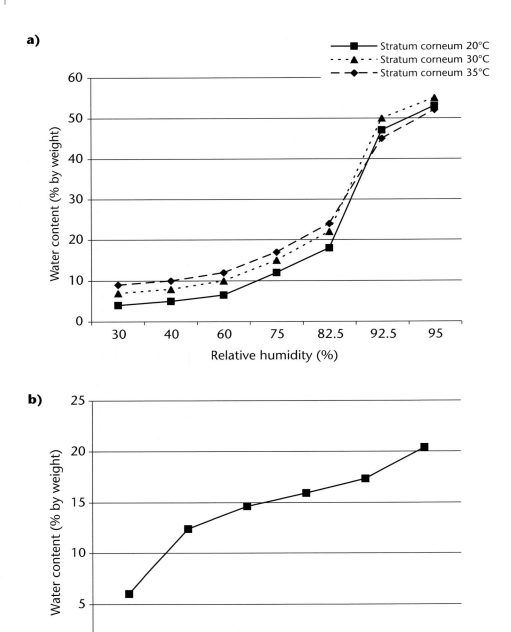

Figure 3.13 **a)** Water content of skin scale (as stratum corneum) in relation to temperature and humidity. Based on data of Spencer *et al.* (1975); **b)** water content of yeast at different humidities. Based on data of Arlian (1977).

functional implications. Feeding by dust mites requires a certain amount of saliva to wet the food and start the process of maceration prior to ingestion. Alberti and Coons (1999) interpreted one of the four cell units in the supracoxal glands as an acinous salivary gland, whereas Brody *et al.* (1972) suggested the salivary glands are adjacent to the supracoxal glands and are large masses compared with the two-cell unit in the supracoxal glands (see section 3.5.2a). If the only salivary glands in the mites open through the pore of the supracoxal glands, it would be impossible for mites to feed effectively below their CEA at all because the pore would be blocked with a precipitate of salt. In order for dust mites to be able to feed below the CEA, there would have to be more than one set of salivary-type glands present (as illustrated by Alberti and Coons, 1999 for a number of non-astigmatid mites, and indicated by the different morphology of the glands observed by Brody *et al.*, 1972). At least one set would have to duct saliva to an area that was independent of the supracoxal glands, presumably in closer proximity to the mouthparts.

e Water balance, changes in metabolic activity and metabolic water

Water is produced from the release of energy from nutrients by aerobic respiration. Like all animals, mites

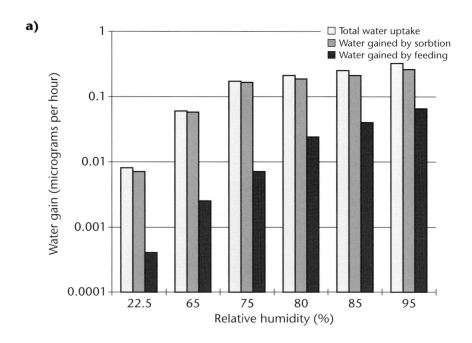

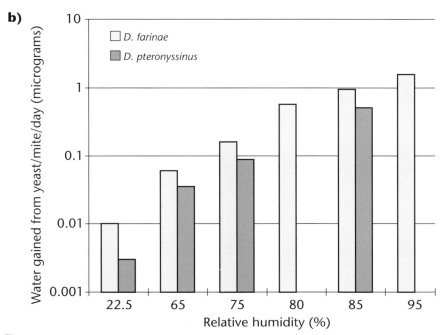

Figure 3.14   **a)** Water gain by sorption compared with water gain by feeding; **b)** rate of water gain from feeding on yeast. Re-drawn from Arlian (1977).

can obtain some metabolic water from oxidation of fats, proteins and carbohydrates (see Table 3.4).

Oxygen consumption by *D. farinae* increased exponentially with increase in temperature (Arlian, 1975b; see Figure 3.15). As the a_v of the ambient air increases, so the rate of water sorption (active + passive components) increases in an exponential fashion (Arlian and Wharton, 1974; see Figure 3.11). This is because increase in a_v leads to a decrease in the pressure against which water has to be moved: the difference in pressure between the a_v of the air and the a_w of the body of the mite is lessened. It takes almost 30 times as much energy to actively sorb water at 0.7 a_v than at 0.98 a_v, but the overall difference this makes to total oxygen consumption at the basal metabolic rate is only in the order of 0.1% (Arlian, 1975b). To actively sorb water at the rate of around 0.1 µg per hour, as estimated by Arlian and Wharton (1974), costs the mite about 1.6×10^{-7} Joules per hour, about 0.1% of the total energy generated by the mite at 0.75 a_v and 25°C. As mites are moved from an a_v at or above their CEA to desiccating conditions, they eventually become motionless and there is a large reduction in oxygen consumption and metabolic rate (Arlian, 1975b): at 0.75 a_v and 25°C mites consumed 0.0091 µl of oxygen per hour. After 6–22 hours exposure to 0.145 a_v, mites consumed 0.0044 µl of oxygen per hour.

The amount of water gained through metabolism is, according to Arlian (1972), insignificant. The example given is as follows. After 1–6 hours exposure to 0.148 a_v, dehydrating mites consumed 0.011 µl of oxygen per hour, while hydrated mites held at 0.75 a_v consumed 0.008 µl of oxygen per hour. Assuming one microlitre of oxygen yields 0.005 calories, or 0.021 Joules, then 2.31×10^{-4} and 1.68×10^{-4} Joules are liberated that would require oxidation of 0.011 and 0.014 µg of carbohydrate per hour. At the CEA this amounts to 0.55 µg of dry weight in 50 hours, contributing 0.31 µg of water.

3.6 Water balance and survival at low or fluctuating humidity

De Boer and Kuller (1995b) demonstrated experimentally that *Dermatophagoides pteronyssinus* could still reproduce when exposed for 3 weeks to only 3 hours per day of air at humidities above the CEH. A brief daily period of raised humidity would seem to be adequate to replenish water lost during the intervening period when humidity was below the CEH. This capacity for rapid water gain following exposure to moist conditions adds an extra dimension to the remarkable water-balance abilities of dust mites, and may be of crucial importance for the maintenance of mite populations in conditions where ambient humidity is extremely low. Indeed, De Boer and Kuller (1995b) suggest that the water vapour that is released every day, when meals are prepared in the kitchen (or when showers or baths are taken), may be of vital importance for the survival of many mite populations.

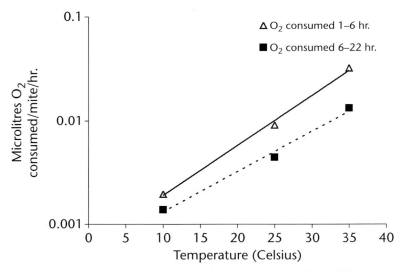

Figure 3.15 Oxygen consumption by *Dermatophagoides farinae*. Re-drawn after Arlian (1975b).

Table 3.4 Amounts of water and energy formed from the oxidation of various food sources (modified from Schmidt-Nielsen, 1975).

	Grams water per gram food oxidised	Metabolic heat value kJ g⁻¹	Grams of water per kJ
Lipid	1.07	39.3	0.027
Starch	0.56	17.6	0.032
Protein	0.45	18.2	0.025

The eggshells of *D. pteronyssinus* are porous to water. Colloff (1987a) observed changes in visible surface area, a surrogate measurement of change in egg volume, in eggs of laboratory-bred and wild populations of *D. pteronyssinus*, fed upon house dust or yeast, incubated in microclimatic conditions designed to mimic the fluctuating humidity and temperature on occupied and vacant beds (15°C, 65% RH for 16 h/day and 30°C, 75% RH for 8 h/day). The amount to reduction in visible surface area, and hence water loss, was dependent upon diet and egg size, not upon the origin of the cultures. Females fed upon yeast produced larger eggs that lost water more rapidly than the eggs of females fed on house dust (see Figure 3.16).

3.7 Conclusions

Domestic mites show considerable variety in their means of avoiding water loss and being able to survive relatively dry conditions. Mites are the smallest terrestrial arthropods, and among the smaller terrestrial invertebrates. Many terrestrial mites are confined to living in relatively moist conditions, such as soil or leaf litter, where water loss is not an issue. By a combination of physiological, behavioural and reproductive modifications, domestic

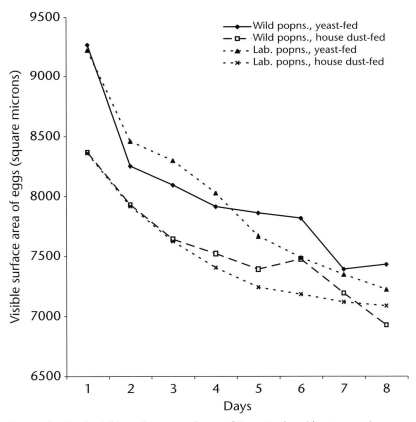

Figure 3.16 Mean reduction in visible surface area of eggs of *Dermatophagoides pteronyssinus* exposed to fluctuating conditions of 15°C, 65% RH for 16 h/day and 30°C, 75% RH for 8 h/day. (Re-drawn after Colloff, 1987a).

mites are able to survive and reproduce in conditions that would prove lethal to many other arthropods. Foremost among these adaptations are:

- the capacity to extract water from unsaturated air;
- the control of water loss by crowding together and shutting down feeding and defaecation;
- the regulation of oviposition water loss through variation of fecundity;
- the extraction of metabolic water from food;
- the absence of a tracheal system or any other body surfaces that are prone to water loss through transpiration;
- the presence in the life cycle of one or more stages that are highly resistant to water loss;
- the capacity to detect and exploit small variations in humidity on a microhabitat scale.

It is because of these adaptations that domestic mites are able to cohabit with humans in the relatively dry climate of the home.

4. Ecology

*In studying the ecology of an animal, we seek to answer the questions:
Why does this animal inhabit so much and no more of the earth? Why
is it abundant in some parts of its distribution and rare in others? Why
is it sometimes abundant and sometimes rare? These are all problems
of distribution and abundance.*

H.G. Andrewartha and L.C. Birch, 1954

4.1 House dust and the genesis of a detrital ecosystem

The idea of house dust as an ecosystem was first presented by van Bronswijk (1972a) to describe the basic house dust food web consisting of skin scales, microorganisms and mites. The diversity of bacteria, fungi, mites and insects that have been recorded from house dust, together with their temperature and humidity preferences were elaborated in a later paper (van Bronswijk, 1979), but consideration of the characteristics of house dust as a discrete ecosystem were constrained to some generalisations about the trophic interactions between various organisms. However, there is a great deal more to ecosystems than simply figuring out who eats whom.

4.1.1 Characteristics of ecosystems

The vast majority of plants and animals live within ecosystems that are ancient, complex, slow to evolve and mature, and that are spatially vast. When we think of examples of ecosystems we tend to conjure up images of wetlands, the soil, freshwater lakes, mangrove swamps, rainforests, prairie grasslands, temperate woodlands and Arctic tundra, though some of these may be better described as biomes (see section 4.3.2). The notion that our beds and

carpets – the textile-covered insides of our houses – merit the same status of ecological organisation as a coral reef seems, at first, alien, almost absurd. Just for starters, the diversity of plants and animals that inhabit a coral reef is countless times greater than the denizens of our mattresses. But diversity is not everything when it comes to ecosystems. One only has to think of the variety of life along a few hundred metres of rocky seashore compared with what one could expect to observe over the same distance of sandy desert, even accounting for the burrowing habit of many desert animals. Comparable with the notion of house dust as an ecosystem, our skin has been regarded as worthy of ecosystem status (Marples, 1965; Andrews, 1976). And the diversity of life on (and in) our own integument, as captivatingly described by Mary Marples in her classic work, *The Ecology of the Human Skin*, is considerably greater than one might expect. Bacterial community composition of the human forearm was recently characterised by Z. Gao *et al.* (2007) using molecular methods. Some 42 years after Marples' book, these authors comment on how little is known about microbial diversity of the skin.

If ecosystems are not defined by biological diversity alone, then what are their characteristics? In fact, finding a unified definition of an ecosystem, like so many other definitions of ecological entities, is not easy. There is usually a mention of the relationship between its living and non-living components. The non-living (abiotic) components are usually represented as a series of resources for exploitation by the living (biotic) components, such as habitat and nutrient sources for a tier of decomposer organisms which, in turn, provide food for predators, leading us to concepts of food webs and communities. Also there is usually some mention of the underlying notion about an ecosystem being self-contained and abiding.

A major ecosystem characteristic is that it is a stable unit generated from the interaction between the biotic community and the abiotic environment. The abiotic component determines the groups of organisms that are able to live there and these organisms may modify the abiotic community in order to allow for a pattern of succession of other organisms to colonise and develop. From an ecologist's point of view, the ecosystem contains two lower levels of organisation: communities of different species, and populations of each species making up the community.

4.1.2 Detrital ecosystems

What human homes, soil, deep sea sediments and estuarine mudflats all have in common is that they are ecosystems based on detritus – the organic debris derived from animals and plant remains, mixed with inorganic constituents derived originally from weathered rock. The principle is the same regardless of the ecosystem; only the constituents differ in nature and origin. And if one is tempted to question the occurrence of weathered rock in house dust ecosystems, one only has to observe the amount of soil and grit that can be beaten out of a hall runner carpet. Marples (1965) came to think of human skin as analogous to an ecosystem simply through applying the principles of classical ecology to the skin and its inhabitants. This refreshingly novel approach is made perfectly clear in the introduction to her book:

The approach to the skin and its inhabitants adopted in the following chapters is a deliberate distortion of the concept of an ecosystem but is based on analogies with the soil. Where the soil receives its non-living component from the parent rock below the surface, the epidermis is constantly receiving increments of keratinised tissue from the proliferating and differentiating epidermal layers. Both soil and skin are permeated with dilute aqueous solutions and both are penetrated by gases derived from the deeper layers and from the air. Both soil and skin support a mixed community of living organisms adapted for life in the various microhabitats.

However, Marples is hesitant to regard the skin as a 'true' ecosystem because she asserts that all interactions between the inhabitants and the substrate represent only the effects of living organisms on each other – there is no abiotic component for the biotic component to interact with. In fact, at least to my mind, there is a clearly identifiable abiotic component, without which there would be no house dust mites or house dust ecosystem. It is the flattened sheets of dead cells and their associated lipids that make up the outermost layer of the skin, the stratum corneum.

When shed from the surface of the skin, human skin scales become the major source of detrital material in the home. The fine dust that accumulates there, most visibly on the smooth surfaces of furniture, as any houseproud finger swept across the surface will reveal, is light grey in colour and is a mixture of skin scales and textile fibres. Freshly shed skin scales deposited onto a hard surface may not at first sight appear to offer a particularly attractive resource for living organisms. The scales when shed will carry with them the viruses, bacteria and yeasts that are part of the normal flora of human skin. These organisms will experience a significant reduction in temperature and humidity from the warm, moist conditions they have been used to on the human body and some will die. Other organisms, especially saprophytic fungi, that are able to withstand the drier, cooler climate on the surfaces onto which the scales have been shed will colonise and start to use the food resources – lipids, proteins, carbohydrates and so forth – that the skin scales consist of. Thus we can envisage a pattern of succession in the flora of shed skin scales. The presence of colonising fungi on the scales then opens up opportunities for fungal-feeding animals to colonise also, and this is where the mites come in.

The detrital habitat represented by skin scales shed onto a hard surface is a pretty hostile one. It is simply too desiccating to support much life. Were interiors of human habitations to be comprised mostly of such surfaces, there would be a very limited diversity of organisms living on them and they would have to possess well-developed mechanisms for coping with the problem of water balance. What made all the difference in the genesis of the detrital ecosystem within human homes was the use of insulative material for human clothing, first in the form of untanned animal hides and vegetation, and later the use of textiles.

4.1.3 Domestic textiles

Without skin scales there would probably be no dust mites. But without textiles for skin scales to accumulate in there would be no dust mites either. Woven fabrics play a crucial role not only in providing a matrix in which skin scales collect, but they buffer the aggregated detritus, and the organisms which feed on it, against low humidity and temperature (see section 4.6.2). The

development of weaving technology and the production of fabrics for domestic use was such an important event in the evolution of the relationship between domestic mites and humans. For this reason it is worth exploring briefly the history and archaeology of textiles (see Table 4.1). For dates, I have used years before present, or simply 'years ago' rather than years BC. I have not attempted to track down the primary reference sources for many of the archaeological finds. Rather, I have cited the secondary source wherein I first read about the discovery. It is important to note at this point that we know virtually nothing about the mites associated with ancient textiles, simply because researchers have not thought to look for them, nor question why they may possibly be of any significance.

Table 4.1 Summary of major archaeological finds and events in the history of domestic textiles (YBP = years before present).

YBP	Period	Event/find	Location	Reference
500		Rise of oriental rug weaving and trade with Europe. Depictions of oriental rugs in European art	Asia Minor, Persia	Bennett, 1988
1000		Seljuk rug weaving	Asia Minor	Bennett, 1988
1700		Rug fragments from Sassanian Dynasty, Persia	Dura–Europas (Qal'at as Salihiya), Syria	Bennett, 1988
2500	Iron Age	The Pazyryk carpet – earliest knotted pile rug	Pazyryk, Altai region, southern Siberia	Harris, 1993; Ford, 1989; Opie, 1992
2700	Late Bronze Age	Carpet patterns decorated on stone paving slabs in Assyrian royal palaces	Nineveh, Iraq	Ford, 1989
3000	Middle Bronze Age	'The Mound People': extensive finds of woven clothing, rugs, cloaks and blankets in tombs	Egtved, Skrydstrup, Muldbjerg, Borum Eshøj, Trindhøj, Denmark	Glob, 1974
5000	Neolithic	Earliest complete garment: elaborate linen shirt from First Dynasty Egyptian tomb	Tarkhan, Egypt	McDowell in Harris, 1993; see also Barber, 1994
7000	Neolithic	Vertical warp-weighted looms used to weave elaborate, patterned, coloured textiles	Lower Danube valley, Hungary	Barber, 1994
8500	Neolithic	Remains of horizontal and vertical looms	Jarmo, Iraq; Çatal Hüyük, Turkey	Barber, 1994
10 000	Mesolithic	Burial couch covered in layers of linen and wool cloth; first evidence of dyed fabric	Gordion, Central Anatolia, Turkey	McDowell in Harris, 1993
17 000	Upper Palaeolithic	Oldest preserved fibre artefact – cord twisted from two-ply vegetable fibre strings	Lascaux, France	Barber, 1994
22 000	Upper Palaeolithic	Bone Venus figures wearing skirts woven from twisted strings	Lespuge, France; Gagarino, Ukraine	Barber, 1994
26 000	Upper Palaeolithic (Old Stone Age)	Imprints of woven fabric on baked clay	Dolní Vestonice-Pavlov, Southern Moravia, Czech Republic	Gamble, 1999

I could have chosen to leave out the history of textiles altogether, but I consider that the story of the genesis of the house dust ecosystem would be glaringly incomplete without it. I hope my brief account here may serve to stimulate some future investigation.

Earliest archaeological records of textiles and weaving technology date from the late Palaeolithic period (26 000 years ago) and consist of impressions of woven fabric on baked clays. These fabrics were probably coarse plain-weaves made of plant fibres. Barber (1994) suggested that some of these earliest weavings – string skirts – had a symbolic function and were worn by women to signify that they were of child-bearing age (see Figure 4.1d). These skirts, which were depicted on bone and stone 'Venus' figurines of the late Palaeolithic 'Gravettian' period, have persisted as part of various eastern European folk costumes up to the present day.

The production of textiles was well advanced in the agricultural societies of Neolithic Europe 8000 to 4000 years ago. Patterned and coloured fabrics were produced on vertical, warp-weighted looms using sophisticated weaving structures such as twills and multiple weft shoots. This latter technique is evidence of weaving as a communal activity. More than one weft thread is inserted through the warps by different weavers sitting on opposite sides of the cloth, passing the weft bobbins to each other. A massive, coarse, brown, woollen cloak dating from about 3300 years ago was excavated at Trindhøj in Denmark and was woven in such a fashion. Such a garment could have provided expansive habitat for late Stone Age dust mites.

By 3000 years ago, piled woollen rugs were probably being produced in the Mediterranean region, based on references to their use in the works of Homer (Ford, 1989). There are dozens of beautifully illustrated books on the subject of oriental rugs and carpets, varying from the coffee table variety to scholarly treatises. Pick one at random and, if it has a section on the historical origins of carpets as most of them do, you will almost certainly find a picture of the celebrated Pazyryk carpet (see Figure 4.1a). Its origins and age are controversial. Its colours (predominantly cochineal red, yellow and light blue), motifs and fineness of knotting (about 260 symmetric knots per square inch, or about 3780 per square decimetre) are of exceptional quality, with borders of diagonal crosses interspersed with rows of gryphons, horsemen and what are quite clearly fallow deer, herded nose-to-tail around the central field of repeated, bordered,

diagonal crosses. The fallow deer are recognisable by the broad, palmate antlers, only found on males, and the spotted patterning of the coats is typical of their summer pelage. The Pazyryk carpet is a breathtaking piece of design, the animal borders wrought with life and symbolism. The burial mound in the Altai Mountains of southern Siberia in which it was found preserved in permafrost, has been dated to about 2500 years ago (Artamonov, 1969). Some scholars believe the carpet to be the work of craftsmen belonging to semi-nomadic central Asian tribes, distantly related to the Scythians (Bennett, 1988). Others assert that the rug was an import from Persian weavers to the south, possibly produced in a workshop because of its fineness and evenness. Such authors cite the uncanny likeness in its geometrical motifs to those on relief carvings on stone floor flags from the doorways of palaces at ancient sites in Assyria (Ford, 1989; Opie, 1992). The mixture of central Asian tribal motifs and elements of Persian design is the source of the debate about its origins. Interestingly, the original range of fallow deer was the Mediterranean region, and from Asia Minor to Iran, as well as north Africa and Ethiopia (Nowak and Paradiso, 1983). This snippet of ungulate biogeography would support the theory of a southerly source for the rug.

Whatever the origins of the Pazyryk carpet, the rendering of wool into piled carpets – functional products of artistic expression and craftsmanship – was well advanced at the time of its production. Thus rug manufacture probably developed long before the Pazyryk carpet was woven, perhaps as much as a thousand years earlier. Hirsch (1991) gives some examples of some ancient Egyptian linen bedcovers with a symmetrically knotted pile weave, like those of many rugs, dating from 3400 years before present (shown in Figures 4.1b, c). With the development of piled weaving, the opportunities for colonisation of human dwellings by dust mites were greatly enhanced.

The history of oriental carpets can be found depicted in the work of European painters such as Holbein, van Eyck, Lotto and Bellini. Several of these paintings date from the 15th century and show richly patterned Turkish rugs being used as altar cloths and on tables, as well as floor coverings (see Mills, 1991). These carpets were the luxury possessions of the wealthy and the powerful, as indeed they still are today. A glance at the portrait of Charles Howard, First Earl of Nottingham, painted by Daniel Mytens ca. 1620

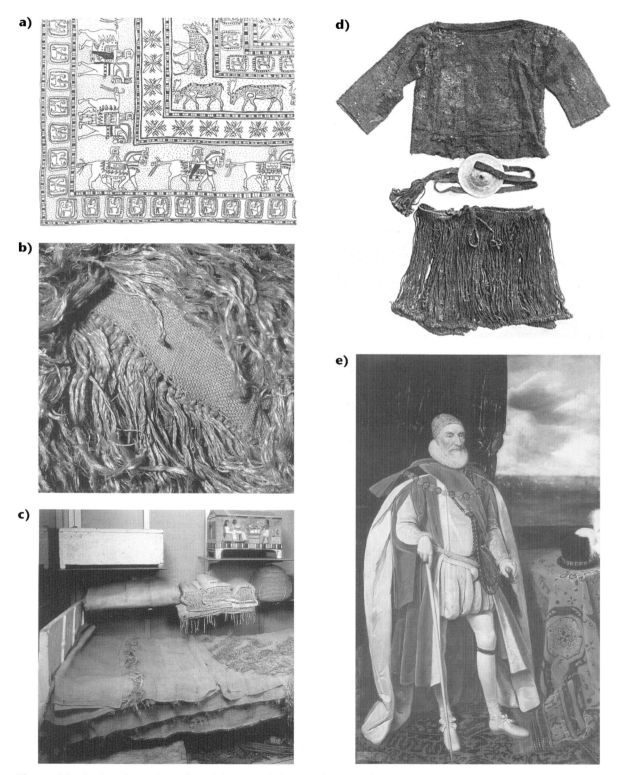

Figure 4.1 Ancient domestic textiles. **a)** A corner of the Pazyryk carpet, from permafrost in a 2500-year-old Siberian tomb; **b, c)** some of the oldest piled textiles: bedcovers from the tomb of Kha, Deir el Medineh, Egypt, 3400 years before present, Turin Museum; **b)** detail of one of the bedcovers from Deir el Medineh. The cloth is linen with a knotted pile; **d)** 3000-year-old costume of a girl, Egtvedt, Denmark. It includes a version of one of the earliest recorded garments, a string skirt, possibly used to symbolise the wearer was of childbearing age; **e)** portrait of Charles Howard, First Earl of Nottingham, by Daniel Mytens c. 1620, National Maritime Museum, Greenwich. Note the Turkish rug covering the table.

(see Figure 4.1e) shows how the artist has used the rich fabric of the carpets to enhance our perception of the status and wealth of the ageing aristocrat.

While the homes of the rich and powerful of 15th century Europe may have been decorated with hand-knotted carpets, the floors of the poor were of beaten earth, covered with straw or rushes, probably little different from Neolithic dwellings. The habitat of any medieval domestic mites in such homes would have been stored food, bedding and clothes. A potent reminder of the capacity of clothes to harbour dust mites is the enormous numbers of live specimens found by Bischoff *et al.* (1992a, b) and Tovey *et al.* (1995) in contemporary wool and cotton clothing.

4.1.4 Beds and bedding

The earliest known artefacts of beds and bedding are Egyptian (shown in Figure 4.1), as mentioned above (section 4.1.3). Carpets and rugs are used not only as floor coverings but as bedding in many tribal nomadic societies, especially those from central Asia with strong weaving traditions such as the Kurds, Turkoman and Beluchi. Unfortunately, almost nothing is known of the mite fauna either of ancient bedding or that of present-day nomads. Research on mites associated with the tents and textiles of the few remaining nomadic tribes would provide an interesting and challenging field of study, not least because the very existence of this lifestyle is threatened, either through policies of settlement in villages and towns or through declining pasture quality and availability. Furthermore, the transition from nomadism to settlement provides a unique opportunity to investigate the nature and extent of changes in exposure to domestic allergens that these people undergo. Such investigations would include issues like urbanisation, exposure to Western cultural practices, changes in diet, social and economic status and living conditions. These are exactly the kind of variables currently under investigation in large-scale epidemiological studies of asthma and allergy in various parts of the world (see Chapter 8).

Van Bronswijk (1981) gives an account of the history of house dust in the Netherlands and the conditions in the oldest Dutch town houses, including the widespread use of cupboard beds for protection against the cold in the draughty buildings. The beds were lined with layers of straw and Bog Myrtle (*Myrica gale*), a plant with insecticidal properties, perhaps used against fleas and bedbugs. These beds, which could be completely enclosed, were cleaned once a year in Spring.

Eventually, Myrtle would have been replaced by mattresses of horsehair, straw or feathers.

Straw and feathers are good habitats for mites. Straw, hay and other forage is often infected with fungal mycelia, which provide a food source for various acarid and glycyphagid mites and their populations can become immense – two million mites per kilo in one-year-old stored hay was recorded by Hallas and Gudmundsson (1987). Dekker (1928) highlighted the association between *Glycyphagus domesticus* and asthma, drawing particular attention to the deleterious effects of pillows filled with oat chaff. The plumage of birds is notorious for the diversity and abundance of its acarine inhabitants (Gaud and Ateyo, 1996), and feather mites have been shown to be allergenic (Colloff *et al.*, 1997). One of the earliest and most celebrated reports of asthma associated with bedding involved John Hamilton (c. 1511–1571), Archbishop of St. Andrews in the Kingdom of Fife, Scotland. Hamilton was a chronic asthmatic who eventually secured the services of the renowned Italian physician, Gerolamo Cardano (also known as Gerome Cardin). After an extensive period of observation, Cardano recommended the covering of the Archbishop's feather mattress with tightly woven silk, the use of a leather-covered pillow, as well as dietary changes, and the patient's asthma improved. This case is often cited as the first recorded instance of allergen avoidance, but, as Ellul-Micallef (1997) points out, Cardano believed that the feather bed was the cause of 'heating of the spine': the concept of allergens was not developed for another 400 years, until Sir Charles Blackley's work on pollens and hay fever was published in 1873.

4.1.5 Early agriculture and the development of food storage

Humans, like budgerigars, live on a staple diet of seeds. Cereal grains, be they rice, wheat, sorghum, millet, maize, rye, barley or oats, provide more than 70% of the daily calorie intake for people in developing countries. In the developed nations, cereals are fed to livestock to be converted into milk and meat, thus providing a more varied range of foods. If one is what one eats, then truly all flesh is grass.

Various cereals started to be domesticated and cultivated in different parts of the world from about 10 000 years ago. In the low-lying lands of Mesopotamia between the Tigris and Euphrates rivers in what is now Iraq, fields were rendered fertile by alluvial deposits and were cultivated and irrigated.

Animal husbandry developed to provide the muscle power for the emerging agriculture. Grains were harvested and stored as buffers against crop failure and famine. Surpluses were traded for other goods and commodities, facilitating the development of writing and the keeping of records and accounts. Permanent settlements were established, populations increased, cities with Old Testament names such as Jericho, Babylon, Ur and Nineveh were born. The same thing was going on in ancient Egypt along the Nile Valley, whose population lived on a staple diet of bread made from emmer wheat and beer brewed from barley, and who established a network of granaries throughout Upper and Lower Egypt, together with an extensive, ever-present fauna of insect pests of stored grain (Levinson and Levinson, 1994). Among the stored products insects known from archaeological remains are species of dermestid beetles known to be the hosts of phoretic acaroid and glycyphagoid mites (see section 4.1.7).

Wherever food and agricultural produce is stored, there is a risk of insect and mite infestations. The origin of stored products mites and insects are most likely natural habitats, including soil, leaf litter, nests of birds and mammals, fruits and seeds. These origins have been explored for insects by Levinson (1984) and Levinson and Levinson (1994), and for mites by OConnor (1979, 1982a, b; 1994) (see also section 1.4.4a).

The investigation by Griffiths (1960) of stored products mites in grassland and stacks of straw, hay and corn in open fields indicates another means by which mites could have been introduced into human dwellings: through their presence in straw used as roofing material or as floor coverings.

4.1.6 Nests of birds and mammals – the origin of the domestic acarofauna

Many mite and insect species are found in the nests of insects, birds and mammals. An extensive list of such associations is given by Zachvatkin (1941), Woodroffe (1953, 1954) and by Wasylik (1959, 1963). The majority of this fauna are scavengers feeding on faeces, the fungi colonising faeces, food remnants of their hosts, including stored food, as well as skin scales, hair and feathers. There is quite a high degree of overlap between the mite fauna of nests and that of food stores and human homes (refer to Table 4.2). This is no coincidence. Several mites are phoretic on stored products insects which are also found in nests. The bird and mammal fauna associated with humans is not coincidental either. Again, it is food and shelter that provides the attraction, and that food is grain. Some of the more important and widespread anthropophilic birds and mammals are seed-eaters: sparrows, pigeons, rats and mice.

One of the big attractions of nests of birds and mammals for insects and mites is that they provide a concentration of potential food materials that are high in nitrogen. Soil and litter, a major habitat of the oribatid mite ancestors of domestic mites, provide relatively poor food. Nitrogen concentrations are low,

Table 4.2 Non-domestic habitats of various domestic mites.

Species	Habitat	References
Acarus siro	Nests of birds and bees; soil, haystacks,[1] grassland	Baker *et al.*, 1976; Hughes, 1976; Sheals, 1956; Griffiths, 1960
Tyrophagus longior	Nests of birds, haystacks, grassland	Baker *et al.*, 1976; Griffiths, 1960
Tyrophagus putrescentiae	Soil, haystacks, grassland	Sheals, 1956; Griffiths, 1960
Aleuroglyphus ovatus	Nests of small mammals	Hughes, 1976
Dermatophagoides pteronyssinus	Nests of birds	Fain, 1966a, b
Dermatophagoides farinae	Nests of small mammals	Fain, 1967
Dermatophagoides sp. (as *Mealia* sp.)	Nests of birds	Woodroffe, 1953
Glycyphagus domesticus	Nests of birds and bees	Hughes, 1976; Woodroffe, 1953
Lepidoglyphus destructor	Soil, haystacks	Zachvatkin, 1941; Sheals, 1956; Griffiths, 1960
Glycycometus geniculatus	Nests of birds	Woodroffe, 1954
Chortoglyphus arcuatus	Nests of sparrows	Wasylik, 1959
Cheyletus eruditus	Nests of birds, haystacks	Woodroffe, 1953; Griffiths, 1960

[1]in open fields.

as are polysaccharides, and most of the carbon is locked up in the form of cellulose. Unless an organism possesses cellulolytic abilities, about the only things worth eating in the soil are other animals (live or dead) or fungi, both of which have higher levels of sugars and essential amino acids than does detritus. Without a nitrogen-rich diet at certain crucial times (such as just before moulting), many arthropods are unable to complete their life cycles.

OConnor (1979) examined the natural ecological distribution of the genera of mites that included stored products species and those found in human dwellings. He came up with four major categories:

1. those associated with specific resources such as fruit, meat, carrion or dung which are not widely distributed in space or time (12 genera);
2. those associated with widespread field resources (six genera);
3. those found in nests of mammals (12 genera); and
4. those in nests of birds (five genera).

Of these, OConnor considered that the category of mites associated with specific resources to be phylogenetically basal, in that they retain the ancestral astigmatid life history. Those associated with field resources were phylogenetically derived (possibly from the mammal nest group) because few astigmatid mites are generally and widely distributed in natural habitats. With regard to mites in mammal nests, those that are specialised for living and feeding on stored grain probably evolved in the food storage chambers of rodent burrows long before the advent of agriculture.

While some of the habitat distinctions may not be hard and fast, either because we know so little about the habitat preferences of so many taxa or because the mites are associated with more than one habitat, OConnor distinguished certain patterns related to life histories. Those mites that have invaded stored products and houses seem better able than their counterparts in natural habitats to tolerate more variable environmental conditions. This has led to a reduced dependency on the hypopial stage for survival of deteriorating conditions and dispersal to better ones. Thus the hypopus may be modified or absent from the life cycle of many mites associated with stored products and human dwellings, such as the Pyroglyphidae. The remaining genera remain restricted to those stored products that most closely resemble their natural habitats.

4.1.7 Mites in archaeological remains

The investigation of archaeological material for evidence of the origins of the association between domestic mites and humans is impeded by some formidable confounders. First, there is the scarcity of source material – ancient human remains that are suitable for study are not unearthed every day, nor is permission to work on them given lightly. Second, the material may only remain suitable for study for a short period; usually only when first discovered and most certainly before it is curated and cleaned. This is because the greater the time elapsed from first discovery, the higher the likelihood of contamination from a contemporary source, such as the clothing of archaeologists. Third, mites found on human remains may not just be post-discovery contaminants. They may be phoretic associates of carrion-feeding insects, transported post-mortem, and therefore will be younger than the remains from which they came.

Kliks (1988) found a variety of unidentified astigmatid mites from gastrointestinal contents and faecal remains of three mummies, one each from Peru, the Aleutian Islands and caves in Kentucky. Radovsky (1970) recovered mites from a coprolite, or sub-fossil human stool, taken from a crevice in Lovelock Cave, Nevada. These were phoretic hypopi of a *Myanoetus* species. Also, a single mite was found in faecal material dissected from the apparently intact pelvis of human mummified remains from Pyramid Lake, Nevada. The specimen turned out to be the hypopus of a new species of the astigmatid mite genus *Lardoglyphus* (later named *L. radovskyi* Baker, 1990). Hides, skins and dried animal remains have their own mite fauna. *Lardoglyphus* spp. (family Acaridae) are particularly associated with such materials and some of them are phoretic on the larvae of the hide beetles of the genus *Dermestes* (Hughes, 1956; 1976), which are serious pests of stored products, as well as being found in nests of mammals, birds and human homes. Baker (1990) found another new species, *Lardoglyphus robustisetosus*, from the gut contents of a human mummy from Chile, indicating the likelihood that the mites were ingested with food. The gut contained dried meat or pemmican, consistent with the high-protein requirements of *Lardoglyphus* species. Pemmican, and its African equivalent, biltong, are among the earliest-known high protein stored foods. Guerra *et al.* (2003) found a variety of mites in human and animal coprolites from a rock shelter at Ferna do Estrago, in Pernambuco, Brazil. The site was occupied sporadically by hunter-gatherers from 8500–11 000 years before present, and was later used as a cemetery (1600–1860 years BP). An unidentified acarid species and the stored products species *Suidasia pontifica* were found in human coprolites, along with oribatids and a gamasid. *Blomia tropicalis* (Echymyopodidae) and the *Hirstia passericola* (Pyroglyphidae) were found in coprolites from unidentified members of the cat family. Although individual coprolites appear not to have been radio-carbon dated, this is the best evidence yet of allergenic pyroglyphid, acaroid and glycyphagoid mites associated with human habitation that predates agricultural settlement.

In a much more recent example, Mehl (1998) mentions that archaeological material dated to around 1500 AD from excavations of the Archbishop's Palace at Nidaros Cathedral, Trondheim contained large numbers of *Euroglyphus maynei* (now quite rare in Norway) and the stored products species, *Acarus siro*.

4.1.8 Mites in communities of indigenous peoples

Extant indigenous peoples are often studied by anthropologists in order to infer something about the human societies of prehistory. Therefore, investigations of the domestic mites associated with tribal peoples are of particular interest in any attempt to understand the evolution of the association between these mites and humans.

a Papua New Guinea

A well-known example of the relationship between dust mite exposure and apparent increase in prevalence of asthma is the work carried out in the 1970s and 1980s in Papua New Guinea. In the 1970s at Degei village (near Lufa, Eastern Highlands, see Figure 4.2) mites were virtually absent from houses of tribespeople (Anderson and Cunnington, 1974) and only 0.2–0.3% of the adult population had asthma (Anderson, 1974). The houses had earthen floors, a central fireplace and a raised woven bamboo sleeping platform around the fire, similar to those in Figure 4.3. Anderson and Cunnington (1974) found low mite numbers in 'low cost' public service houses with wood or concrete floors, fibro-cement walls and corrugated iron roofs, but much higher numbers in 'high cost' homes occupied by Australian medical staff. However, they only took 20 samples altogether and it is impossible to draw any conclusions regarding mite population density of three very different types of houses based on so few samples.

Even though the degree of textile cover within homes was lowest in traditional huts, higher in the 'low cost' houses, and highest in the 'high cost' dwellings, there was no clear correlation with abundance of mites. Nor were there any clear differences in species diversity or composition between the homes.

Green et al. (1982) compared mite numbers in traditional houses at Baiyer River (Western Highlands), where asthma prevalence was about 0.1% of adults (Woolcock, 1972), with homes at Waisa (Eastern Highlands) where asthma had increased sharply (to 2% of adults; Woolcock et al., 1981). Some 11 dust samples were taken from blankets and 23 from floors from Baiyer River, on two occasions in 1975 and 1978. Some 34 samples from blankets and 67 from floors were taken from Waisa in 1978 and 1980. High numbers of *Dermatophagoides* spp. were found in blankets at both Baiyer River and Waisa. Moderately high densities of *Tyrophagus putrescentiae* were found in floor samples from Baiyer River. This was attributed to the floor litter of houses, which consisted of expectorated, masticated sugar cane, plus grass and leaves from bedding. Species diversity appears to be relatively high, especially of glycyphagoids and acaroids. In addition to *Tyrophagus* and *Dermatophagoides*, members of the genera *Cheyletus*, *Acarus*, *Glycyphagus*, *Lepidoglyphus*, *Blomia* and tarsonemids were recorded.

Dowse et al. (1985) reported on mites from homes of people belonging to the South Fore Language Group, south of Okapa (Eastern Highlands Province; see Figure 4.2), and from the same area of the Eastern Highlands studied by Green et al. (1982). Asthma around Okapa was as high as 7.3% of the adult population, among the highest reported worldwide, and was associated with allergy to dust mites (Woolcock et al. 1981; Turner et al., 1988). The major difference from areas of low asthma prevalence was that most of the residents used a single cotton blanket as bedding. Numbers of *Dermatophagoides* spp. in these seldom-washed blankets (n = 32) were very high (1380 per gram of dust), compared with about 27 per gram from floors near the doors (n = 60). Dowse et al. (1985) concluded that the increasing acquisition of cotton blankets during the 1960s and 1970s had provided the necessary habitat needed for dust mites to live and breed in, and mite populations reached densities comparable with those found in city dwellings in developed countries.

Turner et al. (1988) reported on the dust mite fauna of homes in the Asaro Valley, the same area

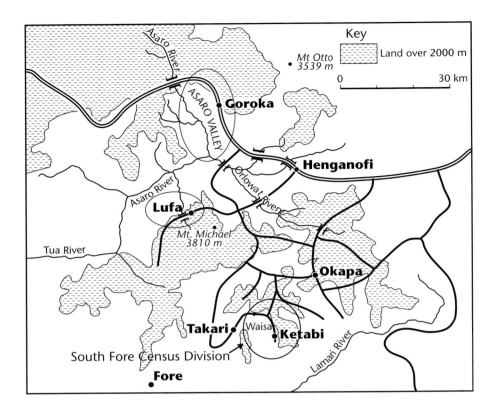

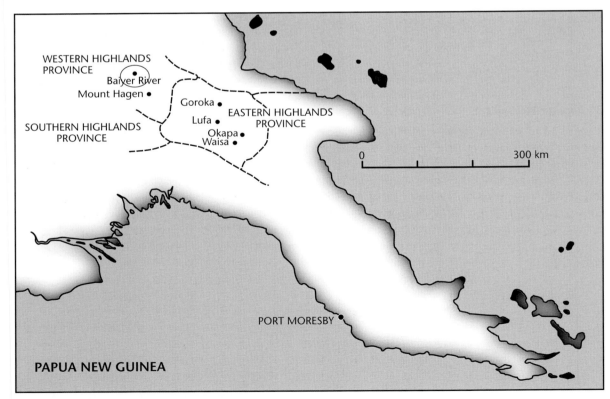

Figure 4.2 Maps of Papua New Guinea showing Eastern Highlands Region where most of the research on dust mites and the prevalence of asthma has been done (bottom); detail of Okapa District and Asaro Valley (top). Shaded area is land over 2000 m. Ellipses are the major study areas.

in which Anderson and Cunnington (1974) had worked (see Figure 4.2). They found housing conditions in the Okapa District and Asaro Valley appeared comparable, but the density of *Dermatophagoides* spp. in blankets in the Asaro Valley (n = 25) was only 283/g compared with 1371/g from blankets from Okapa. Asthma prevalence in the Asaro Valley remained around 0.3% – markedly lower than Okapa – which was attributed to the lower mite population density.

The researchers did not really compare like with like. Anderson and Cunnington (1974) not only had too few samples but the ones at Lufa were taken using a portable 12V vacuum cleaner and a more powerful mains vacuum cleaner was used at Goroka. Green *et al.* (1982), Dowse *et al.* (1985) and Turner *et al.* (1988) shook blankets into large plastic bags, whereas floor samples were obtained by sweeping. The lack of standardised collecting and the different densities of blanket and floor dust makes it impossible to compare mite populations in a meaningful way. Further, the mite data appear to have been duplicated across publications. For example, the mite population densities reported by Dowse *et al.* (1985) are more or less the same as reported by Green *et al.* (1982). The impression gained from the paper by Turner *et al.* (1988) is that it was intended as a follow-up study from Dowse *et al.* (1985) to determine if mite numbers in Okapa really were dramatically higher than elsewhere. However, the datasets for blankets in the two studies are almost identical. It is unclear whether the Asaro samples were taken concurrently with those from Okapa, whether the Asaro samples were taken at a later date or whether there were two separate sets of Okapa samples, one lot taken around 1981, and another taken later (and that coincidentally yielded mite population densities differing by less than 1% from each other). The mean difference in mite numbers in blankets between Asaro and Okapa is well within the range of seasonal fluctuations in various studies around the world (see Chapter 5).

Between 1964 and 1989 there were about 20 publications on dust mites and the epidemiology of asthma in the Highlands of Papua New Guinea. Yet it is not easy to determine which papers stemmed from which piece of fieldwork: who did it and when. The uncertainty regarding the mite population data in blankets calls into question whether there really had been an increase in mite allergen exposure at Okapa compared with Asaro that was causally linked

to a higher prevalence of asthma. Perhaps the most valuable ecological finding was that exposure to dust mites in a tribal society living in traditional dwellings was dependent on the introduction of a suitable textile habitat in which the mites could live and breed (Green *et al.*, 1982). These authors comment that traditional clothing, consisting of bark and leaf belts and genital coverings, has been gradually replaced by Western-style clothing since the advent of Europeans in the Highlands. Most houses, especially in the Eastern Highlands, contained an assortment of old clothing, and cotton blankets are used widely (see Figure 4.3). In contrast to pyroglyphid mites, significant acarid mite populations developed in vegetable matter that had been introduced into homes and left to decay. The use of vegetation as bedding predated the use of textiles and one can infer that *Tyrophagus putrescentiae* may represent a more ancient source of mite allergen exposure than *Dermatophagoides* spp.

b Colombia

The State of Vaupés is in south-eastern Colombia on the border with Brazil. The river Vaupés flows through the region and on, through rainforest and a series of rapids and waterfalls, into the Rio Negro that joins the Amazon at Manaus, 1500 km to the east. The poor navigability of the rivers ensured this isolated territory was virtually unknown by the outside world until the 1850s. It was visited by the English naturalist Alfred Russell Wallace, who established his base at Mitú and collected the local flora and fauna. The people of this region have developed a detailed knowledge of the botany and pharmacology of local hallucinogenic plants, based around a shamanic cult of jaguars and jaguar spirits. The ethnology of these people was described by Riechel-Dolmatoff (1975) in his fascinating book, *The Shaman and the Jaguar*.

Sánchez-Medina *et al.* (1993) studied allergy to dust mites and dust mite exposure among Indians in the Mitú region. They found mean Der p 1 levels in mattresses, hammocks and floors of 3.2, 0.8 and 0.05 µg/g respectively, and that sensitisation to mite allergens was common: 52% of the 82 volunteers gave positive skin test reactions to an extract of *D. pteronyssinus*. Some 19.5% were sensitised to *Blomia tropicalis*. This work has been published only as abstracts. Additional information from Walter Trudeau and Enrique Fernandez-Caldas (*pers. comm.*) concerns the mites found

Figure 4.3 **a)** The Lamari Valley, Mabotasa, Okapa, with terrain typical of the Eastern Highlands of Papua New Guinea; **b)** a hamlet near Keikwambi, Simbari, Eastern Highlands. Note the proximity of the cloud base to the houses, indicating a humid climate; **c)** interior of a Hewa men's house, Lipalipa, Southern Highlands Province. People and pigs sleep naked around a central hearth; **d)** interior of a men's house, Baktamanmin, near Olsobip, Western Province. The occupants are wearing Western-style dress. Note the man at the rear of the hut wrapped in a cotton blanket; **e)** women in mourning, Mount Hagen, Western Highlands. Ashes have been rubbed into the skin and clothes. They are wearing cotton blankets modified into *billum* bags and head-dresses. Photographs by Yvon Perouse, courtesy of Geoffrey M. Clarke.

in woven hammocks. Samples were taken using a vacuum cleaner powered by a portable electrical generator. Pyroglyphid mites were found, as well as large numbers of glycyphagoid mites. The adoption of mattresses by the Mitú people as an alternative to traditional hammocks has parallels with the increasing use of cotton blankets by the tribespeople of the Eastern Highlands Province of Papua New Guinea. In both cases the introduction of novel textiles into traditional dwellings provides increased habitat availability for dust mites, and hence increased risk of exposure of the human population to dust mite allergens.

c Australia

Veale *et al.* (1996) investigated Der p 1 levels in homes of rural Aboriginal communities in Australia. Two communities were on Cape York Peninsula, Queensland (designated CY1 and CY2), and two were in Central Australia (designated CA3 and CA4). Der p 1 levels in the Cape York samples were 12 and 15 μg g^{-1} of fine dust, whereas they were less than 0.05 μg g^{-1} in the Central Australian samples. This can be explained by the moist tropical climate of Cape York compared with the arid climate of Central Australia. Despite the large difference in allergen concentrations, the prevalence of positive skin-prick test reactions to dust mites in CA4 was similar to that in CY1 and CY2, although this was attributed to cross-reactivity with the scabies mite, *Sarcoptes scabiei*. There were differences in housing quality between the communities. In CY1, the housing conditions were relatively good, whereas in CA4, the standard of housing was extremely poor. Studies on asthma and IgE antibody responses to mite allergens in a remote Aboriginal community in the Kimberley found quite high levels of Der p 1 exposure (average 6 μg g^{-1}) and markedly different IgE responses from an urban population in Perth (Hales *et al.* 2007) but lower prevalence of asthma and skin-prick test positivity to dust mite (Bremner *et al.*, 1998).

Of the three indigenous populations examined in this section, the Australian Aboriginal communities have been most influenced by Western intervention and the data are the least informative about the evolution of domestic mite–human relationships. Traditional nomadic and semi-nomadic, hunter-gatherer lifestyles are virtually a thing of the past, replaced by settlement in Western-style housing. Exposure to domestic mites in Aboriginal communities living traditional lifestyles would have been virtually non-existent because of the absence of permanent dwellings, textiles, and the reliance upon fresh-killed or gathered food rather than stored products derived from agricultural production. The Papua New Guinea and Colombia studies are thus the only examples we have of how introduction of domestic textiles within an indigenous society are able to provide the basic habitat requirements for dust mites. Further study of the ethnobiology of dust mites offers the prospect of a fascinating insight into the co-evolution of mites and humans.

4.2 Historical records of mites in houses

4.2.1 1650–1750: Leeuwenhoek, Hooke and Baker

The presence of mites in houses and stored food has been documented in ancient Greek writings (Oudemans, 1926), but illustrations of these mites started to appear only during the 17th century, due to advances in optical technology. The study of mites has been closely associated with developments in microscopy. August Hauptmann (1657) provides a crude woodcut of what has been interpreted as the first illustration of a scabies mite (Hoeppli, 1959, p. 25, pl. 2, Figure 5), but the combination of ovoid body form (rather than a sub-spherical one of *Sarcoptes scabiei*), the very long hysterosomal setae (see section 1.4.3) and three pairs of legs indicates more likely a depiction of a free-living, larval acaroid or glycyphagoid mite, possibly a species of *Tyrophagus*, *Glycyphagus* or *Lepidoglyphus* (see Figure 4.4e).

Antony van Leeuwenhoek (1632–1723) was a pioneer microscopist, founder of microbiology and inventor of a lens grinding technique that allowed him to construct simple microscopes, capable of a magnifying power of up to 270 times (see Figure 4.4b). From 1673 until his death he sent some 200 letters, written in Dutch, to the Royal Society in London describing his researches. In 1693 he wrote to the Royal Society describing the reproductive biology of a species of acarid mite (probably a *Tyrophagus* sp.), found in his house (see Figure 4.4a; Hoole, 1807, Vol. 2, p. 170). Leeuwenhoek made a series of careful and exhaustive observations on these mites, which suggests that they were abundant in his house, and he could easily find them, especially on cheese and other foodstuffs (Heniger, 1976). In fact, Ford (1981a, b; 1985), in his account of the discovery of some of Leeuwenhoek's microscopic preparations in the archives of

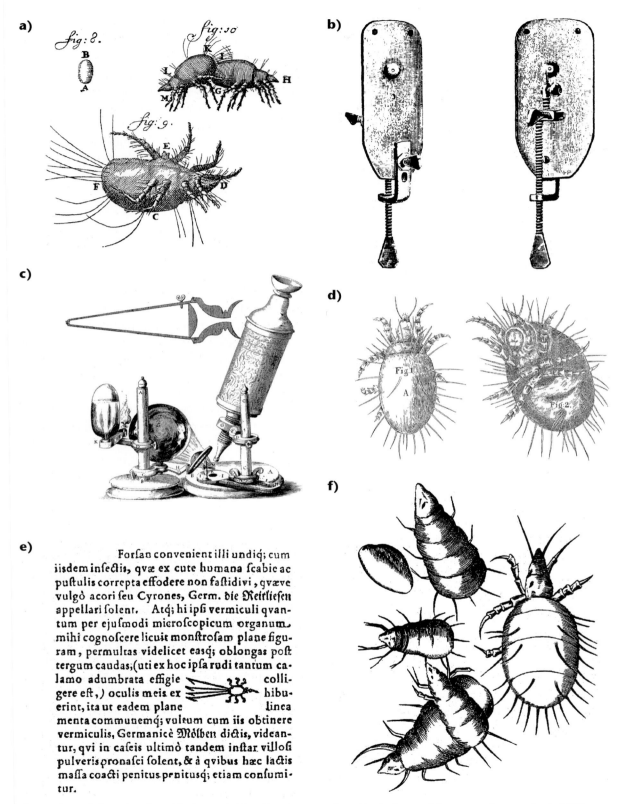

a)

fig: 8.

B

A

fig: 10

K

L

L

M

G

H

fig: 9.

E

F

D

C

b)

P

c)

d)

Fig 1

A

Fig 2

e)

Forſan convenient illi undiq̃; cum iisdem infectis, qvæ ex cute humana ſcabie ac puſtulis correpta effodere non faſtidivi, qvæve vulgò acori ſeu Cyrones, Germ. die Reitlieſen appellari ſolent. Atq̃; hi ipſi vermiculi qvantum per ejuſmodi microſcopicum organum mihi cognoſcere licuit monſtroſam plane figuram, permultas videlicet easq̃; oblongas poſt tergum caudas;(uti ex hoc ipſa rudi tantum calamo adumbrata effigie colligere eſt,) oculis meis ex hibuerint, ita ut eadem plane linea menta communemq̃; vultum cum iis obtinere vermiculis, Germanicè Mölben dictis, videantur, qvi in caſeis ultimò tandem inſtar villoſi pulveris pronaſci ſolent, & à qvibus hæc lactis maſſa coacti penitus penitusq̃; etiam conſumitur.

f)

Figure 4.4 **a)** A species of acarid mite, probably a *Tyrophagus* sp., found in the house of Antony van Leeuwenhoek, 1693; **b)** engravings of Leeuwenhoek's microscope by John Mayall (1885) from Ford (1985); **c)** Hooke's microscope, from an engraving in *Micrographia*; **d)** an acarid mite from Robert Hooke's *Micrographia* (1665); **e)** a woodcut of what may be a *Tyrophagus*, *Glycyphagus* or *Lepidoglyphus*, by August Hauptmann (1657); **f)** an acarid, '*Acarus casei*' from Griendel von Ach (1687).

the Royal Society, mentions that he found and photographed mite fragments on some sections of bovine optic nerve tissue and that Leeuwenhoek had noted the infestation of mites on his animal specimens before he prepared the tissue for study. This find represents the earliest known specimens of domestic mites still extant.

Robert Hooke (1635–1703) was a natural philosopher and architect. In 1662 he became curator of experiments of the Royal Society, and later secretary. In 1665 he published *Micrographia*, an account of his microscopic investigations. There is a fine engraving of an acarid mite in this volume (see Figure 4.4d), with an account of his observations. The following commentary on this acarid indicates the widely held belief of the time in the theory of spontaneous generation; that small, 'primitive' organisms (including mites and moulds) arose spontaneously from putrefying matter.

I have observed it to be resident almost on all kinds of substances that are mouldy, or putrefying, and have seen it very nimbly meshing through the thickets of mould, and sometimes to lye dormant underneath them; and 'tis not unlikely, but that it may feed on that vegetable substance, spontaneous vegetables seeming a food proper enough for spontaneous animals.

Henry Baker in his popular manual, *The Microscope Made Easy*, first published in London in 1742, made this observation on the alternate action of the chelicerae of stored products mites during feeding.

Mites are most voracious animals; for they devour not only cheese, but likewise all sorts of dried fish or flesh, dried fruits, grain of all sorts, and almost everything besides that has a certain degree of moisture without being over-wet: nay, they may often be observed preying on each other. In eating, they thrust one jaw forward and the other backward alternately, whereby they appear to grind their food; and after they have done feeding, they seem to munch and chew the cud.

With the rise in popularity of microscopy as a pastime in the 18th and 19th centuries, stored products mites became a common object of curiosity for the amateur microscopist. Their dramatic morphology and small size continue to capture the public imagination and inspire a degree of fascination and horror combined.

4.2.2 The 19th century

The 19th century saw the development of acarology as a scientific discipline, partly as a result of developments in the science and technology of microscopy. The microscope was the New Technology of the day. It facilitated the study of the natural world and fuelled the massive rise in popularity of natural history, first as an amateur pastime, later as a professional academic endeavour (Allen, 1976). As microscopes became more obtainable and their optical quality improved, they became immensely popular and the true diversity and ubiquity of mites began to be appreciated. Mites were good subjects for amateur microscopists – easy to find, spectacular to look at – perfect objects of curiosity for the Victorian parlour (Butler *et al.*, ca. 1986).

Throughout the 19th century there developed a small but immensely productive group of acarologists in Europe who were concentrating their attentions on the systematics and biology of free-living astigmatid mites, many of which were known as pests of stored products and housing. The major players included Albert Davidson Michael in Britain, who is better know for his work on oribatids, and published his two-volume work on the British Tyroglyphidae between 1901 and 1903. Tyroglyphidae, then known commonly as 'Cheese Mites' is a taxon no longer recognised, but which consisted of large chunks of what we currently refer to as the Glycyphagoidea, Acaroidea and Anoetoidea, as well as various other acarine dribs and drabs. Michael's book still remains one of the most insightful and informative sources on the biology and life history of acaroid and glycyphagoid mites (see Figure 4.5a). In Italy, Antonio Berlese and Giovanni Canestrini between them established much of the systematic basis of the Astigmata; while Paul Kramer in Germany and Philippe Mégnin in France contributed to understandings of the life history and development of acaroids and glycyphagoids, sorting out the nature and function of the hypopus as the phoretic, dispersive deutonymphal stage of the life cycle (see further details in Chapter 5). The connection between glycyphagid mites as inhabitants both of nests of small mammals and human dwellings dates from this period. Michael (1901) had investigated nests of moles and described the life history stages of several glycyphagid species. He mentions massive domestic infestations of these mites, singling out *Glycyphagus domesticus* in rush-covered chairs and quotes the following letter from

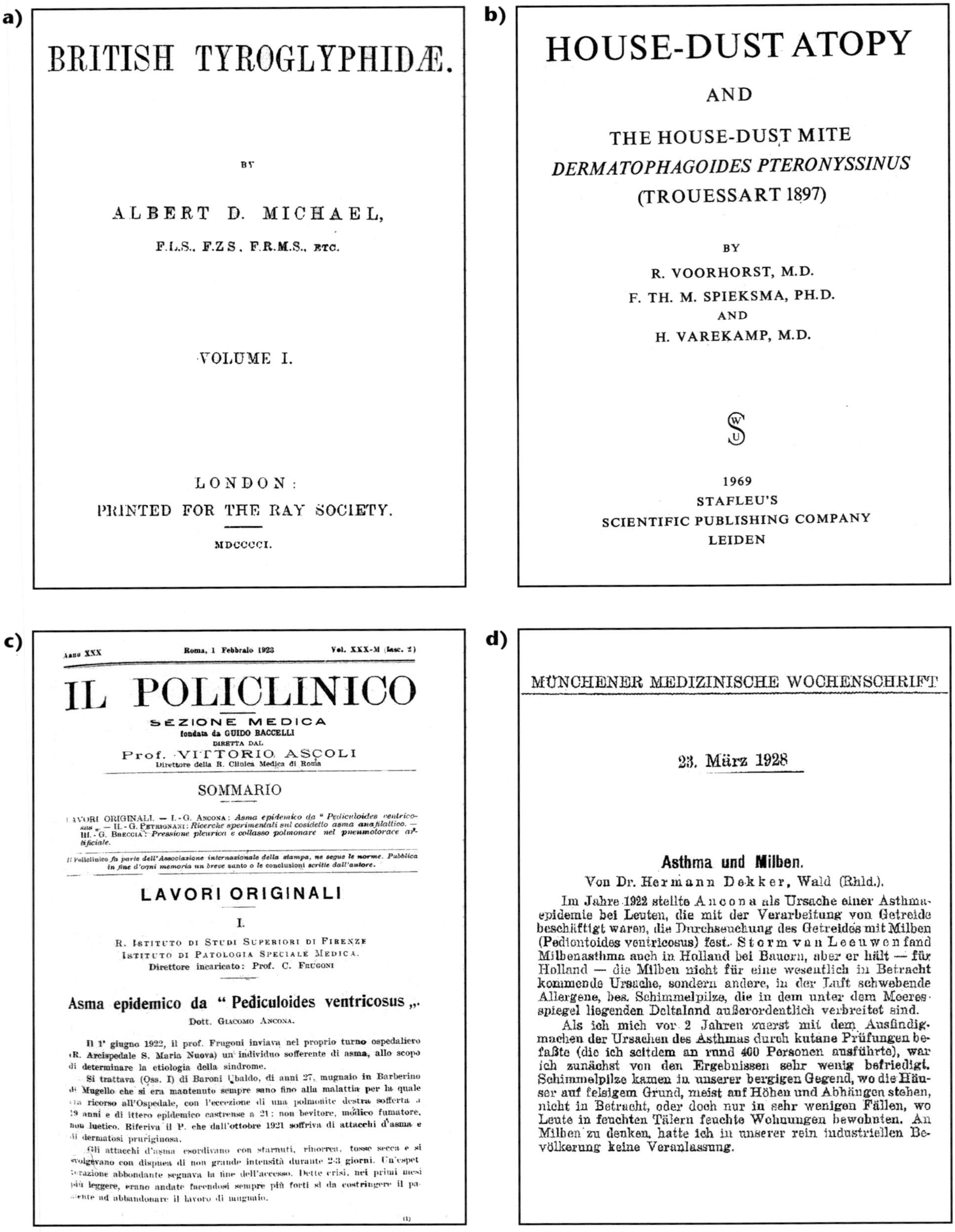

a)

BRITISH TYROGLYPHIDÆ.

BY

ALBERT D. MICHAEL,

F.L.S., F.Z.S., F.R.M.S., ETC.

·VOLUME I.

LONDON:
PRINTED FOR THE RAY SOCIETY.

MDCCCCI.

b)

HOUSE-DUST ATOPY

AND

THE HOUSE-DUST MITE
DERMATOPHAGOIDES PTERONYSSINUS
(TROUESSART 1897)

BY

R. VOORHORST, M.D.

F. TH. M. SPIEKSMA, PH.D.
AND
H. VAREKAMP, M.D.

1969
STAFLEU'S
SCIENTIFIC PUBLISHING COMPANY
LEIDEN

c)

Anno XXX — Roma, 1 Febbraio 1923 — Vol. XXX-M fasc. 2

IL POLICLINICO

SEZIONE MEDICA
fondata da GUIDO BACCELLI
DIRETTA DAL
Prof. VITTORIO ASCOLI
Direttore della R. Clinica Medica di Roma

SOMMARIO

LAVORI ORIGINALI. — I. - G. ANCONA: *Asma epidemico da " Pediculoides ventricosus "*. — II. - G. PETRGNANI: *Ricerche sperimentali sul cosidetto asma anafilattico.* — III. - G. BRECCIA: *Pressione pleurica e collasso polmonare nel pneumotorace artificiale.*

Il Policlinico fa parte dell'Associazione internazionale della stampa, ne segue le norme. Pubblica in fine d'ogni memoria un breve sunto o le conclusioni scritte dall'autore.

LAVORI ORIGINALI

I.

R. ISTITUTO DI STUDI SUPERIORI DI FIRENZE
ISTITUTO DI PATOLOGIA SPECIALE MEDICA.
Direttore incaricato: Prof. C. FRUGONI

Asma epidemico da " Pediculoides ventricosus „.

Dott. GIACOMO ANCONA.

Il 1° giugno 1922, il prof. Frugoni inviava nel proprio turno ospedaliero (R. Arcispedale S. Maria Nuova) un individuo sofferente di asma, allo scopo di determinare la etiologia della sindrome.

Si trattava (Oss. I) di Baroni Ubaldo, di anni 27, mugnaio in Barberino di Mugello che si era mantenuto sempre sano fino alla malattia per la quale era ricorso all'Ospedale, con l'eccezione di una polmonite destra sofferta a 19 anni e di ittero epidemico castrense a 21: non bevitore, medico fumatore, non luetico. Riferiva il P. che dall'ottobre 1921 soffriva di attacchi d'asma e di dermatosi pruriginosa.

Gli attacchi d'asma esordivano con starnuti, rinorrea, tosse secca e si svolgevano con dispnea di non grande intensità durante 2-3 giorni. Un'espettorazione abbondante segnava la fine dell'accesso. Dette crisi, nei primi mesi più leggere, erano andate facendosi sempre più forti sì da costringere il paziente ad abbandonare il lavoro di mugnaio.

d)

MÜNCHENER MEDIZINISCHE WOCHENSCHRIFT

23. März 1928

Asthma und Milben.

Von Dr. Hermann Dekker, Wald (Rhld.).

Im Jahre 1922 stellte Ancona als Ursache einer Asthmaepidemie bei Leuten, die mit der Verarbeitung von Getreide beschäftigt waren, die Durchseuchung des Getreides mit Milben (Pediontoides ventricosus) fest. Storm van Leeuwen fand Milbenasthma auch in Holland bei Bauern, aber er hält — für Holland — die Milben nicht für eine wesentlich in Betracht kommende Ursache, sondern andere, in der Luft schwebende Allergene, bes. Schimmelpilze, die in dem unter dem Meeresspiegel liegenden Deltaland außerordentlich verbreitet sind.

Als ich mich vor 2 Jahren zuerst mit dem Ausfindigmachen der Ursachen des Asthmas durch kutane Prüfungen befaßte (die ich seitdem an rund 400 Personen ausführte), war ich zunächst von den Ergebnissen sehr wenig befriedigt. Schimmelpilze kamen in unserer bergigen Gegend, wo die Häuser auf felsigem Grund, meist auf Höhen und Abhängen stehen, nicht in Betracht, oder doch nur in sehr wenigen Fällen, wo Leute in feuchten Tälern feuchte Wohnungen bewohnten. An Milben zu denken, hatte ich in unserer rein industriellen Bevölkerung keine Veranlassung.

Figure 4.5 Some landmark 19th and 20th century acarological works relating to domestic mites. **a)** Albert Davidson Michael's (1901) *British Tyroglyphidae* title page; **b)** Voorhorst and co-workers, 1969; **c)** title page of Ancona's (1923) paper describing epidemic asthma among grain workers caused by *Pyemotes ventricosus*; **d)** Dekker's (1928) paper on asthma and mites.

the popular journal, *Hardwicke's Science Gossip* (1880, p. 262 in Michael, 1901):

> *Would some reader kindly give me some information regarding the best means of eradicating from household furniture a mite which made its appearance a few months ago in myriads in a bedroom, and has now spread over the whole house? The furniture has been exposed to sulphurous acid fumes, saturated with a solution of carbolic acid, corrosive sublimate turpentine, acetic acid, etc., but although much reduced in numbers the family is still in a flourishing condition.*

In parallel with work on the Tyroglyphidae, E.L. Trouessart was working on the systematics and biology of feather mites, then included in the family Sarcoptidae along with parasitic mites of mammals, which were summarised in the monograph by Canestrini and Kramer (1899). *Dermatophagoides pteronyssinus*, described by Trouessart as *Mealia pteronyssina* (Trouessart, in Berlese, 1897a), and later found to be synonymous with *Dermatophagoides scheremetewski* from the skin of patients with dermatitis in Moscow by Bogdanoff (1864) (see Appendix 1), is mentioned in this work, placed in the subfamily Tyroglyphinae along with most of what are now recognised as the major genera of the acarid and glycyphagid mites.

4.2.3 Twentieth century observers

It has been known for centuries that exposure to house dust can precipitate the onset of wheezing in people with asthma (Ellul-Micallef, 1997). The search for a causative substance only began after Coca and Cooke (1922) clarified that asthma attacks were associated with exposure to allergens and Kern (1921) suggested the existence of a distinct allergen in house dust. This period heralded the beginning of dust sampling in homes, methods for reducing exposure to dust (Storm van Leeuwen, 1924), and investigations into factors affecting the frequency of skin test positivity to house dust extract among people with atopy, such as climate, elevation and soil type (all factors that influence domestic humidity). The first indication of the involvement of mites with asthma was made in 1923 by Ancona, who described 'epidemic asthma' among Italian villagers who were handling sacks containing grain heavily infested with the tarsonemid mite *Pyemotes ventricosus* (Figure 4.5c). Storm van Leeuwen

et al. (1924) described the case history of a farmer with asthma who had inhaled dust from oats which had been heavily infested with the storage mites *Acarus siro* and *Glycyphagus* sp., and suggested that mites may be the source of allergens in house dust.

Dekker (1928) first reported mites in dust from homes of asthmatics, having examined samples of dust brushed from under the beds of his patients. He considered mites to be a very important cause of asthma; responsible, in his estimation, for about 60% of his cases. He recommended intensive cleaning to get rid of the mites and he noticed marked clinical improvement in many patients who carried out this procedure. Posse (1946) also mentioned mites in house dust as a possible cause of asthma.

Between the 1930s and early 1960s, publications on mites found in houses focus almost exclusively on stored products species (such as Hora, 1934; Hughes, 1948; Solomon, 1945, 1946a, b; 1961). The motivation for this research was the importance of ensuring against spoilage of scarce food resources during World War II and the post-war years of food rationing, which ended in the UK as late as 1954.

The discovery of the role of dust mites as a major source of allergens can be attributed to two research teams, working simultaneously and independently, on opposite sides of the world: one in Leiden, the Netherlands; the other in Tokyo, Japan.

Oshima (1964), in an investigation of pruritus in schoolchildren, found *Dermatophagoides* spp. were present in significant numbers in the floor dust of Yokohama schools. This work was stimulated, in part, by the work of Sasa (1950, 1951), who investigated the association between *Dermatophagoides* spp. and suspected cases of human urinary and pulmonary acariasis (see section 8.3.4f), including the description of a *Dermatophagoides* sp. 'from a female patient with bronchial asthma' admitted to the University of Tokyo Hospital (Sasa, 1950). Oshima (1967) notes that he learned of the ecological findings of the Leiden group after the publication of his paper in December, 1964. However, Miyamoto *et al.* (1968), in their work on the allergenicity of *Dermatophagoides farinae* and its role in asthma, state their research was 'stimulated by R. Voorhorst and his co-workers' (Voorhorst *et al.*, 1964; 1967).

Research on dust mite ecology commenced in Europe in 1962 when Spieksma and Spieksma-Boezeman investigated the fauna of dust in houses in Leiden. Though Dekker (1928) had reported his investigations of acarid and glycyphagid mites in dust swept from under the beds of asthma patients, and, as Voorhorst *et al.* (1969, p. 86) mention, the most frequent species of mites considered to occur in houses were acarids and glycyphagids, there had been no previous attempt to study the distribution and abundance of mites in homes using quantitative and reproducible sampling. A major stimulus for the Leiden group was that they knew that the house dust allergen was different from the allergens produced by glycyphagid and acarid mites, and, since an implicit part of their theory was that mites were the source of the house dust allergen (Voorhorst *et al.*, 1964), they reasoned that house dust must contain other, then unknown, mite taxa. A central tenet of their theory is that human exposure to dust mites is commonplace. In order to take the theory forward, the Spieksmas had to be able to assess how many homes contained mites and how abundant the mites were in each home, as well as which species were present. However, their first investigations concerned only one house, where they examined population densities at three-week intervals from May to November. One of the first problems the Spieksmas faced was to develop appropriate techniques to sample the dust and extract the mites from it. The only previous studies that were remotely similar to theirs were on the ecology of free-living mites of soils and leaf litter, and their earliest methods in house dust mite ecology were adapted from these (F.Th.M. Spieksma, 1995, *pers. comm.*), and later modified. They went on to examine population dynamics in three houses in Leiden for a period of a year (1964–1965), and then the fauna of 150 homes where they soon recognised a correlation between mite population density and degree of domestic damp.

The reports of the Leiden group's ecological work (Spieksma and Spieksma-Boezeman, 1967; Spieksma, 1967; Voorhorst *et al.*, 1969; see also Figure 4.5) showed that dust mites were present in every one of the 150 homes they sampled, and that of the 9209 specimens they isolated, *Dermatophagoides pteronyssinus* was present in every home and constituted 88% of the total number of pyroglyphid mites. *Euroglyphus maynei* was found in 53% of homes and constituted 11% of the total pyroglyphids whereas *Dermatophagoides farinae* was present in only 2% of homes and constituted only 1% of the pyroglyphids. The numbers of mites they found are relatively low when compared with figures from numerous other subsequent reports from around the world, and are due to the extraction method that they used.

In the mid-1960s there were two other ecological investigations on dust mites under way. First, in Belgium, where Fain, who had provided taxonomic advice for Spieksma and co-workers, sampled mites in 20 homes in Brussels, Antwerp, Malines, Louvain and La Louvière and found a similar faunal composition and frequency of occurrence as the Leiden Group. Second, Oshima and co-workers in Japan had examined dust mites as a possible cause of papular urticaria outbreaks in schools.

The work of the Leiden and Tokyo research teams provided the stimulus for others to investigate their local dust mite fauna. An era of survey work began. Most researchers sought to answer the same set of questions: how many mites are present in homes and of which species, and how many homes have mites in them? By the end of 1970, surveys ranging from a faunistic analysis of a few samples, to full-scale quantitative investigations of many dwellings, had been published from Australia (Trinca *et al.*, 1969; Domrow, 1970), Barbados (Fain, 1966a), Belgium (Fain, 1965, 1966a, b), Brazil (Fain, 1966b, 1967a, b), Denmark (Haarløv and Alani, 1970), Eire (Spieksma, 1967), England (Brown and Filer, 1968; Cunnington and Gregory, 1968; Maunsell *et al.*, 1968), Germany (Dekker, 1928), Japan (Oshima, 1964; 1967; 1970; Miyamoto *et al.*, 1970), The Netherlands (Voorhorst *et al.*, 1964; 1967; 1969; Spieksma, 1967; Spieksma and Spieksma-Boezeman, 1967; Fain, 1967a), Norway (Spieksma, 1967), Switzerland (Voorhorst *et al.*, 1964; Rufli, 1970), Taiwan (Oshima, 1970), USA (Fain, 1967a; Larson *et al.*, 1969; Mitchell *et al.*, 1969; Sharp and Haramoto, 1970) and Zaire (Fain, 1966a, b; 1967). By 1980 surveys had been reported from some 53 countries, and to date over 250 surveys have been published from over 70 countries (see section 4.9.1).

Most of these surveys represent limited, snapshot views of the house dust fauna: a picture of what is present in relatively few homes in any location, at one particular time. They consist of lists of species, with numbers of mites found, more often than not expressed per unit weight of dust, the proportion of the total haul of mites that each species represented, and the frequency of occurrence of each species. My survey of dust mites in Glasgow (Colloff, 1987c) represents a fairly typical example of the genre; their greatest value being in their contribution to a collective dataset. Largely, these surveys have supported the earlier findings of the ubiquity and abundance of dust mites, but as single studies they tell us little else. When they are databased and analysed (see section 4.9.1) we see a more complete picture of the characteristics of the house dust fauna at regional and global scales.

By the end of the 1960s, people were examining factors associated with the distribution and abundance of dust mites and recognised that domestic microhabitats were not homogeneous. The need for comparisons of mite populations in beds and bedding, upholstered furniture and carpets was recognised early on (Maunsell *et al.*, 1968). Sesay and Dobson (1972) and Dusbabek (1975) took the investigation of such habitats further by examining distribution of mites in different layers of beds and bedding and Blythe (1976) reported detailed observations on how the surface topography of mattresses and local accumulations of dust around seams and buttons was associated with the distribution of mites. Fain (1966a, b) reported greater population densities of mites in old, damp homes, as did Varekamp *et al.* (1966), and the importance of domestic humidity and temperature was soon well established.

A more epidemiologically mediated series of variables started to be investigated in the 1970s. Comparisons of homes of asthmatics versus non-asthmatic volunteers, presence or absence of companion animals, central heating, double-glazing, gas fires, house plants, numbers of people occupying each dwelling, age of beds and carpets, height of apartments above ground, soil type, climate, elevation, proximity to canals, underground water courses and drainage systems, housing construction and design, domestic hygiene standards: all these variables have been examined, at one time or another in relation to dust mite population densities or allergen levels.

4.2.4 Has the species composition of the domestic mite fauna changed?

The numerous reports of mites from European houses during the period 1650 to 1928 refer almost exclusively to acaroid and glycyphagoid species. Pyroglyphids in European houses are first mentioned only in 1964. While it seems highly unlikely that pyroglyphids were absent from houses prior to this period, it is certainly possible that acaroids and glycyphagoids were the more dominant components of the fauna. On the other hand, perhaps glycyphagoids and acaroids were often mentioned simply because they are more noticeable than pyroglyphids because of their larger size and their capacity to develop massive, obvious populations very rapidly. The conditions in homes prior to the 20th century would have been conducive to acaroid

and glycyphagoid population growth. Of particular importance would have been the use of rushes and straw and other vegetation as floor coverings and bedding, and the lack of storage and preservation systems that prevented contamination of foodstuffs with moulds and arthropods (especially comestibles such as cheeses, bread, flour, dried fruit, meat and fish). The trend towards improvements in domestic hygiene and food storage (such as refrigeration) and increased use of textiles in interior furnishings from the 19th century onwards are most likely to have led to declines in the distribution and abundance of glycyphagoids and acaroids and increases in pyroglyphids. There is a certain similarity in this scenario with the data of Green *et al.* (1982) that indicates that *Tyrophagus putrescentiae* predated *Dermatophagoides* spp. as the major source of domestic mite allergens in traditional houses in Papua New Guinea.

There are two retrospective studies of what appear to be changes in species composition. Halmai (1984) found an increase in relative abundance, dominance and frequency of occurrence of *D. farinae* between 1969 and 1984 in 434 dust samples from 300 sites in Hungary. The change in composition was at the expense of *D. pteronyssinus*, and seemed to be associated with dust samples from apartment blocks and samples from flats with carpeted floors. However, more samples came from these sources between 1976 and 1984 than between 1969 and 1976, so the claim of a change in species composition may be an artefact of changes in the proportion of sample type during the 15-year period. Petrova and Zheltikova (2000) examined a total of 340 Moscow apartments between 1983–1985, 1989–1991 and 1993–1997. There was an increase in frequency of occurrence of pyroglyphid mites from 52–59% in the mid-1980s to 78% in the early 1990s, and relative dominance fell from 94% to 78%, mainly due to a 50% decrease in the abundance of *D. farinae*. There was a steady increase in frequency of storage mites, but perhaps the most striking finding was that mean population density increased five-fold during the 14-year period.

4.3 Ecological concepts – distribution, abundance, biodiversity, communities and ecosystems

The iteration of ecological terminology here may seem rather pedestrian but, bearing in mind that the study of dust mites has taken place with rather fleeting, sideways glances at mainstream ecological theory and practice, it may be useful to outline some basic definitions. Ecology is the scientific study of the distribution and abundance of organisms. Krebs (1972) also included the concepts of relationships between different organisms and between organisms and their habitat, and stated that ecology is the scientific study of the interactions that determine the distribution and abundance of organisms; in other words, the study of where organisms are found, what they are, how many are present, and how they interact with each other and their environment.

4.3.1 Distribution and abundance

Distribution, the arrangement of individual organisms in space and time, is closely linked with their abundance (the numbers of organisms present in that unit of space and time), because the simplest measurement of distribution is the recording of the presence or absence of a species, and the determination of patterns of distribution (regular or random or clustered) involves the counting of the individuals present. Distribution and abundance are not the same thing. This may seem an obvious point, but they have occasionally been mistaken as synonyms by house dust mite researchers. They are independent variables but often show related trends. Often, as the pattern of distribution of a species changes, so does its abundance. Distribution of house dust mite species has been assessed on a regional level (that is in prescribed geographical or topographical areas); in different homes, in different habitats within homes (bedding, carpets, upholstered furniture) and in different parts of a single habitat. The frequency of occurrence of a species represents the proportion of the total number of samples in which that species is present (usually expressed as a percentage). Frequency of occurrence is also referred to in the literature as 'occurrence' or, using epidemiological terminology, as 'incidence' or 'prevalence'. The live individuals of a particular species that occupy a particular area or volume at a particular time make up a population.

The measurement of abundance is usually referred to as the density of a species per sample unit ('mite concentration' and 'mite counts' have been used instead of density by a few researchers). Population density can be estimated by *absolute estimates*, whereby all individuals in a unit of habitat are counted; *relative estimates*, which represent the number of individuals encountered per unit sampling effort; and *population indices*, which are relative estimates of population size based on the amounts of waste products (e.g. guanine in faeces or faecal allergens). Most dust sampling provides a relative estimate of the size of the population, not a measurement of all individuals present. To obtain an accurate, reproducible relative estimate, the individuals in each of the sampling units must be counted as exactly as possible, the size of the unit must be constant, representative of the whole area under study and appropriate for the size of the organism and its distribution. Relative density is the number of individuals of a particular species expressed as a percentage of the total organisms in a sample. Relative density is also referred to as 'dominance', i.e. 'in the sample *E. maynei* has a dominance of 25% and *D. pteronyssinus* 70%'.

The presence or absence of a species can be due to factors relating to dispersal, behaviour, other species (predation, parasitism, competition, disease), as well as physical and chemical factors (such as temperature, water and oxygen). These factors are of relevance also for the explanation of abundance – the interrelationship between immigration, natality, mortality and emigration. Figure 4.6 shows the relative species abundance and frequency curve for dust mites from 74 homes in Glasgow, Scotland (Colloff, 1987c). It is fairly typical of the pattern for house dust mite surveys. Some useful insights into the ecology of particular species, and consequences for allergen exposure, can be gained from a simple analysis of the data. Three species, *Dermatophagoides pteronyssinus*, *Euroglyphus maynei* and *Glycyphagus domesticus* account for 95% of individuals found. *D. pteronyssinus* is by far the most abundant species and occurs in all homes and is responsible for the bulk of allergen exposure. *E. maynei* is present in 32% of homes and is likely to contribute to a significant level of allergen exposure such that it cannot be ignored. *Glycyphagus domesticus* is present in 14% of homes but is more abundant than *E. maynei*, suggesting that large populations of this mite develop in relatively few homes. There must be something different about these homes to support high populations of this species (in fact, they tend to be very damp). The residents of these homes are likely to receive a different pattern of allergen exposure than others in the survey.

4.3.2 Populations, communities and ecosystems

Krebs (1972) refers to populations, communities and ecosystems as the three levels of integration in ecology. Each entity has a discrete set of characteristics.

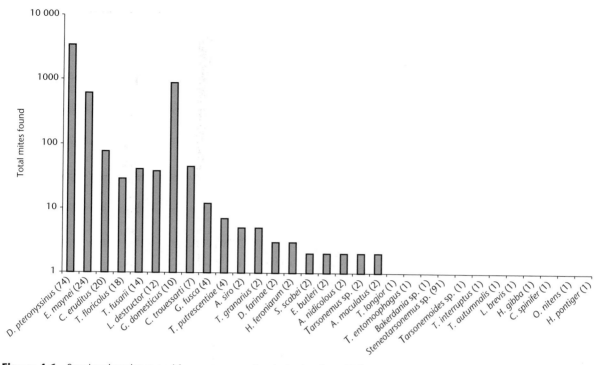

Figure 4.6 Species abundance and frequency curve for dust mites from 74 homes in Glasgow, Scotland (data from Colloff, 1987c), arranged according to their frequency of occurrence (figures in brackets).

Populations have a density, but not species diversity: this is a community attribute. Ecosystems have functional diversity, that is a community of different species that do similar 'jobs' on which the ecosystem depends for its integrity. An ecosystem, then, is a community of organisms and their physical environment that interact as an ecological unit. A biome is a biogeographical region, characterised by distinctive communities of plants (for terrestrial biomes), such as tundra, tropical grassland and savannah; or animals (for marine biomes). It may represent a cluster of different ecosystems, each with its unique community of organisms interacting with each other and their environment as an ecological unit (Lincoln *et al.*, 1998). In extending these definitions to house dust, a single house would represent an ecosystem, whereas the totality of human dwellings represents a biome. I am unconvinced this comparison is useful, and prefer to regard house dust as a fragmented, global ecosystem, operating at a series of nested scales.

4.3.3 Spatial heterogeneity, focality and niche theory

The parts of a dwelling that exhibit the basic requirements for dust mites to reproduce and maintain viable populations will tend to be homogeneous in respect of key habitat suitability determinants. The more

homogeneity of microhabitat characteristics throughout the home, the greater the risk that home will contain high dust mite populations and allergens that constitute a risk factor for asthma. Each of these factors will have a particular distribution within any home. The presence of one factor does not automatically assume that dust mites can live in that home. More correctly, they are like a series of overlays, or three-dimensional Venn diagrams, with the areas of overlap representing the most suitable habitats for mite survival (as shown in Figure 4.7). A similar set of overlapping 3-D spatial conditions will exist for cockroaches, moulds and any other indoor allergen-producing organisms, in fact for all organisms. Brenner (1991a, b) refers to this concept as 'focality': the phenomenon of a population being focused by constraining biotic and abiotic factors (see Figure 4.8). Implicit in focality is that survival of the population is predictable, as is spatial distribution and the characteristics of the principal foci determine optimal conditions for the population.

Focality is a very similar concept to ecological niche theory. The term 'niche' has several interpretations and definitions in ecological usage, varying in complexity and vagueness. It is derived from the French *nicher*, meaning to make a nest. In general usage it has often been used to mean the set of conditions for an

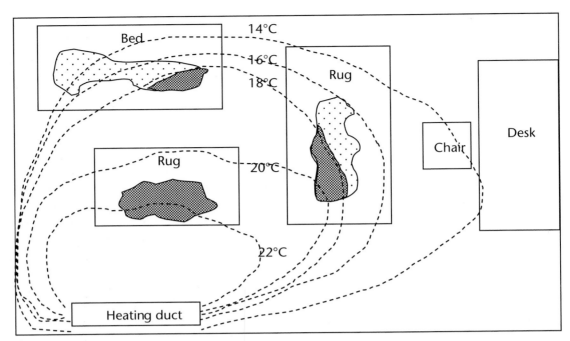

Figure 4.7 Diagram illustrating the concept of focality, based on three niche parameters within a bedroom that collectively determine where dust mite populations can increase: i) availability of habitat (the bed and rugs); ii) availability of food resources; iii) availability of optimum temperature for reproduction. Shaded areas represent areas where all three favourable parameters overlap.

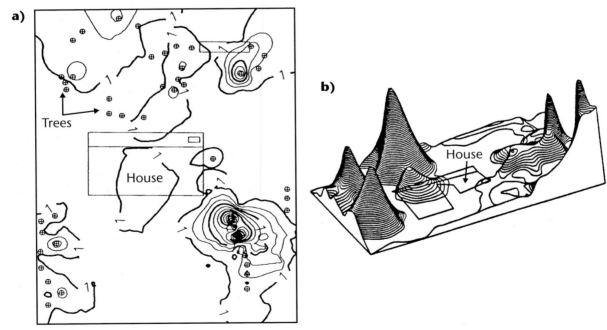

Figure 4.8 Focality and spatial patterns. **a)** 2-dimensional contour map of cockroach population density around a house, modified from Brenner (1991a); **b)** 3-dimensional contour map of distribution and population density of smokeybrown cockroaches around a house, modified from Brenner (1991b).

individual to thrive (as in having found one's niche). This is very similar to early ecological definitions that used the term niche to describe the role of an organism in its community but different from that meaning a subdivision of the habitat (see Krebs, 1972, p. 231 for a brief history of the use of the term). Hutchinson (1958) redefined the term to mean the space represented by the sum of the optima of all the environmental variables that allow the species to survive and reproduce. He referred to this *n*-dimensional hypervolume as the *fundamental niche* of a species. There are some rather basic practical difficulties with the idea of the fundamental niche: the infinite number of dimensions defies the determination of the fundamental niche for any organism!

Nevertheless, niche theory, or focality, is a useful concept in the study of allergenic organisms because it provides a conceptual framework for prediction of population density, and hence allergen concentration and exposure (Brenner, 1993), as well as the systematic manipulation of the microhabitat to make it hostile to allergenic organisms.

4.3.4 Biological diversity, species richness and heterogeneity

Biological diversity, or biodiversity, tends to mean different things to different people and a good, standard definition is hard to arrive at. Often it is used in a general sense to mean simply 'the variety of life'. The term encompasses concepts of genetic diversity (i.e. diversity of genes within a species), species diversity (i.e. species-richness and species function within communities and ecosystems) and ecosystem diversity (i.e. diversity in ecosystem processes and organisation). Species diversity is of most relevance to house dust mite ecology. It comprises species richness (numbers of species in a locality, habitat or ecosystem) and functional diversity (the 'job' that species do for the ecosystem to operate).

Diversity is the measurement of how many different species inhabit a particular place. In house dust mite ecology, species-diversity is used in its simplest form, species-richness, i.e. the number of all species present in a sample (rather than, more correctly, only those which are genuine residents of house dust). Too few samples underestimate species diversity because rare species may not be taken. Too large samples overestimate the importance of rarer species. Over 140 species of mites have been isolated from house dust (van Bronswijk, 1981). Of these, most have been recorded very occasionally and are either accidental introductions from outdoors, plant feeders introduced with house plants or parasites of birds, rodents and domestic pets. The major taxa resident in house dust have been covered in the keys in Chapter 1. Nevertheless, there are several obscure species that are true house dust dwellers. Their occurrence, distribution and abundance have probably been underestimated because they have been either ignored or

confused with more common species. A case in point is the two species described by Oshima (1979): *Calvolia domicola*, found in 29 of 126 dust samples in Tokyo, and *Chortoglyphus longior*, found in 54% of Tokyo dust samples; probably confused with the better-known *Chortoglyphus arcuatus*.

Species richness in a single home is rarely greater than 10 species. In a sample of dust from a carpet, it would be rare to find more than five species of true dust dwellers represented by multiple specimens. Much of the faunistic literature obscures the species-richness within individual dust samples because the abundance and diversity data from all the dust samples tend to be lumped together under one long species list.

Heterogeneity is a measure of species richness combined with relative abundance (see Krebs, 1989, 2001). A problem with counting the numbers of species in a dust sample is that it does not discriminate between rare species and common ones. Typically in house dust samples from temperate latitudes, one species might account for almost all individuals present, although there may be a total of four species in a dust sample. In the tropics, there is more likely to be a more even abundance of three or more major species, with some rare ones in addition. The latter pattern represents a more heterogeneous, or even, community than the former. More species are present and they are more equally abundant.

Species richness and heterogeneity as measures of diversity have been somewhat ignored by dust mite ecologists, probably because diversity is never particularly high. This is a little odd when one considers that each mite species carries its own set of unique allergens. Dust containing several species in a relatively heterogeneous community represents the potential for exposure of patients to a far higher diversity of allergens than dust containing one or two species. In ecology, low species richness is often an indicator of a pioneering community, i.e. one which is relatively young in ecological time, and has yet to develop a stable community structure. There is some evidence to suggest that the dust mite communities remain almost indefinitely in this pioneering stage because of the constraints imposed on them by the domestic microclimate. An equally ecologically plausible hypothesis is that house dust is too structurally simple a habitat to support more than a few species of mites.

Krebs (1989) gives methods for estimating species diversity. A common one is Simpson's index (Simpson, 1949), a non-parametric probability estimate, based on the premise that diversity is inversely proportional to the probability that two individuals picked at random will be of the same species. When used as a measure of diversity, Simpson's index (D) is usually expressed as the complement (i.e. 1– the index), thus:

$$D = \begin{pmatrix} \text{Probability of} \\ \text{randomly selecting} \\ \text{two individuals} \\ \text{that are different} \\ \text{species} \end{pmatrix} = 1 - \begin{pmatrix} \text{Probability of} \\ \text{randomly selecting} \\ \text{two individuals} \\ \text{that are the same} \\ \text{species} \end{pmatrix}$$

$$\text{i.e. } D = 1 - \sum(p_i)^2 \qquad (4.1)$$

where $1 - D$ = complement of Simpson's index of diversity; p_i = proportion of individuals of species i in the community. Simpson's index ranges from 0 (low diversity, single species) to nearly 1 ($1 - 1/s$), where s = the number of species in the sample. The reciprocal, $1/D$, varies from 1 to s. This form is most useful as a measure of the number of common species required to generate the observed heterogeneity. I have used it here to compare diversity of mites from dust samples from Glasgow, Scotland (Colloff, 1987); Perth and Bunbury, Western Australia (Colloff et al., 1991) and Cartagena, Colombia (Fernández–Caldas et al., 1993; see Figure 4.9). Geometric mean diversity per sample was significantly higher in beds and floors from Cartagena (2.13 and 2.03 respectively) than from Western Australia (beds 1.23; floors 1.16) and Glasgow (beds 1.2; floors 1.07). This follows the trend for diversity to be greater in tropical latitudes than temperate ones. The most frequently occurring and abundant members of the mite community at Cartagena were *D. pteronyssinus*, *Blomia tropicalis*, *Chortoglyphus arcuatus*, *Pyroglyphus africanus* and *Cheyletus malaccensis*, whereas in Western Australia they were *D. pteronyssinus*, *Euroglyphus maynei* and *Lepidoglyphus destructor*, and in Glasgow *D. pteronyssinus* and *Euroglyphus maynei* only.

4.3.5 Functional diversity: communities and food webs

Communities consist of populations of species living within a particular area or habitat. Characteristics of communities include species diversity and functional diversity, structure (i.e. degree of complexity of their spatial arrangement), dominance (i.e. major species versus lesser species, in terms of their effect on the rest of the community); abundance, trophic relationships and other species-species interactions.

Communities are dynamic. As they become more species-rich, so the number of potential interactions between species increases. There are patterns of succession and colonisation as some species die out and others move in, often driven by changes in quality and quantity of available food resources. Communities are subject to disturbance and perturbation, resulting in alteration of their characteristics and composition, and they vary over time and space in their stability and persistence.

Energy and nutrients are transferred within ecosystems when one organism eats another. Food chains consist of primary producers (plants that produce food by photosynthesis), primary consumers (animals that feed on the plants) and secondary consumers (animals that feed on primary consumers). Chains are linked to form webs (see Figure 4.10a). Detrital ecosystems lack primary producers. The basal group of organisms are the detritivores, the organisms that feed on the dead and decaying organic matter that comprises the detrital ecosystem, its habitats and is the main source of nutrients and energy. In house dust, the basal group is relatively poorly known and consists primarily of bacteria and fungi, although some of the mites feed directly on detritus also.

a Functional groups – the components of food webs

Groups of organisms with similar food sources and feeding biology are called functional groups or trophic groups, referring to their ecological 'job' of feeding and resource use in which they indulge, such as predation, detritivory or ectoparasitism. The term 'guild' is often used as a synonym of functional group, but guild has often been used to identify particular taxonomic groups of organisms (such as birds or mammals) that compete for a common resource (see Walter and Proctor, 1999, for more detail and definitions). The functional group concept implies no competition between members or no taxonomic constraints on membership. The major functional groups of arthropods follow Walter and Proctor (1999).

Saprophagous microorganisms

These are organisms that feed on dead organic material of animal and plant origin and are known as saprophages, also sometimes called detritivores. Bacteria in house dust are likely to be derived from skin scales, outdoor air and the human gut and respiratory tract. The bacteria that belong to the normal skin flora

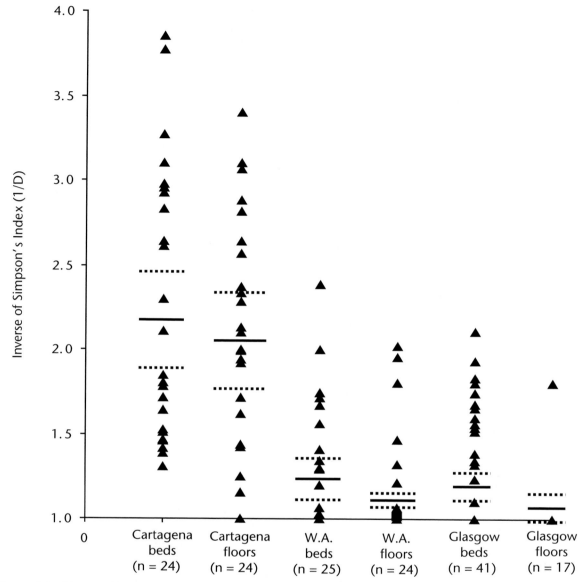

Figure 4.9 Diversity of live mites in samples of house dust in beds and on floors at Cartagena, Colombia (from dataset of Fernández-Caldas *et al.*, 1993); Bunbury and Perth, Western Australia (from dataset of Colloff *et al.*, 1991) and Glasgow, Scotland (from dataset of Colloff, 1987c).

include cluster-forming Gram-positive cocci such as *Staphylococcus aureus* and *S. albus*, chain-forming Gram-positive cocci such as *Streptococcus* spp., Gram-positive rods such as *Corynebacterium* spp. and Actinomycetes such as *Mycobacterium* spp. Oh *et al.* (1986) isolated various bacteria from mite guts including *Bacillus* spp., *Staphylococcus* spp., bacilli and corynebacteria. Ushijima *et al.* (1983) and Kato *et al.* (1983) added members of the genera *Achromobacter*, *Flavobacterium*, *Pseudomonas*, *Proprionibacterium*, *Serratia*, *Citrobacter* and *Enterobacter* to the list of bacterial

isolates from dust mites. *Candida albicans* and *Pityrosporum* spp. are the best-known yeasts associated with the normal skin flora, and yeasts have been isolated from indoor air by Hunter *et al.* (1988), from dust by Andersen (1985), and from dust mites by Oh *et al.* (1986). The mould flora of house dust is much better known than the bacteria, and the literature on occurrence of fungi within houses and other buildings is extensive (see van de Lustgraaf, 1977; 1978a; van Bronswijk, 1981, her chapter 7; Andersen, 1985; van Bronswijk *et al.*, 1986 and Hunter *et al.*, 1988),

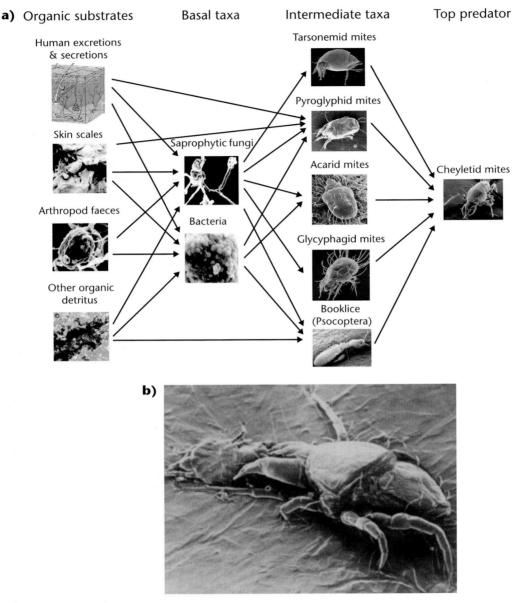

a) Organic substrates   Basal taxa   Intermediate taxa   Top predator

Human excretions & secretions

Skin scales

Saprophytic fungi

Arthropod faeces

Bacteria

Other organic detritus

Tarsonemid mites

Pyroglyphid mites

Acarid mites

Cheyletid mites

Glycyphagid mites

Booklice (Psocoptera)

b)

Figure 4.10 Major components of the detrital food web within house dust. **a)** There are 13 interactions in this web of eight taxa (not including links with organic substrates). There are 8² or 64 possible interactions, so the degree of connectivity is 13/64 or 0.2. Linkage density is 13/8 or 1.63; **b)** a predatory cheyletid mite with its prey (from Mumcuoglu, 1988, Figure 4).

though good information on the ecology of moulds in house dust is harder to find. Members of *Acremonium, Alternaria, Aspergillus, Chrysosporium, Cladosporium, Eurotium, Fusarium, Mucor, Penicillium, Rhizopus, Scropulariopsis, Stachybotrys, Trichoderma, UROcla-dium, Wallemia* and several other genera are found in houses. Of these, species of *Aspergillus, Chrysosporium, Penicillium* and *Wallemia* are among the more significant residents of house dust.

Grazing-browsing microbivore-detritivore arthropods

It would be virtually impossible for mites in house dust to feed on organic matter without ingesting the associated microorganisms. Norton (1985) acknowledged this in relation to soil mites and suggested that saprophagy/detritivory evolved as a way of accessing microbial food within the organic matter; what Walter and Proctor (1999) refer to as 'eating the pie crust to

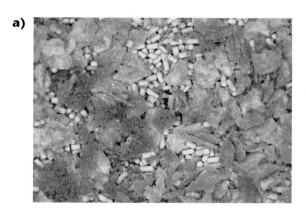 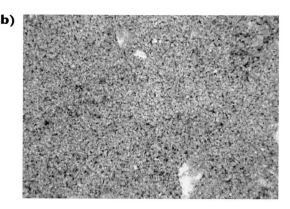

Figure 4.11 The difference in texture between new and old cultures of domestic mites. **a)** Fresh culture of wheat germ, baker's yeast and *Tyrophagus putrescentiae;* **b)** old culture, consisting almost entirely of dead bodies and faecal pellets of *T. putrescentiae.*

get at the filling'. This group includes the pyroglyphids, acaroids and glycyphagoids.

Piercing-sucking microbivore arthropods

Tarsonemid mites are members of the Prostigmata, which are all fluid-feeders with piercing chelicerae for gaining entry to fungal hyphae. The tarsonemids in house dust include the fungal-feeding genera *Tarsonemus* and *Steneotarsonemus.* The biology of *Tarsonemus granarius* was reviewed by Lindquist (1972).

Predators of arthropods

The major predator of house dust mites are the cheyletid mites, especially *Cheyletus* spp. These large mites trap smaller mites between their raptorial palps, inject their styliform chelicerae and suck out body fluids (Wharton and Arlian, 1972; see Figure 4.10b). They are probably not a major constraint on most house dust mite populations because they are not present in every home and, where they are present, their abundance appears to be much lower than one would expect in a predator–prey population equilibrium. It is possible they are more sensitive to water-loss than pyroglyphid mites, though this has not been demonstrated.

b   Transfer of nutrients and energy – trophic interactions

Fungal-mite interactions

Trophic interactions can affect the physical nature of house dust. Conversion of skin scales and fungi into mite faecal pellets represents an example of what has been called 'ecosystem engineering'. The particulate nature of the habitat is altered from one consisting of flakes of skin scales and tubular fungal hyphae into a mix of these plus spheroid faecal pellets. The striking difference in texture in new and old cultures of dust mites is well known to anyone who has grown mites in the laboratory (see Figure 4.11). The process of comminution and faecal pellet production greatly increases the surface area of organic detritus for colonisation by microbes. Fungal spores are ingested by dust mites, pass through the gut undamaged, are concentrated in the faecal pellet and are then able to germinate after it has been voided, producing hyphae and fruiting bodies (see section 2.2.4d). From a fungal viewpoint, a faecal pellet represents a self-contained package of resources, including nutrients in the form of undigested food, leftover enzymes, the remains of digestive cells, and also some residual moisture. Processing and turnover of nutrients in faecal pellets by fungi would lead to an increase in fungal biomass and, in turn, increases in dust mite populations due to feeding on fungi. If house dust fungi use mite faecal pellets as a food source, this may partly explain turnover and degradation of allergens and seasonal variation in allergen concentrations. Grazing by soil mites on fungi can stimulate growth of hyphae (Moore *et al.* 1988), and the potential for fungi to pre-digest skin scales and render them more palatable to mites has already been examined in Chapter 2, as have the implications for the chemical ecology of mites.

Mite–mite interactions: predators and prey

The main trophic interaction is the consumption of mites by *Cheyletus* spp.: a predator–prey interaction. Barker (1991) found *C. eruditus* consumed an average of 0.7 of a prey mite (*Lepidoglyphus destructor*) per day at 25°C. Wharton and Arlian (1972b) describe how *C. aversor* ambushed and grasped specimens of *D. farinae*. After about 20 seconds the prey became paralysed, suggesting the possible involvement of a toxin, allowing it to attack prey much larger than itself. Body fluids were removed over a period of 15–30 min. The capture success rate was only two attempts out of 28 (7%), suggesting that *D. farinae* has some defensive, repugnatorial capability, possibly due to release of chemicals from the lateral opisthosomal glands (see section 2.9.2), or through mechanical protection provided by the long dorsal opisthosomal setae of *D. farinae*. The ambush mode adopted by *C. aversor* indicates that prey density correlates with likelihood of encounters and success rates.

c   Competition

Wharton (1973) studied the effects of mixing populations of *D. pteronyssinus* and *D. farinae*. I have observed that introducing *D. pteronyssinus* into cultures of *E. maynei* leads to the decline in numbers of the latter species but not the former. The reasons for this are not known. Also, van Bronswijk *et al.* (1971) found that the addition of *Acarus* sp. and, separately, *Glycyphagus* sp. to cultures of *Dermatophagoides* sp. suppressed growth of *Dermatophagoides* sp. populations only. This may indicate a trend that faster-growing species inhibit slower-growing ones. Arlian *et al.* (1998a) examined population growth in mixed cultures of *D. pteronyssinus*, *D. farinae* and *E. maynei* and found some evidence of an inhibitory effect of *E. maynei* on the other species.

4.3.6   Colonisation and succession

How long does it take for dust mites to colonise a newly built home or a new mattress or carpet? No mites were found on the floors of new houses prior to occupation but mites were found as soon as the houses were occupied (van Bronswijk, 1974). Most of the mites were dead, suggesting they had been introduced in that state on the clothing of the occupants. Mite populations had become established almost a year after occupation. Van der Hoeven *et al.* (1992) and van der Hoeven *et al.* (1995) found that in newly occupied homes (0–24 months) that had new fitted carpets at the start of occupation, there was no relationship between occupation time and mite allergen concentrations, and the numbers of pyroglyphid mites showed no statistically significant increase over time (see Figure 4.12a). Allergen levels correlated positively with the presence of pet dogs. Most of the mites recovered were dead, indicating that

they were probably introduced to the new homes and were not part of an endogenous population.

Mites are introduced into new homes in furniture and bedding from the previous residence and on clothing. It is rare for people to move into a newly built home that they furnish entirely with new items. In Denver, Colorado, an area with a dry, continental climate, dust mite populations were relatively low (ca. 40 per gram of dust), except in houses containing furniture that had been imported from areas with more humid climates and relatively high mite population densities (coastal California and Texas, Tennessee and Germany; Moyer *et al.* 1985). In these items of furniture, mean mite population density varied from 100–360 per gram, but populations declined over the following two years until they approached the background densities typical for Denver (see Figure 4.12b). This is an important finding because it represents just about the only empirical 'experimental' data to indicate that particular localities have an intrinsic mite population 'carrying capacity' which is defined by regional-scale climatic variables.

One of the characteristics of colonisation and succession is the sequential pattern of changes in populations of colonising species in relation to each other. This is a well-known phenomenon in forensic entomology, whereby the phases of colonisation of a corpse by different functional groups of insects are so well characterised that they can, if ambient temperature is known, be used as a reasonably accurate means of dating the time of death (Smith, 1986). The only data on domestic mites that comes close to this is the study by Ouchi *et al.* (1976) who examined population patterns of mites over a 14-week period in rooms of newly built apartments in Tokyo (see Figure 4.13). *Dermatophagoides* spp. had not established stable populations in either straw mattresses or carpets after 15 weeks, suggesting that the majority of mites were introduced to the home with the occupants, and subsequently died out. *Tyrophagus* spp. populations soared in the straw mattresses (see Figures 4.13a, c), commencing during week 5 in the occupied bedroom, and week 7 in the vacant one. *Tyrophagus* spp. are fungal feeders (Sinha, 1964; Smrž and Catská, 1987), and it is likely that their increase in population density coincided with an increase in fungal populations within the straw filling of the mattresses. This likelihood is further supported by the increase of *Tarsonemus* spp. (see Figures 4.4a, c), which feed exclusively on the liquid contents of fungal hyphae. *Cheyletus* spp., predators

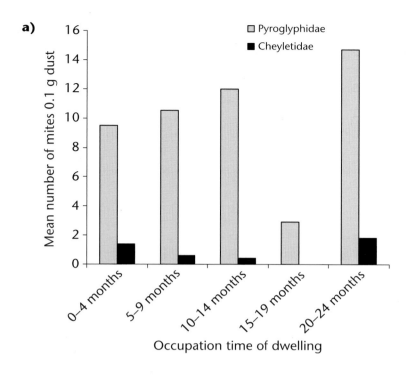

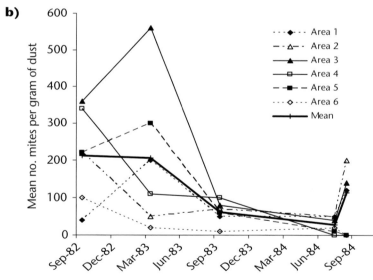

Figure 4.12 Colonisation of newly occupied homes by dust mites. **a)** Graph drawn from data in Table 2 of van der Hoeven *et al.* (1995); **b)** changes in mean mite populations samples in six areas in four homes in Denver, Colorado, that contained items of furniture newly introduced from localities outside Denver (from data in Table 2 of Moyer *et al.*, 1985).

on the other species, also increased in line with the populations of the fungivores. By weeks 14–15, populations of the two fungivores had begun to decline, possibly indicating a change in food resources, microclimate or some other variable, such as competition.

The colonisation of new homes and furniture by dust mites requires that there be adequate habitat, food and microclimate. It is highly likely that most attempts at colonisation are unsuccessful. The accumulation of skin scales in amounts likely to sustain fungi and mites will probably take longer than a few weeks, and the appropriate timescale for studying colonisation is, as van Bronswijk (1974) suggested, likely more than a year. The event documented by Ouchi *et al.* (1976) is better described as a rapid infestation of *Tyrophagus* spp. than a successional sequence of domestic mites.

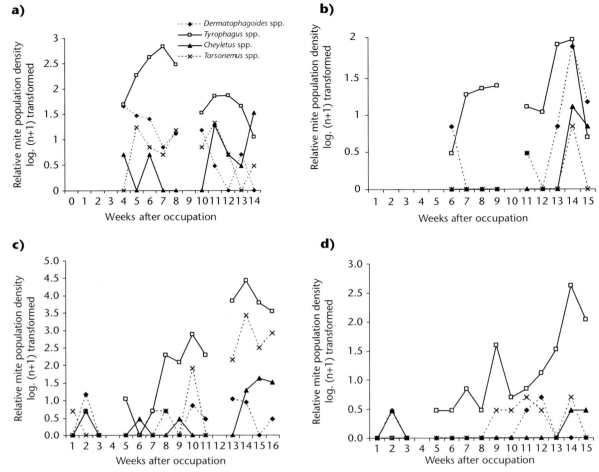

Figure 4.13 Colonisation of four species of dust mites in rooms in a newly built apartment in Tokyo. **a)** Straw mattress in occupied bedroom; **b**) carpet in occupied living room; **c)** straw mattress in vacant bedroom; **d)** wooden floor in vacant room. Graphs drawn from data in Tables 1–4 of Ouchi *et al.* (1976).

4.3.7 Behaviour and dispersal in relation to distribution and abundance

Distribution may be limited by the behaviour of an organism when it is selecting its habitat. The organism chooses to remain in the habitat or not, depending on whether it receives sensory information that tells it the habitat is suitable. It is difficult to study behaviour of house dust mites *in vivo* because of their small size in relation to that of their habitat, but a number of observations have been made on mites *in vitro* that are relevant to habitat-selection, notably the response to humidity gradients. When exposed to the heat given off by the lamp of a stereobinocular microscope, cultures of *D. pteronyssinus* grown in Petri-dishes move to the side of the dish and form small clusters and heaps (the junction between the side and the base of the dish is where humidity is highest). This clustering effect has been suggested to be a consequence of the release of alarm pheromones secreted from the lateral opisthosomal glands (Kuwahara *et al.*, 1980) in response to decreased humidity. The mite that released the pheromones becomes the nucleus of the cluster. Clustering as a behavioural response to reduced relative humidity works because the air spaces between the closely packed mites hold a larger amount of water vapour than the air surrounding the cluster. The mites, having reached the cluster, become quiescent and this helps to reduce water loss by reducing metabolic activity. The response to humidity and heat can be exploited by researchers: clustering behaviour in cultures indicates the humidity of the vessel is too low and should be increased, and also the artificial inducement of clustering using a hot lamp (rather than a cool fibre-optic light source, usually used for examination of mites) can be used to separate mites from their culture medium.

Mollet and Robinson (1995) and Mollett (1996) undertook mark-recapture experiments with live *Dermatophagoides farinae* dyed with Sudan Red 7B. They found that cultures of marked mites released onto a sofa were dispersed from the upholstery onto clothing and then to several other parts of the house and the family car within a 10-day period. It would appear that mites are readily mobile between microhabitats within the home and that clothing provides the vehicle.

4.4 Spatial scales in dust mite ecology

Spatial scales relevant to the distribution and abundance of dust mites can also be thought of in terms of at least four levels of integration. These levels have certain, but imperfect, parallels with the levels of ecological integration – biomes, ecosystems, communities and populations. They provide a practical basis for investigating a series of different, but related phenomena specific to house dust ecology (Colloff, 1998; see Figure 4.14). Temporal scales in dust mite ecology are dealt with in Chapter 5. There is a similar spatial scale for climate (see Linacre and Geerts, 1997, Table 1.1). The relevant climate scale is indicated for each of the categories below in Figure 4.14.

- *The microhabitat scale*, i.e. variation between textile habitats within homes. This scale encompasses the nature of the microhabitat; its microclimate and (diurnal) variation, its physicochemical characteristics, its other residents and their interactions with dust mites.
- *The macrohabitat scale*, i.e. variation between homes. This scale encompasses factors such as housing construction and design and associated biotic factors; its variation in microclimate (daily-seasonally);
- *The regional scale*, i.e. variation between geographical localities in different parts of the world. This scale encompasses regional differences in mite populations brought about by the effects of mesoclimatic variables such as differences in temperature and humidity between localities within the region;
- *The global scale*, i.e. the integration of regional patterns into a continental/global picture, with differences in distribution and abundance influenced by synoptic or global climate variables such as seasonal or annual variation in temperature and rainfall.

For those inclined to fine gradations in classifications, there may be a case for dividing the microhabitat scale up into 'within microhabitats' (i.e. variation within a mattress or carpets) and between microhabitats (i.e. variation between different beds in the home). A similar case could be made for an extra category between macrohabitat scale and regional scale to include variation between suburbs or districts within a city. This latter category is of more relevance to epidemiological considerations. However, for practical purposes, I will stick with four categories.

The features of the house dust ecosystem that make it different from most other ecosystems are that it is highly fragmented and many of its constituent species are found in other ecosystems. There is no spatial continuum of house dust habitat between one house and another. This does not mean that dust mites cannot and do not survive outside the home. Of course they do, because this is where they came from originally (see section 4.6; Table 4.2). However, the unique physical and biological characteristics of the house dust ecosystem mean that from the perspective of a dust mite, each home is an island. This raises the question of whether the house dust ecosystem exists as a self-contained unit within each dwelling or whether it is represented by the entire global archipelago of domestic dwellings.

From an ecological viewpoint, houses are not entirely isolated and self-contained. Humans are highly mobile, and there is evidence of migration of mites in furniture (Moyer *et al.*, 1985) and clothing. When I moved from Scotland to Australia I brought with me in my mattresses and upholstery many thousands of Glaswegian dust mites to settle in Canberra. Some of these would have dispersed, using me as the means of transport, to other Canberran houses. This sporadic mixing, or hybridisation, of geographically isolated populations is likely to have some significant effects on the genetics and evolution of dust mites. However, success of hybridisation events depends on the divergence of the hybridisation gene pools. If two populations are highly divergent genetically, the result is likely to be hybrid depression. If the two populations are not highly divergent, the outcome is more likely to be hybrid vigour. One cannot really generalise about whether an influx to the gene pool of a population will increase fitness, or the likelihood of a beneficial trait becoming fixed.

The insularity of different dwellings can be defined, in part, by the degree of human traffic between them

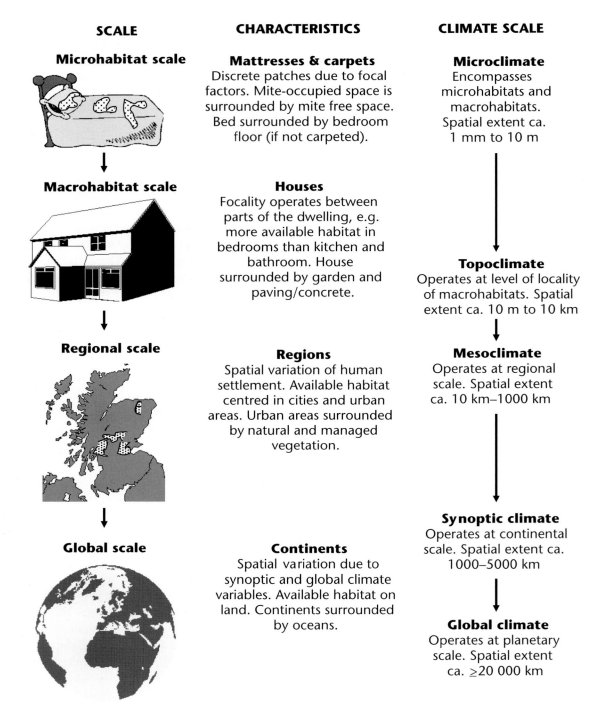

SCALE	CHARACTERISTICS	CLIMATE SCALE

Microhabitat scale

Mattresses & carpets
Discrete patches due to focal factors. Mite-occupied space is surrounded by mite free space. Bed surrounded by bedroom floor (if not carpeted).

Microclimate
Encompasses microhabitats and macrohabitats. Spatial extent ca. 1 mm to 10 m

Macrohabitat scale

Houses
Focality operates between parts of the dwelling, e.g. more available habitat in bedrooms than kitchen and bathroom. House surrounded by garden and paving/concrete.

Topoclimate
Operates at level of locality of macrohabitats. Spatial extent ca. 10 m to 10 km

Regional scale

Regions
Spatial variation of human settlement. Available habitat centred in cities and urban areas. Urban areas surrounded by natural and managed vegetation.

Mesoclimate
Operates at regional scale. Spatial extent ca. 10 km–1000 km

Synoptic climate
Operates at continental scale. Spatial extent ca. 1000–5000 km

Global scale

Continents
Spatial variation due to synoptic and global climate variables. Available habitat on land. Continents surrounded by oceans.

Global climate
Operates at planetary scale. Spatial extent ca. ≥20 000 km

Figure 4.14 Spatial scales in house dust mite ecology in relation to fragmentation of habitat.

and I am therefore inclined to think of the spatial limits of the house dust ecosystem as not confined by four walls, but by a series of more complex constraints, including housing density, geographical boundaries of rural and urban settlements and human social and behavioural interactions.

Houses are pieces in the rather patchy global mosaic of human settlement that contains the house dust ecosystem. The house dust ecosystem has slightly different physical and biological characteristics depending on regional location, just as individual patches of mangrove in coastal central New South

Wales are fragments of an ecosystem (some would say biome), and differ in species diversity, community structure, productivity and biological history from the mangroves of the Florida Keys. Despite their differences, each is immediately recognisable as a mangrove ecosystem.

The idea of a single, global, house dust ecosystem, consisting of a series of subsystems, and differing regionally in biophysical characteristics is, perhaps, a more useful concept than one whereby each house is viewed as containing its own, separate, dust ecosystem. This is because it encompasses the concept of differential fragmentation. This means there is more connectedness in terms of human traffic between some dwellings than others. For example, Feldman-Muhsam *et al.* (1985) found much higher mite population densities in sofas than beds in Israeli houses, which they ascribed to the residents spending several hours daily in socialising. The same effect was found in homes on agricultural collectives (kibbutzim and moshavim) by Mumcuoglu *et al.* (1999). There is likely to be greater human traffic, with accompanying mites, between homes of societies that place strong emphasis on hospitality, kinship and community – typical of the Mediterranean region and the Middle East – than in those societies that are characterised by small family size, few occupants per dwelling, and relatively low social and community interaction.

4.5 Climate

Each of the spatial scales in dust mite ecology involves a consideration of climatic variables, particularly temperature and atmospheric water content and the interaction between the two. Much of the rest of this chapter is concerned with showing how the climate variables of each of the spatial scales in dust mite ecology are influenced by the climate variables of the next largest, as well as the innate conditions operating within each scale. For example, the macrohabitat of dust mites is influenced by the microclimate of the home, which is determined by the topoclimate of the locality (influenced by elevation, distance from the coast and topography), and by characteristics of the home, such as its aspect in relation to solar radiation, insulation characteristics and heating, cooling and ventilation systems. The relative importance of these two groups of factors at each scale provides some useful insights in understanding patterns of spatial variation of dust mite populations.

Temperature and humidity *together* are the major factors that influence the distribution and abundance of dust mites and other terrestrial arthropods. Even if there is an abundance of food, if the microclimate causes greater body water loss than gain, an arthropod cannot survive. Several researchers have sought to isolate the effects of temperature and humidity, ascribing a greater importance to humidity than to temperature. Sometimes they have ignored temperature altogether. But because the drying power of the air is dependent on its temperature, humidity is always a dependent variable.

4.5.1 Temperature

Temperature (T) is a measure of heat energy, which is a measure of the speed of molecular movement. At higher temperatures, molecules in dust mites move faster, so chemical metabolic reactions occur more rapidly, and development and reproduction is completed in less time than at low temperatures. The temperature of the air influences the rate of movement of water molecules, the pressure they exert and the rates of evaporation and condensation (see the following section 4.5.2).

4.5.2 Water and water vapour

The various ways of expressing the amount of water vapour in the air are interrelated and can be derived from each other either by calculation or by using a psychrometric chart (see Linacre and Geerts, 1997, Figure 6.6; Leupen and Varekamp, 1966, Figure 1; van Lynden-van Nes *et al.* 1996, Figure 2.2; Arlian, 1992, Figure 3; Cunningham, 1998, Figure 1). Some basic points about water vapour need to be borne in mind. Air is a mixture of gases with relatively large distances between molecules. Air can accommodate a lot of water vapour, but acts neither as a diluent (i.e. water vapour is not 'dissolved' in air), nor does it 'hold' water. The pressure of water vapour above a volume of water is dependent on the temperature of the water, not on the presence of other gases. Mixed gases behave independently of one another.

a Water vapour pressure and vapour pressure deficit

Under equilibrium between a water surface and the air (i.e. when the number of molecules escaping from the surface balances the number captured), the pressure exerted by the water molecules is called the *saturation vapour pressure* (e_s). Although this might imply this is the vapour pressure when a volume of air is holding as many water molecules as it can, this concept is completely wrong. It is the vapour pressure of the air over water at which evaporation is equal to condensation, and does not relate to the rest of the air. In other words, if more molecules escape than are captured, water evaporates. If more are captured than escape, water is added by condensation. As temperature increases, so does e_s because water molecules are more active and evaporate. As more water molecules are present in the air above the surface, the greater the chances of condensation – collision between airborne water molecules and the water surface – until eventually the rate of evaporation balances the rate of condensation. At this point, the air is saturated. Thus e_s increases with temperature. The rate of increase is given in a psychrometric table, see *Smithsonian Meteorological Tables* (List, 1949; Table 94, *Saturation Pressure Over Water*), or it can be approximated for a given temperature (T) using the following formula (Linacre, 1992):

$$e_s = 6.1 + 0.27\ T + 0.034\ T^2 \qquad (4.2)$$

Saturation vapour pressure (and water vapour pressure and vapour pressure deficit) is measured in hectopascals (hPa: 1 hectoPascal = 1 millibar). From the table of saturation pressure over water, it can be seen that e_s increases exponentially with temperature, roughly doubling with every increase in temperature of 10°C: 6.1 hPa at 0°C, 12.3 hPa at 10°C, 23.4 hPa at 20°C, and 42.4 hPa at 20°C. Note that these values are considerably lower than *atmospheric pressure*, which may typically be around 1000 hPa at sea level, and represents the pressure that the weight of the atmosphere exerts on a square metre of the Earth's surface (1000 hPa = 100 000 kg per square metre).

Vapour pressure (e) is the pressure exerted by the water molecules in a given volume of *unsaturated* air at a given temperature. It is indirectly proportional to the total water molecules in that volume of air. Vapour pressure can be calculated from temperature and relative humidity using equation 4.6. Vapour pressure (and dewpoint, see below) decrease exponentially with increased elevation, so using it as the basis for a comparison of water vapour at places of markedly different altitude creates a bias (see section 4.8.1a).

Vapour pressure deficit (D), also known as the saturation deficit, is the difference between the saturation vapour pressure and vapour pressure of the air at a given temperature.

$$D = (e_s - e) \qquad (4.3)$$

It is proportional to the net evaporation rate (see below) and can be derived from the mixing ratio (see section 4.5.2e). The term ($e_s - e$) refers to saturation vapour pressure at the temperature of the *air*, whereas the same term in Dalton's equation for the rate of evaporation refers to saturation vapour pressure at the temperature of the *water surface* and is referred to as the vapour pressure *difference*. Often the air and water temperatures are similar enough to relate vapour pressure deficit to evaporation. Dalton's equation for evaporation rate (E, in mm per day) is:

$$E = K.u(e_s - e) \qquad (4.4)$$

where u is wind speed in metres per second and K is a constant based on wind turbulence, surface roughness and the height at which wind speed is measured, roughly approximating to 0.2 (when vapour pressures are expressed in hectopascals), but varying inversely with wind speed so that $K.u$ is around 0.5 mm per day. The relevance of this to dust mite ecology and water balance relates to differences in evaporation rates indoors and outdoors. Outside air is more likely to be drier than indoor air at the same vapour pressure deficit because it is more likely to be affected by the drying power of wind. It follows that air within homes in hot climates will be dried considerably by the use of electrical fans.

b   Relative humidity

Relative humidity (RH%, or U) is the ratio of vapour pressure (e) to saturated vapour pressure (e_s), i.e. it is the ratio, expressed as a percentage, of the amount of water vapour that is present in the air to the amount that could be present at a given temperature:

$$U = 100\ \frac{e}{e_s} \qquad (4.5)$$

or:

$$U = 100\ \frac{U.e_s}{100} \qquad (4.6)$$

Note that the vapour pressure deficit represents the *difference* between saturation vapour pressure and actual vapour pressure, not a *ratio* as with relative humidity. To get an approximation of U for a locality, for example a climate station for which there are no recorded measurements of relative humidity, the following formula can be used (Linacre, 1992):

$$U = 100 - 4\ (T - T_d) \qquad (4.7)$$

where T_d is the dewpoint temperature. Relative humidity is a poor index of the amount of moisture in the air because saturation vapour pressure varies with temperature, even though the quantity of moisture in the air does not change. As temperature increases, so does saturation vapour pressure, so at the same vapour pressure but higher temperature, relative humidity will be lower. This means that a relative humidity value does not indicate the moisture content of the air unless accompanied by the temperature of the air. However, it should be noted that all measures of atmospheric water content are temperature-dependent. As an example, the mean daily temperature in Melbourne for the month of January is 19.9°C and the mean vapour pressure is 12.9 hPa. At 19.9°C the saturation vapour pressure over water is 23.2 hPa, so the relative humidity is $(12.9/23.2) \times 100 = 55.5\%$. At the same vapour pressure and a degree warmer, the relative humidity is $(12.9/24.7) \times 100 = 52.2\%$.

Relative humidity varies far more over the course of a day than other measures of atmospheric water. This causes problems in using RH% for deriving estimates of vapour pressure or dewpoint. Data from climate stations are usually given as values collected at 9 am and 3 pm, but temperature is usually given as the daily maximum and minimum values, which probably occurred at other times of day (e.g. midday and before dawn), and are likely to yield very different estimates than would the corresponding 9 am and 3 pm temperature data. Another disadvantage of using relative humidity, especially in regional climate data (see section 4.8), is that finding the average of relative humidity values is not straightforward. One cannot simply add the 9 am and 3 pm RH% readings for a particular climate station and divide by two to get a daily mean. Hourly relative humidity readings, each accurate to within 5%, give a 24-hour average quite different from the mean of daily maximum and minimum values (Kalma, 1968). In an example given by Linacre (1992), a volume of air at 10°C with a relative humidity of 50% (i.e. a vapour pressure of 6.2 hPa), mixed with an equal volume at 30°C and 75% (i.e. 31.8 hPa), will have a temperature of 20°C, a vapour pressure of 19 hPa (the average of 6.2 and 31.8) and thus a relative humidity of 85%. This is 22% greater than 63%, i.e. the average of 50 and 75 RH. The discrepancy is because relative humidity is a ratio and cannot be treated as if it is a scalar quantity by averaging. The solution is to average the vapour pressure values instead.

A good example of the conceptual deceptiveness of relative humidity as a measure of atmospheric water vapour, and the effect of its dependence on temperature, is given by Edney (1977). Consider a scorpion's burrow with the air in the bottom at 15°C and 90% RH. The vapour pressure of the air will be 90×17.04 hPa (i.e. e_s at 15°C) ÷ 100 = 15.3 hPa. If the air outside the burrow is at 28°C and only 50% RH, i.e. 18.9 hPa, water vapour will move from the area of high pressure to the area of low pressure and into the cooler burrow, thus maintaining high humidity. This phenomenon is of more general importance in relation to the ecology of terrestrial arthropods. It is easy to find parallel scenarios in relation to microhabitats of dust mites, for example the microclimate measurements in mattresses (see section 4.6.3a).

Despite these problems, relative humidity has been used widely as an index of the water content of the air in relation to the water balance of dust mites and their population dynamics (refer to Chapters 3 and 5).

c Absolute humidity

Absolute humidity (AH or d_v) is the weight of water (in grams) per cubic metre of air. It is derived from the specific humidity (which is approximately the same as the mixing ratio (see below) and the density of the air. Values for the density (in grams) of saturated aqueous vapour (p_w) are published in *Smithsonian Meteorological Tables* (List, 1949, Table 108, *Density of Pure Water Vapour at Saturation Over Water*). This is the table to use for converting between absolute and relative humidities, *not* Table 94! The relationship between AH and vapour pressure is:

$$d_v = 217e/T_k \qquad (4.8)$$

where T_k is temperature in Kelvin. Thus, air with a vapour pressure of 10 hPa and a temperature of 10°C (= 283 K) has an absolute humidity of 7.7 g m^{-3}. If the air was saturated, the saturation vapour pressure at 10°C would be 12.3 hPa, so the absolute humidity would be 9.4 g m^{-3}. Relative humidity would be 81.9% (from $7.7/9.4 \times 100$ or $10/12.3 \times 100$). The relationship between absolute humidity and relative humidity is:

$$d_v = p_w \, \text{RH\%} \, /100 \qquad (4.9)$$

For the Melbourne example, at 19.9°C the weight of saturated aqueous vapour (p_w) is 17.2 g m^{-3} and mean daily absolute humidity in January is $(17.2 \times 55.5)/100$ = 9.55 g m^{-3}.

d Dewpoint temperature

Dewpoint temperature (T_d), or dewpoint, is the temperature to which air must be cooled for the vapour pressure of that air to become equal to the saturation vapour pressure. In other words, it is the air temperature required for dew to form. Linacre and Geerts (1997) explain it as the temperature of a can of drink from the refrigerator when the dew just disappears from the surface, and give the analogy with a sponge. A sponge full of water represents saturated air and a part-full sponge represents unsaturated air. If the part-full sponge is squeezed, its volume gets smaller, just as if the air is cooled it can hold less water vapour. When the sponge is squeezed to the point that it is full of water, this is analogous with the dewpoint. Additional squeezing causes water to leak out, corresponding to dew.

Dewpoint can be approximated from a table or diagram of empirical daily maximum and minimum temperatures (Linacre, 1992, Figure 3.10; Linacre and Geerts, Table 6.1) or by deriving it from vapour pressure and relative humidity. For the Melbourne example, the air has a vapour pressure of 12.9 hPa and a temperature of 19.9°C. Therefore, the dewpoint is the temperature at which the saturation vapour pressure equals 12.9 hPa. From Table 94 of *Smithsonian Meteorological Tables* (List, 1949), this is 10.7°C.

The monthly mean dewpoint temperature of a locality can be estimated from climate station observations of mean maximum and minimum temperatures and elevation (Linacre, 1992):

$$T_d = 10.9 + 0.63\,T - 0.53\,R_d - 0.35\,R_{ann} - 0.0023\,h \tag{4.10}$$

where T is mean temperature, R_d is the mean daily range of temperature and R_{ann} is the difference between mean temperatures of the hottest and coldest months and h is elevation (metres).

Dewpoint is measured in Celsius so it is easy to compare with the air temperature to determine proximity to saturation. It is quite a good index of the water content of the atmosphere because it has a conceptually clear physical basis: the drier the air, the lower the dewpoint and the greater the range between daily minimum and maximum temperatures.

e Mixing ratio

Mixing ratio (r) is the ratio of the mass of water vapour and the mass of dry air in a given volume of air (in grams per kilogram) and is directly related to vapour pressure (e) and atmospheric pressure (p) so that:

$$r = 622\,e/p \tag{4.11}$$

or

$$e = r.p\,/622 \tag{4.12}$$

The factor 622 is a product of the ratio of the molecular weights of water and air. Thus, at an atmospheric pressure of 1010 hPa and a vapour pressure of 10 hPa the air has a mixing ratio of 6.16 g/kg.

If the mixing ratio of the air at a given temperature is known, the vapour pressure deficit (D) can be calculated. For example, if $T = 10$°C and $r = 6.16$, then vapour pressure (e) = $6.16 \times 1010/622 = 10$ hPa. At 10°C, e_s is 12.3 hPa, so D is $12.3 - 10 = 2.3$ hPa.

4.5.3 Evaporation

Evaporation is the conversion of a substance from a liquid to a gaseous state. In the case of water, evaporation of water from the oceans controls the amount of water in the atmosphere and therefore influences atmospheric humidity. Evaporation rate (E) at terrestrial localities represents a proxy measure of the dryness of the air. It can be estimated using the formula developed by Linacre (1977), which requires only mean daily maximum and minimum temperature, the altitude and latitude of the site:

$$E = \frac{700T_m/(100 - A) + 15(T - T_d)}{(80 - T)} \; (\text{mm day}^{-1}) \tag{4.13}$$

where $T_m = T + 0.006h$, and h is elevation (metres); T is mean temperature, A is latitude (decimal degrees) and T_d is mean dewpoint temperature. Monthly mean values of $(T - T_d)$ can be estimated from the following empirical formula, providing precipitation is at least 5 mm and $(T - T_d)$ is at least 4°C:

$$(T - T_d) = 0.0023h + 0.37\,T + 0.53\,R_d + 0.35\,R_{ann} - 10.9°C \tag{4.14}$$

where R_d is the mean daily range of temperature and R_{ann} is the difference between mean temperatures of the hottest and coldest months.

4.6 The microhabitat scale

The major mite taxa of homes are also found in non-synanthropic habitats. Pyroglyphids live in birds nests, acarids in soil and plant litter and are also associated with insects and small mammals, and glycyphagids live in mammal nests (see section 4.1.6; Table 4.2). No species is specific to human habitation. It follows that houses have features in common with other habitats that make them suitable for mites. For nests and pelage of birds and mammals these characteristics are more obvious than for soil and plant litter. There is adequate heat and moisture, generated by a

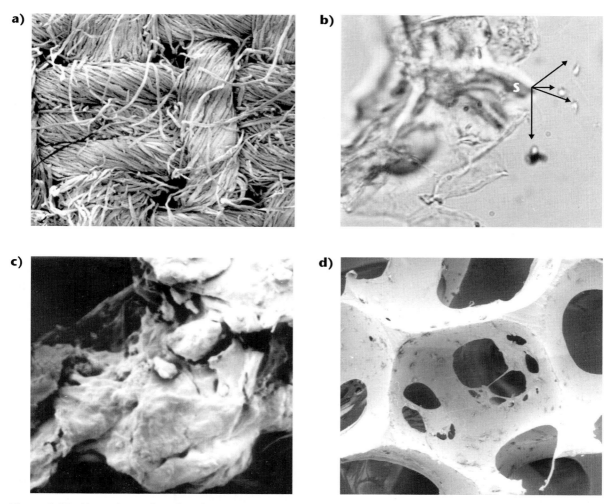

Figure 4.15 House dust and textiles. **a)** Mattress fabric, scanning electron micrograph; **b)** skin scales from mattress dust, light micrograph. Arrows: human spermatozoa (s); **c)** mattress dust, scanning electron micrograph; **d)** synthetic foam from pillow, scanning electron micrograph.

warm-blooded 'host', and there is adequate food: remnants of the host's meals, plus host excretions and secretions (faeces, skin scales, oils) and the fungi and bacteria that colonise these.

4.6.1 The physical environment of the microhabitat

a The nature of house dust

Dust mites live in microhabitats that are dominated by the three great structural macromolecules: keratin, cellulose and chitin. This domain of fibres, particles and scales has a random, chaotic appearance under the scanning electron microscope (see Figure 4.15), yet it is sufficiently structurally complex to provide food and dwelling for a simple community. The microhabitats of dust mites in human homes are found within textiles, including clothing, soft toys and upholstered furniture. The weave or pile of fabrics is an excellent trap for the accumulation of dust particles (shown in Figure 4.15). House dust consists of shed human skin scales (the source of keratin), fungal hyphae (from whence the chitin, as well as the cuticles of the mites themselves) and fibrous material from fabrics (the source of the cellulose). Other components include various mineral particles derived from soil and building materials, fibres from wood and paper, fragments of foam rubber and plastic, food particles (cereal and breadcrumbs, lipid droplets from fried food, and so forth). A quick examination of about a gram of carpet dust shaken out of the bag of my vacuum cleaner revealed the following components, identifiable with the aid of a hand lens, a penknife and a gas lighter: chocolate flakes, sunflower seed, peanut husk, fingernail parings, charcoal, spaghetti fragments, paper, grass seed, bark, leaf fragments, clay, sand, mica, rust flakes, human hair, toast crumbs, walnut kernel, skin scales, nylon fibres, wool, cotton, splinters of wood, paint flakes, insect cuticle, fragment of rice grain, ash and cheese rind. Many of the organic components are potential food sources for fungi and mites. Van Bronswijk (1981), in her book *House Dust Biology*, devotes a chapter to the composition of house dust and its characteristics.

b Spatial and temporal differences in composition of house dust

The striking difference in composition of dust from beds and carpets is immediately apparent when one is extracting mites from dust suspended in lactic acid with lignin pink (see Figure 4.16; Chapter 6). Dust from beds consists almost entirely of skin scales, of relatively uniform size range and shape, plus a few fabric fibres. Carpets tend to contain a far higher diversity of ingredients, including heavier, large particulate material. The constituents of house dust may have a bearing on its suitability as a microhabitat for dust mites from the point of view of its status as a source of food, not only for dust mites but for fungi

also. The water content of individual components will influence not only the amount of dietary water the mites receive but also the water-holding capacity of the dust and hence the humidity within it.

It is likely that house dust has changed in composition over time. The horse was the standard means of transport until motor vehicles became widespread. Up until the 1950s horses were a common sight in many European countries. In London by 1900, there were an estimated 50 000 horses providing, literally, the horse-power to pull Hackney Carriages, trams, carts, coaches and cabs. The horses would have deposited several hundred tonnes of manure daily onto the streets of the capital. The manure was collected by dung carts and deposited into enormous piles (London's Transport Museum, 2004: www.ltmuseum.co.uk). It was removed by farmers who carted hay into the city for use as fodder. Crossing sweepers earned tips from gentlefolk pedestrians, as Lady Charlotte Bonham Carter mentions:

> But with all the horse traffic, there was an awful amount of dirt on the streets, some of them were in a dreadful state. There were crossing sweepers, rather oldish men, and if one gave them a coin they would be very pleased to sweep a path across the street in front of one.

(Quoted by Weightman and Humphries, 1984.)

a)
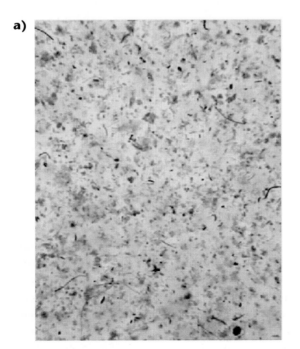

b)
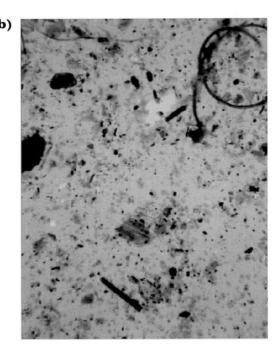

Figure 4.16 Appearance of dust samples suspended in lactic acid, dyed with lignin pink, under a stereobinocular microscope. **a)** Mattress dust; **b)** carpet dust.

a)

b)

Figure 4.17 **a)** Horse dung as an ingredient of 19th century house dust. The volume of horse dung on metropolitan streets was so considerable that crossing sweepers were employed to clear a path for the long and voluminous skirts of Victorian ladies crossing the street, ca. 1860; **b)** a Victorian terrace house in London. The majority of the facing wall area consists of large sash-cord windows, providing ample access to airborne dust from the street.

Van Bronswijk (1981, p. 190) mentions that in 19th century Holland:

Street dust in the towns consisted mainly of powdered horse dung, and penetrated the houses to become part of the dust.

The voluminous floor-length skirts of women of the period would have provided a means for the transfer of dung from the street to the interior of the house (see Figure 4.17).

Decomposing horse dung in the streets would have an accompanying flora of saprophytic fungi and actinomycetes, as well as bacteria and their endotoxins, capable of stimulating innate immunity and suppressing development of atopic allergies, but also increasing the severity of asthma (discussed in more detail in Chapter 8). Large quantities of skin scales and sweat would be generated by the horses and some scales would have wound up in house dust. An editorial in *The Lancet* of the 19th October, 1861 (Anonymous, 1861) considered that horse dung in house dust was a cause of ill health, quoting correspondent J.H.E. of the *City Press*:

The dust of the house is a poison, a small portion only of which is generated within the house. We see it, but we rarely smell or taste its poison. The larger portion of the dust of a London house comes from without – is blown in at the windows, and brought in on the shoes and garments, especially of the females; and the more sumptuously dressed, the larger is the lot they sweep up and bring in-doors. This London dust is chiefly composed of horse-dung, dried and highly comminuted; this it is that in wet weather causes the loathsome smell – e.g. of Cheapside. This nasty material, floating about in our sitting- and bed-rooms, penetrating mouth, nose and eyes, is kept for constant use in large quantities in the carpets, of which dust there must be a daily importation of considerable quantity. The people of the Continent are wiser: they have hard, clean-swept floors, little or no carpet.

Other components that have been subject to change over time include soil and synthetic materials. Nylon fibres and synthetic furniture foam are now a significant component of house dust. Soil particles are likely to be far less common components in house dust of modern towns and cities than in the past, simply because there are now far fewer unpaved roads. In many parts of the world, roads remain unpaved and floors of houses are made of compacted earth. Samples of house dust I took from homes in suburban Bangalore, India, contained large quantities of wind-blown soil. The dust from beds was of a red-brownish colour, rather than the usual light grey. In arid countries, large volumes of topsoil become airborne during dust storms, and this soil finds its way into houses. The significance of soil particles in house dust is that the clay minerals contained within soil have the capacity to bind proteins and other organic compounds onto their surfaces, rendering them more likely to remain in settled dust than become airborne. The nature and extent of binding of mite allergens to clay minerals in house dust has not been studied.

c Microhabitats and spatial heterogeneity

Dust mites are found in beds, carpets, pillows and upholstery. If the domestic climate is moist, they are also found in mould growth on walls (Kort, 1990). All these habitats are a continuum, viewed as different entities only by us humans. To dust mites, the main constraints on habitation are temperature, humidity and food. Doubtless, there are many other, currently unknown requirements. Dividing habitats into furniture and furnishing types is convenient, and most researchers who have looked at the distribution and abundance of mites within homes have done so. However, this approach tends to ignore the spatial heterogeneity of factors that determine the suitability of the habitat for growth and survival of dust mite populations. In other words, all parts of a mattress, or a carpet, are not equally suitable for dust mites.

4.6.2 Microclimate

In biological terminology, microclimate usually refers to the temperature, humidity and air pressure within the habitat of a particular group of organisms. This is slightly different from the meteorological usage, whereby microclimate refers to the climate within tens of metres of the ground (Geiger, 1957) in contrast to mesoclimate, the term used to describe the climate of particular regions (shown in Figure 4.14).

Work on the physiology of water balance of dust mites was well under way by the mid-1970s in Wharton's laboratory at Ohio State University (see Chapter 3), stimulated by the finding a decade earlier of the key role of water as a factor limiting the distribution and abundance of dust mites (see Voorhorst et al. 1969, p. 138 et seq.). Despite this important clue, there seems to have been little uptake or application

of ecophysiological research in relation to epidemiology of mite allergen exposure and disease. Most of the factors investigated to date that are associated with variation in allergen concentrations between homes relate, directly or indirectly, to domestic microclimate (for more detail see Chapter 8). Relatively few epidemiological studies have included ecologically relevant measurements of temperature and humidity in homes, partly because such measurements are not easy to make.

Domestic microclimate as a factor in dust mite ecology was studied by the Leiden group (Varekamp et al., 1966; Leupen and Varekamp, 1966). Geiger (1957) details the links between outdoor microclimate, indoor climate and macroclimate and cites some earlier works on the microclimate of houses, much of it in relation to public health and hygiene. More recently, Korsgaard and co-workers (Korsgaard, 1982, 1983a, 1991, 1998; Harving et al., 1993; Emenius et al., 2000) have greatly extended research on the relationship between indoor humidity, ventilation and mite population densities, including establishing a risk threshold of indoor absolute humidity of 7 g kg^{-1} for mite exposure. Below this threshold, mite population density is likely to be significantly lower than above it.

4.6.3 The climate of the microhabitat

After the discovery of the association between damp homes and high dust mite population densities, studies began on measurement of temperature and humidity within homes. Leupen and Varekamp (1966) set the scene by examining the climatological conditions within homes that permit the growth of mite populations. They defined and charted the normal indoor climate for humans, mites and fungi in homes in the Netherlands, noting that although they overlap, they are very different (see Figure 4.18a). What is immediately apparent from this diagram is that while the boundaries for fungi and mites represent the limits of population growth, that for humans represents the 'comfort climate', i.e. the average optimal indoor climate (defined by place: the Netherlands; and time: the 1960s). Comfort climate varies widely between people, people of different ages (compare the susceptibility to cold of children and their grandparents) and human populations. Furthermore, it has probably changed over time. I suspect the comfort temperature range of the Netherlands in the 1960s was probably slightly lower, and with a broader range, than it is today. These factors

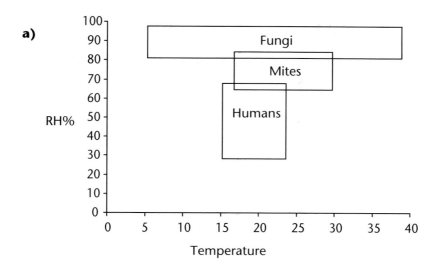

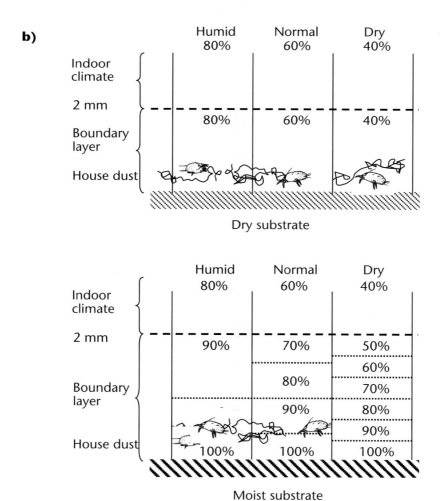

Figure 4.18 a) Microclimatic 'comfort zones' for mites, fungi and humans; **b)** conceptual model of the differences in relative humidity above dry and moist substrates. Re-drawn from Leupen and Varekamp (1966, Figures 2 and 6).

provide important clues to explaining variation in climate between homes.

a Microclimate of beds and carpets

Koekkoek and van Bronswijk (1972) measured humidity in three beds hourly for 24 hours. Relative humidity rose after the bed was occupied, independent of temperature, indicating the generation of moisture by the occupant of the bed. Hughes and Maunsell (1973) recorded temperature and humidity inside and outside a bed, together with fluctuations in mite population density. The single bed, at A.M. Hughes' cottage in Sussex, England, was monitored for a year. They showed that temperature and humidity increased during the late spring and summer and was mirrored by an increase in mite population density. When the bed was occupied, relative humidity fell but reached equilibrium with the surrounding air soon after it was vacated. This study demonstrates that fluctuations in temperature and humidity in beds operate on a diurnal scale, as well as a seasonal one, and are correlated with seasonal variation in mite populations. Dusbabek (1975, 1979) also investigated the relationship between bedroom microclimate and seasonal variation in mite population dynamics.

I examined temperature and humidity within 9×0.24 m^2 quadrats in my double mattress (Colloff, 1988a) in order to discover whether there were differences in diversity and abundance of mite species in different parts of the bed. Highest diurnal fluctuations

in microclimate were in the immediate vicinity of where I was sleeping (on the left hand side of the mattress; column C in Figure 4.22). The centre of the bed (quadrat B2) was relatively hot and dry and this was where fewest mites were found. Most were found at the warm, moist foot end. The head end was the coolest and most humid with least fluctuation, buffered by the presence of pillows. Large fluctuations in hourly mean relative humidity were due to increase in skin temperature followed by bouts of sweating, especially in the last three hours of sleep. The measurement of microclimate in the bed suffered some limitations, not least sleeping on a large metal thermohygrograph probe. The probe was too large to measure anything other than surface conditions on the mattress. It did not accurately reflect microclimate at the microhabitat scale that mites were exposed to. The thermohygrograph was coupled to a pen recorder that had the habit of jamming in the middle of the night. De Boer and van der Geest (1990) inserted 'sword probes', similar to the one I used, into different depths in mattresses during a study on the control of dust mites with electric blankets. When the blankets were in use, there was a gradient effect of decrease in relative humidity and increase in temperature from the top to the bottom of the mattress.

The difficulty of accurate psychrometric measurement was overcome by Cunningham (1998, 1999), who developed a humidity sensor small enough to use within carpet pile or layers of bedding (see Figure 4.19).

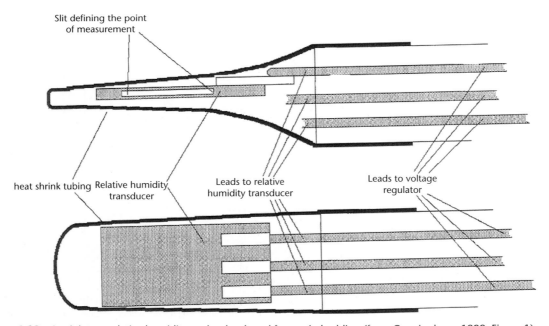

Figure 4.19 A miniature relative humidity probe developed for use in bedding (from Cunningham, 1999, Figure 1).

The sensor can detect humidity differences at the scale of a few millimetres. As in previous studies, Cunningham (1998) found relative humidity fell as soon as beds were occupied, and there was a decrease in the magnitude of the fall with increasing distance from the occupant. The humidity at the base of a carpet pile could be 30% RH higher than the room air (see Figure 4.20). In the absence of any moisture source, vapour pressure of the carpet tracked that of the air in the room. Where a source of moisture (and heat) was present, there was a rise in vapour pressure and temperature, but not necessarily in relative humidity (Figure 4.20). Cunningham (1998) stressed that flows of heat and moisture are determined by temperature and vapour pressure gradients, but the moisture content of fabrics is determined by relative humidity. Therefore, neither absolute humidity nor relative humidity alone determines conditions for survival of dust mites, since neither is independent of temperature.

b Microclimate and the boundary layer

The thin film of air within a millimetre or so of a solid surface is known in biological terminology as the boundary layer. Meteorologists use the term to mean the same thing (Geiger, 1957, p. 57 *et seq.*), but also use it to refer to the layer of air within several metres of the surface of the Earth. The boundary layer only exists when there are differences between the water content or temperature of its underlying surface and the air above. It may disappear if disturbed by turbulence. The boundary layer has some important thermal and hygroscopic properties different from the air further away from the surface. In this zone, heat

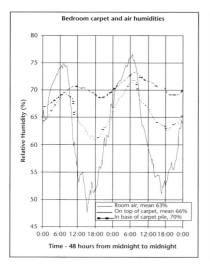
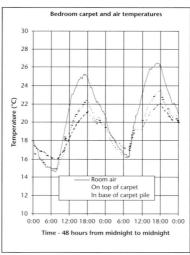
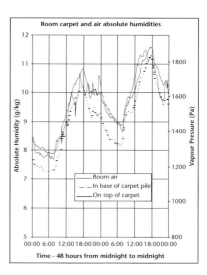
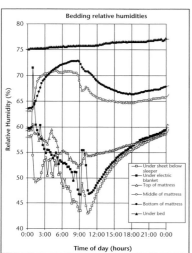
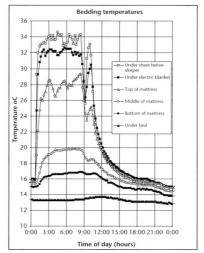
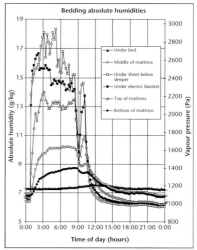

Figure 4.20 Psychrometric data from bedding and carpets using the miniature relative humidity probe to measure humidity and bead-type thermistors to measure temperature (from Cunningham, 1998).

movement or transfer is mainly by molecular heat conduction. Transmission of heat from the surface into the boundary layer takes place along a very steep temperature gradient over a very short distance. A porous surface such as a plaster wall takes up water from unsaturated air, and, if sufficient water vapour is present in the air, additional water can be absorbed by capillarity. Water in the air beyond the boundary layer is always below saturation, so there is evaporation of water from the wall into the boundary layer which becomes saturated. Dust mites are at most only about 0.35 mm long and about 0.15 mm from dorsal to ventral surface. The highest that mites get from the surface is about 0.3 mm when a pair is in copula, so they exist entirely within the boundary layer.

c Microclimate models

Leupen and Varekamp (1966) constructed a model of the boundary layer in contact with both moist and dry substrates and predicted the humidity gradients above moist substrates would reduce by 25–50% over a distance of 2 mm, depending on the humidity of the air of the room (see Figure 4.18b). Their model was confirmed experimentally by Cunningham (1998) who found the boundary layer was considerably moister than the air of the room. This finding may explain why attempts to control dust mites by lowering the humidity of indoor air have met with mixed success. The assumption has been that if one lowers room air to humidities at which mites lose body water, then they will desiccate. The important point here is that if one wants to control dust mites by lowering indoor air humidity, the size of the reduction required could be considerably greater than has been suggested.

The boundary layer effect has important consequences for dust mites and their allergens. If mites are living above an absorptive surface, and if water vapour is evaporating from that surface, the humidity in the boundary layer will always be more favourable to the mites than in the air above the boundary layer. If the temperature of the surface is warmer than the air above the boundary layer, then there will be differences in pressure and air turbulence compared with the air beyond it. Mites and allergen-containing faecal pellets will tend to be held within the boundary layer.

Cunningham et al. (2004) developed a predictive model of the psychrometric conditions in bedding and carpets. Such models are useful for integrating two very complicated phenomena: temperature/humidity and mite population dynamics. What Cunningham's work shows elegantly is the artificiality of laboratory studies on mite population growth and water loss in response to constant temperature and humidity (see Chapters 3 and 5). We knew that in vitro studies were artificial by their very nature anyway, so what makes this any different? We have sufficiently accurate information to start to imitate in the laboratory the psychrometric conditions now known to be found in microhabitats in the home. In order for us to get a more accurate assessment of mite life history parameters, the conditions we need to replicate are ones where both temperature and relative humidity fluctuate over 24 hours. So far, only the effects of fluctuations in humidity have been considered (de Boer and Kuller, 1995b; de Boer et al., 1998; Arlian et al. 1999b).

4.6.4 Distribution and abundance of mites at microhabitat scale

Most estimates of mite abundance are relative, i.e. they represent the number of mites per unit sampling effort, for example in dust collected from 0.25 m^2 sampled for 2 minutes. An absolute population estimate is the number of mites per unit habitat, i.e. all the mites present in an area of 0.25 m^2. Relative estimates may represent only 0.5–5% of total mites present (Hay, 1995), so absolute population estimates give a valuable insight into mite distribution and abundance within microhabitats. With dust mites, the logistic constraints of conducting absolute population estimates have limited their use to the microhabitat scale. They also tend to be destructive, requiring the dissection and removal of parts of the microhabitat. Such studies have been done with mattresses by Mulla et al. (1975), de Boer and van der Geest (1990) and Hay (1995).

a Distribution and abundance within beds

Vertical distribution

Cunnington (cited by Blythe, 1976) and van Bronswijk (1973) dissected old mattresses to discover mites and skin scales were confined to within a few millimetres of the surface. Mulla et al. (1975) dissected an inner-sprung mattress and counted all the mites per unit area in the cloth mattress cover and the underlying cotton fibre-fill of the upper surface. Most mites were present within 5 mm of the surface (see Figure 4.21). The centre of this type of mattress consists of air spaces and springs, so does not afford suitable habitat. De Boer and Kuller (1995a) tested the habitat

material (see Figure 4.7). Wall-to-wall carpets are the norm in all rooms, except the kitchen and bathroom, in many modern Western homes, although there are aesthetic and cultural modifiers here. Scandinavian homes tend to have wooden floors with rugs, and homes in Mediterranean countries have floors of terracotta tiles, stone, timber or terrazzo. Reduction in habitat through removal of textiles has been a practical means of controlling dust mite populations (discussed in Chapter 9).

There are relatively few published studies where the population density of mites of beds has been compared with that of carpets (Sesay and Dobson 1972; Pearson and Cunnington 1973; Arlian et al., 1978, Korsgaard, 1979; Keil, 1983; Colloff, 1987, 1991; Andrews et al., 1992, 1995, Crane et al., 1995; Chew et al., 1999a). From these, the mean ratio of mites in beds to mites in carpets is 2.7 (range 0.6–5.9). In other words, the population density of mites is almost three times higher in beds (mean 1884 per gram) than in carpets (mean 601 per gram). However, this is based on population density per unit weight, not the estimated total mite population of the home. In a three-bedroom home with fitted carpets, the surface area of carpets is in the order of 50–100 m^2, whereas the surface area of beds (allowing for two double beds and single) is only about 10–15 m^2. If one ignores the difference in density between dust of beds and dust of carpets (see Figure 4.16) and assumes proportionality between density of mites per unit weight and mites per unit area, the contribution of fitted carpets to the total mite population of the home is, at a very crude estimate, of the order of 1.6–2.1 times that of beds.

4.7.3 Distribution and abundance between homes

a Furnishings

Abbott et al. (1981) found significant differences in mite numbers between foam, kapok and inner-spring mattresses (see Figure 4.28a), whereas Bigliocchi and Maroli (1995) found no differences between wool and inner-spring mattresses. Mulla et al. (1975) found mattresses of less than five years of age had significantly fewer mites, as did Hegarty (1988; see Figure 4.28b).

There is little information on mites in bedroom carpets and carpets elsewhere in the home. Where bedroom and lounge room carpets have been sampled in the same study (Blythe et al., 1974; Arlian et al.,

1982; Korsgaard, 1982; Bigliocchi and Maroli, 1995; Chew et al., 1999a) the population density is not significantly different.

In the studies by Suto et al. (1991a, b; 1992a, b; 1993), bedroom and living room floors in apartment blocks were covered in tatami and/or carpet, and were of wood or vinyl in traditional homes. Tatami is traditional straw matting (see Figure 4.27a). Yoshikawa (1980) observes that when demand for tatami is particularly high (for example, in response to increased housing construction) straw is sometimes used before it has been fully dried. In tatami with residual moisture, acarid mites flourish as do the predatory cheyletids that feed on them. These have been found to bite humans (Yoshikawa, 1985, 1987; Yoshikawa et al., 1983). Suto et al. (1993) found more D. farinae than D. pteronyssinus in carpets; suggesting carpet was relatively dry compared with tatami because of the tolerance of lower humidity in D. farinae (see Chapter 3). Uchikoshi et al. (1982) found the same thing in Wakayama. Suto et al. (1993) found that on carpeted floors, which have low thermal conductivity and humidity control capacity, and tend to be drier than tatami, D. farinae tended to be numerically dominant, whereas D. pteronyssinus dominated on tatami. Thermal conductivity is the rate of transfer of heat along a material by conduction, measured in Watts per cubic metre per Kelvin. Carpet has low thermal conduction; that of tatami and wood flooring is high.

In Israel, twice as many mites were found in sofas (couches) than beds, and this increased to five and eight times in the dry climates of Beer Sheva and Jerusalem, respectively (Feldman-Muhsam, 1985). The development of large populations in sofas was attributed to the fact that residents spent several hours a day sitting on the sofa, entertaining guests, watching TV, resting and reading. Mattresses are often hung over balconies and aired in the sun, whereas sofas are rarely cleaned. Arlian et al. (1982) and Mumcuoglu et al. (1999) found higher mite densities in heavily used upholstered furniture than in beds.

b Age of homes

The relationship between the age of homes and mite population density is often confounded because older homes tend to be of different design and construction than younger homes. Further, the interiors of older homes are frequently heavily modified from their original condition. I found no difference

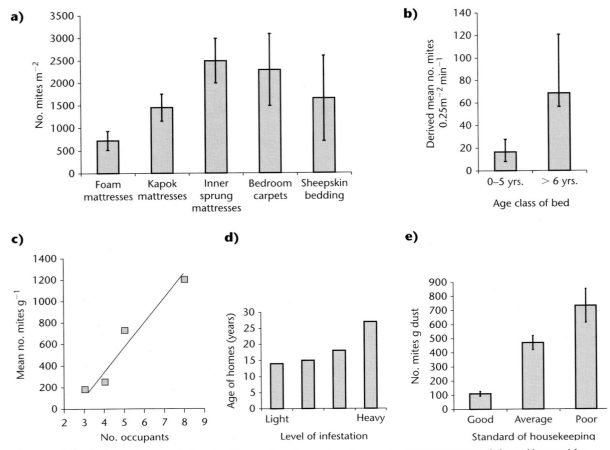

Figure 4.28 **a)** Mean mite population densities on three types of mattresses, bedroom carpets and sheepskins used for infant bedding (drawn from Table 1 of Abbott *et al.*, 1981); **b)** number of mites in relation to the age of beds (± 95% confidence intervals) (drawn from data by Hegarty, 1988); **c)** relationship between mean mite population density and numbers of occupants of homes (re-drawn from Arlian *et al.*, 1978); **d)** relationship between age of home and level of mite infestation (light = 10 mites per ml of dust; heavy = 50 mites per ml of dust; re-drawn from Mulla *et al.*, 1975); **e)** relationship between housekeeping quality and mite population density ± standard error (from Arlian *et al.*, 1978).

in mite abundance in late Victorian Glaswegian tenement flats compared with 1930s–1950s detached or semi-detached houses and 1970s–1980s concrete apartments. However, Mulla *et al.* (1975) found the mean age of homes with heavy mite infestations was 27 years, and the mean for lighter infestations was 14 years (shown in Figure 4.28d).

c Social and economic status and density of occupancy

Despite the lack of relationship between mite abundance to either age of the dwelling or construction type in Glasgow (see above), there was a relationship between mite abundance and the social class of the occupants of these homes. People in the higher social classes 1–2, as defined by the UK Socio-economic Classification (NS-SEC; Office for National Statistics, 2004)

had almost half the mean number of mites in their beds than those in class 3, and five times fewer than those in classes 4–5 (see Figure 4.29). A similar, less-marked trend was present for carpets: classes 1–2 had two-thirds the number of class 3 and almost half the number of classes 4–5 (Colloff, unpublished data). The probable explanation for this is that people in lower socio-economic classes have relatively older mattresses and carpets. Older mattresses and carpets tend to have higher allergen levels (discussed in Chapter 8).

The number of occupants in homes was found to be positively correlated with average mite population densities by Mulla *et al.* (1975) and Arlian *et al.* (1978; see Figure 4.28c). Suto *et al.* (1991b, 1992) and Sakaki and Suto (1994) found that family size, together with the number of rooms in the home, were the most

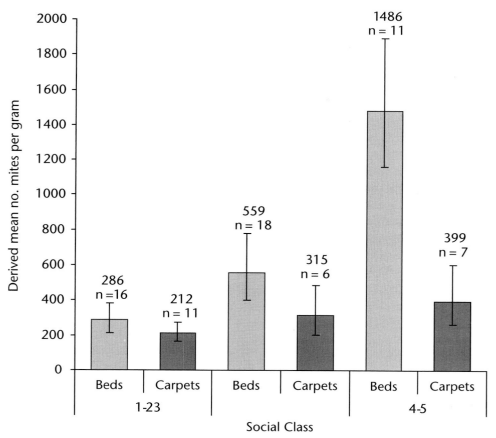

Figure 4.29 Geometric mean (derived from $\log_{10}$-transformed data) number of mites per gram of dust from beds according to socio-economic class of the occupants (± derived standard error), where 1 is 'high' and 5 is 'low' according to the UK socio-economic classification. (Colloff, unpublished data).

important factors associated with prevalence of mites in traditional wooden homes in Nagoya, Japan.

d   Housekeeping

Homes with poor housekeeping (cleaning) practices have been found to have significantly higher mite population densities than homes with good practices (Mulla *et al.*, 1975; Arlian *et al.*, 1978; Bigliocci *et al.*, 1996; see Figures 4.28d, e). Sesay and Dobson (1972) found a mean of 1.4 times as many mites in beds from homes with poor standards of housekeeping than clean homes.

e   Height of dwellings above ground level

Although Suto *et al.* (1992) found a trend towards a negative correlation between the height of apartments above ground level and mite population densities, it was not statistically significant (regression analysis of data in their Table 2: $P = 0.43$). The authors claim that *D. pteronyssinus* tended to be the dominant over

D. farinae in apartments on floors below the third (their Table 5), but again the relationship was not statistically significant when analysed by regression of their whole dataset ($P = 0.292$). There was a negative correlation between species diversity and height of the apartment above ground (their Figure 5).

4.7.4   Habitats other than houses

Pyroglyphid mites have been studied in a few habitats outside the home. These include city pavements (Samsinak and Vobrazkova, 1985), trains (Colloff, 1987d; Green *et al.*, 1992; Custovic *et al.*, 1994; Uehara *et al.*, 2000), ships (King *et al.*, 1989), an airport lounge and a plane (Green *et al.*, 1992), cars (Mollet, 1996; Neal *et al.*, 2002) and they have been found on clothing (Hewitt *et al.*, 1973; Bischoff and Fischer, 1990; Tovey *et al.*, 1995; Neal *et al.*, 2002). Mites and allergens are present in hospitals (Sarsfield, 1974; Green *et al.*, 1992; Babe *et al.*, 1995), nursing homes (Vyszenski-Moher *et al.*, 1986), schools (Oshima, 1964; Vobrazkova *et al.*, 1985; Dybendal *et al.*, 1989a, b; Dornelas de Andrade *et al.*, 1995), and occasionally in offices (Colloff, unpublished data) and cinemas (Green *et al.*, 1992; Konishi and Uehara, 1999). The initial populations in non-domestic buildings most probably have been established after dispersal from clothing. The most obvious manner in which large numbers of mites are dispersed from home to home is in upholstered furniture and carpets. The introduction of an infested second-hand sofa, for example, greatly increases the total mite population in the home. Mites have relatively low capacity for dispersal (vagility), and their walking pace at optimum temperature and humidity is in the order of a few centimetres per minute. They require transport to disperse any appreciable distance. They cannot colonise by walking from home to home, not just because of their low vagility, but because the outdoor environment is relatively hostile, so the mite population of any one home is effectively geographically isolated from that of others. However, it is likely that there are substantial dispersals from home-to-home at irregular intervals in bedding and furniture.

4.8   The regional scale

At the regional scale, climate, latitude and topography (including elevation and distance from maritime or riparian influences and continentality) interact in determining the distribution and abundance of dust mites. At this scale, the effect of indoor microclimate on dust mite populations is heavily moderated by macroclimate, or *topoclimate* – the climate that is applicable to regional scale (refer to Figure 4.14). As Spieksma (1973) succinctly put it:

> …*One can understand that wet places in houses are the most important cause for the presence of mites; that a cold climate, which makes people light their stoves, causes a low air humidity indoors, which hampers the growth of mite populations.*

It was at the regional scale that the first clues to the epidemiology of mite-mediated asthma were worked out by Storm van Leeuwen (1924). He found that asthma due to house dust was more prevalent at low elevations (altitudes), during autumn, and that symptoms improved when patients moved to high elevations. These factors pointed to macroclimate as a key variable in determining degree of exposure to allergens, and suggested the allergens had a biological source. Spieksma *et al.* (1971) found a mean of 0.2 mites per gram of dust in 17 homes above 1200 m in alpine Switzerland; 1.2 g^{-1} in 19 Swiss homes below 1200 m, 2.4 g^{-1} in 60 dry homes in Leiden, the Netherlands (−2 m) and 8.2 g^{-1} in 60 damp homes in Leiden. Mumcuoglu (1975) found a marked negative association between elevation and mite population density at 29 towns in Switzerland over a vertical range of about 1500 m (see Figure 4.30). Vervloet *et al.* (1982) found that mean mite populations in Briançon (1365 m) were very low (2 g^{-1}) and that mite-hypersensitive asthmatic children underwent a mean decrease of 40% of their *D. pteronyssinus*-specific IgE antibody values after a 9-month stay in the town.

Inverse correlations between elevation and mite population density were confirmed by Ordman (1971) in South Africa; Lang and Mulla (1977a) in California; Amoli and Cunnington (1977) and Sepasgosarian and Mumcuoglu (1979) in Iran; Lascaud (1978) in Grenoble, France; Gómez *et al.* (1981) in Northern Spain, Vervloet *et al.* (1982) in Marseilles and Briançon, France; and Kalpaklioglu *et al.* (2004) in Turkey. Rijkaert *et al.* (1981) studied the abundance and frequency of pyroglyphid mites in different European climatic regions. The region with highest abundance had a subtropical climate, followed by regions with temperate maritime climates and then those with Mediterranean climates. This sequence follows a humidity gradient. The most continental town (Monchengladbach) had the lowest mite abundance. The most maritime (Amsterdam) had the highest.

The results of these studies formed the raw material for another of the widespread oversimplifications of house dust mite biology, that 'mite populations decrease with elevation, because the higher you go, the drier it gets'. This may be the case in some temperate regions, but a glance at a global relief map and a global annual precipitation map shows that the wettest places on earth are tropical and at high elevation, including Cherrapunji, in Meghalaya State, India (1313 m) with over 10 m of rainfall per year; most of the highlands of Papua New Guinea; large areas of South-East Asia above 1000 m, and the north-eastern Andean slopes of Venezuela, Colombia, Ecuador and Peru. Moreover, Sanchez-Medina and Sanchez-Gutierrez (1973) and Charlet *et al.* (1979) found relatively high dust mite populations, regardless of elevation, in contrasting climatic zones in Colombia.

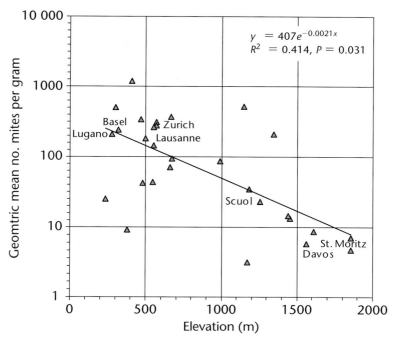

Figure 4.30 Relationship between abundance of dust mites and elevation in a temperate, continental region. Each data point represents geometric mean abundance of mites at one of 29 towns in Switzerland (see Appendix 3 for details). (Data from Mumcuoglu, 1975; see also Voorhorst *et al.*, 1969, their figure on p. 95.)

4.8.1 Regional climate

a Elevation, latitude and distance from the coast

There are a few important globally applicable relationships between variables that affect climate, such as elevation, latitude and distance from the coast. Most of them can be found in standard climatology texts. They include the decrease in temperature with increase in latitude, and elevation – the so-called *lapse rate* of ca. 1°C decrease per increase of 166.67 m, or 0.006°C m^{-1} (Holdridge, 1947). Even this varies seasonally and with latitude. For example, the rate for a set of Colombian climate monitoring stations is 1°C per 181.8 m or 0.0055°C m^{-1} (Mulligan, 2000). Linacre and Geerts (1997, their Figure 3.5) quote 0.0042°C m^{-1} for locations in Africa and the Americas. Temperature decreases with increase in latitude, but not in a linear fashion. The difference between the equator and the tropics is minor. Between 35–40°N or S it is about 0.7°C per degree of latitude; between 45–55°N or S it is about 1.5°C. The relationship between temperature and distance from the coast is not straightforward. In Australia, temperature increases with distance inland during summer because the coasts are cooled by sea breezes. On large continental landmasses such as Canada and Siberia,

summer temperatures may be about the same on the coasts and interiors, but the interiors are considerably colder than the coasts in winter.

Humidity tends to decrease with increasing latitude because of lower temperatures and vapour pressures. Humidity decreases with distance inland, because vapour pressures at the coast are high due to evaporation from sea surfaces. An obvious exception is the Amazon Basin. Increase in elevation is also associated with decrease in humidity because of greater distance inland and lower temperatures and because there is an exponential decrease in vapour pressure with increased elevation (it varies approximately to the power of 3 of the atmospheric pressure: about 30% lower at 1000 m than at sea level). Linacre and Geerts (1997, their Figure 6.8) showed that an increase in elevation of 1000 m in South America, between 30–40°S, had the same effect on dewpoint as an increase in latitude of 8 degrees.

The global climate system is complex and dynamic, and the climate of any particular place depends on the following factors:

- the hemisphere it is in, which determines the warm season outside the tropics of Cancer and Capricorn, or the wet season within them;

- the latitude, which controls day length, annual temperature and range, rainfall and prevailing winds;
- elevation (altitude), which affects the amount of UV radiation, rainfall, mean temperature and range, as well as the location upwind or downwind of a mountain range, which affects temperature and rainfall;
- ocean circulation, which affects sea surface temperatures and climate conditions of the nearest coastline;
- distance downwind of an ocean, which affects temperature range and rainfall; and
- local topography, which influences temperature and local airflow.

b   The Köppen climate classification

Probably of value in attempting to understand the climatic characteristics of the locations in these regional scale studies is a basic appreciation of the Köppen climate classification (Köppen, 1931; most physical geography and climatology textbooks have comprehensive coverage of the topic. For a schematic summary, see *The Times Atlas of the World*, 8th ed. [1990], pl. 3). The Köppen system involves categorising climates according to vegetation types, because natural vegetation is a good indicator of climate. Natural vegetation boundaries, such as the limits of rainforest distribution, represent the boundaries of the climate classes. The characteristics of the relevant Köppen classes are explained for each regional study, but the main categories are:

- moist tropical climates with mean temperatures of all months above 18°C;
- dry climates with low rainfall for most of the year;
- moist mid-latitude climates with mild winters;
- moist mid-latitude climates with cold winters;
- polar climates with very cold summers and winters.

c   Limitations of climate data

There are few studies on distribution and abundance of dust mites with data from sufficient locations to draw inferences about patterns at a regional scale. Even if there are enough, there may not be matching climate statistics available at or near the locations where the mites were sampled. For many meteorological ground stations the only measurement data to be had are monthly means of precipitation (rainfall, snow, hail) and daily maximum and minimum temperature. It is often not possible to get consistent empirical measurements of humidity variables. Precipitation is not a very good proxy measure of humidity because the factors that affect the amount of water that evaporates into the atmosphere (water infiltration to the soil, runoff to drains and waterways and temperature) vary from place to place. For example, in cities with good drainage infrastructure, runoff into storm drains is quite rapid, leaving only a small amount of total rainfall available for evaporation and resulting in only slight, transitory increases in atmospheric humidity.

If there is an absence of empirical humidity measurements for locations of interest, one solution is to use an estimate using formulae such as those developed by Linacre (1977) for estimating evaporation (Equations 4.13 and 4.14). I have used this approach to examine the association between evaporation and mite distribution and abundance in the regional examples given below.

4.8.2   Regional scale examples of effects of climate on mite distribution and abundance

Factors affecting climate, such as distance from the coast, altitude and latitude, are often co-variables, so it makes no sense at the regional scale to treat them as unrelated. I have chosen examples of regional scale studies of dust mites that can be analysed to illustrate the interaction of climate variables and their explanatory power concerning the distribution, abundance and frequency of occurrence of dust mites. I have attempted to select examples where the regional data are derived from either a single study or a small series in which comparable approaches were used. I have avoided combining data from many unrelated papers. There are several regional scale studies that I have not included either because it was difficult to find relevant climate data for most locations, or there were limitations of the mite abundance data (e.g. it was semi-quantitative; expressed as ranges), or there were insufficient numbers of sites, or because the regions were climatically and topographically similar to other, better-documented examples. These studies include Algeria (Louadi and Robaux, 1992), Chile (Casanueva and Artigas, 1985); France (Lascaud, 1978); Iran (Sepasgosarian and Mumcuoglu 1979), India (Modak *et al.*, 1991); Japan (Miyamoto *et al.*, 1970); Spain (Portus and Gomez 1976; Gomez *et al.*, 1981a, b); Thailand (Wongsathayong and Lakshana 1972; Malainual *et al.*, 1995), Turkey (Kalpaklioglu *et al.*,

2004) and the USA (Arlian *et al.*, 1992). Localities, mite abundance and frequency of occurrence for sites from these studies are presented in Appendix 3. The study by Mumcuoglu (1975) on the relationship between altitude and dust mite population density in Switzerland has already been mentioned.

a Israel and Gaza

Israel is small in area (21 000 km²), extending in latitude from 29°30'N to 33°20'N. The population is ca. 4.7 million. It is topographically diverse and encompasses four main climatic regions:

1. the coastal Mediterranean plain;
2. the foothills of Carmel, Judea and the Negev;
3. the mountains of Upper Galilee and Judea; and
4. the Jordan Valley, linking the Sea of Galilee with the Dead Sea; much of the area below sea level.

Large differences in temperature, humidity and rainfall exist over just a few kilometres, reflecting a rapid rise in elevation from the Mediterranean coast to ca. 800 m in the Judean Mountains at Jerusalem, some 50 km to the east (see Figure 4.31).

The Köppen climate classes in Israel are Mediterranean (Csa), corresponding to the coastal plain and the foothills, and semi-arid low-latitude (BSh),

corresponding to the Judean Mountains and the Jordan Valley of the interior. Csa climates have dry, warm summers with the warmest month above 22°C and wet, mild winters with the coolest month between 0°C and 18°C. Mean rainfall of the wettest month is at least three times that of the driest month. Csa homoclimes include Beirut, Marrakech, Perth (Western Australia) and Rome. BSh climates have hot dry summers with a moderate annual range, and winter rainfall. BSh homoclimes include Tehran and Alexandria.

Feldman-Muhsam *et al.* (1985) examined 291 dust samples from five localities, using repeat sampling from December to August. Some 91% of the samples were positive for mites. Highest mean abundance of dust mites was at Bat Yam, the lowest at Jerusalem, and there was a positive correlation between abundance and mean monthly relative humidity of each locality. The dataset I used to examine the association between evaporation and mite distribution and abundance included the five localities studied by Feldman-Muhsam *et al.* (1985) plus Tel Aviv (Kivity *et al.*, 1993) and Gaza (Mumcuoglu *et al.*, 1994) (see Figure 4.31). Climate data for these localities are shown in Figure 4.32. The estimated mean annual evaporation rates for the localities are, lowest to highest, Gaza < Tel Aviv < Nes Ziyyona < Jerusalem < Be'er Sheva < Nir Dawid (shown in Figure 4.32c), and are strongly

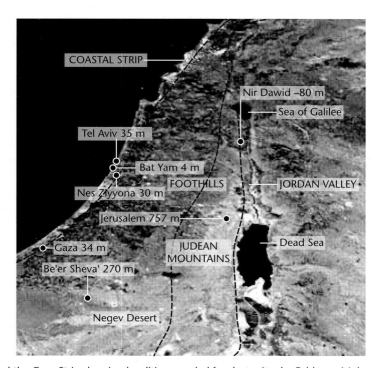

Figure 4.31 Israel and the Gaza Strip showing localities sampled for dust mites by Feldman-Muhsam *et al.* (1985), Kivity *et al.* (1993) and Mumcuoglu *et al.* (1994).

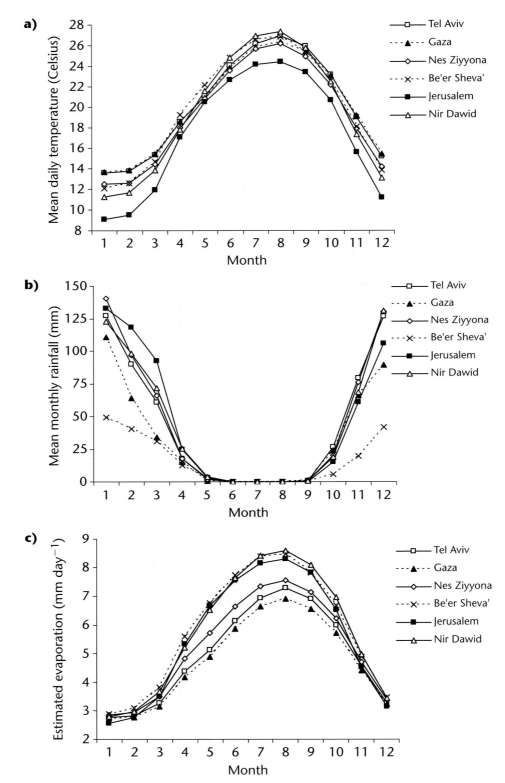

Figure 4.32 a) Mean daily temperature; **b)** mean monthly rainfall; **c)** estimated rate of evaporation of Israeli localities shown in Figure 4.30. Proxy localities: Tel Aviv for Bat Yam, Bet Dagan for Nes Ziyyona, Kefar Yehoshua for Nir Dawid. Data kindly provided by Israel Meterorology Service.

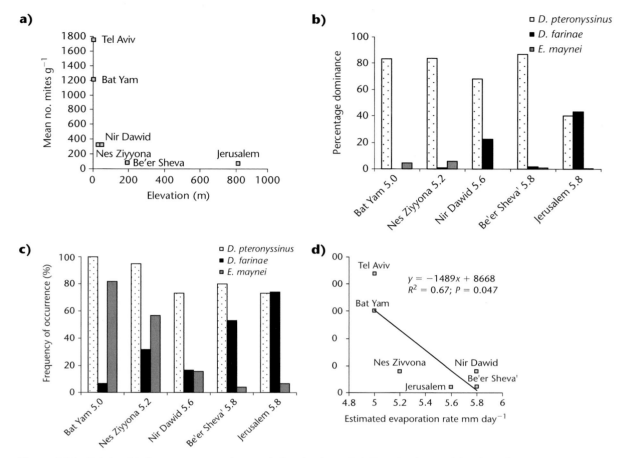

Figure 4.33 Relationship between mean mite population density and **a)** Elevation for Israeli localities shown in Figure 4.30; **b)** frequency of occurrence; **c)** percentage dominance of three major dust mite species at localities studied by Feldman-Muhsam *et al.* (1985), ranked from lowest to highest by estimated rate of evaporation; and **d)** relationship between mean mite abundance and estimated evaporation rate.

positively correlated with distance from the coast ($F = 44.29$; $p = 0.003$). Mite abundance, corrected for the outlier, Gaza, is significantly negatively correlated with estimated evaporation rate ($F = 8.02$, $p = 0.047$; see Figure 4.33d), but not elevation (see Figure 4.33a) or distance from the coast (although there were clear trends), mean annual temperature or rainfall. Frequency of occurrence and dominance of *Dermatophagoides pteronyssinus* and *Euroglyphus maynei* declined significantly with increasing estimated evaporation rate, whereas the reverse pattern was apparent for *D. farinae* (see Figures 4.33b, c).

The abundance of dust mites in Israel is strongly influenced by the complex climate system of the eastern Mediterranean, whereby humidity and temperature are seasonally highly variable and strongly modified by proximity to the coast, elevation and the effect of mountain ranges. The study by Feldman-Muhsam *et al.* (1985) is one of very few that has examined regional *and* seasonal variation in mite population parameters and greatly enhances the explanatory power of the survey.

In contrast with the urban study of Feldman-Muhsam *et al.* (1985), Mumcuoglu *et al.* (1999) focus on regional and seasonal variation in mite populations in rural agricultural communities – kibbutzim and moshavim. Frequency of occurrence (97%), mean population density range (84–2053 g^{-1}), relative dominance of major species (80% *D. pteronyssinus*), and species composition were similar to homes from urban centres. The authors found the lower the temperature and the higher the minimum relative humidity of localities, the greater the mite population.

b   Colombia

Colombia covers an area of ca. 1.14 mn km² with a population of 45.3 million, and straddles the equator from 4°S to 12°N. It is separated into three regions, the Pacific Coastal Lowlands, the Andean Region (consisting of the Eastern, Central and Western Cordilleras of the Andes mountains) and the Eastern Plains, through which flow the many tributaries of the Orinoco and Amazon rivers that have their sources on the slopes of the Eastern Cordilleras (see Figure 4.34). Some 98% of the population lives in the Andean Region at elevations above 1000 m. The Köppen climate classes of those parts of Colombia relevant to the studies on mites, mainly the central and western parts, are tropical wet and dry (savanna; Aw) and dry semi-arid (steppe; Bsh) on the Carribean coast, tropical wet and dry (Aw) and tropical wet (Af) on the Pacific coast, and mid-latitude dry (Cw) and polar (E) in the Cordilleras.

Studies on the dust mites of Colombia have been more intensive than those of any other country. The findings of this research are collected in a monograph by Mulla and Sánchez-Medina (1980). Sánchez-Medina and Sánchez-Gutierrez (1973) examined the species diversity of seven climatic zones, covering localities ranging in elevation from sea level to 2900 m. They determined that highest mite population densities were found at localities below 1500 m. Although large mite populations could be found at high elevations, above 1500 m the mite populations appeared to be less dense. However, no such evidence of a cut-off point was found in a subsequent study by Sánchez-Medina and Zarante (1996), who sampled nine localities of elevations from 1300 to 5820 m over a 10-month period and found correlation between elevation and population density. The highest, La Calera (5820 m) had a mean population of ca. 230 mites per gram of dust over this period; similar to that of Viso Elias (1425 m; 242 mites per gram), Garagoa (1900 m; 237 mites per gram); Obonuco (2527 m; 211 mites per gram); Choconta (2655 m; 263 mites per gram) and Charta (2963 m; 288 mites per gram). The uncanny lack of variation in mite population density over the 10-month period at some of these locations almost defies belief, especially when compared with the quite considerable seasonal variation recorded from Bogotá, Fusagasugá and Girardot by Charlet et al. (1979).

Charlet et al. (1977) examined the relationship between elevation and mite population density at 57 localities (Figure 4.34c). They claim that population densities were highest above 1400 m. Unfortunately

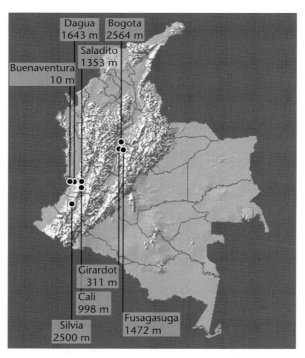

Figure 4.34   Colombia showing localities sampled for dust mites by Charlet *et al.* (1977a, 1979).

they only took one dust sample (from under beds) per locality, so there is no estimate of variation, and the most useful information from this dataset is the occurrence of particular species.

The dataset I used to examine the association between evaporation and mite distribution and abundance came from eight localities studied by Charlet *et al.* (1977a, 1979; Figure 4.34). Climate data for these localities are shown in Figure 4.35. The estimated mean annual evaporation rates for the localities show little seasonal or geographical variation and are highest in Girardot (6.2 mm day^{-1}) and lowest in Fusagasugá (3.6 mm day^{-1}). There was no relationship between mite abundance and altitude (see Figure 4.36a), even when data for Cali (an outlier) were removed. Nor was there any relationship between mite abundance and estimated evaporation rate (see Figure 4.36c).

The Colombian example contradicts the reports that high elevation is associated with lower humidities and fewer mites (see section 4.7 above). There is no significant reduction in abundance with humidity, or increase in evaporation, with increase in altitude. This is not surprising since these sites are in the wet tropics. Annual rainfall varied from 920 mm at Bogota to 7.2 m at Buenaventura. Most reports on dust mites that link high altitude with low humidity are from Europe where the pattern of climate, and the variables that influence it, are very different from that of tropical South America.

c California

California has an area of 404 000 km^2 and a population 35.9 million. It has a number of topographical features in common with Israel, including a long, narrow western coastline and coastal plain, and a central region of higher ground rising to mountains in the east. Like Israel's Jordan Valley and Dead Sea, it has, in the Colorado Desert, a significant inland area of salt lakes and flats below sea level. Köppen climate classes of California are, like Israel, Mediterranean (Csa), corresponding to the coastal strip and foothills, Continental Cool Summer (Db) in the mountains and Dry Semi-arid (Bs) in the southern and eastern interior (shown in Figure 4.37). Climate data are given in Figure 4.38.

The pattern of distribution of *Dermatophagoides* spp. in California was found to be related to gross topography (Furumizo, 1973, 1975b). Of 393 samples from 45 of 58 counties, 152 (39%) were positive for *Dermatophagoides* spp. *D. pteronyssinus* occurred in 68% of positive samples, *D. farinae* in 21% and *D. microceras* in 3%. *D. pteronyssinus* occurs

more frequently in the coastal areas, but both it and *D. farinae* are equally prevalent in the central region. The eastern region was virtually free of mites, probably because it is too dry for them (see Figure 4.39b). Mulla *et al.* (1975) found *D. farinae* occurred in only 30% of homes within 8 km of the coast of Orange County, whereas further inland they infested 65% of homes.

Distribution and abundance of *Dermatophagoides* spp. were studied by Lang and Mulla (1977a) between August and December, 1975 in four climatic zones in southern California: the coast, an inland valley, desert and mountains (see Figures 4.39a–c). They found 93% of homes in Orange County, −3–6 km from the coast (elevation ~13 m), contained mites and *D. pteronyssinus* was the dominant species (78%). At Riverside (elevation 284 m), located inland in a valley, 60% of homes contained mites and *D. farinae* was dominant (67%). Some 54% of homes contained dead mites at Indio (elevation −3 m), an inland desert location, and 27% of homes at Lake Arrowhead (elevation 1582 m) in the San Gabriel mountains, but no live mites were found at either location. Indoor temperature was highest in the desert (mean 28°C, range 18–40°C), lowest in the mountains (mean 20°C, range 13–28°C). There was a direct negative correlation between distance from the coast and mite abundance and frequency of occurrence (see Figure 4.39a).

The dataset I used to examine the association between evaporation and mite distribution and abundance included the four localities studied by Lang and Mulla (1977a), plus Los Angeles and San Diego, studied by Arlian *et al.* (1992). Data from this latter study (number of mites per gram of dust) were converted into number of mites per ml of suspended fine dust (the units that Lang and Mulla used) by dividing by 5.5, the average millilitre-equivalent of a gram of fine dust (Colloff, unpublished). Climate data for these localities are shown in Figure 4.38. There was a significant positive association between altitude and estimated evaporation rates ($F = 12.11$; $p = 0.025$), but not between evaporation rate (or altitude) and mean mite abundance, although a trend was present ($F = 2.75$; $p = 0.17$; see Figures 4.39a, d).

Seasonal variation in populations of *Dermatophagoides* spp. at Orange County (coast) and Riverside (inland valley) by Lang and Mulla (1978) confirmed *D. farinae* as more abundant inland and *D. pteronyssinus* more abundant in coastal locations. Seasonal population optima varied between locations and species.

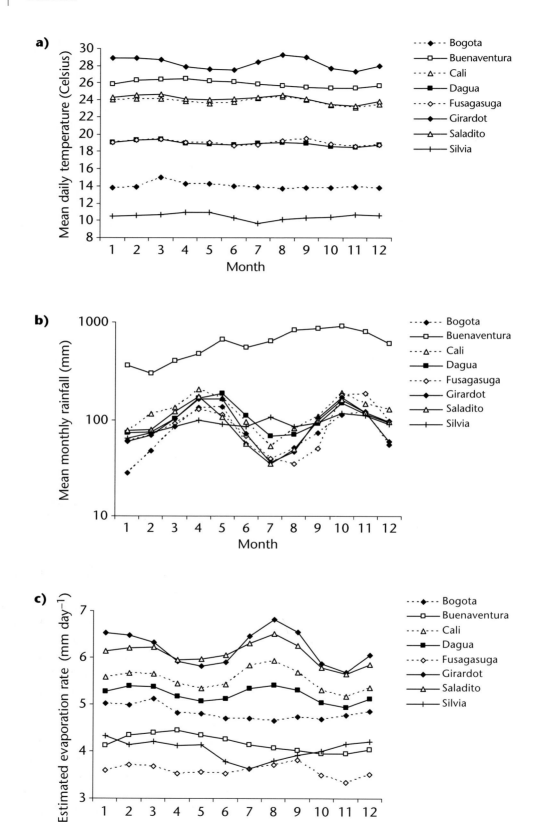

Figure 4.35 a) Mean monthly temperature; **b)** rainfall; **c)** estimated rate of evaporation of Colombian localities shown in Figure 4.34. Data kindly provided by Peter Jones, Centro Internacional de Agricultura Tropical, Cali, Colombia.

mean annual relative humidity (data from Ordman, 1971, Table 1) shows statistically significant positive relationships (8 am: $F = 6.5$, $p = 0.014$; 2 pm: $F = 25.5$, $p = 7.0 \times 10^{-6}$; Figure 4.45c: 2 pm data. 8 am data not shown). Reinforcing the point that elevation is a proxy for humidity, there is a strongly statistical negative relationship between elevation of localities and mean annual relative humidity 8 am ($F = 25.08$; $p = 8.2 \times 10^{6}$) and 2 pm ($F = 47.2$; $p = 1.3 \times 10^{-8}$; Figure 4.30d: 2 pm data. 8 am data not shown). The mite population densities recorded by Ordman (1971) are quite low. The mean for the whole dataset is only 8.3 per gram of dust (range 0–86). Ordman probably used the method of Spieksma and Spieksma-Boezeman (1967), one of few available at the time, which is known to have relatively low efficiency for extracting mites from dust (see Chapter 6).

The dataset I used to examine the association between evaporation and mite distribution and abundance consisted of 16 localities, a subset of those studied by Ordman (1971) for which mean monthly temperature and rainfall data were available on the South African Weather Service website (shown in Figure 4.44). Regression of mite population density against estimated evaporation rate shows a highly statistically significant negative relationship ($F = 15.08$; $p = 0.0017$; Figure 4.45e).

4.8.3 Conclusions from regional scale examples

Rainfall alone is a relatively poor predictor of mite abundance except in Australia where the pattern of rainfall is strongly linked to continentality. Distance from the coast is better, altitude is better still, but evaporation rate is the best predictor in the absence of reliable, consistent data on atmospheric humidity. Mite abundance generally shows a negative association with estimated evaporation rate except in Colombia, where rainfall is high and evaporation rates are relatively low. The relationship does not reach statistical significance for California although the trend is there. But for Israel, Australia and South Africa the relationship is statistically significant. It seems that large, continental regions show clearest climate effects on mite population density and occurrence. Regions with sharply increasing elevations near the coast, or with inland rifts, show less well-defined patterns.

But how does the nature of the relationship between mite abundance and estimated evaporation rates compare between each of the regional studies? One of the constraints of directly comparing the data between the studies is the regional difference in rainfall. To 'correct' for rainfall it is possible to

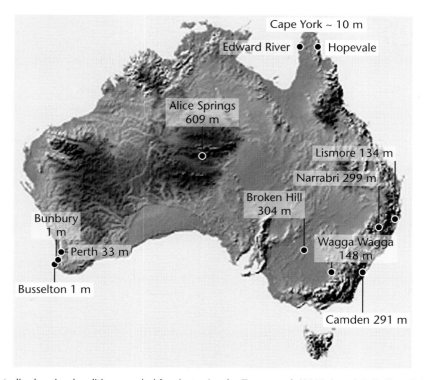

Figure 4.40 Australia showing localities sampled for dust mites by Tovey *et al.* (2000a) and Colloff *et al.* (1991).

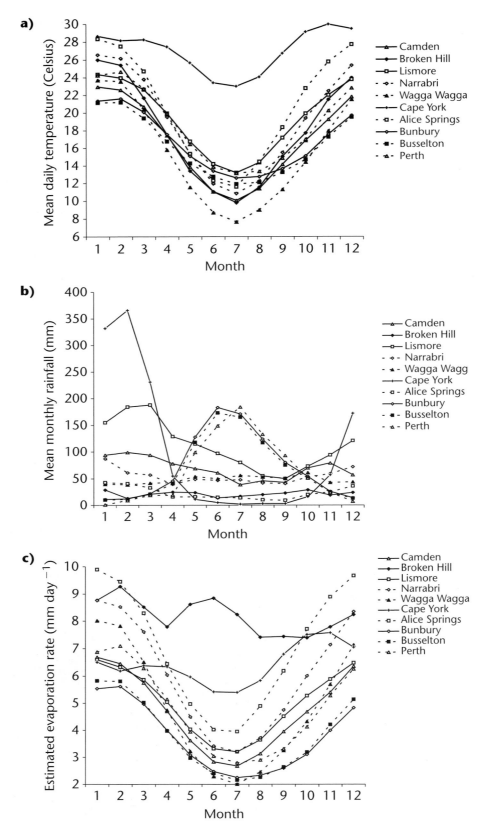

Figure 4.41 **a)** Mean monthly temperature; **b)** rainfall, **c)** estimated rate of evaporation of Australian localities shown in Figure 4.40. Data from Australian Bureau of Meteorology website, www.bom.gov.au.

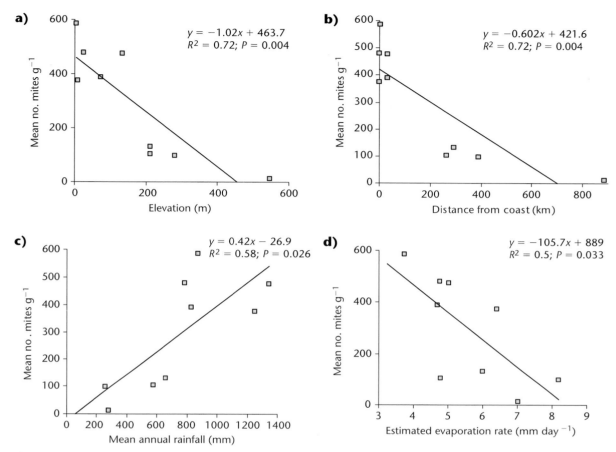

Figure 4.42 **a)** Relationship between abundance of dust mites and **a)** Elevation; **b)** distance from coast; **c)** rainfall; and **d)** evaporation of Australian sites shown in Figure 4.40. One outlier removed (Busselton).

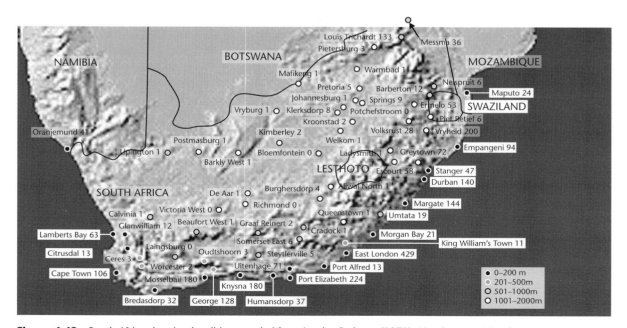

Figure 4.43 South Africa showing localities sampled for mites by Ordman (1971). Numbers next to place names = mean no. mites per 5 g of dust.

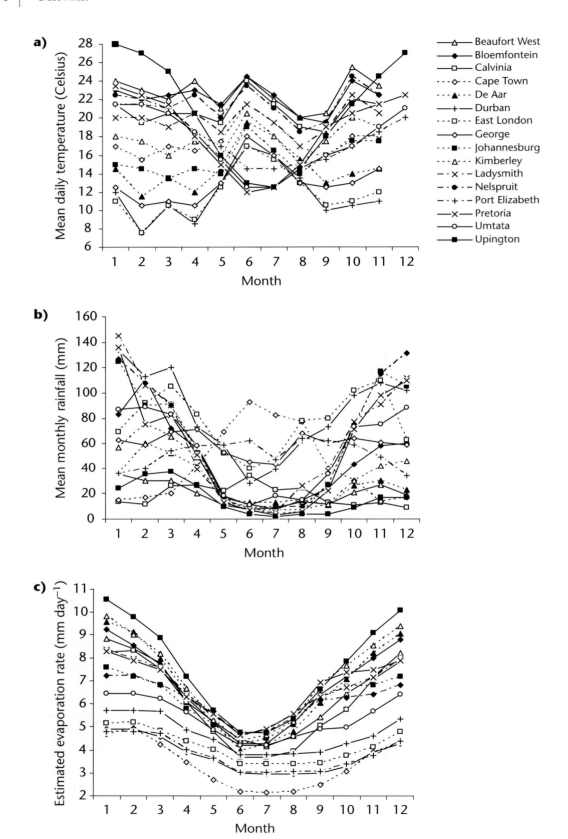

Figure 4.44 **a)** Mean monthly temperature; **b)** rainfall; and **c)** estimated rate of evaporation of 16 of the South African localities shown in Figure 4.43. Data from South African Weather Service website, http://www.weathersa.co.za.

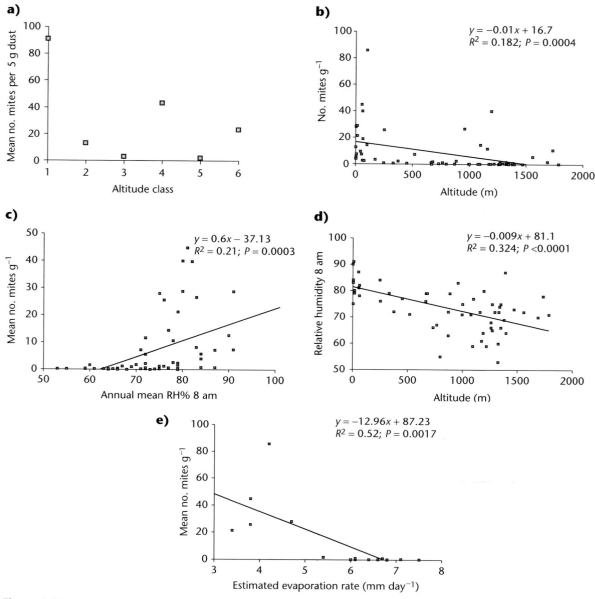

Figure 4.45 **a)** Mean mite population density v. altitude classes of South African localities; **b)** altitude of each locality v. mite population density; **c)** mean relative humidity 2 pm of localities v. mite population density (one outlier removed: East London); **d)** mean relative humidity 2 pm v. altitude of localities; **e)** estimated evaporation rate of 15 localities v. mite population density. Data from Potter *et al.* (1996) and Ordman (1971).

use *aridity*, which represents the ratio of rainfall to the estimated evaporation rate, $P/E_{o\ est}$ (both in mm day^{-1}). Thus a climate is arid when the ratio is low. Figure 4.46 shows mite abundance plotted against P/E_{o} for each of the regional studies, with lines of best fit. The slopes of the lines generally steepen with increasing mean distance of localities from the equator: Colombia: mean latitude of localities 3.78 decimal degrees, slope of line of best fit 4°; Australia: 26.8° latitude v. 20° slope; South Africa:

31° latitude v. 37° slope; Israel: 31.9° latitude v. 35° slope; California: 33.5° latitude v. 57° slope. However, the trend is not quite statistically significant with data from only five regional studies. A likely explanation for the trend in Figure 4.46 relates to greater seasonality in climatic variables with increasing distance from the equator and the effect this has on mite abundance. This relationship is examined in more detail with regard to seasonal population dynamics of dust mites in Chapter 5. Additional data, especially from regional

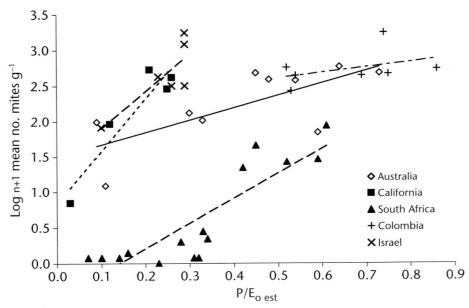

Figure 4.46 Relationship between mite abundance and aridity (the ratio of rainfall, *P*, to the estimated evaporation rate, $E_{o\,est}$, both in mm day^{-1}) for each locality in the regional studies. Statistics: Australia: $y = 1.744x + 1.49$, $R^2 = 0.51$, $P = 0.02$; California: $y = 7.41x + 0.83$, $R^2 = 0.88$, $P = 0.0185$; South Africa: $y = 3.53x - 0.5$, $R^2 = 0.77$, $P < 0.0001$; Colombia: $y = 0.74x + 2.24$, $R^2 = 0.38$, $P = 0.45$; Israel: $y = 5.22x + 1.38$, $R^2 = 0.67$, $P = 0.046$.

studies with mean latitudes of localities between 4 and 25 N or S, and >35 N or S, may clarify the nature of this relationship.

4.9 The global scale

Water balance capabilities, reproductive potential and population dynamics largely determine whether a particular species of mite is capable of successfully colonising and establishing itself in any particular part of the world. This implies that an analysis of global patterns of distribution of dust mites will represent a reflection of the biological characteristics of the mites. The pattern of distribution of different species has significant implications for the diagnosis and management of mite-mediated allergic disease. Many mite allergens are species-specific and knowing precisely which mites one can expect to find where is useful for accurate application of immunodiagnostic technology based on detection of allergen-specific human IgE antibodies. In regard to disease management, seasonal mite population maxima vary dramatically according to species and locations, and this can be important in deciding the best timing for mite control, or predicting periods of high allergen exposure when symptoms may be at their worst.

4.9.1 The biogeography of dust mites

Although both mites and people live permanently in extreme environments (see Figure 4.47), the majority of the human population of the world live in cities located on or near the coast, at elevations of less than 100 m above sea level. Naturally enough, this is where the bulk of surveys of distribution and abundance of dust mites have been done. Also, the epidemiology of allergic asthma has been studied mainly within developed countries in temperate latitudes. For example, the number of surveys of dust mites done in Europe, North America, Australia, New Zealand and Japan accounts for about 70% of the total. This geographical and socioeconomic bias makes it difficult to compare the characteristics of the dust mite fauna in different parts of the world.

After the Second International Workshop on Dust Mite Allergens and Asthma in 1990, I began compiling a database of the global distribution and abundance of the most important species of mites found in dust (Appendix 3). The data show which common species have been recorded from which country and how frequently species occur globally and regionally.

The distribution maps are based on published records of mites in domestic premises. Records from hospitals, day-care centres, retirement homes, schools,

Figure 4.47 Global distribution and abundance of dust mites. Extreme localities where dust mites have been recorded:
a) Highest locality: Huaraz, Peru, 3672 m above sea level; **b)** lowest locality: Nir Dawid, Israel, 111 m below sea level;
c) most northerly locality: Sør-Varanger, Norway, 69°73′N; **d)** most southerly locality: Punta Arenas, Chile, 53°00′S.

offices, shops and farm buildings have not been included. The maps (Figure 4.48) are of the following species: *Dermatophagoides pteronyssinus, D. farinae, D. microceras, Euroglyphus maynei, Malayoglyphus intermedius, Hirstia domicola, Lepidoglyphus destructor, Glycyphagus domesticus, Blomia tropicalis, Chortoglyphus arcuatus* and *Gohieria fusca*. The species complexes *Tyrophagus putrescentiae* and *Acarus siro* are included, even though identifications of *Tyrophagus* spp. and *Acarus* spp. are likely to be the least reliable for any of the more common Astigmata found in house dust. Members of these genera are difficult to identify accurately, as they may contain groups of sibling species and species-complexes (Fan and Zhang, 2007). Records of *Cheyletus* spp. are not mapped but are included in Appendix 3.

There is information in the database of 970 records from 812 named localities from 86 countries. Several of the localities are so close that a single dot on the map covers more than one site.

The Alexandria Digital Library Gazetteer Server (University of California, Santa Barbara) version 3.2: http://middleware.alexandria.ucsb.edu/client/gaz/adl/index.jsp was used to check place names, locations, latitudes and longitudes. Elevations were checked using Global Gazetteer version 2.1: http://www.fallingrain.com/world/. It is worth noting that elevations can vary within major cities (especially if they are in montane regions), and published sources of elevations may differ.

Many sites are from European countries, but with little or no published data from Austria, Ireland, Iceland and parts of central and eastern Europe. Data from China, Canada, large parts of the USA, the former Soviet Union and most of Africa are sketchy. Countrywide, or state/regionwide surveys are available for Algeria, Alsace-Lorraine, Brazil (reviewed by Binotti *et al.*, 2001), California, Chile, Colombia, India, Italy, Japan, Mexico, New Zealand, Norway, South Africa, Switzerland and Thailand, accounting for well over 50 publications.

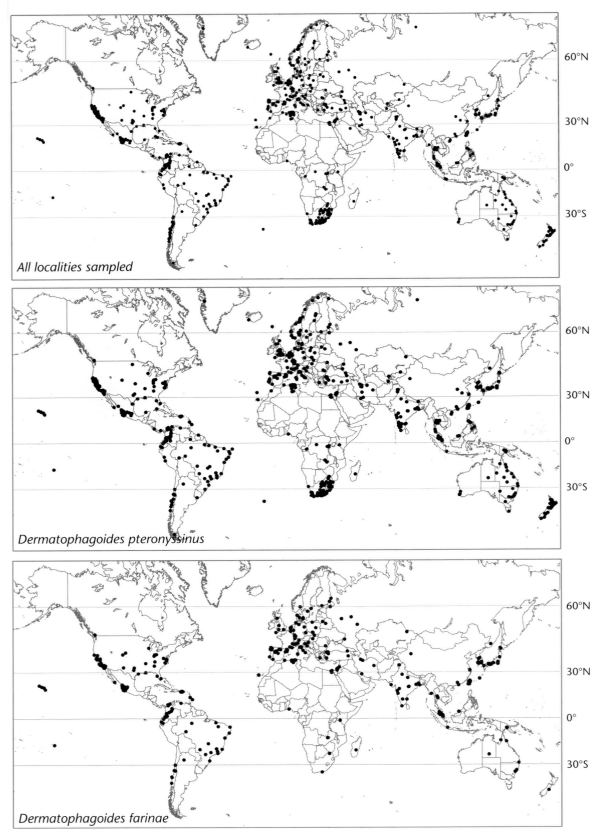

All localities sampled

Dermatophagoides pteronyssinus

Dermatophagoides farinae

Figure 4.48 Global distribution maps of 13 major dust mites species.

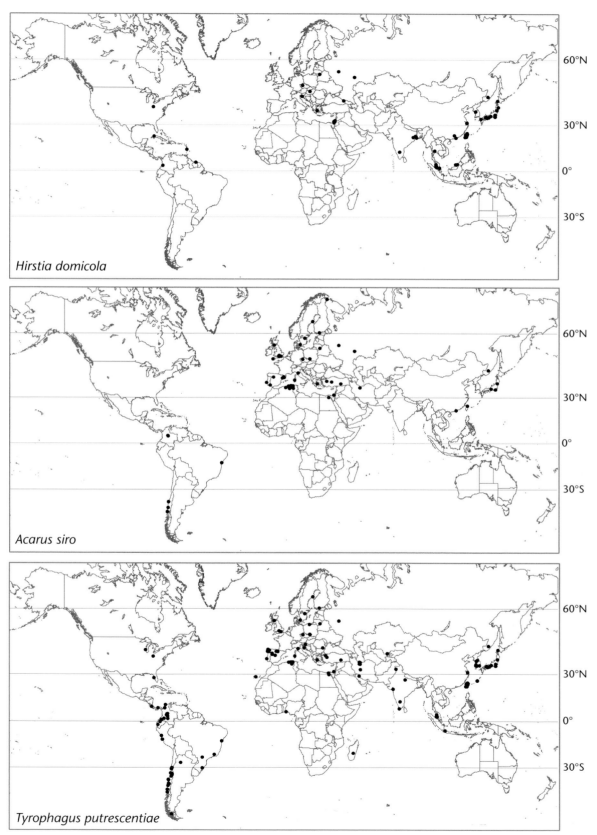

Hirstia domicola

Acarus siro

Tyrophagus putrescentiae

Figure 4.48 Continued

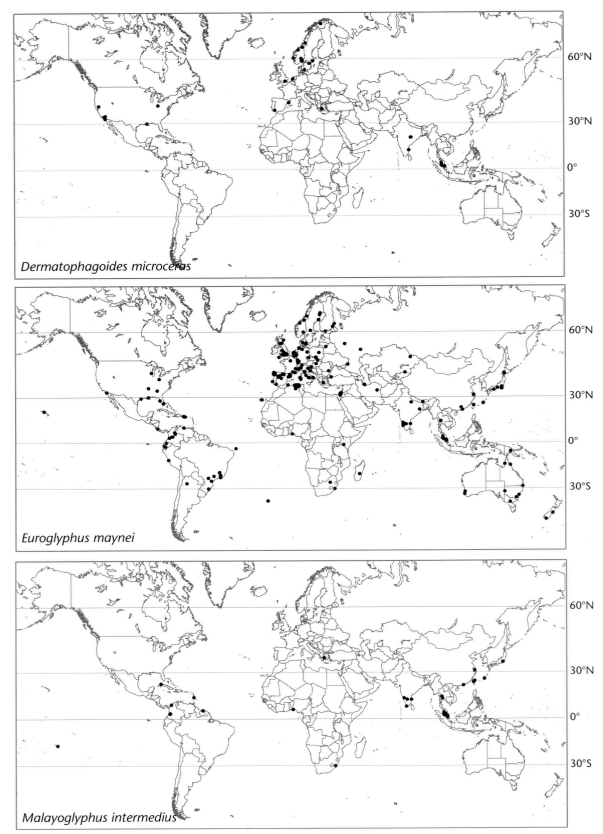

Dermatophagoides microceras

Euroglyphus maynei

Malayoglyphus intermedius

Figure 4.48 Continued

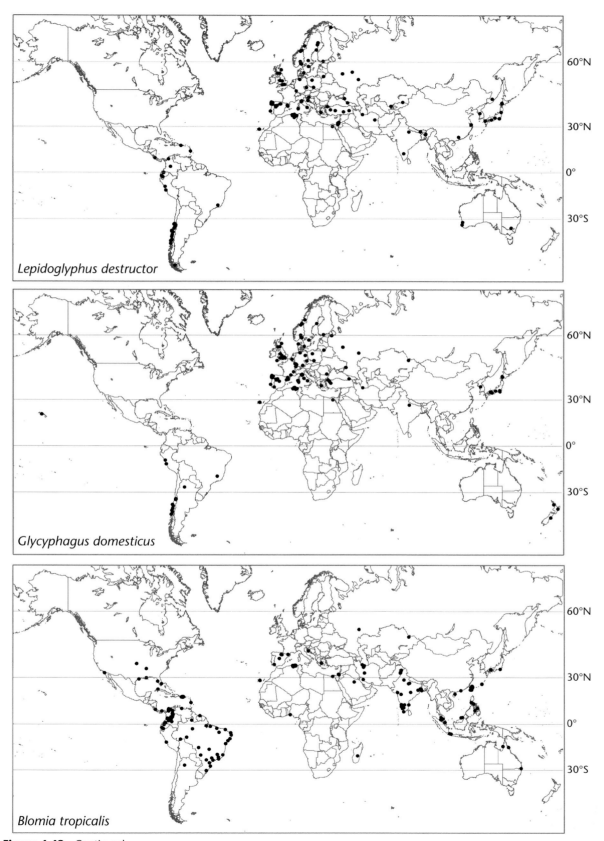

Figure 4.48 Continued

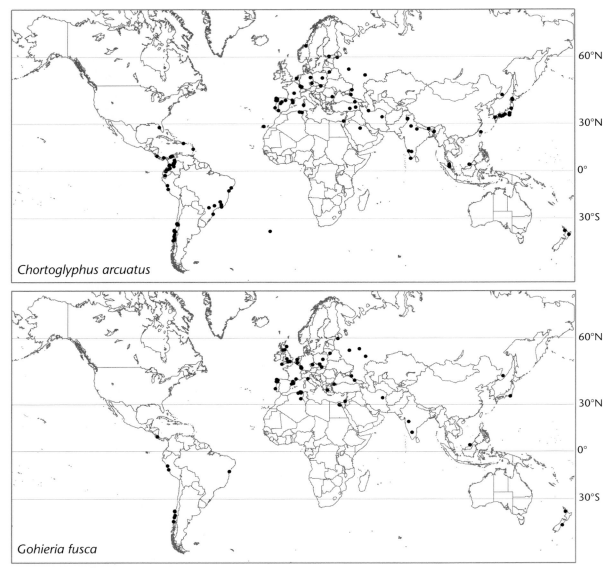

Figure 4.48 Continued

A preliminary analysis of climate data from the localities where the mites were found was done by Anne Bourne, CSIRO Entomology, Brisbane. Median, 1st and 3rd quartiles were calculated for a set of climatic variables from the locations where each species occurred. These were monthly maximum and minimum temperatures, mean relative humidity at 9 am and 3 pm and monthly rainfall. Data were taken from the CLIMEX Met. data file. This file contains relative humidity data for many (though by no means all) localities where mite species have been recorded, and it is possible, at the global scale at least, to use relative humidity data rather than estimated evaporation rate as a measure of atmospheric moisture, at least for now. CLIMEX

is a computer program that enables the prediction of the global distribution and abundance of organisms, based on known geographical distribution and data on population growth (Sutherst and Maywald, 1985). Monthly values for Southern Hemisphere locations were moved by six months so that the summer and winter data from both hemispheres coincided.

Not all locations had the corresponding climatic data and, where possible, proxy data from nearby localities were used; otherwise locations were omitted from the analyses. Numbers of localities with climatic data are recorded for each species below. The median values have then been tabled and graphed (Table 4.3, 4.4, Figure 4.49).

Table 4.3 Means of latitude and elevation of localities where 12 major dust mite species have been recorded. Species are sorted by decreasing affinity for tropical localities, estimated by Median Latitude North (ND = no data: *Malayoglyphus intermedius*, *Dermatophagoides microceras* and *Hirstia domicola* have been recorded almost entirely from Northern Hemisphere localities).

Species	Latitude North (decimal degrees)			Latitude South (decimal degrees)			Elevation (m)		
	1st quartile	Median	3rd quartile	1st quartile	Median	3rd quartile	1st quartile	Median	3rd quartile
Blomia tropicalis	1.0	6.4	22.2	ND	ND	ND	9	34	218
Malayoglyphus intermedius	3.1	4.4	13.8	ND	ND	ND	3	33	45
Cheyletus eruditus	7	34.4	48.6	33	35.1	38.8	16	130	279
Chortoglyphus arcuatus	6.9	33.2	41.5	22.8	35.0	37.0	9	41	142
Dermatophagoides farinae	22.5	35.3	45.2	13.7	21.3	33.0	8	45	216
D. microceras	21.2	41.3	59.3	ND	ND	ND	45	94	192
D. pteronyssinus	13.1	35.0	38.7	17.2	33.0	38.7	9	46	171
Euroglyphus maynei	29.0	41.2	51.5	19.9	27.5	33.5	11	61	220
Glycyphagus domesticus	40.7	45.2	54.6	33.8	36.9	39.7	12	45	138
Gohieria fusca	38.0	45.0	52.5	33.0	36.8	37.9	12	95	214
Hirstia domicola	22.3	31.5	34.8	ND	ND	ND	6	20	71
Lepidoglyphus destructor	33.9	43.3	55.8	30.1	36.3	37.8	12	65	145

a Dermatophagoides pteronyssinus

The distribution of *D. pteronyssinus* is cosmopolitan and there are 827 records (85% occurrence) worldwide. This is the most frequent species recorded, and is the only mite species present at all localities representing extremes of altitude and latitude, as well as at the wettest locality (Buenaventura, Colombia, mean annual precipitation 7392 mm), the driest (Touggourt, Algeria, mean annual precipitation 60 mm) and the most continental (Semipalatinsk, Khazakstan, 2146 km from the nearest coast). Climatic analysis was based on data from 452 localities.

b Dermatophagoides farinae

This species is the second most frequently occurring species worldwide, considerably less so than *D.*

Table 4.4 Medians of monthly medians of climatic data from localities where 11 major dust mite species have been recorded.

Species	Max. monthly temp. (°C)	Min. monthly temp. (°C)	Monthly rainfall (mm)	% relative humidity 9 am	% relative humidity 3 pm
Malayoglyphus intermedius	31.1	22.8	137.0	83.0	68.0
Blomia tropicalis	30.0	20.3	96.5	85.0	66.7
Hirstia domicola	21.1	11.4	98.0	83.0	60.5
Chortoglyphus arcuatus	17.5	8.1	77.0	84.5	64.8
Dermatophagoides pteronyssinus	18.4	8.9	66.0	82.0	65.0
D. farinae	18.9	9.4	66.0	82.0	64.0
Euroglyphus maynei	17.2	7.8	62.5	82.0	63.0
D. microceras	15.9	5.9	58.0	84.0	64.5
Lepidoglyphus destructor	16.7	6.7	56.0	82.0	65.0
Gohieria fusca	16.5	6.3	53.0	83.0	64.8
Glycyphagus domesticus	16.1	6.3	56.0	83.0	64.0

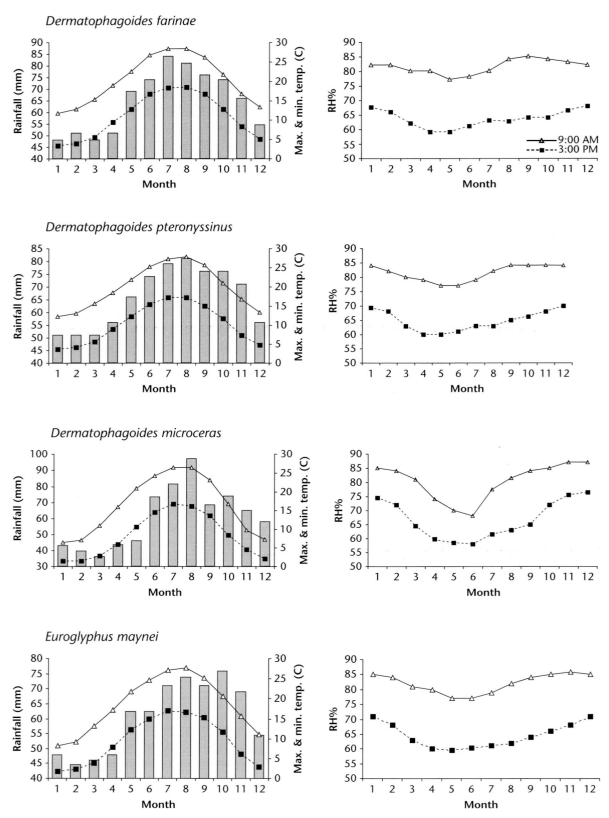

Figure 4.49 Median maximum and minimum daily temperature (Celsius, expressed monthly), median monthly rainfall (mm) and maximum and minimum daily relative humidity (expressed monthly) at locations where each major dust mite species have been recorded.

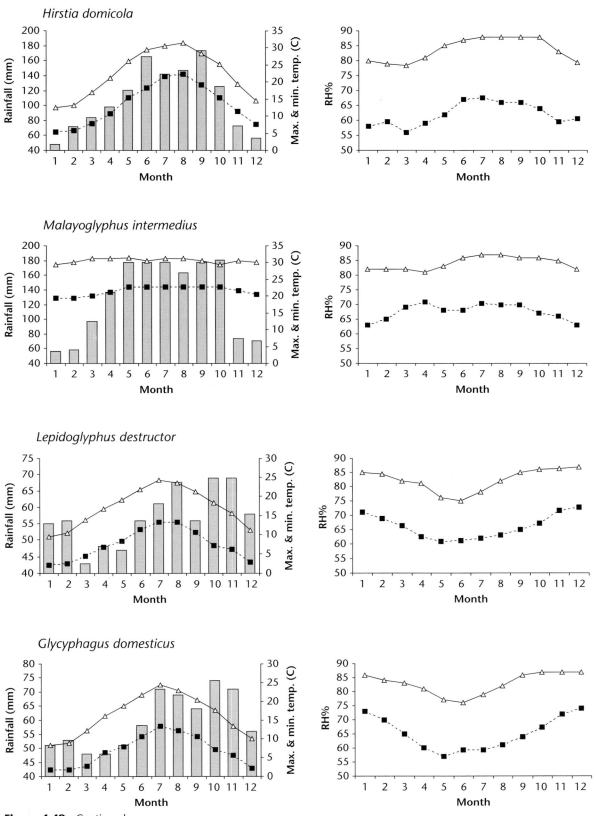

Hirstia domicola

Malayoglyphus intermedius

Lepidoglyphus destructor

Glycyphagus domesticus

Figure 4.49 Continued

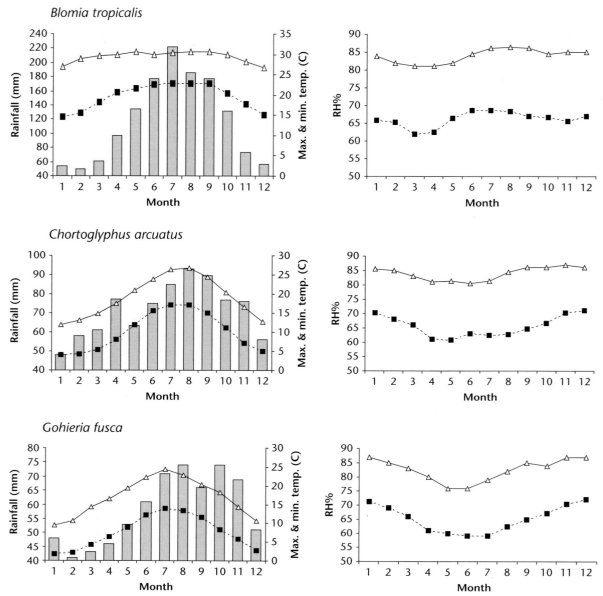

Figure 4.49 Continued

pteronyssinus, with 453 records (47% occurrence). It is most frequently found in the USA, Japan and continental Europe, but is considerably rarer than *D. pteronyssinus* in Australia, UK and Latin America. Climatic analysis was based on data from 259 localities. Despite its cosmopolitan distribution, median maximum summer temperatures at localities where this species are found are rarely below 18°C and minimum summer temperatures rarely below 10°C. There is also high seasonal variation in rainfall.

c   Dermatophagoides microceras

It is almost certain that this species has been widely misidentified as its sibling species, *D. farinae*

(Cunnington *et al.*, 1987) and that it is more commonly found than indicated on the distribution map. This means the frequency of *D. farinae* is even lower than that recorded here. This has clinical implications because it has been known for some time that each of them produces species-specific allergens. Morphologically, they are very similar, but can be discriminated with practice.

There are 35 records (4% occurrence) and climatic analysis was based on data from all the localities. *D. microceras* has been recorded from the Northern Hemisphere only, mostly from high latitude sites, the exceptions being localities in India and Malaysia. Median rainfall at the localities it is found is relatively

Table 4.5 Means of abundance, dominance and occurrence of 12 major dust mite species from all localities where the species have been recorded.

Species	Median occurrence (%)	Geometric mean abundance (No mites g^{-1})	Median dominance (%)
Dermatophagoides pteronyssinus	87.5	186.2	58.1
Blomia tropicalis	73.1	50.1	17.2
Euroglyphus maynei	59.8	45.5	8.4
Dermatophagoides farinae	38.0	62.5	9.1
Lepidoglyphus destructor	32.8	4.1	1.3
Cheyletus eruditus	26.0	6.1	1.1
Glycyphagus domesticus	22.6	5.2	1.9
Hirstia domicola	16.0	4.6	0.6
Malayoglyphus intermedius	13.7	52.1	1.0
Dermatophagoides microceras	13.2	8.4	0.9
Gohieria fusca	9.8	0.8	0.7
Chortoglyphus arcuatus	8.0	3.4	5.2

low, slightly higher in summer months. There is a large range of most of the climatic variables.

d Euroglyphus maynei

This species is the third most frequent species globally, with 261 records (27% occurrence). There was no major difference between temperate and tropical/subtropical regions. Most records are from coastal localities or areas of high rainfall. There are relatively few records from continental interiors. Climatic analysis was based on data from all localities. This species is found at high latitude in conditions of low temperature and rainfall.

e Malayoglyphus intermedius

This is a Northern Hemisphere mite (there are only two records from the Southern Hemisphere – Durban and Tahiti) which prefers tropical, hot, very wet places at low elevations, which have little annual temperature variation. Most records are from tropical and subtropical Asia. This species has probably been under-reported and confused with *Dermatophagoides* spp. There are 41 records (4% occurrence).

f Hirstia domicola

This species is mostly from the Northern Hemisphere and is particularly frequent in Japan and South-East Asia. There are 74 records (8% occurrence). Climatic analysis was based on data from 37 localities. There is strong seasonal variation in temperature and rainfall data at the localities at which *H. domicola* is found.

g Lepidoglyphus destructor

This is a species of high latitude localities with low temperatures and rainfall, both in Northern and Southern hemispheres. There is very little seasonal

variation in rainfall at the localities where it is found. There are 154 records (16% occurrence).

h Glycyphagus domesticus

This is another high latitude species, found at localities with low temperatures and rainfall, with little seasonal variation in rainfall. Its distribution is mainly Palaearctic: Europe, Russia and Japan accounting for the bulk of the records. It does not appear to have been recorded from house dust in North America, Africa, Australia or South-East Asia, and this may be due to confusion with *Lepidoglyphus destructor* or *Blomia tropicalis* which it resembles superficially. There are 140 records (14% occurrence). Climatic analysis was based on data from 82 localities.

i Blomia tropicalis

One of the major findings of this mapping and databasing exercise is just how common and widespread *Blomia tropicalis* appears to be. It has a predominantly tropical distribution, 30° either side of the equator, at low elevation sites, which are hot and wet, with high summer rainfall. Although this species is found throughout Asia and the southern USA, there is a general impression that *B. tropicalis* is a species of Latin America. There are 192 records (20% occurrence). Climatic analysis was based on data from 89 localities.

With this genus there is a high likelihood of confusion over identification. There are six described species of *Blomia*, all very similar in appearance, and of dubious validity. Three species, *B. tijbodas*, *B. kulagini* and *B. tropicalis* have been recorded from house dust. It may be that some records of *B. tropicalis* are actually one of these other species. Alternatively, others believe there is only one valid species in the genus, *Blomia tropicalis*.

j Chortoglyphus arcuatus

There are no major differences in frequency between temperate and tropical localities. This is a widespread species. There are 162 records (17% occurrence). Climatic analysis was based on data from 97 localities. It is found at localities with fairly high seasonal variation in rainfall, some of which have most rain during the winter months.

k Gohieria fusca

This is a high latitude species. Most records are from Europe, though it has been recorded from the tropics. There are 82 records (9% occurrence). Climatic analysis was based on data from 45 localities. It is found at sites of relatively high elevation.

l Cheyletus spp.

There are 91 records (9% occurrence) for *Cheyletus eruditus*, but 89 for *Cheyletus* spp., many of which were not identified to species. Climatic analysis was not done for *Cheyletus* spp.

m Acarus siro and Tyrophagus putrescentiae species complexes

There are 61 records (6% occurrence) for *Acarus siro* and a further 40 for other *Acarus* spp. There are 150 records for *Tyrophagus putrescentiae* (16% occurrence) and a further 42 for other *Tyrophagus* spp. Both *A. siro* and *T. putrescentiae* are cosmopolitan. Climatic analysis was not done.

Several other species have not been mapped because there were either too few records to interpret and/or their known distribution was not cosmopolitan. These include *Suidasia medanensis* and *Aleuroglyphus ovatus* (nine records each), *Dermatophagoides evansi*, *D. siboney*, *Austroglycyphagus malaysiensis*, *Sturnophagoides brasiliensis* and *Hirstia chelidonis*). They crop up occasionally in house dust and may represent species of regional or local significance (see below).

4.9.2 Conclusions

The major findings of this preliminary analysis of the global distribution and frequency of occurrence of dust mites are:

- The countries with highest recorded diversity of dust mites are India (12 spp.), Japan (11 spp.), Greece (11 spp.), Colombia (10 spp.), Czech Republic, Israel, Spain and USA (9 spp.), Algeria, Malaysia, Iran, Italy, Portugal, Poland, Russia and Switzerland (8 spp.). The least biodiverse were Congo, Ireland, Kuwait, Morocco, Namibia, Philippines, Romania, Rwanda, Syria, Uruguay and Uzbekistan (1 sp.), Barbados, Belarus, Canada, Mexico and Zambia (2 spp.), Burma, Egypt, Indonesia, Kenya and Tahiti (3 spp.). These figures reflect relative research effort rather than biological variables.
- *Blomia tropicalis* is found more frequently than previously suspected, and in the tropics is second only to *D. pteronyssinus*.
- *Lepidoglyphus destructor* and *Glycyphagus domesticus* are essentially of temperate distribution and the niche they would probably occupy in the tropics is occupied by *Blomia tropicalis*.
- *D. microceras* is most probably highly under-reported.
- Countries with high dust mite biodiversity are either in warmer, wetter regions or the mites collected in the surveys were identified by people with an acarological training. Those with lower diversity are either in cold, dry or hot dry regions or the mites have been identified by people without an acarological training.
- Several regional species assemblages exist (all contain *D. pteronyssinus* and *D. farinae* unless indicated otherwise):
  - Singapore, Malaysia and Indonesia, consisting of *Austroglycyphagus malaysiensis*, *Blomia tropicalis*, *Sturnophagoides brasiliensis* and *Tyrophagus putrescentiae*;
  - tropical and subtropical Central and South America, consisting of *Blomia tropicalis*, *Suidasia medanensis* and *Aleuroglyphus ovatus*;
  - Europe, consisting of *D. microceras*, *Euroglyphus maynei*, *Lepidoglyphus destructor* and *Glycyphagus domesticus*;
  - Japan, South Korea and China, including *Hirstia domicola*, *Chortoglyphus arcuatus* and *Tyrophagus putrescentiae*;
  - Caribbean, consisting of *Dermatophagoides siboney* and *Blomia tropicalis*;
  - Russia, Baltic States, Poland and Bulgaria, consisting of *Dermatophagoides evansi*, *Gymnoglyphus longior*, *Hirstia chelidonis*, *Lepidoglyphus destructor*, *Glycyphagus domesticus*, *Gohieria fusca* and *Chortoglyphus arcuatus*.

The consequences of regional species assemblages for allergic sensitisation and epidemiology are discussed in Chapter 8.

4.10 Integrating spatial and temporal scales in dust mite ecology

4.10.1 Migration and metapopulations

The size of a population is defined by the net total of births minus deaths, plus the net total of immigration minus emigration (see Chapter 5). Traditional ecological approaches have concentrated on phenomena associated with births and deaths, such as age structure of populations, fecundity and mortality rates and population growth. Metapopulation ecology (see Hanski and Gilpin, 1997) focuses on the phenomena associated with spatially fragmented populations, including migration. A metapopulation is a set of local populations linked together by migration or dispersal. Local populations may become extinct or become re-established following extinction, which means patterns of breeding are highly contingent upon space and time. Thus, the metapopulation model differs from the traditional model of population ecology which assumes equal likelihood of interaction between individuals in a population (panmictic population).

This all means that the patterns of distribution of the local populations and the characteristics of migration between them can have profound effects on the genetics and evolution of organisms. Local populations are constantly being created, added to, growing, shrinking and becoming extinct. So the gene pool of any population that exists as part of a metapopulation is highly dynamic as a result of changes in the membership of the population (see Figures 4.50a, b). In relation to dust mites and allergy, high rates of gene flow and genetic variation could have important clinical consequences relating to molecular polymorphism of allergens and the immune response to them; rates of mite population growth and levels of allergen exposure; genetic resistance to acaricides and resilience to other control measures.

The size and distribution of local populations may fluctuate considerably, but the metapopulation is much more stable. Local population fluctuations are relevant to dust mites at microhabitat to macrohabitat scales, and probably regional scales too. As someone installs central heating or air conditioning, replaces a carpet or installs new bedding and furniture within a home, local populations of mites may become extinct or be substantially reduced. The introduction of a second-hand sofa or mattress into a home may massively increase the mite population of that home and open up prospects for interbreeding and genetic mixing. These are all microhabitat/macrohabitat-scale events, but the extinction or reduction of populations caused by seasonal decrease in humidity and increase in temperature could operate at regional scale.

We have already seen that populations of cosmopolitan dust mite species are subdivided at all the spatial scales detailed here. We also know there is evidence for migration of mites between microhabitats within the home (Mollet and Robinson, 1995; Mollet, 1996). Migration between homes takes place on clothing and in furniture, and large-scale global movement of human populations, accompanied by their goods and chattels, is almost certain to involve co-migration of dust mite populations. If we examine patterns of human migration over the last 150 years or so, we start to get an idea of how dust mite populations from one part of the world have spread to other parts (see Figures 4.51a, b). The pattern in the map for 1850–1950 and the map for the 1970s and 1980s is essentially the same, showing movement from regions that were economically disadvantaged at the time, to countries where prospects were better and had a high demand for migrant labour. We have no data on the genetic and evolutionary consequences of migrant dust mite populations interbreeding with pre-existing populations for the simple reason that there have been no studies done on dust mite population genetics.

Metapopulation approaches may not be an exactly applicable model to the spatial scales of dust mites that I have outlined here. However, they represent a useful approximation and underline the importance of viewing dust mite populations as patchy, fragmented and dynamic; subject to migration and extinction events, rather than as continuous, closed and relatively static.

4.10.2 Spatial and temporal scales and dust mite ecology – some concluding remarks

- Fine-scale processes can be averaged and become constants at a larger scale. One example of this is the high degree of variation in mite population density within habitats versus the relatively predictable population densities that exist at the regional scale.
- Processes that were constant at fine scale become variable or patchy at larger scale. For example, habitat suitability at microhabitat scale is relatively even, but at regional scale, macroclimate variables impose far more complex patterns.

- New interactions and patterns emerge at different scales. The effects of climate on dust mite populations are hard to identify and quantify at microhabitat and macrohabitat scales. At regional and global scales they become dominant variables.
- The scale of observation influences the description of the pattern. For example, the definition of the abundance or rarity of a species is scale-dependent (see Figure 4.50), and is likely to show greater variability at smaller than larger spatio-temporal scales.

- Perceptual scale bias on the part of researchers is ever-present. This process involves attempting to look at 'big picture' questions using 'small picture' methods. Sometimes bias can be quantified in order to minimise errors, but often it is not even recognised.
- There is no single natural scale at which ecological phenomena should be studied. There is a series of biological phenomena that clearly links the different scales. Ignoring these links means we will probably get things wrong.

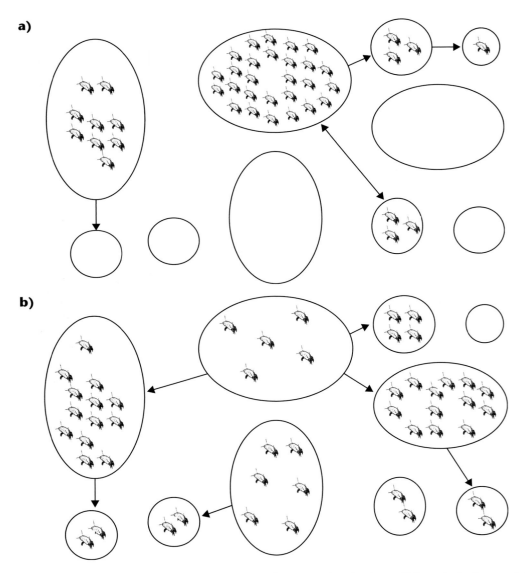

Figure 4.50 Metapopulation dynamics model for dust mites illustrating **a)** extinction and **b)** recolonisation events in local, panmictic populations. Note the total, or metapopulation, density remains the same, and no single local population represents the source of the other populations.

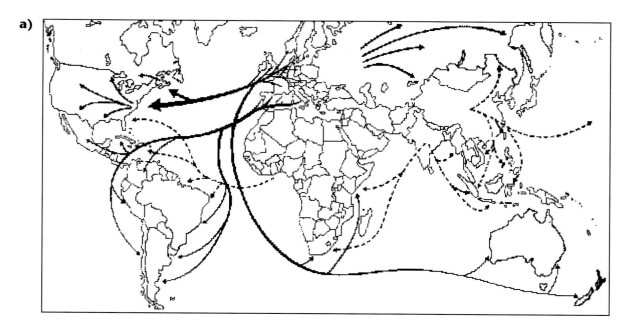

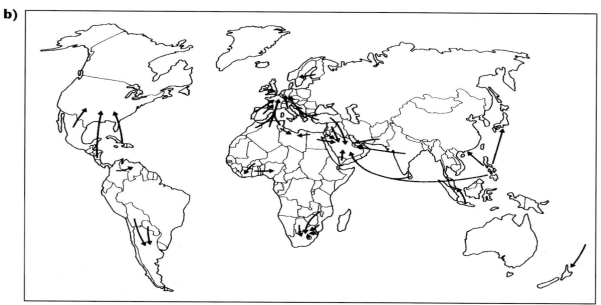

Figure 4.51 a) Major international migration routes, 1850–1950; **b)** major international routes of migrant workers to regional labour markets, 1970s and 1980s (from Jones, 1990).

5. Development, life histories and population dynamics

Dead men may envy living mites in cheese,
Or good germs even. Microbes have their joys,
And subdivide, and never come to death.

Wilfred Owen, A Terre.
In: Blunden, E. (Ed.) The Poems of Wilfred Owen
Chatto & Windus (1931)

5.1 Introduction

The life history of any animal concerns the various details of its reproduction and development. These include characteristics of the life cycle, known as life history traits, such as size at birth, age and size at maturity, longevity, mortality, fecundity, number and sex ratio of offspring, as well as the responses of these variables to factors such as diet, temperature and humidity. Most of these life history traits represent key variables in demography, or the study of populations. Combined, these traits are expressed in ways that affect evolutionary fitness, with inevitable trade-offs between reproduction and survival. A good introduction to the interaction of these factors in relation to the way certain life histories have evolved is provided by Stearns (1992).

Comparative life history biology of pyroglyphid, acaroid and glycyphagoid mites can give some insights into their biological 'success', and their relative importance as allergen-producing organisms. By 'success', I mean simply their global distribution and abundance and their capacity to persist following changes in environmental conditions.

This chapter highlights the differences in life history traits between these major groups of domestic mites, the relationship between these traits and population dynamics, and the likely consequences for allergen exposure, mite control and avoidance. This approach is useful – not least because several of the species considered herein are found within the same habitats, the same homes or the same geographical regions, although there is a paucity of information on the population consequences of their interactions.

We have already seen a simple example of how life history traits may affect the trade-off between reproduction and survival. In Chapter 3, I explored the association between water loss, the size of females and their eggs, and their rates of fecundity. It emerged there were four categories:

1. small mites that lay a few, large eggs (for example *Euroglyphus maynei*);
2. medium-sized mites that lay a few, medium-sized eggs (such as *Dermatophagoides* spp.);
3. large mites that lay moderate numbers of medium-sized eggs (Glycyphagoidea);

4. very large mites that lay large numbers of eggs that are small relative to the size of the female (Acaroidea).

I suggested that a major constraint on the size and number of eggs produced was the amount of water that has to be acquired by the mother, allocated to the eggs and then expelled from the mother's body upon oviposition. Thus small mites with large eggs lay fewer, less often and over a shorter period than large mites with small eggs. As well as water, the egg contains yolk proteins that require energy-expensive synthesis by the mother from the raw materials of digestion. A major determinate of egg size is the proportion of the mother's dietary intake that can be invested in the egg (a process known as 'bestowal' or 'allocation').

Are there trade-offs between egg size and the survival of the offspring? Do mites from bigger eggs that contain more yolk resources have lower mortality? Or do larger eggs take longer to develop and have therefore extended vulnerability to lethal events like desiccation compared with small eggs? These sorts of questions will be examined in this chapter. They deal with trade-offs, the interaction of reproductive traits (numbers of eggs, development time, oviposition period, mortality and so on) and how these characteristics shape the seasonal rises and falls in the size of mite populations which, in turn, influence the amounts of allergens available to which people can become exposed.

5.2 Stages in the life cycle

Pyroglyphid mites have six stages in the life cycle: egg, prelarva, larva, protonymph, tritonymph and adult (see Figure 5.1.). Glycyphagoid and acaroid mites do not have a prelarva but may have an additional nymphal stage, the deutonymph, in the form of a motile or immotile hypopus. The larval and nymphal stages each have periods of activity and feeding followed by quiescent periods prior to moulting.

5.2.1 The egg and the embryo

All the astigmatid mites found in house dust are well under a millimetre in length as adults, which means their eggs are considerably smaller. The eggs tend to be white in colour, oval and elongate. When laid they may be coated in a thin layer of sticky mucus which adheres them to the substratum. The eggs of *Dermatophagoides farinae* are ca. 160–180 μm long, 70–90 μm broad (Walzl, 1988), while those of *D. pteronyssinus*

are ca. 150 μm long by 60 μm broad (see Figure 5.2a) and those of *Euroglyphus maynei* ca. 120 μm long by 55 μm broad (Colloff, 1991b).

The eggs of *Tyrophagus* spp. are smooth when they start to develop but then become heavily ornamented with bands of tubercular microsculpture that protrude from the surface of the shell (Hughes, 1976, her Figure 56). This microsculpture is quite unlike that of other domestic mites which tend to have relatively smooth shells. Scanning electron microscopy reveals the shell (or chorion) of *T. putrescentiae* consists of a complex series of domed exochorionic structures with basal openings (Callaini and Mazzini, 1984). These structures appear to be able to retain a moist layer of air near the surface of the egg in order to reduce water loss, and I suggest that in *Tyrophagus* the egg is the stage in the life cycle that is adapted for survival of desiccating conditions. The females of *Lepidoglyphus destructor* habitually cover most of their eggs with fragments of food and debris (Barker, 1983). This too may be a method for reducing water loss during embryonic development.

Embryological investigations of mites are relatively few and fragmentary. The study by Hafiz (1935) for *Cheyletus eruditus* is one of the more detailed. The following account of the development from egg to larva is based on the study by Walzl (1988) of *Dermatophagoides farinae* (shown in Figures 5.2, 5.3). Embryonic development of *D. farinae* at 75% RH, 25°C takes 170–180 hours, with the completion of the blastula about 20 hours after oviposition (see Figures 5.2b, 5.3b). By 36–40 hours differentiation of the blastoderm is evident, with the germ band and prosomal limb buds present on the ventral surface and extra-embryonic ectoderm present dorsally (see Figures 5.2c, 5.3c). By 48 hours the embryo has closed except for an oval region of extra-embryonic ectoderm in the posterior portion of the embryo (see Figure 5.3d). The cheliceral limb buds form first, then those of the palps, followed by the three pairs of legs. By 74–76 hours a pair of median protrusions appear posteriorly behind the buds representing the third pair of legs (see Figure 5.3e). These protrusions represent the terminal prosomal and initial opisthosomal segment.

5.2.2 The prelarval and larval stages

Pyroglyphid mites, but not glycyphagoids or acaroids, have a vestigial prelarval stage, consisting of a thin membrane that develops inside the eggshell. By about 100 hours after the start of blastulation, a pair

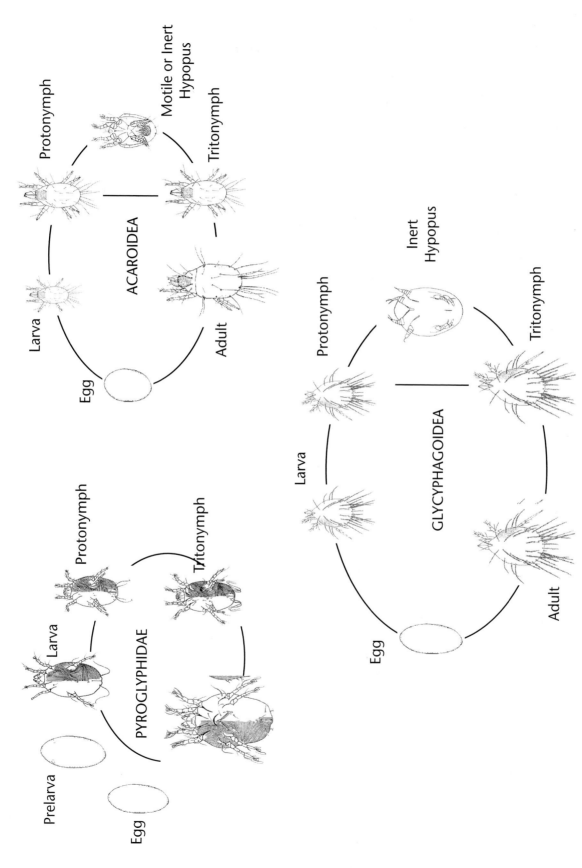

Figure 5.1 Stages in the life cycles of pyroglyphid, acaroid and glycyphagoid mites.

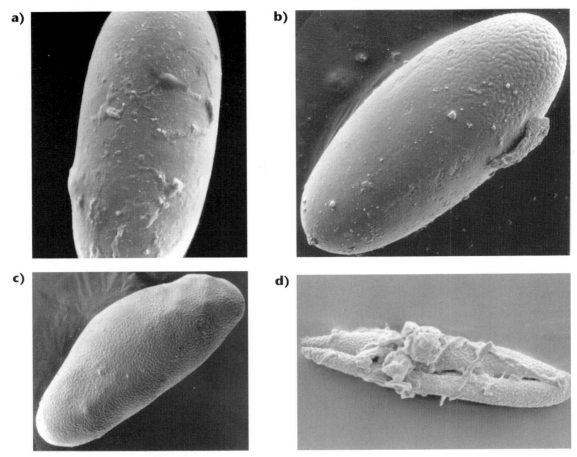

Figure 5.2 Eggs of *Dermatophagoides pteronyssinus*. **a)** Newly laid with layer of mucus; **b)** egg at blastodermal stage with polar change in microsculpture of the cuticle; **c)** at mid/late embryonic stage, with limb buds evident through the cuticle; **d)** cuticle after the larva has hatched, showing dorsoventral line of excision.

of hemispherical egg teeth with pointed tips have developed near where the chelicerae will eventually be, indicating the completed sclerotisation of the prelarval phase (Walzl, 1988). The larva then develops within the prelarval integument. Fain (1977) found females of *Dermatophagoides farinae, D. pteronyssinus, Euroglyphus maynei* and *Sturnophagoides brasiliensis* that contained eggs in which prelarval or larval development was evident. This suggests these species are ovoviviparous, i.e. the offspring develop within the mother prior to oviposition. However, pyroglyphids are probably normally oviparous (i.e. development occurs after oviposition) but some eggs are not oviposited for some reason, and larval development continues regardless. Fain and Hérin (1978) reported larval development in dead females of *Lepidoglyphus destructor*.

The six-legged larva has a characteristic posture within the egg with its first two pairs of legs folded ventrally and pointing posteriorly and the third pair pointing anteriorly (see Figure 5.3f). The eggshell is split longitudinally (see Figure 5.2d) by the egg teeth and the larva emerges, leaving the prelarval integument, along with the egg teeth, inside the shell (Walzl, 1988).

5.2.3 The nymphal stages

a The protonymph

Protonymphs are obligate stages in the life cycle of astigmatid mites. They have four pairs of legs and one pair of genital papillae. In crowded cultures of *Dermatophagoides farinae* and *D. pteronyssinus*, Ellingsen (1974, 1975) observed immobile (quiescent) protonymphs, which represent a facultative stage that appears in response to high population densities. The stage may have a similar function as the non-phoretic hypopi in the

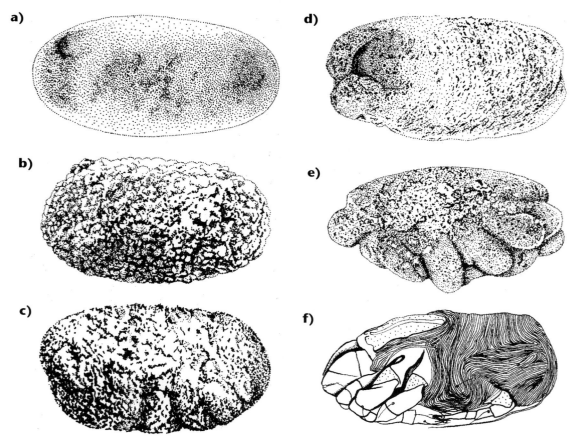

Figure 5.3 The embryonic development of *Dermatophagoides farinae*. **a)** Undifferentiated egg; **b)** uniform blastoderm completed, ca. 20 hours after oviposition; **c)** blastoderm differentiated into germ band and limb buds on ventral side and ectoderm on dorsal side, 36–40 hours; **d)** complete dorsal closure of embryo except for posterior plate of extra-embryonic ectoderm, 48 hours; **e)** limb buds, last prosomal and first opisthosomal segments developed, 74 hours; **f)** hexapod larva fully developed, ca. 100 hours. (Re-drawn from Walzl, 1988.)

Glycyphagidae (Wharton, 1976; see below). Oxygen consumption by quiescent protonymphs is nearly 30 times less than active protonymphs (Ellingsen, 1978) and body water exchange is nearly 140 times slower (Arlian and Wharton, 1974; Ellingsen, 1975), suggesting metabolism is slowed and they represent a survival stage.

b The hypopus

The hypopus (plural: hypopi or hypopodes) is equivalent to the deutonymphal, or second nymphal stage. Its role in the life cycle is the survival of unfavourable conditions and dispersal. In the Pyroglyphidae, the Sarcoptidae (*Sarcoptes* spp. – the scabies mites) and Psoroptidae (*Psoroptes* spp. – the mange mites), the hypopial stage is absent, and protonymphs moult into tritonymphs (see Figure 5.1). In the Acaroidea and Glycyphagoidea, protonymphs may also moult

into tritonymphs or they may go through the hypopial stage. Hypopi come in two types: motile forms (see Figures 5.4a–d) which are dispersed on insects (phoretic dispersal), and inert forms which are a survival stage, but are light enough to be dispersed on air currents (Alberti and Coons, 1999; Figures 5.4e–g).

Within the Acaridae, *Acarus siro* has a facultative motile hypopus. This stage attaches to larger arthropods using a set of suckers in the anal region (shown in Figures 5.4c, d). It is dispersed by hitch-hiking on insects, a process known as phoresy (Binns, 1982). Thus motile hypopi are also occasionally known as phoretomorphs. The occurrence of motile hypopi in populations of *Acarus siro* is rare, but frequency and abundance increases in sub-optimum conditions (Hughes, 1976). Non-hypopial-forming strains exist. *Acarus immobilis* and *A. gracilis* have an inert

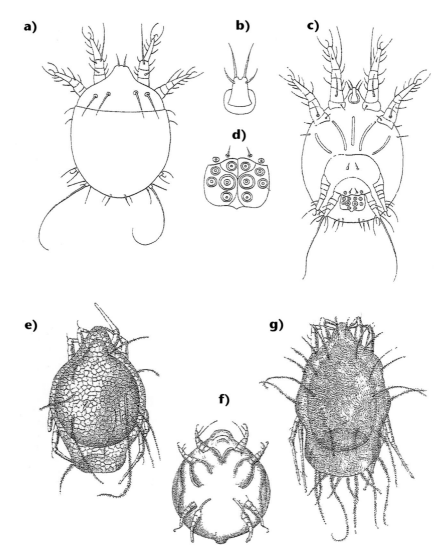

Figure 5.4 a–d) The active hypopus of an acarid mite, *Acarus farris* (from Michael, 1903, pl. 39, Figures 1–4, as *Tyroglyphus longior*, but cf. Griffiths (1964) for details of misidentification). **a)** Dorsal surface; **b)** vestigial gnathosoma; **c)** ventral surface; **d)** anal sucker plate; **e–g)** inert hypopodes of glycyphagid mites (from Michael, 1901, pl. 8, Figures 11, 12, 17); **e)** tessellated protonymphal cuticle of *Lepidoglyphus destructor*; and **g)** striated cuticle of *Glycyphagus domesticus* containing **f)** the inert hypopus, within which the tritonymph develops.

hypopus, which looks like a reduced form of the active hypopus without the anal sucker plate and with fewer setae on the legs and body. Hypopi have been recorded from only one species of *Tyrophagus*. Of the Glycyphagoidea, *Glycyphagus domesticus* and *Lepidoglyphus destructor* have immotile hypopi only. They are different from those of *Acarus* spp., being enclosed within the protonymphal cuticle. In *Lepidoglyphus* and *Glycyphagus* spp. the protonymphal cuticle enclosing the hypopus has a tessellated microsculpture, quite different from that

of the normal protonymphal cuticle (shown in Figures 5.4e, g). The tritonymph may develop within the inert hypopus, within the protonymphal exuvium (Alberti and Coons, 1999, their Figure 17). Hypopi have highly reduced, non-functional mouthparts and show varying degrees of regression of the gut. In the immotile hypopi the internal organs degenerate into a mass of undifferentiated cells and tissue. Only the nervous system remains intact (Hora, 1934; Hughes and Hughes, 1939). The processes of moulting and encapsulation of the hypopus in the protonymphal cuticle of *L. destructor* and *G. domesticus* were studied by Wallace (1960). Hypopi are not known from the genera *Blomia*, *Chortoglyphus* and *Gohieria*.

The existence of the hypopial stage caused considerable confusion for acarologists in the 19th century, so much so that it became known as 'The Hypopus Question' (Michael, 1884). Various theories were advanced (summarised by Michael, 1901), mostly that hypopi were adults of previously unknown taxa. Mégnin (1876), working with what he referred to as *Tyroglyphus mycophagus* (either *Caloglyphus mycophagus* or *C. berlesei*), surmised that the hypopus is 'the "cuirassed" [that is armoured, sclerotised] heteromorphous, adventitious nymph of *Tyroglyphus* entrusted with the preservation and distribution of the species under adverse circumstances'. Even though Mégnin's conclusion is basically correct, it is not the whole story because there is more than one kind of hypopus. Michael (1884) realised that *Caloglyphus* protonymphs moulted to become hypopi, regardless of whether conditions were adverse or not. The hypopi dispersed by attaching themselves by their anal sucker plates to insects. However, not all protonymphs became hypopi. Some moulted and became tritonymphs.

Michael (1901) was also aware that there were three morphologically distinct forms of hypopus, which he describes: those with posterior sucker plates that disperse on insects; others, which he referred to as the Homopus type, which cling to hairs of mammals by means of a bilobed posterior structure (but without suckers); and the third, the rudimentary, quiescent form of certain *Glycyphagus* spp. He appreciated the economic importance of hypopi, as the dispersive stage of stored products mites:

> *The hypopial stage is biologically far the most interesting portion of the life history of the Tyroglyphidae; it is also of considerable commercial importance, because it is the existence of this remarkable provision that enables many of the most destructive species to spread themselves almost all over the world as they do.*

The evolutionary significance of glycyphagid hypopi (Knülle, 1987, 1991, 1995, 2003; Corente and Knülle, 2003) is that hypopi can survive extreme and prolonged dry conditions, as well as being able to disperse to new habitats. Therefore, hypopi represent an ecologically flexible, risk-reduction strategy against fluctuations in conditions in temporary, patchy habitats in which many of these mites live.

c The tritonymph

Tritonymphs are obligate stages in the life cycle of astigmatid mites. They have four pairs of legs and two pairs of genital papillae, compared with one pair in the protonymphs. More leg and body setae appear in the tritonymphal stage. In pyroglyphid tritonymphs that are ready to moult, the adult cuticle can be observed through the tritonymphal cuticle. I have observed adult males 'guarding' inert tritonymphs. All of them turned out to be females, and the males were waiting for them to moult and emerge in order to mate with them.

5.2.4 The adults

Reproductive morphology and function has been covered in Chapter 2. Female pyroglyphids, acaroids and glycyphagoids tend to be larger than their corresponding males, and tend to live longer.

a Sex ratios

Pyroglyphids, acaroids and glycyphagoids are obligate sexual reproducers. The sex ratio in pyroglyphids is typically close to 1:1 (Hodgson, 1976a), but it can vary considerably in some species. An average of 61% of adults of *Lepidoglyphus destructor* were females (range 14–92%; Chmielewski, 1987), and an average of 56% of adults of *Glycyphagus domesticus* were females (range 20–71%; Chmielewski, 1988).

b Andropolymorphism

Some species in the genera *Dermatophagoides*, *Hughesiella* and *Sturnophagoides* are andropolymorphic (Woodring, 1969). The heteromorphic form of the males has the first or third pair of legs greatly enlarged (refer to Chapter 1 for more detail). The adaptive significance of male polymorphism has been studied in the acaroid mites *Caloglyphus berlesei* and *Rhizoglyphus robinii* (Radwan, 1993, 1994). The thickened third pair of legs is used to attack and kill other males (both homomorphic and heteromorphic forms) by puncturing the cuticle. These 'fighter' males appear to be produced in response to low population densities, a circumstance under which it might be hard to find a mate. Poor diet inhibited the appearance of fighters of both species (Radwan, 1994). The trait is not heritable in *C. berlesei*, but *R. robinii* fighters were more likely to have been sired by fighter fathers (Radwan, 1994). In small colonies of *C. berlesei*, fighters were more reproductively successful, mainly because they were able to kill all the other males and monopolise the females, but in larger colonies non-fighter males were more reproductively successful because the fighters tended to be killed by rival males more frequently than non-fighters (Radwan, 1993). The occurrence of the fighter phenotype seems to be a response to conditions where there is a good chance of eliminating rivals without being killed, thereby improving mating success and conferring a selective advantage on these individuals.

c Sperm competition

The observation by Furumizo (1975a) that males and females of *D. farinae* copulate four or five times with different partners suggests that dust mites may exhibit some form of sperm competition. After intromittent transfer, pyroglyphid sperm remains in the female receptaculum seminis until required to fertilise an egg. Male pyroglyphids are not known to remove the sperm of the previous mate of their partners, but in *Caloglyphus berlesei*, the sperm from the last male to mate with the female fertilises 86% of her eggs (Radwan, 1991a), and does so because he is able to displace the sperm of previous males (Radwan, 1991b). Males of *Rhizoglyphus robinii* that produced larger sperm of even size were more reproductively successful than those that produced smaller sperm more variable in size, and reproductive success was independent of the number of sperm per ejaculate, duration of copulation or male body size (Radwan, 1996).

5.3 Why population dynamics is of practical importance

If asked whether it was more effective to control mites when the population was at its highest or lowest density, most people, including many dust mite researchers, would reply: 'When at its maximum, of course – the more mites you kill, the better!' Intuitively, this answer seems to be rational, practical and correct, but it is flawed because it neglects the fact that dust mites are living, reproducing organisms and their populations tend to increase exponentially. The counter-intuitive approach is that control is more likely to be effective when population density is low, but most effective for mites with populations that fluctuate seasonally, at the time of year just before exponential growth. A female *Dermatophagoides* may lay 30 or more eggs in a month, of which 80% survive to adulthood. Half of these will become females (assuming a 1:1 sex ratio) so by the

end of the first month she produces 12 new females (that is 80% of 30 ÷ 2), each capable of generating 12 females. By the end of the second month there would be 144 females. At the end of the third, 1728, and after 6 months, 35.8 million. If a control method achieves a 90% kill of a month-old population there would be one female left. In a 6-month-old population there would be 3.6 million, capable of producing 17.2 million females by the end of the following month. Although the control method seemed to be highly effective at first sight, when put into practice it was a catastrophic failure because the capacity for the dust mite population to regenerate was not taken into account.

This example is both extreme and simplistic (it does not factor in deaths – *Dermatophagoides* spp. only live for about 6 weeks). In nature, a single female dust mite does not give rise to such enormous populations after so few months. Nor do populations continue to grow exponentially. But it serves to emphasise that it is vital to be able to understand the linkage between dust mite control and how populations work in order for dust mite control to be successful. Hitherto, population dynamics have been ignored by most researchers involved in dust mite control. Considering that the control of dust mites effects drastic change on birth and death rates – two of the most basic population parameters – such an omission is remarkable. Furthermore, seasonal fluctuation in mite population density is the underlying cause of seasonal increases in allergen exposure suffered by people with dust mite hypersensitivity. Population dynamics impact directly on allergen exposure, epidemiology and clinical management.

The following sections deal with life tables, the intrinsic rate of natural increase and age structure. These are all elements of classical demographic theory, as developed by Lotka (1922) and extended by Birch (1948): so-called 'stable theory', because it deals with demography of populations that have stable age distributions. I have drawn mainly upon Birch (1948), Andrewartha and Birch (1954), Carey (1993, 2001) and Krebs (2001) as a basis for these sections. The chapter on demography by Stearns (1992) is also useful. Similar accounts can be found in most ecology textbooks.

Carey and Krainacker (1988) point out that many of the concepts and techniques of classical demography are particularly appropriate for the study of mite populations because mites have simple life cycles with overlapping generations, their ages can be estimated from their morphologically unique life cycle stages, they are easy to rear and their populations grow rapidly. Krebs (2001) observed that these demographic techniques can be used to investigate why organisms evolve one type of life cycle rather than another, and Stearns (1992) referred to demography as the key to life history theory. A great deal of the present book is a tale of eight domestic mite species belonging to three major taxa: *Dermatophagoides farinae*, *D. pteronyssinus* and *Euroglyphus maynei* (the Pyroglyphidae), *Lepidoglyphus destructor*, *Glycyphagus domesticus* and *Blomia tropicalis* (the Glycyphagoidea), *Acarus siro* and *Tyrophagus putrescentiae* (the Acaridae). The comparative demography of these mites allows us to better understand how their different life histories have evolved, as well as their patterns of distribution and abundance.

Populations are complex entities. They change in size as a result of many interacting variables, but the basic parameters that are responsible for changes in population growth centre around the balance between the number of individuals being added to the population (that is being born or migrating into it) and the number leaving the population (that is dying or emigrating). For domestic mites, most of the available data on birth and death rates comes from laboratory studies (shown in Tables 5.1, 5.2). The vast majority of these deal with the effects of two variables – temperature and humidity. A few studies deal with the effects of different diets.

Life history parameters of domestic mites are so plastic over a large range of temperatures, humidities and diets that the laboratory approach has been regarded as the only way of determining them with any degree of accuracy. Typically, this is done by rearing mites at constant conditions. The only variable likely to be considered other than humidity and temperature is diet. Of course in human dwellings conditions of temperature and humidity are anything but constant, and vary diurnally, seasonally and spatially, so laboratory-derived data is quite artificial. For those species that occur in stored products, temperatures and humidities in warehouses and granaries are kept reasonably constant and the data is more applicable. Many of the laboratory studies on population parameters were done in order to determine the physical conditions of storage required to inhibit growth of populations of stored products mites (Solomon, 1962; Cunnington, 1965, 1985) because manipulation of temperature is one of the more straightforward methods of control.

Although there are lots of studies on seasonal fluctuations in population size of dust mites in their natural habitats (see below), they are almost entirely descriptive. There are no published studies that have collected empirical data on seasonal fluctuations in population density and modelled them using demographic techniques. Hence, laboratory-derived data on population growth at constant conditions is virtually all the information that is available for demographic analyses. However, this kind of data is especially useful for comparing population performance of different species.

5.4 Demography

Demography is the mathematics of populations – the statistics of births and deaths. The size of a population (X) at any given time can be expressed as:

$$X = (I + B) - (E + D)$$

where B is the total number of live births, I is the total number of individuals migrating into that population, D is the total number of deaths and E is the number of emigrants.

5.4.1 Migration

For dust mite populations, there is virtually nothing known about the migrant component. Unlike, say, waterbirds on wetlands, where several populations may exist with free migration between them, dust mite populations are geographically isolated from each other. There is almost no continuum of habitable space for dust mites between one home and its neighbours. Periodic events such as bringing in a second-hand sofa are likely to have an effect on the size of the mite population, but otherwise there is little intermingling of populations from one home with those of another. Two situations where that may prove to be an exception are the presence of feral dust mite populations in nests of birds and the presence of mites in clothing. Dust mites are known to inhabit nests of anthropophilic birds such as swallows, swifts, martins, sparrows and pigeons; indeed it is from this source that it has been hypothesised that dust mite populations in homes originated (discussed in Chapter 1). Nests within attics or under eaves could conceivably provide a source of immigrant mites, although this has not been investigated. The evidence that there may be immigration of mites on clothing of visitors to the home is more plausible because of the densities

Table 5.1 Publications containing life history data on domestic mites, based on laboratory studies.

Spp. & reference	RH% range	T°C range	Egg duration (days)	Immatures (days)	Adult female	Total longevity	Pre-oviposition	Oviposition	Mean fecundity	Eggs/female/day	Egg mortality (%)	Juvenile mortality
Dermatophagoides pteronyssinus												
Pike *et al.*, 2005	75	23	+	+	+	+		+	+	+	+	+
Arlian *et al.*, 1990	75–80	16–35	+	+	+	+	+	+	+	+	+	
Matsumoto *et al.*, 1986	61–86	25	+	+	+	+	+	+	+	+	+	+
Blythe, 1976	75–80	25		+								
Colloff, 1987b	75–80	25	+									
Dobson, 1979	60–100	20–35	+	+	+	+						
Gamal-Eddin *et al.*, 1983a–d	75	15–40	+	+	+	+		+	+	+	+	+
Hart & Fain, 1988			+				+	+	+	+		
Ho & Nadchatram, 1984	75	23–37	+	+	+	+					+	+
Saleh *et al.*, 1991	75	25	+	+	+	+	+	+	+	+	+	
Spieksma, 1967	75–80	25	+	+	+	+		+	+	+		
Dermatophagoides farinae												
Arlian & Dippold, 1996	75	16–35	+	+	+	+		+	+	+		+
Matsumoto *et al.*, 1986	61–86	25	+	+	+	+	+	+	+	+	+	+
Furumizo, 1975a	75	16–32	+	+	+	+		+	+	+	+	
Hart & Fain, 1988	75	25	+				+	+	+	+		
Gamal-Eddin *et al.*, 1983a–d	75	15–40	+	+	+	+	+	+	+	+	+	+
Euroglyphus maynei												
Taylor, 1975	60–80	25–30	+	+	+	+	+	+	+	+	+	
Colloff, 1992a	60–80	25–30	+	+	+	+	+	+	+	+	+	+
Hart & Fain, 1988	75	25	+				+	+	+	+		
Tyrophagus putrescentiae												
Boczek, 1974	85	20					+	+	+			
Czajkowska & Kropczynska, 1991	85	25	+									+
Eraky, 1995a, b	80–90	18–26	+	+	+	+	+	+	+	+	+	+
Hart, 1990	75–100	12.5	+	+			+	+	+	+	+	+
Liu *et al.*, 2006	73–87	13–30	+	+			+					
Rivard, 1961a, b	70–100	20–30	+	+	+	+	+	+	+	+	+	+
Sanchez-Ramos & Castanera, 2001, 2005	90	19–34	+	+	+	+	+	+	+	+	+	+
Acarus siro												
Davis & Brown, 1969	70, 90	15	+	+	+	+	+	+	+	+	+	+
Chmielewski, 1995	85	20	+	+	+	+	+	+	+	+	+	+
Cunnington, 1985	65–90	5–30	+		+			+	+	+		
Davis & Brown, 1969	70–90	15–20	+	+								
Emekçi & Toros, 1989	70–90	10–25	+	+	+	+	+	+	+	+	+	
Fejt & Zdarkova, 2001	70–90	18–20	+	+	+	+		+	+	+		
Glycyphagus domesticus												
Barker, 1968	70–100	12–24	+	+			+	+	+		+	+
Chmielewski, 1988	65–95	5–25	+	+	+	+	+	+	+	+	+	+
Hart, 1990	75	25	+				+	+	+	+		
Hora, 1934	90	25	+	+								
Lepidoglyphus destructor												
Barker, 1983	75	14–25	+	+			+	+	+	+	+	+
Chmielewski, 1987	65–95	5–25	+	+	+	+	+	+	+	+	+	+
Blomia tropicalis												
Mariana *et al.*, 1996	75	25	+	+	+		+	+	+	+	+	+

Table 5.2 Comparison of some basic life history parameters of domestic mites at or near optimum temperature and humidity for population growth.

Spp.	Optimum temperature	Optimum RH%	Egg duration (days)	Immatures (days)	Adult female	Pre-oviposition period (days)	Oviposition period (days)	Mean fecundity (no. eggs)	Eggs/female/day	Egg mortality (%)	Total juvenile mortality (%)	Reference
Pyroglyphidae												
Dermatopha-goides farinae	27	75	7.1	15.8	24.2	ND	19.8	50	2.5	14	ND	Furumizo, 1975a
D. pteronyssinus	23	75	8.1	25.9	31.2	4.3	23.3	68	2.8	14	ND	Arlian *et al.*, 1990
Euroglyphus maynei	25	75	5.0	21.0	19.0	3.1	13.8	15	1.1	17	61	Taylor, 1975
Glycyphagoidea												
Glycyphagus domesticus	25	85	4.4	17.4	12.6	3.1	9.2	26	2.8	68	68	Chmielewski, 1988
Lepidoglyphus destructor	25	85	4.1	12.2	17.5	3.2	28.8	141	4.9	49	69	Chmielewski, 1987
Blomia tropicalis	25	75	5.7	14.0	57.5	2.8	16.5	28	1.7	19	40	Mariana *et al.*, 1996
Acaroidea												
Acarus siro	25	90	4.3	4.6	25.5	1.1	18.3	315	17.2	14	ND	Emekçi & Toros, 1989
Tyrophagus putrescentiae	25	90	3.8	5.6	40.0	1.6	21.5	502	24.0	1	5	Sanchez-Ramos & Castanera, 2001, 2005

involved. Bischoff *et al.* (1992a) found between 800 and 4000 mites per garment on pullovers, sweatshirts and trousers.

In any event, the direction and magnitude of migration events have not been quantified, so the basis of dust mite population dynamics currently rests solely on the analysis and quantification of births and deaths.

5.4.2 Mortality and natality

a Mortality and the life table

Mortality in the demographic sense means the rate of death per unit time of a group of organisms, the cohort. Mortality may be highest among young individuals or older ones, or it may be constant throughout life. The rates of mortality are expressed as a series of terms, presented in a life table, which summarises the pattern of mortality of the cohort. Table 5.3 is the life table for *Euroglyphus maynei*, then smallest of our species in the water-loss versus size-of-females example

(see Figure 3.6), and *Tyrophagus putrescentiae*, the largest. The temperature and humidity combinations used for comparison are at, or close to, those representing optima for population growth. We do not have to set up a table with every day of the life listed as a row. If we know the duration and the mortality of each stage in the life cycle – the type of data available from the literature (Table 5.1) – we can group rows according to stages. This is called an abridged life table. It is also a single-decrement table, meaning death is treated as a single category and not split into multiple decrements based on cause of death. The components and how they are calculated in Table 5.3 are as follows (see Carey, 1993, p. 19; Carey, 2001, his Table 1):

Column A in Table 5.3 is the stage in the life cycle; n is the average duration of the stage, in days (column B);
x is the age interval of each stage (column C), e.g. the larva of *E. maynei*, which lasts 7 days,

Table 5.3 Abridged cohort single-decrement life table based on mean development times and survivorship of *Euroglyphus maynei* at 25°C, 75% RH (from Colloff, 1992a, based on data of Taylor, 1975) and *Tyrophagus putrescentiae* at 25°C, 90% RH (data from Rivard, 1961a, b).

Row	Column A	B — Duration in days, n	C — Age interval, x to $x + n$ (days)	D — Number alive, N_x	E — Proportion surviving, l_x	F — Period survival, p_x	G — Period mortality, q_x	H — Frequency of deaths, d_x	I — Proportion of days lived in the age interval, L_x	J — Total number of days lived beyond age x, T_x	K — Expectation of life (days), e_x
	Euroglyphus maynei										
1	Egg	5	0–5	85	1.000	0.753	0.247	0.247	4.382	18.587	18.587
2	Larva	7	5–12	64	0.753	0.688	0.313	0.235	4.447	14.205	18.866
3	Protonymph	6	12–18	44	0.518	0.795	0.205	0.106	2.788	9.758	18.850
4	Tritonymph	8	18–26	35	0.412	0.743	0.257	0.106	2.871	6.969	16.926
5	Pre-reproductive period	3	26–29	26	0.306	0.923	0.077	0.024	0.882	4.099	13.400
6	Reproductive period	14	29–43	24	0.282	0.549	0.451	0.127	3.061	3.216	11.392
7	Post-reproductive period	2	43–45	13	0.155	0.000	1.000	0.155	0.155	0.155	1.000
	Final value:	45			0.000			1.000	18.587		
	Tyrophagus putrescentiae										
8	Egg	6	0–6	200	1.000	0.670	0.330	0.330	5.010	10.900	10.900
9	Larva	3	6–9	134	0.670	0.843	0.157	0.105	1.853	5.890	8.791
10	Protonymph	2	9–11	113	0.565	1.000	0.000	0.000	1.130	4.038	7.146
11	Tritonymph	2	11–13	113	0.565	0.903	0.097	0.055	1.075	2.908	5.146
12	Pre-reproductive period	2	13–15	102	0.510	0.108	0.892	0.455	0.565	1.833	3.593
13	Reproductive period	24	15–39	11	0.055	0.818	0.182	0.010	1.200	1.268	23.045
14	Post-reproductive period	3	39–42	9	0.045	0.000	1.000	0.045	0.068	0.068	1.500
	Final value:	42			0			1.000	10.900		

starts being a larva at the end of day $x = 5$ and finishes at the end of day $x + n = 12$;

N_x is the number alive at age x, the beginning of each age interval, based on the size of the cohort (column D);

l_x is the proportion of the cohort still alive at age x (column E); cell D4 divided by D2 = 44/85 = 0.518 (cell E4):

$$l_x = \frac{N_x}{N_0}$$

p_x is the proportion of the cohort still alive at age x (column F) that survive through the period $x + n$; E4/E3 = F3 = 0.688:

$$p_x = \frac{l_{x+n}}{l_x}$$

q_x is the per capita rate of mortality within the period x to $x + n$ (column G); 1 – F3 = 0.205 (G4):

$$q_x = 1 - \frac{l_{x+n}}{l_x}$$

d_x is the proportion of the original cohort that die in the period x to $x + n$ (column H); E3 − E4 = 0.106 (H4):

$$d_x = l_x - l_{x+n}$$

L_x is the per capita proportion of life lived in the period x to $x + n$, assuming that individuals that die in that period do so at its mid-point (Column H); B4 × (E4 − (1/2 × H4)) = 2.788 (I4):

$$L_x = \frac{1}{2}(l_x + l_{x+n})$$

T_x is the total number of days lived beyond age x (Column J); I4 + I5 + I6 + I7 + I8 = 13.176 (J4) and I5 + I6 + I7 + I8 = 10.388 (J5):

$$T_x = \sum_{y=x}^{\omega} L_x$$

e_x is the expected number of additional days an individual aged x will live (column K); J4/E4 = 25.454 (K4):

$$e_x = \frac{1}{2} + \frac{l_{x+1} + l_{x+2} + \cdots + l_{\omega}}{l_x}$$

Table 5.3 provides some useful points of comparison between the two species. Both have highest mortality as immatures (see Figure 5.5) – only a third of all newborns of *E. maynei* survive to adulthood, compared with half of those of *T. putrescentiae* – but mortality during the reproductive period exacts a massive toll on *T. putrescentiae* with over 80% of the females dying during this period, compared with only half of the females of *E. maynei*. The highest probability of death for *T. putrescentiae* occurs during the egg and larval stages, whereas for *E. maynei* it is as a larva or a tritonymph.

b Natality, net reproductive rate and generation time

Natality, or fertility, is the birth rate, the number of eggs produced per female per unit time. By including natality data in the life table (which now becomes a life and fertility table) we can derive the balance between deaths and births. Only the reproductive period needs to be included. The relevant statistic is m_x, the number of female eggs produced per surviving adult female per day at time x. It is half the total number of eggs laid per female per day, assuming an equal sex ratio. By multiplying l_x by m_x and summing the product we get the total number of female offspring for the cohort. This is the net reproductive rate, R_0, which is the multiplication rate per generation (i.e. the period between the birth of parents and the birth of their offspring).

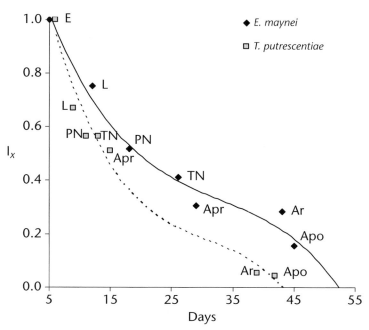

Figure 5.5 Survivorship curves of *Euroglyphus maynei* at 25°C, 75% RH and *Tyrophagus putrescentiae* at 25°C, 90% RH, based on data in Table 5.3. E = egg; L = larva; PN = protonymph; TN = tritonymph; Apr = adult, pre-reproductive period; Ar = adult, reproductive period; Apo = adult, post-reproductive period.

Table 5.4 is the life and fertility table for *E. maynei* at 25°C, 75% RH, and $R_0 = 2.298$, so the population of *Euroglyphus maynei* at 25°C, 75% RH will multiply 2.298 times in each generation. R_0 is not a particularly useful comparative statistic on its own because one also needs to know the generation time for the population concerned. Generation time (T) is the average age of parenthood or, put another way, the average period between birth of parents and birth of offspring and can be estimated approximately from

$$T = \frac{\Sigma x l_x m_x}{R_0}$$

For the data in Table 5.4,

$$T = \frac{75.84}{2.298} = 33 \text{ days}$$

So at 25°C, 75% RH the population of *Euroglyphus maynei* will multiply 2.3 times in 33 days.

5.4.3 The intrinsic and finite rates of natural increase

For the calculation of *r* we need to know the duration of the immature stages and their survival. Ideally we also need the life and fertility table of the adult female. Putting together such a table is not a trivial task. Davis and Brown (1969) derived empirical measurements of adult survival and age-specific fertility of *Acarus siro* which required making lots of rigorous observations every two days for up to five weeks. Are there any short cuts here? Can we use any estimates based on total adult mortality and fertility?

It seems obvious from the life and fertility data in Table 5.4 that in order to calculate the rate of increase in the population we need to know the age distribution (age structure), the age-specific mortality rates (l_x) and age-specific fertility rates (m_x) because the survival and natality vary so much with age. For example, all the females that have started laying eggs within the first two days contribute over a third of the total value of $l_x m_x$, whereas those in the last two days of the reproductive period contribute less than a tenth. In fact, we do not need to know the age structure of the population because if a population is subject to a constant schedule of mortality and natality rates, Lotka (1922) showed it will approach a stable age distribution and increase according to the equation

$$dN/dt = rN, \text{ or } N_t = N_0 e^{rt}$$

Table 5.4 Cohort life and fertility table based on mean development times, survivorship and fecundity of *Euroglyphus maynei* at 25°C, 75% RH, used to calculate value of *r* (from Colloff (1992a) based on data of Taylor (1975)).

Age in days (x)	Proportion surviving (l_x)	No. ♀ offspring per female aged x per day (m_x)	$l_x m_x$	$x l_x m_x$	$e^{-0.0253x}$	$e^{-0.0253x} l_x m_x$
29	0.28	1.65	0.462	13.398	0.4801	0.2218
30	0.27	1.35	0.365	10.935	0.4681	0.1706
31	0.26	1.00	0.260	8.060	0.4564	0.1187
32	0.25	0.75	0.188	6.000	0.4450	0.0834
33	0.24	0.70	0.168	5.544	0.4339	0.0729
34	0.24	0.60	0.144	4.896	0.4231	0.0609
35	0.23	0.60	0.138	4.830	0.4125	0.0569
36	0.22	0.50	0.110	3.960	0.4022	0.0442
37	0.21	0.45	0.095	3.497	0.3922	0.0371
38	0.20	0.45	0.090	3.420	0.3824	0.0344
39	0.19	0.40	0.076	2.964	0.3728	0.0283
40	0.18	0.40	0.072	2.880	0.3635	0.0262
41	0.18	0.40	0.072	2.952	0.3544	0.0255
42	0.17	0.35	0.060	2.499	0.3456	0.0206
Final value:		9.60	R_0 2.298	75.835		1.0016

and Brown (1969), they are really quite similar, despite the differences in microclimate and vastly greater growth rates of *A. siro*. At 15°C and 70% RH there are 41% eggs, 46% immatures and 13% adults, and at 15°C and 90% RH there are 51% eggs, 38% immatures and 11% adults.

In a laboratory culture, or in natural populations, it is much more practical to estimate stage structure than age structure. If the average age of each stage is known, then Table 5.5 can be constructed using age intervals that correspond to stages. By classifying individuals by stage rather than age, Lefkovitch (1965) used matrix models to make population projections. Caswell (2001, Chapter 4) gives details on the construction of stage-classified matrix models.

Carey (1982) made the point that stable stage structure of spider mites provides meaningful information on field populations and that field populations are often at, or close to, the stable stage distribution because they are able to increase rapidly and stable stage is approached quickly. For spider mites, stable stage distribution was approximately 66% eggs, 26% immatures and 8% adults. Field data on stable stage distribution of *Dermatophagoides pteronyssinus* showed the most frequent pattern found was immatures dominant, then eggs, then adults; similar to the pattern for *E. maynei* in Figure 5.6a, indicating that *D. pteronyssinus* also tends to be at stable stage distribution in the field (Colloff, 1992a; Figure 5.6b).

5.5 Life history traits and demographic parameters

5.5.1 Age and size at maturity – effects on fecundity and mortality

Life cycles consist of pre-reproductive and reproductive periods. The relative duration of these periods can have a major effect on fitness, because the pressures and trade-offs that operate on immature individuals are different from those for adults. For example, during the life cycle mortality rates tend to be highest for immature mites and they may be more susceptible to dehydration than adults. Intuitively, there should be clear benefits to becoming sexually mature early in the life cycle. Less time spent as juveniles means lower likelihood of mortality and a higher likelihood of reaching sexual maturity and achieving successful reproduction. Individuals that mature early will produce more offspring sooner. The cost of early maturity may be that individuals are smaller than if they matured later, and smaller individuals may be less successful at mating, produce fewer offspring, or the offspring may have a lower chance of survival.

In fact, the relationship between age at maturity and juvenile mortality is not quite as straightforward as this. Although there is a clear positive correlation for *D. farinae* and *D. pteronyssinus* (mites that mature early have lowest mortality; see Figure 5.7c), with the acarid *Tyrophagus putrescentiae* there is a rather loose negative correlation (mites that mature early tend to have higher mortality than ones that mature late; see Figure 5.7a). With the glycyphagids *Glycyphagus domesticus* and *Lepidoglyphus destructor*, there is no relationship at all. Average juvenile mortality is uniformly high (65–80%), regardless of the age at maturity (see Figure 5.7b). The same pattern was found for egg mortality and age at maturity as for juvenile mortality (data not shown). The negative relationship, or lack of relationship, between juvenile and egg mortality and development time in acarids and glycyphagids makes sense for species that live in field habitats (OConnor, 1982a; Table 4.2), where they will encounter a wide range of temperatures and may have to endure cold for extended periods. In general, the lower the temperature the longer the development times. If mortality was highest when age at maturity was greatest, the capacity for populations to survive low temperatures may be less than if mortality were negatively correlated with development time, or independent.

The other side of the mortality story is the relationship between development time and fecundity. In this case, the pattern is the opposite of the mortality one. *T. putrescentiae* shows no relationship (see Figure 5.7d), whereas for *G. domesticus* and *L. destructor* there is a significant negative correlation: mites that mature early have higher fecundity than ones that mature late (shown in Figure 5.7e), as is the case with the two pyroglyphid species (see Figure 5.7f).

Large parts of the juvenile life cycle of astigmatid mites are spent quiescent, waiting to moult (see Figure 5.8). The egg stage is inactive, so are the pre-moult stages of the larva and the two nymphal stages. The duration of the life cycle spent inactive, at any given temperature and humidity, can have some consequences for population dynamics, partly because the risk of mortality is much lower for quiescent than active mites because the main predator, cheyletid mites, are ambush assailants that remain static and attack in response to movement (Wharton and Arlian, 1972b). Also quiescent immatures have

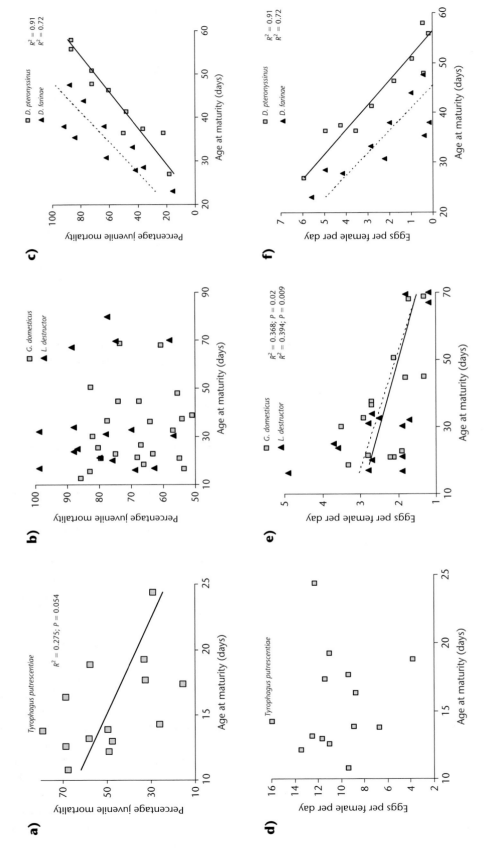

Figure 5.7 Relationship between age at maturity (total mean longevity of all the juvenile stages) and total mean juvenile mortality and fecundity respectively for **a)** and **d)** *Tyrophagus putrescentiae* (data from Rivard et al., 1961a, b); **b)** and **e)** *Glycyphagus domesticus* (data from Chmielewski, 1988); and *Lepidoglyphus destructor* (data from Chmielewski, 1987); **c)** and **f)** *Dermatophagoides farinae* and *D. pteronyssinus* (data from Gamal-Eddin et al., 1983a–d), over a range of temperature and humidity combinations.

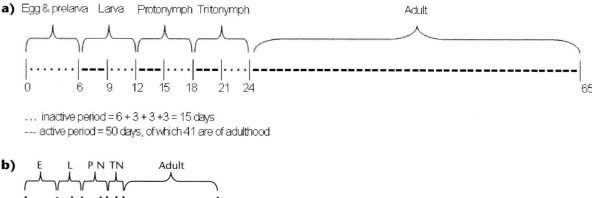

... inactive period = 6 + 3 + 3 +3 = 15 days
--- active period = 50 days, of which 41 are of adulthood

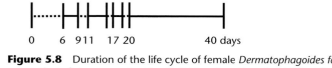

Figure 5.8 Duration of the life cycle of female *Dermatophagoides farinae* at different temperatures and 75% RH, showing proportion of stages spent active (solid lines) and inactive (dotted lines). E = egg; L = larva; PN = Protonymph; TN = Tritonymph. **a)** Mean duration at 21°C; inactive period = 14 + 5 + 5 + 5 = 29 days; **b)** mean duration at 32°C; inactive period = 6 + 2 + 1 + 1 = 10 days. (Data from Furumizo, 1975a.)

behavioural and physiological adaptations to limit water loss and survive desiccation. Figure 5.8 shows the mean duration of each stage of *Dermatophagoides farinae* at 21°C and at 32°C. A quarter of the life cycle is spent inactive, regardless of temperature, but at 32°C the inactive period is three times shorter than it is at 21°C. The period spent immotile has implications for the estimation of population density using the 'Heat Escape Method' (Bischoff and Fischer, 1990; Chapter 6), which depends on the mites being active. If a substantial proportion of the population is immotile, then trapping will be less efficient than it would be otherwise, and estimates of population size are likely to be biased.

Age at maturity (development time) in relation to rearing temperature follows a pattern similar to that for oviposition water loss (Figure 5.9; see also Chapter 3). The mites with the largest females (Acaridae and Glycyphagidae) are fastest to mature and the smallest (Pyroglyphidae) are the slowest. Figure 5.10 shows the inverse power relationship between size of females and age at maturity. Adult female size and population doubling time at comparable temperatures and humidities follow an inverse exponential relationship (Figure 5.11), with the largest mites having the shortest doubling time, the smallest mites the longest, and the medium-sized mites in between. The relationship between female size and finite rate of increase is a positive linear one – big mites have the fastest rate of increase. This is the same pattern as for fecundity and development time (Table 3.3; Figure 5.9) and water loss during oviposition.

5.5.2 Differences in life history traits between dust mites

In summary, big mites lay more eggs and develop more rapidly than small mites. I suggest they are able to do so because they are less constrained by water loss. For *Acarus*, *Glycyphagus* and *Lepidoglyphus* when humidity falls to desiccating conditions, or food resources deteriorate, these mites are able to produce a surviving and dispersing stage, the hypopus. Although production of hypopi by *Tyrophagus* spp. is rare and is known from only one species, members of the genus are extremely fecund, producing between 100 and 700 eggs (Fan and Zhang, 2007). It is the egg that probably represents the survival stage (see section 5.2.1 above), so *Tyrophagus putrescentiae* does not need to produce hypopi. The eggs may also be dispersive – they are smaller than the hypopi of glycyphagids which are thought to be dispersed on air currents. Of the Pyroglyphidae, which do not produce a hypopus, *Dermatophagoides farinae* at least is able to tolerate low humidity as a quiescent protonymph (see section 3.3.4). Furthermore, pyroglyphids tend to have a longer adult life span than the Acaridae and Glycyphagidae, and are better able to withstand low humidity as adults, recovering from desiccation after relatively short exposure to favourable conditions.

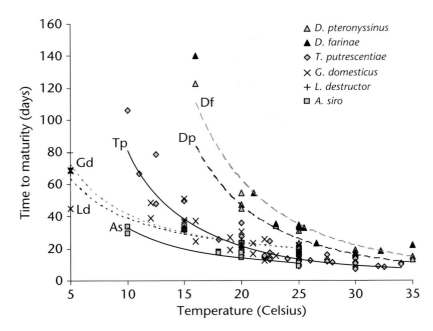

Figure 5.9 Relationship between development time (duration of period, egg to adult) and temperature for six species of domestic mites. Lines of best fit are described by the following power equations: *D. pteronyssinus* (Dp): $y = 114512x^{-2.61}$, $R^2 = 0.782$; *D. farinae* (Df): $y = 141661x^{-2.581}$, $R^2 = 0.888$; *T. putrescentiae* (Tp): $y = 6427x^{-1.902}$, $R^2 = 0.853$; *G. domesticus* (Gd): $y = 201.62x^{-0.717}$, $R^2 = 0.6947$; *L. destructor* (Ld): $y = 267.02x^{-0.81}$, $R^2 = 0.909$; *Acarus siro* (As): $y = 526x^{-1.202}$, $R^2 = 0.925$.

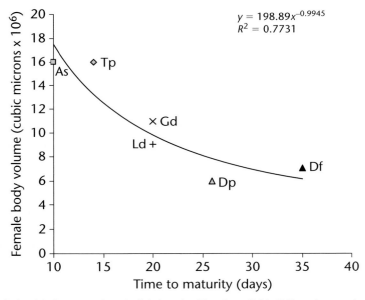

Figure 5.10 The relationship between size of adult females (data from Table 3.3) and mean development time (at 25°C) of six species of domestic mites. Note the clustering of the Acaridae (*A. siro, T. putrescentiae*), Glycyphagidae (*L. destructor, G. domesticus*) and Pyroglyphidae (*Dermatophagoides* spp.). (Legend as for Figure 5.9.)

5.6 Factors affecting population dynamics

This section deals with comparative demography of domestic mites, mostly based on data from laboratory studies at constant conditions of temperature and humidity.

5.6.1 Temperature and humidity

a Effects on stages in the life cycle

We have already seen the effects of temperature on the development time of different species (Figures 5.8, 5.9). The most comprehensive studies of the effects

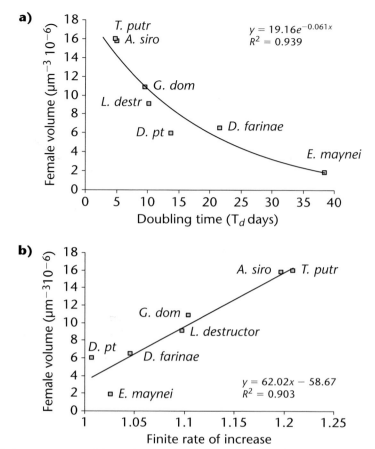

Figure 5.11 Population parameters of domestic mites in relation to size (female volume, Table 3.3). **a)** Doubling time (T_d); **b)** finite rate of increase (λ).

of temperature and humidity on the demographic parameters of dust mites are by Gamal-Eddin *et al.* (1983a–d) and Matsumoto *et al.* (1986) for *D. farinae* and *D. pteronyssinus*. These observations were made at 50–95% RH and 25°C and at 15–40°C and 75% RH, and included six variables relating to longevity, mortality and fecundity. Given so many variables, a single parameter would be of value in order to make sense of the data. R.C. Fisher (1938) developed an Index of Suitability (*I*) (not to be confused with Fisher's Index developed by the statistician R.A. Fisher), where:

$$I = \frac{NV}{L + T}$$

and *N* = number of eggs laid, *V* = egg viability (i.e. the inverse of egg mortality), *L* = duration of the oviposition period and *T* = duration of the egg stage. This index relates to oviposition and hatching, and the most favourable conditions are those that produce the highest

values. It provides a utilitarian comparison of egg laying performance and survival and was used by Cunnington (1985) for determination of the physical limits for complete development of *Acarus siro* at 27 combinations of temperature and humidity. The Index of Suitability is useful because it does not require detailed mortality and longevity data for immatures and adults, which is often missing from laboratory studies on life history parameters. Figure 5.12 shows values of the Index of Suitability (*I*) for two pyroglyphid species, two acarids and two glycyphagids. The data used for the pyroglyphids by Gamal-Eddin *et al.* (1983a–d) was done at temperatures from 15–40°C, but only at 75% RH, and at 50–95% RH, but only at 25°C, rather than the full matrix of 60 combinations, so it is presented differently from the others. It shows the optimum for *D. farinae* is 75% RH, 22°C and for *D. pteronyssinus* is slightly higher at 80% RH, 25°C. Because *I* does not consider mortality data, it tends to be better at predicting the conditions at which populations will not grow, or grow comparatively

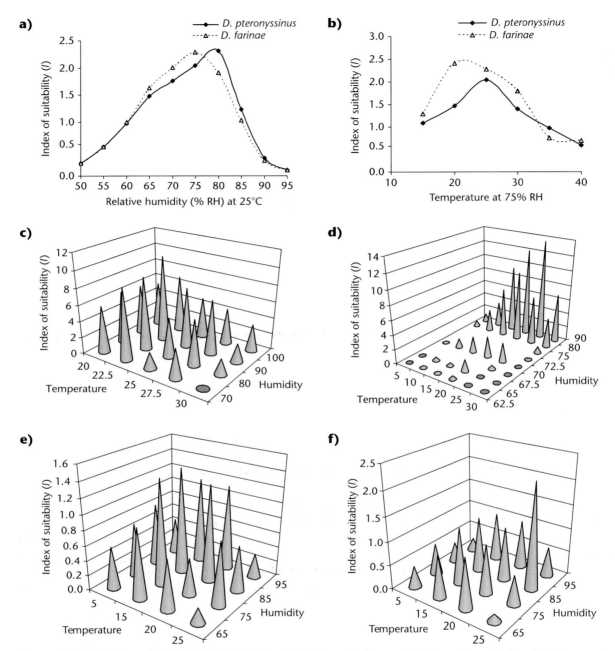

Figure 5.12 The Index of Suitability (*I*) of R.C. Fisher (1938) for egg laying and hatching of *Dermatophagoides farinae* and *D. pteronyssinus* at **a)** a range of humidities and constant temperature (25°C) and **b)** a range of temperatures and constant humidity (75% RH) (data from Gamal-Eddin *et al.*, 1983a–d); **c)** *Tyrophagus putrescentiae* (data from Rivard, 1961a, b); **d)** *Acarus siro* (data from Cunnington, 1985); **e)** *Glycyphagus domesticus* (data from Chmielewski, 1988); and **f)** *Lepidoglyphus destructor* (data from Chmielewski, 1987).

slowly, than the conditions at which growth is optimal. Of the optimum temperatures and humidities for the species in Table 5.6, the highest value of *I* corresponded with the temperature and humidity combination at which λ and T_d would predict most rapid population increase for *Lepidoglyphus destructor*, *Acarus siro* and the pyroglyphids, but not *Glycyphagus domesticus* or

Tyrophagus putrescentiae. Another suitability index based on immature development times and their mortality was developed by Howe (1971).

Teasing apart the effects of temperature and humidity is not straightforward, and there are many examples of the interaction between the two variables on life history parameters. Colloff (1987b) and Biddulph *et al.* (2007)

Table 5.6 Temperature and humidity optima for population growth of domestic mites based on various life history parameters.

Taxon	T°C	RH%	Reference	Index of suitability (I)	Finite rate of increase (κ)	Doubling time (T_d)
Pyroglyphidae						
Euroglyphus maynei	25	75	This chapter	0.6	1.026	39.2
Dermatophagoides farinae	26.5	75	Furumizo, 1975	1.7	1.051	21.6
D. pteronyssinus	25	75	Matsumoto et al., 1986	2.1	1.041	24.3
Glycyphagidae						
Glycyphagus domesticus	25	85	Chmielewski, 1988	0.6	1.123	8.1
Lepidoglyphus destructor	25	85	Chmielewski, 1987	2.2	1.134	7.5
Acaridae						
Acarus siro	28	85	Aspaly et al., 2007	13.6	1.2	5.1
Tyrophagus putrescentiae	30	90	Sanchez-Ramos & Castanera, 2005	10.4	1.9	1.1

provide examples of the interaction for *D. pteronyssinus*. The publication by Stratil *et al.* (1980) on the innate capacity for increase of *L. destructor* represents probably the most comprehensive demographic dataset on the effects of temperature and humidity. The effect of humidity (60–80%) on survivorship (l_x) of *E. maynei* (see Figure 5.13) shows rates of mortality decrease with increasing humidity, especially among immatures, so that at 80% RH mortality is almost constant throughout life, but that low humidities prolong adult life. In essence, temperature is the major driver of population growth because of the thermodynamics of physiological reactions of poikilothermic animals (Frazier *et al.*, 2006). Growth will be slower at lower temperatures, even if humidity is optimal. But humidity acts as a constraining factor on population growth at optimum temperatures (see effects of fluctuating conditions below), especially if an animal is having to expend energy on balancing its body water at the expense of reproduction.

b Effects of fluctuating conditions

Fluctuating conditions of humidity at a given temperature have the effect of prolonging development times (de Boer *et al.*, 1998; Arlian *et al.*, 1998b, 1999b; Pike *et al.*, 2005; see Figure 5.14a). Each stage in the life cycle of *D. farinae* was extended, with the adult and nymphal stages doubling or trebling their duration when exposed to 4 hours per day of 75% RH and 20 hours at 35% RH,

though the egg stage was affected least (Arlian *et al.*, 1999b). The most important finding of this study was that *D. farinae* can complete its life cycle, though over a longer time course, even though humidity is sub-optimal for the majority of the time. de Boer *et al.* (1998) found partially dehydrated *D. pteronyssinus* were able to re-hydrate and survive when moist conditions were provided for only 90 min. per day. Oviposition still occurred when dehydrating conditions prevailed most of the time, though fecundity rates were much reduced (shown in Figure 5.14b).

It seems that dust mite populations are capable of surviving cool, dry conditions for prolonged periods, with brief respite periods for rehydration, though de Boer *et al.* (1998) found one culture survived for two and a half months at constant conditions of 44% RH and 16°C, indicating that *D. pteronyssinus* is able to overwinter in an active state and recommence population growth when conditions become more favourable (de Boer and Kuller, 1995c, 1997). Active overwintering by *D. pteronyssinus* contrasts with *D. farinae*, which overwinter as quiescent protonymphs (Suto and Sakaki, 1990; Sakaki and Suto, 1991).

c Effects on laboratory populations

Populations of *D. farinae* in laboratory cultures grew most rapidly to highest density at 25°C and 60% RH (Waki and Matsumoto, 1973a; see Figure 5.15a). At

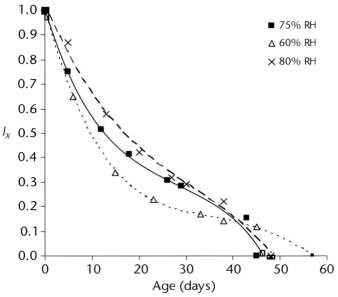

Figure 5.13 Effects of different relative humidities at 25°C on stage-specific survivorship of *Euroglyphus maynei* (from Colloff, 1992a, based on data from Taylor, 1975).

the same temperature and 57% RH, population growth during the exponential phase was similar, but densities tended to be lower, with a higher frequency of fluctuations, and at 75% density was lowest. At 30°C and both 57% and 60% RH, population densities fluctuated at markedly lower levels than at 25°C. The cultures were quite large. They were maintained in conical flasks with 50 g of culture medium, and some would have contained tens or hundreds of thousands of mites. The pattern of growth in these cultures showed a classic exponential growth phase followed by a period of chaotic fluctuation, though with an underlying mean and equilibrium. Equilibrium was reached in 8–10 weeks. When lines of best fit are added to the population growth data in Figure 5.15a (describing 3rd- or 4th-order polynomial models), it can be seen that they follow a basic sigmoid or logistic curve, indicating growth in a limited environment, rather than geometric or exponential growth typical of an unlimited environment (see Figure 5.15b). These patterns suggest that in laboratory cultures, within a relatively short period, dust mite populations exhibit characteristics typical of wild populations (see below).

5.6.2 Diet

Females of *Dermatophagoides pteronyssinus* fed on house dust and freeze-dried human semen produced twice as many eggs as mites fed on house dust alone

(Colloff, 1988b). Dried semen is common in mattresses of sexually active people and represents potentially high-quality food for mites, containing substantial amounts of protein, sugar, phospholipids and nucleic acids. However, what this study indicated was not so much the high nutritional quality of dried human semen, but the relatively low quality of the diet of house dust. Most of the laboratory studies listed in Table 5.1 that have been used to derive population growth rates have used high-quality diets rich in lipids, carbohydrates, protein and vitamins such as yeast, dried fish food, powdered bovine liver, dried milk and wheat germ. There are no studies that critically examine differences in population growth of pyroglyphid species on such high quality diets with something approaching their natural diet, though Spieksma (1976) successfully grew dust mites on animal skin scales.

Koren and Eckhard (1995) found that population growth of *D. pteronyssinus* increased with higher protein content of the house dust; the optimal level being at least 11%. Similarly, Matsumoto (1975) and de Saint Georges-Gridelet (1984) found that about 6% fat content of skin scales was optimal for population growth of *D. farinae* (Chapter 2). Furthermore, Koren and Eckhard (1995) found that dust from mattresses had significantly higher protein content (because of the high proportion of keratin-containing skin scales) and, when used as a laboratory

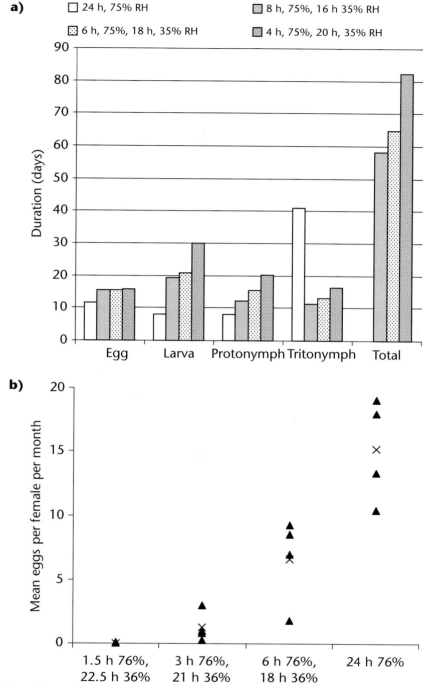

Figure 5.14 a) Effect of fluctuating conditions of relative humidity at 21°C on development times of *Dermatophagoides farinae* (data from Arlian *et al.*, 1999b); **b)** fecundity of *D. pteronyssinus* at fluctuating humidities at 16°C (× = mean; data from de Boer *et al.*, 1998).

culture medium at controlled temperature and humidity, supported higher mite populations than dust from carpets which had a low protein content. It seems therefore unlikely that all house dust is of equal nutritional quality and that within a home there is a fair amount of spatial variability. There has been a tendency to assume that food is never a limiting factor to dust mites under natural conditions, that it is always super-abundant and that dust mites never go short. These assumptions may well prove unfounded, and

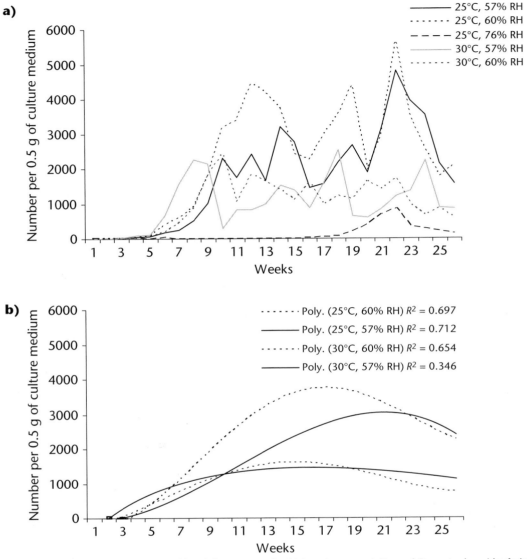

Figure 5.15 **a)** Effects of temperature and humidity on growth of laboratory populations of *Dermatophagoides farinae* reared on dried yeast and fish meal over a period of 31 weeks. No population growth was recorded at 25°C, 40% RH and 30°C, 40% and 75% RH; **b)** the same data (with the values omitted for the 25°C, 76% RH treatment) with lines of best fit to a 3rd- or 4th-order polynomial model, showing logistic growth. (Data from Waki and Matsumoto, 1973a.)

there could well be circumstances under which poor food quality represents a limiting factor for growth of dust mite populations.

Laboratory studies by Ree *et al.* (1997a) showed highest population growth rates by 8 weeks, for both *D. farinae* and *D. pteronyssinus*, on a diet of yeast and fish food, intermediate growth on fish food only, and lowest growth on yeast only (see Figure 5.16). Waki and Matsumoto (1973b) measured population growth of *D. farinae* on diets based on yeast and dried fish to which they added varying amounts and types of lipid and protein. Highest population

growth was achieved on diets containing 5% lipid in the form of oil or lard, or protein in the form of bean flour or egg albumin.

5.6.3 Predation, competition and crowding

a Predation

Cheyletus spp. are the only candidate predators for dust mites. These mites are used to control stored products mites in commercial premises (see Pekár and Žd'árková, 2004 for many references). Predator–prey interactions of *Cheyletus* spp. have been studied by Gause *et al.*

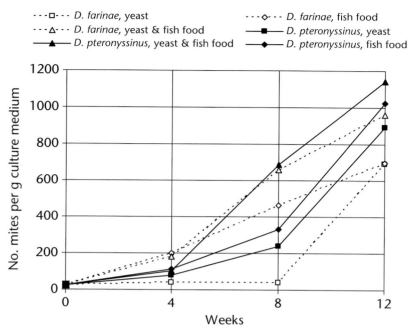

Figure 5.16 Effect of diet on population growth of dust mites. (Data from Ree *et al.*, 1997a.)

(1936), Bereen (1984) and Pekár and Žd'árková (2004). Life history parameters were determined by Bereen and Metwally (1984) and Barker (1991).

There is little information to gauge how important *Cheyletus* spp. might be in regulating dust mite populations under natural conditions, except in relation to some general statistics on their distribution and abundance. From the data on global distribution (see Appendix 3) *Cheyletus* spp. occur in about 124 localities (about 17% of the total). Abundance is usually low in relation to abundance of major dust mite species (as is typical of ratios in abundance between predators and their prey), and frequency of occurrence within homes at any particular locality is also well under 20%. So in the majority of homes, in most parts of the world, there is probably no regulatory effect on dust mite populations from predation.

b Competition

Dermatophagoides farinae and *D. pteronyssinus* are often found together in homes in North America and Europe, sometimes with *Euroglyphus maynei* or other species. Where multiple species are present, one is usually numerically dominant, and this has been ascribed to differences in microclimatic requirements; the classic example being the distribution of *D. farinae* and *D. pteronyssinus* in California (Lang and Mulla, 1977a, b, 1978; Chapter 4). Within 3 months *D. farinae* always became the dominant species in cultures that

commenced with equal numbers of both species, even though population growth of *D. pteronyssinus* is fastest (Arlian *et al.*, 1998a). This could be due to subtle differences in life history parameters (such as greater longevity of female *D. farinae*), or that 3 months was insufficient for *D. pteronyssinus* to become dominant. Or it could be due to *D. farinae* being able to outcompete *D. pteronyssinus* in some way. When *Euroglyphus maynei* was added to cultures, *D. pteronyssinus* and *D. farinae* showed slower growth rates than in single-species cultures, indicating an inhibiting effect of this species, even though *E. maynei* populations declined. There is lots of scope for clarifying these relationships further.

c Density dependent effects

Anyone who has grown dust mites in the laboratory will be familiar with the necessity to make subcultures on a regular basis in order to stop the cultures becoming too crowded and dying. The processes that regulate birth and death rates in relation to population size are called density dependent effects. They can be negative, whereby deaths exceed births and the population declines, or positive (so-called 'inverse density dependence') whereby the population is up-regulated. McGregor and Peterson (2000) detected inverse density dependence in cultures of mites that were newly inoculated into a 38–75 μm sieved fraction of medium (consisting mostly of faecal pellets from which eggs had been excluded) taken from a mature culture. Mite populations with higher proportions of faecal pellets in their culture medium grew significantly faster than those with lower proportions or the negative control, and this result was consistent over 10 replications. The mechanism responsible for this effect is not clear, though beneficial effects resulting from coprophagy is one possibility (see Chapter 2). This example serves to demonstrate that population growth can be regulated by modifications that dust mites make to their microhabitat. Extrapolating these results to natural populations suggests that growth will be faster in habitats that previously contained high populations that have died out than in habitats that did not. In natural populations, rapid population growth to high densities occurs in summer and autumn, followed by a crash in late autumn or winter and gradual increase the following late spring/early summer (see below). Could this growth phase be more rapid, all other factors being equal, if the previous population peak was particularly dense?

5.6.4 Differences between laboratory and natural populations

Most of the life history studies in Table 5.1 were done with cultures of mites that have been kept in the laboratory, at constant temperature and humidity on relatively high-quality food, sometimes for many years and countless generations of mites. A laboratory culture might have started with a few hundred or thousand founder individuals, forming the genetic basis for all subsequent generations. It is common practice for researchers in different parts of the world to exchange cultures among themselves and these may get incorporated into other lab cultures over time. By comparison, populations in the wild are subject to diurnal and seasonal fluctuations in microclimate, and food quality may vary in time and space. Wild populations have some limited opportunity for mixing and out-crossing, but considerably more so than for laboratory populations. With species that are globally distributed and exist as metapopulations, it is likely that a population of *Dermatophagoides pteronyssinus* from Belgium will be genotypically and phenotypically different from a population from Brazil. Genetic and phenotypic differences between laboratory and wild populations, and between wild populations from different parts of

the world, could potentially manifest themselves as differences in life history parameters and population growth rates, tolerance to extremes of temperature and humidity, and also allergen polymorphisms (discussed further in Chapter 7).

Colloff (1987a) exposed eggs from wild and laboratory populations to diurnal hygrothermic fluctuations to simulate those in a mattress (16 h. at 15°C, 60% RH, 8 h. at 30°C, 75% RH). Wild eggs had more rapid development times and lower mortality due to dehydration than eggs from laboratory cultures. The result suggests that wild eggs are better able to withstand these fluctuations because that is what they normally get exposed to, whereas laboratory cultures have become acclimatised to constant conditions. Colloff (1987b) found that at constant, favourable conditions (30°C, 80% RH) laboratory populations had shorter egg development times and lower egg mortality, but that at cooler, dryer conditions (20°C, 60% RH) wild populations did better. This finding was extended by Hart *et al.* (2007) who found that at favourable conditions (25°C, 75% RH) wild mites were less fecund, with a trend to longer egg development times and pre-oviposition period, than laboratory mites, but at less favourable conditions (25°C, 64% RH) wild mites had better reproductive performance than laboratory mites.

The message is that if life history data based on laboratory populations are used to parameterise population growth models, those models will give inaccurate predictions of growth of natural populations, as was suggested by Pekár and Žd'árková (2004) in relation to the effects of laboratory diet. Laboratory studies of life history parameters show a great deal of variation for the same species (*D. pteronyssinus*) kept at the same conditions (23–25°C, 72–76% RH; Table 5.7). Some parameters varied more than others, with oviposition rates and duration of the oviposition period being especially variable in relation to diet. The poorest performance was on wheat bran, the best was on diets containing powdered liver.

5.7 Seasonal dynamics of natural populations and allergens

5.7.1 Mites

Seasonal population dynamics of house dust mites are of importance because they are related to the amounts of allergens that are produced, and when allergen exposure is likely to be highest. This begs the

Table 5.7 Variation in life history parameters of *Dermatophagoides pteronyssinus* from laboratory cultures at 23–25°C, 72–80% RH. (Where stage duration is given based on eventual gender (such as egg duration; males, females) I have quoted data for females.)

Reference	RH%	T°C	Egg development (days)	Larva (days)	Protonymph (days)	Tritonymph (days)	Total immatures (days)	Adult female (days)	Longevity (instars & female, days)	Pre-oviposition period (days)	Oviposition period (days)	Mean total eggs per female	Oviposition per female per day	Egg mortality (%)	Juvenile mortality (%)	Diet
Arlian et al., 1990	75	23	8.1	10.4	6.9	8.3	34.0	31.2	65.2	4.3	23.3	68.4	2.8	14.0		Yeast, animal protein
Ho & Nadchatram, 1984	75	23–37	5.7	8.8	8.0	9.5	33.0	23.0	56.0					19.0		Dried milk, cereal, liver, yeast, vitamin B
Pike et al., 2005	75	23	8.8	11.8			51.1	100.8	151.9							Yeast, wheat germ
Blythe, 1976	70–74	25	7.0	10.0	9.0	9.5	33.0									No data
Colloff, 1987b	75	25	4.7											11.0		Yeast
Hart & Fain, 1988	75	25					14.3			9.0	33.9	58.2	1.8			Fishmeal, yeast, beard shavings
Saleh et al., 1991	75	25	7.5	4.1	3.5	9.6	24.6	32.9	57.5	3.1	16.2	7.4	0.5	10.8		Wheat bran
Saleh et al., 1991	75	25	7.3	6.3	5.7	14.4	37.6	47.1	84.7	4.3	22.3	9.4	0.4	7.5		Wheat bran, yeast
Hart et al., 2007	75	25	3.5				13.3	45.2	58.5	2.2	35.5	100.0	3	0.0	0.0	Liver, yeast
Hart et al., 2007	75	25	3.3				16.4	39.9	56.3	2.6	27.0	79.7	3.2	0.0	0.0	Skin scales, mattress dust
Matsumoto et al., 1986	76	25	6.2	10.7	8.6	10.7	36.3	72.0	108.3	3.5	40.6	76.2	2.1	20.0	35.7	Not known
Spieksma, 1967	75–80	25	6.0	6.0	6.0	6.0	18.0	100.0	118.0		45.0	30.0	0.7			Skin scales, fishmeal
Gamal-Eddin et al., 1983a–d	75–80	25	8.3	3.0			36.5	51.5	88.0		46.0	123.0	2.7	10.0	22.0	Skin scales, bovine liver, yeast
Dobson, 1979	80	25	7.5	10.5	9.4	17.8	45.2	38.7	83.9							Beard shavings, yeast

question as to whether periods of higher risk of allergen exposure can be predicted.

Seasonal rises in house dust mite population density tend to be associated with changing weather conditions, usually an increase in mean monthly humidity and temperature, and many of the studies on seasonal variation include correlations with local climate data (see Table 5.8). Most of these studies lasted up to a year, but it is possible to gauge between-year variation in the timing of population maxima and minima from those studies that ran for about 20 months or longer (Dusbabek, 1975; Dar and Gupta, 1979; Murray and Zuk, 1979; Arlian *et al.*, 1982, 1983; Solarz, 1997). Murray and Zuk (1979) found no consistency in seasonal maxima neither for month nor density. The maximum in their second year (1978) occurred in July and in 1977 it occurred in October. Population density of the 1978 maximum was nearly five times greater than the 1977 one. Arlian *et al.* (1982) recorded population maxima in August in the first and second years and June in the third year with mean maximum density about 3 times that of the second year and 2.5 times that of the first. Murray and Zuk (1979) and Arlian *et al.* (1982) showed the fluctuations in mite density were synchronous with fluctuations in mean indoor relative humidity, coinciding with seasonal use of indoor climate regulation (heating and cooling).

If we plot out the seasonal fluctuations in mite populations from all the studies in Table 5.8, bearing in mind they were done in different years and in different parts of the world (see Figure 5.17), the graph looks fairly chaotic. But on closer inspection, there is usually one major peak per year and one major trough. There are some peaks and troughs from different studies that coincide, and populations are usually on the decline between mid-winter and early spring. If we eliminate the years and line up all the studies according to the month they commenced (correcting by 6 months for the Southern Hemisphere studies, eliminating those studies where n = <3 homes), we can then examine patterns of seasonal fluctuation in more detail. Figure 5.18 shows the result, with mean monthly abundance highest between September and October. If we calculate the ratio of mite abundance for each month to that during the month when abundance was lowest, we get the fold-increase (a crude measure of the maximum rate of population growth during the exponential phase) in abundance. The mean fold-increase shows a similar pattern as mean abundance except for a second peak in August. So although at any

particular locality, in any particular year, there may be variation in timing of seasonal population maxima, on average we see a consistent pattern. Populations are lowest during late winter, commence their exponential growth phase during spring, are growing most rapidly during late summer, peak in autumn and then decline sharply during early winter.

So what is the effect of geographical location? When mite numbers during the months of maximum and minimum population densities are plotted against the latitude of the locality there is no obvious trend (see Figure 5.19a). But when mite numbers during the month of maximum population are divided by the numbers in the month prior to the maximum to give a fold-increase, there is a strongly positive correlation (shown in Figure 5.19b). In other words, the higher the latitude, the greater the rate of growth during the exponential phase of population increase, even though there is still a lot of variation in fold-difference between 40° and 50°. The correlation with latitude is more marked when the coefficient of variation of abundance is used (see Figure 5.19c). We calculate the mean and standard deviation of abundance for each month for the sampling period of each study and then divide the standard deviation by the mean to get the coefficient of variation. It is a particularly useful statistic when comparing datasets with broadly different means. What it shows is that there is a likelihood of a consistently greater variation in population size throughout the year with distance from the equator. This accords with greater seasonal variation in temperature at higher latitudes. Putting the data together with that on timing of seasonal maxima, it means that localities at latitudes from about 35–55°N will show relatively greater increases in mite population densities during autumn than localities from 5–35°N.

5.7.2 Allergens

There are not enough studies on seasonal variation of allergen concentrations to examine the association between latitude and seasonal maxima and minima, as for dust mite populations. Much of the data is based on samples taken only every 3 or 6 months from many different homes (for example Kalra *et al.*, 1992; Chang-Yeung *et al.*, 1995a) or localities (Lintner and Brame, 1993), rather than repeat monthly sampling in the same homes as for most studies of mite populations. Of those studies with monthly or bi-monthly data (Platts-Mills *et al.*, 1987; Lau *et al.*, 1990;

Table 5.8 Studies on seasonal variation of dust mite populations, giving locality details, the period during which each study was done, the timing of seasonal population maxima and minima, and whether local climate data (temperature and humidity) were included. *Duration (max. to min.) = the duration (in months) between the timing of the population maximum and the minimum.

Locality	Latitude (decimal)	Years	Duration (months)	Month of maximum population	Month of minimum population	Duration (max. to min.)*	Floors (F) or beds (B)	No. homes	Climate data?	Reference
Leiden	52.15	1964–1965	13	Sep	Mar	6	F	3	✗	Spieksma & Spieksma Boezeman, 1967
Davos	46.8	1967–1968	12	Oct	Feb	4	F	4	✗	Spieksma et al., 1971
Basel	47.58	1967–1968	12	Sep	Nov	2	F	3	✗	Spieksma et al., 1971
Waikiki	21.28	1968–1969	8	Jan	Jul	6	ND	5	✗	Sharp & Haramoto, 1970
Manoa	21.28	1968–1969	8	Oct	Sep	1	ND	5	✗	Sharp & Haramoto, 1970
Groesbeek	51.08	1968–1969	12	Nov	May	6	F	3	✓	Van Bronswijk et al., 1971
Brisbane	27.4	1969–1970	12	Feb	Jun–Jul	4	F	4	✓	Domrow, 1970
Groesbeek	51.78	1970–1971	13	Jul	Jan–Mar	6	B	3	✓	Van Bronswijk, 1973
Knoxville	35.97	1970–1971	12	Sep	Feb	5	F	15	✓	Shamiyeh et al., 1973
Rudgwick	51.1	1970–1971	11	Jun	Mar–Apr	9	B	1	✓	Hughes & Maunsell, 1973
Riverside	34	1971–1972	14	Jun	Feb	8	B	4	✓	Furumizo, 1978
Groesbeek	51.78	1971–1972	8	Jul	Dec	5	F	3	✗	Van Bronswijk, 1974
Prague	50.08	1972–1973	20	Jul	Jan	6	B	1	✓	Dusbabek, 1975
Basel	47.58	1972–1973	12	Sep	Mar	6	BF	32	✓	Mumcuoglu, 1975
Delhi	28.67	1972–1974	24	Aug	Feb–Mar	6	BF	15	✗	Dar & Gupta, 1979
Barcelona	41.35	1972–1974	17	Oct	Mar	5	ND	6	✓	Portus & Blasco, 1977
Prague	50.08	1972–1973	13	Jul	Nov	4	B	1	✓	Dusbabek, 1979
Prague	50.08	1972–1973	13	Jul	Mar	8	B	1	✓	Dusbabek, 1978
Grenoble 200 m	45.18	1973–1974	13	Sep	Dec	3	B	4	✓	Lascaud, 1979
Grenoble 700 m	45.18	1973–1974	13	Mar	Nov	8	B	3	✓	Lascaud, 1978
Grenoble 1200 m	45.18	1973–1974	13	Jun	Nov	5	B	3	✓	Lascaud, 1978
Silistra	44.1	1974–1975	12	Jul	Mar	8	ND	ND	✗	Todorov, 1979
Silistra	44.1	1974–1975	12	Jun	Jan	7	ND	ND	✗	Todorov, 1979
Tokyo	34.75	1974–1976	14	Jul	Jan	6	F	2	✓	Miyamoto & Ouchi, 1976
Brasilia	15.78	1975–1976	12	Mar	Sep	6	B	ND	✓	Cardoso et al., 1979
Bogota	4.37	1975–1976	10	Oct	Dec	2	B	11	✓	Charlet et al., 1978
Reus	41.17	1975–1976	12	Oct	Jul	9	ND	5	✓	Gomez et al., 1981b
Puigcerda	42.43	1975–1976	12	Oct	Mar	4	ND	6	✓	Gomez et al., 1981b
L'Ametlla	41.83	1976–1977	12	Aug	Feb	6	ND	6	✓	Gomez et al., 1981b
Bogota	4.37	1976–1977	7	Sep	Oct	1	B	11	✓	Charlet et al., 1979
Fusagasuga	4.3	1976–1977	7	Oct	Feb	4	B	4	✓	Charlet et al., 1979

(continued)

Table 5.8 (Continued)

Locality	Latitude (decimal)	Years	Duration (months)	Month of maximum population	Month of minimum population	Duration (max. to min.)*	Floors (F) or beds (B)	No. homes	Climate data?	Reference
Girardot	4.3	1976–1977	7	Jan	Sep	8	B	5	✓	Charlet *et al.*, 1979
Groesbeek	51.78	1976–1977	13	Jul	Jan–Mar	7	B	1	✓	Lustgraaf, 1978b
Vancouver	49.33	1977–1979	24	Oct, Jul	Dec, Jan	6	B	3	✓	Murray & Zuk, 1979
Dayton	39.75	1977–1979	27	Jun–Aug	Feb–Mar	7	BF	19	✓	Arlian *et al.*, 1982, 1983
Wakayama	34.21	1977–1979	18	Sep	Mar, Dec	6	BF	55	✓	Uchikoshi *et al.*, 1982
Bangalore	12.97	1978–1980	17	Sep, Dec	Apr	7	B	12	✓	Ranganath & Channa-Basavanna, 1988
Thisted	56.97	1980	12	Oct	May	7	B	1	✓	Hallas & Korsgaard, 1983
Tokyo	34.75	1981–1982	12	Jul	Dec	5	ND	6	✗	Yoshikawa *et al.*, 1982
Tokyo	34.75	1981–1982	12	Jul	Nov	4	ND	6	✗	Yoshikawa *et al.*, 1982
Saitama	36.42	1981–1982	12	Aug	Aug	12	ND	6	✗	Takaoka & Okada, 1984
Sosnowiec	53	1984–1985	24	May, Oct	Dec, Mar	7	B	2	✓	Solarz, 1997
Aurang-abad	19.87	1984–1985	18	Aug–Sep	May	9	B	ND	✓	Tilak & Jogdand, 1989
Shanghai	31.17	1984–1985	12	May	Dec	7	ND	ND	✓	Cai & Wen, 1989
La Paz, Mexico	24.17	1985–1986	13	Dec	Jun	6	BF	17	✓	Servin & Tejas, 1991
Taipei	25.03	1986–1987	12	Oct	Jul	9	B	61	✓	Chang & Hsieh, 1989
Mie Prefecture	32.82	1989–1990	12	Jul	Nov	4	BF	14	✗	Matsuoka *et al.*, 1995
Okinawa	26.53	1990–1991	12	May	Jan	8	ND	11	✓	Toma *et al.*, 1993
Pal-machim	32.00	1995–1996	12	Jun	Apr	10	BF	2	✓	Mumcuoglu *et al.*, 1999
Zova	31.78	1995–1996	12	Jun	Jan	7	BF	2	✓	Mumcuoglu *et al.*, 1999
Taichung	24.09	1998–1999	11	July	Jan	6	BF	8	✓	Sun & Lue, 2000
Dayton	39.75	1998–1999	18	July	Mar	7	F	71	✓	Arlian *et al.*, 2001
Ponteve-dra	42.42	2002	12	May	Jan	4	B	1	✗	Boquete *et al.*, 2006
La Coruña	43.37	2002	12	Jun	Jan	7	B	1	✗	Boquete *et al.*, 2006
Orense	42.33	2002	12	Oct	Jan	9	B	1	✗	Boquete *et al.*, 2006
Lugo	43.00	2002	12	Aug	Mar	5	B	1	✗	Boquete *et al.*, 2006
Mansoura	30.13	2004–2005	12	Spring	Winter	ND	BF	3	✗	El-Shazly *et al.*, 2006

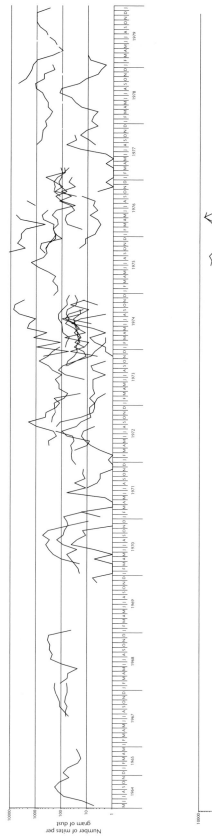

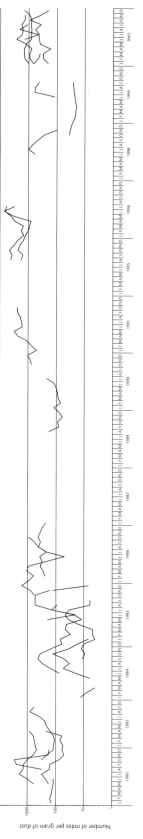

Figure 5.17 Seasonal variation in mite abundance: data from all studies, 1964–2002. The two Southern Hemisphere studies (Domrow, 1970 in Brisbane and Cardoso et al., 1979 in Brasilia) were adjusted by 6 months to match with Northern Hemisphere seasons.

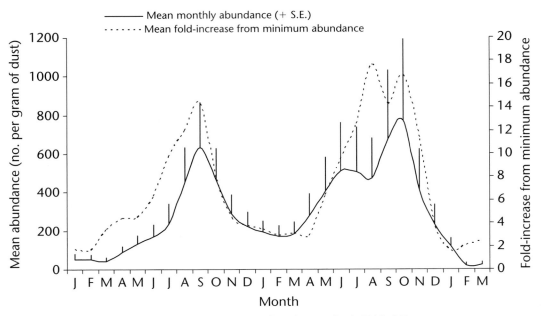

Figure 5.18 Seasonal fluctuations of dust mite populations, based on studies in Table 5.7.

Miyazawa *et al.*, 1996; Garrett *et al.*, 1998; G.L. Chew *et al.*, 1999; Sidenius *et al.*, 2002a), seasonal maxima were between late autumn and early winter. Minima were between late spring and early summer. Of the studies that investigated seasonal variation of both allergen concentrations and population density of mites, Platts-Mills *et al.* (1987) found a slight lag of allergen maxima of up to 8 weeks, whereas Sidenius *et al.* (2002), with observations taken bi-monthly, found mite and allergen fluctuations were more or less synchronous.

Crisafulli *et al.* (2007) analysed seasonal variation of Der p 1 from repeat-sampling of beds of children over a 7-year period (1997–2004) in relation to indoor and outdoor climate in Sydney. This is by far the most comprehensive information available on seasonal fluctuations of allergens. Der p 1 was highest in mid-autumn to winter (May–June) and lowest in mid-summer (January; see Figure 5.20). This pattern is similar to that of van Strien *et al.* (2002), based on a large dataset from the Netherlands, though their minimum falls slightly earlier. Sydney has a temperate climate (mean monthly temperature range ca. 10–23°C) with about 800 mm of rainfall distributed fairly evenly throughout the year. Fluctuations in allergen lagged behind relative humidity (indoors and outdoors) typically by about 2 months. It is likely that increases in humidity and temperature in late summer (February–March) stimulate

mite population growth and that allergen levels peak around 8 weeks thereafter. As mite populations decline, stimulated by falling temperatures and humidities, allergen concentrations tend to persist until reduced, presumably, by microbial decomposition.

One of the most striking characteristics of the Sydney Der p 1 data is the mean seasonal maxima are only two-to-three times higher than the minima. Although allergen concentrations fluctuate, they are basically stable between years. The grand mean for the dataset (interpolated from Figure 1 of Crisafulli *et al.*, 2007) is 7.3 µg g^{-1} (range 2.2–17), and for over three-quarters of the time mean allergen levels are between 4 and 9 g^{-1}. For those studies where repeated measurements were done in the same homes, Platts-Mills *et al.* (1987) recorded a difference between seasonal maxima and minima of about eight-fold for Charlottesville, Virginia. The difference for Berlin (Lau *et al.*, 1990) and Tokyo (Miyazawa *et al.*, 1996) was four-fold, and was about two-fold for Cartagena, Colombia (Fernández-Caldas *et al.*, 1996), the La Trobe Valley, Australia (Garrett *et al.*, 1998), Boston (G.L. Chew *et al.*, 1999) and Copenhagen (Sidenius *et al.*, 2002). Markedly higher variation was recorded from Taipei by Li *et al.* (1994), though mean fold-changes per home were not detailed. The fold-difference in seasonal maxima and minima of ca. two to eight for allergens is within the same range as fold-differences

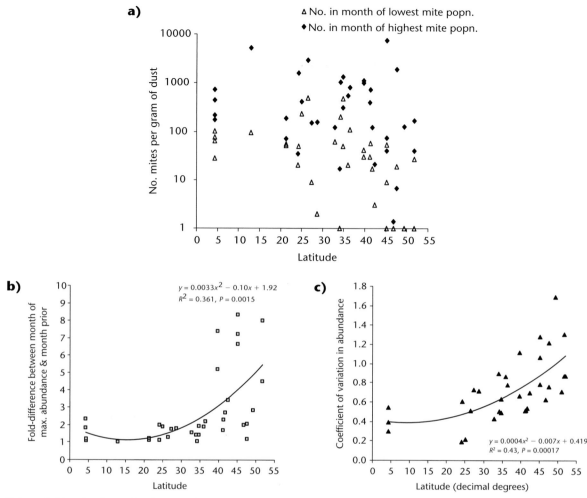

Figure 5.19 **a)** Relationship between latitude and seasonal maxima and minima of mite abundance; **b)** relationship between latitude and the fold-difference between number of mites during month of maximum abundance and number in the month prior; **c)** relationship between latitude and the coefficient of variation in seasonal mite population density from studies of seasonal variation. Lines of best fit for b) and c) are 2nd-order polynomials.

for mite population density during the period of exponential growth (shown in Figure 5.19b).

The data on seasonal variation, together with the regional-scale examples in Chapter 4, start to suggest that, within certain constraints, particular localities may have characteristic (and therefore predictable) mite population densities and allergen concentrations. This issue is explored further in Chapter 8.

5.8 Population models

Any mathematical formula used to describe the behaviour of a population represents a population model, including the equations for stable age distribution and intrinsic rate of natural increase detailed herein. Models represent an explanation for a particular phenomenon or observation that allow a hypothesis to be

refined or tested, but they do not represent a test in themselves. Models are particularly useful in helping explain how complex variables might interact, but they are invariably simplifications of reality. For example, Figure 5.5 shows two neat 3rd-order polynomial curves fitted to the empirically derived survivorship data for *Euroglyphus maynei* and *Tyrophagus putrescentiae*. They are models that describe mortality as most rapid early in life and that a greater proportion of the population of *E. maynei* survives longer than *T. putrescentiae*. But because the curves are fitted to the entire dataset for each species, they do a very poor job of detecting the subtle differences of stage-specific mortality: that there is no mortality between the protonymphal and tritonymphal stage and low mortality between the reproductive and post-reproductive stages for *T.*

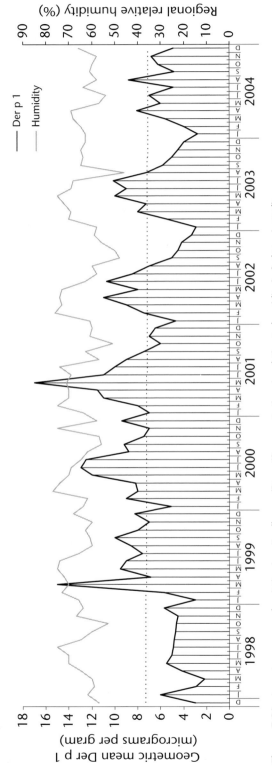

Figure 5.20 Long-term seasonal variation in mite allergen (Der p 1) concentrations in relation to humidity in Sydney, Australia. (Re-drawn from data by Crisafulli et al., 2007.) The dotted line is the geometric mean Der p 1 concentration, 7.3 µg g⁻¹.

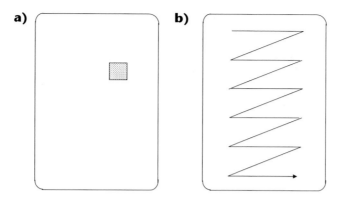

Figure 6.1 Sampling dust from beds. A fixed quadrat **(a)** versus a zig-zag **(b)** sampling pattern; with the fixed quadrat and repeated sampling, the area is likely to be depleted of dust with each sampling, causing an artefact when mite population density is expressed as numbers per gram of dust. With the latter sampling approach and repeated sampling, the same area is unlikely to be covered twice.

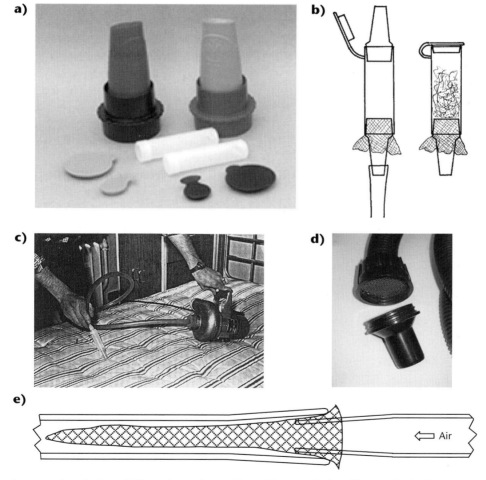

Figure 6.2 Dust sampling devices. **a)** Mitest dust collector (Martin Chapman, Indoor Biotechnologies, http://www.inbio.com). This device is a plastic sleeve containing a dust collector. It fits onto the hose or wand of a vacuum cleaner. After dust has been collected, the bottom of the collector can be capped; **b)** sampling device developed by Sesay and Dobson (1972). A cappable tube is attached to a plastic hose leading to a portable vacuum pump; **c)** an aquarium tube connector is used to hold in place a 25 µm mesh nylon cloth filter that catches the dust; **d)** a simple dust trap on the hose of a vacuum cleaner. The steel mesh supports a piece of tissue paper or nylon cloth mesh on which the dust sample is trapped; **e)** the sampling device developed by Tovey (refer to Sercombe *et al.*, 2005), consisting of a nylon mesh bag inserted into the nozzle of a vacuum sampler tube.

sleeve that can be attached to the end of a vacuum cleaner hose, with a mesh of some description to catch the dust. Typically, after the dust has been collected, the collection tube can be capped, removed and a fresh tube inserted. The simplest device is a piece of nylon gauze inserted into the vacuum cleaner tube.

6.1.2 Comparisons of vacuum sampling and brushing

Brushing of fabrics has been used primarily in places where mains electricity is unreliable or unavailable. Vigorous shaking of items of bedding or clothing into plastic bags has also been used (Hewitt *et al.*, 1973; Dowse *et al.*, 1985).

Abbott *et al.* (1981) took samples from adjacent areas of carpets and mattresses and found the geometric mean number of mites per m^{-2} of mattress was 95 times less, and of carpets 24 times less, when estimated by brushing than by vacuuming. Choice of collecting technique can influence estimates of species-diversity as well as abundance. Stenius and Cunnington (1972) found brushing of mattresses yielded species of Pyroglyphidae, Acaridae, Glycyphagidae and Cheyletidae while vacuuming yielded only the pyroglyphid species *D. pteronyssinus*, *D. farinae* and *E. maynei*. From the above, it is clear that it is not valid to compare quantitative data from studies in which vacuum cleaning was used with data from studies in which brushing was used. The different collecting methods produce markedly different results. Therefore, attempting to compare abundances of mites from studies that have used different sampling techniques may almost be meaningless. Indeed, Blythe *et al.* (1974) stated:

> An examination of the structure of the mite populations revealed by the two methods [vacuuming and brushing] brings to light serious discrepancies ... Compared with vacuum cleaner samples, the brushed samples showed a pronounced bias in favour of the larger instars.

6.1.3 Trapping

Methods of sampling live mites within fabrics were first developed by Bischoff and co-workers in Mainz, Germany, as part of their research program in dust mite ecology and their need for reliable, reproducible methods to estimate in the laboratory the effects of domestic acaricides on mortality of mites seeded into pieces of carpet of standardised size (Bischoff *et al.*, 1986a; Bischoff, 1988). There have been various subsequent modifications by other researchers and trapping techniques have been used in field studies as an alternative to vacuum sampling. Relative performances of the mobility and heat escape methods were evaluated by Brown (1994).

a The heat escape method

With the heat escape method, the underside of a carpet or piece of fabric is heated slowly on a hotplate and the mites are driven away from the source of heat, due to reduced humidity, and are trapped on a piece of adhesive plastic weighted down on the upper surface of the fabric. The plastic is then peeled off and the mites counted directly under a stereo binocular microscope without further preparative steps. In pieces of carpets seeded with mites, 65% were recovered. The technique is limited to use on fabrics that can be manipulated from both sides and for this reason has been confined to use in laboratory studies. Bischoff and Fischer (1990) found their techniques gave considerably higher estimates of mite density than were achieved by vacuum sampling: in one instance the number of mites extracted from an overcoat was 150 times that achieved by vacuuming. Also they found the heat escape method recovered almost 19 000 mites from a jacket a year after the garment had been dry-cleaned, whereas vacuuming recovered only 50 mites; about the same as had been present before dry-cleaning. The heat escape method also appears to be a far more accurate way of estimating rates of mite re-colonisation. A year after mattresses were treated with an acaricide, Bischoff and Fischer (1990) found mite populations of between 5.5 and 8% of the size of pre-treatment populations, whereas vacuum sampling indicated that no re-colonisation had taken place.

b The mobility test

The mobility test is similar to the heat escape method except no heat is used, so the adhesive plastic is left in situ for 24 hours. Bischoff *et al.* (1986a) and Bischoff and Fischer (1990) recovered 8–30% of mites from mite-seeded fabrics by using the mobility test.

c Heat attractant trapping

Tovey (*pers. comm.*) has developed a modification of the heat escape method to allow collection of mites from fabrics that cannot be manipulated from both sides, such as fitted carpets and mattresses. A hot water bottle or plastic bag is filled with warm water (ca. 40°C) (the bags inside wine boxes are particularly useful for this) and is placed over a piece of adhesive tape 100 cm². The water bag is covered and the 'trap' left for up to 12 hours, usually overnight. With this method the mites move towards the heat source, rather than away from it, because it is mild enough not to decrease humidity substantially but warm enough to increase their mobility.

6.1.4 Comparison of trapping and vacuum sampling

There has been only limited direct comparison of the efficiencies and biases of vacuum sampling and trapping (van Bronswijk, 1984; Bischoff and van Bronswijk, 1986). The most obvious difference is that vacuum sampling collects live and dead mites, eggs and faecal pellets, as well as dust particles, whereas live trapping techniques have been specifically designed to sample only live, mobile mites. Thus live eggs, and the immobile stages in the life cycle will tend not to be collected. What are the consequences for the efficiency of live trapping?

The extraction efficiency of the heat escape method has been estimated at around 65% (Bischoff et al., 1992a, b), using live mites seeded into carpet squares. A growing mite population has a preponderance of immatures in it (see Chapter 5). For example, a growing population of 100 mites may comprise 25 eggs, 35 larvae, 30 nymphs and 10 adults. Bearing in mind that trapping methods depend for their success on the mites being active and moving about, it is possible to estimate the proportion of the population that cannot be trapped at any one time from the numbers of each stage of the life cycle in the total population and the proportion of each life cycle stage that is spent immotile. All of the egg stage is spent immotile, and half each of the larval and both nymphal stages in the pre-moult periods, whereas adulthood is spent active. Thus the number of immotile mites for the population would be 25 + 17.5 + 15 = 57.5. So the trappable component is only 42.5 mites. Therefore, with an efficiency of 65% for active mites, the total population extraction efficiency is only 27.6%. The important additional consequence of this example is that the efficiency of the trapping method will vary according to the 'stage structure' of the population in any home. It will be more efficient in homes that have a higher proportion of adults in the population because it is biased against

stages in the life cycle that are inactive. This example demonstrates that the true measure of extraction efficiency of trapping methods, relevant to the field rather than the laboratory, cannot be gauged by measuring the proportion of active mites which are trapped following seeding test carpet squares with known numbers of active mites. Extraction efficiency should reflect the catchability of the total population present in their natural environment.

It would appear, at first sight, that trapping techniques may provide more accurate estimations of the size of the live mite population than vacuum sampling. However, there are a number of problems of extraction efficiency and extraction bias that need to be addressed. It also needs to be tested, rather than presented as an act of faith, whether the extraction efficiency of trapping methods really does represent a significant improvement over vacuum cleaning.

6.1.5 Sampling total mite populations

So far I have dealt with methods for making 'absolute' estimates of partial mite populations. Hay (1995) developed a method for estimating total mite populations. The method was not developed primarily for routine use because it involves taking cores of mattress material and is therefore destructive. Rather, the intention was to compare the total mite populations present with the estimates made by vacuum sampling. Plugs of material were cut out of the mattress, separated into layers and the material teased apart under lactic acid, and the mites removed and counted. Using this technique estimates of 8200–26 800 mites m^{-2} were obtained compared with 3–46 m^{-2} by vacuum cleaning. Core sampling recovered a significantly higher proportion of eggs than did vacuuming.

6.1.6 Partial and total estimates of mite populations

If vacuum sampling estimates only some 0.5–5% of all the mites present, should not a correction factor be applied to samples taken by vacuum cleaning in order to give a more accurate estimate of total mite populations? It would certainly be informative if researchers estimated and reported the efficiency of their sampling devices as a matter of best practice. But the fact remains that vacuum cleaning is the quickest, easiest and most practical method of sampling mites. The heat escape method is not possible, because the heating device has to be placed under the fabric. The coring method involves cutting chunks out of mattresses and is similarly impractical. Tovey's heat attractant trapping method may prove an alternative to vacuuming, but it requires a repeat visit to collect the sample and is therefore potentially more time consuming than the single visit required for a vacuum sample.

6.2 Dimensions of the samples and units of measurement

With vacuum sampling, various dimensions have been employed. For example, Carswell et al. (1982) used a 600 cm² quadrat for sampling mattresses; the first sample taken from the area below (not beneath) the pillow, repeat samples taken from 'equivalent' areas. Korsgaard (1983a, b, c) sampled the whole upper area of the mattress in one minute, while Colloff (1987c) used a zig-zag sampling pattern which scanned the whole upper surface of the mattress, but removed mites from a total of 0.25 m² of it (calculated from the diameter of the sampling nozzle multiplied by the length of the sampling sweeps). This latter method allows repeat sampling from the same general area but with a minimum risk of artefactual reduction in mite density because exactly the same area within the general area is never exactly covered twice (see Figure 6.1).

Dust from different substrata varies greatly in physical density: that from floors contains particles of grit and sand, while that from mattresses consists mostly of skin scales, which have a very low density. These differences can result in misleading data if numbers of mites from dust samples from mattresses and carpets, expressed per unit weight, are compared. To avoid this problem, Furumizo (1975c) developed a technique called volumetric washing (discussed further in section 6.3 below). Numbers of mites are expressed per unit volume (ml) of sieved dust that has been allowed to settle in ethanol in a measuring cylinder.

Can samples be taken uniformly and accurately from the very different surfaces of a mattress, a carpet and an upholstered chair and the results compared in a meaningful way? The problem is not an easy one to solve. The topography of each of these objects bears no relation to the other. It is likely that the microclimate in each will differ and thus also the distribution and abundance of mites. The easy option is to sample as much of the surface area as possible within a set time and express the numbers of mites or the numbers of species per unit weight. Is this strategy valid?

The topography of a mattress is, on first appearance, deceptively simple: a rectangular object, consisting of springs and stuffing materials, or foam rubber, covered with fabric, with seams and buttons at various intervals. But where does one take the sample – from the sides, the horizontal surfaces, or both? If samples are taken from the top and the mattress is turned over at regular intervals by the owner, what effect would this have on mite populations, and would data from samples from mattresses that are regularly turned be comparable with those from non-turned mattresses? How does one take into account the seams and buttons? Blythe (1976) showed there were greater population densities of mites in areas immediately adjacent to seams than elsewhere on the mattress. These sorts of questions have, to a large extent, been evaded or ignored by house dust mite researchers. Mostly they have chosen sampling units as the weight of house dust recovered by vacuum cleaning, not because it is known to be the most appropriate method scientifically, but because it requires the least effort to standardise.

The argument on whether unit weight or unit area is more appropriate for the expression of mite densities or allergen concentrations in dust samples taken by vacuuming is crucial to the practice of house dust mite ecology. It rests on whether or not the numbers of mites extracted from the dust by vacuuming are dependent on or related to the quantity of dust present in the substrate. In other words, is there any equivalence between the number of dust particles removed by vacuuming and the number of mites or quantity of allergen? The simple answer is that for dead mites and faecal pellets there is some relationship. They are, essentially, components of the dust itself and are likely to be removed with similar ease as any other component. But for live mites there is less relationship: they have sucker-like pulvilli at the ends of their legs which provide them with a good grip on the substrate, making them more difficult to remove than dead mites.

Table 6.1 summarises the advantages and disadvantages of using unit weight and unit area. Of these points, the most significant concerns the use of unit weight for data derived from repeated sampling during the course of a field trial or a clinical trial of house dust mite control. Effective control of house dust mites requires a very high killing efficiency followed by thorough, preferably repeated, vacuum cleaning to remove as much of the allergen pool of dead mites and faecal pellets as possible. If

Table 6.1 Advantages and disadvantages of expressing population density of mites by unit weight and by unit area.

Unit weight (no. mites per gram of dust)	Unit area (no. mites m^{-2})
Easy to use when sampling objects with different topography, e.g. furniture v. carpets	Requires the area to be sampled to be determined, e.g. by use of a quadrat or a set length of sweep of a sampling device
Comparison of data from substrates containing dust of different densities gives misleading results	Data from substrates with different densities of dust is more directly comparable
Use not valid for repeated measures sampling because dust is removed on each sampling occasion	Valid for use in repeated measures sampling
Easy to compare data from different studies: has been used in the overwhelming majority of studies	Few studies have used unit area, and it is difficult to convert from one to the other
Can only be used for samples collected by vacuum cleaning	Can be used for samples taken by any method

data are expressed per unit weight and the density of total mites or allergen concentration compared with baseline values taken prior to commencement of control methods, this will tend to underestimate the reductions because of the amount of dust removed by vacuum cleaning. An example of this effect can be seen from a re-examination of the data gathered during a clinical trial of mite control in which liquid nitrogen was used as a freezing agent to kill mites (Dorward *et al.*, 1988). To compare reductions of mite density per unit weight and per unit area, total mite density from each of four bi-weekly samplings of mattresses following treatment was expressed as a ratio of the pre-treatment density and the mean rate of reduction per mattress estimated from regression analysis, as done by Colloff *et al.* (1989). The grand mean reduction of mite density per gram of dust after 8 weeks was 1.8-fold (range 1.01–4.9) while that of mite density per m^2 was 5.5-fold (range 1.8–13.8). A mite control trial that involves addition of material to the substrate in order to kill mites, for example, wet powders containing acaricides, may also result in a biased estimation of efficacy of reduction. Any residual material not removed by vacuum cleaning may have the effect of 'diluting' mite density or allergen concentration if density is expressed in unit weight. It is only common sense to recommend the data should be expressed per unit area for studies involving repeated measurements of mite population density or allergen concentration in the same home, or the substrate. Such studies involve alterations to the dust load by replicate vacuum cleaning and dust sampling.

Most dust mite surveys have given numbers of mites per unit weight (see Appendix 3; Chapter 8), which makes unit weight the easiest and most convenient

way to compare population densities, for the same habitat type, between different published studies. Comparing numbers of mites in bed dust in one study with those in floor dust in another is not likely to be very informative.

But if expressing numbers of mites or allergen concentrations per unit weight of dust is less appropriate than per unit area, why do we see statistically significant correlations between values expressed in this way and other factors under investigation? For mite densities the answer lies partly in the fact that differential determination of the numbers of live mites and the numbers of dead mites has rarely been performed. The efficiency of collection of dead mites is much higher than that of live mites because dead mites have no active method of clinging to the substrate. They are probably extracted with similar efficiency with other similar-sized particulate non-living material in the dust. Takaoka and Okada (1984) showed statistically significant positive correlations between density of total mites (dead and live) per unit weight and density of mites per unit area, thus demonstrating an area–weight relationship (shown in Figure 6.3).

6.3 Extraction of mites from dust samples

Many techniques have been developed for the removal of mites from house dust. These were reviewed by Wharton (1976); Gridelet de Saint Georges (1975) and van Bronswijk *et al.* (1978). The basic types are flotation, suspension and heat extraction. The first two are the most widely used.

6.3.1 Flotation

Flotation involves the exploitation of the difference in density of mites from the aqueous media in which

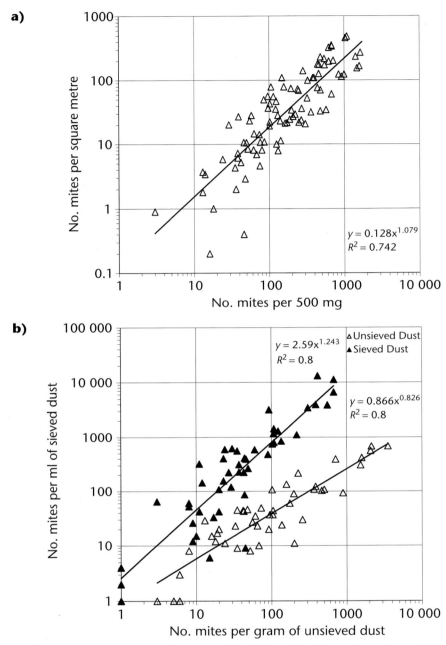

Figure 6.3 **a)** Relationship between density of mites per unit weight of fine dust and per unit area (data from Takaoka and Okada, 1984); **b)** relationship between mites per gram in unsieved (coarse) dust and mites per millilitre of sieved (fine) dust (data from Massey *et al.*, 1988).

they are submerged. The mites float to the surface, become concentrated there, and can be aliquotted off. The major disadvantage of flotation techniques is that they have relatively low extraction efficiency.

The first method for extraction of mites from dust was developed by Spieksma and Spieksma-Boezeman (1967) and involved flotation in lactic acid assisted by centrifugation. It has a relatively low extraction

efficiency. Miyamoto and Ouchi (1976) developed a similar method.

Mites have been extracted in saturated sodium chloride solution (Sasa *et al.*, 1970; Shamiyeh *et al.*, 1971; van Bronswijk, 1973; Nourrit *et al.*, 1975; Fain and Hart, 1986; Hart and Fain, 1987; Sakaki *et al.*, 1991). Despite claims of efficiencies of up to 97%, only three of the samples examined by

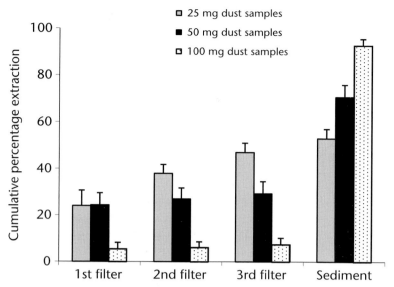

Figure 6.4 Determination of the percentage efficiency (mean ± S.E.) of a saturated NaCl flotation technique for the extraction of mites from dust samples (n = 3 for each sample class; Colloff, unpublished data).

Hart and Fain (1987) were house dust. The rest were mammal nest material, debris in the bottom of museum jars containing pickled mammals and laboratory cultures of mites – none of which have the particulate characters of house dust. Using their technique, which involves exploiting differences in densities between ethanol and saturated sodium chloride solution, I found a geometric mean extraction efficiency for 33 samples of house dust of only 27% (range 3–88%; see Figure 6.4), that the bulk of mites remained in the sediment, and that the proportion extracted increased with increased weight of dust samples. The technique was most efficient for samples that contained little or no fibrous material, but mites got trapped in the dust and remained in the sediment or stuck to the walls of the vessel when the supernatant was poured off. Additionally, quantities of fine dust particles floated to the surface along with the mites. The main advantage of flotation is that it is relatively rapid. Mites are separated from the dust automatically, although a long preparation time may be required. Hart and Fain's (1987) method requires the dust sample be left to stand for 12–24 hours in ethanol before it can be processed further.

Volumetric washing (Furumizo, 1975c) combines aspects of flotation and suspension. The method is apparently highly efficient (96–98% mite recovery). Numbers of mites are expressed per unit volume (ml)

of dust that has been sieved free of coarse material and has then been allowed to settle in ethanol in a measuring cylinder. Although it has several advantages, this technique has not become widely adopted, partly because it involves over 20 separate steps to extract the mites and because it is not straightforward to compare numbers of mites expressed in ml with numbers expressed in grams of dust. I have used an approximate conversion figure for sieved dust (based on 32 samples from beds) of 1 ml to 5.5 g. Massey *et al.* (1988) show a log.-linear relationship between numbers of mites per ml and per gram of coarse dust (shown in Figure 6.3b).

Oshima (1964) isolated mites from dust using a separating mixture of ether and tetrachloride. Maunsell *et al.* (1968) used dichloromethane. A method based on flotation at the interface of methylated spirits or ethanol and kerosene was developed by Thind and Griffiths (1979), with subsequent modifications (Thind and Wallace, 1984; Thind, 2000; Thind and Dunn, 2003). Methods using organic solvents obviously have higher occupational health and safety risks compared with methods using sodium chloride solution.

6.3.2 Suspension

Suspension extraction involves placing the dust sample in lactic acid, ethanol or sodium chloride solution, usually in a glass Petri dish, wetting and dispersing the particles, staining and – in the case of lactic acid suspensions – heating them to macerate the mites. Mites are removed under a stereo binocular microscope. The disadvantages are that it can be time-consuming, especially if large numbers of mites are present, although there are ways to minimise this, for example by pouring the suspension into a fresh Petri dish and examining the film remaining in the first dish. The mites are removed and the process repeated with another fresh dish until all the fluid has been examined. This method is quicker and more reliable than examining all the sample in one dish – as done by Gridelet and Lebrun (1973) and Gridelet de Saint Georges (1975) – there are far fewer particles per unit area in a film of fluid than in a full dish and the mites are much more easily visible. Suspension in viscous fluids such as lactic acid is easier than in alcohol because the particles move around less. The major advantages of suspension techniques are that the entire sample can be examined by eye and, if done correctly, a high extraction efficiency achieved. Also a good picture can be gained of the composition of the dust sample including the size and consistency of particles, the presence of different types of fibres and other components such as feathers and pollen. The suspension technique results in few losses of mites due to transfer from different vessels because the contents of each vessel are examined thoroughly after every transfer.

Arlian *et al.* (1982) described a modified suspension technique whereby 50 mg dust samples were suspended in saturated NaCl with detergent, wet-sieved onto 45 μm mesh and stained with crystal violet, which stains the dust particles but not the mites. This technique has the advantage over lactic acid suspension that live mites are not killed and a very accurate assessment of live and dead mites can be made. Natahura (1989) came up with a similar method, using methylene blue as a stain.

6.3.3 Heat extraction

Heat extraction is a technique borrowed from the Tullgren funnel extraction method, widely used in soil biology, where the sample of dust is placed under a light bulb and the live mites migrate from the dust as the sample starts to lose water (van Bronswijk, 1981; Figure 13.7). It will only extract active stages in the life cycle and is most useful for removing live mites in order to form starter populations for in vitro cultures.

6.3.4 Extraction efficiency

Each procedure differs in the efficiency with which mites are extracted and there is generally an inverse relationship between the amount of 'hands on' time and effort that the method requires and the extraction efficiency. It is desirable to determine the extraction efficiency (number of mites found divided by the number actually present in the sample) of a technique at the beginning of the study. Variation in extraction efficiency between operators can be considerable. However, operators tend to increase efficiency with practice. Additionally, extraction efficiency varies according to species and stages in the life cycle of the mites (shown in Figure 6.5). The extraction efficiencies for intact mites of the same stage were always higher than for damaged mites and efficiency increases with the size of the mites. There was no significant correlation between extraction efficiency and the relative abundance of each stage. Smaller instars and damaged specimens were recovered less efficiently than larger instars and intact specimens. Experimental errors of this sort have obvious implications for population ecology, especially studies of age-structure of populations. It was surprising to find the extraction efficiencies of the very small *Tarsonemus* sp. were similar to those of the much larger *D. pteronyssinus*. This was probably due to an ability to discriminate between the markedly different morphology of tarsonemid mites and pyroglyphid mites and override the factor of size with that of morphology while searching for specimens.

6.4 Mounting, counting and identification of mites

Ideally all mites extracted from the dust samples should be washed, cleared and mounted on microscope slides for identification under a compound microscope. Some researchers have simply counted mites when examining the sample under a stereo binocular microscope (e.g. Walshaw and Evans, 1987); others have taken subsamples of mites for identification (Tovey *et al.*, 1975; Korsgaard, 1982). These approaches have several disadvantages. It is valuable to make permanent preparations of all mites extracted from dust so that the identity of specimens can be checked or re-examined following new developments such as the discovery of concomitant populations of *D. farinae* and its sibling species *D. microceras* in Europe (Cunnington *et al.*, 1987). Permanent

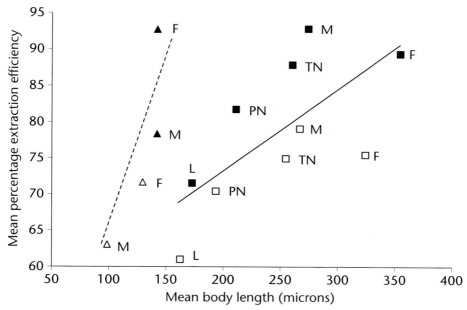

Figure 6.5 Relationship between mean extraction efficiency of *Tarsonemus* sp. (triangles) and *Dermatophagoides pteronyssinus* (squares) and mean body length of specimens (n = 10 specimens of each stage of each taxon). Closed symbols = intact (live) specimens; open symbols = damaged (dead) specimens. L = larvae, PN = protonymphs, TN = tritonymphs, M = males, F = females. (Modified from Colloff, 1991d.)

preparations also allow for subsequent, more detailed analysis of population structure or some other factor at a later date.

6.4.1 Mounting mites on microscope slides

Mounting mites from dust samples on microscope slides one-by-one is extremely tedious. To avoid this, Colloff (1989a) developed a mass-transfer method which allows 200–300 mites to be mounted in a single operation (see Figure 6.6). The mites are frozen in water in a watch-glass (Figure 6.6a) and transferred in ice to a microslide (Figure 6.6b) that has been coated with silicone solution (e.g. 'Sigmacote') and allowed to dry. The water is evaporated off on a hotplate (Figures 6.6c, d) and mountant and cover slip are added in the usual way (Figures 6.6e–h). Aqueous mounting media such as the various modifications of Berlese's gum-chloral medium, e.g. de Faure's or Hoyer's media (see section 6.5.1 and Krantz, 1978, for recipes) have been widely used. They have the advantage that mites can easily be remounted at a later date (refer to Fain, 1980, for method). However, gum chloral-based media have a major disadvantage in that they tend to dry out if the cover slip is not ringed. Also, there are a number of other significant problems associated with these

media, especially damage of specimens due to the crystallisation of some components of the medium, as detailed by Upton (1993). Media soluble only in organic solvents, such as Euparal or Canada Balsam, involve rather lengthy preparation stages and are not particularly suitable for mites. They also involve the use of xylene as a solvent, which is undesirable for occupational health and safety reasons. I have achieved good results using polyvinyl alcohol in lactophenol, a modification of Heinze PVA. The original recipe (Heinze, 1952), and the modification of it by Boudreaux and Dosse (1963), contain errors regarding the proportions of the ingredients which render it less than satisfactory. However, the recipe presented in section 6.5.7 is the same as that used at the Central Science Laboratories, UK, and works well.

6.4.2 Counting and identification

Counting of mites is best done at the same time as identifying them. Correct identification of mites is vital. This requires some expertise in basic acarology, but is not as impossible as some clinical researchers imagine. Keys for adults of all pyroglyphid species are presented in Chapter 1 and by Fain *et al.* (1988, 1990). Additionally, there are various courses in acarology available (e.g. at the Acarology Laboratory

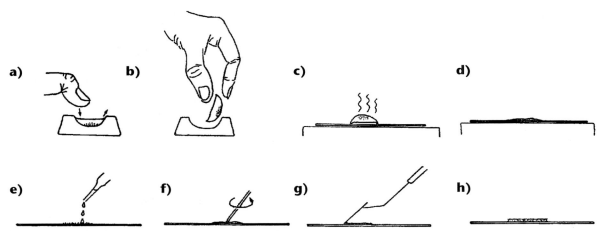

Figure 6.6 The frozen block technique for mounting dust mites on a microslide (modified from Colloff, 1989a). See text for details.

at Ohio State University) that can provide that expertise in a relatively short space of time to persons with little or no previous experience. The determination of immature pyroglyphids to species is more difficult. Only immatures of *D. pteronyssinus*, *D. farinae* and *E. maynei* have been described adequately (Mumcuoglu, 1976). The identification of different stages in the life cycle is relatively easy and can provide important data on the age-structure of populations. Likewise the determination of damaged and intact specimens, using the criteria of Arlian *et al.* (1982), is a useful means of estimating the numbers of mites that were alive or dead at the time of sampling. House dust mite ecology is unusual in that it is concerned not only with the living population but with dead mites – they are part of the allergen pool to which persons with atopy are exposed.

6.5 Laboratory testing of mite control methods

6.5.1 The test substrate

The sampling units should consist of individual pieces of carpet, rather than a single piece of carpet marked into sampling units. The reasons for this are:

- it eliminates the risk of migration of mites from one sampling unit to another;
- the sampling units are easier to handle and seed with live mites;
- the other sampling units are not disturbed during sampling of mite population densities units when sampling;

- the sampling units are easier and more controllable to incubate at a standard temperature and humidity;
- it allows for convenient use of the heat escape method of mite sampling.

The carpet pieces need to be checked with the manufacturer to determine whether they have been treated with a mothproofing insecticide, as this could also have a negative effect on mite populations. If the mothproofing insecticide is a synthetic pyrethroid, it can be denatured by exposure to UV light.

Even if the carpet pieces are new, they can contain mites, depending on where they were stored post-manufacture and for how long. It is thus vital to freeze the carpet squares to kill any live mites, at least to −30°C overnight. The carpet pieces should then be vacuum cleaned thoroughly and the vacuum samples examined to ensure that no live mites are present.

6.5.2 Temperature and humidity control

A major confounding variable in these sorts of experiments is the effect of temperature and humidity on the reproductive rate and mortality of the dust mite population. It is therefore essential that temperature and humidity are controlled constantly and precisely with as little variation as possible. This requires careful monitoring with an electronic thermohygrograph. A climate-controlled room with humidification facilities and shelving for the carpet pieces, as used by Bischoff *et al.* (1986b) is the ideal set-up for such a study. Temperature should be 25 ± 2°C and humidity 75 ± 2% RH. Alternatively, if a climate-controlled room is not available, humidity-controlled incubators can be used, using the methods for humidity control outlined below (see section 6.6.3).

It is important to equilibrate carpet squares at the required temperature and humidity for at least 24 hours prior to seeding the squares with mites. It is also important to ensure the cultures have been grown at the same temperature and humidity at which they will be exposed in the carpet. If not, the mites will also require equilibration or their reproductive rate and mortality may change after they have been seeded into the carpet.

6.5.3 Seeding the carpet squares with mites

The number of mites seeded into the carpet needs to be known as accurately as possible and not guessed at. The purpose of studies on the efficacy of dust mite control methods is to determine if the control method has an effect on the size of the dust mite population. If the size of the starting population is not known, or based on an estimate of undetermined accuracy, then the study is virtually invalid.

The easiest way of estimating numbers is to take a series of aliquots of mite culture of known weight and count the live mites, calculating the mean and standard deviation, until an acceptably small range is achieved. Researchers can then define, with precise statistical confidence, how many mites are contained within a given piece of carpet at the outset of the experiment. This work is time consuming but essential. It needs to be done shortly before the culture is to be used otherwise the density of mites in the culture may have changed due to water loss by the time they are seeded.

If the mites are sieved from the culture medium prior to seeding (which is the easiest way of getting a high density of live mite material), then a food source needs to be introduced or the mites in the control carpet pieces will die of starvation after seeding, thus confounding the experiment. A suitable culture medium for dust mites (there are many) should be equilibrated at the appropriate test microclimate at least 24 hours before use. Alternatively, mites can be left within their culture medium, though this makes for greater variation (and greater difficulty) when trying to count live mites prior to seeding them.

There are no ideal starting population densities prescribed, but there need to be sufficient mites per unit area so that they will come into contact with each other frequently enough in order to mate. Below a certain population density, the frequency of mating encounters will diminish dramatically and the rate of

population increase decline sharply, thus confounding the experiment. As a minimum, something in the order of 10–20 mites cm^{-2} would probably allow for sufficient mating encounters. The mites and the culture medium should be distributed evenly over the surface of the carpet squares and gently brushed in and left for at least 3 days to settle and disperse within the carpet pieces.

6.5.4 Monitoring population growth prior to the mite control treatment

Adequate numbers of carpet pieces need to be used to allow for sufficient replication of treatments and controls. Three pieces for each of treatment and control is probably too few, 20 is too many. Sufficient replicates should be included to allow for pieces that have to be discarded due to lack of mite population increase. It is important to monitor the live mite populations in each piece of carpet prior to assigning the pieces to treatment and control groups. At least three samplings would be ideal. In this way an estimate is gained of the reproductive performance of the mites since the time they were seeded. Any carpet pieces in which the population has declined should be discarded because if it were assigned to the treatment group, it would be impossible to determine whether subsequent decline in mite population density was due to the acaricide or to a continuance of the pre-treatment decline.

6.5.5 Mite control treatment

If the treatment is an acaricide, care should be taken to avoid contaminating the control pieces, either through spray drift or storage of control carpet pieces in close proximity to treatment pieces.

6.5.6 Estimating the size of the mite population after treatment

There are two main types of method for doing this: the heat escape method and vacuum sampling followed by suspension extraction (discussed in sections 6.1 and 6.3 above).

a The heat escape method

The underside of the carpet squares is heated on a hotplate at 80°C for 15–30 min. Live mites, if present, move away from the heat source and are captured on strips of transparent adhesive tape laid over the top surface of the carpet square. I have used 3 × 100 cm^2 strips of tape per square, which were removed after heating and examined under a stereo binocular microscope for the presence of live mites.

b Vacuuming and suspension extraction

Carpets are vacuum sampled at the recommended rate of 1 m² in 2 min. using a filter system or similar to trap the mites (see section 6.1.1). The material adhering to the filter is removed, weighed and the mites recovered, weighed and extracted as follows. The sample is suspended in 50 ml saturated sodium chloride solution, with a wetting agent added. The suspension is then stirred and filtered through a sieve of mesh size small enough to retain mites and their eggs, and stained on the sieve using 1% aqueous crystal violet which stains everything except the mites. After rinsing, the contents of the sieve are transferred, with water, to a 9 cm diameter glass Petri dish. Active, live mites can be seen on the meniscus or below the surface and detected by movement under a stereo binocular microscope.

c Post-treatment monitoring of mite populations

Sampling for live mites should be repeated at sufficient intervals commensurate with the life cycle of the mite at the microclimate at which it has been incubated (ca. 3 weeks), e.g. post-treatment and then 2-weekly on four occasions for treatments and controls.

d Statistical treatment

Appropriate statistical methods for comparing rates of change in mite population density for treatments and controls are essential. It is important to remember that the type of study design described above represents a repeated measures design. Thus only statistics appropriate for repeated measures should be used.

6.5.7 Recipe for Heinze polyvinyl alcohol (PVA) mounting medium

This recipe is from the Central Science Laboratories, UK.

  10 g polyvinyl alcohol (PVA);
  40–60 ml distilled water;
  10 ml glycerol;
  25 ml of a 1.5% solution (weight/volume) of phenol in distilled water (e.g. 375 mg phenol in 25 ml distilled water);
  100 g chloral hydrate;
  35 ml lactic acid.

1. Place PVA powder in a beaker and add water gradually, stirring, while heating the beaker in a water bath on a magnetic hotplate/stirrer at just below boiling point.
2. Add lactic acid and stir.
3. Add glycerol and stir.
4. Mix 1.5% phenol solution and place in a beaker in a water bath on a magnetic hotplate/stirrer. Add chloral hydrate and stir until dissolved.
5. Add chloral hydrate-phenol solution to the PVA solution when the latter is lukewarm.
6. Filter and store in a brown glass bottle to prevent oxidation by UV light.

Note: To dissolve hardened PVA on slide mounts, e.g. for demounting or remounting of slides, use warm lactophenol or, even better, warm 60% aqueous glycerol.

Chloral hydrate is a controlled drug in many countries and may require a licence for research use. An alternative to PVA is a ready-made mounting medium called CMC-10, which is available from Masters Chemical Company Inc., Elk Grove, IL, USA.

6.6 Culturing house dust mites in the laboratory

House dust mites are, by and large, easy to grow in the laboratory providing that sufficient care and attention is devoted to selection of an adequate diet, proper maintenance of microclimate in the culture vessels, prevention of contamination by fungi, cross-contamination with any other mite species, and regular subculturing to prevent overcrowding. There are several reviews and methods papers (Woodring, 1963; Boczek, 1964; Solomon and Cunnington, 1964; Voorhorst *et al.*, 1969; Sasa *et al.*, 1970; Hall *et al.*, 1971; Araujo-Fontaine *et al.*, 1972; Miyamoto *et al.*, 1975; Wharton, 1976; Mumcuoglu, 1977a; James and Mulla, 1978; van Bronswijk, 1981; Castagnoli *et al.*, 1989; Parkinson, 1992).

6.6.1 Basic equipment

The basic equipment requirements are a plastic or glass desiccator, a stereo binocular microscope with a cold light source, a humidity meter and thermometer, suitable culture vessels (shown in Figure 6.7) and a starter culture of the species required. The choice of food may be determined by the final use to which the mites are to be put. Those to be used for making allergen extracts (such as making polyclonal antibodies or doing western blotting) ideally need to be reared on a diet of low intrinsic allergenicity.

Figure 6.7 Mite culturing equipment. **a)** Desiccator with sodium chloride sludge in the base to control humidity, Kilner jar with NaCl sludge and rack to hold individual Robertson cells; **b)** detail of Perspex rack and individual Robertson cells.

6.6.2 Starter cultures

a   Starter cultures from existing laboratory cultures

Starter cultures of various species may be obtainable commercially from manufacturers of diagnostic allergology products. The Central Science Laboratories, UK, has supplied mite cultures commercially. These cultures were developed over 45 years, at what was the Pest Infestation Laboratory at Slough, UK. This laboratory built up and maintained a unique collection of stored products mites and dust mites (Parkinson, 1992). The collection used to number at least 36 species, but at the time of writing, it is no longer clear from the CSL website whether such a service is still offered. The cultures were accumulated as a result of research on the biology and population dynamics of stored products pests such as *Acarus siro* by Solomon (1945, 1946a, 1962) and Cunnington (1965) and *Tyrophagus putrescentiae* by Cunnington (1969). The development of methods to grow these mites in the laboratory (Solomon and Cunnington, 1964) went hand-in-hand with this research and forms the basis of what we know about culturing free-living astigmatid mites today. Other likely sources of mite cultures are research groups.

b   Developing a culture from house dust samples

If starter cultures are not available, researchers can prospect for live material in dust taken from houses. Take a series of dust samples from a range of bedding and carpets in different homes and incubate them in the desiccators and see what grows. It may take between a few days to several months before there are noticeable numbers of mites present. When the population is established, gradually add culture medium to replace the dust.

6.6.3 Microclimate

Dermatophagoides pteronyssinus, D. farinae, Euroglyphus maynei and several of the glycyphagoid species such as *Blomia tropicalis* or *Lepidoglyphus destructor* can be maintained with good population growth at room temperature (ca. 18–22°C) and 75% RH. Population growth can be speeded by increasing temperature to 28–33°C and upping the relative humidity slightly (up to 80%). Humidities in excess of 80% are difficult to maintain mites in because mould growth is encouraged. In fact the easiest way of controlling mould growth is to keep humidity constant at 75% and provide the cultures with weekly aeration and shaking, preferably in a laminar flow hood. Shaking helps prevent the build-up of a mat of fungal mycelia in the culture medium.

The easiest way of maintaining humidity inside the plastic desiccator is by using one of the mixtures of saturated salt solutions recommended by Winston and Bates (1960). Saturated aqueous sodium chloride solution, enough to cover the bottom of the desiccator, is mixed with solid sodium chloride to make a sludge. This will keep the humidity at a steady 75% RH (± 1%) over a very wide temperature range. NaCl mixed 1:1 with KCl will give an RH% around 70–71 between 20 and 30°C, $(NH_4)_2SO_4$ maintains humidity around 80% RH between 10 and 40°C. For further details, consult the reference. Other methods include use of mineral acids or strong alkalis (Solomon, 1951) and pose various risks in relation to occupational health and safety. Another method, safer than strong acids and alkalis, is the use of glycerol solutions (Miner and Dalton, 1953).

6.6.4 Culture vessels

Typically, glass tubes (ca. 2 cm diameter × 5 cm long), 9 cm diameter plastic Petri dishes and cell-culture bottles have been used. Glass tubes are also normally stoppered with non-absorbent cotton wool. This allows gas exchange between the inside and outside of the tube, but is relatively insecure and mites can escape. Also, the plug becomes full of mites eventually and these may fall off the plug onto the bench when sub-culturing, thus increasing the opportunities for cross-contaminating cultures. Petri dishes are easy, cheap and secure. Some petroleum jelly or similar grease (high vacuum grease as used for laboratory vacuum pumps and the like is very good) is loaded into a plastic syringe and run round the inside edge of the lid. This produces an escape-proof seal – mites get tangled in the grease. However, the trade-off for good security is poor gas exchange and the cultures have to be aired regularly to prevent CO_2 build-up. This is especially important if using live yeast as a culture medium. Another consequence of reduced gas exchange is that humidity in the culture vessels may come out of equilibrium with that in the desiccators, again, especially if the culture medium is hygroscopic.

Cell culture bottles that have a microporous filter incorporated into the lid allow gas exchange and are relatively mite-proof. Additionally, the small opening to the flask makes it more difficult to get access to the cultures (e.g. if subculturing or adding food material) than does a Petri dish. Doubtless, there are other containers that could be used to circumvent some of these problems.

Small numbers of mites can be raised and observed in modified Robertson cells (Robertson, 1944; see Figure 6.7b). These are black Perspex cells with a central concave chamber milled into them. The lid is a glass microscope slide held in place with a hinge of double-sided adhesive tape. The original Robertson cells had a black filter paper base to provide gas exchange. In the modified versions shown in Figure 6.7b, it was found there was sufficient gas exchange between the top of the cell and the underside of the lid and a filter paper base was unnecessary. A Kilner jar with a Perspex rack can be used to store several cells. Humidity is controlled as for a larger culture vessel, using saturated solutions of inorganic salts.

6.6.5 Food

Mites live in their food; they lay their eggs in it, they defecate in it and they eventually die in it. The particle size of the culture medium can be important in determining population growth rates. As a rule, the smaller the particle size, the poorer the growth.

Powdery media tend to encourage more fungal growth than granular media. This can be avoided by adding glass beads to help break up any conglomeration when the culture is shaken.

De-fatted human skin scales, derived from hair collected from a barber's shop, de-fatted in acetone (Spieksma, 1969), ox liver powder, baker's yeast granules, fish food flakes or dried Daphnia, dog biscuits, various cereal preparations including wheat germ, cultures of fungi, and house dust have all been used for culturing dust mites. Yeast and dried Daphnia 1:1 mix is very good for rapid growth of many species. Yeast granules alone give good growth for several species but yeast contains a streptavidin-binding factor, so if the mite extracts are to be used in immunochemical procedures involving streptavidin-biotin binding reactions, the mites must be cleared of culture medium by sieving and/or starving the mites. Axenic culture of house dust mites on chemically defined diets have not been tried but may offer the best prospect for media of low intrinsic allergenicity.

The recent findings of Barber *et al.* (1996) and Sánchez-Monge *et al.* (1996) that cereals contain a series of inhibitors of certain *Dermatophagoides* spp. allergens has some important implications for the choice of culture media in which dust mites are grown. It would appear that use of wheat germ is not advisable in mite cultures that are grown for the purpose of allergen production.

Mumcuoglu (1977a) attempted to rear various mites on simple sugars and achieved some success with *Tyrophagus putrescentiae*, but not *Dermatophagoides pteronyssinus*. Other artificial diets are detailed by Rodriguez and Lasheen (1971), Rodriguez and Blake (1979) and Singh (1977).

6.6.6 Subculturing

Subculturing needs to be done when the culture is starting to get excessively crowded. As a general rule, this can be gauged by looking at the culture medium with the naked eye and seeing if any movement can be detected. If the medium appears to be heaving or shimmering slightly, remove about two-thirds of the contents and replace with fresh culture medium. Be sure to keep the stock culture medium in its own desiccator (with no mite cultures) and at the same temperature and humidity as the mite cultures. Baker's yeast granules that have been stored at dry conditions can absorb a lot of water from the air when placed in humid conditions and may lower the humidity of the culture vessel dramatically and lethally. Dust mites can survive a relatively short duration at low humidity (Arlian, 1992) and will recover. Gross effects of water loss from mites involve shrinkage of the body: the mites appear shrivelled and flattened dorsoventrally, instead of ovoid. If they appear flattened instead of tumescent then they have probably lost a sizeable quantity of body water and should be moved to a humidity above 75% RH immediately.

Slowing of activity may indicate all is not well with the culture. Normally the mites are moving constantly, couples can be observed in copula and eggs are present. Absence of any of the above may indicate humidity is too low or too high and, as a consequence, the mites are suffering the effects of toxic fungal metabolites.

6.6.7 Isolating mites from the culture medium

This can be done either by picking out individuals with a fine brush or needle or, if many mites are required, by driving them out of the culture medium using a hot light source. When inspecting mites in culture, fibre optic light sources should always be used because the light is relatively cool and will not desiccate them. A hotter light source directed at the middle of the culture will drive mites out to the edge of the vessel where they can be scooped up with a brush. For even larger scale isolation, several cultures can be pooled and sieved through a series of Endecott test sieves (or similar) with meshes of different sizes (e.g. 1 mm, 500 μm, 250 μm, 100 μm) on a sieve-shaker to obtain a mite-rich fraction. The 100 μm mesh will result in a faecal pellet-rich fraction (pellets are ca. 10–50 μm in diameter), though it will also contain some eggs and larvae. Making a faecal pellet-rich fraction using meshes of 50 μm or below is usually not practical because the mesh gets clogged up with pellets, with very little yield. Shimomura *et al.* (1982) described a method of collecting each developmental stage from a culture using sieving.

Having generated a mite-rich fraction, the procedure of Lind and Løwenstein (1983) can be used. In short, the mites from the mite-rich sieve fragment are mixed with saturated sodium chloride in a separating funnel. Funnels must be large enough so that the passage in the tap does not become blocked when drawing off liquid/mites. The culture material will fall to the bottom of the funnel and the mites will float to the top. Repeat twice. This will provide clean mites free of culture medium. Arlian *et al.* (1979b) describe a similar method for separating mites from culture medium using an ethanol suspension. Automatic devices have been described for separating large quantities of mites from the culture medium by Woodring (1968) and Stepien and Rodriguez (1972).

Starving the mites will clear their guts of faecal material and food. This may be important if the culture medium is allergenic, but bear in mind that the faecal pellets and bodies of the mites have different allergen repertoires, e.g. Der p 1 and Der p 3 are mainly found in the faeces but Der p 2 is found mainly in mite bodies. Starving can be done by removing the mites from the culture medium using a hot light and transferring them to a clean vessel without culture medium for 24 hours. Repeating the flotation process will separate faeces from bodies, but the faeces may not then be suitable for allergen investigation because of the high water solubility of group 1 allergens.

6.6.8 Laboratory hygiene and safety

Good laboratory hygiene will keep cultures free of major fungal infection and of cross-contamination. Any cultures found to contain more than one species of mites or with visible fungal growth should be discarded into a bleach bucket. Specific details for keeping cultures sterile are detailed by Parkinson (1992).

Culturing dust mites renders the laboratory worker at risk of becoming sensitised to them. The rule is to ensure that procedures are done in a laminar flow hood or fume cupboard, and that a gown or lab coat, gloves, goggles and a mask are worn. As little skin as possible should be exposed, and culturing operations should be done in a specially designated room away from other people who may not be wearing protective clothing.

7. Dust mite allergens

Biologists work very close to the frontier between bewilderment and understanding. Biology is complex, messy and richly various, like real life; it travels faster nowadays than physics or chemistry (which is just as well, because it has so much farther to go), and it travels nearer to the ground.

Peter Medawar,
Pluto's Republic (1982),
Oxford University Press, Oxford.

7.1 Why do dust mites produce allergens?

There can be few tasks in biological research messier than emptying a vacuum cleaner bag of dust, sieving it and preparing an aqueous suspension in order to make allergen extracts or remove mites. The fine grey powder gets everywhere and is not very wettable. It sticks to glassware, and the smell of old house dust and decomposing skin scales is like a mixture of the rancid fat of dirty kitchens and the mustiness of neglected libraries. In the 1920s and 1930s it was thought house dust contained a single allergen. Making aqueous extracts of house dust would have been a standard method. By the 1950s, it was becoming clear that house dust contained a complex mixture of allergens (see Voorhorst *et al.*, 1969). By the late 1960s dust mites were known to be a major component of this mix. By the early 1970s it had been determined that the faecal pellets of dust mites contained allergens. Halmai and Alexander (1971) isolated faecal pellets from the bodies of mites using a micromanipulator and used them in skin-prick tests to demonstrate their allergenicity. In 1981, Tovey *et al.* discovered that mite faeces were a major source of the dust mite allergen Der p 1. We now know of a large

variety of allergens, most characterised to the last amino acid, many for which their original biological function is known, and several with their tertiary structure modelled.

Most allergens are biochemically active molecules and include enzymes, enzyme inhibitors, and proteins involved in molecular transport, regulation and cell and tissue structure. Reviews of mite allergens include those by Platts-Mills and Chapman (1987); International Workshop Report (1988); Platts-Mills *et al.* (1989b); Arlian (1991); Arruda and Chapman (1992); Stewart, Bird, Krska *et al.* (1992); Stewart (1994); Stewart and Thompson (1996); Colloff and Stewart (1997); Stewart and McWilliam (2001); Arlian (2002); Thomas *et al.* (2002); Kawamoto *et al.* (2002b); Stewart and Robinson (2003); and Nuttall *et al.* (2006). Most of this work deals with allergens produced by the pyroglyphid mites *Dermatophagoides pteronyssinus*, *D. farinae*, *D. microceras* and *Euroglyphus maynei*. More recently, data on storage mites such as *Lepidoglyphus destructor* and *Blomia tropicalis* have emerged, as well as the parasitic mites *Sarcoptes scabiei* and *Psoroptes ovis* which are related to the Pyroglyphidae. It is now clear that each of the major groups of dust contains several faecally derived allergens, some of which have been shown to be hydrolases corresponding to those commonly associated with digestion in both vertebrates and non-arachnid invertebrates.

The enzymatic nature of many allergens begs the question, 'What are the most abundant sources of enzymes in dust mites?' The answer is the digestive system. So what facets of digestion are responsible for the production of large amounts of allergens? The digestive system of dust mites appears to be rather inefficient by mammalian standards, and is partially intracellular (see Chapter 2). Relatively large amounts of enzymes, which in an organism that digests extracellularly would be recycled and partly reclaimed, are voided in the faecal pellet, so they are the most abundant proteins derived from dust mites present in the domestic environment. We begin to get a clue as to why the digestive enzymes of dust mites are important allergens.

In general, the spectrum of digestive enzymes detected in an organism indicates the degree of trophic specialisation, and there are some data suggesting that this may be the case with mites (Bowman, 1984). The enzymes (and, therefore, allergens) detected in faecal extracts may reflect the diet of the mite (Colloff, 1994a). For pyroglyphid mites, diet includes protein, lipid and carbohydrate derived from sources such as human skin scales, bacteria and fungi, whereas in storage mites such nutrients may be derived from seed proteins and carbohydrates as well as microorganisms. There may be quantitative and qualitative differences in the spectrum of enzymes produced by mites occupying different ecological niches and, ultimately, recognised as allergenic by susceptible individuals. For example, skin contains a number of structural proteins usually resistant to conventional proteases found in the digestive tracts of both invertebrates and vertebrates. These include collagen and keratin, and it is possible that the mite Der p 1 cysteine protease evolved to act in concert with a reducing environment in the mite gut in order to digest such proteins. Collagen and keratin are not present in grain, suggesting that a similar enzyme will not be prominent in the faeces of storage mites.

Mites are assumed to digest bacteria and fungi, and the most likely taxa to be consumed will be the common commensals such as *Staphylococcus*, *Streptococcus*, Mycobacteria, Propionibacteria and *Aspergillus* species. Whether mites with different diets such as the stored products mites possess the same range of enzymes (and, therefore, allergens) as those

produced by the pyroglyphid mites is not yet clear despite an overall qualitative similarity in the spectrum of enzymes found in whole mite extracts. A more definitive answer must await a detailed examination of mite-free faecal extracts. Despite this, interspecies similarities with regard to physicochemical and biochemical properties of allergens are present which account for the observed immunological cross-reactivity. However, the absence of cross-reactivity does not preclude the presence of a particular allergen in a mite species since functionally equivalent proteins may exist.

With new techniques in molecular biology and bioinformatics over the last decade, whole allergens have been sequenced (see Table 7.1) and tertiary structures modelled (e.g. Topham *et al.*, 1994; Mueller *et al.*, 1997; Derewenda *et al.*, 2002). Development and growth of gene and protein databases and bioinformatics tools greatly facilitated comparisons of the primary structure of allergens with other proteins of known function. Allergenic identity of different allergen groups in different mite species and complete or partial amino acid sequence data are available for over a hundred mite allergens (see Table 7.1).

As I show in this chapter, the focus on allergens as enzymes associated with mite faecal pellets (which when they become airborne are regarded as a primary vehicle for sensitisation) is central to an ageing paradigm that is in need of revision. Several major allergen groups are proteins involved in intracellular regulation and processing (e.g. groups 13, 14, 16 and 17) and are not associated with the gut at all. Others are components of muscle tissue (groups 10 and 11). Almost half of all known allergen groups are not associated with faecal pellets of dust mites (see Table 7.2). The fact that atopic patients become sensitised to these allergens means that they must be exposed not only to faecal pellets but also to particles of cell and muscle tissue. These particles, especially cellular debris, are likely to be considerably smaller than the 15–20 µm-diameter faecal pellets. Potentially, the nature of the exposure to those allergens not associated with the digestive system is quite different to that of the faecal hydrolases, which, in any event, are also carried on particles much smaller than faecal pellets (De Lucca *et al.*, 1999). The epidemiological implications of differential allergen exposure in relation to particle size are examined in more detail in Chapter 8 (see section 8.4.4).

In this chapter I have not gone into any detail on the mechanisms involved in the immune responses by

Table 7.1 Dust mite allergens for which full or partial amino acid sequences have been published and/or databased. Short N-terminal amino acid sequences (ca. ≤ 30 residues) are not included. ND = no data; *Databases: GenBank, EMBL, UniProt/SwissProt. GenBank numbers are entry/locus number (not accession numbers); (I) = incomplete sequence; N/A = not applicable (e.g. not databased). Der f = *Dermatophagoides farinae*; Der p = *D. pteronyssinus*; Der s = *D. siboney*; Eur m = *Euroglyphus maynei*; Pso o = *Psoroptes ovis*; Sar s = *Sarcoptes scabiei*; Aca s = *Acarus siro*; Ale o = *Aleuroglyphus ovatus*; Sui m = *Suidasia medanensis*; Tyr p = *Tyrophagus putrescentiae*; Gly d = *Glycyphagus domesticus*; Lep d = *Lepidoglyphus destructor*; Blo t = *Blomia tropicalis*.

Allergen name	No. amino acids (mature protein)	Calculated molecular weight (kDa)	Predicted isoelectric point	Predicted charge	N-glycosylation sites	NCBI *Entrez* accession no.*	References
Group 1							
Der f 1	223	25.191	6.14	−1.5	1	N/A	Dilworth *et al.*, 1991
	223	25.148	6.14	−1.5	1	ABO34946	Yasuhara *et al.*, 2001
	146 (I)	16.852	8.21	6.0	1	AF194431	Park *et al.*, unpublished
	107 (I)	12.277	8.21	3.5	0	AF194432	Park *et al.*, unpublished
	210 (I)	23.548	6.38	−0.5	1	AF285763	Hao *et al.*, unpublished
	196 (I)	22.037	6.77	1.0	1	DQ185509	Piboonpocanun *et al.*, unpublished
Der p 1	78 (I)	8.304	5.49	−1.5	1	DEPMHDA	Thomas *et al.*, 1988
	222	25.394	6.61	0.5	1	N/A	Chua *et al.*, 1988
	210 (I)	23.509	6.5	0.0	1	DPDERPIG	Kent *et al.*, 1992
	222	25.019	5.92	−2.0	1	DPU11695	Chua *et al.*, 1993
	117 (I)	13.421	5.5	−1.5	1	AY947536	Lucentini *et al.*, unpublished
	133 (I)	14.965	4.91	−4.5	1	AF145247	Park *et al.*, unpublished
	222	24.992	5.92	−2.0		DQ185508	Piboonpocanun *et al.*, unpublished
Eur m 1	217 (I)	24.375	6.61	0.5	1	EMB X60073	Kent *et al.*, 1992
	223	25.129	6.38	−0.5	1	AF047610	Smith *et al.*, 1999
Pso o 1	184	20.278	8.03	3.5	1	AF495854	Lee *et al.*, 2002
Sar s 1	226	25.326	9.1	12.5	0	AY525148	Holt *et al.*, 2004
Blo t 1	221	25.126	8.36	9.5	0	AF277840	Mora *et al.*, 2003
	239	27.232	6.0	−1.5	0	AY291322	Chew *et al.*, unpublished
Group 2							
Der f 2	129	14.021	6.9	1.0	0	N/A	Trudinger *et al.*, 1991
	129	14.044	6.9	1.0	0	DEPDER1	Yuuki *et al.*, 1991
	129	14.076	6.9	1.0	0	DEPDER2	Yuuki *et al.*, 1991
	129	14.034	6.9	1.0	0	DEPDER3	Yuuki *et al.*, 1991
	129	14.063	6.5	0.0	0	AB195580	Tsuuki *et al.*, unpublished
	129	14.081	7.0	1.5	0	AJ862836	Nandy *et al.*, 2003
	129	14.045	6.9	1.0	0	AJ862837	Nandy *et al.*, 2003
	158	17.628	7.8	5.0	0	AF346905	Hao *et al.*, unpublished
	129	14.035	6.1	−1.0	0	AY066008	Jin *et al.*, unpublished
	129	14.123	7.9	3.0	0	DQ185511	Piboonpocanun *et al.*, unpublished
	129	14.044	6.9	1.0	0	S70378	Okuhira, 1994
Der p 2	129	14.121	7.0	1.5	0	AF276239	Chua *et al.*, 1990a, b
Der p 2	129	14.008	7.0	1.5	0	N/A	Thomas & Chua, 1995
	129	14.088	7.4	2.5	0	DQ185510	Piboonpocanun *et al.*, unpublished
Der s 2	129	14.043	6.9	1.0	0	DQ367850	Jorge *et al.*, unpublished
Eur m 2	129	14.086	6.5	0.0	0	AF047613	Smith *et al.*, 1999
Pso o 2	126	13.468	6.5	0.0	0	AF187083	Temeyer *et al.*, 2002
Ale o 2	129	13.876	6.9	1.5	1	AY497898	Ramjan & Chew, unpublished
Sui m 2	126	13.493	7.1	2.0	0	AY497879	Reginald *et al.*, unpublished
Tyr p 2	128	13.54	8.1	3.5	0	EMB Y12690	Eriksson *et al.*, 1998
Gly d 2	128	13.79	4.7	−5.0	1	GDO249864	Gafvelin *et al.*, 2001
Gly d 2.02	125	13.366	8.4	4.0	0	GDO272216	Gafvelin *et al.*, 2001
Gly d 2.03	125	13.183	7.8	.02	0	AY288673	Chew *et al.*, unpublished
Lep d 2	125	13.108	6.5	0.0	0	N/A	Varela *et al.*, 1994
Lep d 2.0201b	125	13.108	6.5	0.0	0	LDRNAALL	Schmidt *et al.*, 1995
Lep d 2.0201a	125	13.147	6.9	1.0	0	LDLEPD1A	Schmidt *et al.*, 1995
Lep d 2.0202	125	13.175	6.9	1.0	0	AJ487973	Kaiser *et al.*, 2003
Lep d 2.0102	125	13.138	6.5	0.0	0	AJ487972	Kaiser *et al.*, 2003
Lep d 2.013	125	13.183	7.8	2.0	0	AY288142	Chew *et al.*, unpublished
Lep d 2.017	123	12.998	7.2	1.0	0	AY288143	Chew *et al.*, unpublished

continued

Table 7.1 Continued

Allergen name	No. amino acids (mature protein)	Calculated molecular weight (kDa)	Predicted isoelectric point	Predicted charge	N-glycosylation sites	NCBI Entrez accession no.*	References
Lep d 2.023	125	13.183	7.8	2.0	0	AY288144	Chew et al., unpublished
Lep d 2.024	125	13.183	7.2	1.0	0	AY288145	Chew et al., unpublished
Lep d 2.025	125	13.169	7.2	1.0	0	AY288146	Chew et al., unpublished
Lep d 2.031	125	13.125	8.1	3.0	0	AY288147	Chew et al., unpublished
Lep d 2.035	125	13.183	7.8	2.0	0	AY288148	Chew et al., unpublished
Lep d 2.039	125	13.183	7.8	2.0	0	AY288149	Chew et al., unpublished
Lep d 2.042	125	13.184	7.8	2.0	0	AY288150	Chew et al., unpublished
Blo t 2	124	13.528	6.9	1.0	0	AY288139	Chew et al., unpublished
	127	13.530	6.7	0.5	0	AY288140	Chew et al., unpublished
	126	13.518	6.3	−0.5	0	AY288141	Chew et al., unpublished
Group 3							
Der f 3	232	24.954	5.7	−2.0	1	DFU54781	Smith & Thomas, 1996
	232	24.914	5.3	−3.0	1	DEPPDF3	Nishiyama et al., 1995
Der p 3	129 (I)	14.595	5.5	−2.5	1	AF145248	Park et al., unpublished
	229	24.987	8.0	4.0	1	DPU11719	Smith et al., 1994
Eur m 3	232	25.031	6.1	−1.0	1	AF047615	Smith et al., unpublished
Sar s 3	231	25.77	8.1	5.0	2	AY333071	Holt et al., 2003
Gly d 3	219	22.734	7.4	2.0	1	AY288674	Chew et al., unpublished
Lep d 3	219	22.499	7.4	2.0	1	AY291571	Chew et al., unpublished
Blo t 3	231	23.787	8.35	5.0	0	AY090091	Flores et al., 2003
Blo t 3	231	23.824	8.8	7.0	0	N/A	Cheong et al., 2003
	231	23.843	8.1	4.0	0	AY291323	Chew et al., unpublished
Group 4							
Der p 4	496	57.15	7.2	8.5	1	AF144060	Mills et al., 1999
Eur m 4	496	57.341	7.4	10.5	1	AF144061	Mills et al., 1999
Blo t 4	484	55.055	7.02	7.5	0	AY291324	Chew et al., unpublished
Group 5							
Der f 5	113	13.614	5.6	−2.5	1	BAE45865	Tsukui et al., unpublished
Der p 5	113	13.587	4.9	−4.5	0	CAA35692	Tovey et al., 1989
	113	13.587	4.9	−4.5	0	S76340	Lin et al., 1994
	113	13.529	5.1	−3.5	0	P14004	Lin et al., 1994
	117	14.178	5.0	−3.5	0	DQ354124	Weghofer et al., unpublished
Gly d 5.01	239	27.517	5.0	−8	0	AY288675	Chew et al., unpublished
Gly d 5.02	233	26.406	5.1	−6.5	1	AY288676	Chew et al., unpublished
Lep d 5	110	12.550	6.4	−0.5	0	Q9U5P2	Eriksson et al., 2001
5.02	115	13.057	6.4	−0.5	0	AY288151	Chew et al., unpublished
5.04	115	13.057	6.4	−0.5	0	AY288152	Chew et al., unpublished
Blo t 5	72 (I)	8.373 (I)	N/A	N/A	0	AAB49396	Caraballo et al., 1996
	114	13.497	4.9	5.0	1	O96870	Arruda et al., 1995, 1997a
Group 6							
Der f 6	230	25.034	6.1	−1.0	0	AF125187	Kawamoto et al., 1999
Der p 6	231	24.885	5.3	−3.0	0	N/A	Bennett & Thomas 1996
Blo t 6	230	24.743	5.7	−2.0	0	AY291325	Chew et al., unpublished
Group 7							
Der f 7	196	21.863	4.7	−10.0	1	S80655	Shen et al., 1995b
Der p 7	198	22.179	4.6	−11.0	1	DPU37044	Shen et al., 1993
Gly d 7	200	22.198	9.7	6.5	1	AY288677	Chew et al., unpublished
Lep d 7	196	21.951	7.9	3.0	0	AJ271058	Eriksson et al., 2001
Blo t 7	177 (I)	N/A	N/A	N/A	1	AY291326	Chew et al., unpublished
Group 8							
Der p 8	219	25.589	6.8	0.5	2	S75286	O'Neill et al., 1994a
	219	25.668	8.5	4.0	2	AY825938	Dougall et al., 2005
	217	25.621	9.0	4.5	1	AY825939	Dougall et al., 2005

continued

Table 7.1 Continued

Allergen name	No. amino acids (mature protein)	Calculated molecular weight (kDa)	Predicted isoelectric point	Predicted charge	N-glycosylation sites	NCBI *Entrez* accession no.*	References
Pso o 8	219	25.944	7.8	3.5	1	AF078684	Lee *et al.*, 1999
Sar s 8	219	25.841	8.3	2.5	1	AY825933	Dougall *et al.*, 2005
Gly d 8	214	24.763	9.0	5.5	3	AY288679	Chew *et al.*, unpublished
Lep d 8	214	24.793	9.0	5.5	3	AY291572	Chew *et al.*, unpublished
Blo t 8	236	27.630	4.8	−8.0	1	AY283284	Chew *et al.*, unpublished
Group 9							
Der f 9	132 (I)	14.374	9.5	5.0	0	AF194430	Park *et al.*, unpublished
Der p 9	220	23.756	6.8	0.5	2	AF409110	King *et al.*, unpublished
	224	23.766	9.0	8.0	1	AY211952	Bi *et al.*, unpublished
	228	24.485	8.5	5.0	1	AF409111	Mathaba *et al.*, unpublished
Blo t 9	224	23.598	8.6	6.5	0	AY291327	Chew *et al.*, unpublished
Group 10							
Der f 10	284	32.955	4.4	−23.0	0	DEPMAG44	Aki *et al.*, 1995
Der p 10	284	32.901	4.5	−22.5	0	DPY14906	Asturias *et al.*, 1998
	284	32.973	4.5	−22.5	0	AF016278	Smith *et al.*, unpublished
	281	32.553	4.5	−22.5	0	DQ247970	Lucentini *et al.*, unpublished
Pso o 10	284	32.914	4.5	−20.5	0	AM114276	Nisbet *et al.*, 2006a
Tyr p 10	284	32.955	4.4	−27.5	1	AY623832	Lee *et al.*, unpublished
Gly d 10	284	32.674	4.3	−27.5	0	AY288684	Chew *et al.*, unpublished
Lep d 10	284	32.949	4.4	−23.0	0	AJ250096	Saarne *et al.*, 2003
Blo t 10	284	33.003	4.5	−22.5	0	N/A	Yi *et al.*, 2002
Der g 10	284	32.674	4.3	−27.5	2	AM167555	Nisbet *et al.*, 2006b
Group 11							
Der f 11	692 (I)	81.372	6.0	−8.0	2	AF352244	Tsai *et al.*, 1998
Der p 11	875	102.417	5.6	−16.5	4	AY189697	Lee *et al.*, 2004
Pso o 11	875	102.647	5.5	−16.5	3	AM114275	Nisbet *et al.*, 2006a
Sar s 11	679 (I)	79.784	5.8	−9.5	2	DQ131648	Zheng *et al.*, unpublished
	876	102.454	5.7	−14.0	4	AF317670	Mattsson *et al.*, 2001.
Blo t 11	875	102.028	5.7	−14.0	5	AF525465	Ramos *et al.*, 2001
Group 12							
Lep d 12	124	14.055	6.1	−3.5	0	AY293744	Chew *et al.*, unpublished
Blo t 12	125	14.337	5.6	−6.5	0	BTU27479	Puerta *et al.*, 1996a
Group 13							
Der f 13	132	14.980	6.8	0.0	0	2A0AA	Chan *et al.*, 2006
Aca s 13	130	14.297	5.3	−1.0	1	ASI6774	Eriksson *et al.*, 1999
Tyr p 13	132	14.573	6.9	0.0	1	AY710432	Jeong *et al.*, 2005a
Gly d 13	132	14.82	7.0	0.0	0	AY288679	Chew *et al.*, unpublished 2003
Lep d 13	132	14.723	7.0	0.0	0	AJ250279	Eriksson *et al.*, 2001
Blo t 13	131	14.8	5.4	−1.0	0	BTU58106	Caraballo *et al.*, 1997
	130	14.814	5.4	−1.0	0	AY283286	Chew *et al.*, unpublished
	130	14.814	5.4	−1.0	0	AY283287	Chew *et al.*, unpublished
	131	14.862	5.3	−1.5	0	AY283294	Chew *et al.*, unpublished
Group 14							
Der f 14	341 (I)	39.668	7.5	4.5	0	DEPMAG	Aki *et al.*, 1994a
Der f 14	349 (I)	40.545	7.6	6.0	1	D17686	Fujikawa *et al.*, 1996
Der p 14	1788	205.737	7.9	28.5	1	AF373221	Epton *et al.*,1999, 2001a
Eur m 14	1650	189.541	8.2	31.5	1	AF149827	Epton *et al.*, 1999
Sar s 14	115 (I)	13.174	9.2	4.5	2	BM176880	Harumal *et al.*, 2003
	330 (I)					AF462196	Fischer *et al.*, 2003
	719 (I)					DQ109676	Ljunggren *et al.*, 2006
Blo t 14	338 (I)	39.353	9.6	9.0	1	AY090092	Flores *et al.*, unpublished
Group 15							
Der f 15	535	61.111	5.3	−10.5	1	AF178772	McCall *et al.*, 2001
Der p 15	538	61.413	5.1	−13.0	1	DQ078740	O'Neill *et al.*, 2006

continued

Table 7.1 Continued

Allergen name	No. amino acids (mature protein)	Calculated molecular weight (kDa)	Predicted isoelectric point	Predicted charge	N-glycosylation sites	NCBI *Entrez* accession no.*	References
Der p 15s	512	58.868	5.1	−12.0	1	DQ078741	O'Neill *et al.*, 2006
Group 16							
Der f 16	480	55.131	6.2	−2.5	3	AF465625	Kawamoto *et al.*, 2002a
Group 18							
Der f 18	437	49.468	6.0	−7.5	2	AYO93656	Weber *et al.*, 2003
Der p 18	437	49.227	5.8	−8.5	2	DQ078739	O'Neill *et al.*, 2006
Blo t 18	437	49.231	5.8	−8.0	2	AY291330	Chew *et al.*, unpublished
Group 19							
Blo t 19	70	7.226	8.4	5.0	0	N/A	Nge & Chua, unpublished
Group 20							
Der p 20	356	40.477	8.5	7.5	0	N/A	Thomas, unpublished
Group 21							
Der p 21	125	14.909	5.0	−3.5	0	ABC73706	Weghofer *et al.*, unpublished
Blo t 21	129	14.944	5.6	−2.5	0	AY800348	Gao *et al.*, 2007
Other allergens							
D. farinae							
Alt a homologue	490	54.197	5.4	−5.5	2	AY283296	Chew *et al.*, unpublished
Mag 29 (HSP 70)	145 (I)	15.594	4.4	−10.5	0	DEPMAG29	Aki *et al.*, 1994b
CPW1 (cathepsin F)	127	14.057	4.5	−5.5	2	AF145249	Park *et al.*, unpublished
D. pteronyssinus							
13.8 kDa enzyme	134	14.064	8.6	8.0	0	AF409109	Mathaba *et al.*, 2002
T. putrescentiae							
α-tubulin	450	50.037	4.8	−15.5	1	AY986760	Jeong *et al.*, 2005b
S. medanensis							
Profilin	130	14.127	6.5	0.0	1	AY803196	Reginald *et al.*, unpublished
G. domesticus							
Jun a 3 (thaumatin)	213	23.096	6.6	1.0	0	AY288680	Chew *et al.*, unpublished
Mal d 3 (lipid transfer protein)	90	9.476	7.3	2.0	0	AY288681	Chew *et al.*, unpublished
Mala s 6 (cyclophilin)	197	21.609	8.7	4.0	1	AY288682	Chew *et al.*, unpublished
Enolase homologue	362	39.639	5.5	−4.5	4	AY288683	Chew *et al.*, unpublished
L. destructor							
α-tubulin	450	50.038	4.8	−15.5	1	AJ428050	Saarne *et al.*, 2003
Cystatin	98	11.293	5.5	−2.5	0	AJ428051	Kaiser, unpublished
B. tropicalis							
Mag 29 (HSP 70)	650	71.495	5.2	−9.0	5	AY291333	Chew *et al.*, unpublished
Aldehyde dehydrogenase	416	45.241	5.6	−3.0	2	AY291328	Chew *et al.*, unpublished
Gal d 1 (IGF-binding protein)	248	27.828	4.4	−12.5	1	AY291331	Chew *et al.*, unpublished
Profilin	130	14.127	6.5	0.0	0	AY291334	Chew *et al.*, unpublished
Alt a 6 (ribosomal protein P2)	114	11.674	4.3	−8.0		AY291329	Chew *et al.*, unpublished
Hexosaminidase	341	39.432	5.3	−6.0	3	AY291332	Chew *et al.*, unpublished
Ves m 1 (lipase)	260	28.941	7.2	5.0	1	AY291335	Chew *et al.*, unpublished
Cathepsin D homologue	384 / 384	42.516 / 42.574	7.6 / 7.6	5.5 / 6.5	4 / 4	AY792951 / AY792952	Chew *et al.*, unpublished / Chew *et al.*, unpublished

*some are nucleotide and some are protein database numbers

atopic patients to mite allergens. The literature on this topic is vast and complex and its inclusion would greatly increase the length of this chapter, probably to no great end. This book is mainly about the biology of dust mites. I have tried therefore to remain focused on the biological role of allergens within the mites, with some consideration of those properties that render them allergenic.

Table 7.2 Functions of mite allergen groups. Loc. = location within mites; G = gut digestive cells/faecal pellets; M = muscle cells; I = intracellular (other than midgut cells). Cysteine residues: refers to number of conserved residues in aligned sequences (see Figures 7.9–7.28). Unpublished references refer to entries of sequence data on databases, see Table 7.1.

Group	Mite spp.	Identity/function	Loc.	References
1	Df, Dm, Dp, Ds, Em, Po, Ss, Bt	Cysteine peptidase	G	Chapman & Platts-Mills, 1980; Stewart & Turner, 1980; Lind, 1986b; Stewart & Fisher, 1986; Chua *et al.*, 1988; Kent *et al.*, 1992; Ferrándiz *et al.*,1995b; Smith *et al.*, 1999; Mora *et al.*, 2003
2	Df, Dp, Ds, Em, Po, Ao, Tp, Sm, Gd, Ld, Bt	Lipid binding protein	I	Heymann *et al.*, 1986, 1989; Yuuki *et al.*, 1991; Trudinger *et al.*, 1991; Valera *et al.*, 1994; Schmidt *et al.*, 1995; Eriksson *et al.*, 1998; Gafvelin *et al.*, 2001; Temeyer *et al.*, 2002; Kaiser *et al.*, 2003; Ichikawa *et al.*, 2005; Keber *et al.*, 2005
3	Df, Dp, Ds, Em, Ss, Gd, Ld, Bt	Trypsin	G	Heymann *et al.*, 1989; Stewart *et al.*, 1989, 1992, 1994; Ando *et al.*,1993; Smith *et al.*, 1994; Nishiyama *et al.*, 1995; Smith & Thomas, 1996; Ferrándiz *et al.*, 1997; Flores *et al.*, 2003; Holt *et al.*, 2003; Cheong *et al.*, 2003
4	Dp, Em, Bt	Alpha-amylase	G	Lake *et al.*, 1991; Mills *et al.*, 1999
5	Df, Dp, Gd, Ld, Bt	Structural protein		Tovey *et al.*, 1989; Lin *et al.*, 1994; O'Neill *et al.*, 1994b; Caraballo *et al.*, 1996; Arruda *et al.*, 1997a; Eriksson *et al.*, 2001; Liaw *et al.*, 2001; Kuo *et al.*, 2003; Yi *et al.*, 2004
6	Df, Dp, Bt	Chymotrypsin	G	Yasueda *et al.*, 1993; King *et al.*, 1996; Bennett & Thomas 1996; Kawamoto *et al.*, 1999
7	Df, Dp, Gd, Ld, Bt	Function unknown		Shen *et al.*, 1993, 1995a, 1995b, 1996; Eriksson *et al.*, 2001
8	Dp, Po, Ss, Gd, Ld, Bt	Glutathione-s-transferase	?I	O'Neill *et al.*, 1994; Dougall *et al.*, 2005; Lee *et al.*, 1999; C. Huang *et al.*, 2006
9	Df, Dp, Bt	Collagenase	G	King *et al.*,1996
10	Df, Dp, Po, Tp, Gd, Ld, Dg	Tropomyosin	M	Aki *et al.*, 1995; Asturias *et al.*, 1998; Yi *et al.*, 2002; Saare *et al.*, 2003; Huntley *et al.*, 2004; Nisbet *et al.*, 2006a, b
11	Df, Dp, Po, Ss, Bt	Paramyosin	M	Tsai *et al.*, 1998; Ramos *et al.*, 2001; Mattson *et al.*, 2001; Huntley *et al.*, 2004; Lee *et al.*, 2004; Teo *et al.*, 2006
12	Ld, Bt	Contains peritrophin A domain: chitin binding	I	Puerta *et al.*, 1996a
13	Df, As, Tp, Gd, Ld, Bt	Fatty acid-binding protein	I	Caraballo *et al.*, 1997; Eriksson *et al.*, 2001; Jeong *et al.*, 2005a; Chan *et al.*, 2006
14	Df, Dp, Em, Po, Ss, Bt	Vitellogenin: egg yolk storage protein	I	Aki *et al.*, 1994a; Fujikawa *et al.*, 1996; Mattsson *et al.*, 1999; Epton *et al.*, 1999, 2001a; Harumal *et al.*, 2003; Huntley *et al.*, 2004; Ljundgren *et al.*, 2005
15	Df, Dp	Family 18 chitinase	I	McCall *et al.*, 2001; O'Neill *et al.*, 2006
16	Df	Gelsolin: actin-binding	I	Kawamoto *et al.*, 2002a
17	Df	Calcium-binding protein	I	Tategaki *et al.*, 2000
18	Df, Dp, Bt	Chitinase	I	Weber *et al.*, 2003; O'Neill *et al.*, 2006
19	Bt	Anti-microbial peptide	G	Thomas *et al.*, 2002
20	Dp	Arginine kinase	?I	Thomas, unpublished
21	Bt, Dp	Structural protein	G	Gao *et al.*, 2007; Weghofer, unpublished
Dpt4	Dp	Lipoprotein		Stewart *et al.*, 1983
13.8 kDa bacteriolytic enzyme	Dpt, Df	P60 hydrolase		Mathaba *et al.*, 2002

continued

Table 7.2 Continued

Group	Mite spp.	Identity/function	Loc.	References
Major allergen 1	Ss			Fischer *et al.,* 2003
Alpha-tubulin	Tp, Ld	Alpha-tubulin	I	Jeong, 2005b; Saarne *et al.,* 2003
Mag 29	Df, Bt	Heat shock protein 70	I	Aki *et al.,* 1994b
CPW1	Df	Cathepsin F homologue		Park *et al.,* unpublished
39 kDa allergen	Ld	Function unknown		Ansotegui *et al.,* 1991
79/93 kDa allergen	Ld	Function unknown		
Cystatin	Ld	Cystatin A		Kaiser *et al.,* unpublished
Aldehyde dehydrogenase	Df, Bt	Aldehyde dehydrogenase		Chew *et al.,* unpublished
Gal d 1 homologue	Bt	Ovomucoid allergen; insulin growth factor-binding protein		Chew *et al.,* unpublished
Enolase	Gd	Phosphopyruvate hydratase		Chew *et al.,* unpublished
Mal d 3 homologue	Gd	Type 1 non-specific lipid transfer protein		Chew *et al.,* unpublished
Mala s 6 homologue	Gd	Peptidylprolyl isomerase B		Chew *et al.,* unpublished
Jun a 3 homologue	Gd	Thaumatin homologue		Chew *et al.,* unpublished
BTP1	Bt			Chew *et al.,* unpublished
Profilin	Sm, Bt	Plant pan-allergen		Chew *et al.,* unpublished
Hexosaminidase	Bt	Beta-*n*-acetyl hexosaminidase		Chew *et al.,* unpublished
Alt a 6 homologue	Bt	Ribosomal protein P2		Chew *et al.,* unpublished
Ves m 1 homologue	Bt	Lipase		Chew *et al.,* unpublished
Cathepsin D homologue	Bt	Cathepsin D homologue		Chew *et al.,* unpublished

Ao = *Aleuroglyphus ovatus*; As = *Acarus siro*; Bt = *Blomia tropicalis*; Df = *Dermatophagoides farinae*; Dm = *D. microceras*; Dp = *D. pteronyssinus*; Ds = *D. siboney*; Em = *Euroglyphus maynei*; Gd = *Glycyphagus domesticus*; Ld = *Lepidoglyphus destructor*; Po = *Psoroptes ovatus*; Sm = *Suidasia medanensis*; Ss = *Sarcoptes scabiei*; Tp = *Tyrophagus putrescentiae*; Dg = *Dermanyssus gallinae*.

7.2 Which mites produce clinically important allergens?

Probably any mite, or indeed virtually any organism, is capable of producing proteins and peptides to which humans could become allergic, given the right circumstances. In fact, only a few hundred organisms, out of millions, are known to have allergenic properties. The ones that tend to cause problems are ones that humans are likely to come into contact with on a regular basis. There has always been an intimate association between surveys of mites in homes and assessment of mite allergenicity. The mites that were investigated for their allergenicity were the ones that were recorded repeatedly in houses. The establishment of the distribution and abundance of species provided the basis for investigations of their allergenicity. For example, some 35 species of mites had been recorded from house dust in Japan, with *D. farinae* known to be a source of allergens similar to those isolated from house dust, having a mean relative abundance of only 4% (Oshima, 1970). This led Miyamoto *et al.* (1969) to conclude that in order to account for the clinical significance of allergens of *D. farinae*, they must cross-react with those of more abundant mites. They did, with those of *D. pteronyssinus* which had a relative mean abundance of about 30%. Later, when the high prevalence and abundance of *Blomia tropicalis* was recognised in the Caribbean and Latin America, patterns of sensitisation and cross-reactivity of this and other dust mites, such as *D. siboney*, began to be investigated (Puerta *et al.*, 1991; Ferrándiz *et al.*, 1995a, b).

What follows is a summary of research on those superfamilies, families and genera of mites known to cause allergies (see Table 7.3).

7.2.1 Pyroglyphidae

a Dermatophagoides

The allergens of members of this genus are by far the most thoroughly investigated. Sensitisation to *D. farinae* (then called *D. culinae*, a junior synonym of *D. farinae*, see Appendix 1) using skin-prick testing with aqueous extracts of the mites was reported by Pepys *et al.* (1968), Spieksma and Voorhorst (1969) and to *D. pteronyssinus* by Spieksma and Voorhorst (1969) and Miyamoto *et al.* (1969). In the 1970s and 1980s, scores of publications detailing allergic reactions to *D. pteronyssinus* and *D. farinae* were published. There are too many to mention here, but the relevant abstracts can be found in the *Review of Applied Entomology (Series B, Medical and Veterinary)*, volumes 58–78 (1970–1990).

During the 1970s there were several reports of trials of hyposensitisation of allergic patients with crude extracts of *Dermatophagoides* spp. (Maunsell *et al.*, 1971; Gaddie *et al.*, 1976; Gabriel *et al.*, 1977) as well as measurements of anti-*Dermatophagoides* IgE antibodies for diagnosis of mite allergy, for epidemiological studies (Turner *et al.*, 1975) and for monitoring the course of immunotherapy (Nakamura and Yoshida, 1977). It became apparent that the crude extracts of dust mites contained a mixture of allergens, and research turned to attempts at isolation and purification (Ishii *et al.*, 1973; Biliotti *et al.*, 1975; Baldo *et al.*, 1977; Nakagawa *et al.*, 1977;

Chapman and Platts-Mills, 1978; Kabasawa and Ishii, 1979).

The first major allergen, now known as Der p 1, was purified and characterised independently by Stewart and Turner (1980) and Chapman and Platts-Mills (1980), shortly followed by Der f 1 by Dandeu *et al.* (1982), Der p 2 (Lind, 1985) and Der f 2 (Yasueda *et al.*, 1986). These allergens were given informal names before current nomenclature was introduced (see Table 7.4). There followed a period of intense research activity involving further physicochemical characterisation of the allergens, production of monoclonal antibodies and mapping of epitopes, characterisation of further allergens and defining their effects on the immune system. Most of this research has been summarised in the proceedings of the first two international workshops on dust mite allergens and asthma (International Workshop Report, 1988; Platts-Mills *et al.*, 1992).

Extensive cross-reactivity between *D. pteronyssinus* and *D. farinae* has long been known (Miyamoto *et al.*, 1969; Dasgupta and Cunliffe, 1970). Platts-Mills *et al.* (1986a) found that a major allergen of *D. pteronyssinus* (Der p 1) possessed both species-specific epitopes and ones that cross-reacted with those of Der f 1 of *D. farinae*. Further, Der m 1 from *D. microceras* and Der s 1 from *D. siboney* show close physicochemical and immunological similarities with Der p 1 and Der f 1 (Lind, 1986a, b; Ferrándiz, 1997; Ferrándiz *et al.*, 1998).

b Euroglyphus

There is increasing recognition that *Euroglyphus maynei* is common in house dust throughout the world and is a major source of mite-derived allergens in house dust (reviewed by Colloff, 1991c). Voorhorst *et al.* (1969) first reported that extracts of *E. maynei* produced positive skin-prick test reactions and that *E. maynei* and *D. pteronyssinus* contained the same or a similar allergen. Platts-Mills *et al.* (1986a) found that extracts of *E. maynei* incubated with anti-Der p 1 sera inhibited more than 90% of Der p 1 binding activity, indicating that *E. maynei* produces an allergen which cross-reacts with Der p 1. Mumcuoglu (1977b, c) found *E. maynei* extracts elicited the most frequent and severe positive scratch test reactions of nine mite species and noted some cross-reactivity between *D. pteronyssinus*, *D. farinae* and *E. maynei* as well as species-specific allergens. Using the Schultz-Dale test of smooth-muscle reaction of guinea pigs, Mumcuoglu (1977c) found that *E. maynei* extracts evoked positive responses only in those animals that had been

Table 7.3 Genera of mites known or suspected of producing allergens. Abbreviations for evidence of the presence of allergens in mites: S = skin test or RAST with mite extract; C = clinical evidence; P = purification and/or amino acid sequencing of allergens. See text for references.

Order, family and genus	Major habitat	Evidence of allergens	Most important setting for exposure
ASTIGMATA			
Pyroglyphidae			
Dermatophagoides	House dust	S,C,P	Domestic
Euroglyphus	House dust	S,C,P	Domestic
Malayoglyphus	House dust	S	Domestic
Sturnophagoides	House dust/nests of birds	S	Domestic
Acaridae			
Acarus	Stored products	S,C,P	Occupational, esp. bakers and farmers
Tyrophagus	Stored products	S,C,P	Occupational
Aleuroglyphus	Stored products	S,P	Occupational
Suidasia	House dust/stored products	S,C,P	Domestic/occupational
Thyreophagus	Stored products	S,C	Occupational
Glycyphagidae			
Glycyphagus	House dust/stored products	S,P	Domestic/occupational
Lepidoglyphus	House dust/stored products	S,C,P	Domestic/occupational
Gohieria	House dust/stored products	S	Domestic/occupational
Austroglycyphagus	House dust/nests of birds and mammals	P	Domestic
Hemisarcoptidae			
Hemisarcoptes	Predator of agricultural pests	S,C	Occupational
Echymyopodidae			
Blomia	House dust/stored products	S,C,P	Domestic/occupational
Chortoglyphidae			
Chortoglyphus	House dust/stored products	S	Domestic/occupational
Sarcoptidae			
Sarcoptes	Ectoparasite of mammals	S,C,P	Domestic
Psoroptidae			
Psoroptes	Ectoparasite of mammals	S,C,P	Occupational
PROSTIGMATA			
Cheyletidae			
Cheyletus	House dust/stored products	S,C	Domestic/occupational
Pyemotidae			
Pyemotes	Stored products	S,C	Occupational
Tetranychidae			
Panonychus	Pest of horticultural crops	S,C	Occupational
Tetranychus	Pest of horticultural crops	S,C	Occupational
MESOSTIGMATA			
Dermanyssus gallinae	Nests of birds in houses	C,P	Domestic/occupational
Phytoseiulus persimilis	Predator of phytophagous mites	S,C	Occupational: greenhouse workers
Hypoaspis miles	Predator of phytophagous mites	S,C	Occupational: greenhouse workers
Amblyseius cucumeris	Predator of phytophagous mites	S,C	Occupational: greenhouse workers

Table 7.4 Synonyms of informal names of mite allergens.

Informal or original name	Formal name/new designation	Reference
Ag 11	Der f 1	Le Mao *et al.*, 1981; Dandeu *et al.*, 1982
Bt-M	Blo t 5	Caraballo *et al.*, 1996
DF1	Der f 1	Yasueda *et al.*, 1986
DF2	Der f 2	Yasueda *et al.*, 1986
DF5	Der f 6	Yasueda *et al.*, 1993
Df6	Der f 1	Lind, 1986b
Df 11	Der f 1	Le Mao *et al.*, 1985
Df642	Der f 11	Tsai *et al.*, 1998
Dm6	Der m 1	Lind, 1986b
Dp1	Der p 1	Yasueda *et al.*, 1989b
DP2	Der p 2	Yasueda *et al.*, 1989b
DP5	Der p 6	Yasueda *et al.*, 1993
λ Dp 15	Der p 8	O'Neill *et al.*, 1994a
Dp42	Der p 1	Lind, 1985, 1986b
Dpt 12	Der p 1	Stewart, 1982; Stewart & Turner, 1980
Dpt 22	Der p 2	Chua *et al.*, 1988
DpX	Der p 2	Lind, 1985
Lep d 1	Lep d 2	Varela *et al.*, 1994; Schmidt *et al.*, 1995
M-177	Der f 14	Fujikawa *et al.*, 1998
Mag 1	Fragment of Der f 14	Aki *et al.*, 1994a
Mag 3	Fragment of Der f 14	Fujikawa *et al.*, 1996
Mag 15	Der f 16	Tategaki *et al.*, 2000; Kawamoto *et al.*, 2002a
Mag 44	Der f 10	Aki *et al.*, 1995
Mag 50	Der f 17	Tategaki *et al.*, 2000
MSA1 (not ASA1)	Sar s 14	Mattsson *et al.*, 1999
Me1	Der f 2	Haida *et al.*, 1985
Me2	Der f 1	Yamashita *et al.*, 1989
P_1	Der p 1	Chapman & Platts-Mills, 1980
Po16k	Pso o 2	Temeyer *et al.*, 2002
Ssag1	Sar s 14	Harumal *et al.*, 2003

sensitised with *D. pteronyssinus* extract, and not those sensitised with extracts of the stored products mites *Tyrophagus putrescentiae*, *Chortoglyphus arcuatus* and *Lepidoglyphus destructor*. The likelihood of common cross-reacting allergens between *E. maynei* and *D. pteronyssinus* was reinforced by the finding that *D. pteronyssinus*, which was the numerically dominant species in 71% of homes and was present in 89% of homes, gave positive skin tests in 75% of the patients, while *E. maynei*, which caused 85% of positive skin tests, was numerically dominant in only 17% of homes and was present in 37% (Mumcuoglu, 1977b). A similar lack of correlation between skin-prick test

positivity and frequency of occurrence of *E. maynei* was noted by Charpin *et al.* (1986). In marked contrast to Mumcuoglu's findings, Nannelli *et al.* (1983) found only 0.6% positive skin-prick tests to *E. maynei* in atopic individuals.

Van Hage-Hamsten and Johansson (1989) investigated the cross-reactivity of extracts *of E. maynei* with those of *Acarus siro*, *Lepidoglyphus destructor* and *D. pteronyssinus* using sera from Swedish farmers. They found positive reactions to *E. maynei* in 4.5% of sera that were mite-positive by RAST (radioallergosorbent test), compared with 5.2% for *D. pteronyssinus*. Some 8% of farmers had IgE against *E. maynei* only, indicating *E. maynei* has species-specific allergens. This was confirmed by RAST inhibition studies whereby *D. pteronyssinus* extract failed to inhibit *E. maynei*-positive sera from binding to paper discs coated with *E. maynei* extract. An allergen seemingly homologous with Der p 1 was found to be present in *E. maynei*-rich dust. Van Hage-Hamsten and Johansson (1989) found moderate cross-reactivity with *D. pteronyssinus* but none with *Lepidglyphus destructor* or *Acarus siro*. Although *E. maynei* possesses a group 1 allergen homologue, as evidenced by Kent *et al.* (1992), who reported 85% amino acid sequence homology with Der p1 and Der f 1, Arruda and Chapman (1992) found no evidence for *E. maynei* group 2 allergens. However, a 14 kDa allergen was identified by immunoblotting by Colloff *et al.* (1992a) and group 2 allergens have been purified (Smith *et al.*, 1999). Major *E. maynei*-specific IgE-binding components of 16, 38 and 62 kDa have been identified by immunoblotting, as well as components common to *D. pteronyssinus* of 14, 24, 97 and 175 kDa using sera from atopic people who had domestic exposure to both species.

c Malayoglyphus and Sturnophagoides

In a survey of dust mites in Singapore, Chew *et al.* (1999a) found *Malayoglyphus intermedius* and *Sturnophagoides brasiliensis* in 10–31% and 42–84% of dust samples. They found 50–70% of atopics were sensitised to these species. A more detailed study on frequency of sensitisation found 72% of patients with allergic asthma or rhinitis were skin-prick test positive to *S. brasiliensis* (Chew *et al.*, 1999b), fifth in rank frequency after *Blomia tropicalis*, *Dermatophaogides farinae*, *D. pteronyssinus* and *Austroglycyphagus malaysiensis*.

7.2.2 Acaroidea

a Acarus

The cosmopolitan genus *Acarus* contains some of the most important pests of stored products (Hughes, 1976), although its members occur in a wide range of habitats including house dust and agricultural soils, grasslands, poultry houses and birds' nests. Exposure to these mites can therefore occur in domestic, agricultural or occupational settings. Sensitisation to *A. farris* and *A. siro* was reported among farmers by Ingram *et al.* (1979) and grain workers by Blainey *et al.* (1989). Allergy to *Acarus siro* was detailed among asthmatics by Maunsell *et al.* (1968) and by Pepys *et al.* (1968); among asthmatic children versus non

asthmatic controls by Morrison Smith *et al.* (1969); atopic outpatients attending an allergy clinic by Spieksma and Voorhorst (1969); cheese-makers by Molina *et al.* (1975, 1977); patients with asthma or rhinitis by Wraith *et al.* (1979); farmers by van Hage-Hamsten *et al.* (1987); asthmatic children by Boner *et al.* (1989); and bakers by Revsbech and Dueholm (1990). *Acarus siro* has been implicated in cases of anaphylaxis following ingestion of contaminated food in France (Dutau, 2002). Cross-reactivity of *A. siro* with dust mites and other storage mites was investigated by Mumcuoglu (1977b), Griffin *et al.* (1989) and Johansson *et al.* (1994). Aca s 13, a fatty-acid binding protein, shows significant sequence homology with other group 13 allergens and the recombinant protein was bound by 23% of patients who were RAST-positive to *A. siro* extract (Eriksson *et al.*, 1999).

b Tyrophagus

Exposure to species of *Tyrophagus* can, like *Acarus*, occur in domestic and occupational circumstances. Farmers, grain handlers, bakers, cheese-makers and small-goods processors have become sensitised (summarised by Eriksson *et al.*, 1998), mostly to what has been reported as *Tyrophagus putrescentiae*, although occupational exposure is also likely to *T. longior*, *T. nieswanderi*, *T. palmarum* and *T. perniciosus*, species that are significant pests of stored products (Hughes, 1976). Czernecki and Kraus (1978) reported an outbreak of dermatitis in a butcher's family, exposed to *T. dimidiatus* growing on mouldy bacon.

Sensitisation to *Tyrophagus putrescentiae* was reported by Spieksma and Voorhorst (1969); Miyamoto *et al.* (1969); Araujo-Fontaine *et al.* (1974); Green and Woolcock (1978); Ingram *et al.* (1979); Blainey *et al.* (1989); and Heyraud *et al.* (1989). Allergenic properties were investigated by Arlian *et al.* (1984a). Cross-reactivity of *Tyrophagus* spp. with dust mites and other storage mites was investigated by Arlian *et al.* (1984b); Griffin *et al.* (1989); and Johansson *et al.* (1994). Rufli (1970) reported skin-prick test cross-reactivity between *D. pteronyssinus* and *T. putrescentiae*, as did Mumcuoglu (1977b). The group 2 allergen was characterised from *T. putrescentiae* by Eriksson *et al.* (1998).

c Aleuroglyphus

The only allergenically important species in this genus, *Aleuroglyphus ovatus*, is found in stored products, mammal nests, farms and human dwellings. Sensitisation to *A. ovatus* was reported by Miyamoto *et al.* (1969) and Puerta *et al.* (1993). Prevalence of specific IgE against this species was reported by Silton *et al.* (1991). *Aleuroglyphus* sp. has been implicated in cases of anaphylaxis following ingestion of contaminated food in France (Dutau, 2002).

d Suidasia

The only known species of this genus of allergenic significance is *Suidasia medanensis*. It is probably a commensal species associated with bees (Hughes, 1976), but also crops up in stored food and houses. Sensitisation to *S. medanensis* was reported by Miyamoto *et al.* (1969). Some 73% of asthmatic patients in Cartagena, Colombia had positive IgE reactivity to *S. mendanensis*, and cross-reactivity was demonstrated between *S. medanensis*, *Blomia tropicalis* and *Dermatophagoides farinae* (Puerta *et al.*, 2005). An un-named *Suidasia* sp. was found in large numbers in flour eaten by Venezuelan patients who had suffered anaphylaxis (Sánchez-Borges *et al.*, 1997, 2001) and *Suidasia nesbitti* has been implicated in cases of anaphylaxis following ingestion of contaminated food in France (Dutau, 2002).

e Rhizoglyphus

Rhizoglyphus contains several species that are pests of tubers and corms such as potatoes, onions and tulip bulbs. Occupational atopic dermatitis caused by *Rhizoglyphus* sp. was reported by Pigatto *et al.* (1991).

f Thyreophagus

Exposure of German farmers to *Thyreophagus* sp. was documented by Musken *et al.* (2003 and earlier papers cited therein). *Thyreophagus entomophagus* has been implicated in cases of anaphylaxis following ingestion of contaminated food in France (Dutau, 2002).

g Hemisarcoptes

This genus contains several egg predators of scale insects. Arlian *et al.* (1999a) found *H. cooremani* to be allergenic and, using immunoblotting, identified two putative allergens with molecular weights of 16 and 19 kDa.

7.2.3 Glycyphagoidea

a Glycyphagus

Sensitisation to *Glycyphagus domesticus* has been reported by Maunsell *et al.* (1968); Pepys *et al.* (1968); Voorhorst *et al.* (1969); Ingram *et al.* (1979); van Hage-Hamsten *et al.* (1987); Blainey *et al.* (1989); and to *Glycyphagus privatus* by Boner *et al.* (1989). Cross-reactivity

of *G. domesticus* and *G. privatus* with dust mites and other storage mites was investigated by Mumcuoglu (1977b). Some 11 allergens from *G. domesticus* have been characterised and sequenced (see Tables 7.1, 7.2).

b Lepidoglyphus

Lepidoglyphus destructor is one of the major species responsible for both occupational allergies due to exposure to contaminated stored products mite and exposure within the home. Sensitisation to this species has been reported by Spieksma and Voorhorst (1969); Araujo-Fontaine *et al.* (1974); Ingram *et al.* (1979); van Hage-Hamsten *et al.* (1987); Boner *et al.* (1989); Blainey *et al.* (1989); and Heyraud *et al.* (1989). Allergen components of an extract of *L. destructor* were investigated by immunoblotting by Johansson *et al.* (1988). Cross-reactivity with dust mites and other storage mites was investigated by Mumcuoglu (1977b); Griffin *et al.* (1989); and Johansson *et al.* (1991). Monoclonal antibodies to *L. destructor* allergens were produced by Ventas *et al.* (1991) and Härfast *et al.* (1992). A 39 kD allergen was identified by Ansotegui *et al.* (1991), and cDNA analysis of Lep d 1 was reported by Schmidt *et al.* (1995). To date, about 12 allergens from *L. destructor* have been characterised and sequenced (see Tables 7.1, 7.2).

c Gohieria

Gohieria fusca is a common stored-products species found in house dust worldwide. Cross-reactivity of *G. fusca* with dust mites and other storage mites was investigated by Mumcuoglu (1977b). Sensitisation to this species has been reported by Boner *et al.* (1989).

d Blomia

Blomia kulagini is probably the same species as *Blomia tropicalis* (see Chapter 1). However, they have been treated as separate species in studies of allergens. *B. tropicalis* is widely distributed in the tropics and subtropics (see Chapter 4). Sensitisation to *Blomia* was reported by Miyamoto *et al.* (1969) using skin-prick tests. A high prevalence of sensitisation to this mite occurs in tropical countries (Chew *et al.*, 1999b; Kuo *et al.*, 1999, 2003). Multiple, species-specific allergens were detected by Arlian *et al.* (1993). Immune responses to *B. kulagini* were investigated by van Hage-Hamsten *et al.* (1990a, b). Positive nasal challenge responses to an extract of *B. tropicalis* were recorded by Stanaland *et al.* (1996). Allergic cross-reactivity with *Lepidoglyphus destructor* and clinical significance were investigated by van Hage-Hamsten *et al.* (1990a) and Puerta *et al.* (1991), and cross-reactivity of Blo t 5 with Der p 5 by Kuo *et al.* (2003). Arruda *et al.* (1995) purified and identified the allergen Blo t 5; Arruda *et al.* (1997a) identified Der p 5 as a major sensitising allergen. Y. Gao *et al.* (2007) purified Blo t 21. Some 24 examples of allergens from *B. tropicalis* have been characterised and sequenced (Tables 7.1, 7.2), and the solution structure of Blo t 5 has been detailed by Naik *et al.* (2008; see Figure 7.6b).

e Chortoglyphus

Sensitisation to *Chortoglyphus arcuatus* was reported by Miyamoto *et al.* (1969) and Puerta *et al.* (1993). Cross-reactivity of *C. arcuatus* with dust mites and other storage mites was investigated by Mumcuoglu (1977b) and Garcia-Robaina *et al.* (1998).

f Austroglycyphagus malaysiensis

In a survey of dust mites in Singapore, Chew *et al.* (1999a) found *Austroglycyphagus malaysiensis* in 20–54% of dust samples and 50–70% of atopics were sensitised to this species. A more detailed study on frequency of sensitisation found 72% of patients with allergic asthma or rhinitis were skin-prick test positive to *S. brasiliensis* (Chew *et al.*, 1999b), the fourth most important mite in frequency of allergenicity for this patient group.

7.2.4 Psoroptoidea

a Sarcoptes

The genus *Sarcoptes* contains *S. scabiei*, the causative agent of scabies in humans and other mammals. Infestation with *S. scabiei* causes an allergic skin reaction, lesions and itching. Falk and Bolle (1980a, b) found scabies infection stimulated production of IgE antibodies that cross-reacted with those of *Dermatophagoides* and that *S. scabiei* also produced specific allergens, as did Arlian *et al.* (1988). *S. scabiei* homologues of *Dermatophagoides* allergens have been identified (groups 1, 3, 8 and 11; Table 7.2).

b Psoroptes

Psoroptes is a genus that includes several species that cause mange in mammals, including *P. ovis*, the agent of sheep scab. Psoroptic mange is a form of allergic dermatitis affecting livestock. It is a fatal disease of considerable economic significance that requires stock to be quarantined if there is any likelihood they are infected. Cross-reactivity between *Psoroptes* spp. and *D. pteronyssinus* allergens was detailed by Stewart and Fisher (1986). A search for potential vaccines led to the identification, cloning and sequencing of group 2 allergens (Pruett, 1999; Temeyer *et al.*, 2002). Homologues of group 10, 11 and 14 allergens have been identified by Huntly *et al.* (2004) and Nisbet *et al.* (2006a).

7.2.5 Cheyletidae

a Cheyletus

This genus is a member of the Prostigmata. *Cheyletus* spp. have piercing-sucking mouthparts (see Chapter 2, Figure 2.1) and are usually predators of other mites (Wharton and Arlian, 1972b). Sensitisation to *C. malaccensis* was documented by Morita *et al.* (1975). *C. eruditus* and *C. malaccensis* and a member of a related genus, *Chelacaropsis*, have been recorded as biting humans, feeding on tissue fluid (Yoshikawa, 1980, 1987). These mites cause itchy papular skin lesions (Yoshikawa, 1980, 1985; Yoshikawa *et al.*, 1983; Yamada *et al.*, 1988). Liquid-feeding mites do not produce faecal pellets, so the source of their allergens is likely to be the salivary glands. Over half of a group of asthmatics skin-prick tested with an extract of *Cheyletus* were found to be positive to it, and extracts of house dust and a *Dermatophagoides* sp. (Pérez Lozano, 1979). Some preliminary isolation and characterisation of the allergens of *Cheyletus eruditus* using immunoblotting was reported by Musken *et al.* (1996), including a possible dust mite group 4 allergen.

7.2.6 Pyemotidae

a Pyemotes

This genus, also a member of the Prostigmata and having piercing-sucking mouthparts, was identified as the cause of asthmatic and allergic symptoms in grain workers by numerous authors from the mid-19th century onwards (reviewed by Alexander, 1984). *Pyemotes ventricosus* causes allergic symptoms of what is known as 'Grain itch' and was identified as the agent responsible for epidemic allergic asthma among grain workers by Ancona (1923; Figure 4.5d).

7.2.7 Tetranychoidea

Tetranychoid mites include the plant-feeding spider mites and false spider mites. Occupational allergy to these mites is usually associated with fruit growing and harvesting. Bernecker (1970) found that half the asthmatics who were skin-prick test positive to *Dermatophagoides pteronyssinus* were also positive to *Panonychus ulmi*, and two-thirds were also positive to *Tetranychus urticae*. Rufli (1970) also reported on skin-prick test cross-reactivity between *D. pteronyssinus* and *T. urticae*. Michel *et al.* (1977) reported asthma, conjunctivitis and dermatitis apparently due to *Panonychus ulmi* among apple growers; Reunala *et al.* (1983) reported similar symptoms, plus skin-prick test and RAST results, in two greenhouse workers exposed to the two-spotted spider mite, *Tetranychus urticae*. Other reports of allergy to spider mites include those of Kroidl *et al.* (1992), Kim *et al.* (1999), Lee *et al.* (2000) and Kronqvist *et al.* (2005).

7.2.8 Trombiculidae

Chiggers, also known as itch mites, are the parasitic larvae of trombiculid mites, and include species that transmit rickettsial diseases of mammals. Over 20 species have been reported as responsible for itchy papular eruptions in humans (Wharton and Fuller, 1952).

7.2.9 Mesostigmatid mites

Most mesostigmatid mites are predators or parasites, with piercing chelicerate mouthparts. Alexander (1984) reviewed skin eruptions caused by mesostigmatid mites. Bernecker (1970) found that a fifth of the asthmatics who were skin-prick test positive to *Dermatophagoides pteronyssinus* were also positive to *Dermanyssus gallinae*. Rufli (1970) also reported on skin-prick test cross-reactivity between these two species. Frenken (1962) and Sexton and Haynes (1975) reported allergic symptoms due to bites of *Dermanyssus gallinae*, and Kronqvist *et al.* (2005) documented IgE sensitisation and related asthma and rhinoconjunctivitis by greenhouse workers to *Phytoseiulus persimilis* and *Hypoaspis miles*, two biocontrol agents for the red spider mite, *Tetranychus urticae*. De Jong *et al.* (2004) reported on a group of bell pepper growers of whom 23% had occupational allergy to *Amblyseius cucumeris*, a predatory mite and biocontrol agent. A group 10 tropomyosin allergen from the poultry mite *Dermanyssus gallinae* (Der g 10) has been cloned, sequenced and databased (Nisbet *et al.*, 2006b).

7.3 Localisation of allergens within the mites

Long before it was known that mite allergens were mostly digestive enzymes, various investigations were made to localise within the mites the sites of synthesis and storage of the allergens. The first attempt was that of Halmai and Alexander (1971) that indicated that the isolated faeces were allergenic. Mumcuoglu and Rufli (1979) used indirect immunofluorescence microscopy of sections of dust mites to detect sites of allergen localisation. They used polyclonal rabbit anti-sera raised against a partially purified extract of *Dermatophagoides pteronyssinus* to screen some 1500 sections and found that fluorescence was most intense in the posterior midgut, anterior hindgut and cuticular regions. During preparation the mite extract was heated to 80°C for 4 hours, followed by reduction by boiling for an unspecified period. This procedure would have virtually removed all of the binding activity of the heat-labile Der p 1 (Lombardero *et al.*, 1990) and the resulting antiserum that was used for immunolabelling would have been directed against a high proportion of heat-stable allergen epitopes such as those of Der p 2. Additionally, the mites were fixed in 25% glutaraldehyde, which may denature the allergen and thus interfere with allergen-antibody binding. Nevertheless, the results greatly strengthened previous indications that the allergens of dust mites were associated with the digestive system. Group 2 allergens were subsequently localised by Jeong *et al.* (2002).

Stewart and Turner (1980) demonstrated that fluorescent-labelled polyclonal anti-Der p 1 antibody stained the gut wall of *Dermatophagoides pteronyssinus*, and Thompson and Carswell (1988) estimated rates of incorporation of radiolabelled culture medium into mite tissues. They suggested the long time course (>14 days) for the process was indirect evidence that Der p 1 is a protein synthesised by the mite and secreted into the digestive system rather than a by-product of digestion, in which case a far shorter time course would have been anticipated. Subsequently, Thomas *et al.* (1991) localised Der p 1 in the gut of *D. pteronyssinus* using immunogold labelling. Tovey and Baldo (1990) used fluorescent-labelled rabbit polyclonal antibodies raised against whole mite extract and against Der p 1, as well as a murine monoclonal anti-Der p 1 antibody to isolate the sites of deposition of the allergen in fresh frozen sections of *D. pteronyssinus*. With the polyclonal anti-whole mite extract, fluorescent antibody binding was most intense in the gut wall and gut contents as well as in the gnathosoma and cuticular-subcuticular regions – a similar pattern to that demonstrated by Mumcuoglu and Rufli (1979). With the polyclonal anti-Der p 1 antibody, fluorescence was localised to the anterior midgut and its contents and the posterior midgut. The same pattern was observed with the monoclonal anti-Der p 1 with additional labelling in the oesophageal region. Tovey and Baldo (1990) suggested that Der p 1 is secreted from the wall of the anterior midgut and becomes incorporated into the faecal pellet. Fluorescent-labelled soybean lectin (specific for D-galactose) bound to the gut wall, contents and the peritrophic membrane, whereas wheat germ agglutinin (specific for N-acetyl-D-glucosamine) was more generally distributed, with strong fluorescence from the walls of the anterior midgut. Rees *et al.* (1992) used polyclonal monospecific sheep anti-Der p 1 antibody to localise Der p 1 to the gnathosoma, faecal pellets and midgut, visualised by peroxidase staining, in paraffin sections of mites.

Van Hage-Hamsten *et al.* (1992a) used murine monoclonal antibodies raised against an extract of *Lepidoglyphus destructor* to localise the 39 kDa allergen in frozen sections of the mite using immunoperoxidase staining. They found labelling of the gut, and faecal pellets as well as cuticular and subcuticular tissues, though sera from *L. destructor*-allergic patients only bound to the gut and faecal pellets. Subsequently van Hage-Hamsten *et al.* (1995) localised Lep d 2 and a 79 and 93 kDa complex in *Lepidoglyphus destructor* and visualised them using confocal microscopy. Anti-Lep d 2 monoclonals bound most intensely to the gnathosomal region, with fainter staining in the gut and faecal pellets, with no staining of the cuticle, whereas anti-79/93 kDa antibodies bound to the gnathosomal region and the cuticle but not the faecal pellets. Despite all of the above work, we are not much further forward in understanding precisely where the sites of synthesis of mite allergens are located. However, this can be postulated by a closer examination of ultrastructural data together with a consideration of the enzymic properties of the allergens (see section 2.2.3a).

Der f 15, a major allergen for atopic dogs, has been identified as a chitinase. It has been localised mainly in the chitin-lined parts of the gut of *D. farinae* – the oesophagus and hindgut – but not within faecal pellets (McCall *et al.*, 2001). The same pattern of localisation was found for Der f 18, which also shows homology with chitinases (Weber *et al.*, 2003).

Blo t 21 is not an enzyme but a small alpha-helical protein of unknown function related to groups 5 and 7 allergens. Gao *et al.* (2007) found it was present in the gut and faecal pellets of *Blomia tropicalis*. Most intense staining could be seen along the epithelium of the midgut caecae as well as in the semi-digested food balls.

The regions of the body that have stained positive for allergen in these studies, namely the buccal-pharyngeal-oesophageal region, the anterior and posterior midgut and the cuticular and subcuticular tissues, indicate the presence of four groups of allergen-enzymes. First, a group of salivary-gland associated carbohydrases which may well be group 4 allergens; second, the cysteine proteinase group 1 allergens associated with the midgut and faecal pellets and probably localised within midgut digestive cells; third, an enzyme associated with moulting, most probably localised within epidermally derived haemocytes associated with ecdysis (see section 2.5.1); and fourth, some chitinases associated with the anterior midgut.

7.4 Groups of mite allergens and their classification

7.4.1 Allergen nomenclature

The International Union of Immunological Societies (IUIS) Subcommittee on Allergen Nomenclature and Standardisation decided that allergens should be abbreviated using the first three letters of the generic name and the first letter of the specific name of the organism from which they are derived (Marsh *et al.*, 1987; King *et al.*, 1994). These are written in Roman characters with a capital for the first letter of the generic abbreviation, hence Der p. There follows a number, indicating the group to which the allergen belongs (Der p 1, Der p 2), which historically has reflected the chronology of their purification and characterisation, but now is intended to reflect functional biochemical and amino acid sequence similarities of allergens within a major taxon (i.e. all mite group 3 allergens are trypsins, but group 3 allergens of cats are cystatins). Before this the letters were written in italics and the numbers in Roman numerals, hence *Der p* I, *Der p* II. Prior to the IUIS standardisation, allergens were given informal names, which can be confusing for the reader when dealing with older literature. A list of commonly used informal names and their current synonyms is given in Table 7.4.

7.4.2 Classification of mite allergens

Now that the amino acid sequences of over 20 groups of mite allergens have been databased, it has become possible to explore similarities in primary and secondary protein structure using molecular bioinformatics tools such as Basic Local Alignment Search Tool (BLAST) searches, sequence alignment and peptide statistics programs. These tools have been particularly important in providing comparative data on proteins of known function with similar amino acid sequences to mite allergens.

Mite allergens have been divided into groups on the basis of physicochemical properties and functions where these are known or can be deduced. Thus, group 3 allergens are all trypsins; group 4, amylase. This section deals with the higher order classification of allergen groups, to attempt to determine what other groups of allergens the group 3 trypsins and group 4 amylases are most similar to and place them in 'families'. The purpose of such classifications is that they might be of predictive value in determining, for example, the common molecular and structural features of the groups that make them allergenic.

The allergen classification detailed by Aalberse (2000) covered allergens from all sources, not just mites, and was based on secondary protein structure. It consisted of four families:

- those with anti-parallel β-strands including the immunoglobulin fold family (mite group 2) and serine proteases (groups 3, 6 and 9);
- anti-parallel β-sheets closely associated with α-helices (lipocalin – mite group 13);

- α and β structures that are not closely associated one with the other, including mite group 1 allergens; and
- α-helical lipid transfer proteins, e.g. Fel d 1 chain 1.

A new classification of mite allergens described in this chapter in Table 7.5 is based on:

- comparisons of amino acid compositions;
- secondary structure as represented by topological maps;
- tertiary structure of allergens or other homologous proteins;
- known functional similarities;
- amino acid sequence homology; and
- shared conserved structural domains (Table 7.7).

The outline of the classification is given in Table 7.5. It includes six 'families': the peptidases (mite allergen groups 1, 3, 6 and 9), glycosidases (groups 4, 12, 15 and 18), transferases (groups 8 and 20), small alpha-helical proteins (groups 5, 7 and 21), muscle proteins (10 and 11) and the lipid-binding proteins (2, 13 and 14). This leaves groups 16, 17 and 19 unclassified. Group 16 are gelsolins, involved in actin binding and capping. Amino acid sequences of group 17 allergen have not been published or databased as far as I can discover. Group 19 is represented only by Blo t 19 which is very small (7 kDa) as mite allergens go, and about which not much is known, other than its amino

Table 7.5 Classification of dust mite allergen groups into families with shared characteristics.

Family and groups	Most frequent amino acids	Least frequent amino acids	Mean % aromatic amino acids	Mean % polar amino acids	Mean % charged amino acids	Tertiary structure
Peptidases 1, 3, 6, 9	Ile, Tyr, Trp	Phe, Glu	10.6	45.1	21.6	Globular, in two halves; central cleft containing active site
Glycosidases 4, 12, 15, 18	His, Thr, Trp, Tyr	Ala, Arg	16.7	47.9	28.7	4, 15 and 18: globular $(\alpha/\beta)_8$ barrel
Transferases 8, 20	Met, Tyr, Phe, Leu	Cys, Ser	13.7	46.8	28.8	Globular, two domains separated by a surface cleft
Small alpha-helical proteins 5, 7, 21	Met, Glu, Lys, Ile	Cys, Trp	10	47.3	36.6	5 and 21: coiled-coil alpha-helical bundle
Muscle proteins 10, 11	Glu, Gln, Arg	Cys, Gly Phe, Pro, Trp	4.2	62.3	43.4	Elongated alpha-helix
Lipid-binding proteins 2, 13, 14	Lys, Val, Asp, Ile	Trp, Pro	8.0	33.9	30.3	2 and 13: globular beta-barrels

(a) The peptidases

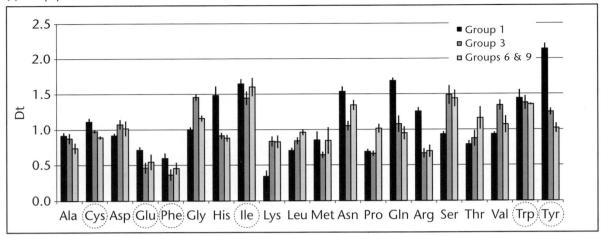

(b) The glycosidases

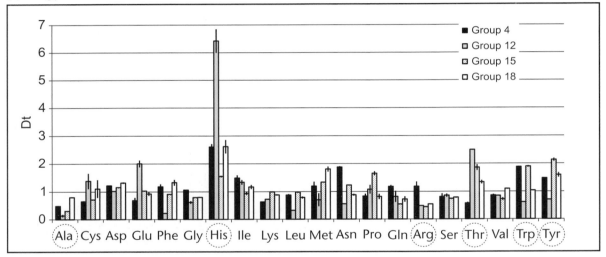

(c) The transferases

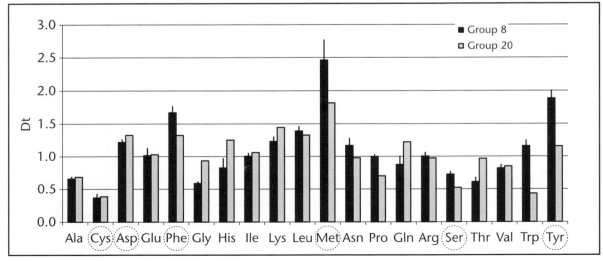

Figure 7.1 Amino acid composition of the various groups of dust mite allergen sequences in Table 7.1, represented as the transformed Dayhoff statistic (D_t), where D_t = molar percentage of each amino acid divided by the Dayhoff statistic (D), which is the relative occurrence of an amino acid per 1000 residues, normalised to occurrence per 100 residues. Figures represent means ± standard error. All calculations were done using the mature protein sequence only. Dotted circles highlight the signature patterns.

(d) The small alpha-helical proteins

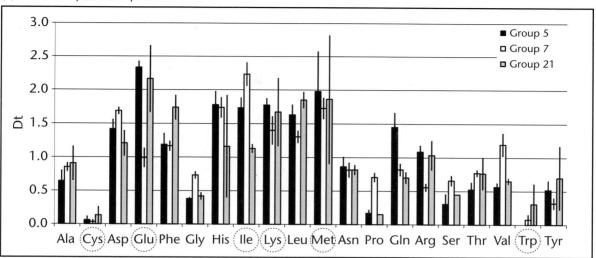

(e) The muscle proteins

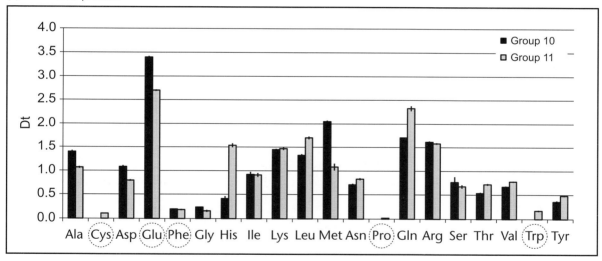

(f) The lipid-binding proteins

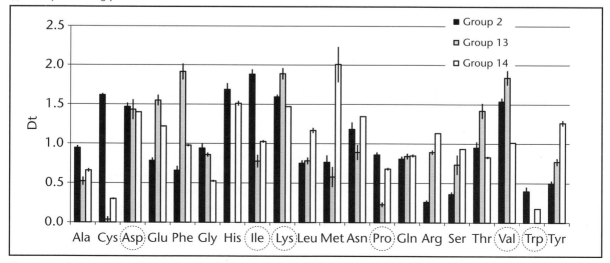

Figure 7.1 Continued

acid sequence, and its possible anti-microbial function (deduced from amino acid sequence comparisons).

Amino acid compositions represent the relative proportion of the different amino acids that make up the protein. Direct comparisons of proportions tend to have relatively high levels of variation, so to eliminate 'noise' I used the transformed Dayhoff value, D_t. This is the molar percentage of each amino acid divided by the Dayhoff statistic (D), which is the relative occurrence of an amino acid per 1000 residues, normalised to occurrence per 100 residues. Figure 7.1 shows D_t values for each of the six allergen groups. I used the program PEPSTATS (Australian National Genomic Information Service, http://www.angis.org.au/).

Secondary structure, as represented by topological maps of the proteins, is simply a method of predicting the arrangement of the structural elements in a linear map (Figure 7.5), detailing how much and which parts of the molecule are made up of alpha-helices, beta-strands and random coils. I used the Network Protein Sequence Analysis (Combet *et al.*, 2000), http://npsa-pbil.ibcp.fr; 3-D jigsaw (http://www.bmm.icnet.uk); and PeptideStructure (GCG). Some secondary structure predictions were found to be of limited use because they gave predictions contrary to x-ray crystallography-based tertiary structure. I have used them sparingly. For details of tertiary structure of allergens and related proteins I used the RCSB Protein Data Bank (www.pdb.org).

Known functional similarities are based either on direct experimental evidence of functional attributes of the allergen molecule (e.g. group 3 allergens shown to have trypsin activity by artificial substrate specificity) or are inferred from sequence comparisons with proteins of known function. Much of the experimental evidence for digestive enzyme-allergens has been covered in the section on digestion in Chapter 2 (see section 2.2.3).

Amino acid sequence homology (see Figures 7.10–7.27; Table 7.3) was determined using BLAST protein-protein searches (Altschul *et al.*, 1997). Amino acid sequences were aligned using sequence alignment software such as MUSCLE (Edgar, 2004; http://phylogenomics.berkeley.edu/) and T-COFFEE (Notre-dame *et al.*, 2000; http://tcoffee.vital-it.ch/cgi-bin/Tcoffee/tcoffee_cgi/index.cgi). Shared structural domains were determined using conserved domain searches in the NCBI database (Marchler-Bauer and Bryant, 2004) and Pfam, the Protein Families Database (http://www.sanger.ac.uk/Software/Pfam/).

7.4.3 The peptidases

This includes all allergens that are peptidase enzymes (EC 3.4), including groups 1 (cysteine proteinases), 3 (trypsins), 6 (chymotrypsins) and 9 (collagenases) (see Table 7.5). Characteristic of the amino acid composition of the peptidases (shown in Figure 7.1a) are high frequencies of isoleucine, tyrosine and tryptophan (mean transformed Dayhoff value, D_t of 1.6, 1.5 and 1.4 respectively), and the lowest are glutamate and phenylalanine (D_t 0.6 and 0.5). The peptidases also have a relatively high cysteine content (D_t 1.0) with six conserved residues. Predicted secondary structures indicate proteins with a series of 3–6 short alpha-helices (18–24% of the molecule), and 6–12 beta-strands (20–25%) with no clear patterns of association. About 53–60% of the molecule is made up of random coil.

Allergen groups 3, 6 and 9 share trypsin-like serine protease domains and are members of the chymotrypsin family S1. These are secretory proteins, synthesised with a signal peptide which targets it to the secretory pathway, and a pro-enzyme which prevents activation until the enzyme is required. Proteolytic activation occurs extracellularly, or within cellular vesicles or other storage organelles (Rawlings and Barrett, 1994). Group 1 allergens have peptidase C1 domains (papain family cysteine protease). Cysteine protease activity of group 1 allergens was confirmed experimentally by Stewart *et al.* (1991), as was serine protease activity for group 3 (Stewart, Ward *et al.*, 1992), group 6 (Yasueda *et al.*, 1993) and group 9 allergens (King *et al.*, 1996).

There is mounting evidence that the enzymatic activity of some mite allergens plays a significant role in enhancing their allergenicity (e.g. Hewitt *et al.*, 1995). The effects of dust mite peptidases on the initiation and enhancement of allergic responses have been reviewed by Robinson *et al.* (1997), Caughey and Nadel (1997), Shakib *et al.* (1998), Sharma *et al.* (2003) and Reed and Kito (2004), and it appears mite peptidases may augment or facilitate the action of endogenous peptidase (such as mast cell tryptase and elastase) in allergic inflammatory reactions, or act as proteolytic triggers of inflammation in their own right. The important point here is that not all mite allergens need to be proteases for such mechanisms to be important, because protease allergens can facilitate the allergenicity of presentation of non-enzymic allergens. Some of the effects of mite proteases are as follows:

- Proteoloytic activity of Der p 1, 3, 6 and 9 causes disruption of the lung epithelium, allowing the

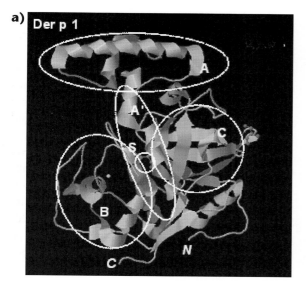
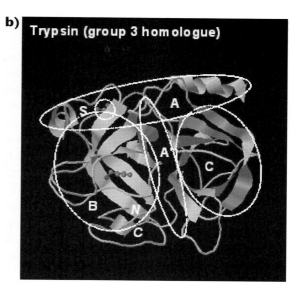

Figure 7.2 Tertiary structures of members of the peptidase family of mite allergens (groups 1, 3, 6 and 9) and related proteins. **a)** Der p 1 in monomeric form with pro-region attached (circled area, A and A') (Meno *et al.*, 2005; RCSB Protein Data Bank entry (PDB) 1XKG). The left-hand region is predominantly alpha-helix, the right-hand region beta-strands. The cleft-like region between them contains the catalytic site (S) and in the inactivated form of the molecule is covered by the A' region of the pro-enzyme. **b)** Trypsin (group 3 allergens) (Leiros *et al.*, 2004; PDB 1UTN). A is the alpha-helical region, B and C are the left-hand and right-hand beta-strand regions, A' is the central cleft between the two and S is the active site.

allergens access to underlying antigen-presenting cells (Herbert *et al.*, 1995; Wan *et al.*, 1999, 2001).

- Der p 1 can cleave immunomodulators such as CD23, the human low affinity IgE receptor, and CD25, the alpha-subunit of the human interleukin 2 receptor (reviewed by Shakib *et al.*, 1998).
- Gough *et al.* (1999) found enzymically active Der p 1 tripled the amount of Der p 1-specific IgE produced by mice compared with inactivated control Der p 1. Active Der p 1 also enhanced the IgE response to ovalbumin (Gough *et al.*, 2001). This adjuvant effect was IgE-specific. There was no similar effect on IgG antibody production.
- The proteoloytic activity of Der p 1, 3 and 9 is responsible for triggering the release of pro-inflammatory mediators such as interleukins. Der p 3 can switch on protease-activated receptors (PARs) on respiratory epithelial cells resulting in the release of interleukin-8 (IL-8). Der p 1 can also induce IL-8 but via a different mechanism (Adam *et al.*, 2006).
- *D. farinae* serine protease (probably Der f 3) has kallikrein activity, causing liberation of kinins from kininogens (Takahashi *et al.*, 1990). Kinins are important inflammatory mediators.
- Der p 3 and Der f 3 were found to activate the complement system (Mauro *et al.*, 1997).

a Group 1 allergens

The group 1 allergens are polymorphic, 25 kDa, acidic/neutral proteins. They belong to the cysteine group of proteolytic enzymes which include the mammalian enzymes cathepsin B and H, and the plant enzymes actinidin and papain. Group 1 allergens are recognised by at least 70% of mite allergic individuals. They have been identified in *Dermatophagoides pteronyssinus*, *D. farinae*, *D. microceras*, *D. siboney*, *Euroglyphus maynei* and *Blomia tropicalis* (Chapman and Platts-Mills, 1980; Stewart and Turner, 1980; Lind, 1986b; Chua *et al.*, 1988; Kent *et al.*, 1992; Ferrándiz *et al.*, 1995b; Mora *et al.*, 2003), as well as in the mange mites *Psoroptes cuniculi* and *P. ovis* (Stewart and Fisher, 1986; Lee *et al.*, 2002) and the scabies mite *Sarcoptes scabiei* (Holt *et al.*, 2004). The allergens are found in both whole body and faecal extracts, and are synthesised by cells lining the gastrointestinal tract (Tovey and Baldo, 1990; see section 7.3).

Group 1 allergens have not been isolated from *Acarus*, *Tyrophagus*, *Glycyphagus* and *Lepidoglyphus* spp. and it seems unlikely that these mites possess them or they would have been detected by now. One reason might be the diet of storage mites (seed storage proteins and the moulds that live on them) is a rich source of protease inhibitors. Barber *et al.* (1996) found cereal prolamins inactivated Der p 1. It may be that stored products mites use alternative proteinases that are not inhibited by compounds in grain or moulds.

The amino acid sequences of Group 1 allergens from the pyroglyphids (*Dermatophagoides* spp. and *Euroglyphus maynei*) comprise a signal peptide of 18/19 residues in *Dermatophagoides* spp., followed by a pro-peptide of 79/80 residues and the mature protein of 222–223 residues (Chua *et al.*, 1988; Dilworth *et al.*, 1991; Kent *et al.*, 1992; Table 7.1; Figure 7.9). Other cysteine proteases also have transient pre- and pro-form intermediates (Chua *et al.*, 1988; Dilworth *et al.*, 1991). The enzyme is produced in an inactive form and becomes active following cleavage of the pro-peptide by auto-catalysis. A similar situation probably applies to Pso o 1 (Lee *et al.*, 2002), as well as for other peptidase allergens.

The crystal structure of the pro-enzymic form of Der p 1 (Meno *et al.*, 2005; RCSB Protein Data Bank entry: 1XKG) shows the active site is located in a large surface cleft. The pro-peptide, which consists of four alpha-helices, is folded in such a way that it covers the cleft and active site, rendering the enzyme inactive. The mature Der p 1 is a globular molecule folded into two domains separated by the surface cleft. The left-hand domain (consisting of the N-terminal half of the mature protein: residues 21–116 in Figure 7.9) is predominantly alpha-helix and the right-hand domain (the C-terminal part: residues 117–223, Figure 7.9) is mostly beta-sheet. This overall structure is concordant with the model of the 3-D structure of Der p 1 by Topham *et al.* (1994) and is illustrated in Figure 7.2a. The active site consists of Q29 and C35 on the left and H171 and N190 on the right.

de Halleux *et al.* (2006) showed the mature, active form of Der p 1 undergoes dimerisation at pH 8 but is monomeric at pH 7.5. X-ray crystallography showed the monomers bound via H73 on the first momomer to D147, D149 and Y166 on the second. N169 binds to its counterpart on the other monomer. This region would be blocked by the pro-enzyme, so only the mature protein can dimerise. The authors suggest that the concentration of Der p 1 in faecal pellets is sufficiently high to maintain dimerism under natural conditions and infer that the dimer is likely to be more allergenic than the monomer. Interestingly the midgut of dust mites, where digestive hydrolysis takes place, is slightly acid and Der p 1 would likely exist there as a monomer. But the hindgut, where faecal pellets accumulate, is neutral or slightly alkaline, allowing for dimers to form.

Meno *et al.* (2005) and de Halleux *et al.* (2006) found that Der p 1 contains a metal binding site near the catalytic site, though its biological function is not known.

The pyroglyphid Group 1 allergens have the highest sequence identity (ca. 78% between the three allergens). Pso o 1 has ca. 58% identity with the pyroglyphid allergens, with which it shares a potential glycosylation site at N53, and only 42% with Sar s 1. Blo t 1 has ca. 31% identity with the pyroglyphids and 29% each with Sar s 1 and Pso 1. Sar s 1 has 34% identity with the pyroglyphids. Highest homology is in the region of the active site and the cysteine residues (Figure 7.9). The pairing pattern of cysteines in disulphide bridge formation was identified by Memo *et al.* (2005) as C4–C118, C32–C72 and C66–C104. C118 is substituted by asparagine in Blo t 1. Also H73, the crucial residue for dimerisation in Der p 1, is substituted for aspartate in Sar s 1, glycine in Pso o 1 and glutamate in Blo t 1.

Der p 1 has only one gene (W.A. Smith *et al.*, 2001), but since isoallergens exist, they must be alleles. By contrast, the scabies mite, *Sarcoptes scabiei*, has at least 10 group 1 genes, with signal peptides of 19–26 amino acids and mature proteins of 225–251 amino acids (Holt *et al.*, 2004). Some of the genes have mutations that would render the protein inactive. A similar multi-gene family exists for Sar s 3 (see below). This phenomenon is thought to be an adaptation to parasitism involving evasion of the host immune response (Holt *et al.*, 2003).

b Group 3 allergens

The group 3 allergens are trypsin-like serine protease enzymes with calculated molecular weights of 23–25 kDa, although on SDS-PAGE gels they migrate as a 30 kDa band. They have been isolated from bodies and faecal pellets of *Dermatophagoides farinae*, *D. pteronyssinus*, *D. siboney*, *Euroglyphus maynei*, *Sarcoptes scabiei*, *Glycyphagus domesticus*, *Lepidoglyphus destructor* and *Blomia tropicalis* (see Tables 7.1 and 7.3).

The group 3 allergens have a 15–20 residue signal peptide followed by a pro-enzyme sequence of 9–26 residues and a mature protein of 219–232 residues (Figure 7.11). Recombinant Der f 3 expressed with the intact pro-sequence was enzymatically inactive, but became active on removal of the pro-sequence (Nishiyama *et al.*, 1995). This is similar to the situation with group 1 allergens where the pro-enzyme prevents enzymatic activity by occluding the active site. The group 3 allergens belong to the S1 family of serine proteases, as do group 6 and 9 allergens. All of them possess the catalytic residues H41, D85 and S185 characteristic of the S1 chymotrypsin family of serine proteases, as well as a series of conserved sequences associated with the active sites (Figure 7.2b; Rawlings and Barrett, 1994; their Figure 2).

The group 3 allergens are polymorphic and in Der p 3 are allelic, since the allergen is encoded by a single gene (Smith and Thomas, 1996). *Sarcoptes scabiei* has a multi-gene family for group 3 allergens (Holt *et al.*, 2003).

Pyroglyphid and *Blomia* group 3 allergens are not glycosylated. Gly d 2 and Lep d 2 have a single putative glycosylation site and Sar s 3 has two sites. Conserved cysteine residues are paired to form disulphide bonds as in trypsin (numbering for Der f 3; Figure 7.11): C26–C42, C152–C169 and C181–C207. In the pyroglyphid allergens there is also an unpaired conserved cysteine (C12). Amino acid sequence homology with Der f 3 shows 82% identity with Eur m 3, 80% with Der p 3, 49% with Blo t 3, and 45% with trypsin from the mosquito *Anopheles gambiae*. Sar s 3 shows ca. 43–44% identity with the pyroglyphid allergens.

The IgE binding frequency of group 3 allergens is relatively high (mean 67%, range 40–100%; Table 7.6). Members of group 3 are considered to be major allergens. The enzymatic activity of serine protease allergens may enhance their allergenicity though initiating non-allergic inflammatory responses, as with group 1 enzyme-allergens. Der p 3 and Der p 9 activated the receptor PAF$_2$ on lung epithelia which triggers the release of the pro-inflammatory cytokines GM-CSF and eotaxin (Sun *et al.*, 2001). Der p 3 and Der f 3 were found to activate the complement system, cleaving the complement components C3 and C5 to produce the anaphylatoxins C3a and C5a (Mauro *et al.*, 1997).

c Group 6 allergens

The group 6 allergens are 25 kDa serine proteases that have chymotrypsin activity based on their substrate affinity (Yasueda *et al.*, 1993). They are known from *Dermatophagoides farinae*, *D. pteronyssinus* and *Blomia tropicalis*. Group 6 allergens appear to be recognised by a markedly lower proportion of mite allergic individuals than the other peptidases (31–65%; Table 7.6; Yasueda *et al.*, 1993; King *et al.*, 1996; Bennett and Thomas, 1996; Kawamoto *et al.*, 1999).

Der 6 allergens have no N-glycosylation sites on the mature protein and consist of 16–17 hydrophobic residues, indicating a signal peptide (Kawamoto *et al.*, 1999), followed by a 32–34 pro-enzyme and a 230–231 mature protein. The active sites and conserved residues around the active sites are the same as in groups 3 and 9: typical of S1 peptidases. The six conserved cysteines are paired similarly. Der p 6 and Der f 6 have 75% identity. They share about 35–37% sequence identity with the group 3 allergens and about 27–30% with group 9 (see Figures 7.2, 7.14, 7.17). There is little cross-reactivity between groups 3 and 6 (Yasueda *et al.*, 1993). Bennett and Thomas (1996) identified serine 178 as conserved in all chymotrypsins and essential for substrate specificity. In Blo t 6 (GenBank accession no. AY291325, Chew *et al.*, unpublished) this serine is substituted by a glutamine.

d Group 9 allergens

The group 9 allergens migrate to positions around 27–30 kDa on SDS-PAGE and have a calculated molecular weight of 23.8–24.5 kDa (Table 7.1). They are serine proteases with collagenase activity and have been isolated and characterised from *Dermatophagoides farinae*, *D. pteronyssinus* and *Blomia tropicalis*. Collagenase activity was also detected in *Euroglyphus maynei* (Stewart *et al.*, 1994). Der p 9 was isolated from faeces-enriched extracts and was found to cleave collagen (King *et al.*, 1996). It showed IgE-binding to 92% of sera from mite-allergic patients, and had limited cross-reactivity with Der p 3 but not Der p 6. The relatively high concentration of Der p 9 isolated from the faeces (2.8% w/w) suggests group 9 allergens are not only important digestive enzymes but are likely to be present in substantial concentrations in homes. This may account for the high IgE-binding frequency (Table 7.6). The proteolytic activity of Der p 9 (along with Der p 3) activated PAF$_2$ to induce release of the cytokines GM-CSF and eotaxin (Sun *et al.*, 2001).

Der p 9 has a signal peptide of 16 residues (based on Der p 3), a pro-enzyme region of nine residues and a mature protein of 219 residues. A BLAST search showed the Der p 9 sequence in Figure 7.17 had 70% sequence identity with Der p 3 and 61% with Der f 3. The active sites and conserved residues around the active sites are the same as in groups 3 and 6. There are six conserved cysteines and no N-glycosylation sites. In Der p 9, the equivalent of serine 178, conserved in all chymotrypsins and essential for substrate specificity, is aspartate 166, although it is still serine in the Der f 9 and Blo t 9 sequences.

Table 7.6 IgE binding frequency to enzyme allergens and non-enzyme allergens. * = sera from children.

Enzyme allergens. Native [n] or recombinant [r]	% frequency of IgE response	Raw frequency data	Reference	Non-enzyme allergens. Native [n] or recombinant [r]	% frequency of IgE response	Raw frequency data	Reference
nDer f 1	79	18/23	Ando et al., 1993	nDer f 2	83	19/23	Ando et al., 1993
nDer f 1	89	78/88	Yasueda et al., 1993	nDer p 2	93	82/88	Yasueda et al., 1993
nDer f 1	90	28/31	Aki et al., 1995	nDer f 2	93	82/88	Yasueda et al., 1993
nDer p 1	100	51/51	Stewart et al., 1992	nDer p 2	100	35/35	King et al., 1996
nDer p 1	90	79/88	Yasueda et al., 1993	rDer p 2	100	10/10	Weghofer et al., 2005
nDer p 1	96	34/35	King et al., 1996	rDer p 2	88	36/41	Shen et al., 1996
nDer p 1	100	10/10	Weghofer et al., 2005	rDer p 2	90	27/30	Shen et al., 1993
nDer p 1	93	28/30	Shen et al., 1993	rDer p 2	82	14/17	Chua et al., 1990b
rBlo t 1	62	17/27	Mora et al., 2003	nGly d 2	94	16/17	Gafvelin et al., 2001
nDer f 3	70	16/23	Ando et al., 1993	rGly d 2	94	16/17	Gafvelin et al., 2001
nDer f 3	42	37/88	Yasueda et al., 1993	nLep d 2	95	21/22	Ventas et al., 1991
nDer p 3	100	55/55	Stewart et al., 1992	rLep d 2	59		Kronqvist et al., 2000 (cf. Eriksson et al., 2001)
nDer p 3	51	45/88	Yasueda et al., 1993	nDer f 2	74	23/31	Aki et al., 1995
nDer p 3	97	34/35	King et al., 1996	rDer p 2	92	22/24	Tsai et al., 1998
rBlo t 3	51	23/45	Cheong et al., 2003	Der p 5	52	13/25	Lin et al., 1994
nDer p 4	46	12/27	Lake et al., 1991	Lep d 5	9	4/45	Eriksson et al., 2001
nDer p 4	25	5/20*	Lake et al., 1991	Der p 5	50	5/10	Weghofer et al., 2005
rDer p 4	30	3/10	Mills et al., 1999	Der p 5	55	21/38	Tovey et al., 1989
nDer p 4	50	5/10	Weghofer et al., 2005	Blo t 5	47	47/100	Caraballo et al., 1996
nDer f 6	31	27/88	Yasueda et al., 1993	Lep d 7	62	28/45	Eriksson et al., 2001
rDer f 6	40	15/38	Kawamoto et al., 1999	Der p 7	30	3/10	Weghofer et al., 2005
nDer p 6	39	34/88	Yasueda et al., 1993	Der p 7	46	19/41	Shen et al., 1996
nDer p 6	65	24/35	King et al., 1996	rDer p 7	53	16/30	Shen et al., 1993
nDer p 8	96	53/55	C. Huang et al., 2006	Der p 10	0	0/10	Weghofer et al., 2005
nDer p 8	75	15/20	C. Huang et al., 2006	rDer p 10	6	4/71	Asturias et al., 1998
nDer p 8	65	11/17	C. Huang et al., 2006	nDer f 10	81	25/31	Aki et al., 1995
rDer p 8	84	46/55	C. Huang et al., 2006	rBlo t 10	29	27/93	Yi et al., 2002
rDer p 8	20	2/10	Weghofer et al., 2005	rBlo t 10	20	7/35	Yi et al., 2002
rDer p 8	40	77/193	O'Neill et al., 1994	rBlo t 11	52	33/63	Ramos et al., 2001
nDer p 9	92	32/35	King et al., 1996	nDer f 11	88	21/24	Tsai et al., 1998
rDer p 15	70	19/27	O'Neill et al., 2006	rBlo t 11	5	1/22	Teo et al., 2006
nDer f 18	54	13/24	Weber et al., 2003	nBlo t 11	77	17/22	Teo et al., 2006
rDer p 18	63	17/27	O'Neill et al., 2006	rBlo t 11	50	11/22	Teo et al., 2006
				rDer p 11	50	50/100	Lee et al., 2004
				rBlo t 12	50	16/32	Puerta et al., 1996a
				Lep d 13	13	6/45	Eriksson et al., 2001
				rAca s 13	23	3/13	Eriksson et al., 1999
				rBlo t 13	11		Caraballo et al., 1997
				Der p 14	0	0/10	Weghofer et al., 2005
				rDer f 14	30	13/43	Aki et al., 1994a
				nDer f 14	70	16/23	Fujikawa et al., 1996
				rDer f 14	39	9/23	Fujikawa et al., 1996
				rDer f 16	35	ND	Tategaki et al., 2000
				rDer f 16	47	8/17	Kawamoto et al., 2002a
				rDer f 17	35	ND	Tategaki et al., 2000
				rBlo t 19	10	ND	Thomas et al., 2002
				rHSP70	10	4/41	Aki et al., 1994a

7.4.4 The glycosidases

This group of allergens belongs to the enzyme class EC 3.2, the glycosidases (Table 7.5; cf also Chapter 2, 2.2.3b). It includes allergen group 4 (alpha-amylase) and groups 15 and 18 (chitinases). Blo t 12 is a chitin-binding protein possessing a peritrophin A domain. It is included here because it shows some sequence homology with the chitinases though it is probably not an enzyme.

With the exception of group 12 (146 residues) this family consists of comparatively large molecules as mite allergens go (437–555 residues). With regard to their amino acid composition, the highest frequency is histidine (mean D_t 3.3), then threonine (1.6), tyrosine (1.5) and tryptophan (1.4). The lowest is

alanine (0.4) then arginine (0.7). The glycosidases have a relatively high cysteine content (mean D_t 1.0), a feature they share with the peptidases (Figure 7.1b).

Alpha-amylases and chitinases share basically the same tertiary structure, an $(\alpha/\beta)_8$ barrel which is detailed below. The tertiary structure of the group 12 allergens are not known, but they are too small to be an $(\alpha/\beta)_8$ barrel. The secondary structure prediction indicates two major N-terminal alpha-helices and six major beta-strands.

a Group 4 allergens

The group 4 allergens are alpha-amylases (Lake *et al.*, 1991), migrating on SDS-PAGE at 60 kDa, and with a calculated molecular weight of 55–57 kDa. They have

a)

b)

c)

d)

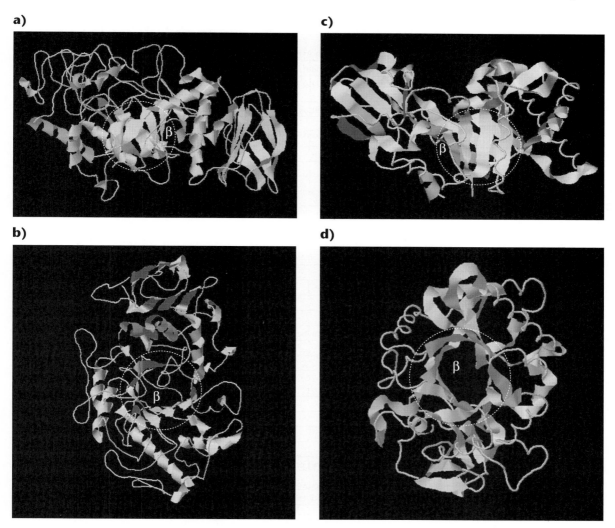

Figure 7.3 Tertiary structures of proteins related to members of the glycosidase family of mite allergens (groups 4, 12, 15 and 18). **a)** Lateral view; and **b)** end-on view of alpha-amylase homologue (mealworm, PDB 1CLV) of group 4 allergens showing an $(\alpha/\beta)_8$ barrel structure (β), which Janeček (1997) referred to as 'this charming fold'. **c)** Lateral view; and **d)** end-on view of chitinase homologue (human, PDB 1HKK) of groups 15 and 18.

Table 7.7 Structural domains identified within dust mite allergen molecules, cf. SDAP – Structural Database of Allergenic Proteins (http://fermi.utmb.edu/SDAP/index.html) and Pfam, the Protein Families Database (http://www.sanger.ac.uk/Software/Pfam/).

Allergen group	Domain/family	Examples of non-mite allergens with same domain
1	Peptidase C1 (Papain family cysteine protease)	Kiwi fruit: Act c 1; pineapple: Ana c 2; papaya: Car p 1
2	1) ML: MD-2-related lipid recognition domain 2) Immunoglobulin E-set domain	None
3	Tryp-SPc: trypsin-like serine protease	Bee: Api m 7
4	1) Alpha-amylase, catalytic domain 2) Alpha-amylase, C-terminal all-beta domain	Fungi: Asp o 21
5	No conserved domains detected	None
6	Tryp-SPc: trypsin-like serine protease	None
7	No conserved domains detected	None
8	1) Glutathione S-transferase, N-terminal domain 2) Glutathione S-transferase, C-terminal domain	Cockroach: Bla g 5 Fungi: Pen c 24
9	Tryp-SPc: trypsin-like serine protease	None
10	Tropomyosin	Nematode: Ani s 3; insects: Chi k 10, Lep s 1, Per a 7; Crustacea: Pen a 1, Cha f 1; molluscs: Cra g 1, Per v 1, Hal d 1, Mim n 1
11	1) Myosin tail 2) Smc, chromosome segregation ATPases	Nematode: Ani s 2
12	Chitin binding Peritrophin-A domain	None
13	Lipocalin/cytosolic fatty-acid binding protein family	Cow: Bos d 2; horse: Equ c 1; dog: Can f 1, Can f 2; cat: Fel d 4; mouse: Mus m 1; rat: Rat n 1
14	LPD-N Lipoprotein N-terminal domain	Chicken: Gal d vitellogenin
15	1) Glycosyl hydrolases family 18 2) Chitin binding peritrophin-A domain	None None
16	GEL: 4 gelsolin homology domains	None
17	Calcium-binding EF hand	None
18	Glyco 18 domain	None
19	No conserved domains detected	None
20	1) ATP guanido phospotransferase, N-terminal domain 2) ATP guanido phospotransferase, C-terminal catalytic domain 3) Actin-binding domain	None
21	No conserved domains detected	None
Aldehyde dehydrogenase	Aldedh, Aldehyde dehydrogenase	Fungi: Alt a 10
13.8 kDa allergen	NLPC_P60	
Lipase homologue	Pancreatic lipase-like	Wasp: Ves m 1
Hexosaminidase	Glycosyl hydrolase family 20, catalytic domain	Plants: Ole e 10
Cathepsin F	Peptidase_C1A	None
Cathepsin D	Eukaryotic aspartyl protease	Insects: Bla g 2; fungi: Asp f 10
Enolase	Enolase	Fungi: Alt a 6, Alt a 11, Asp f22, Cla h 6, Pen c 22; plants: Hev b 9
Cyclophilin	cyclophilin_ABH_like	Fungi: Asp f 11, Mala s 6; plants: Bet v 7
Thaumatin	THN, Thaumatin family	Plant: Act c 2, Jun a 3, Mal d 2
Cystatin A	Cystatin domain	Cat: Fel d 3
Heat shock protein 70	HSP70	Fungi: Alt a 3
IGFP	1) IB, insulin growth factor-binding protein homologues 2) KAZAL, Kazal type serine protease inhibitors 3) IGcam, immunoglobulin cell adhesion molecule	Chicken: Gal d 1
Alpha-tubulin	alpha_tubulin	None
Profilin	PROF, Profilin	Plants: Amb a 8, Ara h 5, Cyn d 12, Hev b 8, Mal d 4, Ole e 2, Phl p 12
Lipid transfer protein	AAI_LTSS; Alpha Amylase Inhibitor	Plants: Amb a 6, Gly m 1, Mal d 3
Ribosomal protein P2	RPP1A, Ribosomal protein	None

been isolated and characterised from *Dermatophagoides farinae*, *D. pteronyssinus*, *Euroglyphus maynei* and *Blomia tropicalis* (Tables 7.2, 7.3; Lake *et al.*, 1991; Mills *et al.*, 1999; Chew *et al.*, unpublished), and alpha-amylases have been detected in several stored products species (Bowman, 1984; Stewart *et al.*, 1991; cf. 2.2.3b).

The group 4 allergens have a 22–25 residue signal sequence followed by a 496 residue mature protein (484 for Blo t 4; Figure 7.12). Blo t 4 has no N-glycosylation sites. Der p 4 has one at N10, Eur m 4 has one at N147. There are eight conserved cysteines that form disulphide bonds.

Alpha-amylases from animals are highly conserved along the entire sequence (Janeček, 1997). They have been used as 'molecular clocks' to estimate the timing of evolutionary divergence of animal groups (Hickey *et al.*, 1987). Amino acid sequences of Der p 4 and Eur m 4 were 94% identical, and Der p 4 shows 55–60% identity with insect alpha-amylases such as those from the flies *Drosophila* spp. and *Bibio marci* and the beetle *Tenebrio molitor*. Several arthropods have multiple genes for amylases, and isoforms of mite amylases have been noted (Bowman and Lessiter, 1985; Lake *et al.*, 1991). Cereals contain amylase inhibitors, which are important occupational allergens (Sánchez-Monge *et al.*, 1996). In stored products mites, multi-genes for amylase isozymes may be an evolutionary response to dietary inhibitors.

The three-dimensional structure of amylase consists of an $(\alpha/\beta)_8$ barrel – similar but not identical to a TIM barrel – consisting of eight parallel beta-strands forming an inner cylindrical beta-sheet, surrounded by eight alpha-helices, thus the barrel consists of eight repeated alpha/beta units (Janeček, 1997; Figure 7.3).

Group 4 allergens have intermediate to low frequency of IgE binding (mean 38%; Table 7.6). Frequency of IgE binding to a panel of sera from children was only 25%, compared with 46% with sera from adults (Lake *et al.*, 1991). This may reflect the relative abundance of this allergen: it was isolated from spent mite growth medium in far lower concentrations than were serine proteases (Lake *et al.*, 1991; King *et al.*, 1996), though Mills *et al.*, (1999) pointed out that levels could potentially vary with differential gene expression.

b   Group 12 allergens

This group consists of the 14 kDa Blo t 12 (Puerta *et al.*, 1996a) and Lep d 12 (AY293744; Chew *et al.*, unpublished; Table 7.1). The composition of the 123–125 amino acid residues of the mature proteins

fits best with the chitinase groups 15 and 18 (Figure 7.1b), although the mean transformed Dayhoff value for histidine in Blo t 12 is markedly higher than for the others.

A BLAST search revealed some limited homology with a hypothetical protein from the fruit fly *Drosophila melanogaster*, a peritrophic membrane binding protein from the cabbage looper moth *Trichoplusia ni* and Der f 15 (Figure 7.20). The peritrophic membrane (PM) chitin binding protein, CBP1 from *Trichoplusia ni*, isolated by Wang *et al.* (2004), is much larger than Der p 12, containing 1171 amino acid residues (124 kDa). The sequence similarity is most apparent in the C-terminal end of the CBP1 protein, predominantly in the cysteine-rich, chitin-binding, peritrophin-A domain which has a general consensus signal $CX_{13-20}CX_{5-6}CX_{9-19}CX_{10-14}CX_{4-14}C$ (Tellam *et al.*, 1999). The terminal cysteine appears to be absent from the Blo t 12 amino acid sequence entered in Genbank (BTU27479). However, if the nucleotide sequence for Blo t 12 is run through a translation program (e.g. *Translate* [GCG]), there is an extra Cys residue at position 151 ($N_{146}XLINC_{151}$). The peritrophin-A domain, underlined for CBP1 in Figure 7.20, corresponds to domain no. 11 of Wang *et al.* (2004; their Figure 1).

The high cysteine content indicates that a key structural property of the peritrophin-A domain is resistance to proteases in the arthropod midgut. It is involved in the binding of PM proteins to chitin fibrils and forming the structural basis of the peritrophic envelope (see Chapter 2). It appears probable that Blo t 12 is involved in this process.

Puerta *et al.* (1996a) found recombinant Blo t 12 bound IgE in sera of patients allergic to *Blomia tropicalis* with a frequency of 50% (Table 7.6).

c   Group 15 allergens

Group 15 allergens have been characterised from *Dermatophagoides farinae* (McCall *et al.*, 2001) and *D. pteronyssinus* (O' Neill *et al.*, 2006). Der f 15 is a protein of 63.2 kDa and 555 amino acids which on SDS-PAGE has an apparent molecular weight of 98/109 kDa due to extensive O-glycosylation (almost 50% is carbohydrate) along a proline/serine/threonine-rich O-glycosylation domain of 84 residues (McCall *et al.*, 2001). It shows homology with insect chitinases which belong to family 18 of the glucohydrolase superfamily (Figure 7.23). The highly conserved sequence YXFDG-LDLDWEYP is typical of insect chitinases (Kramer and Muthukrishnan, 1997; Weber *et al.*, 2003; Merzendorfer and Zimoch, 2003) and is consistent with the active site signature of the family 18 chitinases, including the glutamate residue (E) that is essential for catalysis (Figure 7.23). As well as this catalytic domain, family 18 chitinases possess a PEST-region (proline [P]/glutamate [E]/serine [S]/threonine [T]-rich) associated with O-glycosylation and a cysteine-rich region (Kramer and Muthukrishnan, 1997).

Manduca sexta chitinase was the first insect chitinase to be cloned and sequenced. The enzyme is expressed during moulting and is up-regulated by ecdysteroid and down-regulated by juvenile hormone (Kramer *et al.*, 1993). The sequence is aligned with Der p 15 and Der f 15 in Figure 7.23 and shows 35% identity. Der p 15 and Der f 15 have ca. 90% identity. Der f 15 also shows some homology with Blo t 12 (Puerta *et al.*, 1996a) and Der f 18. Like Der f 18, Der f 15 was found to be a major allergen for atopic dogs sensitised to dust mites, with 80% frequency of IgE-binding of sera. O' Neill *et al.* (2006) found 70% IgE-binding frequency with mite-sensitive human sera. Prior to this, the only published data on IgE-binding frequency of group 15 allergens was a 2-D immunoblot study by Le Mao *et al.* (1998). The amino acid sequence of part of one of the proteins (AFEPHGYLLTAAV) shows 100% identity with Der f 15 (residues 166–178). The amino acid sequences for two isoforms of Der p 15 have been published (O' Neill *et al.*, 2006; Table 7.1).

d Group 18 allergens

Group 18 allergens are known from *Dermatophagoides farinae*, *D. pteronyssinus* and *Blomia tropicalis* (Table 7.2). Der f 18 is a 60 kDa (50 kDa for the mature protein) homologue of chitinase (Weber *et al.*, 2003). Some 54% of human sera that were positive to extracts of *D. farinae* were bound by Der f 18, and some 57–77% of canine sera. The amino acid sequence for Der p 18 has been published (O'Neill *et al.*, 2006; Table 7.1) and shows a high degree of homology with Der f 18 (Figure 7.25). Both allergens consist of a 25 amino acid signal sequence and a 437 amino acid mature protein. A BLAST search of the Der p 18 sequence showed highest homology to other invertebrate chitinases. Homology with the group 15 allergens (cf. above) is only about 30%; highest in the regions of the conserved cysteine residues and the two chitinase catalytic domains (Figures 7.23, 7.25).

7.4.5 The transferases

The transferases belong to EC group 2 and include allergen groups 8 (glutathione transferase) and 20 (arginine kinase). The amino acid composition of the transferase mite allergen family is fairly uniform between the two groups and quite different from the other families (Figure 7.1c; Table 7.5). It has relatively high frequencies of methionine, tyrosine, phenylalanine and leucine (mean Dt of 2.1, 1.5, 1.5 and 1.4 respectively). Lowest mean frequencies are for cysteine (0.4) and serine (0.6).

a)

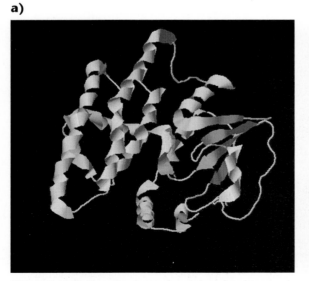

b)

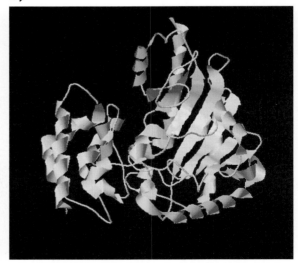

Figure 7.4 Tertiary structures of proteins related to members of the transferase family of mite allergens (groups 8 and 20). **a)** Glutathione s-transferase (human, PDB 1EEM); **b)** arginine kinase (PDB 1P52).

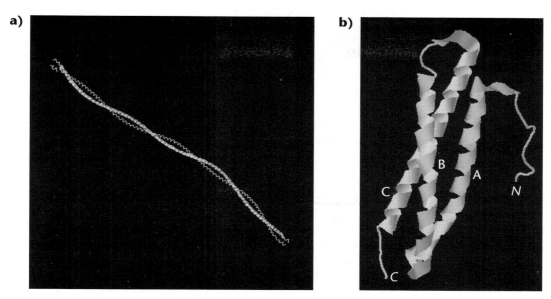

Figure 7.6 Tertiary structures of helical allergens. **a)** Tropomyosin – a homologue, the muscle protein family of mite allergens (groups 10 and 11); **b)** a member of the small alpha-helix protein family (groups 5,7 and 21), Blo t 5 (PDB 2JMH), showing three coiled-coil triple helices (A–C).

copy gene, and has an alpha-helical structure like the other group 5 allergens. Positive skin-prick tests to Blo t 21 were found in 60–95% of patients with asthma and/or rhinitis, and there was little cross-reactivity between Blo t 21 and Blo t 5. The localisation of Der p 21 and its possible function in the gut and faecal pellets has been mentioned above (see Group 5 allergens).

7.4.7   The muscle proteins

Included here are the allergen groups 10 (tropomyosin) and 11 (paramyosin). Characteristic of their amino acid composition are high frequencies of glutamate, glutamine, arginine and methionine (mean Dt of 3.1, 2.0, 1.6 and 1.6 respectively) (Figure 7.1e). Lowest mean frequencies are for cysteine (0.1) and proline (0.01). The tertiary structure of these molecules is a departure from the typical small, globular structures encountered so far. These are long, strand-like alpha-helices.

a   Group 10 allergens

Group 10 allergens are known from *Dermatophagoides pteronyssinus*, *D. farinae*, *Psoroptes ovis*, *Tyrophagus putrescentiae*, *Glycyphagus domesticus*, *Lepidoglyphus destructor* and *Dermanyssus gallinae* (Table 7.2). This is another allergen group (along with group 20) for which there is homology with crustacean allergens. Tropomyosins are muscle-binding proteins involved in the regulation of the contraction of actin. They are highly conserved and are found in all animals. Tropomyosin allergens from crustaceans, molluscs and insects all cross-react with those of dust mites (Reese *et al.*, 1999; Jeong *et al.*, 2006), and they are regarded as so-called pan-allergens (Reese *et al.*, 1999).

Group 10 allergens, e.g. Blo t 10 (Yi *et al.*, 2002) and Der p 10 (Asturias *et al.*, 1998), have a derived molecular weight of about 33 kDa (Table 7.1) consisting of a 15 residue signal peptide and a mature protein of 285 residues. On SDS-PAGE, tropomyosins appear as a band at about 37 kDa. There are no glycosylation sites and no conserved cysteines, consistent with its tertiary structure: a dimeric coiled-coil, consisting entirely of alpha-helix (Figure 7.6a).

Sequence identity within the seven mite tropomyosins in Figure 7.18 is 77%, higher than for any other allergen except paramyosin (cf. below). IgE-binding frequency was reported as 81% for nDer f 10 (Aki *et al.*, 1995). Group 10 allergens can be abundant in nature and they probably can be regarded as major allergens. Recombinant allergens show much lower binding, e.g. 13% for Lep d 10 (Saarne *et al.*, 2003) (Table 7.6).

b   Group 11 allergens

Group 11 allergens are paramyosins, another ubiquitous group of muscle proteins, identified from *Dermatophagoides pteronyssinus*, *D. farinae*, *Psoroptes ovis*, *Sarcoptes scabiei* and *Blomia tropicalis*

(Table 7.2). Like tropomyosin they are highly conserved and cross-reactive (Figure 7.19).

They are large molecules, with a derived molecular weight of 102 kDa (ca. 98 kDa on SDS-PAGE gels) and show 42–80% IgE binding frequency (Tsai *et al.*, 1998, 2005; Ramos *et al.*, 2001; Lee *et al.*, 2004). Like group 10 allergens, IgE reactivity to the recombinant form of Blo t 11 was significantly lower than to the natural form (Teo *et al.*, 2006).

7.4.8 The lipid-binding proteins

Included here are the allergen groups 2, 13 (fatty acid-binding proteins) and 14 (vitellogenin). The amino acid composition of the lipid-binding allergens is somewhat less concordant than it is for other mite allergen families. They have lysine and asparagine at consistently high frequencies (mean D_t of 1.7 and 1.5 respectively), followed by aspartate (1.4) and isoleucine (1.2). Lowest frequencies are for tryptophan (0.2) and proline (0.6). Tryptophan is absent from group 13 allergens. Similarly, cysteine occurs at low frequency in groups 13 and 14 but is common in group 2. Histidine is common in groups 2 and 14 but absent in group 13 (Table 7.5; Figure 7.1f).

Tertiary structure based on X-ray crystallography shows groups 2 and 13 allergens consist of beta-barrels (Figure 7.7a, b). Group 14 is much larger and more complex, and is likely to present allergenically as a series of subunits (cf. below).

a Group 2 allergens

The group 2 allergens are polymorphic, neutral to basic 14 kDa non-glycosylated allergens recognised by the majority of mite allergic individuals (Tables 7.1 and 7.6). They are probably involved in lipid binding and transport, although their precise function remains a mystery. They appear not to be enzymes. Group 2 allergens have been identified in a higher diversity of mites than any other allergen, from *Dermatophagoides pteronyssinus*, *D. farinae*, *D. siboney*, *Euroglyphus maynei*, *Psoroptes ovis*, *Aleuroglyphus ovatus*, *Tyrophagus putrescentiae*, *Suidasia medanensis*, *Glycyphagus domesticus*, *Lepidoglyphus destructor* (previously designated Lep d 1) and *Blomia tropicalis*.

The complete amino acid sequences of group 2 allergens (the sequence of Der s 2 is only partial) indicate the allergens are synthesised as preproteins with signal peptides of 16–17 residues and mature proteins of 125–129 residues (Yuuki *et al.*, 1991; Chua

et al., 1990a; Trudinger *et al.*, 1991; Varela *et al.*, 1994; Schmidt *et al.*, 1995). Group 2 allergens are quite highly conserved (Figure 7.10). Homologous sequences are not concentrated around active sites and cysteine residues as they are in group 1 allergens but are evenly distributed along the molecule. Differences that reflect the phylogenetic relationships of the mite species are evident, and the sequences in Figure 7.10 have been arranged according to families or superfamilies to emphasise this characteristic.

The tertiary structure of Der p 2 and Der f 2 have been modelled using nuclear magnetic resonance spectroscopy (NMR; Mueller *et al.*, 1998; Ichikawa *et al.*, 2005). The crystal structure of Der p 2 and Der f 2 was derived empirically by X-ray diffraction (Derewenda *et al.*, 2002; Johannessen *et al.*, 2005) and found to be very similar. The NMR models differ from the crystal structures in not showing a central cavity. The crystal structure shows a cylindrical molecule containing 10 beta-strands and a short apha-helix. It shows a characteristic immunoglobulin-like fold comprising two anti-parallel beta-sheets, one consisting of three beta-strands, the other of five, overlying each other at an angle of about 30° to form a beta-sandwich or beta-barrel (Figure 7.7a, b). Between the sheets is a large central cavity lined with hydrophobic and aromatic residues, suggesting a lipid-binding function.

The cysteine-cysteine disulphide bond pairings of Der f 2 are C8–C119, C21–C27 and C73–C78 (Nishiyama *et al.*, 1993). This arrangement is likely to be conserved in all group 2 allergens as well as in the 16 kDa mammalian epididymal secretory proteins, human HE1 (Kirchoff *et al.*, 1996) and its porcine homologue (Okamura *et al.*, 1999), which have ca. 25–35% sequence identity with Der p 2. Thomas and Chua (1995) suggested that the group 2 allergens are secretions of the male mite reproductive system. But it is unlikely they are confined to males. Localisation of group 2 allergens in thin sections of mites shows they were mainly in the gnathosomal region, and no concordance with the male reproductive system (section 7.5 below). Males constitute only about 10–15% of natural dust mite populations (Colloff, 1992a; Chapter 5) and would have to be working overtime to produce the amounts of group 2 allergens present in house dust (which are comparable to the levels of group 1 allergens; Chapter 8).

HE1 is the same as Niemann-Pick C2 protein (NPC2; Friedland *et al.*, 2003), involved in cholesterol binding and transport via the low-density lipoprotein

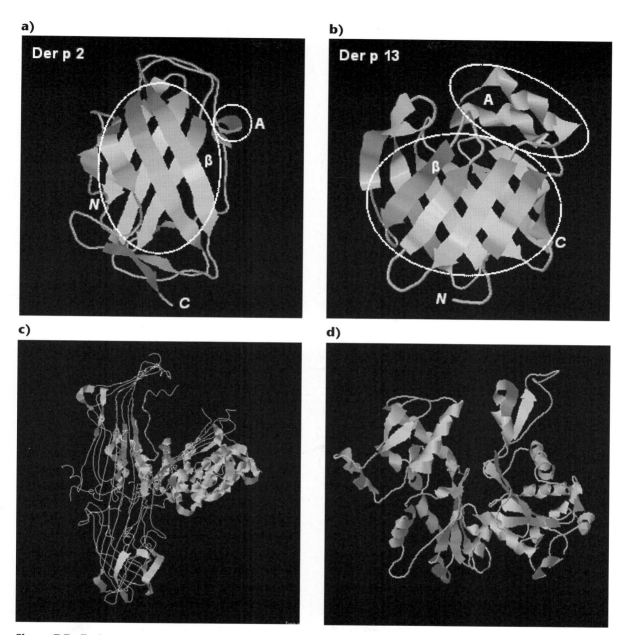

Figure 7.7 Tertiary structures of members of the lipid-binding family of mite allergens (groups 2, 13, 14 and 16) and related proteins. **a)** Der p 2 (PDB 1KTJ) showing a beta-barrel of two anti-parallel beta-sheets, one consisting of three beta-strands, the other of five. A = singly-turn alpha-helix; **b)** Der p 13 (PDB 2A0A) showing a beta-barrel (β) consisting of two sets of five beta-strands. A = helix-turn-helix motif; **c)** vitellogenin homologue of group 14 allergens; **d)** gelsolin homologue of group 16 allergens.

pathway. Absence of NPC2 in humans causes Niemann-Pick disease, a hereditary, fatal neurodegenerative disease which involves the accumulation in mutant cells of cholesterol-filled lysosomes and subsequent cell death. When superimposed on each other, the crystal structures of NPC2 and Der p 2 are quite similar (Friedland *et al.*, 2003, their Figure 5b) although the hydrophobic core differs in some important respects.

There is some sequence and structural homology between group 2 allergens and the human Rho-specific guanine dissociation inhibitor, RhoGDI (Derewenda *et al.*, 2002) and MD-2 that binds to lipopolysaccharides and the lipophilic GM2-activator protein (Keber *et al.*, 2005; Johannessen *et al.*, 2005). These are lipid-binding molecules belonging to the ML family of proteins involved in lipid recognition, metabolism and transport. A sequence homology search by Mueller *et al.* (1998) revealed 26% identity with the protein esr16 (which had 36% sequence identity with HE1) from tracheal epithelial cells of a moth, the tobacco hornworm, *Manduca sexta* (Mészáros and Morton, 1996a). Levels of mRNA for the esr16 gene increase prior to moulting and the expression of the esr16 gene can be regulated with ecdysteroids, the moulting hormones (Mészáros and Morton, 1996a, b). The residues that form the lipid-binding cavity in Der p 2 are conserved in esr16 (Mueller *et al.*, 1998). Ecdysteroids are lipid hormones found in arthropods with very similar structure to cholesterol (which does not function as a hormone in arthropods). It may be that ecdysteroids as well as dietary cholesterol are the natural lipid ligands of group 2 allergens. The binding of hormones such as ecdysone is associated with regulation of moulting (Chapter 2). Although the site of ecdysteroid production in dust mites is not known, it is interesting that Lep d 2 was localised in the gnathosomal region (van Hage-Hamsten *et al.*, 1995), corresponding with the location of the tracheal epithelial cells in *Manduca sexta*.

b   Group 13 allergens

Group 13 allergens have not been isolated from one pyroglyphid species, *Dermatophagoides farinae*, as well as from *Acarus siro*, *Tyrophagus putrescentiae*, *Glycyphagus domesticus*, *Lepidoglyphus destructor* and *Blomia tropicalis*. All these allergens have been fully sequenced (Table 7.1). These allergens are 15–17 kDa proteins showing high homology with fatty acid-binding proteins (FABP) belonging to the lipocalin family. Lipocalins are involved in the binding of small hydrophobic molecules such as fatty acids, glycerol, steroids, arachidonic acid, histamine and many other molecules with critical cellular physiological function (Flower, 1996). Group 13 allergens show relatively high sequence homology with retinoic acid binding proteins of decapod Crustacea (Gu *et al.*, 2002) and insects (Mansfield *et al.*, 1998).

Amino acid sequences of group 13 indicate an absence of conserved cysteine residues, signal or propeptide sequences, and only Aca s 13 and Tyr p 13 have potential glycosylation sites. The degree of sequence homology is extremely high, with residue identities of between 58 and 82% (Figure 7.21), but the degree of allergenicity is very low: only 10–23% of sera from patients allergic to the relevant mite showed IgE binding activity (Table 7.6).

The tertiary structure of Der f 13, modelled using NMR spectroscopy (Chan *et al.*, 2006), shows an N-terminal beta-strand followed by a helix-turn-helix motif and nine anti-parallel beta-strands forming a beta-barrel, with the two alpha-helices at one end of the barrel (Figure 7.7b). This structure is very similar to vertebrate FABPs such as human brain FABP (Balendiran *et al.*, 2000). It is broadly similar to group 2 allergens which also appear to be involved in lipid binding.

c   Group 14 allergens

Group 14 allergens have been found in *Dermatophagoides farinae*, *D. pteronyssinus*, *Euroglyphus maynei*, *Psoroptes ovis*, *Sarcotes scabiei* and *Blomia tropicalis* (Tables 7.1, 7.2). Complete amino acid sequences are available for Eur m 14 (M-177; Epton *et al.*, 1999) and Der p 14 (Epton *et al.*, 1999, 2001a, b). They are 190–206 kDa homologues of vitellogenin and/or the related protein apolipophorinin. Mag 1 (Aki *et al.*, 1994a) and Mag 3 (Fujikawa *et al.*, 1996) are fragments of Der f 14 (Epton *et al.*, 1999), as is Ssag1 (Harumal *et al.*, 2003) and MSA1 (Mattsson *et al.*, 1999). MSA1 (Mattsson *et al.*, 1999) is different from the antigen ASA1 (Ljunggren *et al.*, 2006) even though it is this latter paper that is cited against the GenBank entry for MSA1 (DQ109676). Fujikawa *et al.* (1996) reported that Mag 3 is not the same as the large lipoprotein allergen Dpt4 (Stewart *et al.*, 1983). IgE-binding frequency of Group 14 allergens was 30–39% for rDer f 14 (Aki *et al.*, 1994; Fujikawa *et al.*, 1996) and 70% for nDer f 14 (Fujikawa *et al.*, 1996).

Vitellogenins are large (200–700 kDa), complex carotenoid lipoproteins that constitute the egg yolk of invertebrates and vertebrates. They can be heavily glycosylated and phosphorylated, and may bind with lipids, metals and vitamins. In athropods they are expressed in the invertebrate equivalent of the liver – the fat body in insects, or the hepatopancreas in crustaceans (as well as the ovaries) – secreted into the haemolymph, taken in to the ovary and processed further (Chen *et al.*, 1997; Tsang *et al.*, 2003). Consistent with its role as an egg storage protein secreted into the haemolymph, Harumal *et al.* (2003) showed immunohistochemical localisation of Sar s 14 in internal organs, cuticle and in eggs. Fujikawa *et al.* (1996) found Mag 3 localised in the oesophagus, gut and other internal organs.

Group 14 allergens show relatively high homology (Figure 7.22). Der p 14 shows next highest identity (20–22%) with the N-terminal regions of several vitellogenins from prawns and shrimps (Chen *et al.*, 1997; Tsang *et al.*, 2003; Avarre *et al.*, 2003; Phiriyangkul and Utarabhand, 2006). Although sequence identity is relatively low, it is consistent and evenly distributed along the Der p 14 molecule. Crustacean vitellogenin is encoded by multiple genes (Tsang *et al.*, 2003), present in females and males but only expressed by females. Gravid females constitute only about 10% of individuals in dust mite populations, and eggs about 30% (Colloff, 1992a; Chapter 5). Even accounting for eggs containing fully developed larvae, somewhere between a third and a half of the individuals in a mite population are likely to contain some group 14 allergens. In eggs at an early stage of development, the bulk of the protein present is likely to be vitellogenin, so this allergen is likely to be present in relatively large amounts in natural populations. However, as Epton *et al.* (1999) point out, it is likely to be underestimated in aqueous allergen extracts because it is not water soluble.

Cleavage sites are present in vitellogenins corresponding to the consensus motif RXRR or RXKR which are recognised by furin convertases. These subtilisin-like enzymes cleave the protein into subunits (vitellins) prior to secretion (Chen and Raikhel, 1996). Two groups of cleavage site motifs are highlighted on the sequences in Figure 7.22 at positions 835 and 1137 (Der p 14 numbering), which do not correspond to those of crustaceans (Avarre *et al.*, 2003) or insects (Chen *et al.*, 1997). Cleavage of Der p 14 at these points would result in three subunits of 95, 36 and 76 kDa (fragments I, II and III). Group 14 allergens are large lipoproteins and seem at first sight to be rather unlikely candidates for allergens. But it appears they could be exposed to the human immune system in smaller, subunit forms. This fits with the observation by Fujikawa *et al.* (1998) that Der f 14 broke down on storage and that the fragments were more allergenic than the whole protein. The fragments they obtained on SDS-PAGE were 144, 125 and 52 kDa, which corresponds with M177 minus fragment II, M177 minus fragment III, and M 177 minus fragments I and II, adjusting for differences between derived molecular weight of Der p 14 (206 kDa) and SDS-PAGE molecular weight of Der f 14 (177 kDa). Epton *et al.* (2001a, b) also found western blotting of mite eggs with antisera to a rDer p 14 fragment showed three major bands, consistent with subunit cleavage.

7.4.9 Unclassified allergens

a Group 16 allergens

Der f 16 (Mag 15; Tategaki *et al.*, 2000) is a 53 kDa gelsolin-like protein that bound IgE from allergic patients with a frequency of 47% (Kawamoto *et al.*, 2002b). Gelsolins (Figure 7.7d) are actin-binding proteins that sever and cap actin filaments and are involved in gel-to-sol (hence the name) transformations within the cell, allowing for cell movement (Silacci *et al.*, 2004). The process is regulated by Ca^{++} and the gelsolin molecule has several calcium-binding domains. Der f 16 showed ca. 34–38% amino acid sequence homology with crustacean, insect and human gelsolins (Kawamoto *et al.*, 2002a).

The amino acid composition of Der f 16 (Figure 7.24) shows some similarity with the methionine-rich alpha-helical proteins (groups 5, 7 and 21), but topology mapping indicates only 31% of the molecule is alpha-helix, compared with >50% for the other allergens in this group (Figure 7.5).

b Group 17 allergens

Der f 17 (a.k.a. Mag50) is a 30 kDa calcium-binding protein homologue, deduced on the basis of the allergen possessing Ca^{++}-binding capacity and a helix-loop-helix (EF-hand) motif (Tategaki *et al.*, 2000; Kawamoto *et al.*, 2002b). The full amino acid sequence has not been published and I have been unable to find a database record. Tategaki *et al.* (2000) stress the importance of the Ca^{++}-binding component of the EF-hand motif in IgE-binding activity, but the partial amino acid sequence shows no homology with other EF-hand Ca^{++}-binding allergens such as fish parvalbumins or the pollen allergens Bet v 4 and Ole e 3. Der f 17 bound IgE from mite-allergic patients with a frequency of 35%.

c Group 19 allergens

Blo t 19 is a 7 kDa peptide (Nge and Chua, IUIS database). Few details about this allergen have been published. The amino acid sequence is given in Figure 7.26 courtesy of Cheong Nge (pers. comm., 2006). The recombinant allergen was cited by Thomas *et al.* (2002) as having only 10% frequency of IgE-binding activity.

A BLAST search revealed 76% homology between Blo t 19 and the antibacterial factor, ASABF-a from the nematode, *Ascaris suum*. ASABF (*Ascaris suum* antibacterial factor) peptides are induced as part of the immune response to the presence of bacteria in the pseudocoelom of the nematode (Pillai *et al.*, 2003). Blot 19 has limited hology with CSaβ-type antimicrobial peptides, including insect defensins (Zhang and Kato, 2003). Blo t 19 has eight conserved cysteine residues, most of which are part of an invertebrate defensin consensus (Figure 7.26; Dimarcq et al., 1998). Like the group 12 allergens, the high cysteine content suggests that a key structural property of Blo t 19 is resistance to proteases in the midgut, which is one of the most obvious sites for induction of an immune response against bacteria. Zhang and Kato (2003) detail the cysteine-pairing for ASABF (C1–C5, C2–C6, C3–C7, C4–C8), whereby disulphide bridges are formed, indicating a tightly folded, highly protease-resistant molecule.

7.4.10 Other mite allergens not included in the IUIS classification

There are many other allergens that have been isolated but not assigned to groups 1–21. For some, especially those that are known only from database entries, the evidence of their allergenicity may be based only on amino acid sequence homology with known allergens. For others, data on frequency of IgE-binding has been published.

a Oxidoreductases

Oxyreductases (EC 1.2) act on the aldehyde or oxo-group of donors and include aldehyde dehydrogenase (EC 1.2.1.3) which catalyses the oxidation of aldehydes to carboxylic acids in the presence of NAD or NADP as a co-factor. Two aldehyde dehydrogenase (ALDH) homologue allergens have been databased, from Blomia tropicalis (AY291328; Chew et al., unpublished) and Dermatophagoides farinae (AY283296, Chew et al., unpublished; Figure 7.28). Other aldehyde dehydrogenases include the mould allergens Alt a 10 from Alternaria alternata and Cla h 3 from Cladosporium herbarium. The mite allergens have a conserved cysteine (Cys230), found in all aldehyde dehydrogenases with catalytic activity (Perozich et al., 1999).

b Hydrolases

A 13.8 kDa bacteriolytic enzyme was isolated and characterised from Dermatophagoides pteronyssinus (Mathaba et al., 2002). The protein shows homology with the P60 family of bacterial cell wall-associated hydrolases. The signal peptide is eukaryotic rather than bacterial, suggesting the protein may be derived from bacteria within the mites and incorporated into the mite genome via horizontal gene transfer. The allergenicity of this molecule has not been determined.

A Blomia tropicalis protein (AY291335; Chew et al., unpublished), showing homology with the allergen Ves m 1 (Hoffman, 1994) from the wasp Vespula maculifrons, is a lipase-like enzyme homologue similar to pancreatic triglyceride lipase and phospholipase A1 (EC 3.1.1.4). Lipases are esterases that hydrolyze long-chain acyl-triglycerides into di- and monoglycerides, glycerol, and free fatty acids. The active site of a lipase contains a catalytic triad consisting of Ser-His-Asp/Glu.

A putative allergen from Blomia tropicalis (AY291332; Chew et al., unpublished) shows homology with beta-N-hexosaminidase (EC 3.2.1.52). This enzyme catalyses the hydrolysis of terminal non-reducing acetylhexosamine residues in N-acetyl-beta-D-hexosaminides. The enzyme has the same active site as for N-glucosaminidase and N-galactosaminidase activities and therefore could be involved in moulting via the hydrolysis of N-acetylglucosamine residues of chitin.

A cathepsin F (EC 3.4.22.41) homologue has been identified from Dermatophagoides farinae (CPW1 cysteine proteinase AF145249; Park et al., unpublished; Table 7.1).

The cathepsin D homologue allergen and its isoform from Blomia tropicalis (EC 3.4.23.5; isoform AY792952; Chew et al., unpublished) show high homology with other arthropod aspartic proteases, including that from the tick Haemaphysalis longicornis (Boldbaatar et al., 2006; Figure 7.29).

Carbon-oxygen lyases (EC 4.2): Enolase, phosphopyruvate hydratase (aka 2-phospho-D-glycerate hydrolase) EC 4.2.1.11. Enolases are homodimeric enzymes that catalyse the reversible dehydration of 2-phospho-D-glycerate to phosphoenolpyruvate as part of the glycolytic and gluconeogenesis pathways. The reaction is facilitated by the presence of metal ions. Thus these enzymes are involved in CHO transport and metabolism. They include Glycyphagus domesticus enolase homologue AY288683 (Chew et al., unpublished).

cis-trans-Isomerases (EC 5.2): Peptidylprolyl isomerase B (Cyclophilin; EC 5.2.1.8). These enzymes are involved in protein folding and assembly and include Glycyphagus domesticus Mala s 6 homologue (AY288682; Chew et al., unpublished).

c Enzyme inhibitors

Thaumatins are protease inhibitors, functioning as anti-fungal agents. Glycyphagus domesticus Jun a

3 homologue (AY288680; Chew *et al.*, unpublished) appears to be a thaumatin.

Cystatin A (Stefin A; Cystatin AS) is a cysteine protease inhibitor. These proteins regulate endogenous protease activity and protect against exogenous protease activity. *Lepidoglyphus destructor* cystatin is claimed to be IgE-binding (AJ428051; Kaiser *et al.*, unpublished).

Mite heat shock protein 70 homologues (Figure 7.30) include *Dermatophagoides farinae* Mag29 (DEPMAG29; Aki *et al.*, 1994b) and *Blomia tropicalis* Mag29 (AY291333; Chew *et al.*, unpublished). Heat shock proteins inhibit the release of lysosomal proteases that trigger cell death, and are expressed in cells stressed by heat, radiation or toxins (Nylandsted *et al.*, 2004). They are highly conserved and found in bacteria, plants and animals. The *B. tropicalis* protein showed over 90% homology with a 70 kDa heat shock cognate protein (HSC70) from the midge *Chironomus tentans* (Figure 7.30; Karouna-Renier *et al.*, 2003). A recombinant Mag29 fragment from *D. farinae* showed a positive IgE-binding frequency (by immunoblotting) of 4 out of 41 sera (9.8%) from mite-allergic patients (Aki *et al.*, 1994b). The putative allergenicity of Mag29 from *Blomia tropicalis* is presumably based on its amino acid sequence homology with Mag29 from *D. farinae*.

d Insulin growth factor–binding protein homologue

These proteins show homology with:

- high affinity binding partners of insulin-like growth factors;
- Kazal-type serine protease inhibitors and follistatin-like domains;
- the immunoglobulin domain cell adhesion molecule (cam) subfamily – members are components of neural cell adhesion molecules (N-CAM L1), Fasciclin II and the insect immune protein hemolin (Sun *et al.*, 1990).

A putative IGF-BP allergen is the Blo t homologue of allergen Gal d 1 (AY291331; Chew *et al.*, unpublished).

e Cytoskeletal proteins

Alpha-tubulins are involved in cytoskeletal microtubule formation. There is published experimental evidence of IgE binding activity to alpha-tubulins of *Lepidoglyphus destructor* (Saarne *et al.*, 2003) and *Tyrophagus putrescentiae* (Jeong *et al.*, 2005b). These are highly conserved molecules (Figure 7.31). The amino acid sequence of the *L. destructor* alpha-tubulin shows >90% identity with those from mouse, chicken and *Xenopus laevis*.

Profilin binds actin monomers, membrane polyphosphoinositides and poly-L-proline. Profilin can inhibit actin polymerisation into F-actin by binding to monomeric actin (G-actin) and terminal F-actin subunits, but – as a regulator of the cytoskeleton – it may also promote actin polymerisation. It plays a role in the assembly of branched actin filament networks. Profilins are considered as pan-allergens in plants, and profilins have been identified from *Blomia tropicalis* (AY291334; Chew *et al.*, unpublished) and *Suidasia medanensis* (AY803196; Reginald *et al.*, unpublished). The mite profilins (Figure 7.32) show ca. 38–40% sequence homology with profilin allergens from plants such as those from brazil nut, banana, mango and pineapple.

f Proteins involved in cell maintenance

Type 1 non-specific lipid transfer proteins (LTPs) can enhance *in vitro* transfer of phospholipids between membranes and can bind acyl chains. *Glycyphagus domesticus* Mal d 3 homologue appears to be an LTP (AY288681; Chew *et al.*, unpublished).

Ribosomal protein P2 plays an important role in the elongation step of protein synthesis. P1 and P2 exist as dimers at the large ribosomal subunit. P2 belongs to the ribosomal protein L12P family and is involved in translation, ribosomal biogenesis and structure. *Blomia tropicalis* Blo t homologue of allergen Alt a 6 appears to be a P2 protein (AY291329; Chew *et al.*, unpublished).

g Others

- Allergens of 39 kDa from *D. farinae* (Aki *et al.*, 1994b) and *L. destructor* (Ansotegui *et al.*, 1991), both of unknown function; the latter with IgE-binding activity in eight out of eight sera positive to *L. destructor*.
- A glucoamylase-like allergen (Muzinich *et al.*, 1997) from *D. pteronyssinus*.
- Dpt4, a high molecular weight phospholipid-containing lipoprotein allergen, with an apparent molecular weight of 244 kDa, from *D. pteronyssinus* (Stewart and Turner, 1980b; Stewart *et al.*, 1983).
- An allergen designated Dpt 36 from *D. pteronyssinus* (Stewart *et al.*, 1988).
- Two allergenic components of 79 and 93 kDa of unknown function from *L. destructor*, which

bound IgE from *L. destructor*-positive sera with a frequency of 60% (Olsson *et al.*, 1994).

7.5 Cross-reactivity and sequence polymorphisms of mite allergens

Allergens are said to be cross-reactive when IgE antibodies that are specific to a particular allergen will also bind to another. This implies there is a similarity in the structure and conformation of the part of the allergen molecule to which the antibody molecule binds. Cross-reactivity has some significant diagnostic implications. People who are sensitised to a particular allergen may give a positive skin-prick test reaction to a different allergen (or allergen source) to which they are not exposed, as was found for an urban population who were skin-prick test positive to *D. pteronyssinus*, but also to several storage mites to which they had little or no exposure (Luczynska *et al.*, 1990b).

The cross-reactivity between Der p 2, Der f 2 and Eur m 2 was found to be due to a relatively high degree of amino acid sequence homology, and the surface of the molecule where antibody binding occurs was relatively conserved between the allergens. Lep d 2 and Tyr p 2 have multiple amino acid substitutions across the antigenic surface and do not cross react with the others (A.M. Smith *et al.*, 2001).

This finding is consistent with the greater phylogenetic distance between the pyroglyphid mites than between the glycyphagoid and acaroids mites (Chapter 1, Figure 1.2).

The source of all the allergens that have been purified and characterised has been mites grown in culture. Sometimes these mites have been grown in the laboratory for years, and it is not surprising that their allergens might differ from mites from the wild to which people are naturally exposed. Polymorphisms have been investigated in wild and cultured *Dermatophagoides farinae* by Thomas *et al.* (1992) and Yuuki *et al.* (1997), in *D. pteronyssinus* by Chua *et al.* (1996), W.A. Smith *et al.* (2001), Hales *et al.* (2002) and Nandy *et al.* (2003) as well as in *Lepidoglyphus destructor* by Schmidt *et al.* (1995) and Kaiser *et al.* (2003). The paper by W.A. Smith *et al.* (2001) deals with group 1 allergens, the others with group 2 allergens, and Smith and Thomas (1996) examine polymorphisms of Der p 3.

Group 1 and 2 allergens are highly polymorphic or genetically variable. Other mite allergens have relatively low levels of polymorphism, such as groups 4 and 5, 10 and 11, and 15 and 18. The implications of high levels of polymorphism (allelic variation) are:

- high frequency of IgE binding due to increased opportunity for T-cell receptor binding and increased overall responsiveness of the allergic population;
- IgE binding to local or regional variants (alleles), with implications for accuracy of diagnostics and epidemiological investigations;
- evolution of distinct polymorphisms that vary in their allergenicity – different polymorphisms may activate low-affinity T-cell interactions that drive Th2-type responses.

The nature of the polymorphism differs between group 1 and group 2 allergens. For Der p 1, polymorphism is represented primarily by seemingly random substitutions of amino acids within the allergen molecule (W.A. Smith *et al.*, 2001). There is little evidence that such polymorphism is cumulative and has become fixed within mite populations. For Der p 2, there is a distinct evolutionary pattern, with substitutions focussed on amino acids 40, 47, 111 and 114 (W.A. Smith *et al.*, 2001; Hales *et al.*, 2002). Further, substitutions 47 and 114 usually co-occur. For Lep d 2, substitutions were found at positions 55 and 102. The difference between Der p 2 and Lep d 2 is that in the latter, there is more than one copy of the gene, whereas in the former, there appears to be only a single gene. This is evidenced by the presence of an intron in one Lep d 2 amplicon but not in another (Kaiser *et al.*, 2003). However, unlike Der p 2, no major difference in IgE binding or T-cell responses could be attributed to Lep d 2 polymorphism.

7.6 Conclusions

Why are some proteins allergens and others not? The proteolytically active allergens appear to be able to do a lot of enzymic damage to the human immune system – cleaving IgE receptors, making the lung epithelium leaky, enhancing specific IgE production, releasing inflammatory mediators and generally bringing about the up-regulation of the allergic response. So why can non-enzymes (like group 2) also be major allergens? The answer may lie in the timing of allergen presentation and exposure. If enzymes and non-enzymes are presented to the human immune system at about the same time, then the enzymic activity facilitates the allergic response to the non-enzymic allergens, as was seen in the example of the adjuvant effect of Der p 1 on production of specific IgE to the bystander allergen, ovalbumin (Gough *et al.*, 2001). So the presence in homes of high levels of a diverse cocktail of airborne

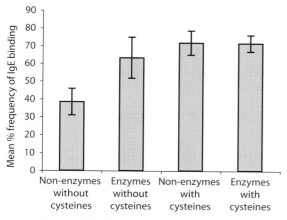
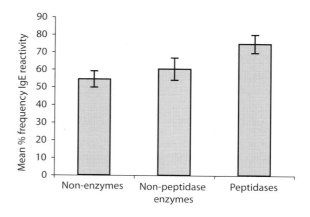

Figure 7.8 IgE binding frequency (± standard error) of major dust mite allergens in relation to their enzymatic function and the presence of conserved cysteine residues. Based on data in Table 7.6.

dust mite allergens, whether borne in faecal pellets, fragments thereof or on other types of particle, is likely to result in co-exposure of the airways to enzymes that pre-condition the immune environment for the bystander non-enzymes.

Whether allergens are major or minor in terms of frequency of IgE responses in a clinical population, may depend not just on the absolute amounts to which people are exposed but whether they are co-presented with enzyme-allergens. To examine this further I plotted frequency of IgE responses to particular allergens against whether the allergens were enzymes or non-enzymes. The enzymes had slightly higher IgE binding frequencies, but the difference was not statistically significant. If the enzyme groups were divided into peptidases and others and the analysis re-run, mean IgE binding to proteases was 75%, significantly higher than the non-peptidase enzymes at 60% and non-enzymes at 54% (Table 7.6; Figure 7.8). There is no IgE-binding data available for allergens 20 and 21. That the peptidase allergens (groups 1, 3, 6 and 9) are on average more likely to elicit IgE responses than non-peptidase enzyme allergens (groups 4, 8, 12, 15, 18) or non-enzyme allergens (groups 2, 5, 7, 10, 11, 13, 14, 16) needs to be qualified. First, the degree of variation between allergen groups within these categories is quite high (e.g. group 6 allergens show much lower frequency of IgE reactivity than other peptidases, group 2 allergens show higher frequency than all other non-enzyme allergens), as it is between mite species (Der p 1: 93–100%, Blo t 1: 62%). Second, the only IgE binding data available for some groups (12,

13, 15, 16 and 19) is from recombinant allergens, which may have lower IgE-binding activity than their native counterparts. Third, the data on IgE binding frequency may not be representative of what happens in the population generally. In cases where the IgE-binding data comes from studies where it has been used to demonstrate the allergenicity of a newly isolated allergen, it may be more representative of what sera happen to be available in the laboratory freezer at the time. Nevertheless, the relationship is intriguing, but it does need to be tested more rigorously.

In a search for common properties of allergens, Furmonaviciene and Shakib (2001) analysed 3-D structures of those allergens showing a common motif, but they really only dealt with group 1 allergens. Chan *et al.* (2006) reckoned that the frequency of charged and polar residues on the allergens is an important factor in determining the allergenicity of the molecule. They found that the non-allergic human homologues of Der f 13 (human FABPs 1–8), Der p 2 (NPC2) and the horse allergen Equ c 1 (lipocalin 9) had significantly fewer charged residues than the allergens, indicating that charged residues play an important role in IgE binding and allergenicity, either because the molecules are more soluble or because charged residues may form more suitable epitopes. Liaw *et al.* (2001) observed that Der p 5 had a high content of charged residues (40%), that non-enzyme allergens such as groups 5, 7, 10, 11 and Mag 29 contain coiled-coil helices consisting of multiple heptad repeats, and that many coiled-coil proteins are known to be strongly antigenic.

Appendix – amino acid sequences

```
                                           ▼
Der f 1 AB034946  -98 : --------MKFVLAIASLLVLSTVYARPASIKTFEEFKKAFNKNYATVEEEEVARKNFLE -47
Der p 1 DPU11695  -96 : --------MKIVLAIASLLALSAVYARPSSIKTFEEYKKAFNKSYATFEDEEAARKNFLE -45
Eur m 1 AF047610 -102 : --KHLSTIMKIILAIASLLVLSAVYARPASIKTFEEFKKAFNKTYATPEKEEVARKNFLE -45
Sar s 1 AY525148 -104 : MNTFCCKFACIVFITLYLDFISIRCDENESINTFDLFKSRFQKTYRSIEDELDAERNFND -42
Pso o 1 AF495854  -92 : -----------LAIASLLVLSVVYAYPSEIRTFEEFKKAFNKHYVTPEAEQEARQNFLA -41


                                                            1  4
Der f 1 AB034946  -46 : SLKYV-EANKGAINHLSDLSLDEFKNRYLMSAEAFEQLKTQFDLNAETSACRINSVNVPS  13
Der p 1 DPU11695  -46 : SVKYV-QSNGGAINHLSDLSLDEFKNRFLMSAEAFEHLKTQFDLNAETNACSIN-GNAPA  12
Eur m 1 AF047610  -46 : SLKYV-ESNKGAINHLSDLSLDEFKNQFLMNANAFEQLKTQFDLNAETYACSINSVSLPS  13
Sar s 1 AY525148  -43 : SLRYVLQTIGTKINAHSDLSGEEFARKFSSNIESIEASNDDY----KPYACDFLPGQYPA  13
Pso o 1 AF495854  -40 : SLEHIEKAGKGRINQFSDMSLEEFKNQYLMSDQAYEALKKEFDLDAGAQACQIGAVNIPN  13
Blo t 1 AF277840      : ----------------------------------------------------------IPA   3


                             32 35                                66
Der f 1 AB034946   14 : ELDLRSLRTVTPIRMQGGCGSCWAFSGVAATESAYLAYRNTS---LDLSEQELVDCASQ-  69
Der p 1 DPU11695   13 : EIDLRQMRTVTPIRMQGGCGSCWAFSGVAATESAYLAYRNQS---LDLAEQELVDCASQ-  68
Eur m 1 AF047610   14 : ELDLRSLRTVTPIRMQGGCGSCWAFSGVASTESAYLAYRNMS---LDLAEQELVDCASQ-  69
Sar s 1 AY525148   14 : QIDLRDIGHLTSIKNQGNCGACWAFATICTIESLLLASKQVSPCKFSLSEQNLIDCASK-  72
Pso o 1 AF495854   14 : EIDLRALGYVTKIKNQVACGSCWAFSGVATVESNYLSYDNVS---LDLSEQELVDCASQ-  69
Blo t 1 AF277840    4 : NFDWRQKTHVNPIRNQGGCGSCWAFAASSVAETLYAIHRHQN---IILSEQELLDCTYHL  60
                                  ▲     ▲


                        *72*                    104          118
Der f 1 AB034946   70 : -------HGCHGDTIPRGIEYIQQNGVVEERSYPYVAREQQCR-RPNSQHYGISNYCQI- 120
Der p 1 DPU11695   69 : -------HGCHGDTIPRGIEYIQHNGVVQESYYRYVAREQSCR-RPNAQRFGISNYCQI- 119
Eur m 1 AF047610   70 : -------NGCHGDTIPRGIEYIQQNGVVQEHYYPYVAREQSCH-RPNAQRYGLKNYCQI- 120
Sar s 1 AY525148   73 : -------RGCDGEKQSTGYRFLQQNGTCESSRYPYVAKVQQCK-HPFGPNYKIRDFCMI- 123
Pso o 1 AF495854   70 : -------HGCGGDTVLNGLRYIQKNGVVEEQSYPYKAREGRCQ-RPNAKRYGIKDLCQI- 120
Blo t 1 AF277840   61 : YDPTYKCHGCQSGMSPEAFKYMKQKGLLEESHYPYKMKLNQCQANARGTRYHVSSYNSLR 120


                                            *  *            *  *
Der f 1 AB034946  121: YPPDVKQIREALTQTHTAIAVIIGIKDLRAFQHYDGRTIIQHDNGYQPNYHAVNIVGYGS 180
Der p 1 DPU11695  120: YPPNVNKIREALAQTHSAIAVIIGIKDLDAFRHYDGRTIIQRDNGYQPNYHAVNIVGYSN 179
Eur m 1 AF047610  121: SPPDSNKIRQALTQTHTAVAVIIGIKDLNAFRHYDGRTIMQHDNGYQPNYHAVNIVGYGP 180
Sar s 1 AY525148  124: HPENPTIVKTVLAKFKASVSTSLQILDMDKFRHYDGKSVIINDFKAKPLNHAVNIVGYGP 183
Pso o 1 AF495854  121: YPPNGDKIRTYIATKQAALSVIIGIRDLDSFRHYDGRTLQSDNGGKRNFHAINIVG--- 177
Blo t 1 AF277840  121: YRAGDQEIQAAIMNHGPVVIYIHGTEA--HFRNLRKGILRGAGYNDAQIDHAVVLVGWGT 178
                                                                    ▲


Der f 1 AB034946  181: TQGVDYWIVRNSWDTTWGDSGYGYFQAGNNLMMIEQYPYVVIM 223
Der p 1 DPU11695  180: AQGVDYWIVRNSWDTNWGDNGYGYFAANIDLMMIEEYPYVVIL 222
Eur m 1 AF047610  181: TQGVDYWIVRNSWDTTWGDNGYGYFAANINLMMIEQYPYVVML 223
Sar s 1 AY525148  184: KFGRQAWYRNSWGTSWGDKGYAYVSQDSSIFGITKRVYVAWL 226
Pso o 1 AF495854  178: --------------------YGYFAAN-------------ASL 184
Blo t 1 AF277840  179: QNGIDYWIVRTSWGTQWGDAGYGFVERHHNSLGINNYPIYASL 221
                                     ▲
```

Figure 7.9 Multiple amino acid sequence alignment of group 1 allergens (cysteine proteinase). **C** = conserved cysteine residues (numbered); **L** = invariant sequences; L = conservative substitutions; __ = signal (pre-) peptide; ▼ = starting position of pro-peptide (Der 1); 1 = starting position of mature protein; **NXS NXT** = N-glycosylation sites (mature protein); ▲ = residues comprising the active site (from Meno *et al.*, 2005); residues involved in dimerisation of Der p 1 (from de Halleux *et al.*, 2006).

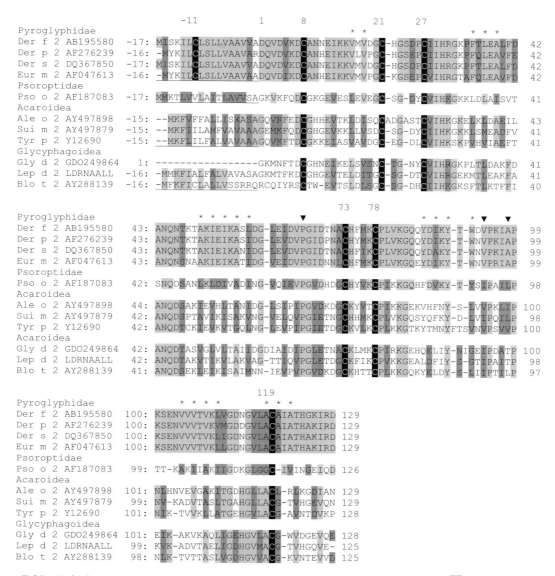

Figure 7.10 Multiple amino acid sequence alignment of group 2 allergens (function unknown). **C** = conserved cysteine residues (numbered); **L** = invariant sequences; **L** conservative substitutions; __ = signal peptide; 1 = starting position of mature protein. ▼ = residues essential to cholesterol binding in NPC2 protein (Friedland *et al.*, 2003); * = hydrophobic and aromatic residues lining central cavity identified by Derewenda *et al.* (2002).

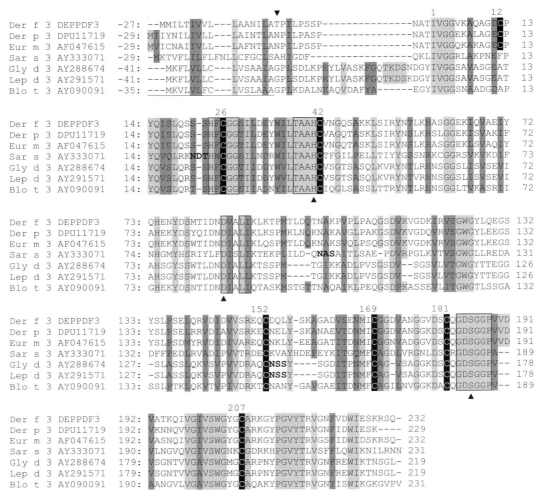

Figure 7.11 Multiple amino acid sequence alignment of group 3 allergens (trypsin). **C** = conserved cysteine residues (numbered against Der f 3); **L** = invariant sequences; L conservative substitutions; __ = signal peptide; ▼ = starting position of pro-peptide; **NXS NXT** = N-glycosylation sites (mature protein, where X = any amino acid except proline); 1 = starting position of mature protein; ▲ = catalytic residues typical of the serine peptidase family S1; boxed sequences = conserved domains of S1 serine peptidases.

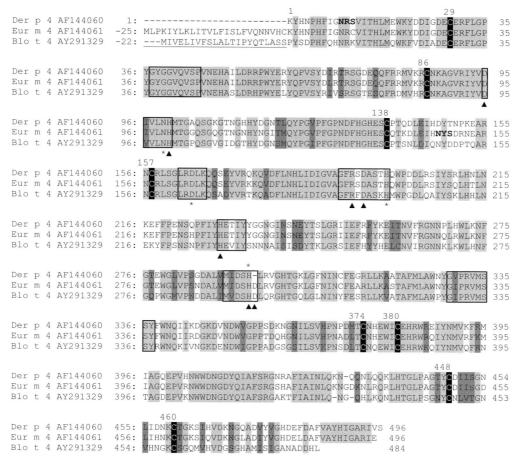

Figure 7.12 Multiple amino acid sequence alignment of group 4 allergens (amylase). **C** = conserved cysteine residues (numbered); **L** = invariant sequences; **L** conservative substitutions; 1 = starting position of mature protein; * = the first position is not the N-terminus of the complete molecule; __ = signal peptide; **NXS NXT** = N-glycosylation sites (mature protein, where X = any amino acid except proline); boxed regions = conserved regions typical of alpha-amylases (see Janeček, 1997); ▲ = catalytic residues; * = calcium-binding residues.

```
                          1
Der f 5  AB195581  -18: ------------------------MKFIIAIAVCTLA-VVCVS----------------
Der p 5  CAA35692  -34: ---------LFLENKDPKPLKKISIMKFIIAFFVATLA-VMTVS----------------
Lep d 5  Q9U5P2
Lep d 5  AY288152  -52: MTGVKTHLQHELKRTDLNFL-----EKFNLDEITAPLN-VLTKELTEVQKHVKAVESDEV   2
Blo t 5  BTU59102  -19: ------------------------MKFAIVLIACFAASVLAQE----------------
Der p 21 ABC73706 -20: ------------------------MKFIITLFAAIVM-AAAVS------------GFIV   2
Blo t 21 AY800348 -16: ------------------------MKFIIALAALIA---VACA---------------

Der f 5  AB195581   3: GEPKKHDYQNEFDFLLMQRIHEQMRKGEEALLHL-HHQINTFEENPTKEMKEQILGEMDT  61
Der p 5  CAA35692   3: GEDKKHDYQNEFDFLLMRIHEQIKKGELALFYL-QEQINHFEEKPTKEMKDKIVAEMDT  61
Lep d 5  Q9U5P2     1: -----DDFRNEFDRLLIHMTEEQFAKLEQALAHL-SHQVTELEKSKSKELKAQILREISI  59
Lep d 5  AY288152   3: AIPNPDEFRNEFDRLLIHMTEEQFAKLEQALAHL-SHQVTELEKSKSKELKAQILREISI  61
Blo t 5  BTU59102   3: HKPKKDDFRNEFDHLLIEQANHAIEKGEHQLYL-QHLQLDELNENKSKELQEKIIRELDV  61
Der p 21 ABC73706  3: GDKKEDEWRMAFDRLMMEELETKIDQVKGLLHL-SEQYKELEKTKSKELKQILRELTI  61
Blo t 21 AY800348   3: LPVSNDNFRHEFDHMIVNTATQRFHEIEKFLLHITTHEVDDLEKTGNKDEKARLLRELTV  62

Der f 5  AB195581   62: IIALIDGVRGVLNRLMKRTDLDIFERYNVEIALKSNEILERDLKKEEQRVKKIE-V---  116
Der p 5  CAA35692   62: IIAMIDGVRGVLDRLMQRKDLDIFEQYNLEMAKKSGDILERDLKKEEARVKKIE-V---  116
Lep d 5  Q9U5P2     62: GLDFIDSAKGHFERELKRADLNLAEKFNFESALSTGAVLHKDLTALATKVKAIETK---  117
Lep d 5  AY28815221 62: GLDFIDSAKGHFERELKRADLNLAEKFNFESALSTGAVLHKDLTALATKVKAIETK---  117
Blo t 5  BTU59102   62: VCAMIEGAQGALERELKRTDLNILERYNYEEAQTLSKILLKDLKVKDIQTQ---  117
Der p 21 ABC73706  62: GENFMKGALKFFEMEAKRTDLNMFERYNYEFALESIKLLIKKLDELAKKVKAVNPDEYY  120
Blo t 21 AY800348   63: SEAFIEGSRGYFQRELKRTDLDLLEKFNFEAALATGDLLLKDLKALQKRVQDSE-----  115
```

Figure 7.13 Multiple amino acid sequence alignment of allergens of groups 5 and 21 (alpha-helical proteins). **L** = invariant sequences; **L** conservative substitutions; 1 = starting position of mature protein; = signal peptide; **NXS NXT** = N-glycosylation sites (mature protein, where X = any amino acid except proline).

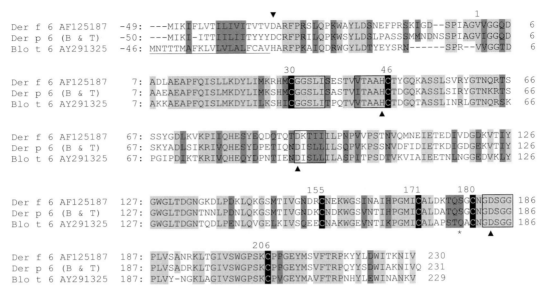

Figure 7.14 Multiple amino acid sequence alignment of group 6 allergens (chymotrypsin). Der p 6 (B & T) = Bennett & Thomas (1996). **C** = conserved cysteine residues (numbered); **L** = invariant sequences; **L** conservative substitutions; __ = signal peptide; ▼ = putative starting position of pro-peptide; 1 = starting position of mature protein; ▲ = catalytic residues typical of the serine peptidase family S1; boxed sequences = conserved domains of S1 serine peptidases; * = serine residue (Der 6) conferring substrate specificity.

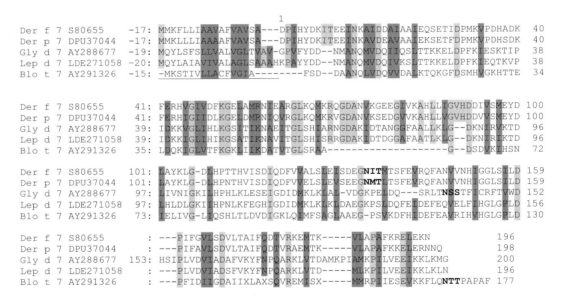

Figure 7.15 Multiple amino acid sequence alignment of group 7 allergens (function unknown). **L** = invariant sequences; **L** conservative substitutions; __ = signal peptide; **NXS NXT** = N-glycosylation sites (mature protein, where X = any amino acid except proline); 1 = starting position of mature protein.

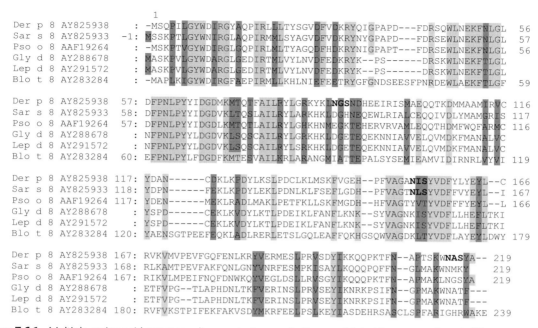

Figure 7.16 Multiple amino acid sequence alignment of group 8 allergens (glutathione-s-transferase). **L** = invariant sequences; **L** conservative substitutions; **NXS NXT** = N-glycosylation sites (mature protein, where X = any amino acid except proline); 1 = starting position of mature protein.

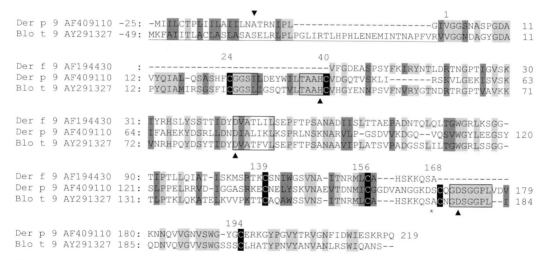

Figure 7.17 Multiple amino acid sequence alignment of group 9 allergens (collagenase). **L** = invariant sequences; **L** conservative substitutions; __ = signal peptide; ▼ = putative starting position of pro-peptide (based on Der p 3); 1 = starting position of mature protein; **C** = conserved cysteine residues (numbered from Der p 9); ▲ = catalytic residues typical of the serine peptidase family S1; boxed sequences = conserved domains of S1 serine peptidases. * = equivalent residue to serine 178, the residue in Der 6 allergens conferring substrate specificity.

```
                                    1
Der f 10 DEPMAG44  -15: FFFVAAKQQQQPSTKMEAIKKKMQAMKLEKDNAIDRAEIAEQKARDANLRAEKSEEEVRA  45
Der p 10 AF016278    1: --------------MEAIKKKMQAMKLEKDNAIDRAEIAEQKARDANLRAEKSEEEVRA  45
Pso o 10 AM114276    1: --------------MEAIKKKMQAMKLEKDNAIDRAEIAEQKARDANLRAEKSEEEVRG  45
Tyr p 10 AY623832    1: --------------MDAIKNKMQAMKLEEDNAIDRAEIAEQKARDANLKSEKTEEEVRA  45
Gly d 10 AY288684    1: --------------MEAIKKKMQAMKLEKDNAIDRAEIAEQKSRDSNLRAEKSEEEVRG  45
Lep d 10 LDE250096   1: --------------MEAIKNKMQAMKLEKDNAIDRAEIAEQKSRDANLRAEKSEEEVRG  45
Der g 10 AM167555    1: --------------MEAIKNKMQAMKLEKDNAADRADVAEQQSKEAVLRAEKAEEEVRG  45


Der f 10 DEPMAG44   46: LQKKIQQIENELDQVQEQLSAANTKLEEKEKALQTAEGDVAALNRRIQLIEEDLERSEER 105
Der p 10 AF016278   46: LQKKIQQIENELDQVQEQLSAANTKLEEKEKALQTAEGDVAALNRRIQLIEEDLERSEER 105
Pso o 10 AM114276   46: LQKKIQQIENELDQVQEQLSAANTKLEEKKKALQTAEGDVAALNRRIQLIEEDLERSEER 105
Tyr p 10 AY623832   46: LQKKIQQIENELDQVQENLTQATTKLEEKEKALQTAEADVAALNRRIQLIEEDLERSEER 105
Gly d 10 AY288684   46: LQKKIQLIENELDQVPESLTRANTKLEEKEKSLPTAEGDVAALNRRIQLIEEDLERSEER 105
Lep d 10 LDE250096  46: LQKKIQQIENELDQVQESLTQANTKLEEKEKSLQTAEGDVAALNRRIQLIEEDLERSEGR 105
Der g 10 AM167555   46: LQKKIQQIENELDQVQEQLATANNSLEEKDKALAAAEGDLAALNRRIQLIEEDLERSEER 105


Der f 10 DEPMAG44  106: LKVATAKLEEASHSADESERMRKMLEHRSITDEERMDGLESQLKEARLMAEDADRKYDEV 165
Der p 10 AF016278  106: LKIATAKLEEASQSADESERMRKMLEHRSITDEERMEGLENQLKEARMMAEDADRKYDEV 165
Pso o 10 AM114276  106: LKIATAKLEEASQSADESERMRKMLEHRSITDEERMDGLENQLKEARMMAEDADRKYDEV 165
Tyr p 10 AY623832  106: LKIATSKLEEASQSADESERMRKMLEHRSITDEERMEGLESQLKEARMMAEDADRKYDEV 165
Gly d 10 AY288684  106: LKIATSKLEEASQSADESERMRKMLEHRSITDEERMEGLESQLKEARMMAEDADRKYDEV 165
Lep d 10 LDE250096 106: LKIATSKLEEASQSADESERMRKMLEHRSITDEERMEGLESQLKEARMMAEDADRKYDEV 165
Der g 10 AM167555  106: LKVATAKLEEASATADESERMRKMLEHRNITDEERMDQLEANLKEAKLMAEDADRKYDEV 165


Der f 10 DEPMAG44  166: ARKLAMVEADLERAEERAETGESKIVELEEELRVVGNNLKSLEVSEEKAQQREEAYEQQI 225
Der p 10 AF016278  166: ARKLAMVEADLERAEERAETGESKIVELEEELRVVGNNLKSLEVSEEKAQQREEAHEQQI 225
Pso o 10 AM114276  166: ARKLAMVEADLERAEERAETGESKIVELEEELRVVGNNLKSLEVSEEKAQQREEAHEQQI 225
Tyr p 10 AY623832  166: ARKLAMVEADLERAEERAETGESKIVELEEELRVVGNNLKSLEVSEEKAQQREEAYEQQI 225
Gly d 10 AY288684  166: ARKLAMVEADLERAEERAETGESKIVELEEELRVVGNNLKSLEVSEEKAQQREEAYEQQI 225
Lep d 10 LDE250096 166: ARKLAMVEADLERAEERAETGESKIVELEEELRVVGNNLKSLEVSEEKAQQREEAYEQQI 225
Der g 10 AM167555  166: ARKLAMVEADLERAEDRAETGETKIVELEEELRVVGNNLKSLEVSEEKALQKEETYEMQI 225


Der f 10 DEPMAG44  226: RIMTTKLKEAEARAEFAERSVQKLQKEVGRLEDELVHEKEKYKSISDELDQTFAEL--- 281
Der p 10 AF016278  226: RIMTTKLKEAEARAEFAERSVQKLQKEVDRLEDELVHEKEKYKSISDELDQTFAELTGY 285
Pso o 10 AM114276  226: RIMTAKLKEAEARAEFAERSVQKLQKEVDRLEDELVHEKEKYKSISDELDQTFAELTGY 285
Tyr p 10 AY623832  226: RIMTAKLKEAEARAEFAERSVQKLQKEVDRLEDELVHEKEKYKSISDELDQTFAELTGY 285
Gly d 10 AY288684  226: RIMTTKLKEAEARAEFAERSVHKLPTEVDRLEDELVHEKEKYLSLSDDLDPSLAERTGC 285
Lep d 10 LDE250096 226: RIMTSKLKEAEARAEFAERSVQKLQKEVDRLEDELVHEKEKYESISDELDQTFAELTGY 285
Der g 10 AM167555  226: RQMTGRLQEAEARAEFAERSVQKLQKEVDRLEDELVQEKEKYKSISDELDQTFAELTGY 285
```

Figure 7.18 Multiple amino acid sequence alignment of group 10 allergens (tropomyosin). L = invariant sequences; L conservative substitutions; __ = signal peptide; 1 = starting position of mature protein.

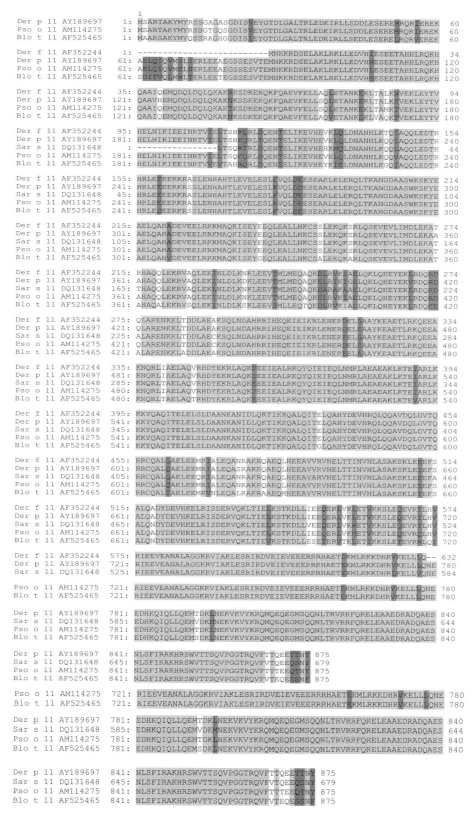

Figure 7.19 Multiple amino acid sequence alignment of group 11 allergens (paramyosin). **L** = invariant sequences; **L** conservative substitutions; 1 = starting position of mature protein.

```
                              1
Blo t 12 BTU27479    -21: MKSVLIFLVAIALFSSANIVSADEEQTTRGRHTEPDDHHEKPTTQCTHEETTSTQHHHEE    39
Lep d 12 AAQ55550    -20: MKSVLIFLVAIALFS-ANIVSAD-EQTTRGRHTEPDDHHEKPTTHATHEETTSTQHHHEE    38
CBP1     AY345124    929: EEAPAICAAEGSSGVLVAHENCNQFYKCANGVPVAFTCSASLLYNPYRGDCDWPSNVECG   988


Blo t 12 BTU27479     40: VVTTQTPHHEEKTTTEETHHSDDLIVHEGGKTYHVVCHEEGPIHIQ---EMCNKYIICSK    96
Lep d 12 AAQ55550     39: -VTTQTPHHEEKTTTEETHHSDDLIVHEGGKTYHVVCHEEGPIPHP---GNVHKYIICSK    94
CBP1     AY345124    989: NRPISVPDDNNVGTSTTTMPDDNQVINDDPSQAPSICAENGSSGVLVAHENCNQYYICSA  1048


Blo t 12 BTU27479     97: SGSLWYITVMPCSIGTKFDPISRNCVLDN         125
Lep d 12 AAQ55550     95: SGSLWYITVMPCSIGTKFDPISRNCVLDN---     123
CBP1     AY345124   1049: GRPV----PMPCSSGLLFNPVNRACDWPQNVVC   1077
```

Figure 7.20 Amino acid sequence alignment of group 12 allergens, with the peritrophic membrane protein CBP1 (chitin binding protein 1) from cabbage looper caterpillar (*Trichplusia ni*). **L** = invariant sequences; **L** conservative substitutions; **C** = conserved cysteine residues; 1 = starting position of mature protein; __ = signal peptide. **CAEN...CSA** = the peritrophin-A domain of CBP1.

```
                       1
Der f 13 2A0AA       1: MASTEGKYKLEKSEKFDEFLDKLGVGFMVKTAAKTLKPTFEVATENDQYIFRSLSTFKNT    60
Tyr p 13 AY710432    1: MVQLNGSYKLEKSDNFDAFLKELGVNFVTRNLAKSASPTVEVIVDGDSYTIKTSSTLKNS    60
Aca s 13 (E, 1999)   1: --QLNGSYKLEKSDNFDAFLKELGLNFVTRNLAKSATPTVEVSVNGDSYTIKTASTLKNT    58
Gly d 13 AAQ54609    1: MANIVGQYKLEKSENFDQFLDKLGVGFLVKTAAKTVKPTLEVAVDGDTYIFRSLSTFKNT    60
Lep d 13 CAB62213    1: MANIAGQYKLDKSENFDQFLDKLGVGFLVKTAAKTVKPTLEVAVDGDTYIFRSLSTFKNT    60
Blo t 13 BTU58106    1: -MPIEGKYKLEKSDNFDKFLDELGVGFMVKTAAKTLKPTLEVDVQGDTYVFRSLSTFKNT    59


Der f 13 2A0AA      61: TEAKFKLGEEFEEDRADGKRVKTVIQKEGDNKFVQTQFGDKEVKIIREFNGDEVVVTASC   120
Tyr p 13 AY710432   61: SEIKFKLGEEFEEDRADGKKVQTSVTKEGDNKLVQVQKGDKPVTIVREFSEEGLTVTATV   120
Aca s 13 (E, 1999)  59: TETSKFLGEEFEEARADGKTVKTVVNKESDTKFVQVQQGDKEVTIVREFSDEGLTVTATV   118
Gly d 13 AAQ54609   61: TEIKFKLGEEFEEDRADGKRVKTVVNKEGDNKFVQTQFGDKEVKVVREFKGDEVEVTASV   120
Lep d 13 CAB62213   61: TEIKFKLGEEFEEDRADGKRVKTVIVKDGDNKFVQTQYGDKEVKVVREFKGDEVEVTASV   120
Blo t 13 BTU58106   60: TEIKFKLGEEFEEDRADGKRVKTVVNKEGDNKFIQTQYGDKEVKIVRDFQGDDVVVTASV   119


Der f 13 2A0AA     121: DGVTSVRTYKRI 132
Tyr p 13 AY710432  121: NGVTSVRFYKRQ 132
Aca s 13 (E, 1999) 119: SGVTSVRFYKRQ 130
Gly d 13 AAQ54609  121: DGVNSVRLYKRL 132
Lep d 13 CAB62213  121: DGVTSVRPYKRA 132
Blo t 13 BTU58106  120: GDVTSVRTYKRI 131
```

Figure 7.21 Multiple amino acid sequence alignment of group 13 allergens (fatty acid binding protein). **L** = invariant sequences; **L** conservative substitutions; 1 = starting position of mature protein; **NXS NXT** = N-glycosylation sites (where X = any amino acid except proline); __ = cytosolic fatty acid-binding protein signature. E, 1999 = Eriksson *et al.*, 1999.

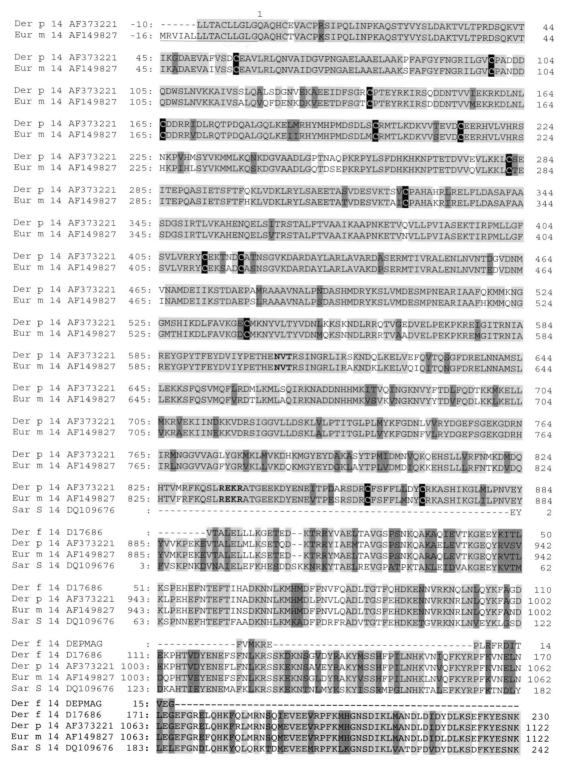

Figure 7.22 Multiple amino acid sequence alignment of group 14 allergens (apolipophorin or vitellogenin). DEPMAG = Mag 1; D17686 = Mag 3. L = invariant sequences; L conservative substitutions; C = conserved cysteine residues; __ = signal peptide; 1 = starting position of mature protein; **NXS NXT** = N-glycosylation sites (mature protein, where X = any amino acid except proline); **RX(K/R)R** = subtilisin cleavage sites.

```
Der f 14 DEPMAG        : -----------------------------------------------------
Der f 14 D17686    231: GTPIELQYKISGKDRSKRAADLGAEDVEGVIDYKNNGSPIDSKMHAHLKMKGNNYGYDSE  290
Der p 14 AF373221 1123: GTPIELQYKVSGKDRSKRAAEMNAEDVEGVIDYKNGSPIDSKMHAHLVKGNNYGYDSE 1182
Eur m 14 AF149827 1123: GTPIELQYKVSGKDRSKRAAELGAEDVEGVIDYKNNGSPIDSKMHAHLKAKGNHYEYDSE 1182
Sar S 14 DQ109676  243: GTPMELQYNLKGKDRSKRAAEKNQEEIEGKIDYKNNGSPIDSKMNANLQAWGNQYAYESE  302

Der f 14 DEPMAG        : -----------------------------------------------------
Der f 14 D17686    291: LKQTQPQQYEGKITLSKNDKKIFINHKSEMTKPT------------------------
Der p 14 AF373221 1183: LKQTEPQQYEGKMTLSKNDKKIFITHKTEMTKPTSTFLLKTDADVSYSESDMKKHYHMEF 1242
Eur m 14 AF149827 1183: LKQTQPQQYEGKITMSKNDKKIFINHKSEMTKPTNTFHLKTDADVSYSDSEMKKHYQMEF 1242
Sar S 14 DQ109676  303: LKQVEPQRYEGKITMSKNDKKIFITHKDEMAKPTDTFHLKSEAEVTFSDSEDKKNYFVEL  362

Der f 14 DEPMAG        : -----------------------------------------------------
Der f 14 D17686        : -----------------------------------------------------
Der p 14 AF373221 1243: KKENDIYTLRSTVERDGQLFYENYLTVHKGGKLNLNYRRNDRKILLDLDNALSPREGTMK 1302
Eur m 14 AF149827 1243: KKENDIYTLRSTVERDGQLFYENYLTIHKGGKLNLNYRRNDRKILLDLDNALSPREGTMK 1302
Sar S 14 DQ109676  363: KKDKDLYSMKSNVKRNNEIFYENNMDLEKNGKMNWYYKRNDRTWNMDLDNAFNPRDGTMK  422

Der f 14 DEPMAG     51: ------------------------NENAYIKNGKLHLSLMDPSTLSLVTKADGKIDMT    51
Der f 14 D17686        : -----------------------------------------------------
Der p 14 AF373221 1303: LNIKDREYNFVLKRDPMRYRDITVEGNENAYVKNGKLHLSLIDPSTLSLVTKADGQIDMT 1362
Eur m 14 AF149827 1303: LNIKDREYNFALKRDPLRYRDITVEGNENAYVKHGKLHLSLMDPSTLSLVTKADGKIDMT 1362
Sar S 14 DQ109676  423: LQVKDRIYDIKLKREPFRYGDLHIEGNENPLIKKGDLHMSLVDPLTLNVLTKNDGIVDMT  482
Blo t 14 AY090092      : -----------MDRYFKYITLKVDGNENALIKNGKAHLSIMDPTTLNLVTKANSNVDFS    49

Der f 14 DEPMAG     52: VDLISPVTKRASLKIDSKKYNLFHEGELSASIVNPRLSWHQYTKRDSREYKSDVELSLRS   111
Der f 14 D17686        : --------------------NTFH----------------------LKTDADVSYSD   339
Der p 14 AF373221 1363: VDLISPITKRASLKVDSKKYNLFHEGELSASLVNPRLSWHQYTKRDSREYKTDVDLSLRS 1422
Eur m 14 AF149827 1363: VDLISPVTKRASLKIDSKKYNLFHEGELSASLMNPRLSWHQYTKRDSREYKSDVDLSLRS 1422
Sar S 14 DQ109676  483: LDLVSPNTKKAALKINSKKYDLDHDGEITVSIFNPRMTWKHHTRKGDMELNIDADITRKG   542
Blo t 14 AY090092   50: MDLFASINKKVALKIDSPKYNFFHDGDIDLSIVNRRLLWKSLTKKDDREYKFNADIARKG   109

Der f 14 DEPMAG    112: SDIALKITMPDYNSKIHYSRQGDQINMDIDGTLIEGHAQGTIREGKIHIKGRQTDFEIES   171
Der f 14 D17686    340: SDMK------------------------------------------------------
Der p 14 AF373221 1423: SDIAVKITMPDYNSKIHYSRQNDQISMDIDGTLIEGHAKGTIKEGKIHIKGRQSDFEIES 1482
Eur m 14 AF149827 1423: SDIALKITMPDYNSKIHYSRQNDQLNLDIDGTLIEGHAQGTVKEGKIHIKGKQSDFEIES 1482
Sar S 14 DQ109676  543: SLITYSRKEPDDSTKVRYSRQGNQVSMEVDSKLIEGHANGTLTDGKIHVKGRESDFEIES   602
Blo t 14 AY090092  110: SMISLTKVTPDRTSSVQYSRNGEKIEVNIDTEYLEGKVEGDRFSGKIVLKNKQNDYELES   169
                                          ***
Der f 14 DEPMAG    172: NYRYEDGKLIIEPVKSENGKLEGVLSRKVPSHLTLETPRVKMNMKYDRYAPVKVFKLDYD   231
Der f 14 D17686        : -----------------------------------------------------
Der p 14 AF373221 1483: NYRYEDGKMLIEPVKSENGKLEGVLSRKVPSHLTLETPRVKMNMQYDRHSPVKMFKLDYD 1542
Eur m 14 AF149827 1483: NYRYEDGKVLIEPVKSENGKLEGVLSRKVPSHLTLETPRVKMNMQYDRHAPVKMFKLDYD 1542
Sar S 14 DQ109676  603: TYKVEDGKLMIEPTKTQNGKLEGLLSRKVPSHLVLETPRVKMNMKYDRFAPVKILKLDYD   662
Blo t 14 AY090092  170: TYKRENGRLVIESVNGKNAKMEAVFSRKEPSKFVLETPNTKAKIDMDLTAPVKTFKLDFD   229

Der f 14 DEPMAG    232: GIHFEKHTDIEYEPGVRYKIIGNGKLKDDGRHYSIDVQGIPRKAFNLDADLMDFKLKVSK   291
Der f 14 D17686        : ------------------------------KHYQME-----------------------
Der p 14 AF373221 1543: GIHLEKHTDLLYEPGVQYKIIGNGKIKDDGSHYSIDTQGPRKAFKLDADMMNFKLNVNK 1602
Eur m 14 AF149827 1543: GIHFEKHTDIQYEPGVRYTIVGNGKLKDDGSHYSIDVQGKPRKAFKLDADMMNFKLKVDK 1602
Sar S 14 DQ109676  663: GLNYEKHIDAEYEPSNHYKYFTDGKSKRSGKGYSIKLDGKPKKALKVDVDMPDFKFNVNK   722
Blo t 14 AY090092  230: NPRYQKKIDASMEPKSKFKYSSYSNQKNEKKERKIEIDGVHMKEFNVDIDFPDFKFKVKQ   289

Der f 14 DEPMAG    292: PEDSNKAQFSYTFNEYTETEEYEFDPHRAYYVNWLSSIRKYIQN----------------   291
Der p 14 AF373221 1603: PEDSNKAQFSYTFNDYTETEEYEFDPHRAYYINWISSIRKYIQNLEGVLSRKVPSHLTLE 1662
Eur m 14 AF149827 1603: PEDSNKAQFSYTFNDYTETEEYEFDPHRAYYVNWLSSIRKYIQT---------------
Sar S 14 DQ109676  723: PEDSNKAQFSYTFNDYTETEEYEFDPHRAYILNWARAIRQYLQT---------------
Blo t 14 AY090092  290: PESSKKVEFSYTFNNYTETEEYDFDPHKAYLVNWVNALRQYVQT---------------

Der f 14 DEPMAG        : -----------------------------------------------------
Der p 14 AF373221 1663: TPRVKMNMQYDRHAPVKMFKLDYDGIHFEKHTDIQYEPGVRYTIVGNGKLKDDGSHYSID 1722
Eur m 14 AF149827      : -----------------------------------------------------
Sar S 14 DQ109676      : -----------------------------------------------------
Blo t 14 AY090092      : -----------------------------------------------------

Der f 14 DEPMAG        : -----------------------------------------------------
Der p 14 AF373221 1723: VQGKPRKAFKLDADMMNFKLKVDKPEDSNKAQFSYTFNDYTETEEYEFDPHRAYYVNWLS 1782
Eur m 14 AF149827      : -----------------------------------------------------
Sar S 14 DQ109676      : -----------------------------------------------------
Blo t 14 AY090092      : -----------------------------------------------------

Der f 14 DEPMAG        : --FIVEDN   341
Der p 14 AF373221 1783: SIFIVEDH  1790
Eur m 14 AF149827      : --FIVWDH  1652
Sar S 14 DQ109676      : --FIVE--   770
Blo t 14 AY090092      : --FVVQN-   338
```

Figure 7.22 Continued

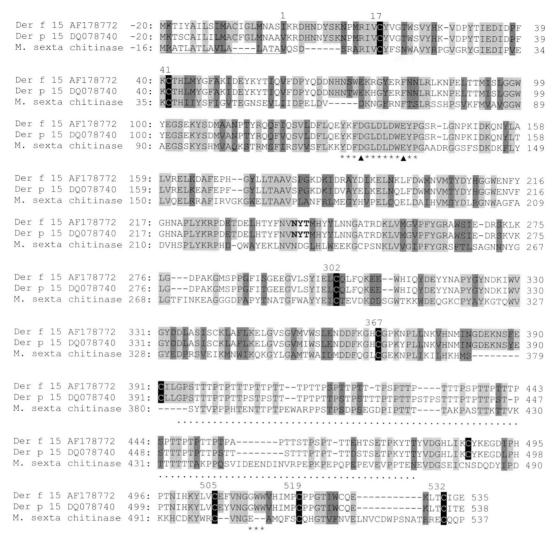

Figure 7.23 Multiple amino acid sequence alignment of group 15 allergens with chitinase from the moth *Manduca sexta* (sequence from Kramer and Muthukrishnan, 1997). **L** = invariant sequences; **L** conservative substitutions; **NXS NXT** = *N*-glycosylation sites (mature protein); ___ = signal peptide; 1 = starting position of mature protein; **C** = conserved cysteine residues (numbered); *** = conserved sequences typical of arthropod chitinase domains; ▲ = catalytic residues; ..... = proline/serine/threonine-rich *O*-glycosylation domain.

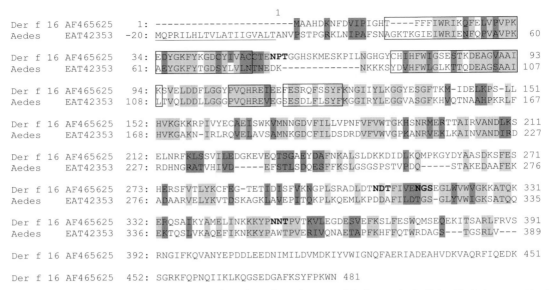

Figure 7.24 Amino acid sequence alignment of Der f 16, the only group 16 allergen (gelsolin) with *Aedes aegypti* gelsolin precursor (EAT42353). **L** = invariant sequences; **L** conservative substitutions; __ = signal peptide; 1 = starting position of mature protein; **NXS NXT** = *N*-glycosylation sites (mature protein); boxed = gelsolin homology domain.

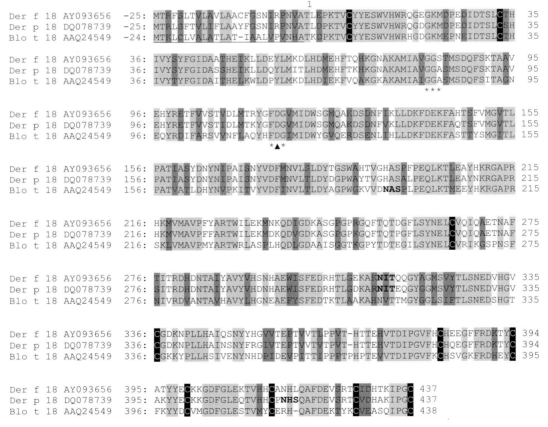

Figure 7.25 Multiple amino acid sequence alignment of group 18 allergens (chitinase). **L** = invariant sequences; **L** conservative substitutions; **NXS NXT** = *N*-glycosylation sites (mature protein); __ = signal peptide; 1 = starting position of mature protein; **C** = conserved cysteine residues; *** = conserved sequences typical of arthropod chitinases; ▲ = catalytic residues.

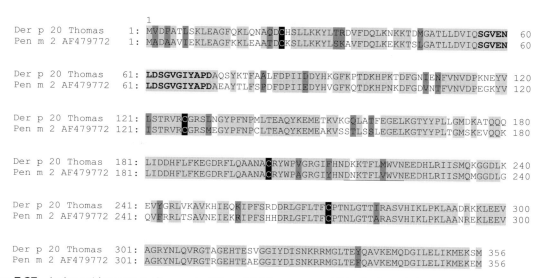

Figure 7.26 Amino acid sequence alignment of Blo t 19, the only known group 19 allergen, with the antibacterial factor ASABF- α from the roundworm, *Ascaris suum*. **C** = conserved cysteine residues; **L** = invariant sequences; L = conservative substitutions; __ = signal peptide; 1 = starting position of mature protein; CXXXC = invertebrate defensin consensus; N & C = Nge and Chua, unpublished data; Nge, personal communication.

```
Der p 20 Thomas     1:  MVDPATLSKLEAGFQKLQNAQDCHSLLKKYLTRDVFDQLKNKKTDMGATLLDVIQSGVEN  60
Pen m 2 AF479772    1:  MADAAVIEKLEAGFKKLEAATDCKSLLKKYLSKAVFDQLKEKKTSLGATLLDVIQSGVEN  60

Der p 20 Thomas    61:  LDSGVGIYAPDAQSYKTFAALFDPIIDDYHKGFKPTDKHPKTDFGNIENFVNVDPKNEYV 120
Pen m 2 AF479772   61:  LDSGVGIYAPDAEAYTLFSPDFDPIIEDYHVGFKQTDKHPNKDFGDVNTFVNVDPEGKYV 120

Der p 20 Thomas   121:  ISTRVRCGRSLNGYPFNPMLTEAQYKEMETKVKGQLATFEGELKGTYYPLLGMDKATQQQ 180
Pen m 2 AF479772  121:  ISTRVRCGRSMEGYPFNPCLTEAQYKEMEAKVSSTLSSLEGELKGTYYPLTGMSKEVQQK 180

Der p 20 Thomas   181:  LIDDHFLFKEGDRFLQAANACRYWPVGRGIFHNDKKTFLMWVNEEDHLRIISMQKGGDLK 240
Pen m 2 AF479772  181:  LIDDHFLFKEGDRFLQAANACRYWPAGRGIYHNDNKTFLVWVNEEDHLRIISMQMGGDLG 240

Der p 20 Thomas   241:  EVYGRLVKAVKHIEQKIPFSRDDRLGFLTFCPTNLGTTIRASVHIKLPKLAADRKKLEEV 300
Pen m 2 AF479772  241:  QVFRRLTSAVNEIEKRIPFSHHDRLGFLTFCPTNLGTTARASVHIKLPKLAANREKLEEV 300

Der p 20 Thomas   301:  AGRYNLQVRGTAGEHTESVGGIYDISNKRRMGLTEYQAVKEMQDGILELIKMEKSM 356
Pen m 2 AF479772  301:  AGKYNLQVRGTRGEHTEAEGGIYDISNKRRMGLTEFQAVKEMQDGILELIKMEKEM 356
```

Figure 7.27 Amino acid sequence alignment of Der p 20 (Thomas, unpublished), the only known group 20 allergen, with the arginine kinase allergen, Pen m 2, from the prawn, *Penaeus monodon* (Yu *et al.*, 2003). **C** = conserved cysteine residues; **L** = invariant sequences; L = conservative substitutions; 1 = starting position of mature protein; **SGV...APD** = active site; __ = putative actin-binding domain.

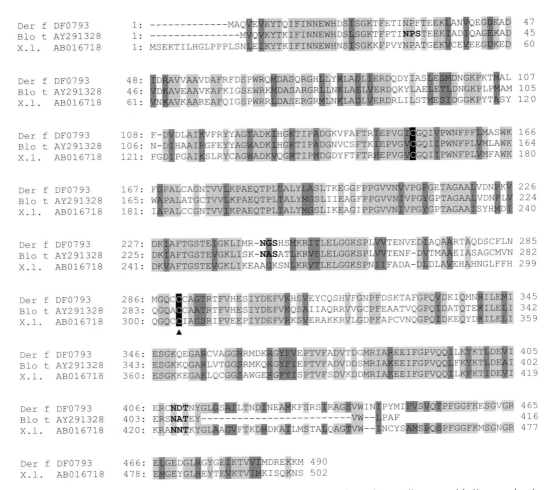

Figure 7.28 Amino acid sequence alignment of aldehyde dehydrogenase homologue allergens with *Xenopus laevis* aldehyde dehydrogenase. **C** = conserved cysteine residues; **L** = invariant sequences; ▲ = catalytic cysteine; **L** conservative substitutions; **NXS NXT** = N-glycosylation sites (mature protein).

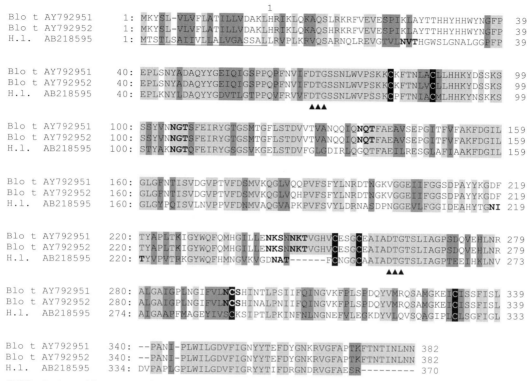

Figure 7.29 Amino acid sequence alignment of cathepsin D homologue allergens with *Haemaphysalis longicornis* aspartic protease (Boldbaatar *et al.*, 2006). **C** = conserved cysteine residues; **L** = invariant sequences; **L** conservative substitutions; 1 = starting position of mature protein; __ = signal peptide; ▲ = catalytic residues of aspartic proteases.

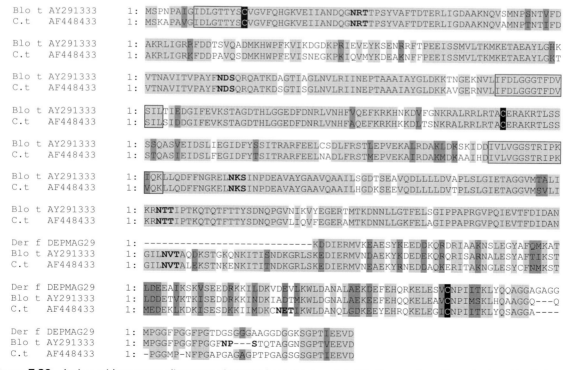

Figure 7.30 Amino acid sequence alignment of Mag29 heat shock protein 70 allergens from *Dermatophagoides farinae* and *Blomia tropicalis* with heat shock cognate protein 70 from the midge, *Chironomus tentans* (Karouna-Renier *et al.*, 2003). **C** = conserved cysteine residues; **L** = invariant sequences; **L** conservative substitutions; **NXS NXT** = N-glycosylation sites (mature protein); boxed residues = heat shock protein 70 family signatures.

```
Lep d  AJ428050   1: MRECISVHVGQAGVQIGNACWELYCLEHGIQPDGQMPSDKTIGTGDDSFNTFFSETGSGK  60
Tyr p  AY986760   1: MRECISVHVGQAGVQIGNACWELYCLEHGIQPDGQMPSDKTIGTGDDSFNTFFSETGSGK  60

Lep d  AJ428050  61: HVPRAVYVDLEPTVVDEVRTGTYRQLFHPEQLITGKEDAANNYARGHYTIGKEIVDLVLD 120
Tyr p  AY986760  61: HVPRAVYVDLEPTVVDEVRTGTYRQLFHPEQLITGKEDAANNYARGHYTIGKEIVDVVLD 120

Lep d  AJ428050 121: RIRKLADQCTGLQGFLIFHSFGGGTGSGFTSLLMERLSVDYGKKSKLEFAVYPAPQVSTA 180
Tyr p  AY986760 121: RIRKLSDQCTGLQGFLIFHSFGGGTGSGFTSLLMERLSVDYGKKSKLEFAVYPAPQVSTA 180

Lep d  AJ428050 181: VVEPYNSILTTHTTLEHSDCAFMVDNEAIYDICRRNLDIERPTYTNLNRLIGQIVSSITA 240
Tyr p  AY986760 181: VVEPYNSILTTHTTLEHSDCAFMVDNEAIYDICRRNLDIERPTYTNLNRLIGQIVSSITA 240

Lep d  AJ428050 241: SLRFDGALNVDLTEFQTNLVPYPRIHFPLVTYAPVISSEKAYHEQLTVSEITNTCFEPAN 300
Tyr p  AY986760 241: SLRFDGALNVELTEFQTNLVPYPRIHFPLVTYSPVISAEKAYHEQLTVAEITNTCFEPQN 300

Lep d  AJ428050 301: QMVKCDPRHGKYMACCLLYRGDVVPKDVNAAIAAIKTKRSIQFVDWCPTGFKVGINYQPP 360
Tyr p  AY986760 301: QMVKCDPRHGKYMACCLLYRGDVVPKDVNAAIAGIKTKRSIQFVDWCPTGFKVGINYQPP 360

Lep d  AJ428050 361: TVVPGGDLAKVQRAVCMLSNTTAIAEAWARLDHKFDLMYAKRAFVHWYVGEGMEEGEFSE 400
Tyr p  AY986760 361: TVVPGGDLAKVQRAVCMLSNTTAIAEAWARLDHKFDLMYAKRAFVHWYVGEGMEEGEFSE 400

Lep d  AJ428050 401: AREDLAALEKDYEEVGLDSTEADDTAGEEF 430
Tyr p  AY986760 401: AREDLAALEKDYEEVGLDSTEAEGGDGEEF 430
```

Figure 7.31 Amino acid sequence alignment of alpha-tubulin allergens from *Lepidoglyphus destructor* and *Tyrophagus putrescentiae*. L = invariant sequences; L conservative substitutions.

```
Blo t  AY291334   1: MSWQSYVDNQICQHVECRLAVIAG-LDGSVWAKFEKDIPKQVSQQELKTIADAIRTNPNS 59
Sui m  AY803196   1: MSWQSYVDNQICQHVDCSLAVIASNQDGAIWAQFERE-NQSISPNELKTIAETIRQNPAG 59

Blo t  AY291334  60: FLEGGIHLGGEKYICIQADNSLVRGRKGSSALCIVATNTCLLAAATVDGFPPGQLNNVVE 109
Sui m  AY803196  60: FLDNGIHIGGSKYICIQADNTLVRGRKGSSALCIVATNTCLLIAATVDGFPPGQLNNVIE 109

Blo t  AY291334 110: KLGDYLKANNY 120
Sui m  AY803196 110: KLGDYLRSNNY 120
```

Figure 7.32 Amino acid sequence alignment of profilin allergens from *Blomia tropicalis* and *Suidasia medanensis*. C = conserved cysteine residues; L = invariant sequences; L conservative substitutions; **NXS NXT** = N-glycosylation sites (mature protein).

8. Allergy and epidemiology

Say what the use, were finer optics given,
To inspect a mite not comprehend the heaven?
Or touch, if trembling alive all o'er,
To smart and agonise at every pore.

Alexander Pope, 1733,
An Essay on Man.

8.1 Introduction

This chapter deals with the allergic disorders associated with dust mite allergens, principally allergic asthma, atopic eczema and allergic perennial rhinitis. The main theme is the natural history of allergen exposure, concentrating on allergen levels and mite population densities in relation to temporal and spatial variation in the prevalence of these disorders, particularly atopic asthma. I do not review epidemiology of asthma or allergy more generally. This topic has been investigated in two large, long-term, international studies of asthma and allergies in adults and children respectively: the European Community Respiratory Health Survey (ECRHS; Burney *et al.*, 1994; European Community Respiratory Health Survey, 1996; Janson *et al.*, 2001; Sunyer *et al.*, 2004; www.ecrhs.org) and the International Study of Asthma and Allergies in Childhood (ISAAC; International Study of Asthma and Allergies in Childhood Steering Committee, 1998; Asher *et al.*, 2006; Pearce *et al.*, 2007; Ait-Khaled *et al.*, 2007; http://isaac.auckland.ac.nz). Several 'environmental' factors have been examined in these studies, including effects of climate (Verlato *et al.*, 2002; Weiland *et al.*, 2004), economic status (Stewart *et al.*,

2001), diet (Ellwood *et al.*, 2001) and mite allergen exposure (Zock *et al.*, 2006). Although there were differences in absolute prevalence values between the two studies, in general the patterns of prevalence were similar: low in Eastern Europe and higher in Western Europe with a strong northwest-southeast gradient, and highest in English-speaking countries (Pearce *et al.*, 2000).

The story of the relationship between dust mites, allergy and asthma is complicated by the fact that atopy, the genetic predisposition to make IgE antibodies to common allergens, may or may not be associated with the presence of disease, and that asthma, eczema and rhinitis can be provoked by agents other than allergens and may not be associated with atopy. Therefore, the disease states of relevance to this chapter are those combining allergen-specific IgE antibodies and/or a positive skin-prick test to dust mite and other common airborne allergens; where there is a family history of atopy; and where patients have active asthma, eczema and rhinoconjunctivitis. Nomenclature of allergy and allergic diseases was revised by Johansson *et al.* (2001) and Figure 8.1 indicates the overlap between atopy and allergic disease

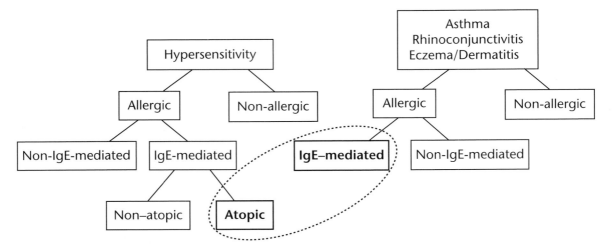

Figure 8.1 Interaction between categories of hypersensitivity and immune response (left) and disease (right) in relation to allergy, IgE antibody isotype and atopy. Dotted line: interface between atopy and disease relevant to mite allergens. Modified from Johansson *et al.* (2001).

based on this classification. For practical purposes in epidemiological studies, atopy has been defined by a positive skin-prick test to common airborne allergens such as those of dust mites, cats, dogs, pollens, cockroaches and various moulds. If the proportion of people with asthma who have specific IgE or positive skin-prick tests is known, the population attributable fraction of asthma due to atopy can be calculated (Pearce *et al.*, 1999, 2000; Sunyer *et al.*, 2004; Weinmayr *et al.*, 2007).

8.2 Diseases associated with dust mites

Over the years, dust mites have been implicated in several diseases and disorders, mostly with an allergic basis. The evidence for the associations ranges from very high to possible for most of the allergic conditions, and low to non-existent for the non-allergic ones.

8.2.1 Allergic diseases

a Allergic asthma

Asthma is probably not a single disease. It is hard to characterise unambiguously, but it involves wheeze, shortness of breath, airway narrowing and inflammation. However, the wide variety of provoking agents and the variable time course of symptoms in childhood and adulthood point towards the existence of different types of asthma – so-called phenotypes. The recognition and classification of different phenotypes has been around for many years, and non-allergic asthma and allergic asthma is one of the better known broad classifications. But it is not clearly understood

whether these phenotypes represent clinical manifestations of different underlying diseases or whether they are different stages in the progression of the pathology of a single disease – inflammation of the airways – that presents differently in different people according to their susceptibility to different provoking agents. Classifications of asthma phenotypes are starting to emerge, with improved characterisation derived from large-scale epidemiological studies and clinical trials of patients with particular phenotypes (Wenzel, 2006).

A large proportion of spending on healthcare is devoted to asthma and it accounts for massive loss of time from school and work, with associated productivity losses (Weiss *et al.*, 1992). Asthma has been the subject of billions of dollars worth of research funding involving thousands of researchers, and yet we still have no clear understanding about how to prevent it, despite many developments in our knowledge, despite it being the subject of tens of thousands of research papers and several major textbooks (e.g. Barnes *et al.*, 1997; Naspitz *et al.*, 2001; Gershwin and Albertson, 2001; the British Library Catalogue has over 1500 records of books with the word 'asthma' in the title). What has happened is that we have realised that it is complex, challenging, elusive, and that many of the concepts that were received wisdom 10 years ago have since been discarded or are being re-evaluated.

Asthma is among the most common chronic diseases of childhood, particularly in urbanised, English speaking countries with a high standard of living. While treatable, asthma has increased in prevalence since the 1960s, though some studies indicate a levelling off in the last few years (Asher *et al.*, 2006; Anderson *et al.*, 2007). In the ISAAC study, 4–32% of 6–7-year-olds (mean 14%) and 2–37% of 13–14-year-olds (mean 12%) had wheeze in the previous 12 months. Prevalence was above 20% for 13–14-year-olds in Australia, New Zealand, the UK, Ireland, the USA, Canada, Peru, Costa Rica and Brazil, and below 6% in India, China, Taiwan, Indonesia, Albania, Georgia, Romania, Russia and Greece (International Study of Asthma and Allergies in Childhood Steering Committee, 1998). In a subset of centres, the population attributable fraction of wheeze due to atopic sensitisation was 41% in countries with high annual per-capita gross national income (GNI; range 13–60%) and 20% in countries with low GNI (range 0–94%; Weinmayr *et al.*, 2007). In the ECRHS study, for those countries common to the ISAAC study, adult asthma (diagnosed) also had a higher prevalence in Australia, New Zealand, the UK and the USA (7–12%), and lower prevalence in India and Greece (~3%; European Community Respiratory Health Survey, 1996). The mean population attributable fraction of adult asthma due to atopic sensitisation was 30%, and 18% for sensitisation to dust mites (Sunyer *et al.*, 2004).

b Rhinitis, rhinoconjunctivitis, keratoconjuntivitis and otitis media

Allergic rhinitis means sneezing, runny, blocked or itchy nose. Allergic rhinoconjunctivitis means these symptoms plus watery, inflamed or itchy eyes. Allergic rhinitis has been classified as seasonal, provoked by pollens (what most people would call hay fever) and perennial, triggered by indoor allergens (mites, pets, moulds, cockroaches). More recently it has been classified as intermittent or persistent, and by severity (see Mösges and Klimek, 2007). Allergic rhinitis affects 5–30% of the population, with children aged 6–14 being most affected (Bousquet *et al.*, 2001; Asher *et al.*, 2006). Effects of rhinitis can include sleep impairment and hence poor cognitive function and impaired quality of life (Meltzer, 2007).

Keratoconjunctivitis

This is a rare, chronic severe inflammation of the conjunctiva affecting mostly young boys. It is considered to be an IgE mediated disorder associated with atopy (Frankland and Easty, 1971; Johansson *et al.*, 2001), although not invariably so. Evidence for an association with dust mites is slight, and includes

mite-specific IgE in tears and sera of patients (Sompolinsky *et al.*, 1984) and a seasonal peak in symptom severity corresponding with maximum mite population density in patients' homes (Mumcuoglu *et al.*, 1988).

Secretory otitis media (glue ear)

It has been reported that 20–90% of children with glue ear are sensitised to common inhalent allergens. Nasal allergy is considered to have some causal involvement (Pelikan, 2007). The association between allergens and glue ear was reviewed by Bisgaard and Mygind (1987).

c Atopic eczema and papular urticaria

Atopic eczema (or atopic dermatitis) is a chronic inflammatory skin disease, particularly common in infancy (10–20% prevalence; Sehra *et al.*, 2008). Prevalence of eczema in general in the ISAAC study was 1–20% and increasing (Asher *et al.*, 2006). In the ECRHS study, a mean of 7.1% of adults had eczema in the previous 12 months and 2.4% had eczema attributable to atopy (Harrop *et al.*, 2007). Prevalence of eczema, like asthma, shows marked geographic variation, tending to be high in Western Europe, Australia and New Zealand and low in Eastern Europe, the Mediterranean and South-East Asia (Asher *et al.*, 2006; Harrop *et al.*, 2007). Positive skin-prick tests, elevated levels of serum IgE to at least one airborne allergen, as well as allergen-specific cellular responses, are common findings among patients with atopic eczema (Mitchell *et al.*, 1982; Chapman *et al.*, 1983; Rawle *et al.*, 1984; Reitamo *et al.*, 1986; Tanaka *et al.*, 1989).

Papular urticaria (also known as nettle-rash: the common stinging nettle is *Urtica dioica*) is an itchy eruption consisting of localised non-pigmented papules. Alexander (1972) reported positive intradermal tests using *Dermatophagoides* spp. extract and high densities of dust mites in homes of a third of a group of children with papular urticaria but not in a control group, suggesting that some cases of urticaria may be due to dust mite allergens. Dixit (1973) reported three cases of people with urticaria who were positive by skin-prick test to dust mites and whose condition seemed to be associated with exposure to dust mites.

d Anaphylaxis

Anaphylaxis is a rapid-onset, severe, systemic allergic reaction. Symptoms include respiratory and cardiac failure. This condition can be fatal. Edston and van Hage-Hamsten (2003) report a case of a 47-year-old farmer who was allergic to dust mites and who died of anaphylactic shock. Serum tryptase activity of post-mortem blood was substantially elevated, indicating massive release of this enzyme from mast cells, a characteristic marker of anaphylaxis. Serum anti-*D. pteronyssinus* and *D. farinae*-specific IgE was also elevated, as were levels of house dust mite allergen in the patient's bed. Dutau (2002) reviewed cases of anaphylaxis due to ingestion of food contaminated with mites (see below). Wen *et al.* (2005) ascribed a case of anaphylaxis to consumption of a pancake contaminated with *Blomia freemani*. A similar case was described by Erben *et al.* (1993) involving pancake mix contaminated with *Dermatophagoides farinae*. In a Venezuelan study of several cases of anaphylaxis that occurred after eating flour and wheat products, high levels of mite contamination were found in 25/30 cases, (Sánchez-Borges *et al.*, 1997, 2001). Oral, mite-induced anaphylaxis was reviewed by Sánchez-Borges *et al.* (2005).

Sudden Infant Death Syndrome (SIDS)

A role for anaphylaxis in SIDS caused by mite allergens was proposed by Mulvey (1972), having observed anaphylactic-type reactions among several patients inadvertently given high doses of mite allergen preparations during immunotherapy. Turner *et al.* (1975) considered anaphylaxis induced by allergy to dust mites may be one causative factor in SIDS deaths in Western Australia. Surveys in homes of SIDS patients (Mulvey, 1972) and in nursery bedding (Tovey *et al.*, 1975; Ingham and Ingham, 1976) indicated that infants received exposure to high population densities of mites. Post-mortem levels of serum beta-tryptase (a marker for anaphylaxis) in infants who died from SIDS, compared with children who died from known non-anaphylactic causes, was elevated in one study (Buckley *et al.*, 2001) but no different from the control group in another (Hagan *et al.*, 1998). The role, if any, for mite-mediated anaphylaxis in SIDS remains unclear.

e   Gastrointestinal allergy

There are a few reports of ingestion of pyroglyphid mites causing symptoms of gastrointestinal allergy. Scala (1995) described a case of a 5-year-old girl,

skin-test positive to dust mites, who had suffered persistent vomiting but no respiratory symptoms. Her bedroom was indicative of high dust mite exposure and her symptoms abated after allergen avoidance. Subsequent nasal challenge with *Dermatophagoides* extract induced vomiting. It was suggested that sensitisation of the gut to dust mites, following allergen inhalation and transfer to the gut by saliva deglutition and oesophageal peristalsis, may have resulted in gastroenteric symptoms. Specific IgE to inhalent allergens has been found in intestinal washings of children with atopic eczema, and the enteric mucosa may sometimes be the first tissue to receive exposure to inhalent allergens (Marcucci *et al.*, 1985).

There are several well-documented cases due to ingestion of relatively large quantities of stored products mites in foodstuffs (Matsumoto *et al.*, 1996), and many foods can become contaminated with mites, especially grain and flour, dried meat and fish, and even beer (mites can contaminate malted barley). In parts of the world where flour and dried goods are still sold retail as loose commodities, mite contamination and consumption is common. Cheeses become infested by mites of the genera *Tyrophagus* and *Tyrolichus* (Robertson, 1952) and some have mites introduced onto them intentionally, including Altenburger (colonised by *Tyrolichus casei*), and Cantal from the Auvergne region of France. The mites colonise as the cheeses are maturing and a layer of dead mites develops on the rind, giving the cheeses a distinctive, slightly salty taste. Barber *et al.* (1996) and Sánchez-Monge *et al.* (1996) in their studies on the interactions of *Dermatophagoides* spp. allergens with inhibitors derived from various cereals, assert that infestation of flour with *Dermatophagoides pteronyssinus* is a common event, but although *D. farinae* was first described from samples of flour (Hughes, 1961), *Dermatophagoides* spp. are not considered to be major pests of flour compared with acaroid and glycyphagoid mites.

The oral administration of allergen can induce specific immune tolerance (Cox *et al.*, 2006), and this occurs naturally in the gut as a result of exposure to the many allergens contained in foods. It seems not unreasonable that immune tolerance in the gut may be induced by allergens associated with inhalent particles.

8.2.2 Non-allergic disorders associated with dust mites

a Acariasis

This is a condition in which live mites are deemed to be living in the lungs, gut or the urinogenital system. Usually mites are found during pathological examination of samples of sputum, faeces or urine (Sasa, 1950, 1951; Chen and Fu, 1992; Li *et al.*, 2003). With pulmonary acariasis, mites may have been accidentally inhaled following high occupational exposure involving stored products, whereas intestinal acariasis is likely due to ingestion of contaminated food.

b Kawasaki disease

Kawasaki disease (KD) is a systemic vasculitis of unknown aetiology, involving skin rash and fever, first recognised in the 1960s, and mainly affecting children <5 years old (Tanaka *et al.*, 1976). The role of mites is now all but discounted, and I mention KD for historical interest because of the fuss and flurry it created at the time. *Rickettsia*-like organisms, assumed to be vectored by dust mites, were found in biopsies (Hamashima *et al.*, 1973, 1982; Carter *et al.*, 1976). Some patients had raised IgE antibody (Kusakawa and Heiner, 1976; Furusho *et al.*, 1981; Fumimoto *et al.*, 1982) and prior respiratory disease (Bell *et al.*, 1981), while others did not (Klein *et al.*, 1986). Onset of KD was claimed to be associated with domestic cleaning and mite exposure (Patriarca *et al.*, 1982; Ohga *et al.*, 1983), prompting mite surveys in homes of patients (Ishii *et al.*, 1983; Klein *et al.*, 1986). A new twist emerged when a strain of *Proprionibacterium acnes* was isolated from blood of patients, and mites from their homes. Since mites ingest skin scales and the bacterium causes acne, the finding is unremarkable. But H. Kato *et al.* (1983) found *P. acnes* caused cardio-pathology similar to KD in experimental animals and suggested mites were vectors of *P. acnes*. The theory of microbial aetiology persisted, with *Pseudomonas* spp. (Keren and Wolman, 1984), *Coxiella burnetti* (Lambert *et al.*, 1985) and retroviruses (Shulman and Rowley, 1986) as candidates. So what does cause Kawasaki disease? It is almost certainly not dust mites. Much of the speculation was fuelled by studies published as short letters without peer-review. Those mite researchers involved reported neutral findings (Ishii *et al.*, 1983) or cast doubt upon the role of dust mites (Jordan *et al.*, 1983; Murray *et al.*, 1984).

c Delusions of parasitosis

Delusions of parasitosis is a psychological disorder in which the patient believes that small parasites, insects, mites or worms are living in the skin or in the bodily orifices (Alexander, 1984). It is a very distressing condition of chronic duration, but it is not caused by allergy dust mites, even though some patients may claim otherwise (Woodford, 1980).

d Sick building syndrome

Sick building syndrome is an occupational disease associated with air quality in modern buildings (Apter *et al.*, 1994). There is no general definition, but there is a complex of non-specific symptoms including dizziness, rhinoconjunctivitis, sore throat, headache, fatigue, chest tightness, wheeze, skin dryness and gastrointestinal symptoms. It has been attributed to electromagnetic radiation from office equipment, organic chemicals in cleaning and construction materials, indoor air of low humidity, pollen, and microorganisms in the air-conditioning system. A role for dust mites in sick building syndrome has been suggested because some of the symptoms were assumed to have an allergic basis and because dust mites and their allergens can be detected in carpets and office furniture (Janko *et al.*, 1995). There are no hard data to indicate that dust mites play any role in sick building syndrome.

8.3 Sensitisation and the development of allergy and allergic disorders

8.3.1 Birth cohorts

Early infancy has been identified as the critical period for sensitisation to allergens (e.g. Holt *et al.*, 1990). Several studies have charted the unfolding relationship between allergen exposure, the development of allergy and the expression of allergic disease from birth through childhood. These birth cohort studies involve newborn infants at high risk of development of allergy because of a family history of atopy. In a birth cohort of 67 children with a family history of allergic diseases, Sporik *et al.* (1990) found all but one of the children with active asthma at age 10 had more than 10 µg g^{-1} of Der p 1 in their beds and carpets at age one. The higher the Der p 1 concentration, the earlier was the onset of the first episode of wheeze. In a cohort of over 900 children followed up to age seven, Lau *et al.* (2000) found a linear relationship between the concentrations of Der 1 (Der p 1 + Der f 1) in homes and the proportion of children who developed allergies to mites. By age three, there was a higher proportion of mite-sensitised children who had developed asthma than children who were not sensitised (ca. 30% v. 10%).

More recently, several studies of infants and children have not found linear dose-response relationships between concentrations of allergens and development of allergy or asthma. The Childhood Asthma Prevention Study (CAPS) involved about 600 children in Sydney, Australia, between 1997 and 2004. Half the children received allergen avoidance measures and the other half were a control group (Mihrshahi *et al.*, 2003). Results were analysed at ages five and eight. There were no differences in numbers of children with asthma, eczema or allergy by age eight. Also, children with low or high mite allergen levels in their beds had lower prevalence of asthma, eczema or allergy than those with medium levels of exposure, i.e. the relationship between allergen exposure, sensitisation and disease was a bell-shaped curve (Almqvist *et al.*, 2007; Tovey *et al.*, 2008; Figure 8.2 below). One explanation for this result is that reducing allergen exposure in homes with high levels of mite allergens

also reduces exposure to something in house dust that may be beneficial or protective against the development of allergy.

The Manchester Asthma and Allergy Study (MAAS; Custovic *et al.*, 2002) was based on 251 newborn children with a family history of atopy, randomised to receive stringent allergen avoidance or a control group. The allergen avoidance group had a *greater* prevalence of children with allergy to dust mites at age three, although they had better lung function than children in the control group (Woodcock *et al.*, 2004). Again, this suggests allergen control reduces exposure to some sort of component that is protective against the development of allergy.

A non-linear dose-response between allergen exposure (to cat but not mite) was found by Torrent *et al.* (2007) in a birth cohort study of 1500 children in Spain and the UK (the Asthma Multicentre Infant Cohort Study). They suggested a minimum level of exposure was required to induce sensitisation in an all-or-nothing manner, and above this threshold there was no obvious dose-response.

The PARSIFAL study (Prevention of allergy – risk factors for sensitisation in children related to farming

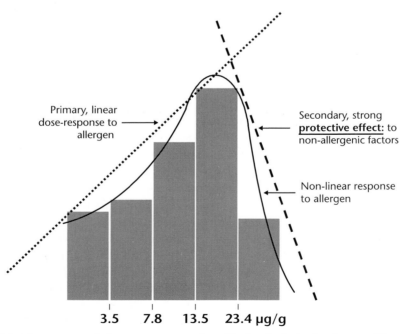

Figure 8.2 Non-linear dose-response between exposure and sensitisation to dust mite allergens. (Figure courtesy of Euan Tovey.) Columns represent relative prevalence of sensitisation to dust mites among 516 children at age five in the CAPS study. Figures on the *x* axis are quintiles of time-weighted average Der p 1 levels (micrograms per gram of dust) in the mattresses of the children. Mattress dust was sampled when children were aged three, six, nine and 12 months, then six-monthly until age five (cf. Almqvist *et al.*, 2007).

and anthroposophic lifestyle) involved 400 children from five European countries and aimed to identify lifestyle factors associated with farming and anthroposophy (the teachings of Rudolph Steiner) that predispose towards lower prevalence of atopy. Farming families have high exposure to bacteria and fungi through contact with soil, crops and animals, and anthroposophists tend to avoid the use of antibiotics and have a diet that includes fermented vegetables containing live lactobacilli (Alm *et al.,* 1999). PARSIFAL was not a birth cohort study, but found, like the CAPS study, a bell-shaped dose-response curve between mite allergen concentrations in mattress dust and prevalence of sensitisation to dust mites (Schram-Bijkerk *et al.,* 2006). The authors suggested that the protective effect at high allergen concentrations was due to exposure to immune modulators such as endotoxins from bacteria as well as soluble beta-glucans and other polysaccharides from fungi.

These studies add to the evidence of a causal relationship between exposure to domestic allergens and development of sensitivity and asthma. They also indicate that the relationship between allergen exposure and disease is not straightforward. The finding in the CAPS and PARSIFAL studies of a non-linear (bell-shaped) relationship between levels of allergen in beds of children and the prevalence of asthma and allergy indicates a primary dose-response and a secondary, seemingly protective effect at high exposure. The data suggest a scenario whereby the development of large populations of dust mites and allergen concentrations is associated with the development of large microbial communities. The removal of dust mite allergens, deemed to be beneficial within a linear dose-response paradigm, appears also to remove some useful compounds derived from microorganisms that modulate the immune system and help prevent the development of allergies and allergic asthma.

There is quite a lot known about bacterial endotoxins and their protective effect on the immune system, including early reports of 'tolerance' following repeated injections of endotoxin in children with asthma (Peterson *et al.,* 1964) and some studies have found a protective effect of endotoxin against sensitisation to allergens (Braun-Farlander *et al.,* 2002; Gehring *et al.,* 2004). But we know almost nothing about the interactions between bacteria, fungi and mites, and the extent to which large mite populations are associated with the development of diverse and abundant microbial communities in house dust. Some basic elements of the interaction between mites and microorganisms were examined in Chapter 2 (Figures 2.7, 2.18). As a general ecological principle, the more diverse and abundant a community in terms of species-richness, functional groups and numbers of individuals present, the greater the prospects for the development of complex interactions between the component organisms. This principle is the basis of research on community and food web ecology (Chapter 4). It is plausible that diverse and abundant communities of mites, fungi and bacteria have far greater levels of complexity than we have considered hitherto, with consequences for our understanding of the interaction between allergen exposure and disease.

8.3.2 The hygiene hypothesis and the microbiota hypothesis

In affluent societies, we have become obsessed with attempting to eradicate 'germs'. Because people tend to be unaware that most microorganisms are not pathogenic, and that they have an immune system designed to give them efficient protection against the ones that are, we spend absurd amounts of money on antibacterial cleaning products, soaps, wipes, swabs and sprays. Dirt and germs have become a collective phobia. Our lack of basic biological knowledge has been exploited and we have been sold the idea that *all* bacteria are harmful. If we do not rigorously attempt to remove them from our kitchen surfaces and bathrooms, according to the advertising messages, then we must be neglectful and slovenly and place our families at risk of gastroenteritis or worse, even though frequent use of cleaning sprays has been associated with an elevated risk of asthma (Zock *et al.,* 2007). In our ignorance, daily we flush thousands of gigalitres of water laden with cleaning chemicals down our sinks and lavatories with virtually no thought to the environmental impact.

Most bacteria are beneficial. Without them, we would not be here, because they drive all the major biogeochemical cycles on the planet. Decomposition and nutrient cycling of nitrogen, phosphorous, carbon and sulphur are essential to plant growth, biodiversity and food production. It has been known for a very long time that consuming certain bacteria stimulates the so-called innate immune system.

The hygiene hypothesis, simply stated, is that allergic diseases can be prevented by infection in early childhood. Declining family size and higher standards of hygiene and cleanliness in recent years have reduced opportunities for cross-infection and resulted in more widespread clinical expression of atopic disease (Strachan, 1989, 2000; see also Rook and Stanford, 1998; Hamilton, 1998). Exposure to a range of viruses, bacteria and fungi – including *Mycobacterium* spp., found in soil and water; lactobacilli in probiotic foods – can confer protection against the development of allergies through exposure to microbial endotoxins, beta-glucans and polysaccharides. The elicitation of cellular immune mechanisms involve two sets of T-cell helper (Th) phenotypes, Th1 associated with diseases caused by infectious agents and Th2, with allergic responses. The theory for the role of infection in early childhood revolves around the notion that a reduction in infections due to increased standards of healthcare and hygiene has led to a failure to elicit Th1-type responses, thus leaving the immune system open to challenge from allergen exposure and Th2-type responses, resulting in greater prevalence of allergy (reviewed by Holt *et al.*, 1999).

Wold (1998) proposed a variation of the hygiene hypothesis called the 'microbiota hypothesis', whereby altered bacterial colonisation of the infant gut resulting in failed induction of immune tolerance is responsible for increased prevalence in allergic sensitisation. The microbiota hypothesis is based on the premise that exposure to lactobacilli, bifidobacteria and mycobacteria skew immune responses to immunoregulation rather than inflammatory responses, and lack of exposure to such bacteria in industrialised societies may be responsible for the increase in allergy. There are strong lines of evidence that link gut microbiota with atopic sensitisation and disease (reviewed by Noverr and Huffnagle, 2005; Penders *et al.*, 2007) and the role of the gut, and the gut-associated lymphoid tissue as an immune organ of relevance to allergy is now starting to be investigated.

8.4 Allergen exposure

Mite allergens are only one of several groups of allergens found within homes, the others being those derived from mammalian pets, fungi and insect pests. In addition there are microorganisms, pollutants such as cigarette smoke, gases from cookers and heaters, volatiles from certain building materials and cleaning products and a host of other compounds capable of affecting the human immune system. The evidence for the causal relationship between exposure to house dust mite allergens and asthma has been steadily amassed since the work of the Leiden group in the 1960s. Voorhorst *et al.* (1964, 1967, 1969) showed that the prevalence of positive skin-prick test reactions to extracts of *Dermatophagoides pteronyssinus* correlated well with atopy and the density of dust mites in homes (which also correlated with the degree of dampness of the homes). Independently, Miyamoto and co-workers (Miyamoto *et al.*, 1968, 1969, 1970; Oshima, 1967) concluded that house dust mites were the most important allergenic component of house dust and showed that the allergenic potency of dust related to the population density of the mites it contained.

Some reviews of the epidemiology of mite-allergic asthma (Sporik and Platts-Mills, 1992; Peat, 1995) have taken the approach of assessing the eight criteria of Bradford Hill (1965) relating to causal associations between disease and environmental factors. These include the strength, consistency, specificity, timing and biological plausibility of the association between disease and risk factor; whether there is a 'dose-response' relationship between factor and disease; whether a cause-and-effect interpretation concords with the known biology of the disease and whether the association can be demonstrated experimentally. All these criteria have been addressed for dust mites and allergic asthma, to a greater or lesser extent, though the dose-response relationship has been the topic of debate (Peat, 1995; Marks *et al.*, 1995b; Platts-Mills *et al.*, 1995; see below).

8.4.1 Mite allergen concentrations as risk factors for allergy and asthma

The First International Workshop on House-Dust Mite Allergy, held in 1987 (Platts-Mills *et al.*, 1989b), established preliminary guidelines for levels of mite allergens in houses that represented a risk factor for allergic disease. The guidelines, based on the very few epidemiological studies available at the time (e.g. Korsgaard, 1982, 1983b), were that 2 μg of *Dermatophagoides* group 1 allergen per gram of reservoir dust from mattresses and carpets (deemed equivalent to 100 mites per gram) be regarded as a risk factor for sensitisation and the development of asthma, and that 10 μg of Der 1 g^{-1} (500 mites per gram) be regarded as a major risk factor for the development of acute asthma in mite-allergic individuals. At the Second International Workshop (Platts-Mills *et al.*, 1992), these threshold levels were regarded as relevant to the development of asthma, based on studies of mite allergen exposure and risk, published after the First International Workshop (Lau *et al.*, 1989; Charpin *et al.*, 1991; Peat *et al.*, 1987, 1993; Sporik *et al.*, 1990; Arruda *et al.*, 1991). The paper by Sporik *et al.* (1990) details a longitudinal birth cohort (and represents a key study linking allergen exposure during infancy to development of asthma), whereas the others were cross-sectional studies of mostly school-age children. The threshold levels have been widely adopted as a reference framework for allergen 'exposure', evidenced by the many publications that present data in terms of frequencies of dust samples in exposure classes (typically, <2, 2–10 and >10 μg g⁻¹), rather than more informative statistics such as a geometric mean or median and a measurement of variation.

The concept of 2 μg and 10 μg Der 1 as exposure risks includes the assumptions that:

- there is a positive dose-response (be it linear, or log.-linear or sigmoidal; Platts-Mills *et al.*, 1995) between exposure, sensitisation and disease;
- mite allergens are a primary risk factor for allergic sensitisation and disease;
- allergen measurement in reservoir dust is an accurate and clinically meaningful measure of exposure;
- a fixed level of exposure is globally applicable; and
- there is equivalence between exposure measured as mite population density and allergen concentrations.

So what is the evidence for these assumptions? We have seen examples of large birth cohort studies where there is a non-linear relationship, or no relationship, between allergen exposure, sensitisation and disease (Woodcock *et al.*, 2004; Schram-Bijkerk *et al.*, 2006; Almqvist *et al.*, 2007; Torrent *et al.*, 2007; Tovey *et al.*, 2008). There is now more evidence from longitudinal studies that the relationship between exposure, sensitisation and disease is complex and multifactorial, involving microorganisms, the gut as well as the lungs as an immune organ, diet, lifestyle, exposure to infectious agents, allergens other than dust mites, including the apparently protective effect of exposure to pets early in life (reviewed by Pearce *et al.*, 2000), as well as other factors (cold air, stress) that can trigger asthma. Little of this was known in 1987, partly because large birth cohort studies had not yet been done.

Most allergens have been measured in settled, or reservoir dust. For risk factor values of allergens in reservoir dust to be relevant to sensitisation and disease, we have to assume that the quantity of allergens released into the air and inhaled is proportional to the amount in settled dust. The problem is that there is no obvious, predictable relationship between allergen concentrations in reservoir dust and the amount of allergen inhaled (O'Meara and Tovey, 2000; see below). There may well be a scaling effect in that reservoir levels in *large* long-term studies represent an *indicator* (i.e. an indirect measure) of airborne exposure, but the indicator signal is not apparent in smaller studies of shorter duration.

Literature records of Der 1 allergen levels in beds indicate the geometric mean concentration of Der 1 is >2 μg at two-thirds of localities around the world and >10 μg at a third (see section 8.7 below and Figure 8.12 therein). In other words, the risk factors of Der 1 exposure for sensitisation and disease fall well within the range of bed Der 1 in most homes in most parts of the world surveyed to date. Levels of 2 μg Der 1 g⁻¹ in beds would be considered low compared with those in Sydney, Strasbourg and Seattle, but very high in Tokelau, Tartu and Turin. Of the studies used to support the 2 μg and 10 μg risk factors, only that by Lau *et al.* (1989) was at a low allergen location (Berlin). The locations of the other studies fall within the top 20 most heavily Der 1-polluted places in the world (Charpin *et al.*, 1991 – Marseilles; Peat *et al.*, 1987 – Sydney; Sporik *et al.*, 1990 – Poole; Arruda *et al.*, 1991 – Sao Paulo). In a case-control study of 74 children, Marks *et al.* (1995b) found homes in Sydney all exceeded the risk levels for allergens and there was no dose-response relationship between exposure and sensitisation or asthma. There was no difference in allergen exposure between children with atopy and non atopics. Children with asthma and house dust mite allergy had slightly but significantly *lower* allergen exposure than those with no asthma or who were not sensitised to dust mites.

One might expect lower thresholds at places where exposure is low. Of studies done at such locations (Southampton, Stockholm, Berlin respectively), Price *et al.* (1990) proposed a sensitisation threshold of 0.5 μg Der 1 g⁻¹ in carpet dust, Wickman *et al.* (1991) found Der 1 in beds of mite-sensitised children was 106 ng g⁻¹ (geometric means calculated from their Figure 1) compared with 34 ng g⁻¹ for atopic children not sensitised and 44 ng g⁻¹ from non-atopic controls, and Wahn *et al.* (1997) found children who were sensitised to dust mite by age three had a median of 0.9 μg

Der 1 g^{-1} in carpet dust compared with 0.2 µg g^{-1} for children who were not sensitised. Peat *et al.* (1995a, b) found where levels of dust mite allergens were low (central Australia), children became sensitised to other, more abundant allergens like *Alternaria* or rye grass. Similarly, Sporik *et al.* (1995) at Los Alamos, New Mexico, found that where mite allergen levels were low, sensitisation was mainly to high levels of cat allergen.

Significant positive correlations have been demonstrated between Der p 1 concentrations and mite density (Tovey *et al.*, 1981; Lind, 1986b; Colloff *et al.*, 1991; Warner *et al.*, 1998; Mumcuoglu *et al.*, 1999; Terra *et al.*, 2004; see section 8.7.1 below), though showing some variation in equivalence to 2 µg of Der 1 g^{-1} = 100 mites g^{-1} and 10 µg of Der 1 g^{-1} = 500 mites g^{-1}. Other studies found no statistical relationship between Der 1 and mite numbers (Wen and Wang, 1988).

In a systematic review of the epidemiological data available up to about 1999 on allergen exposure as a risk factor for asthma, Pearce *et al.* (2000) emphasised the limitations of the cross-sectional population-based studies on asthma prevalence in relation to allergen exposure. Because prevalence is a product of incidence (i.e. the number of new cases per unit time) *and* duration (i.e. how long the disease persists in individuals), a factor such as allergen exposure might prolong duration, thus translating into higher prevalence in the population, even though it might have little or no effect on incidence. Longitudinal studies (if they last long enough) can separate the contributions of a factor to both incidence and duration. Pearce *et al.* (2000) considered the proportion of cases attributable to exposure to mite allergens (the 'population attributable risk'), and found 'allergen exposure is at most a minor risk factor for development of asthma in children'. However, in adults there are positive associations of current mite allergen exposure with current asthma (Gelber *et al.*, 1993; Björnsson *et al.*, 1995; van der Heide *et al.*, 1997c).

In summary, there is some evidence (though it is not consistent) that the level of exposure is linked to the risk of developing sensitisation to dust mites in childhood (though not necessarily other allergens), and that the relevant level of exposure associated with sensitisation is likely to vary geographically. Exposure to dust mite allergens varies enormously in different parts of the world. Where mite allergen levels are high, prevalence of atopic sensitisation to mites tends to be high (e.g. Peat *et al.*, 1996), but the prevalence of asthma and atopic sensitisation *in general* are not closely or consistently linked with house dust mite exposure in infants and children. Trials of allergen avoidance aimed at primary prevention of childhood asthma have generally been disappointing (van Schayck *et al.*, 2007; Chapter 9). However, levels of allergen exposure are associated with asthma in atopic adults who are sensitised to dust mites.

8.4.2 Allergen levels in homes in relation to disease status

We have already seen that there is no consistent evidence that high allergen exposure is related to prevalence of sensitisation to dust mites. It has been assumed that people who have developed mite-allergic asthma have more mites or higher levels of allergens in their homes than the general population (e.g. Sidenius *et al.*, 2002a), but is it true or just another house dust mite myth? Of those surveys involving comparisons of dust mite populations in homes of patients with asthma with healthy control subjects at the same location, no significant difference was found in six examples (Sesay and Dobson, 1972; Ishii *et al.*, 1979; Htut *et al.*, 1991; Hart and Whitehead, 1990; Mumcuoglu *et al.*, 1994; Sun and Lue, 2000); significantly more mites in homes of mite-sensitised asthmatics were found in two (Korsgaard, 1983b; Saha *et al.*, 1994); and significantly more mites in homes of controls in one (Colloff, 1987c). For mite allergens, there was no difference in seven surveys (Call *et al.*, 1992; Pauli *et al.*, 1993; Konishi and Uehara, 1995; van Strien *et al.*, 1995; Wang and Wen, 1997; Sopelete *et al.*, 2000; Scrivener *et al.*, 2001) and significantly higher concentrations in homes of asthmatics in one (Addo-Yobbo *et al.*, 2001). By way of contrast, homes of patients with atopic dermatitis had consistently higher mite populations or allergen concentrations than other homes (Beck and Korsgaard, 1989; Harving *et al.*, 1990; Colloff, 1992c; Holm *et al.*, 1999).

8.4.3 Sampling and measurement of allergens in reservoir dust

The main methods for sampling reservoir dust are brushing or vacuum cleaning, although vacuuming is the method of choice. The consensus view is that important sampling sites are the mattress and bedding because most people are in contact with them for about eight hours per day, but bedroom carpets, living room carpets and upholstered furniture are also important, especially for infants who may spend as much or more time per day in contact with these items than with bedding. Brushing results in markedly lower estimates of mite population density than vacuum cleaning (Abbott *et al.*, 1981), though a higher diversity of species may be collected (Stenius and Cunnington, 1972). It is clear that it is not valid to compare amounts of mites or allergens collected by brushing with those collected by vacuum cleaning (Chapter 6). The power and airflow rate of vacuum cleaners vary considerably and may influence amounts of mite and allergen collected, although there is no consistent evidence that this introduces a systematic bias during monitoring. A turbo head fitted to a vacuum cleaner removed considerably higher quantities of house dust from carpets, but not house dust mites, than did a plain suction nozzle (Wassenaar, 1988b). The type of surface being sampled may influence sampling efficiency. Mulla *et al.* (1975) found 35% of total mites present were removed by the first of three vacuum samplings of a carpet, whereas 80% were removed from a bare floor. Similarly, air samples above synthetic carpets yielded relatively smaller amounts of allergens than above natural ones, possibly due to differences in static electric charge (Price *et al.*, 1990).

There is an issue regarding whether mite allergen concentration should be expressed by unit weight of dust or by unit area sampled. Most researchers have used the former, and most of the available data is expressed this way. That expressed by unit area cannot be compared with anything else in a meaningful way unless both units are used (which is becoming more common), or a conversion factor is given. Custovic *et al.* (1995) showed a high statistical correlation for the two measures with mattress dust (with a conversion factor of approximately $y = x - 1$, i.e. $10\,\mu g\,g^{-1} = 9\,\mu g$ per m^2). Doull *et al.* (1997) found the weight of sampled dust and its Der p 1 content were significantly correlated in samples taken from both planar surfaces (mattresses) and non-planar surfaces (mixed lounge room furniture/carpet samples). Unit weight tends to be easier to standardise than unit area especially when sampling non-planar surfaces like upholstered furniture, but unit weight may be biased by the origin and differential density of the dust (Abbott *et al.*, 1981). Dust from carpets tends to contain a larger proportion of dense particles of sand and grit than dust from mattresses which consists almost entirely of low-density skin scales. Sieving may help to equalise density but is not always guaranteed to remove dense particles (van Leeuwen and Aalberse, 1991). Also, interpretation of trials of allergen avoidance is difficult when unit weight has been used because the size of the dust reservoir may be depleted by intensive vacuum sampling used to remove allergens, potentially leading to artifactual changes in allergen concentration.

The most common method for quantifying domestic allergens is enzyme-linked immunosorbent assay (ELISA), and for group I allergens of *Dermatophagoides* spp. two ELISA systems dominate: one developed in Denmark (Lind, 1986), one in the USA (Luczynska *et al.*, 1989). Both systems yield very similar results with the same samples (see, for example, Brunekreef *et al.*, 2005). The assay involves adsorbing group I-specific monoclonal antibody to wells of a microtitre plate, followed by successive incubations with an extract of the dust specimen and biotinylated group I-specific monoclonal antibody. The reaction is visualised colorimetrically using streptavidin-labelled horseradish peroxidase. Amplified ELISA methods can increase sensitivity of conventional ELISA by around 15-fold (Custovic *et al.*, 1999). Sampling and assay of indoor allergens was reviewed by Luczynska (1997).

Guanine assay is a semi-quantitative proxy measurement of allergen in house dust (Bischoff *et al.*, 1985; van Bronswijk, 1986). Guanine is the major excretory product of arachnids, so is an indicator of the abundance of dust mite faecal pellets and Der 1 concentration. There is some evidence that use of the semi-quantitative version of this test, which can be done simply in homes of patients in about two minutes, may increase patient compliance with allergen avoidance (Ransom *et al.*, 1991). Statistically significant positive correlations have been demonstrated between concentrations of guanine in house dust and both concentrations of allergen Der p 1 (van Bronswijk *et al.*, 1989) and density of house dust mites (van Bronswijk, 1986).

8.4.4 Monitoring airborne allergens and personal exposure

Airborne allergens and personal allergen exposure were the subjects of a review by O'Meara and Tovey (2000), their paper being essential reading on these topics. This section summarises some of its main points.

There is little or no correlation between reservoir allergen concentrations and airborne ones in most studies, yet the relationship between reservoir allergen and the amount that becomes airborne and available for inhalation is critical to understanding the relationship between personal exposure, sensitisation and development of symptoms. Many factors are involved, including the size and shape of allergen-bearing particles, type of allergen, the amount of disturbance that reservoir dust is subject to, the nature of the reservoir source, the duration of exposure and proximity of the person to the source of the allergen. Suspension of allergen-containing particles resulting from simulated walking disturbance indicates an interaction between the mechanical disturbance and aerodynamic effects (Gomes et al., 2007).

Airborne group 1 and 2 mite allergens are found on particles of >5 μm diameter, which may be flakes, fibres or spherical faeces (De Lucca et al., 1999; Custovic et al., 1999). Particle size is an important factor influencing the duration that allergen-bearing particles remain airborne (Swanson et al., 1985; Platts-Mills et al., 1986b; Luczynska et al., 1990a). Reservoirs of allergen may be large, but only a small proportion becomes airborne when disturbed (Tovey et al., 1981) and only a tiny portion of the airborne component is inhaled (see Figure 8.3). There is enormous variation in airborne allergen levels. Allergen is often undetectable in the air in the absence of disturbance but may increase over 1000-fold when reservoir dust is disturbed during activities such as cleaning and vacuuming, bed-making, walking across a carpeted floor

or simply moving around in bed. Sakaguchi et al. (1992) found a mean of ca. 220 pg Der 1 per m³ in the air around sleepers, compared with about 20 pg per m³ in the lounge room and suggested a mean nightly exposure of approximately 0.6 ng. The addition of a mattress cover dropped the airborne level in the bedroom to about 10 pg per m³, indicating the bed was the major source of airborne allergen for the sleeper.

Measurements of airborne allergen can be done using large Petri dishes as settling plates left exposed for up to a week (Tovey et al., 1992; Karlsson et al., 2002). More technically sophisticated sampling involves either static or breathing zone volumetric air samplers, but few airborne particles may be detected in the absence of disturbance of the reservoir dust. In a comparison of sampling of reservoir dust, volumetric air sampling and settling dust (using two methods), there was moderate correlation, except for reservoir dust versus settling on Petri dishes (Tovey et al., 2003).

Direct personal exposure can be measured simply and accurately using intranasal samplers, consisting of a pair of soft plastic sleeves that fit snugly into the nostrils. Each sampler contains an adhesive filter strip onto which particles impact (Graham et al., 2000; Figure 8.4). The device is inserted, and particles collected over 10–30 min. on the filter during normal breathing. The strips are removed and immunoassayed using a modified protein blotting and immunostaining system. Allergen-containing particles are visualised as halo-like bodies, and counted under a

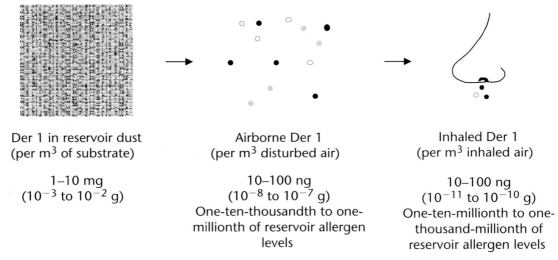

Der 1 in reservoir dust
(per m³ of substrate)

1–10 mg
(10^{-3} to 10^{-2} g)

Airborne Der 1
(per m³ disturbed air)

10–100 ng
(10^{-8} to 10^{-7} g)
One-ten-thousandth to one-millionth of reservoir allergen levels

Inhaled Der 1
(per m³ inhaled air)

10–100 ng
(10^{-11} to 10^{-10} g)
One-ten-millionth to one-thousand-millionth of reservoir allergen levels

Figure 8.3 Relationship between approximate orders of magnitude in amounts of allergen in reservoir dust and amounts inhaled. Amount in reservoir dust per cubic metre based on 10–100 g of dust per square metre of carpet with a pile depth of 1 cm. Inhaled amount based on inhalation rate of 3–6 Der p 1-bearing particles hr^{-1} (Gore et al., 2002) and a minute volume of 10 L.

a)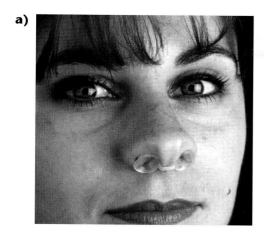

b)

Figure 8.4 Nasal filters for monitoring of personal allergen exposure. **a)** Inserted in nostrils; **b)** showing details of filter (cf. Tovey *et al.*, 2000b for details). (Images courtesy of Euan Tovey.)

compound microscope (Tovey *et al.*, 2000b). Particles without halos are also counted. Those particles with mite allergen represent a relatively small fraction of total particles (Poulos *et al.*, 1999). The sampler collects almost all particles with an aerodynamic equivalent diameter of ≥10 μm (ca. 50% of those of ~5 μm). The immunostaining method is highly reproducible and sensitive down to single particles carrying amounts of allergen in the order of <1 pg. If serum IgE from an atopic patient with multiple sensitisation is used for immunostaining (instead of monoclonal antibody), it will detect all the particles bearing the different allergens that the patient is allergic to.

Assessment of personal exposure to mite allergen using intranasal sampling of 12 volunteers under natural conditions (in bed at home, three samples per night for six nights) found significant correlations between the median number of Der 1 and Der 2-bearing particles and the reservoir allergen levels in the beds (Gore *et al.*, 2002). The median number of allergen-bearing particles inhaled per person during a 30 min. period was three (range 0–79). Based on an estimate of 0.1 ng Der p 1 per faecal pellet (Tovey *et al.*, 1981), this provides a direct measure of nightly exposure to mite allergens of approximately 5 ng of Der p 1 during eight hours of sleep.

8.5 Spatial scales of variability in mite allergens in reservoir dust

Temporal (seasonal) variation in allergen levels and mite populations was covered in Chapter 5. The following sections deal with spatial variation in mite allergen concentrations. The same scaling effects and groupings are applicable as used in Chapter 4, namely

microhabitat (variation within homes), macrohabitat (variation between homes) and variation between regions to provide a global perspective.

Apart from the well-known tendency for beds to contain higher concentrations of allergens than carpets, there is little information on variation between habitats within homes. Tovey (1995) partitioned bedroom carpets in 10 different rooms into squares of 0.25 m² and sampled each square. In one sample the range was 17–85 μg g⁻¹ of Der p 1, with no obvious distribution pattern apart from slightly lower levels near the window. Similarly, Simpson *et al.* (1998) found no consistent pattern of allergen distribution and coefficients of variation around 80–90% for within- and between-room Der p 1 concentrations in carpets. Loan *et al.* (1998) found a mean coefficient of variation around 50% in living rooms and suggested that for large-scale epidemiological studies a single site from the centre of the room, in front of a couch or chair or from a corner, is representative of the whole room. In replicated samples taken at a two-week interval, Marks *et al.* (1995a) found the range for a single sample from beds and floors was accurate to 3.1-fold and 3.5-fold respectively.

8.6 Variability in allergen concentration between homes

When researchers identify factors for increased risk of allergen exposure they usually mean factors associated with differences in allergen concentrations in reservoir dust. Since there is no clear relationship between the amount of allergen in reservoir dust and the risk of exposure, or for that matter between mite populations and allergen production, it is probably best to think of these factors in a more restricted sense as simply indicators of high or low reservoir allergen levels.

Some of these factors were covered in Chapter 4 for mite population densities (including age of homes, socio-economic status of occupants, housekeeping, crowding and height of homes above ground level). Here they are considered in relation to allergen concentrations. The two sections could have been combined, but the rather scant data on mite population densities are mostly from small-scale ecological studies involving only a few factors. Those on allergen concentrations are mostly from much larger population-based epidemiological studies with high statistical power, where many factors were examined simultaneously using multivariate statistical techniques. One of the aims of many of these studies is to attempt to explain the high variation in allergen concentrations between homes (Simpson *et al.*, 2001).

Although risk factors associated with housing are sometimes referred to as determinants, there are few examples of direct causal associations with elevated allergen concentrations. Risk factors are often surrogates for immediate influences on mite population size. For example, presence of visible mould on walls and ceilings is an indicator of elevated humidity which favours mite population growth. The classification herein attempts to make some sort of sense of risk factors and how they may help explain differences in reservoir allergen concentrations between homes. Factors are divided into those associated with:

- housing characteristics, including condition, construction and design;
- human behaviour, social and economic status; and
- the immediate external environment.

In order to obtain estimates of the relative magnitude of allergen concentrations associated with various risk factors, I have focused on epidemiological studies with data from discrete geographic localities, rather than pooled data from several localities where climate influences, even over small distances, may have confounding effects (Kuehr *et al.*, 1994a; Basagaña *et al.*, 2002). For the studies in Table 8.1, statistically significant differences in allergen concentrations are given as geometric means or medians of Der 1 (mostly Der p 1) in µg g^{-1} from floors or beds, based on initial, univariate models. The reason for presenting these data is to contrast the between-study variation in unadjusted means. For the sake of clarity at the expense of completeness, I have not included in the table those studies that give odds ratios in the absence of means or, for the most part, separate models of Der p 1 and Der f 1. Risk factors for the two allergens are not consistent (Kuehr *et al.*, 1994b; Hirsch *et al.*, 1998; Gross *et al.*, 2000; Zock *et al.*, 2006; Fig. 8.5), reflecting different physiological and environmental requirements of *Dermatophagoides farinae* and *D. pteronyssinus*. Most of the information presented in Table 8.1 relates to Der p 1 from discrete geographic localities, whereas much of that for Der p 1 and Der f 1 is pooled from several localities.

Most authors also conducted multivariate linear regression. Factors that might be significant with univariate analyses may not be so following multivariate analyses (i.e. after adjusting for all other factors) and vice-versa. Adjusted, independent risk factors from multivariate analyses are listed in Table 8.2.

8.6.1 Housing characteristics

a Type of home

Apartments or flats tend to have lower Der 1 concentrations than detached or semi-detached houses, but the size of the effect is relatively small (average for statistically significant studies 1.8-fold for floors) and inconsistent (2/6 studies of floors showed statistical significance). This factor is likely to be co-dependent with the age of the home, construction (Mihrshahi *et al.*, 2002) and its elevation above ground (Dharmage *et al.*, 1999), because apartments may tend to be newer and/or higher up and made of brick or concrete rather than timber.

b Age of home and state of repair

Older homes had higher Der 1 concentrations and the effect was greater and more frequent for beds than floors (average 2.1-fold for carpets, 4/9 studies statistically significant; 3-fold for beds, 4/5 studies statistically significant). Varekamp and Voorhorst (1960) found high inter-home variability in the allergenicity of dust samples they used to make extracts for skin-prick testing. Damp homes contained dust that was more allergenic. Older homes in Leiden, the Netherlands, tended to be damper and in poorer repair than newer homes, with 72% of pre-1918 homes having dry rot, compared with only 11% of post-1918 homes. Julge *et al.* (1998) made the point that older apartments in Tartu, Estonia, had damp

Table 8.1 Residential and domestic factors associated with (mainly) elevated exposure to mite allergens, based on univariate analyses. NS = not statistically significant

Factor	Floor				Bed				Reference
	1st value	2nd value	Fold-difference	Value of P	1st value	2nd value	Fold-difference	Value of P	
Type of home									
Apartment v. house	0.26	0.38	1.46	NS					Julge *et al.,* 1998
Apartment v. house	8.70	9.30	1.07	NS					Plácido *et al.,* 1996
Apartment v. detached	9.92	13.10	1.32	NS					Mihrshahi *et al.,* 2002
Non-detached v. detached	12.40	19.10	1.54	<0.05	17.2	21.2	1.2	<0.05	Dharmage *et al.,* 1999
Apartment v. detached	0.83	1.77	2.13	0.04	1.4	1.9	1.4	NS	Chan-Yeung *et al.,* 1995
Apartment v. detached	0.40	0.79	1.98	NS	0.9	0.6	0.7	NS	Chan-Yeung *et al.,* 1995
Age of home									
=10 yr v. >10 yr	1.30	1.40	1.08	NS					Atkinson *et al.,* 1999
<16 yr v. 17–36 yr	2.30	3.80	1.65	NS					Luczynska *et al.,* 1998
<16 yr v. >16 yr	28.20	43.60	1.55	NS					Wickens *et al.,* 2001
<20 yr v. >20 yr	6.40	14.50	2.27	<0.05					Plácido *et al.,* 1996
<25 yr v. =25 yr	0.21	0.44	2.10	<0.001					Julge *et al.,* 1998
=10 yr v. >10 yr	9.41	15.31	1.63	0.001	9.3	18.8	2.0	<0.0001	Mihrshahi *et al.,* 2002
<20 yr v. >20 yr	0.85	1.99	2.34	0.04	0.9	2.5	2.8	0.01	Chan-Yeung *et al.,* 1995
<20 yr v. >20 yr	0.66	0.78	1.18	NS	0.5	0.8	1.4	NS	Chan-Yeung *et al.,* 1995
<16 yr v. >36 yr	13.90	19.10	1.37	NS	8.0	23.5	2.9	<0.05	Dharmage *et al.,* 1999
<15 yr v. >55 yr					0.4	1.6	4.5	<0.001	Simpson *et al.,* 2002
Size of home									
>2 rooms v. =2 rooms	5.10	5.30	1.04	NS					Plácido *et al.,* 1996
Housing construction									
Concrete v. timber	0.22	0.55	2.50	<0.05					Julge *et al.,* 1998
Brick v. weatherboard	1.60	3.02	1.89	NS					Couper *et al.,* 1998
Brick v. weatherboard	12.28	12.39	1.01	NS	12.6	20.7	1.6	<0.0001	Mihrshahi *et al.,* 2002
Other v. brick cladding	10.50	21.70	2.07	<0.01	19.4	28.6	1.5	<0.01	Garrett *et al.,* 1998
Foundation/flooring									
Concrete v. timber	0.82	2.37	2.89	NS					Couper *et al.,* 1998
No sub-floor space v. space	15.40	19.10	1.24	NS	15.4	21.2	1.4	<0.05	Dharmage *et al.,* 1999
Concrete v. timber	9.80	15.80	1.61	0.001	10.2	19.6	1.9	<0.0001	Mihrshahi *et al.,* 2002
Timber v. concrete	12.20	23.30	1.91	<0.01	24.5	23.6	1.0	NS	Garrett *et al.,* 1998
Floor not wood v. wood on stumps					11.2	23.8	2.1	<0.05	Dharmage *et al.,* 1999
Not concrete v. concrete					3.5	7.4	2.1	0.007	Luczynska *et al.,* 1998

continued

Table 8.1   Continued

Factor	Floor				Bed				Reference
	1st value	2nd value	Fold-difference	Value of P	1st value	2nd value	Fold-difference	Value of P	
Presence of damp									
No damp v. damp	0.23	0.51	2.22	<0.05					Julge *et al.*, 1998
No damp v. damp in living room	1.30	4.30	3.31	0.002					Atkinson *et al.*, 1999
No damp v. damp	7.20	20.00	2.78	<0.001					Plácido *et al.*, 1996
No damp v. damp	15.40	23.50	1.53	<0.05	17.2	29.0	1.7	<0.05	Dharmage *et al.*, 1999
Condensation absent v. present	2.80	4.50	1.61	<0.05	3.4	4.7	1.4	NS	Luczynska *et al.*, 1998
Condensation absent v. present					40.0	49.5	1.2	0.05	Wickens *et al.*, 1997
No damp v. damp					44.8	60.0	1.3	0.04	Wickens *et al.*, 1997
No damp v. damp in bedroom					0.5	0.7	1.4	NS	Atkinson *et al.*, 1999
No window condensation v. present					0.8	1.5	2.0	0.001	Simpson *et al.*, 2002
No damp v. damp in bedroom					1.1	4.5	4.1	0.001	Simpson *et al.*, 2002
Presence of mould									
No mould v. mould	1.68	4.21	2.51	0.05					Couper *et al.*, 1998
No mould v. mould	15.40	23.50	1.53	<0.05	17.2	29.0	1.7	<0.05	Dharmage *et al.*, 1999
No mould v. mould	12.14	17.15	1.41	NS	13.7	23.2	1.7	0.01	Mihrshahi *et al.*, 2002
No/slight mould v. heavy mould	15.40	15.70	1.02	NS	20.5	29.5	1.4	<0.05	Garrett *et al.*, 1998
Indoor temperature & humidity									
<18°C v. >18°C	0.11	0.32	2.91	<0.001					Julge *et al.*, 1998
Mattress RH <51% v. >51%					41.3	52.8	1.3	0.03	Wickens *et al.*, 1997
Heating									
Central heating v. wood stove	1.42	2.11	1.49	NS					Couper *et al.*, 1998
No central heating v. with	15.40	21.20	1.38	<0.05	21.2	21.1	1.0	NS	Dharmage *et al.*, 1999
Heating unused week prior v. used	12.66	11.32	0.89	NS	14.1	20.4	1.5	NS	Mihrshahi *et al.*, 2002
Other v. forced air	0.55	2.31	4.20	0.009	1.1	2.3	2.2	NS	Chang-Yeung *et al.*, 1995a
Other v. forced air	0.53	0.75	1.42	NS	1.2	0.6	0.5	NS	Chang-Yeung *et al.*, 1995a
Central heating v. none					1.1	3.2	3.1	<0.001	Simpson *et al.*, 2002
Gas appliances									
Without gas oven v. with	3.00	3.70	1.23	NS					Luczynska *et al.*, 1998

continued

Table 8.1 Continued

Factor	Floor				Bed				Reference
	1st value	2nd value	Fold-difference	Value of P	1st value	2nd value	Fold-difference	Value of P	
Ventilation									
Kitchen extractor fan v. none	2.50	3.90	1.56	NS					Luczynska et al., 1998
Without kitchen extractor fan v. with	12.50	19.30	1.54	<0.05	23.3	19.1	0.8	NS	Dharmage et al., 1999
Open fireplace in lounge v. without	1.10	4.00	3.64	0.03					Luczynska et al., 1998
Curtains and blinds									
Heavy curtains absent v. present	15.40	19.10	1.24	<0.05	19.1	21.2	1.1	NS	Dharmage et al., 1999
Venetian blinds present v. absent	10.10	19.10	1.89	<0.05	20.3	21.2	1.0	NS	Dharmage et al., 1999
Insulative properties									
Carpet =4 mm v. =5 mm	30.10	45.80	1.52	NS					Wickens et al., 2001
Underlay =7 mm v. =8 mm	37.10	40.50	1.09	NS					Wickens et al., 2001
Floor insulated v. not	26.10	42.90	1.64	<0.05					Wickens et al., 2001
Walls insulated v. not	31.30	48.70	1.56	<0.05					Wickens et al., 2001
No double glaze in lounge v. with	1.10	1.60	1.45	0.03					Atkinson et al., 1999
No double glaze in bedroom v. with					0.5	0.5	1.0	NS	Atkinson et al., 1999
No double glaze v. with					1.2	0.9	0.7	NS	Simpson et al., 2002
Age of mattress									
<5 yr v. >5 yr					15.4	23.5	1.5	<0.05	Dharmage et al., 1999
=2 yr v. >2 yr					10.3	15.5	1.5	0.003	Mihrshahi et al., 2002
=1 mo v. >2 mo					0.3	0.5	1.7	<0.001	Atkinson et al., 1999
<1 yr v. >5 yr					0.4	2.6	6.1	<0.001	Simpson et al., 2002
<1 yr v. >1 yr					1.0	4.7	4.7	0.05	Luczynska et al., 1998
Type of mattress									
Inner spring v. foam					14.2	21.4	1.5	NS	Mihrshahi et al., 2002
Inner spring v. foam					50.2	39.3	0.8	NS	Wickens et al., 1997
Foam v. inner spring					13.0	25.0	1.9	<0.01	Garrett et al., 1998
Type of bedding									
Quilt on bed v. not					19.1	48.4	2.5	<0.05	Dharmage et al., 1999
Blankets on Bed					19.1	26.1	1.4	<0.05	Dharmage et al., 1999
Blankets not wool v. wool					12.9	27.2	2.1	<0.0001	Mihrshahi et al., 2002
Blankets not synthetic v. yes					13.2	22.2	1.7	0.001	Mihrshahi et al., 2002

continued

Table 8.1 Continued

Factor	Floor				Bed				Reference
	1st value	2nd value	Fold-difference	Value of P	1st value	2nd value	Fold-difference	Value of P	
No sheepskin on bed v. yes					13.8	23.0	1.7	0.03	Mihrshahi *et al.*, 2002
Synthetic pillow v. not					14.2	16.5	1.2	NS	Mihrshahi *et al.*, 2002
No feather pillow v. yes					14.0	16.2	1.2	NS	Mihrshahi *et al.*, 2002
Blankets not cotton v. yes					14.2	15.4	1.1	NS	Mihrshahi *et al.*, 2002
No synthetic quilt v. yes					13.5	14.9	1.1	NS	Mihrshahi *et al.*, 2002
Feather quilt v. not					13.5	14.7	1.1	NS	Mihrshahi *et al.*, 2002
Floor covering									
No carpet v. carpet	0.91	2.25	2.47	NS					Couper *et al.*, 1998
No carpet v. carpet	5.70	21.20	3.72	<0.05					Dharmage *et al.*, 1999
No carpet v. carpet	9.00	22.30	2.48	0.01					Plácido *et al.*, 1996
Rugs v. no rugs	10.07	12.83	1.27	NS	19.8	13.7	0.7	0.047	Mihrshahi *et al.*, 2002
No carpet v. carpet	0.60	1.40	2.33	0.03	0.5	0.5	1.0	NS	Atkinson *et al.*, 1999
Hard floor v. carpet	4.11	15.10	3.67	<0.0001	18.5	13.9	0.8	NS	Mihrshahi *et al.*, 2002
No carpet v. carpet	3.90	28.10	7.21	<0.0001	19.9	48.1	2.4	0.001	Wickens *et al.*, 1997
Type of carpet									
Other v. wool	15.40	32.10	2.08	<0.05					Dharmage *et al.*, 1999
Synthetic v. wool	9.90	15.70	1.59	<0.01					Garrett *et al.*, 1998
Synthetic v. wool	22.60	44.90	1.99	NS					Wickens *et al.*, 2001
Age of carpet									
Carpet <1 yr v. >1 yr	0.80	5.10	6.38	0.02					Luczynska *et al.*, 1998
Carpet <1 yr v. >5 yr	4.50	20.20	4.49	<0.05					Dharmage *et al.*, 1999
Carpet =1 yr v. >1 yr	19.40	29.60	1.53	NS					Wickens *et al.*, 1997
Carpet =1 yr v. >5 yr	0.25	1.79	7.16	<0.001					Simpson *et al.*, 2002
Number of occupants									
<3 v. =3 children	23.20	29.20	1.26	0.03					Wickens *et al.*, 1997
1–2 v. =3 children	33.50	51.00	1.52	0.05					Wickens *et al.*, 2001
=5 v. =6	1.71	5.29	3.09	0.03					Couper *et al.*, 1998
=4 v. >4	7.00	9.20	1.31	NS					Plácido *et al.*, 1996
<4 v. >4	0.99	1.95	1.97	0.02	1.4	2.1	1.6	NS	Chan-Yeung *et al.*, 1995
<4 v. >4	0.55	0.72	1.31	NS	0.6	0.9	1.6	NS	Chan-Yeung *et al.*, 1995
=3 v. =6	1.10	1.70	1.55	0.02	0.4	0.6	1.5	NS	Atkinson *et al.*, 1999
Income									
>£30 k v. <£10 k					0.8	2.4	3.1	<0.01	Simpson *et al.*, 2002
Smoking									
2 smokers in home v. no smokers	1.50	5.00	3.33	0.007					Luczynska *et al.*, 1998
Smokers in home v. no smokers	0.25	0.33	1.32	NS					Julge *et al.*, 1998
No maternal smoking v. smoking	1.89	2.18	1.15	NS					Couper *et al.*, 1998
Smokers in home v. no smokers	1.00	1.70	1.70	0.001	0.5	0.5	1.0	NS	Atkinson *et al.*, 1999

continued

Table 8.1 Continued

Factor	Floor				Bed				Reference
	1st value	2nd value	Fold-difference	Value of P	1st value	2nd value	Fold-difference	Value of P	
Cleaning									
Vacuuming =weekly v. <weekly	0.47	2.46	5.23	0.007					Couper et al., 1998
Sweep bedroom ever v. never	0.30	2.09	6.97	0.02					Couper et al., 1998
Type of vacuum cleaner									
New type (cylinder) v. old (soft bag)	3.30	6.30	1.91	0.005					Luczynska et al., 1998
>1000 W v. =1000 W	33.50	48.00	1.43	NS					Wickens et al., 2001
Vacuuming =weekly v. <weekly	38.90	82.10	2.11	NS					Wickens et al., 2001
Laundry									
Dried outdoors v. indoors	1.85	8.07	4.36	0.05					Couper et al., 1998
Exposure									
House in windy location v. not	1.78	2.34	1.31	NS					Couper et al., 1998
Height of home off ground									
1st floor home v. ground floor	0.18	0.41	2.28	<0.001					Julge et al., 1998
Lounge room on 1st floor v. ground	0.70	3.70	5.29	<0.001					Luczynska et al., 1998
Lounge over garage v. on ground	25.00	46.00	1.84	<0.001					Wickens et al., 2001
Bedroom on 1st floor v. ground	3.00	7.20	2.40	0.006					Luczynska et al., 1998
=1st floor home v. ground floor	12.00	14.70	1.23	NS					Plácido et al., 1996
Bedroom on 1st floor v. ground	13.90	17.20	1.24	NS	12.5	23.5	1.9	<0.05	Dharmage et al., 1999

associated with poor construction (leakiness), which allows water ingress, but also facilitates natural ventilation. Damp in urban homes in Denmark and Sweden is likely to be associated with condensation resulting from efficient insulation and poor ventilation(Korsgaard et al., 1983a, b; Wickmann et al., 1991; Sundell et al., 1995).

c Housing construction, foundations and flooring

Construction characteristics tend to be co-factors of type of home and age. For example, Garrett et al. (1998) found the higher allergen concentrations associated with older homes was no longer significant after adjusting for construction type (brick cladding) and foundation type. Homes with concrete slab ground floors have been associated with higher allergen concentrations than homes with an under-floor crawl space (Munir et al., 1995; Luczynska et al., 1998). Dharmage et al. (1999) found slab floors associated with lower concentrations of Der 1, as did

Mihrshahi et al. (2002), but in the latter study this effect was not significant after adjusting for other factors because only 22% of older homes had concrete foundations. The type of foundations, as well as other construction factors (brick v. timber), is likely to co-vary with the age of the home, depending on local architectural styles and building codes.

d Temperature and humidity, damp, condensation and mould

High indoor humidity is among the most important risk factors for increased mite exposure (Korsgaard, 1982; 1983a, b; Verhoeff et al., 1995), and the relationship between domestic damp and asthma has long been known (Varekamp, 1925; Storm van Leeuwen, 1927). Emenius et al. (2000) found absence of condensation on double-glazed windows and low indoor water vapour generation (<3 g per m³) were highly predictive of low indoor humidity, and a reasonable predictor of low allergen

Table 8.2 Independent risk factors, derived from multivariate analyses, of factors associated with high Der 1 concentrations in homes.

Factor	Wickens et al., 1997, Wellington	Couper et al., 1998, S. Tasmania	Garrett et al., 1998, La Trobe Valley	Luczynska et al., 1998, Norwich	Atkinson et al., 1999, Ashford	Dharmage et al., 1999, Melbourne	Wickens et al., 2001, Wellington	Mihrshahi et al., 2002, Sydney	Simpson et al., 2002, Manchester	Basagaña et al., 2002, Barcelona	Basagaña et al., 2002, Menorca
Mould/damp/condensation		×	×		×	×			×		
Age of mattress				×	×	×		×	×		
Age/type of carpet	×					×	×		×		
Number of occupants	×	×				×	×				
Smokers in home (-ve association)				×	×					×	×
Indoor RH%	×	×	×								
Age of home						×	×	×	×		
Woollen blankets/underblanket	×		×					×			
Carpet v. hard floor		×				×		×			
Mattress type	×		×							×	
Height of home/room above ground				×			×				
Brick cladding			×								
Open fireplace				×							
Old vacuum cleaner				×							
Presence of sub-floor space						×					
No insulation							×				
Less ventilation/air conditioning										×	
High social class											×

concentrations, but presence of condensation was associated with only 20–30% risk of high indoor humidity. Presence of damp was a consistent predictor of high allergen concentrations; statistically significant in all studies of floors (average 2.3-fold greater) and 5/6 studies of beds (average 2.1-fold). Presence of mould was a good predictor for beds (statistically significant in all studies) but only associated with 1.6-fold greater Der 1 levels, whereas for floors only 2/4 studies showed significantly elevated allergen (2-fold). Garrett *et al.* (1998) found a positive correlation between bed allergen concentrations and mean bedroom relative humidity (though not absolute humidity) over quite a narrow humidity range (48–67% RH), whereas Dharmage *et al.* (1999) found only a weak association over a higher, broader range (mean 60% RH, range 32–80). A weak negative association between temperature

and Der p 1 levels was confounded by humidity in their multiple regression. Temperature is likely to show variable effects in different parts of the world depending on season and climate.

e Heating and ventilation

The reduction of indoor humidity through the use of ventilation systems is associated with lower exposure to dust mites and allergens (Harving *et al.*, 1993, 1994; Wickman *et al.*, 1994a; Arlian *et al.*, 2001). Luczynska *et al.* (1998) found the presence of an open fireplace in the living room was associated with 3.6-fold lower concentrations of Der p 1 in the lounge room. The fireplaces were usually not used (most of the homes that had them were centrally heated), hence they represent a ventilation factor rather than a heating factor. These authors also found homes with gas cookers had higher Der p 1 concentrations in the lounge room.

Central heating was both negatively and positively associated with allergen concentrations (see Table 8.1). This may be due in part to patterns of use in different parts of the world, as well as comparisons of central heating with different types of heating. Centrally heated houses may be less ventilated because there is a tendency to keep windows closed if the heating is on (Dharmage *et al.*, 1999).

f Insulative properties

Wickens *et al.* (2001) found the absence of insulation from floors was the most important factor associated with elevated Der p 1 concentrations in floors. The authors suggested that insulation may buffer against ingress of humid outdoor air. A similar effect was found by van Strien *et al.* (1994), with a 3-fold increase of Der 1 on non-carpeted floors with rugs but no sub-floor insulation. Double glazing was not consistently associated with high Der 1 concentrations.

g Beds and bedding

Bed type

Mosbech *et al.* (1991) found no significant differences in allergen concentrations between different types of mattress, though Garrett *et al.* (1988) found almost double the concentration in inner-spring mattresses compared with foam, whereas Mihrshahi *et al.*

(2002) found almost double the concentrations in foam mattresses, and Wickens *et al.* (2001) found no difference.

Bed age

Older mattresses have consistently higher allergen concentrations than newer ones, in concordance with the finding of increased mite population density with mattress age (see Chapter 4). 'Newer' in most studies was often less than a year old, though Custovic *et al.* (1996) showed significant accumulation of Der p 1 in new mattresses after only four months. The best data, showing a cumulative exponential increase in allergen with age, are those of Simpson *et al.* (2002; see Figure 8.5). Mihrshahi *et al.* (2002) found the age of the bed to be the strongest predictor of high Der p 1 concentrations in beds in Sydney. More dust tends to be extracted from older beds (Kuehr *et al.*, 1994b), indicating the accumulation of skin scales over time.

Bedding type

Mihrshahi *et al.* (2002) found the use of woollen or synthetic blankets, but not cotton ones, was associated with high allergen concentrations. This may be due to frequency of cleaning. Cotton blankets can be washed in a domestic washing machine whereas wool blankets have to be dry-cleaned. Garrett *et al.* (1988) also found

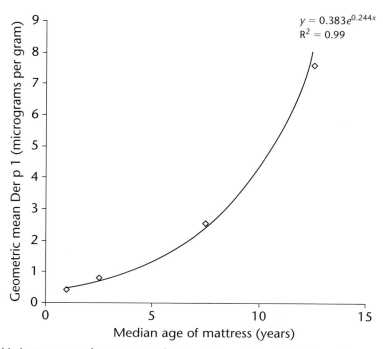

Figure 8.5 Relationship between age of mattresses and concentration of Der p 1 (data from Simpson *et al.*, 2002).

a strong positive association with woollen bedding, including underblankets, as did Kuehr *et al.* (1994b). As suggested by Mihrshahi *et al.* (2002), wool may have particular properties that create a favourable microenvironment for growth of dust mite populations.

h   Carpets

Fitted carpets were consistently associated with high Der 1 levels – almost 4-fold greater than floors with no carpet. Wool carpet had higher concentrations (1.9-fold) than synthetic ones (statistically significant in 2/3 studies). Carpets more than one year old had a mean 6-fold greater allergen concentration than those less than a year old.

8.6.2   Human behaviour and social and economic factors

a   Number of occupants

Homes with higher numbers of occupants tend to have higher Der 1 concentrations in floors (1.9-fold on average; 5/7 studies) but not beds. Higher density of occupancy is related to greater water use (in eastern Australia, currently around 200 L per person per day), and generation of water vapour through activities such as showering, bathing, cooking and laundry may be responsible for higher allergen levels.

b   Social and economic variables

Living in the high poverty part of Boston was associated with lower Der 1 concentrations in homes (based on allergen exposure class frequencies, not geometric means) and higher cockroach allergen levels, but family income was not significantly associated with Der 1 levels after adjusting for home characteristics and ethnicity (Kitch *et al.*, 2000). But high poverty areas were dominated by apartments, which had half the frequency of high mite allergen exposure than houses. In Chapter 4, higher mite populations were found in mattresses in homes of people classified as being of lower social and economic status. This factor has no direct effect on mite populations, but it is likely to be an indicator of mattress age, because people with higher incomes are more likely to replace their beds and furnishings more frequently. Older mattresses tend to have higher allergen concentrations. Simpson *et al.* (2002) found beds of people with annual incomes under £10 000 (A$ 22 500) had 3.1-fold greater Der p 1 concentrations than people earning more than £30 000 (A$ 67 500), but income was not a significant ind-

ependent variable in their multivariate analysis. Chen *et al.* (2007) found no association with income.

c   Smoking

Luczynska *et al.* (1998) found a strong statistically significant negative 'dose-response' relationship with smoking and Der p 1 concentrations: no smokers in home, 5.0 µg g^{-1}; one smoker, 1.9 µg; two smokers, 1.5 µg g^{-1}. Hypotheses for this effect were denaturation of Der p 1 by nicotinic acid or a toxic effect of nicotine on dust mites. Nicotine is a potent insecticide and acaricide (Rodriguez *et al.*, 1979; Eldefrawi *et al.*, 1985). Another possibility is that Der p 1-bearing particles adhere to sticky tar condensates from cigarette smoke that are deposited onto carpets. Atkinson *et al.* (1999) found significantly higher Der p 1 on floors, but not in beds, of non-smokers. Lau *et al.* (1997) also found a significant negative association with smoking, as did Basagaña *et al.* (2002), including a 'dose-response' effect based on number of cigarettes smoked by the occupants. Julge *et al.* (1988) and Couper *et al.* (1999) found similar trends but they were not significant. Simpson *et al.* (2002) found no association and Munir *et al.* (1995) found a positive association, but with pooled data from three geographical regions.

d   Domestic cleaning

Frequency of cleaning has been associated with large differences in allergen concentrations on floors (Couper *et al.*, 1998). Age and type of vacuum cleaner were predictors of Der 1 concentrations in one study where homes with older, leakier, soft-cloth bag ('upright') vacuum cleaners had higher floor allergen levels than homes with modern cylinder-type models (Luczynska *et al.*, 1998).

8.6.3   Environmental factors

a   Proximity to water courses

Maunsell (1951) and Voorhorst *et al.* (1969, their Figure 27) found higher prevalences of sensitisation to dust and atopic asthma with proximity of homes to watercourses. However, in relation to Der 1 concentrations, no increased levels were found within 500 m of a lake or river (Kuehr *et al.*, 1994b; Luczynska *et al.*, 1998).

b   Soil type

Voorhorst *et al.* (1969, pp. 33–35) mention a study by Tissot van Patot (1929) who found that house dust extracts taken from homes built on peaty or mixed soils in the Netherlands gave larger reactions in skin

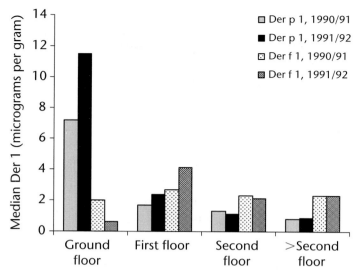

Figure 8.6 Relationship between height above ground of homes (storey level) and Der p 1 and Der f 1 concentrations in beds from two surveys, one year apart, in 1050 homes in Freiburg, Loerrach and Kehl in the Upper Rhine Valley, south-west Germany (data from Kuehr et al., 1994b).

tests than those from homes built on sandy soils. This may relate to the water-holding capacity of soils and a greater likelihood of damp in homes on heavy soils than in homes that are free draining, but this factor has not been investigated further.

c Height of home above the ground

I have included this as an environmental factor because it relates to the interaction between the building and the external climate. Of those studies that show this factor to be of significance, the average magnitude of the effect is 3-fold greater Der 1 concentrations in homes at ground floor level than those at first floor level or higher. The most likely explanation for this effect is that the water vapour content of the air decreases exponentially with distance above the ground, and the drying capacity of the air via wind and turbulence increases (Geiger, 1957). Atmospheric relative humidity of 75% RH at ground level may decrease to 60% RH at 5 m above the ground. The humidity gradient is modified by diurnal fluctuations in heat exchange at the ground surface. Wickens et al. (2001) found significantly lower Der p 1 concentrations in lounge room floors that had a room or garage underneath them. There was no correlation between Der p 1 and relative humidity of carpets and room air, but having a room or garage below the lounge room was highly predictive of low lounge room humidity. Dharmage et al. (1999) found ground floor bedrooms had almost 2-fold higher bed Der p 1, but this effect was confounded by home age in their multivariate analysis. Kuehr et al. (1994b) found ground floor

homes were more humid and had higher concentrations of Der p 1 than homes on higher floors, but that Der f 1 was not associated with the height of the home above ground (Figure 8.6).

d Exposure to wind

Couper et al. (1998) found a 1.3-fold lower mean Der p 1 concentration in homes situated in windy locations than homes in sheltered spots, but the difference was not statistically significant.

8.6.4 Summary – hidden sources of variation?

Factors associated with variation in mite allergen concentrations between homes fall into three general categories that are likely to show interaction and overlap. They are:

1. those that relate to processes of *colonisation and establishment* of mite populations, including habitat availability and development of habitat complexity;
2. those associated with *retention or removal* of allergens;
3. those that are associated, directly or indirectly, with variation in indoor *microclimate*, and influence the rate of growth and population density of mite populations once they become established.

The first category includes age (and possibly type) of mattresses and carpets (including their presence or absence); the second covers cleaning and the third almost all other factors. The nature of the association between allergen concentrations and smoking is unclear and does not fit clearly into any category.

Most of the studies were based in English-speaking countries in temperate latitudes. Four were done in Australia, three in the UK and two in New Zealand. Basagaña *et al.* (2002) observed that most of the risk factors associated with the English-speaking countries lacked an equivalent association in Barcelona and Menorca and were irrelevant. Perhaps the starkest example of such differences is the finding of El Sharif *et al.* (2004) that the most significant independent factor for high allergen exposure in Ramalla, Palestine, was whether or not the home was in a refugee camp. Several studies in Table 8.2 are based on subsets of homes of parents and infants enrolled in longitudinal birth cohort studies. Selection of homes on the basis of their representative housing types might be desirable. Three studies involved random selection of homes, independent of a pre-existing study (Luczynska *et al.*, 1998; Garrett *et al.*, 1998; Simpson *et al.*, 2002). Wickens *et al.* (2001) selected houses to represent a bimodal distribution of Der p 1, based on their previous study (Wickens *et al.*, 1997).

Few authors provide an assessment of the combined effects of risk factors on Der 1 concentrations, after adjustment. Mihrshahi *et al.* (2002) found unheated brick apartments or townhouses less than 10 years old, with concrete foundations, with carpets but no rugs, with inner-spring mattresses less than two years old and bedding that did not include blankets, synthetic or feather quilts or feather pillows, but did have a synthetic pillow would have a Der p 1 concentration of 3.2 µg g^{-1} (compared with a mean of 12.1 µg g^{-1} for beds and floors). When only bedding factors were considered, beds would have a mean of 8.6 µg g^{-1} compared with a mean of 14.3 µg g^{-1}. Based on the data in Tables 8.1 and 8.2, the lowest risk of high allergen concentrations would be in an uncrowded, well-ventilated newer apartment, above ground level, that had hard floors and a newer mattress without blankets.

Risk factors are basically indicators or predictors of a variable – allergen levels or mite population densities. Indicators are most useful when they are simple to assess and predictive of a more direct factor that is difficult or expensive to measure. If the direct factor is easily and accurately measured, there is no need for an indicator. A good example is indoor relative humidity, which is hard to measure in a manner that directly relates to physiological drivers of mite population growth (partly because of the co-dependent effect of temperature). Point-source or 'snapshot' measurements of humidity are of lower utility than repeated monitoring because of large seasonal and diurnal fluctuations and high spatial variation. But in regions with high precipitation, the presence of damp and mould are good indicators of high indoor humidity, while absence of condensation is a good indicator of low indoor humidity.

The remarkable feature of the allergen variation data is that most significant factors are associated with only 2- to 3-fold greater concentrations. Larger differences (4.5 to 7-fold) are associated mostly with presence of carpet or its age. The significant factors explain only a small amount of the between-home variation which is typically 2–4 orders of magnitude (for variation in larger, repeat-sampling datasets, see Marks *et al.*, 1995a, their Figure 1; Kuehr *et al.*, 1994b, their Figure 1; Crisafulli *et al.*, 2007, their Figure 2).

A source of possible additional variation, currently unexplored, is the spatial distribution of survey homes within a town or city and the effect of urban microclimatic variation (Bridgman *et al.*, 1995). Urban heat islands (UHI) are volumes of air over cities that are significantly warmer and drier than over surrounding rural areas (Oke, 1982; Collier, 2006). The largest temperature difference is at night, during summer, and in still air. A UHI is caused partly by the emission of heat through human activities but is mainly due to the surfaces of buildings, roads and roofs warming up during the day and releasing heat at night. Materials like brick, concrete and asphalt absorb and radiate heat far more effectively than air, water or vegetation. Urban areas have lower rates of evaporation and heat loss is reduced because of low turbulence and the shading effects of buildings (Sánchez and Alvarez, 2004). Within a metropolitan area, there can be quite large differences in temperature and humidity between the central business district, high-density inner-city housing, parks and gardens, commercial districts, medium-density suburban housing, low-density outer suburbs and the rural hinterland (e.g. Coutts *et al.*, 2007). Generally, the larger the city, the larger the UHI effect (Oke, 1982). Major modifying factors on urban climate include distance from the coast, sea-breeze circulations and surrounding mountains (Ohashi and Kida, 2004). Studies have yet to be done on the geographic variation in dust mite populations or allergen concentrations within cities in relation to urban climate factors.

8.7 Regional and global variation in dust mite abundance and allergen concentrations

8.7.1 Datasets

Repeated measurements of mite allergens within the same homes over periods ranging from weeks to years have shown a reasonable degree of consistency (Kuehr *et al.*, 1994b; Marks *et al.*, 1995a; Matheson *et al.*, 2003; Antens *et al.*, 2006) though Topp *et al.* (2003), over six years, found low consistency for a group of homes containing low levels of mite allergens. Crisafulli *et al.* (2007) found remarkably consistent fluctuations over a 2- to 3-fold range in a 7-year study of over 1000 homes. In other words, mite allergen levels in homes are sufficiently stable and consistent over periods of several years to use single-measurement samples in epidemiological studies with a reasonable degree of confidence (Antens *et al.*, 2006). The same kind of studies have not been done for dust mite populations, but population growth, based on stable age theory (Chapter 5), predicts that population densities will tend to stabilise and fluctuate around a particular level, constrained by the range of temperature and humidity they are exposed to. At the regional scale, macroclimate, elevation and continentality have major effects on dust mite population densities (Chapter 4). These arguments suggest that allergen concentrations and mite population densities may be characteristic of localities that share similar climatic and geographic features.

In order to examine regional and global-scale variation in mite populations and allergens in homes, I compiled datasets from published records of the abundance of total mites, plus abundance, frequency of occurrence and percentage dominance of the top four species (see Chapter 4, Appendix 2): *Dermatophagoides farinae, D. pteronyssinus, Euroglyphus maynei* and *Blomia tropicalis* (Appendix 3), as well as concentrations of *Dermatophagoides* group 1 allergens in dust from beds and floors (Appendix 4). There was insufficient data to compare some of the other allergenic species of interest, notably *Lepidoglyphus destructor, Glycyphagus domesticus, Acarus siro* and *Tyrophagus putrescentiae*. Most of these species are represented by less than 30 records of abundance or dominance or frequency. This does not imply they are not insignificant, but rather they are often reported lumped together in their family categories of Glycyphagidae and Acaridae.

The dataset on mites is based on almost 23 500 samples from over 9000 homes. That on allergens is based on over 34 000 samples from nearly 28 000 homes. Data were included where a latitude and longitude could be matched to a record. Data pooled from multiple localities were excluded from any analyses (but not from the dataset). Means are given as geometric means wherever possible. Where arithmetic means were presented in the original publications, I have converted them to geometric means using graphs or tables where such data were available, and included 'negative' samples in the calculation. Mite abundance and allergen concentrations were excluded if semi-quantitative, based on 'exposure classes', because it is not possible to accurately calculate a measurement of central tendency from these. Values repeated in multiple publications were not replicated in the dataset. Where abundances or concentrations were presented per unit area they were excluded. The vast majority of data are expressed per unit weight and there is no reliable way of converting or comparing area data. Data from places other than homes (offices, hospitals, nursing homes and public facilities) are not included, nor are those based on samples taken by sweeping rather than vacuuming.

On the principle of *habeas corpus*, I have not attempted any integration of mite abundance and allergen concentration data regarding the mapping of distributions. The physical presence of mite bodies, identified to species, is better evidence that populations persist at a particular locality than the presence of a particular allergen. This does not mean the presence of allergen is not a good indicator of presence of mites. It is, but it is only an indicator and can tell us a lot less about structure, growth or persistence of mite populations. For now, because these data have not been collated previously, it is best the mite abundance data and allergen data be presented separately.

8.7.2 Population densities of mites

The global distribution pattern of total mite abundance (i.e. all species) in beds, floors and mixed samples is shown in Figures 8.7 and 8.8 (see also Appendix 3). Abundance data were presented as classes based on threshold equivalents (100 mites roughly equivalent to $2\,\mu g\,g^{-1}$ and 500 equivalent to $10\,\mu g\,g^{-1}$). The reason for doing so is mainly because these thresholds have been widely used and represent a familiar benchmark for many researchers. Data from 'mixed samples' (i.e. dust samples that included material from both beds and floors, or mean abundances calculated from both floor and dust samples) were mapped together with data from floors (with which they had a similar median and range) in order to increase geographic coverage. About 75 locations were represented by mixed samples only.

Localities with the highest total mite populations in the tropics were Central and South America and the Caribbean (Colombia – Caracas, Tolú; Ecuador – Guayaquil; Barbados, Costa Rica), South-East Asia (Singapore, Kuala Lumpur, Bangkok), Africa (Nairobi, Lagos). In temperate regions, the highest numbers of mites were recorded at Iranian cities on the Caspian Sea, Algerian coastal cities, New Zealand (Wairoa, Wellington), Australia (Melbourne) and Japan (Naha, Wakayama, Osaka, Sendai, Tokyo). Lowest mite populations were at inland places in Australia (Wagga Wagga, Alice Springs), desert oases in Algeria (Biskra, Touggourt, El-Oued, Arris, Batna, Ain-Touta), central Iran (Shiraz, Karaj, Isfahan), far southern Chile (Punta Arenas, Valdivia), alpine Switzerland (Zermatt, Davos, St Moritz), continental northern Europe and Scandinavia (Katowice, Copenhagen, Umeå, Sør-Varanger), central and south-eastern Turkey, and montane continental USA (Denver).

The global pattern of abundance classes in Figures 8.7 and 8.8 shows a high proportion of large dots (>500 mites per gram) in tropical and subtropical latitudes, corresponding with the zone of highest mean annual rainfall and lowest mean fluctuation in temperature, but also at coastal localities (including inland seas like the Caspian), on islands and at places that receive high annual precipitation due to maritime climatic influences such as New Zealand and Japan. Contrastingly, Chile and coastal Peru receive low annual precipitation due to the effects of cold northerly ocean current systems (as does the west coast of southern Africa), and mite abundance is correspondingly low. Smaller dots (=100 mites per gram) tend to be in continental interiors or above 55°N.

The range of abundances based on the distribution curves (see Figure 8.9) indicates for beds 112/167 localities (67%) had more than 100 mites g^{-1}, and 59 (35%) had more than 500 mites g^{-1}. For floors, the figures were 36/115 (31%) and 17 (15%). Median abundance for beds was 243 mites g^{-1} (interquartile range, IQR 62–835) and for floors 14 mites g^{-1} (IQR 2–54, see Table 8.3). Mixed sample data were not included. Globally, *Dermatophagoides pteronyssinus* was by far the most widespread, abundant, frequently occurring and dominant species (dominance being the percentage of *D. pteronyssinus* individuals in the total mite population) in both beds and floors (Table 8.3). Beds had consistently higher densities of mites than floors. The highest recorded mean population density of *D. pteronyssinus* was from Caracas (nearly 13 000 per gram; Hurtado and Parini, 1987). *Dermatophagoides farinae* was the next most abundant and frequently occurring species, but not that much greater than *Blomia tropicalis* and *Euroglyphus maynei*. *D. farinae* was more frequent (though not as abundant) on floors, where it is more likely to be the dominant species than in beds. *Blomia tropicalis* was almost as frequent and abundant as *D. farinae*, and appears to show no marked preference for beds over floors. In tropical South-East Asia (Malaysia, Singapore) and Latin America, this species was either the dominant species or co-dominant with *D. pteronyssinus*, and it had the highest recorded mean population density of any dust mite species on floors: 8250 per gram in Singapore (Chew *et al.*, 1999a). *Euroglyphus maynei* was more abundant and occurred slightly more frequently in beds than on floors. Highest densities (>1000 g^{-1}) were all recorded from the tropics (Nairobi, Kenya; Guayaquil, Ecuador and Caracas, Venezuela) and the highest frequencies of occurrence (>50%) were at coastal localities with warm moist or temperate moist climates.

Mite population densities in beds and on floors were exponentially negatively correlated with latitude (see Figure 8.10). Ordman's (1971) records for floors from South African localities were excluded from Figure 8.10b because they were artificially low; probably due to the use of an early and inefficient extraction technique. As stated, tropical and subtropical localities tend to have the highest numbers of mites. Population densities within 30° of the equator were 2- to 11-fold greater than at 50°N. For beds, based on the regression equation, with every 10-degree increase in latitude from the equator, there was a 1.6-fold decrease in mean population density. For floors, the decrease was 1.5-fold. The significance of this relationship lies not so much with the higher rainfall zones of the tropics, for rainfall may be highly seasonal and many of the world's great deserts lie between the tropics of Capricorn and Cancer. Rather, it is due to the lower amplitude of diurnal and seasonal variation in temperature, especially at tropical coastal localities. A glance at the climate graphs for major cities in the *Times Atlas of the World* shows hardly any seasonal variation in mean monthly temperature for Singapore, Caracas or Bangkok, but at least 3- to 4-fold variation for places at temperate latitudes or inland ones in the

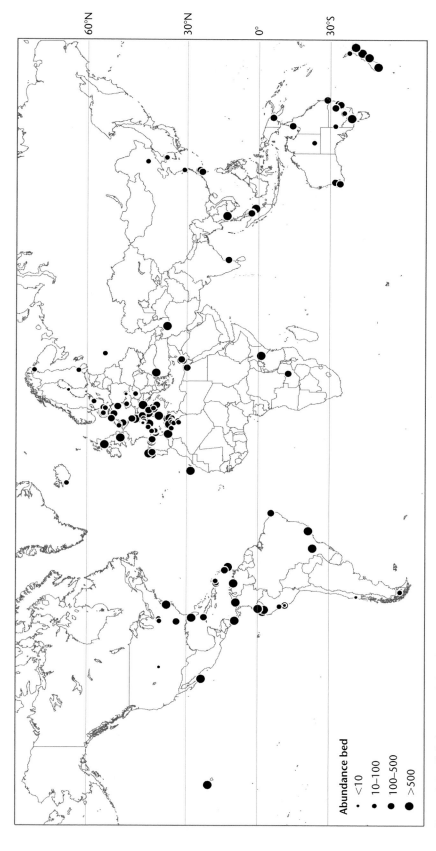

Figure 8.7 Global pattern of abundance (number per gram) of total mites (all spp.) in beds, based on published surveys (Appendix 3).

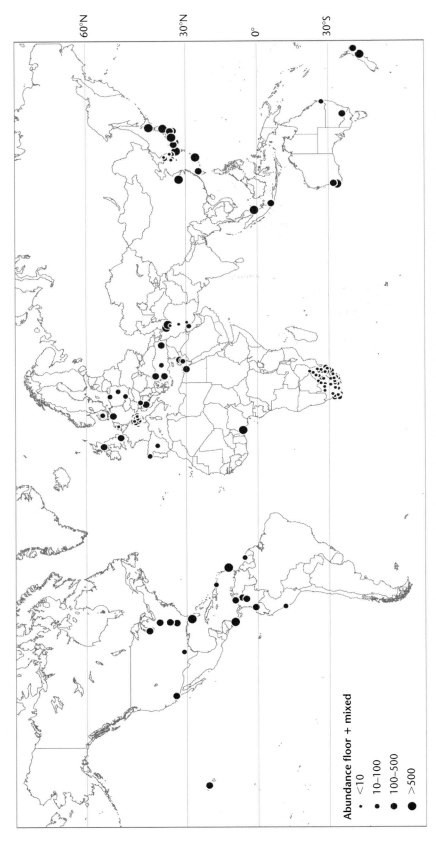

Figure 8.8 Global pattern of abundance (number per gram) of total mites (all spp.) in floors and mixed samples, based on published surveys (Appendix 3).

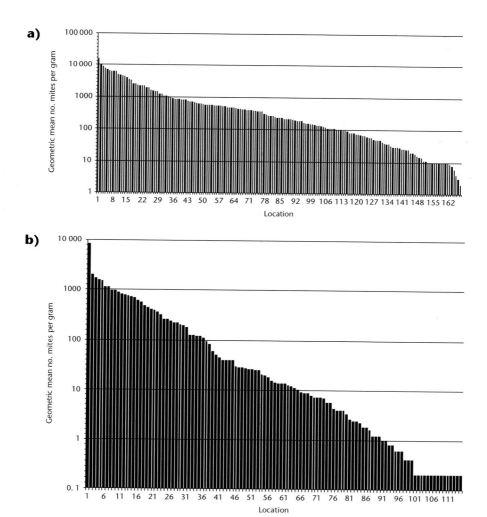

Figure 8.9 Distribution curves of ranked mean total mite abundance. **a)** In beds; **b)** on floors; with localities ordered according to abundance, based on published surveys (Appendix 3).

sub-tropics. The annual temperature range at 40–50°N is in the order of 30–40°C (maximum v. minimum recorded temperatures). More uniform seasonal temperature in the equatorial belt is accompanied by consistent day length and little difference between daytime and night-time temperatures, whereas at 50°N, mid-winter day length may be half that of mid-summer, and outdoor temperature at midday 10°C higher than at midnight. It should come as no surprise that individual global climate variables based on annual means have relatively low explanatory power in relation to mite populations. For example, monthly mean minimum temperature (expressed as an annual mean) is a significant factor (Figure 8.11a), but only part of the story.

Although highly statistically significant, the relationship with latitude explains only a small proportion of the variation in population density. An additional amount of variation can be accounted for according to whether a location is on the coast or inland. Coastal locations had five times the median number of mites in beds and eight times the number in floors than localities that were inland (see Table 8.4). Although precise distances from coasts were not calculated, the pattern on the maps (Figures 8.7, 8.8) suggests there is a tendency for localities that are strongly continental to have lower mite populations. There was no relationship between either elevation of localities above sea level or mean annual rainfall and mite population densities in beds or carpets at the global scale, although with rainfall there appears to be a cut-off point around 270 mm, below which mite populations are uniformly low.

Table 8.3 Median (and interquartile range, IQR) abundance (number per gram), percentage frequency of occurrence and percentage dominance of dust mites in beds and floors, based on published surveys (Appendix 3).

Total mites	Abundance	IQR	n	% frequency	IQR	n	% dominance	IQR	n
Beds	243	62–835	164	99	70–100	109			
Floors	14	2–54	156	100	90–100	31			
D. pteronyssinus									
Beds	191	33–585	104	87	60–98	104	67	35–83	106
Floors	83	4–337	31	100	60–100	14	34	12–69	30
D. farinae									
Beds	33	5–185	73	34	9–69	77	6	2–24	80
Floors	14	3–79	27	50	20–67	9	16	4–70	24
E. maynei									
Beds	18	2–125	68	22	8–43	56	3	1–16	68
Floors	12	4–38	11	17	13–54	10	4	1–13	13
B. tropicalis									
Beds	23	3–115	32	33	19–50	43	8	1–27	38
Floors	17	8–1100	9	95	78–100	6	41	6–73	11

The distribution and abundance of dust mites are dependent on a favourable microclimate in which they can live and reproduce. The mite population density influences allergen levels, human exposure and, to a certain extent, the prevalence and severity of disease. Microclimate within homes is influenced, for at least part of the year, by outdoor climate, even in well-sealed homes. The influence of macroclimate on dust mite populations appears, at the regional and global scales, to override the effects of indoor microclimate. In summary, some of the major lines of evidence for macroclimate effects on dust mite populations are:

- low density mite populations at high altitude in temperate latitudes, but very high densities at high altitude localities in the tropics (Caracas, Bogota, Nairobi) due to seasonally high rainfall and outdoor humidity (mean monthly RH is >81% for Caracas);
- a dust mite fauna dominated numerically by *Dermatophagoides farinae*, with its lower critical equilibrium activity, in the drier parts of Europe and North America, but by *D. pteronyssinus* in the more humid regions;
- highest dust mite population densities recorded from coastal cities with high rainfall and

moderate-to-warm outdoor temperature regimes (e.g. Melbourne, Sydney, Singapore, Tokyo), but low mite population densities from cities with dry climates, e.g. those in continental interiors or high latitudes (e.g. Helsinki, Wagga Wagga, Katowice, Briançon and Denver);

- seasonal fluctuations in outdoor humidity are followed by corresponding fluctuations in dust mite population density and allergen concentrations; and
- where outdoor climate is unfavourable for part of the year (e.g. cold and dry, as in Scandinavia), the frequency of occurrence of dust mites in homes is considerably less than 100% (e.g. Copenhagen, 60%; Uukuniem and Ilomantsi, 32.4%; Oslo, 23.5%; Reykjavik, 10%).

In homes where outdoor climate is favourable for most of the year (e.g. warm-temperate and moist, as in coastal Australia and New Zealand), just about all homes contain dust mites.

8.7.3 Concentrations of allergens

There was a strongly significant correlation between mite population densities and allergen concentrations in beds at the 32 localities where both were measured

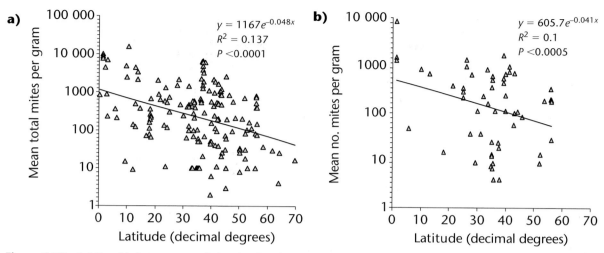

Figure 8.10 Relationship between population density of mites and latitude. **a)** In beds; **b)** on floors; with South African localities of Ordman (1971) excluded. Southern Hemisphere localities were corrected to positive latitudes.

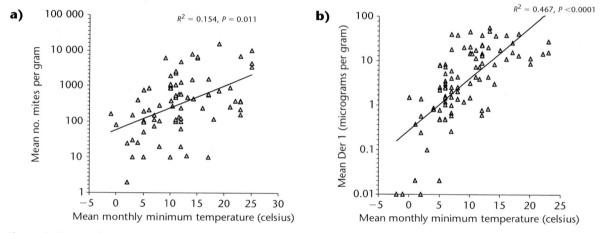

Figure 8.11 Relationship between **a)** population density of mites; and **b)** concentrations of Der 1 in beds and mean monthly minimum temperature, averaged annually.

(see Figure 8.12). There was no significant difference between mites and allergens measured as part of the same study (14 locations) and in different studies (and in different years), lending weight to the hypothesis that mite populations and allergen concentrations are relatively stable at localities with shared climatic and geographic features. Der 1 concentrations of 2 and 10 µg g^{-1} were equivalent to 150 and 450 mites g^{-1} respectively, quite close to the 100 and 500 mites g^{-1} equivalents proposed by Platts-Mills et $al.$ (1989b). There were not enough data to compare mites and allergens on floors.

The geometric mean concentration of Der 1 in beds at 128 localities worldwide was >2 µg at 66% of localities and >10 µg at 35% (Figure 8.13a).

Floors had >2 µg at 49% of localities and >10 µg at 19% (Figure 8.12b). The median value of Der 1 for beds was 4.4 µg g^{-1} (IQR 1.2–14.8) and for floors 1.9 µg g^{-1} (IQR 0.7–7.6; Table 8.5).

The global distribution of Der 1 concentrations in Figures 8.14 and 8.15 shows a similar pattern as for mite abundance. Like the abundance maps, data were presented as classes based on threshold concentration (2 µg g^{-1} and 10 µg g^{-1}). Again, there is a high proportion of large dots (>10 µg Der 1 g^{-1}) in tropical and subtropical latitudes and at coastal localities, and smallest dots (<0.2 µg Der 1 g^{-1}) tend to be in continental interiors or above 55°N. Exceptions are the high level of Der 1 in beds at Moscow (based on paired

Table 8.4 Median (and interquartile range) *Dermatophagoides* group 1 allergen concentrations (micrograms per gram) and total mite population density in coastal and inland localities, based on published surveys (Appendices 3 and 4). For mite population densities in floors, Ordman's (1971) South African records were excluded.

		Median	25th percentile	75th percentile	n
Total mites per gram					
Beds	Coast	546	189	1527	75
	Inland	107	30	428	92
Floors	Coast	383	182	807	32
	Inland	50	13	202	24
Der 1 μg per gram					
Beds	Coast	7.4	1.9	18.9	60
	Inland	3.1	0.8	11.6	68
Floors	Coast	4.4	1.8	13.9	29
	Inland	1.1	0.3	2.5	36

mothers' and children's beds in a small sample of apartments) and Fyn/Viborg in Denmark (from baseline measurements of a clinical trial of allergen avoidance).

The dense cluster of high allergen localities (>10 μg g^{-1}) for beds in northern Spain (Figure 8.14) represent some of the highest Der 1 concentrations recorded in Europe. This has been ascribed to an interaction of indoor risk factors with high humidity and warmer temperatures during winter at coastal localities (Echepichía *et al.*, 1995; Boquete *et al.*, 2006; Zock *et al.*, 2006), reflecting the temperate high-rainfall climate associated with westerly Atlantic low pressure systems. High allergen levels were also found at lower elevation sites in the montane Rioja region, which has a less favourable climate than the coast, though this was considered to be related to higher allergen concentrations in rural than urban homes (Lobera *et al.*, 2000). A group of very low allergen sites (with only 2–23% frequency of Der 1) in Saudi Arabia (Al-Frayh *et al.*, 1997; Figure 8.15) was associated with arid desert conditions in the central Arabian peninsula, but also at one central coastal location (Jeddah). Much higher allergen levels (and 70% frequency) were found at the most southerly town, Abha, which is only 70 km from the coast but almost 2300 m above sea level. Temperature is lower and humidity markedly higher here than at the coast.

As was found for dust mite populations, coastal locations globally had much higher median Der 1 concentrations than inland localities (2.4-fold for beds, 4-fold for floors; see Table 8.4). There was no relationship between allergen concentrations and

either elevation of localities or mean annual rainfall. Allergen concentrations showed a highly significant decrease with distance from the equator (shown in Figure 8.16) and a highly significant positive correlation with mean monthly minimum temperature (Figure 8.11b). The relationship between allergen concentration and temperature is much stronger than for mite population density. Since Der 1 production is a function of population size and rates of feeding and digestive physiology, perhaps populations in colder climates are both smaller and egest faecal pellets at lower rates than those in warmer climates.

8.7.4 Regional and global patterns of allergen diversity and exposure

Previous compilations of records of mite populations and allergen concentrations (Woolcock *et al.*, 1991; Arlian *et al.*, 2002) have been preliminary or of limited geographic coverage because there were far fewer published records available than there are now. The datasets in Appendices 3 and 4 are certainly not complete. There are several publications I have been unable to obtain. Some of these are abstracts of conference proceedings, are published in very hard-to-find periodicals or both. I would be grateful to all readers who can supply copies of publications that have not been cited.

One of the main outcomes of mapping the abundance of dust mites and their allergens has been to highlight the information gaps, the main one being uneven geographic coverage. Quantitative data are harder and more expensive to generate than

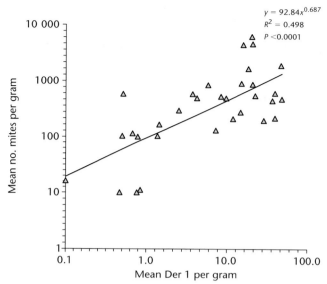

Figure 8.12 Relationship between mean Der 1 concentrations and mean total numbers of mites in beds, based on published surveys.

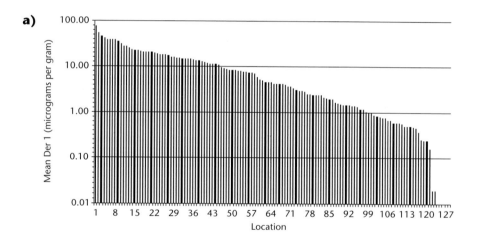

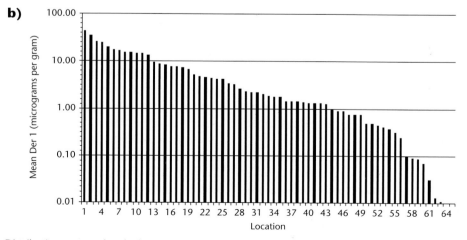

Figure 8.13 Distribution curves of ranked mean Der 1 concentrations. **a)** In beds; **b)** on floors; with localities ordered according to abundance, based on published surveys (Appendix 3).

Table 8.5 Median (and interquartile range, IQR) concentrations and percentage frequency of *Dermatophagoides* group 1 allergens (micrograms per gram), based on published surveys (Appendix 4).

		Concentration	IQR	n	% positive samples	IQR	n
Der 1							
	Beds	4.4	1.2–14.8	128	91	74–100	43
	Floors	1.9	0.7–7.6	65	86	45–100	12
Der p 1							
	Beds	2.8	0.5–13.3	113	85	45–100	53
	Floors	1.9	0.7–6.6	50	97	68–100	21
Der f 1							
	Beds	0.7	0.2–2.2	62	68	26–81	32
	Floors	0.3	0.1–0.8	25	38	15–68	13

point-source species records, and much less is available (compare map in Figures 8.14 and 8.15 with the 'all localities sampled' map in Figure 4.48). Nonetheless, many more published records could have been included in the quantitative maps had they been more geographically precise and/or expressed the data as geometric means.

The reason for compiling these records is, in part, to assist clinicians in making better-informed decisions about patterns of mite allergen exposure in their region, especially regarding appropriate diagnostic skin-prick testing and IgE antibody assays for allergies. The distribution maps in Chapter 4 show which species occur where, but the quantitative maps and appendices provide information on the relative magnitude of exposure of *Dermatophagoides* group 1 allergens and the abundance and frequency of the four most widespread and clinically important species. These species assemblages are examined in more detail below.

a Species assemblages and allergen exposure

The main regional combinations of dust mite species, based on mite community structure and relative dominance, are listed below. These species assemblages are responsible for different global patterns of allergen diversity, exposure and IgE responses.

Dermatophagoides farinae and *D. pteronyssinus* co-occur frequently in most of continental Europe, Russia, the Middle East, Asia, North America and Latin America. Mixed populations are rare in the UK, Norway, Portugal, north-western Spain, North Africa, the Caribbean, Australia and New Zealand. Zock *et al.* (2006) found much higher frequency of occurrence of

Der p 1 in Western Europe and the Atlantic coast (where *Euroglyphus maynei* may also be present), whereas Der f 1 was the more frequent mite allergen on the Mediterranean coast, central and north-eastern Europe and eastern Scandinavia. Relative concentrations of the two allergens show similar distribution patterns and frequencies (frequency of occurrence of both Der f 1 and Der p 1 is strongly correlated with concentration), with Der p 1 dominating in Western and southern Europe (see Figure 8.17). Where total Der 1 concentrations are high (>10 µg g^{-1}), Der p 1 is almost always the dominant allergen. This trend persists worldwide (one exception being southern Brazil) and may be related to the capacity of *D. pteronyssinus* for more rapid population growth than *D. farinae* under optimal conditions (Arlian *et al.*, 1998a). A slightly less pronounced longitudinal trend is present in both Latin America and the USA, with Der p 1 tending to be the dominant allergen on the western sides of the continents and Der f 1 on the east (shown in Figure 8.18).

Dermatophagoides farinae, *D. pteronyssinus* and *D. microceras* co-occur mainly in Scandinavia, though group 1 allergen from *D. microceras* (Der m 1) has also been found with Der p 1 and Der f 1 in Belgium, Austria and the USA (Lind, 1986b; Schwartz *et al.*, 1987). It is likely that the abundance, frequency of occurrence and geographical range of *D. microceras* has been underestimated because it has been confused with its sibling species, *D. farinae* (Cunnington *et al.*, 1987). This finding may be of clinical significance because Lind *et al.* (1987) demonstrated *D. microceras*-specific antibody responses in some patients with atopic asthma. In Norway, where *D. farinae* is rare,

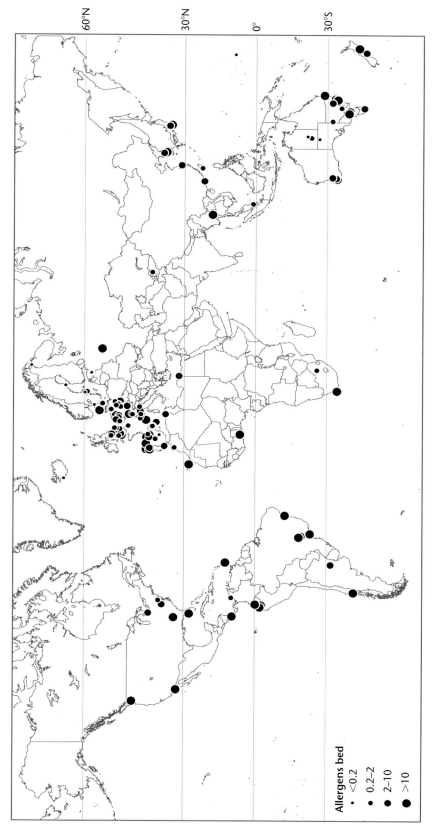

Figure 8.14 Global pattern of *Dermatophagoides* group 1 allergen concentrations (micrograms per gram) in beds, based on published surveys (Appendix 4).

Allergens bed

· <0.2
· 0.2–2
● 2–10
● >10

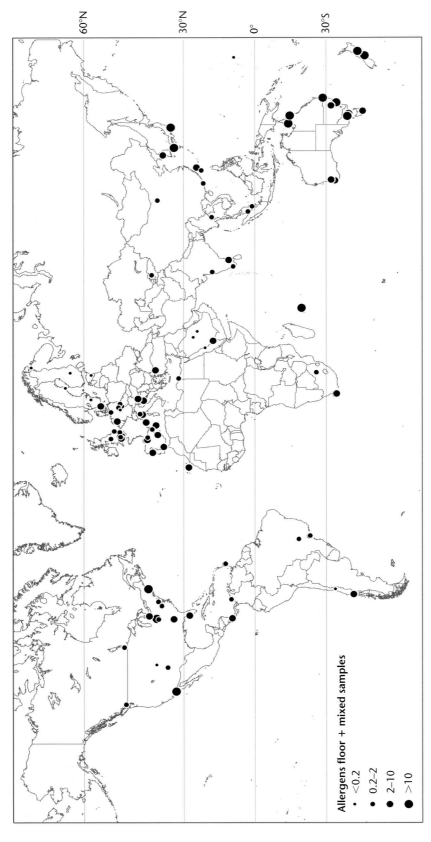

Figure 8.15 Global pattern of *Dermatophagoides* group 1 allergen concentrations (micrograms per gram) on floors, based on published surveys (Appendix 4).

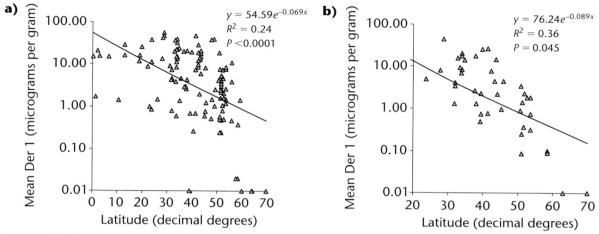

Figure 8.16 Relationship between Der 1 concentrations and latitude. **a)** In beds; **b)** on floors. For floors, tropical desert localities (Saudi Arabia) were removed. Southern Hemisphere localities were corrected to positive latitudes.

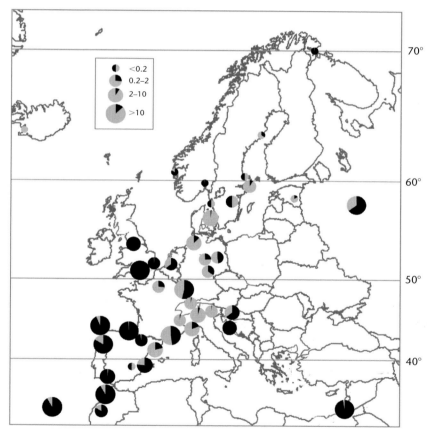

Figure 8.17 Relative concentrations (micrograms per gram) of Der p 1 (black) and Der f 1 (grey) in beds (and some mixed samples) in Europe, North Africa and the Middle East, based on records in Appendix 4.

D. microceras is the second most frequently occurring species after *D. pteronyssinus*. Of 540 homes in Oslo, *D. pteronyssinus* was present in 24% and 9% contained *D. microceras*. Only one specimen of *D. farinae* was found (Mehl, 1998). Der m 1 concentrations at three localities in Sweden ranged from 0.1–932 µg g⁻¹ and represented about 30% of the total group 1 concentration, and Der m 1 was found in about half the homes where group 1 allergens were detected (Warner *et al.*, 1998).

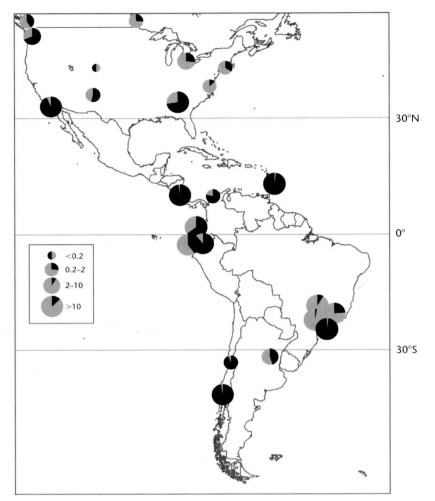

Figure 8.18 Relative concentrations (micrograms per gram) of Der p 1 (black) and Der f 1 (grey) in beds (and some mixed samples) in Latin America, the Caribbean, USA and Canada, based on records in Appendix 4.

D. pteronyssinus and *D. siboney* are found together in Cuba, where *D. siboney* occurs in over 80% of homes and represents ca. 40% of the total mite population (Dusbabek *et al.*, 1982; Cuervo *et al.*, 1983). *Blomia tropicalis* (see below) also commonly co-occurs, but *D. farinae* is rare or absent (Ferrándiz *et al.*, 1996). About 80% of patients with asthma had IgE antibodies to Der s 1 and the species is regarded as an important cause of sensitisation to dust mites in Cuba (Ferrándiz, 1997; Ferrándiz *et al.*, 1995b, 1996). The significance of *D. siboney* is that it is likely to be much more widespread in the Caribbean than just Cuba. It has been found in dust samples from homes in Kingston, Jamaica (Colloff, unpublished) and infrequently (though in substantial numbers) in mattress dust in Puerto Rico and Martinique (Montealegre *et al.*, 1997; Lafosse Marin *et al.*, 2006).

It probably also occurs on mainland South and Central America also, where it may have been misidentified as *D. farinae.*

Blomia tropicalis and *D. pteronyssinus* are found together in the tropics and subtropics, particularly coastal Latin America, the Gulf of Mexico and the Caribbean, West Africa, South-East Asia and northern Australia. At localities where both species are present, rates of co-occurrence are often very high (>80% of samples). *B. tropicalis* tends to occur most frequently nearest to the equator (see Figure 8.19), often in very large numbers, and often as the numerically dominant species. Hence it is a major source of allergen exposure in the tropics, as confirmed by high concentrations of total Blo t allergens, detected by inhibition immunoassay (Puerta *et al.*, 1996b; Zhang *et al.*, 1997), and by ELISA for Blo t 5 (Yi *et al.*, 2004; Lee *et al.*, 2005)

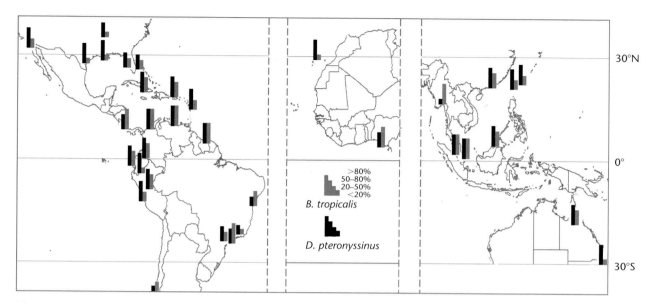

Figure 8.19 Frequency of occurrence of *Dermatophagoides pteronyssinus* and *Blomia tropicalis* at localities where both species are present, based on records in Appendix 4.

there is a high prevalence of sensitisation among atopic asthmatic patients to allergens of both *B. tropicalis* and *D. pteronyssinus* (summarised by Fernández-Caldas and Lockey, 1995, 2004). *D. farinae* and *Euroglyphus maynei* may also co-occur at localities with *B. tropicalis* and *D. pteronyssinus*, but there are insufficient data to get a clear breakdown of rates of co-occurrence within individual homes.

D. pteronyssinus and *E. maynei* are the two major species that co-occur, where *D. farinae* is rare, in the UK, Australia and New Zealand, as well as a couple of locations in Spain (Santiago de Compostella – Agratorres *et al.*, 1999; Santa Cruz, Tenerife – Sanchez-Covisa *et al.*, 1999) and coastal Algeria (Louadi and Robaux, 1992). Population growth of *E. maynei* is slower than most other domestic mites (Chapter 5) and it is rarely the numerically dominant species, usually accounting for less than 25% of the total mites (Colloff, 1991c).

D. pteronyssinus and *Lepidoglyphus destructor* are found together in rural areas and farming communities in Northern Europe. Some of the exposure to *L. destructor* allergens is occupational, relating to farming and the presence of mites in hay, straw and barns (van Hage-Hamsten *et al.*, 1985; Cuthbert *et al.*, 1984; Cuthbert, 1990; Härfast *et al.*, 1996; Radon *et al.*, 2000). In other cases, storage mites, including *L. destructor*, are associated with IgE sensitisation in urban homes (Warner *et al.*, 1999). In a random sample of 540 urban dwellers in Reykjavik, Iceland, 25% had a positive skin-prick test, and 6% had IgE-mediated allergy, to *L. destructor* (Gislason and Gislason, 1999). Over half the population had handled mite-contaminated hay or been exposed to hay dust during the short, intensive, rainy Icelandic hay harvest. For this population, *L. destructor* is possibly a more important source of allergens than *D. pteronyssinus* (Hallas *et al.*, 2004).

Currently our knowledge of regional and global scale variation in allergen concentrations is based solely on data for *Dermatophagoides* group 1 allergens. Group 2 data are available from a few studies but currently insufficient to be able to attempt a synthesis. For mites other than Pyroglyphidae, Lep d 1 and Blo t 5 assays are being used, but again there is as yet insufficient information to draw any inferences.

8.8 Epidemiological implications of variation in allergen concentrations

In regions like Scandinavia where there are relatively low population densities of dust mites, there tend to be far fewer homes that have detectable mites or allergens than in those areas where dust mite populations are high (Wickman, 1993, 1995; Wickman *et al.*, 1991, 1993). Under these circumstances, the prevalence of sensitisation will be only as high as the prevalence of homes with sensitisable levels of allergen in them, which may fall well short of the genetic potential of the population to develop sensitisation to dust mites and clinical symptoms of atopy. In regions where all homes have house dust mites, the prevalence of sensitisation may be close to what the genetic status of the population will allow and the population potential for clinical atopy nears saturation. Two of the most contrasting examples are as follows: Dotterud *et al.* (1995) in a survey of 424 children at Sør Varanger, Northern Norway, found only 20 were sensitised to dust mite (4.7%; current asthma 2.6%) and that 10 of them lived in homes containing detectable mites (mean ca. 190 g^{-1}) whereas 19 randomly selected, non-sensitised controls had no detectable mites in their homes, giving an odds-ratio for sensitisation if mites were present of >20. By way of contrast, Marks *et al.* (1995b) in Sydney, Australia, found 41% of 80 children were sensitised to dust mites (current asthma 15%) and that there was actually a slightly higher Der p 1 concentration in homes of children who were not sensitised to dust mites (64 µg g^{-1}) than in those who were (41 µg g^{-1}). In areas of low dust mite populations, sensitised people receive seasonal variation in exposure and may not have symptoms and bronchial hyperreactivity all year. In areas of high dust mite populations, sensitised people receive continued high dose exposure, enough to maintain bronchial hyperreactivity for most of the time (Peat *et al.*, 1996). Demonstration of a dose-response relationship between allergen exposure, sensitisation and asthma depends on there being sufficient variation in allergen concentration, seasonally or between homes, for a broad range of asthma severity to become manifest. This variation in exposure is likely to be highest in areas that have most homes with few or no mites and a few homes with a great many, rather than in areas where almost all homes have large numbers of mites (Marks, 1998). So, in areas of low dust mite populations, the range of disease severity is likely to be high, from mild to severe; and a dose-response is demonstrable. In areas of high dust mite populations, the range of disease severity is lower because most people with asthma have more severe symptoms and a dose-response is not evident (see Figure 8.20), suggesting there is an upper threshold for exposure beyond which increasing exposure

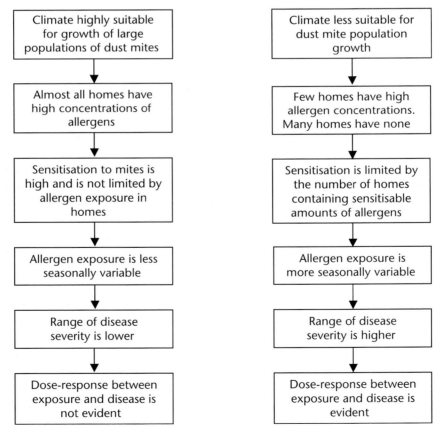

Figure 8.20 Conceptual model of the influence of macroclimate on mite allergen exposure and the prevalence of sensitisation to dust mites and the severity of mite-mediated atopic asthma.

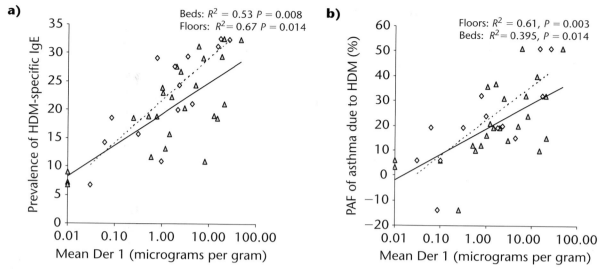

Figure 8.21 Relationship between allergen concentrations in beds (triangles, solid lines) and floors (diamonds, dashed lines) from Appendix 4 and **a)** sensitisation to dust mites based on prevalence of mite-specific IgE; **b)** the population-attributable fraction of asthma due to sensitisation to dust mites from the European Community Respiratory Health Survey (ECRHS; Sunyer *et al.*, 2004).

does not cause further risk of sensitisation or disease. Marks (1998) reckons this level is around 10 μg g⁻¹. It follows that even with allergen avoidance methods that can achieve large reductions in exposure, patients in areas of low exposure are likely to benefit more: an 80% reduction of 10 μg g⁻¹ Der 1 still leaves a concentration that would be considered high for most places in the temperate latitudes of the Northern Hemisphere.

The European Community Respiratory Health Survey included 48 centres in 22 countries. Prevalence of mite-specific IgE has been determined and the attributable fraction of asthma due to sensitisation to mites has been calculated for 36 centres involving almost 13 600 volunteers in 16 countries (Sunyer *et al.*, 2004). Of these, 28 centres had matching data on mite allergen concentrations (shown in Appendix 4), most of which were determined by Zock *et al.* (2006) as part of ECRHS Phase II (though not all of these were used herein). There were statistically significant positive correlations between Der 1 concentrations in both beds and floors and the prevalence of mite-specific IgE as well as the mite-attributable fraction of asthma (see Figure 8.21). At a mean of 10 μg g⁻¹ Der 1, prevalence of mite-specific IgE was 28%, almost twice that at 0.1 μg g⁻¹, and mite-attributable asthma was 35%, over four times as high. These relationships indicate there is probably a log.-linear dose-response relationship between mite allergen exposure and both prevalence of sensitisation and mite-mediated asthma in adults

in this large trans-regional study. The International Study of Asthma and Allergies in Childhood had 28 centres with data on the population-attributable fraction of asthma due to a positive skin-prick test or allergen-specific IgE (Weinmayer *et al.*, 2007). But corresponding allergen concentrations (Appendix 4) were available for only nine of them, so a similar analysis was not attempted.

8.9 Changes in exposure to mite allergens?

It has been suggested there has been an increase in dust mite population density and allergen concentrations in homes since the 1970s (e.g. Sporik *et al.*, 1990; Woolcock *et al.*, 1995). Petrova and Zheltikova recorded an increase in frequency of mites in Moscow apartments between 1983 and 1997, but a decrease in mean abundance. Halmai (1984) claimed there was a significant increase in frequency and abundance of *D. farinae* between 1969 and 1984 in Hungary, but statistical analyses were lacking. Neither of these studies was based on repeated-measures sampling, whereas the longest repeated-measures surveys of allergen concentrations, though more recent, show no increase over 6–8 years (Topp *et al.*, 2003; Antens *et al.*, 2006; Crisafulli *et al.*, 2007).

But current patterns of mite allergen exposure are probably quite different now from what they were 100 years ago or more. Before the advent of improved food storage and hygiene, people ran a daily risk of acquiring a not inconsiderable part of their daily fat and protein intake from the consumption of insect and mite pests of stored products. There is anecdotal evidence that stored products mites were more common in urban homes than they are today (Chapter 4), so probably the spectrum of allergen exposure was more diverse. In the 19th century, airborne particles in cities would have included quite a lot of horse dung, together with its bacterial endotoxins (see Figure 4.17). Insect and mite contaminants of food were a fact of life in Westernised countries until the implementation of food standards and packaging around the 1950s. In many parts of the world where cereals, pulses, dried fish and meat are sold loose, they still are. The regular ingestion by a sizeable proportion of the human population of insects, mites and their frass would represent a marked difference both in route of exposure and allergen spectrum, especially bearing in mind that oral administration of allergen can induce immune tolerance. If people used not to get asthma so much in the past because their immune system was steered towards Th1 responses by increased prevalence of childhood infections, then maybe also there was less asthma around because people naturally desensitised themselves as they ate.

There have been major changes in housing design in the last 40 years. The building of energy-efficient homes (predominantly a phenomenon of developed countries with temperate climates in the Northern Hemisphere), which are insulated, double-glazed, centrally heated, and with low ventilation, was triggered partly by the oil crisis of the 1970s and the need to conserve heating fuel. Fitted carpets have become the standard floor covering in many homes, whereas before the 1960s, linoleum, polished floorboards and rugs were standard. These changes in housing design have resulted in warmer, moister homes and it has been claimed that the more favourable domestic microclimate encouraged the development of large populations of dust mites. This, in turn, has led to a boom in dust mite allergen exposure which has been cited as a possible reason for the recent increase in asthma prevalence. So goes the story. But the data on housing risk factors (see 8.6 above) does not support it. Apparently, people spend more time indoors on average than they did 40 years ago. According to Platts-Mills et al. (1996) human behavioural factors that may contribute to the increase in the prevalence and severity of asthma include poor standards of domestic cleanliness, passive smoking, lack of exercise and obesity. People spend more time in front of the television or computer, less time exercising outdoors or engaged in manual toil; they eat more processed food containing more saturated fat, and less antioxidant-containing fresh fruit and vegetables; and have less contact with soil and less exposure to beneficial bacteria.

These lifestyle changes are linked in part to increasing affluence. Increased prevalence of asthma and atopy has been associated with higher gross national product per capita (Stewart et al., 2001) and with lower intake of cereals, nuts, vegetables and plant starches (Ellwood et al., 2001). But in urban societies it is people on low incomes who tend to eat worst, exercise least and have the highest rates of morbidity and mortality (Syme, 1986). Exposure to a so-called 'modern' or 'Western' lifestyle is frequently invoked as a factor that explains increases in asthma prevalence and is linked to increasing urbanisation (Figure 8.22) and rural depopulation and dietary shifts associated with urban living in developing countries (Weinberg, 1999; Hooper et al., 2008), and to higher social class and income in developed ones (Heinrich et al., 1998). Asthma is of higher prevalence in Westernised countries, but developing countries may have a higher burden of asthma morbidity because they contain a much higher proportion of world population.

There is now widespread acceptance that the temperature of the planet is increasing and that climate change is taking place as a result of anthropogenic greenhouse gas emissions. A great deal of research effort has been devoted to predicting the biological effects of global climate change, focusing on the effects on ecosystem diversity, pests of crops and arthropod vectors of diseases of humans and domesticated animals. The issue is that warming does not automatically equate to higher population densities of mites. On balance, it is too early to say what will happen with dust mite populations, though increased prevalence of asthma has been linked to climate change (Beggs and Bambrick, 2005). In Chapter 4 we saw that the best predictor of abundance of dust mites at the regional scale was the estimated daily rate of evaporation, a site-specific function of temperature, rainfall, elevation and latitude. In most climate change models, predictions of changes in evaporation are much less certain than predictions of changes in temperature. Global average precipitation projections show increases with time as the hydrological cycle is enhanced by global warming, with higher latitudes having increased precipitation, the tropics having least

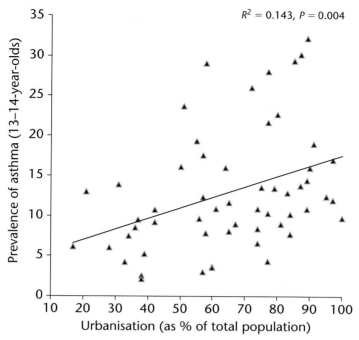

$R^2 = 0.143$, $P = 0.004$

Figure 8.22 The relationship between prevalence of asthma among 13–14-year-old adolescents and urbanisation in 56 countries. Asthma prevalence data from ISAAC Steering Committee (1998), urbanisation from country population data, World Bank Group (http://www.worldbank.org/data/countrydata/countrydata.html).

and the subtropics showing little or no change (Johns *et al.*, 2003). Some regions are predicted to get markedly dryer with reduced average summer rainfall by 2080, including most of Europe (Buonomo *et al.*, 2005), though certain areas (the central and eastern Mediterranean, central Spain, central and eastern Continental Europe and southern Scandinavia) are predicted, paradoxically, to receive a higher frequency of heavy rainfall events despite strong summer drying.

In conclusion, mites are important in asthma and allergies but they are only part of the story. Diet and certain foods, exposure to microorganisms and other allergens, urbanisation, changes in social and economic status and lifestyle all seem to have a role in relation to the prevalence of allergies and allergic diseases. All of this makes mites seem like a smaller part of a much bigger picture than at the time of the First International Workshop on mites and asthma in 1987. This is probably a good thing, for there is no reason to believe that interactions between the environment and human disease should be characterised by simple, linear phenomena governed by only a few factors. Mites need to be seen within the appropriate context of a bigger, more complex set of interacting determinants than we ever imagined 20 years ago.

9. Control of dust mites and allergen avoidance

I took a glass tube somewhat larger than a swan's quill, one end of which I stopped with cork, and after putting into the glass tube some hundreds of mites, I cut a small piece of nutmeg of a size that would fit into it; and I perceived that the mites next to the nutmeg soon died. I then put another piece at the other extremity of the tube where there were many live mites, which also died in a short time.

Antoni van Leeuwenhoek, 1695,
Missive 88, to Antonius Heinsius.
(see Ford, 1985, p. 83.)

9.1 Introduction

In Leeuwenhoek's account of probably the first documented experiment on acaricides, he describes how he discovered the toxic effect was due to a volatile substance evaporating from the nut (Figure 9.1). Incidentally, nutmeg and nutmeg oil have been used traditionally as an insecticide (Norman, 1990; Y. Huang *et al.*, 1997). Nutmeg is only one of several plant-associated acaricides that have been examined for their ability to control domestic mites. Some, like caffeine (Russell *et al.*, 1991), nicotine, phenyl salicylate and azadirachtin (derived from the neem tree, *Azadirachta indica*), have got no further than laboratory or field testing. Others, such as pyrethroids (synthetic analogues of compounds from the pyrethrum daisy) and benzyl benzoate (from Peru balsam) have been deployed in clinical trials. Tannic acid has been used as an allergen denaturant, and tea-tree oil and eucalyptus oil have been used as laundry additives (McDonald and Tovey, 1993; Tovey *et al.*, 2001).

Initial studies on reduction in allergen exposure focused on the removal of patients to low-allergen environments, either in hospitals (Storm van Leeuwen *et al.*, 1927; Platts-Mills *et al.*, 1982), or at high altitude locations (Storm van Leeuwen *et al.*, 1924; Storm van Leeuwen, 1927; Vervloet *et al.*, 1979, 1982) where there were few mites (Voorhorst *et al.*, 1969). More recent high-altitude trials were reviewed by Tovey (1997). Generally patients showed marked improvement of the clinical symptoms of asthma. For example, Platts-Mills *et al.* (1982) showed dramatic reduction in bronchial hyperreactivity to histamine (PC_{20}), indicating the underlying importance of mite allergens in airway constriction and spasm. Bronchial hyperreactivity is a defining feature of allergic asthma. It follows that reduction in hyperreactivity is a key outcome of interventions designed to reduce allergen exposure.

Once it appeared that low-allergen environments were of clinical benefit, a new wave of trials focused on reducing patients' exposure within their homes. Improvement in symptoms of asthma due to domestic

Figure 9.1   Leeuwenhoek's (1695) test set-up to examine the acaricidal effects of nutmeg on mites, consisting of a glass tube ca. 13 inches long, closed at one end (E), containing about 150 000 mites and some pieces of nutmeg (A–B, C–D). (From Hoole, 1798; Plate 10, Figure 1, p. 289 *et seq.*)

cleaning and removal of mites was first reported anecdotally by Dekker (1928), but the first trials began in the 1970s, mostly short term, with children, and using mattress covers, cleaning and washing of bedding as the main interventions (Sarsfield *et al.*, 1974; Burr, *et al.*, 1976, 1980a; Warner, 1978). These trials showed marginal benefits or were unsuccessful. Only that by Burr *et al.* (1980a) included the monitoring of mite populations and it found no reduction in symptoms. In the 1970s, most people would have had no knowledge of the role of mite allergens in asthma and access to little information, and the news that their homes were infested with mites which were making their children sick may have reduced their compliance with the interventions. Indeed, Sarsfield *et al.* (1974) observed that 'great tact' was needed to put across the concept of mite allergy: 'the mother must not be allowed to think that dust is associated with uncleanliness or that she has failed in her role as housewife and mother.' Contrast this with the contemporary perspective, where mite infestation carries no stigma, and most patients view allergen avoidance as a logical procedure for a disease caused by allergen exposure.

By the 1990s, allergen avoidance was being considered potentially useful to prevent the development of asthma in infants at high risk of becoming sensitised (Arshad *et al.*, 1992; Hide, 1996). Reduction of allergen exposure prior to development of allergies is referred to as *primary avoidance*, whereas *secondary avoidance* refers to reduction in exposure after allergic disease has developed. *Tertiary interventions* refer to the removal of patients to low allergen environments.

The results of clinical trials have been mixed, but the notion that allergen avoidance could be used for preventing and managing asthma prompted a burgeoning industry in anti-mite products from the early 1990s. Products were sold direct to the public and used without medical supervision or a diagnosis of mite-induced allergy. Some products have been subject to field and clinical trials, but for many their efficacy remains undemonstrated. Several clinical trials used a particular product as the sole or primary treatment (so called 'mono-faceted' trials), often with disappointing results. Others ('multifaceted' trials) used combinations of control methods deployed more rigorously, and with a tendency for greater success.

Methods for dust mite control and allergen removal and isolation have been reviewed exhaustively (de Saint Georges-Gridelet *et al.*, 1988; Thompson and Stewart, 1989; Colloff, 1989b, 1990; Colloff *et al.*, 1992b; Tovey, 1992; van Asperen, 1993; Tovey, 1997; Tovey and Marks, 1999; Platts-Mills *et al.*, 2000). In the first part of this chapter I detail the various methods that are available for killing dust mites and removing allergens, including the mode of action and investigations on acaricidal or allergen-reducing efficacy. I focus only on those methods that have been tested in field and clinical trials, are generally accessible and reasonably widely used. This excludes the many products and procedures that enjoyed a brief flurry of interest but are either no longer commercially available or were the subject of only one or two studies. I avoid the use of trade names where possible, but where they are used for the sake of clarity, no endorsement of the product should be implied. In the second part of the chapter I examine clinical trials and the clinical efficacy of killing mites and avoiding their allergens. Most clinical trials have been done on groups of patients with allergic asthma, and far fewer with patients with rhinitis or eczema.

Killing dust mites is not the same as allergen avoidance, although they are often assumed to be synonymous. One could kill every mite in a home, but unless the allergens are removed, the allergenic load and exposure risk remains the same. If one removes allergens without killing mites, the allergen load may decrease temporarily, but will then build up again as mites grow, reproduce and defecate. Substantial reductions in mite population density and allergen concentrations are likely to be necessary if a lasting clinical benefit is to be achieved. This means a combination of acaricidal methods (be they chemical or physical) and thorough vacuuming and cleaning (see Table 9.1).

Table 9.1 Summary classification of methods for dust mite control and allergen avoidance (modified from Colloff, 1989b).

Killing of dust mites		
Direct chemical: acaricides	**Indirect chemical: fungicides**	**Physical**
Bioallethrin Benzyl benzoate Disodium octaborate Plant essential oils	Natamycin	**Passive** Central heating Electric blankets Airing Mechanical ventilation Air conditioning Sun-drying **Active** Steam-cleaning Freezing • Liquid nitrogen • Outdoors in cold climate Boil-washing Dry cleaning
Removal/inactivation of allergens		
Removal	**Denaturation**	**Immobilisation**
Dry vacuuming Wet vacuuming Damp dusting Interior redesign Air filtration Dry cleaning	Tannic acid Steam-cleaning Hot-washing Ionisation	Mattress covers
Prevention of mite colonisation of homes		
Design and construction of 'low allergen' housing		
Acaricidal treatment of domestic textiles during manufacture		

9.2 Methods for killing dust mites

9.2.1 Chemical acaricides

Almost all insecticides and acaricides work by mimicking or inhibiting endogenous molecules involved in metabolism, and kill by preventing a chemical reaction or a metabolic pathway from being completed. More detail on the mode of action of insecticides and acaricides is given by Coats (1982) and Kerkut and Gilbert (1985).

A variety of sprays, foams, powders and paints containing acaricides was developed in the 1980s for use on dust mites (Colloff, 1990; Schober *et al.*, 1992). Most were not widely used and are no longer available. Acaricides that have been tested in field or clinical trials include organochlorides (Mitchell *et al.*, 1985), pyrethroids (Geller-Bernstein *et al.*, 1995) and benzyl benzoate (van der Heide *et al.*, 1997a).

Acaricidal and anti-microbial compounds are also added to fabrics and textiles during the manufacture of mattresses, pillows, carpets and clothing (Allanach *et al.*, 1990; Dean, 1993). These compounds are the most widely used domestic acaricides, even though the consumer may not know, or be only faintly aware, they have purchased a textile product that has been so treated. These compounds include permethrin (a synthetic pyrethroid), thiabendazole (a benzimidazole), tributyltin oxide (an organotin), triclosan (a phenol) and silver nanoparticles. Brown (1996) found permethrin-treated carpets were ineffective at preventing colonisation by dust mites. As well as being incorporated into carpets during manufacture, permethrin has been used in impregnated mattress liners and bedding (Cameron, 1997; Cameron and Hill, 2002; Chen and Hsieh, 1996). There appear to be no journal publications of clinical, field or laboratory trials of anti-mite properties of domestic fabrics and textiles treated with either thiabendazole, tributyltin oxide, triclosan or silver particles.

Triorganotin compounds are used as biocides in marine anti-fouling paints, timber treatments, agricultural pesticides, anti-microbials in plastic flooring and wall coverings (Markarian, 2006). Tributyltin is a highly potent endocrine disruptor and immunotoxin.

In mammals it inhibits the proliferation of immature T-cells causing immunosuppression. Its use in anti-fouling paint is the subject of a worldwide ban due to the toxic effects on non-target marine organisms (the International Maritime Organization global ban came into effect on 17 September 2008). Tributyltin has been found in carpet dust in several surveys at concentrations ranging from ca. 0.001–15 mg kg^{-1} (reviewed by Fromme et al., 2005). The widespread presence of tributyltin in house dust indicates that it can become available for inhalation or ingestion. Children are considered to be especially at risk, though there are not enough data at present for a reliable toxicological risk assessment (Fromme et al., 2005).

Natamycin (pimaricin) is an antifungal antibiotic derived originally from the fungus Streptomyces natalensis (Raab, 1971). It is a polyene macrolide with a large lactone ring, similar to macrocyclic lactones like ivermectin. Natamycin is claimed to disrupt the apparent symbiosis between dust mites and fungi (de Saint-Georges Gridelet, 1981b; Lebrun and de Saint Georges-Gridelet, 1984; see also Chapter 4). A randomised, double-blind, controlled trial showed no effect on allergen concentrations or on symptoms of asthma (Reiser et al., 1990). Similar results were found in a controlled trial with patients with atopic dermatitis (Colloff et al., 1989). Another controlled trial showed improved asthma symptom scores and medication use. Leclerq-Foucart et al. (1985) and van de Maele (1983) found improved symptom scores in an uncontrolled trial, but neither study included measurement of mites or allergens. de Saint Georges-Gridelet (1988) found the effective dose was twice that recommended by the manufacturer, and Schober et al. (1992) and Koren (1993) found its fungistatic and acaricidal capacity poor compared with other acaricides. Since then, interest in natamycin as an acaricide has declined and it is no longer in widespread use.

Borate compounds are worth a mention. Boric acid is widely used as an insecticide, fungicide and antiseptic. Disodium octaborate tetrahydrate ($Na_2B_8O_{13}.4H_2O$) has been used effectively in trials on dust mites in carpets, delivered as an aqueous solution via a carpet-cleaning machine (Vyszenski-Moher and Arlian, 2003). Borate insecticides have a long history of use, have low mammalian toxicity and prolonged residual activity. They are used widely as anti-fungal timber treatments and for the control of termites, ants, cat fleas and cockroaches, often with sucrose baits. The insecticidal mode of action is not entirely known, but they are probably gut poisons (Cochran, 1995). In 44 homes where carpets were treated with disodium octaborate, Codina et al. (2003) found a mean reduction from 130 to 3 live mites per gram of dust in 6 months (98%), compared with 170 to 75 in the placebo group (59%).

a Benzyl benzoate and related compounds

The precise mode of action of benzyl benzoate on dust mites is not known. It is described as a gut poison by Kersten et al. (1988). Bischoff et al. (1992c) found it functioned like a contact acaricide at high doses, but at lower doses it proved lethal following ingestion. Kalpaklioğlu et al. (1996) indicated it had contact toxicity at low concentrations. Benzyl benzoate is a very good solvent of lipids. Part of its contact acaricidal function may be that it dissolves cuticular lipids, thus increasing water loss and inducing death by dehydration. On ingestion, benzyl bezoate would break down into benzoic acid which may be responsible for toxic effects on the digestive system.

Benzyl benzoate is the most extensively tested domestic acaricide. Laboratory studies show it to be highly effective (Bischoff et al., 1986b, 1992c; Kalpaklioğlu et al., 1996). However, results are disappointing from both field trials (Burr et al., 1988; Elixmann et al., 1991; Lau-Schadendorf et al., 1991; Ehnert et al., 1992; Kalra et al., 1993; Rebmann et al., 1996) and clinical trials (Bischoff et al., 1986a; Elixman et al., 1988; Kersten et al., 1988; Pauli et al., 1989; Brown and Merrett, 1991; Dietemann et al., 1993; Huss, R.W. et al., 1994; Sette et al., 1994; Jooma et al., 1995; van der Heide et al., 1997a; Kroidl et al., 1998). In the trials analysed herein (see 9.4.2 below; Tables 9.2 and 9.3), the mean of mean percentage reductions in allergen concentrations in the active group was only 30% in clinical trials and 36% in field trials compared with controls of 36% and 7% respectively.

The lack of acaricidal efficacy of benzyl benzoate in vivo has been attributed to the inability of the delivery system (foam or moist powder) to penetrate fabrics, or because the product needs to be applied in higher doses, more frequently and for longer than the manufacturer's instructions (Kalpaklioğlu et al., 1996). Hayden et al. (1992) found a markedly improved reduction in Der 1 in carpets after leaving benzyl benzoate powder in situ for 12 hours rather than 4 hours.

Instructions for use of benzyl benzoate include post-treatment vacuum cleaning to remove dead mites and allergens. In clinical trials this has often been left

Table 9.2 Clinical trials of allergen avoidance used in the re-evaluation herein (see text for details). Abbreviations: AF = air filtration; BB = benzyl benzoate; B-T = benzyl tannate; c/w/a = vacuum cleaning, washing, airing; HEPA c = HEPA vacuum cleaning; MC = mattress covers; TA = tannic acid; ND = no data; BHR = bronchial hyperreactivity; FEV$_1$ = forced expiratory volume in 1 second; PEF = peak expiratory flow.

Reference	n (in analyses)	Adults/children	Duration (months)	Active and control groups	Multifaceted?	Allergen/mite outcome	BHR, active	BHR, control	Other clinical outcomes improved – active group	Other clinical outcomes improved – control group
HDM counts										
Chen & Hsieh, 1996	73	C	12	Permethrin v. placebo v. none	✗	−	ND	ND	−	−
Dorward et al., 1988	18	A	2	Liquid nitrogen, c/w v. none c/w/a, new bedding, no carpet v.	✓	+	+	−	Hours of wheeze	−
Korsgaard, 1983a	46	A	3	none	✓	−	ND	ND	Symptoms	−
Walshaw & Evans, 1986	42	A	12	MC, c/w, no carpet v. none	✓	+	+	−	Drugs, FEV$_1$, PEF, symptoms	−
Der 1 ELISA										
van den Bemt et al., 2004	52	A	2	MC v. placebo MC, BB, c/w v. placebo MC & BB, c/w	✗	+	ND	ND	PEF	−
Carswell et al., 1996	49	C	6		✓	+	−	−	Drugs, FEV$_1$, symptoms	−
Charpin et al., 1990	42	A	3	Bioallethrin v. placebo	✗	+	ND	ND	PEF	−
Cloosterman et al., 1999	157	A	5	BB, c/w, MC v. placebo BB & covers	✓	+	−	−	FEV$_1$	FEV$_1$
Dharmage et al., 2006	30	A	6	MC v. placebo	✗	+	−	−	Quality of life	Quality of life
Dietemann et al., 1993	24	A	12	BB v. placebo	✗	−	ND	ND	FEV$_1$, symptoms	Symptoms
Ehnert et al., 1992	21	C	12	MC, TA (carpets) v. BB v. placebo	✓	+	+	−	ND	ND
Geller-Bernstein et al., 1995	27	C	6	Bioallethrin, c/w v. placebo, c/w	✗	−	ND	ND	−	−
Halken et al., 2003	47	C	12	MC v. placebo	✗	+	+	+	Drug use	−
van der Heide et al., 1997a	40	A	12	BB v. placebo v. MC	✗	+	+	−	−	−
van der Heide et al., 1997b	45	A	6	AF v. placebo AF, MC v. AF, MC	✓	+	+	−	IgE, eosinophils	ND
Htut et al., 2001	23	A	12	Steam v. steam, AF v. none	✓	+	+	−	ND	ND
Huss, R.W. et al., 1994	12	A	12	BB v. placebo	✗	−	−	−	−	−

continued

Table 9.2 Continued

Reference	n (in analysis)	Adults/children	Duration (months)	Active and control groups	Multifaceted?	Allergen/mite outcome	BHR, active	BHR, control	Other clinical outcomes improved – active group	Other clinical outcomes improved – control group
Jooma *et al.*, 1995	60	A	6	B-T v. covers v. none	✗	–	–	–	–	–
Lee, 2003	42	A	1	MC, w/a v. none	✗	–	–	–	Symptoms	–
Luczynska *et al.*, 2003	31	A	12	MC v. placebo	✗	–	ND	ND	–	–
Manjra *et al.*, 1994	59	C	3	BB, detergent v. detergent v. none	✗	–	–	–		
Marks *et al.*, 1994	35	A	6	B-T, MC v. placebo spray	✗	–	–	–	–	–
Morgan *et al.*, 2004	869	C	12	MC, HEPA c, AF, v. home visits	✓	+	ND	ND	Acute visits, symptoms	–
Reiser *et al.*, 1990	46	C	3	N, c v. placebo, c	✗	–	–	–	–	–
Rijssenbeek-N. *et al.*, 2002	30	A, C	12	MC, w v. placebo, w	✗	+	–	–	Quality of life	Quality of life
Thiam *et al.*, 1999	12	C	4	Covers v. HEPA v. none	✗	–	ND	ND	Symptoms	–
de Vries *et al.*, 2007	105	A	24	MC v. placebo	✗	+	ND	ND	–	–
Warner *et al.*, 2000	40	A, C	12	AF, HEPA c, v. AF v. HEPA c v. none	✓	+	–	–	–	–
Williams *et al.*, 2006	34	C	12	MC, c/w/a v. none	✓	+	ND	ND	Functional severity score	–
Woodcock *et al.*, 2003	628	A	12	MC v. placebo	✗	–	ND	ND	PEF	PEFR

to the patients and may be one reason for the relatively poor performance. Benzyl benzoate has been used in multifaceted trials (Jooma *et al.*, 1995; Carswell *et al.*, 1996; Chang *et al.*, 1996; Cloosterman *et al.*, 1999), though it is not possible to separate its effects from the other measures. It has also been used effectively as a laundry additive for washing bedding (McDonald and Tovey, 1993; Vanlaar *et al.*, 2000; see below).

Benzyl benzoate – tannic acid preparations

These products consist of tannic acid to denature the allergens, combined with benzyl alcohol or benzyl benzoate to kill the mites. Laboratory studies showed promising results (Green *et al.*, 1989; Tovey *et al.*, 1992; Hart *et al.*, 1992), but field trials achieved reductions in group 1 allergen in carpets of the active group of only 28% (Warner *et al.*, 1993a) and 31% (Lau

et al., 2002). In a clinical trial, Quek *et al.* (1994) found significant improvement in PC_{20} of active group patients over controls, but did not measure allergens. Kroidl *et al.* (1998) found no significant difference in clinical improvement between active and control groups and Manjra *et al.* (1994) found no clinical benefit and no change in allergen concentrations.

Aerosol sprays seem to be ineffective at delivering active ingredients into mattresses and carpets with a deep pile. This would account for the generally poor field performance of several spray-delivered acaricides and allergen denaturants that were effective under laboratory conditions.

Disinfectants containing benzyl benzoate

Paragerm AK is a mixture of plant essential oil-derived compounds (e.g. phenol salicylate, thymol, terpineol,

Table 9.3 Field trials of allergen avoidance used in the re-evaluation herein (see text for details).

Reference	n (in analysis)	Duration (months)	Active and control groups	Mattresses and/or carpets?	Allergen/mite outcome
HDM counts					
Colloff, 1986	10	2	Liquid N_2 v. none	M	+
Harving et al., 1994	30	24	New MV housing v. none	M, C	+
Massey et al., 1993	33	3	Paragerm, vacuum cleaning v. placebo, vacuum cleaning	M	+
Natuhara et al., 1991	22	1	Covers v. none	M	+
Penaud et al., 1977	30	1	Paragerm v. placebo	M	+
Shibasaki et al., 1996	7	2.5	Liquid N_2 v. none	M	+
Der 1 ELISA					
Arlian et al., 2001	71	17	Dehumidification v. none	C	+
Chew et al., 1996	27	4	Benzyl tannate, washing v. none	M, C	−
Codina et al., 2003	93	6	Disodium octaborate v. placebo v. none	C	+
Colloff et al., 1995	12	N/A	Steam-cleaning v. none	C	+
Custovic et al., 1995b	12	3	Dehumidification v. none	M, C	−
Fletcher et al., 1996	18	12	Mechanical ventilation v. none	M, C	−
Kalra et al., 1993	16	N/A	Liquid N_2 v. none	M, C	−
Lau et al., 2002	22	2	Benzyl tannate, vacuum cleaning v. placebo, vacuum cleaning	C	−
Lau-Schadendorf et al., 1991	22	2	Benzyl benzoate v. placebo	M, C	+
Medina et al., 1994	17	3	Dehumidification v. none	M	+
Mosbech et al., 1988	20	12	Electric blankets v. none	M	+
Niven et al., 1999	20	15	Mechanical ventilation v. none	M, C	−
Olaguibel et al., 1994	48	2	Vacuum cleaning, washing, airing v. none	M	−
Owen et al., 1990	16	3	Mattress covers v. none	M	+
Rebmann et al., 1996	12	18	Benzyl benzoate v. placebo	M	−
Sporik et al., 1998	85	16	Carpet cleaner v. placebo	C	−
Tovey et al., 1992	25	1	Benzyl tannate v. none	M, C	−
Vanlaar et al., 2000	28	2	Mattress covers, washing, benzyl benzoate v. none	M	+
Vichayond et al., 1999	31	6	Mattress covers v. none	M	+
Vojta et al., 2001	22	2	Covers, prof. w. v. covers, dom. w. (bed); steam-clean v. vacuuming (carpet)[1]	M, C	+
Warner et al., 1993a	16	0.5	Benzyl tannate, vacuum cleaning v. vacuum cleaning	C	+

[1]prof. w. = professional washing of bedding; dom. w. = standard domestic washing of bedding.

citrus oil, *Syringa* and *Nardus* oil), chlorophenol, liquid paraffin and benzyl benzoate. It performed well in laboratory trials (Schober *et al.*, 1992), and Penaud *et al.* (1977) halved the mite population in a field trial. Massey *et al.* (1993) achieved a reduction in group 1 allergen of 70%. The only clinical trial (Dutau and Rochiccioli, 1979) was uncontrolled and did not measure mites or allergens.

b Pyrethroids

Pyrethroids disrupt arthropod neurotransmission leading to paralysis and death. The passage of a nerve impulse along an axon is accompanied by a shift in the normal sodium gradient due to the rapid opening of sodium channels and the depolarisation of the axon. It appears that pyrethroids bind to sodium channels preventing them from closing properly, causing the nerve to be rendered continuously depolarised.

The synthetic pyrethroid bioallethrin, synergised with piperonyl butoxide, available as an anti-dust mite spray, has been tested successfully in the laboratory (Schober et al., 1992) and in a field trial by Tafforeau et al. (1988). Clinical trials have been disappointing. One was uncontrolled (Chivato et al., 1993), and the others found no improvement in clinical symptoms or lung function in the active groups compared with controls. Of these trials, mean reductions in mites or allergens in the active and control groups were: 59 and 50% (Geller-Bernstein et al., 1995), 45 and 29% (Chen and Hsieh, 1996) and 93 and 49% (Charpin et al., 1990). The semi-quantitative results of Bahir et al. (1997) show greater reduction in the controls than the active group.

The poor performance of bioallethrin in these trials, due in part to the considerable reductions in the control groups, suggests either that interventions by the control groups were having an effect or the trials were done at times of year when mites and allergens were undergoing seasonal decline. Of those trials for which intervention dates were reported (Charpin et al., 1990; Bahir et al., 1997; Chen and Hsieh, 1996), all commenced when mite populations would have been naturally decreasing.

9.2.2 Drying, heating and freezing

Dehydration and death of dust mites at temperatures above 40°C and at humidities lower than the critical equilibrium water activity for dust mites (Chapter 3), as well as the thermo-labile nature of group 1 and 2 allergens (Cain et al., 1998), suggest that various methods of heating and drying may be useful control measures.

a Reduction of indoor humidity

Korsgaard (1982, 1983a) found indoor absolute humidity above 7 g kg^{-1} of air was associated with increased mite population densities. This threshold value is equivalent to 46–36% RH at a room temperature range of 18–22°C. Field and clinical trials on lowering indoor humidity have mainly used mechanical ventilation or dehumidifiers. Oshima et al. (1972) deployed large bags of silica gel on floors but without

success. Korsgaard (1983c), in a controlled clinical trial, achieved limited reduction of indoor humidity by passive airing of homes. But there was a marked increase in mite numbers in the active and control groups by the end of the trial and no improvement in clinical symptoms of asthma.

Mechanical ventilation systems

Wickman et al. (1994a) found that Swedish houses with mechanical ventilation tended to have humidities below 7 g kg^{-1}, and lower allergen concentrations than homes with natural ventilation. Using a mechanical ventilation system in new Danish apartments, Harving et al. (1994) eradicated mites from 11 of 16 pre-infested mattresses within 12 months. By contrast, Fletcher et al. (1996) found that houses with mechanical ventilation in the north of England showed no reduction in indoor humidity (it remained well above 55% RH) compared with controls, or any reduction in mites or allergens. Mechanical ventilation is not designed specifically to remove water from the air, though it has some drying power and works better in dryer climates than moister ones (Colloff, 1994b). A modified system with a dehumidifier reduced allergen levels in both active and control groups, but with no difference between the groups (Niven et al., 1999). Humidity was only slightly lower in the active group (37% RH, 5 g kg^{-1}, at 18.6°C) than the control group (50% RH, 6.5 g kg^{-1}, at 18.1°C) during the trial. In the clinical trial by Warner et al. (2000) in the UK, there was no significant reduction in mites or allergens in the mechanical ventilation group versus the control. The ventilation group had mean absolute humidities (AH) of ≥7 g kg^{-1} for 6 months of the 9-month trial and control group AHs were only slightly higher than the mechanical ventilation group.

Dehumidifiers

These are usually portable, electrical devices that remove water from the air and store it in a tank from whence it can be discarded. Machines vary a lot in their capacity for dehumidification and there have been no clinical trials. Of the field trials, Yoshikawa et al. (1988) achieved large reductions in mite populations but with only one home in the active group and five in the control group. Medina et al. (1994) achieved a significant reduction in allergen levels (82%) compared with the control group (29%), and a fall to below 50–51% RH by the end of the trial. Custovic et al. (1995b) found a 75% reduction in allergen levels in the active group, but a 64% reduction in the controls. Indoor humidity did not differ between the two groups and was well above 55% RH. Hyndman *et al.* (2000) found some reduction in humidity for their active group, but not enough to achieve a difference in mites or allergens compared with the control. The field trial by Cabrera *et al.* (1995) does not include before-and-after allergen levels and could not be compared with the others.

The previously mentioned trials were done with portable dehumidifiers of fairly low drying capacity. Arlian *et al.* (2001) found carpets in homes in which humidity was kept below 51% RH with a high-efficiency portable dehumidifier (capable of extracting ca. 60 L water per day) for 17 months had significantly less allergen (baseline v. end: 17 to 4 μg g^{-1} Der 1) and live mites (401 to 8 per g of dust) than homes that were not, which showed seasonal peaks of 40–70 μg g^{-1} Der 1 and 500–1000 mites (see Figure 9.2). This study was done in Ohio, USA, which has a humid, temperate climate. Outdoor humidity was 60–80% RH. The 50–51% RH threshold used by Arlian *et al.* (2001) is a bit higher than the 7 g kg^{-1} threshold (30–45% RH at room temperature) of Korsgaard (1982), but it seems to have been an appropriate dividing point for the dataset. Cunningham (1998) found that relative humidity in the immediate environment of dust mites, at the base of a 5 mm carpet pile, was considerably moister than in the air of the room. Thus the magnitude of the reduction in humidity of the air of the home required to kill dust mites is only a secondary variable. What is important is that conditions within the microenvironment of the mite are sufficiently desiccating (Ucci *et al.*, 2007). This may require extra energy expenditure to achieve, at extra cost. Furthermore, it may not be achievable at localities with high outdoor humidities.

b Heating

Attempts at killing mites by heating can be divided into those that aim to heat the home by a few degrees sufficient to cause dehydration, and those that heat fabrics well beyond the upper thermal death point of the mites of ca. 50–55°C (Spieksma, 1967; Kinnaird, 1974; Shibasaki and Takita, 1994). Of the former, Sidenius *et al.* (2002b) found no effect of a 3°C increase in indoor temperature, partly due to seasonal effects, but also because the target temperatures were not achieved. de Boer (2003) found homes with sub-floor heating tended to have fewer mites than homes

without, but not significantly so. de Boer and van der Geest (1990) achieved reductions in mite populations in mattresses with electric blankets of 19–84%, but Shibasaki and Takita (1994), using an electrical heating carpet, found no difference in mite populations between active and control groups.

Steam

As Glass and Needham (2004) point out, there is a difference of ca. 50–60°C between steam-cleaning using super-heated water and commercial carpet 'steam-cleaning', which is basically just a hot water wash. There are no trials of the latter approach – those detailed here used super-heated steam and have shown promising results. A commercially available domestic steam cleaner was used to treat carpet squares which had been seeded with known numbers of dust mites *in vitro* (Colloff *et al.*, 1995). It was able to inject steam at ca. 110°C into the carpet pile, and also appeared to have an allergen-denaturing effect. After regular monitoring for four months, there were no live mites detectable in the treated carpet, whereas mean population density was 740 per m^2 in untreated control carpet. Htut *et al.* (2001) and Vojta *et al.* (2001) also found steam was effective in reducing allergen levels.

Sunlight and ultraviolet exposure

Tovey and Woolcock (1994) recorded a decrease in relative humidity from 76% to 30% in carpets that were exposed to direct sunlight for 6 hours on a summer's day in Sydney, Australia. Temperature increased from a baseline of 25°C around 9 am, peaking at 55°C after 5 hours. Over the first 2.5 hours a massive migration of mites occurred (shown in Figure 9.3). For the final 3 hours no live mites were found and there was a 10-fold increase in dead mites collected, mostly dehydrated in appearance. It appears dehydration of mites in textiles by exposure to intense sunlight is a highly effective method of control. It is also simple, free and safe.

Needham *et al.* (2006) found eggs of *Dermatophagoides farinae* failed to hatch after exposure to 5–15 sec. exposure to ultraviolet light (wavelength 253.7 nm). UV vacuum cleaners and lights for mite control have become available commercially, but as yet there are no published clinical studies of their efficacy.

Hot washing and tumble-drying

McDonald and Tovey (1992) determined by hot washing that the LT$_{50}$ (lethal temperature at which 50% of the test population died) was 49°C. All mites

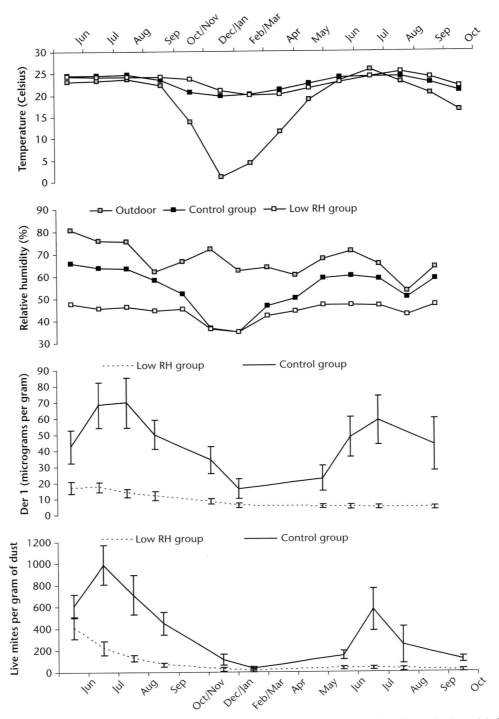

Figure 9.2 Temperature and humidity, allergen concentrations and mite population densities (± standard error) in homes with an indoor humidity <51% by high-efficiency dehumidifiers, compared with control homes. (Re-drawn from the data of Arlian *et al.*, 2001.)

were killed by water temperatures of ≥55°C. Andersen and Rosen (1989) found complete extermination of mites at a wash temperature of 58°C. Autoclaving was found to be highly effective (de Boer, 1990b). Mason *et al.* (1999) achieved a mean decrease in live mites of 99% in duvets in a domestic tumble dryer for one hour at ca. 59°C and <10% RH. The clinical trial by Lee (2003) involved the use of cotton mattress covers which were washed by boiling, followed by 'disinfesting by sunlight ... the

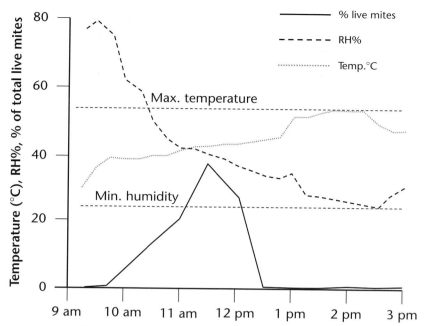

Figure 9.3 Temperatures, humidities and numbers of mites collected during exposure of carpets to direct sunlight. (Re-drawn from Tovey and Woolcock, 1994.)

traditional Korean approach to cleaning bedding'. This achieved a reduction of only 28% of Der 1 in the active group, and no significant clinical improvement, versus 11% in the control group.

c Freezing

In a two-month field trial, liquid nitrogen, combined with intensive vacuum cleaning, reduced live mite numbers by 99%, compared with a reduction of 59% in the control group (Colloff, 1986; Figure 9.4). Kalra *et al.* (1993) found no significant effect of liquid nitrogen on Der p 1 in 10 homes, but did not measure live mites nor undertake rigorous cleaning to remove dead mites post-treatment. The mean reduction in mattress Der p 1 in the active group was 24%, versus an increase of 70% in the control group. Shibasaki *et al.* (1996) used three liquid nitrogen treatments of carpets to achieve a lasting reduction of <6% of the baseline mite population. Following the first two treatments they found mites recolonised within a month. The boiling point of liquid nitrogen is −196°C and as it boils off through the treated fabric, liquid nitrogen has the effect of loosening dust, making it easier to remove by vacuuming. The treatment should only be done by an operator who is fully trained in cryogenic safety handling and the use of liquefied gases. Large reductions in mite densities (95% in the active group versus 39% in the control group), and improved bronchial hyperreactivity, were achieved in a

controlled clinical trial with adult asthmatics (Dorward *et al.*, 1988).

Van Bronswijk and Koekkoek (1972) reported that it took 7–14 days to kill dust mites at −9°C, 1–7 days at −15°C and death was instantaneous at −28°C. Paul and Sinha (1972) found only 15% of *D. farinae* individuals survived 2°C for more than seven days. Dodin and Rak (1993) found the lower thermal death point of all life-cycle stages of *D. pteronyssinus* was −30°C for 5 min. in dry air. Eggs were more resistant than larval and post-larval stages. Tsundona *et al.* (1992) determined the supercooling point of *D. pteronyssinus* as −23°C, below which mites died from formation of ice crystals within their bodies. A temperature of −20°C for 30 min. achieved almost 100% mortality, indicating that a standard domestic freezer could be used for killing mites in relatively small items such as soft toys, pillows and items of clothing that cannot be hot washed.

9.2.3 Domestic redesign

This approach is about reducing available habitat for dust mites and making the remaining habitat more hostile. Carpets can be replaced with vinyl, tiles or bare wooden floors with washable rugs (see Figure 9.5). In some clinical trials this kind of approach, together with other measures, has been associated with clinical improvement of asthma (Walshaw and Evans, 1986; Murray and Ferguson, 1983).

Figure 9.4 Experimental application of liquid nitrogen to a mattress for the control of dust mites.

The selective removal of carpet makes the indoor environment more spatially heterogeneous because it removes not only a textile in which the mites live and lay their eggs, and in which their food is trapped and accumulated, but one which provides insulation and higher humidity than the ambient air of the home (as modelled by Leupen and Varekamp, 1966). Many homes would have wool-polyester carpets in all rooms except kitchens, bathrooms and laundries, typically covering at least 50 m^2 of the floor area. Removing the carpet would cut the available habitat for dust mites to

mattresses and upholstered furniture; typically 5–10 m^2, or 10–20% of the total surface area of carpets, textiles and upholstered furniture.

Reduction and fragmentation of the area of available habitat has a tendency to reduce the survivability of populations of many organisms, though species sensitivity to fragmentation varies greatly (Krebs, 2001, Chapter 19). The rendering of the indoor environment hostile to dust mites through habitat manipulation and fragmentation via the reduction in surface area of carpets and other domestic textiles, combined

a)

b)

Figure 9.5 **a)** A room with bare wooden floors, no curtains or upholstered furniture versus **b)** a room with abundant textile-covered surfaces that provide habitats for dust mites.

with lowering of indoor humidity, offers the starting point for an ecologically sound means of control. If carpets are removed in favour of hard floor surfaces, then these floors still have to be cleaned. Tovey (*pers. comm.*, 2008) found no difference in allergen levels in settled dust above hard floors versus carpets (see Figure 9.6), indicating that allergen is dispersed from elsewhere in the home and that it settles on both hard floors and carpets alike.

de Blay *et al.* (2002) reviewed the role of architecture in the reduction of allergen exposure. They found considerable interest among a large group of architects in working with medical doctors on improved housing design for improvement of indoor air quality. Flechtmann *et al.* (1998) produced a manual on the design and modification of houses for people with allergies.

9.3 Methods for removing or isolating allergens

From the 1920s, when it was recognised that dust was involved in asthma, avoidance of dust has been advocated (Dekker, 1928). But there was no efficient method for dust removal without generating lots of disturbance and airborne allergens until the widespread availability of the domestic vacuum cleaner. A simple but seemingly unappealing approach was to use an oil-based dust retardant on bedding (Wright, 1963); also used by Bowler *et al.* (1985) in a clinical trial, but with no demonstrated clinical benefit.

9.3.1 Vacuum cleaning

Live mites are more difficult to remove by vacuum cleaning than dead ones because they cling on to fabric fibres. The ends of their legs (the pretarsi) are like miniature sink-plungers (see illustrations in the identification keys in Chapter 1). Intensive vacuum cleaning can remove significant amounts of dust from carpets, thereby diminishing the allergen reservoir (Massey and Massey, 1984; Wassenaar, 1988a, b; Kato *et al.*, 1991; Munir *et al.*, 1993; Wickman *et al.*, 1997). Vacuum cleaning is a removal technique only, and can be regarded only as an adjunct to methods that kill mites. Vacuum cleaners in their various shapes and forms are one of the only practical means people have of removing dust from their homes. Prior to the invention of the electric vacuum cleaner in the early 20th century, rugs and carpets had to be taken outside and beaten in order to clean them effectively. Naturally, this is an impractical procedure with fitted carpets. The widespread adoption of fitted carpeting from the 1970s onwards was dependent in part on the existence of the vacuum cleaner as the only practical cleaning method (see Platts-Mills *et al.*, 2000).

Carpet sweeping and beating machines had been in existence since the 1850s and some even had primitive suction devices (see Figure 9.7). The first machines that employed a vacuum to remove dust from fabric were invented around 1902 by Herbert Cecil Booth, an English civil engineer. One of these vacuum cleaners

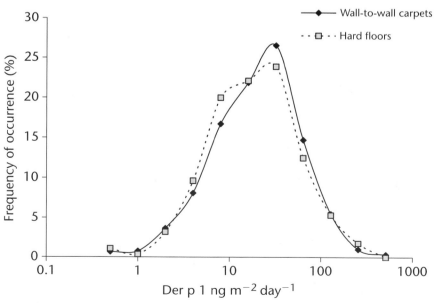

Figure 9.6 Aeroallergen settling in dishes over 50 days in rooms with different floor coverings. (Figure courtesy of Euan Tovey.)

was a very large device which had to be brought to premises, much as commercial carpet cleaners are hired today. Known as 'The Noisy Serpent', it was used to clean the carpet in Westminster Abbey prior to the coronation of King Edward VII. The first electric domestic vacuum cleaner was invented by James Murray Spangler in 1907. An asthmatic, working as a janitor in an Ohio department store, Spangler deduced his manual carpet sweeper was provoking his symptoms. To remedy this, he designed and built a prototype vacuum cleaner and formed the Electric Suction Sweeper Company. Later he sold the patent to his cousin, William Hoover, who took over his business.

The modern vacuum cleaner differs relatively little from the first Hoover models in basic design elements, consisting of a set of brushes powered by an electric motor linked to a fan which creates the vacuum. Dust is sucked up through the fan and deposited into a porous bag. Without the porosity of the bag, the vacuum cleaner will soon overheat. However, this porosity is also a potential route for allergen leakage and recirculation (see below). Different types of exhaust filters have been developed to counteract this problem (Vaughan et al., 1999; Vicentini et al., 2002). A survey by Which magazine (UK; 19 December, 2007) showed relatively little

Manual carpet sweeper, 1890s

Manual bellows vacuum cleaner, 1901

The first electric vacuum cleaner, 1907

Early electric vacuum cleaner, 1920s

1960s cylinder vacuum cleaner

Modern vacuum cleaners, 2007

Figure 9.7 The evolution of the vacuum cleaner, from manual sweepers with roller brushes in the late 19th century to high efficiency modern cleaners with filtration systems.

difference in performance between the basic modern types of machine.

a Dry vacuum cleaning

Many of the early clinical trials relied heavily on vacuum cleaning, mainly because there were few other options other than mattress covers. Sarsfield *et al.* (1974), in an uncontrolled trial on 14 patients, employing mattress covers, vacuuming and other anti-dust measures, found an improvement in symptoms and a reduction in mite densities. However, Burr *et al.* (1980a), who used less stringent measures (no mattress covers) in a controlled trial, found no clinical improvement. Similarly, negative observations were made in a trial by Gillies *et al.* (1987) on 26 children, despite a significant reduction in mite densities. Kato *et al.* (1991), in an uncontrolled clinical trial, found intensive vacuum cleaning reduced mite allergen levels in bedrooms of asthmatic children leading to improved asthma symptom scores. Popplewell *et al.* (2000) found no difference between conventional vacuum cleaners and those with filtration systems in their capacity to reduce levels of Der p 1, but Fel d 1 was significantly lower following filtration vacuum cleaning, leading to clinical improvements in asthma.

Vacuum cleaning can increase the amount of airborne allergens (Sly *et al.*, 1985; Swanson *et al.*, 1989). 'Medical' vacuum cleaners contain filters which prevent escape of particles of >1 µm diameter. Kalra *et al.* (1990) determined that medical vacuum (i.e. with a filter) cleaners vented less respirable, aerosolised allergen than conventional vacuum cleaners, as did Hegarty *et al.* (1995). Vaughan *et al.* (1999) found those vacuum cleaners recommended for allergic subjects leaked lower amounts of cat allergen. Vacuum cleaners with single-thickness bags were leakiest compared with 2- and 3-layer bags, and best performance was from machines with a pre-filter and a HEPA exhaust filter. Vicentini *et al.* (2002) found that vacuum cleaners with either polyethylene or HEPA exhaust filters vented about 30 times less Der p 1 than a vacuum cleaner with no exhaust filter and a double-thickness paper bag.

b Wet vacuum cleaning

Der p 1 is water soluble, as are some of the other allergens of dust mites, and techniques involving wet vacuum cleaning might be expected to increase their removal from carpets. Wassenaar (1988a)

showed that wet vacuum cleaning of a carpet was associated with subsequent increase in population densities of *Dermatophagoides pteronyssinus*, probably due to elevated humidity in the carpet caused by the procedure. Korsgaad and Iversen (1991) found wet vacuuming with carpet shampoo was no more effective than intensive dry vacuum cleaning at reducing live mite populations. Various cleaning products containing anti-mite compounds are available for use with wet vacuuming (Fell *et al.*, 1992). Thompson *et al.* (1991) found one such product reduced both dust and Der p 1 in carpets by 70% and in mattresses by 63% and 84% respectively, an effect which lasted for several weeks. de Boer (1990a) and de Boer *et al.* (1996) found wet vacuuming was inefficient at killing mites, but did achieve significant reductions in allergen levels.

9.3.2 Washing

Washing can remove allergens from fabrics and bedding but only a hot wash (>60°C) will kill mites (Andersen and Rosen, 1989; McDonald and Tovey, 1992). The use of laundry additives such as benzyl benzoate, or essential oils of tea-tree and eucalyptus, have been shown to have a major acaricidal effect at wash temperatures below 55°C (McDonald and Tovey, 1993; Vanlaar *et al.*, 2000) and the use of such additives may overcome the need to use hot washing for mite control in bedding.

There is no information available on the anti-allergen effects of the traditional method of cleaning rugs by beating them outdoors followed by washing and drying in the sun (see Figure 9.8), probably because it is quite labour-intensive. Yet this is the most common carpet cleaning method used worldwide, and is the standard for cleaning hand-knotted oriental rugs (Amini, 1981). Traditional rug cleaning methods may combine beating, washing, airing in the sun or, in cold climates, leaving them out in the snow during winter. Platts-Mills *et al.* (2000) illustrate a municipal platform in Helsinki, Finland, for local residents to wash their rugs in sea water, and washed rugs draped over balconies to dry is still a common sight in rural parts of Central Europe. Families in Palestinian villages expose their mattresses to sunlight almost every day, and this was considered to be an effective control measure for mites (El Sharif *et al.*, 2004).

Figure 9.8 Carpet washing and drying in the sun. (Images courtesy of Euan Tovey.)

9.3.3 Allergen denaturation

a Chemical denaturation

Tannic acid has been used for the chemical denaturation of allergens, either as an aqueous solution or mixed with the acaricide benzyl benzoate (see 9.2.1 above). But tannic acid inhibits Der 1 allergen assays at concentrations of ca. 0.1% w/v, possibly by interfering with the binding of the allergen to the coating monoclonal antibody (Woodfolk *et al.*, 1994, 1995), unless the sample is extracted in 1–5% bovine serum albumin (BSA). Assay inhibition appears to have resulted in an overestimation of the efficacy of tannic acid in denaturing allergens, which has been used as an intervention in five clinical trials (Shapiro *et al.*, 1995; Tan *et al.*, 1996; Ehnert *et al.*, 1992; Jooma *et al.*, 1995; Marks *et al.*, 1994) and six field trials (Green *et al.*, 1989; Tovey *et al.*, 1992; Warner *et al.*, 1993a; Woodfolk *et al.*, 1994, 1995; Lau *et al.*, 2002). Presence of residual tannic acid from prior use of commercial products by patients might pose a small risk of skewing a trial, but apparently tannic acid can be detected in house dust by adding iron (III) chloride solution to an aqueous suspension of dust which turns blue-black in the presence of phenols (Woodfolk *et al.*, 1994). Of those trials in which mite allergens were measured, BSA was added to prevent assay inhibition by Tovey *et al.* (1992), Marks *et al.* (1994) and Woodfolk *et al.* (1994, 1995). For the rest, either there is no mention of the addition of BSA or it was added only at low concentration (<1%). The trials by Green *et al.* (1989) and Warner *et al.* (1993a) involved skin-prick testing with dust extracts from blankets treated with tannic acid, and may also have been subject to inhibition effects.

The uncertainty about inhibition effects does not necessarily invalidate the trials in which tannic acid was used as one of several treatments (Ehnert *et al.*, 1992), or where it was used on carpets but not mattresses (Jooma *et al.*, 1995), but it remains difficult to get a clear picture of the magnitude of allergen reduction for all studies. Of those that used adequate concentrations of BSA to prevent tannic acid inhibition, reduction in allergen levels in the active group tends to be in the range of 30–60%.

As an alternative to tannic acid, which can stain light-coloured carpets, Sevki *et al.* (2006) found a solution of alum (hydrated aluminium potassium sulphate, $KAl[SO_4]_2.12H_2O$), applied at 3–9 g in 60 ml per square metre of carpet, reduced Der p 1 concentrations by 49–95%. Alum is a protein binder and precipitant, used in the dressing of leather. The authors consider Der p 1 was precipitated and absorbed by alum, but did not investigate any possible interference with the Der p 1 ELISA.

b Physical denaturation

Dry heat was found to denature both group 1 and 2 allergens in an unreplicated laboratory study (Cain *et al.*, 1998). Heating at 100°C for 15 min. denatured 97% of a sample of Der f 1, whereas the same period at 120°C was required to denature 94% of the Der p 1. The group 2 allergens were less heat labile: 120°C for 15 min. to denature 86% of Der f 2 and 140°C for 30 min. to denature 92% of Der p 2.

After steam treatment of standardised areas of carpet in a home there was a mean reduction of 87% in Der p 1 concentration in treated areas compared with only 5% in adjacent, untreated control areas (Colloff *et al.*, 1995). The steam was applied at an average temperature of 105°C for 2 min. m^{-2}.

The corona discharge from ionisers has been shown to denature allergens (Goodman and Hughes, 2002, 2004), and ionisers have been used in clinical trials but without much success (Nogrady and Furnass, 1983; Warner et al., 1993b; Johnsen et al., 1997). A product of the ionisation process is ozone, which can exacerbate negative bronchial responses to allergens (Holz et al., 2002).

9.3.4 Barrier covers for mattresses and bedding

Barrier methods encompass a variety of covers for mattresses, box springs, pillows and duvets (referred here collectively as 'mattress covers'). Desirable features of mattress covers have been reviewed by Pearson (1995). Initially these covers were made of plastic or rubber, but more recently, microporous covers have been developed which allow the passage of water vapour while excluding mites and their allergens. The capacity of mites to penetrate different fabrics has been evaluated in detail by Mahakittikun et al. (2003, 2006). Mattress covers should be used in combination with other anti-mite measures. Barrier methods and cleaning methods are not mutually exclusive. Covers may be the most effective method of isolating people from the mites and allergens in their bedding, but unless old mattresses and pillows (that are likely to be heavily infested) are treated, mites will continue to survive beneath the cover.

Studies in which mites were vacuumed from mattress covers (Walshaw and Evans, 1986; Sarsfield et al., 1974) showed a reduction in the order of 30–100-fold compared with those in dust in the mattresses themselves. The total amount of Der p 1 recovered from mattresses coated with polyurethane was about 1% of that from control mattresses (Owen et al., 1990). The majority of studies suggest clinical improvement following use of mattress covers, usually combined with cleaning measures, but only three included a separate control group (Walshaw and Evans, 1986; Murray and Ferguson, 1983; Howarth et al., 1992). Clinical benefit was demonstrated in all three.

The clinical trial by Ehnert et al. (1992) showed significant reduction in bronchial hyper-responsiveness (BHR) to histamine in the mattress cover group, but not in the control group or the benzyl benzoate intervention. The mean reduction in mite allergens over a 12-month period was from 2 to 0.3 µg g^{-1} for the mattress cover group (nearly 99%), whereas the control group experienced an increase from 1.3 to 2.4 µg g^{-1} (up 85%). Schmidt and Gøtsche (2005) were highly critical of this heavily cited study, alleging the increase in PC$_{20}$ in the active group was not statistically significant, the trial had too few patients and was un-blinded. They opined that it was 'disturbing' that seven children in a non-blinded trial could have been so influential for recommendations on allergen avoidance interventions. The data of Ehnert et al. (1992) is more convincing when presented as in Figure 9.9 which shows the strong and statistically significant correlation between improvement of BHR and reduction in group 1 allergen in the active group versus the control.

In a meta-analysis of 33 clinical trials on the efficacy of mattress covers for treatment of asthma, Recer (2004) found that of the 19 trials that reported adequate allergen exposure and BHR data, four showed significant reduction in allergens and BHR compared with the control group, 10 had a decrease in allergens but not BHR and five reported no significant effect on either allergens or BHR. Collectively, the effects of mattress covers showed a modest but non-significant benefit. Mattress covers were the only intervention used in nine trials, of which one showed reduction in allergens and BHR, and were combined with other allergen avoidance measures in the remaining 10 trials, of which three showed reductions in allergens and BHR.

From an experimental design viewpoint, sampling the mattress during the baseline period and the cover surface during the endpoint phase are not really comparable because the samples are taken from different places. Yet the majority of successful trials have involved mattress covers and have used this measurement. From a patient exposure viewpoint, sampling baseline mattress and endpoint cover surface might seem valid, because the aim of the trial is to measure the effect of reduction in allergen exposure on clinical outcomes. However, what is being measured is neither a direct estimate of reduction in allergen exposure nor a change in concentration of mite allergen in the bedroom. This problem is typified by results from trials where the control group has no covers (compared with allocating it a placebo cover), and therefore mattress surface dust is sampled on each occasion, whereas for the intervention group the mattress surface is only sampled at baseline. Owen et al. (1990) found Der p 1 for the control group at the end of the trial

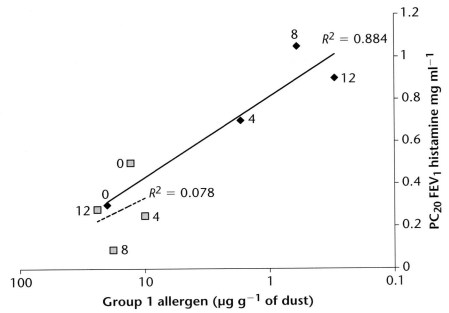

Figure 9.9 Relationship between mean bronchial hyper-responsiveness and allergen concentrations in active group (diamonds) and control group (squares) during a 12-month intervention with mattress covers. Numbers next to data points indicate the month of the trial. (Data from Ehnert *et al.*, 1992.)

was seven times higher than baseline whereas in the intervention group it was 14 times less, suggesting an impressive (though misleading) adjusted decrease of 20-fold.

9.3.5 Air filtration devices

None of the studies on the effects of air filters on symptoms of allergic airways disease have involved pre- and post-treatment monitoring of allergens. Scherr and Peck (1977), in what is essentially a tertiary trial, found a strong trend for high efficiency particulate air (HEPA) filtration units to reduce the incidence and severity of nocturnal asthma attacks in a two-year study conducted at a summer camp for asthmatic children. Bowler *et al.* (1985) in a single-blind crossover trial of 12 adult patients with asthma and rhinitis used electrostatic or mechanical HEPA filters alone and in combination with conventional cleaning methods and oiling of bedding, but found no improvement in symptoms. However, the treatment regimes ran for periods of only 2 weeks. The use of HEPA filters fitted to bed headboards was shown to be associated with clinical improvement in a 4-week crossover study with house dust-sensitive children with asthma (Zwemer and Karibo, 1973). Villaveces *et al.* (1977) and Verrall *et al.* (1988), using similar methods, also reported significant improvement. Mitchell and Elliott (1980) used electrostatic

precipitators in homes of 10 mite-sensitive children, but reported no clinical improvement. In a placebo-controlled crossover study, slight clinical improvement in asthmatic/rhinitic symptoms occurred after 4 weeks' use of HEPA filters in the patients' bedrooms (Reisman *et al.*, 1990). Antonicelli *et al.* (1991) also tested efficacy of HEPA filters.

All studies that found clinical benefits involved the use of HEPA filters. In two studies, HEPA filters were placed in quite close proximity to the faces of sleeping children. A committee consulted about the use of air cleaning devices concluded that although HEPA filters were more effective than electrostatic ones, neither could be recommended in the absence of other forms of environmental control (Nelson *et al.*, 1988). A systematic review of 10 randomised controlled trials of air filtration and asthma found lower total symptom scores and sleep disturbance in the intervention groups, but not medication use or morning peak expiratory flow values (McDonald *et al.*, 2002).

9.4 Clinical trials in homes of patients with allergic asthma

9.4.1 Meta-analyses

In the introduction I mentioned that the results of clinical trials of allergen avoidance to improve symptoms of asthma have been mixed. Many failed because the methods did not reduce allergen exposure sufficiently to achieve clinical benefit. Others were inappropriately designed in various ways (Gøtzsche and Johansen, 2008). Very few trial designs have used comprehensive, multiple methods that are known to consistently reduce allergen exposure and kill mites. Rather, researchers have tended to test the efficacy of a method (or a combination of methods) at reducing allergen exposure, while at the same time measuring any clinical effect. If the control method does not adequately reduce allergen exposure, then the patients are unlikely to improve. From an experimental design viewpoint, this approach leaves the allergen reduction component *and* clinical symptoms as the two major groups of response variables of the intervention, with the clinical variables dependent upon the allergen reduction variables. This design is inherently weaker and more uncertain than the low-allergen environment trial design, where reduction in allergen exposure is certain and of considerable magnitude (Platts-Mills *et al.*, 2000), and the *only* group of response variables is the clinical symptoms.

In a meta-analysis of randomised controlled clinical trials aimed at assessing the effects of reducing exposure to mite allergens on symptoms of mite-allergic asthmatic patients, Gøtzsche and Johansen (2008) concluded that mite control measures cannot be recommended because there were no differences in numbers of patients who improved, their symptom scores, morning peak flow or medication use between the intervention group and the non-intervention group. (Earlier versions of the meta-analysis are by Gøtzsche *et al.*, 1998, 2004.) Other clinical outcome measures included bronchial responsiveness, days sick and number of visits to physicians/hospitals. Of the 42 trials for which data were included in the meta-analysis, significant reduction in amounts of allergen or numbers of mites compared with the control group occurred in 13 trials, was unsuccessful in 20, and was not measured or reported in eight. In other words, two-thirds of the trials did not demonstrate any reduction in mite exposure. It is hardly surprising therefore that there was no clinical benefit to patients, since it is unlikely their symptoms will improve if their exposure to allergens has not fallen. The meta-analysis definitely does *not* show that reducing mite exposure does not improve asthma and there is ample evidence that reducing mite exposure can and does improve asthma, particularly from trials on removing patients to low allergen

environments (reviewed by Tovey, 1997). What the meta-analysis indicates is that improvement in asthma could not be demonstrated using the kinds of trial designs and implementation of mite and allergen reduction and monitoring methods that were evaluated, a point that the authors make. Publication of the 1998 meta-analysis generated some criticism (Strachan, 1998; O'Connor, 2005; correspondence, *British Medical Journal*, 318, 870–871), the main points of which are given below.

- There was a lack of acknowledgement by the authors that clinical improvement is contingent on reduction in allergen exposure, and that most of the trials did not demonstrate this.
- Morning peak expiratory flow rate, one of the clinical parameters selected for the meta-analysis, is too variable to be a good measure of asthma severity at the population level, and that changes in bronchial hyperreactivity are more informative.
- Interventions in the first trials in the late 1970s and early 1980s are known to be ineffective and these studies should not have been included.
- The meta-analysis included several studies in which the 'placebo' treatments or dust sampling protocols could have had a significant effect in reducing allergen levels (e.g. Chang *et al.*, 1996).

The 2004 update of the meta-analysis was skewed by one large (n = 628 patients) study with negative clinical outcomes (that by Woodcock *et al.*, 2003). Another large study with positive clinical outcomes (n = 869 patients; Morgan *et al.*, 2004) was excluded from the 2008 update because 'the study was not blinded and the positive results for these subjective outcomes were obtained through telephone interviews'. Other meta-analyses of allergen avoidance trials include those by Recer (2004; see above) on the efficacy of mattress covers for treatment of asthma; McDonald *et al.* (2002) on air filtration, Sheikh *et al.* (2007) for rhinitis, van Schayck *et al.* (2007) for primary intervention and Macdonald *et al.* (2007) for a mixture of primary and secondary interventions.

9.4.2 Clinical and field trials – a re-evaluation

a Introduction

One of the arguments in favour of trials of mite and allergen control is that they test methods that are likely to be practical, available to the patient and relatively easy to implement, and they do so in a rigorous manner. This is not necessarily borne out by the evidence. The purpose of this section is to have another look at published secondary clinical trials; not as a formal meta-analysis, but to examine patterns and trends relating to the magnitude of the reductions in allergen exposure that were associated with clinical improvements in asthma. In contrast to clinical trials, field trials test the efficacy of the intervention techniques at killing mites and removing allergens, but without any clinical measurements of patients. The same arguments for practicality, availability and implementation apply to field trials as to clinical trials, so field trials form a useful group to compare with clinical trials.

b Methods

For secondary clinical trials, I did a comprehensive literature search using the same search keywords as those of Gøtzsche and Johansen (2008), but with additional citation tracking, i.e. examination of all papers that cited each publication of interest. Abstracts or short letters were excluded because often there was not enough information to be able to accurately assess the details of the trial.

Publications were checked for multiple reporting of the same trial. Inclusion criteria were as follows:

- randomised, controlled trials of allergen avoidance intervention in homes of patients with allergic asthma;
- with parallel control and intervention groups;
- with quantitative measurement of allergens and/ or mites.

These criteria specifically exclude crossover trials, which Gøtzsche and Johansen (2008) included. The different timing of intervention and control periods represent a confounder for measurement of allergens and mites because of seasonal fluctuations in population size, which can occur in a timeframe of weeks. The criteria also exclude semi-quantitative measures of allergens (e.g. guanine) because quantitative measurements were required to calculate percentage reductions in allergens.

I examined the difference in allergen concentration or mite population density at baseline and the end of the trial in active (intervention) and control groups. Before-and-after differences were not corrected for the duration of the trials. Where there was more than one intervention group, I selected either the group with the largest difference from the control or the group with a single type of intervention if there was one. Inclusion of more than one active group would mean having to use the control group data twice, which would skew the control dataset.

For field trials, the selection criteria were:

- controlled trials within homes;
- parallel intervention and control groups;
- quantitative measurement of allergens and/or mites at least at baseline and end of trial;
- trials based on natural populations of mites and concentrations of allergens.

This latter criterion was designed to eliminate colonisation trials (e.g. rates of colonisation of new mattresses and carpets, and studies where beds or carpets were seeded with mite cultures), in order to have a field trial group that was comparable with the clinical trial group.

c Results and discussion

Clinical trials

I identified 30 secondary clinical trials that included measurement of changes in mite numbers (n = 4 trials) or allergen levels (n = 26; see Table 9.2) in mattresses. In 12 trials (40%) the active group showed significantly greater improvement than the control group in one or more measures of asthma (lung function, bronchial challenge tests, symptom and quality of life scores, medication use, hospital emergency visits, number of hours of wheezing, serum IgE or eosinophil counts; shown in Table 9.2). Mite numbers were the only measure of allergen exposure in two of the trials with positive clinical outcomes, and were reduced by 95 and 99% in the active groups and 40 and 47% in the control groups respectively. In the remaining 10 'clinically positive' trials, group 1 allergen levels were reduced from a grand mean of 9.5 to 0.8 μg g^{-1} in the active groups (92%; range 47–99%) compared with 5.8 to 5.5 μg g^{-1} (5%) in the control groups. In the 18 trials with negative clinical outcomes, allergen levels were reduced from 12.0 to 7.5 μg g^{-1} in the active group (38%) compared with 15.4 to 13.3 μg g^{-1} (14%) in the control groups (Figure 9.10). These data suggest that no clinical improvement in asthma occurs without a reduction in allergen levels of at least two-fold: 7 out of 10 'clinically positive' trials had reductions in allergen levels of >80%.

Of the 30 trials, 11 of them also included data on mites and/or allergens in carpets, but only three had clinically positive outcomes, so the data were not considered further.

Clinical trials with mattress covers had the highest grand mean percentage reduction in allergen levels (active groups: 74%; control groups: 5%) and benzyl benzoate had the lowest – no better than the control (active: 30%; control: 36%).

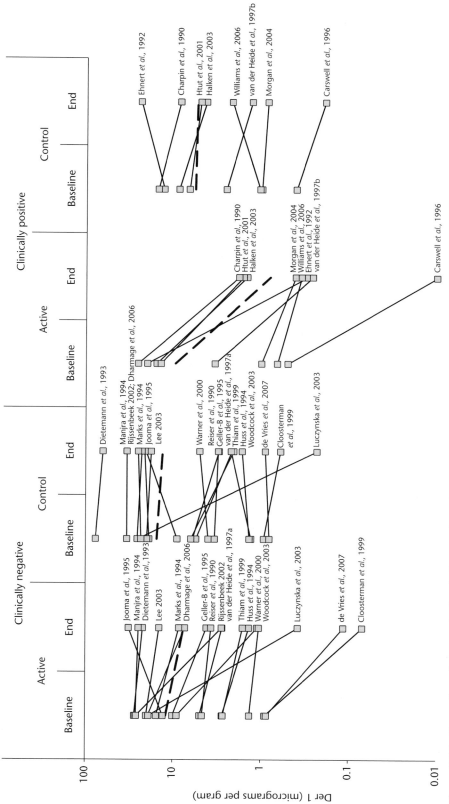

Figure 9.10 Comparison of mean reductions in *Dermatophagoides* group 1 allergen levels in clinical trials of allergen avoidance with significantly improved measures of clinical outcomes and trials in which there was no improvement. Trials are listed in Table 9.2. Dashed line = arithmetic mean of means.

Mono-faceted trials (i.e. with only one main method of control) achieved lower allergen reductions (active: 41%; control: 5%) than multifaceted trials (active: 85%; control: −25%, i.e. an increase). Mono-faceted trials were five times less frequently associated with positive clinical outcomes than negative outcomes, whereas significant clinical improvement was four times more frequent than not in multifaceted trials. Trials with children were more likely to have positive clinical outcomes (five positive v. five negative) than trials with adults (five positive v. 13 negative) and showed higher mean allergen reduction in the active group compared with the control (children, active groups: 79%; control groups: −15%; adults, active groups: 91%, control groups: 34%).

Trials in which the active group had significantly greater improvement than controls in bronchial hyperreactivity, as measured by challenge with histamine or allergen (PC_{20}), showed a mean allergen reduction of 81% (13% for the controls) compared with 44% reduction for the active groups in trials

where there was no improvement in bronchial hyperreactivity (5% for controls).

Field trials

I found 27 field trials with measurement of changes in mite numbers (n = 6) or allergen levels (n = 21; see Table 9.3). Of those that reported allergen levels, 15 had data on allergen levels in mattresses and were comparable with the clinical trial dataset, while 14 had data on allergens in carpets, and were used for a within-field trial comparison with mattresses. Of the 15 trials that included allergen data for mattresses, there was a statistically significantly greater reduction in allergen levels in the active groups compared with controls in seven cases (47%). For these 'positive' trials, group 1 allergen levels were reduced from a grand mean of 31.6 to 5.1 µg g^{-1} in the active groups (84%; range 68–95%) compared with 33.4 to 32.0 µg g^{-1} (4%) in the control groups (shown in Figure 9.11). In the eight trials with no significant difference between active and control groups, mean allergen levels were reduced from 7.0 to 4.1 µg g^{-1}

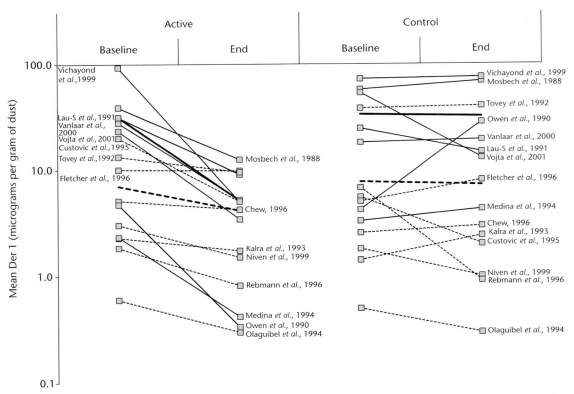

Figure 9.11 Comparison of mean reductions in *Dermatophagoides* group 1 allergen levels in mattresses in field trials of allergen avoidance for which there was a significant difference between active and control groups ('positive trials'; solid lines) and for which there was no difference ('negative trials'; dotted lines). Trials are listed in Table 9.3. Thick solid line = arithmetic mean of means, positive trials; thick dotted line = arithmetic mean of means, negative trials.

(41%; range 87 to −70%) in the active group and 7.7 to 7.2 µg g⁻¹ in the control group (7%).

Reductions in allergen levels in carpets were similar to those for mattresses. For the positive trials, the grand mean fell from 17.3 to 3.4 µg g⁻¹ in the active groups (80%) and 19.9 to 19.2 µg g⁻¹ (4%) in the control groups (see Figure 9.12). In the trials with no difference in allergen reduction between active and control groups, mean allergen levels were reduced from 9.6 to 5.4 µg g⁻¹ (44%) in the active group and 9.2 to 5.2 µg g⁻¹ in the control group (44%).

The magnitude of allergen reduction achieved in 'positive' field trials is similar to 'positive' clinical trials (ca. 80–90%). The frequency of clinical and field trials that gave significantly greater improvements in outcomes (clinical improvement or allergen reduction) in the intervention group was also similar (40–50%).

The conclusion that allergen avoidance strategies do not work and should not be recommended for use by patients with asthma is not supported by these data. The most parsimonious explanation for why about half the trials did not work can be gained from an examination of Figures 9.10–9.12. In most cases, either the reduction in allergen levels in the active groups was too small or, if it was large, so was the fall in allergen levels in the control group (e.g. the trial by Luczynska et al., 2003; Figure 9.10). Possible explanations for the latter situation are either the placebo treatment had an anti-mite effect or the trial was done at a time of year when mite populations and allergen levels were undergoing seasonal decline, or both. The trial by Luczynska et al. (2003) commenced in January, which coincides with the period of declining mite population densities in the Northern Hemisphere (January to March; Figure 9.13; Chapter 5).

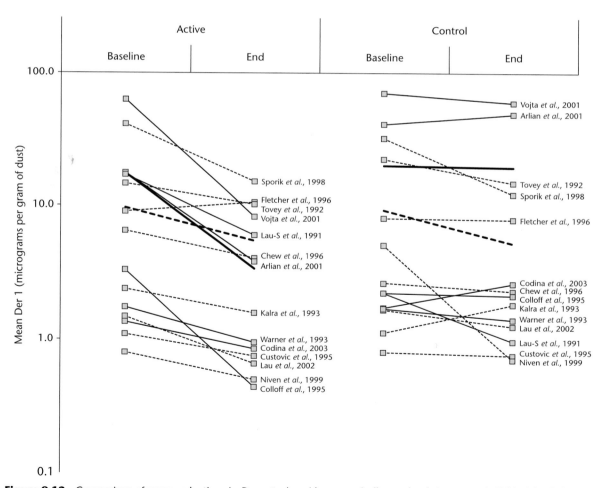

Figure 9.12 Comparison of mean reductions in *Dermatophagoides* group 1 allergen levels in carpets in field trials of allergen avoidance for which there was a significant difference between active and control groups ('positive trials'; solid lines) and for which there was no difference ('negative trials'; dotted lines). Trials are listed in Table 9.3. Thick solid line = arithmetic mean of means, positive trials; thick dotted line = arithmetic mean of means, negative trials.

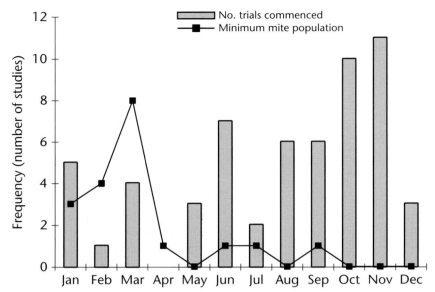

Figure 9.13 Timing of the start of clinical trials in relation to seasonal variation in mite populations (corrected for latitude of trial locations). The histogram indicates the number of trials that commenced during each month. Most trials commenced in late autumn to avoid the pollen season. The line graph shows the number of (Northern Hemisphere) studies on seasonal population fluctuations (Chapter 5) that recorded the lowest mite population density in any particular month.

The re-evaluation of clinical trials presented herein covers a set of studies that were done between 1983 and 2007 and all of them used reservoir dust as a basis for assessment of reduction in allergen concentrations. During this period our understanding of the relationship between reservoir dust and allergen exposure has changed dramatically. Measuring reservoir dust levels is only an indirect assessment of exposure even though almost all data on exposure are collected in this way. Effective allergen avoidance depends on understanding the mechanisms of exposure. We now know that allergens become airborne from a series of reservoirs when those reservoirs are disturbed; that sampling aeroallergens is more representative than sampling reservoir dust; and sampling aeroallergens that are inhaled is the best measure of exposure (O'Meara and Tovey, 2000). So far, no clinical trials have been done that assess the relationship between reduction in allergens in reservoir dust and changes in personal exposure.

An example of the variation in different methods of measuring allergens in dust following the encasing of a mattress is shown in Figure 9.14. Reservoir dust sampling becomes an issue especially with use of mattress covers, because trials invariably use baseline data from the upper mattress surface and compare this with end-of-trial data from the upper surface of the cover. Such a comparison is highly likely to show a dramatic fall in allergen levels by the end of the trial,

unless large amounts of allergen were deposited from the air onto the upper surface of the cover during the trial. There is no *a priori* reason that mattress covers should reduce the level of allergen within the mattress. In fact, levels tend to increase after covers are installed (Carswell *et al.*, 1996; Wickman *et al.*, 1994b). A better approach would be to compare direct measures of exposure like airborne sampling or use of an intranasal personal sampler (Graham *et al.*, 2000; Chapter 8) at the beginning and end of the trial, thus ensuring the sampling of like-with-like and use of a less indirect measure of exposure.

Despite the shortcomings of measuring reduction of allergens in reservoir dust, the data in Figure 9.10 indicate that trials with significantly improved clinical outcomes are associated with a greater reduction in reservoir allergens than those where no clinical improvement was evident.

9.4.3 Other issues with clinical trials

Some authors have reported very low allergen levels compared with other published data for the same geographical location of their trials. For example, levels for mattresses reported by Popplewell *et al.* (2000) are an order of magnitude lower than those of an earlier study by the same research group (Warner *et al.*, 2000) on patients recruited from the same catchment area. Baseline allergen concentrations may affect the outcome of the trial because reductions from low

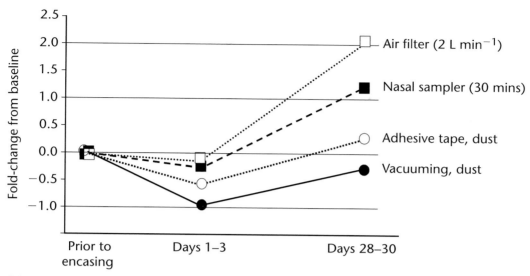

Figure 9.14 Fold-changes in allergen exposure measured by different methods following the encasing of mattresses and bedding with covers. (Data and figure courtesy of Euan Tovey.)

baseline may be less clinically relevant or, conversely, changes from a high allergen baseline may not be large enough to achieve a clinical effect. The only trial where baseline allergen levels were considered as part of the design was that by Luczynska *et al.* (2003), whereby a cut-off of >2 µg g^{-1} (the level at which mite allergen exposure is considered to be a risk for development of asthma; Platts-Mills *et al.*, 1989b) was used as a criterion for inclusion. During screening prior to the trial, the authors found only 23% of the asthma patients in S.E. London who had mite-specific IgE also had exposure to >2 µg g^{-1} Der p 1, and suggested that only a low proportion of mild to moderate asthmatic patients that present to general practitioners or allergy clinics would be likely to benefit from allergen avoidance. The caveat here is that although >2 µg g^{-1} Der p 1 is too high a standard of exposure for this particular group of patients, it does not mean that lower levels (e.g. 1–2 µg g^{-1}) are not clinically relevant to them. What constitutes high allergen exposure is most probably population-specific, because of seasonal and geographic variation in allergen levels (Chapters 5 and 8).

Prior allergen avoidance measures in the home tend not to be assessed (Rijssenbeek-Nouwens *et al.*, 2002 is an exception). I am not aware of any trials where patients were excluded for this reason, yet it should be part of a standardisation protocol because it could lead to a skewing of relative allergen levels in the control or avoidance group if not controlled for. There is a risk of including patients with allergen avoidance

experience in a control group because they may not adhere to a non-intervention protocol. The effects of patients undertaking parts of the intervention (e.g. vacuum cleaning, applying acaricides) reduces standardisation. The Hawthorne Effect is a phenomenon whereby patients or their families respond to the increased attention they receive from their enrolment in a clinical trial by improving their performance (at mite and allergen control measures), regardless of whether they are in the control or intervention group (McCarney *et al.*, 2007).

Many of the trials done in the 1980s and 1990s focused only on reducing exposure to dust mites, even though the patients may also have been sensitised to allergens from pollens, moulds, cockroaches and domestic pets. Some more recent trials have tailored avoidance measures to multiple allergens (e.g. Morgan *et al.* 2004), but there are few data on the clinical consequences and interaction effects of relative reductions in multiple allergens. Measures aimed at reducing exposure to allergens from a single biotic source may be less likely to work on patients with multiple sensitisation. However, Popplewell *et al.* (2000) found reduction in Fel d 1, but no reduction in Der p 1, led to clinical improvement in asthmatics sensitised to both mites and cats.

It is a lot harder than has been assumed to get mite and allergen levels down and even harder to keep them down for extended periods. This becomes more of an issue in trials of longer duration. Htut *et al.* (2001) reported that mattress dust showed a six-fold reduction in Der p 1 after steam-cleaning, then it crept up again: four-fold at 6 months post-treatment and two-fold at 12 months. It took steam-cleaning and installation of a mechanical ventilation system to maintain a three-to-five-fold reduction by 12 months. Woodcock *et al.* (2003) found significantly lower Der p 1 in a group allocated impermeable mattress covers compared with those given permeable placebo covers, but by 12 months post-intervention there was no difference.

9.4.4 Secondary trials of patients with allergic diseases other than asthma

a Rhinitis

Relatively few trials have focused on the influence of reducing house dust mite allergen exposure on symptoms of allergic rhinitis. Of the studies addressing this (Howarth *et al.*, 1992; Bowler *et al.*, 1985; Kersten *et al.*, 1988; Brown and Merrett, 1991; Kneist *et al.*, 1991; Reisman *et al.*, 1990), two contained no monitoring of allergens and cannot be interpreted in relation to reduction in allergen exposure (Bowler *et al.*, 1985; Reisman *et al.*, 1990). The remaining papers contain reports of benefit with a reduction in allergen exposure. However, two of these trials, both with Acarosan treatment of the bedroom mattress, soft furnishings and carpet, which reported a 70% and 60% reduction in symptoms after a 12-week and a 52-week period of observation respectively, were uncontrolled open studies (Kersten *et al.*, 1988; Kneist *et al.*, 1991). Terreehorst *et al.* (2003) reported significant allergen reduction in the active group but no improvement in symptoms.

A meta-analysis of clinical trials of dust mite control for rhinitis (Sheikh *et al.*, 2007) concluded there was only marginal clinical benefit.

b Atopic dermatitis

Platts-Mills *et al.* (1983) described prolonged improvement of atopic dermatitis in six patients who undertook regular bedroom cleaning and/or removal of carpets and soft furniture, three who adopted diets free of milk or eggs and one who used both treatments. A further 20/23 patients improved within 10 days admission to hospital without changing their medical treatment. Hospitalisation for up to 2 weeks resulted in 'dramatic improvement' of 13/16 patients (Platts-Mills *et al.*, 1991). Der p 1 concentrations in beds of 15/16 patients were well above 10 $\mu g\ g^{-1}$, whereas in hospital beds they were less than 0.2 $\mu g\ g^{-1}$. Adinoff *et al.* (1988) reported marked improvement in skin symptoms in 10 patients who moved from

their usual environments. Roberts (1984) found that 6 weeks after intensive cleaning, combined with the use of mattress covers, the clinical scores were improved in 15/18 patients with severe atopic dermatitis. Using similar methods, August (1984) found that after 6 months the clinical scores of 32/37 patients had improved and that overall, by the end of the trial, 60% were clear of lesions or almost clear, 27% had improved partially, 13% were unchanged and none were worse. Complete cessation of itching after 10 days' hospitalisation in a clean room was achieved, but not in patients who stayed in an ordinary ward (Sanda *et al.*, 1992). Time until symptom relapse was nearly 9 months in the clean room group, compared with almost 2 months for the ward group. These studies were uncontrolled and, with one exception (Platts-Mills *et al.*, 1991), did not involve mite or allergen monitoring. The patient groups contained both mite-sensitive adults and children. Case reports of five adults whose homes had been treated with Acarosan showed 70–90% improvement in symptoms (Brown and Merrett, 1991). Colloff *et al.* (1989), in a controlled trial with adults, found lower mite numbers in the groups that had used vacuuming, with or without natamycin, than in the groups that used natamycin or placebo spray without vacuuming, but there was no significant improvement in clinical scores, skin-prick test reactions or total or mite-specific IgE values.

Tan *et al.* (1996) achieved a 98% fall in dust exposure on mattresses using covers, but the severity of eczema decreased in both active and placebo groups. Ricci *et al.* (2000), in a multifaceted, crossover trial with children, achieved a four-fold reduction in allergen levels and improved clinical scores. Holm *et al.* (2001) found mattress covers and placebo covers both reduced allergen levels and that the severity of eczema was improved in both active and placebo groups. Gutgesell *et al.* (2001), using mattress covers and benzyl-tannate, achieved a greater reduction in allergen levels in the active group v. the control, but without any difference in clinical improvement. Oosting *et al.* (2002), using mattress covers, achieved greater than two-fold allergen reduction over the placebo group, but again with no difference in clinical parameters.

The chronic nature of atopic dermatitis, combined with the finding that homes of patients with the disease contain significantly greater population densities of mites than those of other patients with atopy and healthy non-atopics (Beck and Korsgaard, 1989; Colloff, 1992c), suggests that successful mite eradication that translates into clinical improvement will require rigorous multifaceted approaches, including the removal of carpets as well as covering of bedding.

9.4.5 Primary trials – allergen control to prevent children developing asthma

If a causal relationship exists between exposure of infants to domestic allergens and development of allergic symptoms, it follows that providing a newborn baby with a low-allergen environment may prevent or delay the onset of symptoms. This was first investigated by Arshad *et al.* (1992) in a controlled prospective study of infants with a uni- or bi-parental family history of atopy, by combining avoidance of allergenic foods (both directly and via restricted diets for lactating mothers) with reduction in mite allergens using benzyl benzoate, cleaning and new bedding. In the first year 25 (40%) of the infants in the control group

developed one or more allergic disorders compared with eight (13%) in the treatment group, but it was not possible to separate the effects of food allergen avoidance from those of reduced mite allergen exposure. The authors suggest fewer infants may have developed symptoms if avoidance of pets and tobacco smoke had been instigated, and if mite eradication procedures had commenced before the infant was brought home from hospital.

Several studies since have shown disappointing results in relation to the prevention of allergies and allergic disease (reviewed by van Schayck *et al.*, 2007), though multifaceted intervention studies were deemed to have a better chance of success than mono-faceted studies. However, these studies have been invaluable in developing a better understanding of the complex interaction between allergen exposure, development of allergies and appearance of disease (Chapter 8).

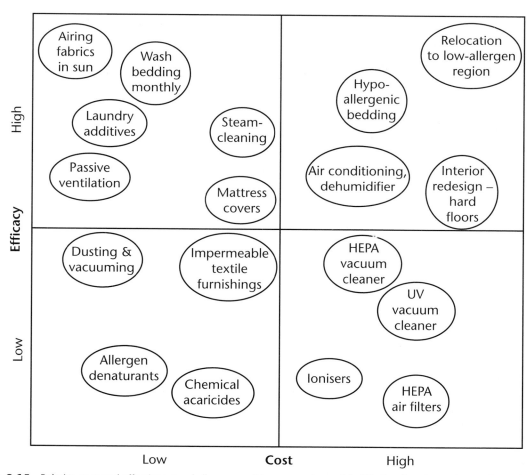

Figure 9.15 Relative cost and effectiveness of allergen avoidance measures. (Modified and re-drawn from a figure by Euan Tovey, *pers. comm.*)

9.5 Integrated approaches to mite and allergen control

Can control methods be combined to provide integrated, long-term strategies for the management of dust mites? Integrated Pest Management (IPM), used in crop protection since the 1950s (Smith and Allen, 1954; Smith *et al.*, 1976), applies ecological principles to the control of pests, aiming to maintain populations at tolerably low levels. IPM emerged in response to the indiscriminate use of chemical pesticides and the development of resistance in pest populations, and often involves use of biological and chemical control, timed to achieve optimal efficacy, and supported by monitoring and adaptive management principles. Biological control of dust mites is not feasible. Their predators are mites of the genus *Cheyletus* which are present in relatively few homes (Chapter 4), are allergenic (Pérez, 1979) and also bite people (Yoshikawa *et al.*, 1983). It would be ridiculous to advocate application of agricultural IPM to a domestic setting, but I do advocate the adoption of a more structured, planned approach to mite control that will be flexible enough to take into account two important facts:

- the indoor environment of no two houses are the same;

- the degree of compliance of patients with a set of instructions for removing indoor allergens is dependent on social and economic factors, including education (K. Huss *et al.*, 1992; de Blay *et al.*, 2003; Williams *et al.*, 2006).

To stand any chance of widespread adoption, mite control strategies must be flexible enough so they can be tailored to suit individual budgets and capabilities, while remaining effective. This requires an understanding of the trade-offs between cost, efficacy and ease of use (Figure 9.15). Few if any methods can achieve the total removal of mites and allergens without prohibitive expense. There are sound ecological reasons, primarily relating to rates of mite re-colonisation, why total eradication strategies should not be pursued. If one accepts this view, then it follows that the maintenance of dust mite populations at levels low enough to ameliorate clinical symptoms is the only alternative.

The primary measurable outcome of an integrated strategy is regarded as the reduction in the size of the population of allergen-producing organisms and of the domestic allergen pool. An integrated approach stems from the following observations:

- control methods have their primary effect upon the mite population and not on the patients: alleviation

Table 9.4 A key for a flexible, simple, cheap, integrated dust mite control strategy, which incorporates synergistic effects (modified from Colloff, 1995). The strategy deals with treatment of the mattress, bedding, including pillows, upholstered furniture and carpets. The idea is to read both parts of each numbered section fully before making a decision and moving to the next option and then to make a note of achievable options and compile a list.

1 – fit microporous mattress covers wherever possible	go to 3
– if covers not affordable, unavailable or cannot be fitted	go to 2
2 – air mattress daily by stripping off bedding, opening windows and doors	go to 3
– if airing not possible	go to 3
3 – hot wash bedding monthly	go to 5
– if bedding cannot be hot washed and mattress not covered	go to 4
4 – add tea-tree oil to cold wash if hot wash unachievable	go to 5
– if not desirable	go to 5
5 – remove carpet from bedrooms	go to 7
– if removal not possible, e.g. floor is rough board or concrete	go to 6
6 – do weekly vacuuming of carpets and upholstered furniture with a HEPA filter vacuum cleaner	go to 7
7 – if HEPA filter cleaner not available, use ordinary vacuum cleaner	go to 8
8 – if ordinary vacuum cleaning too onerous, hire or buy a steam cleaner for upholstered furniture and mattresses	go to 9
9 – if living in climate with hot, sunny summers (ca. 30°C) or cold dry winters (below 0°C overnight), remove upholstered furniture, rugs, mattresses and bedding outside for 12 hr, followed by vacuuming.	

of symptoms is not a consequence of the use of mite control methods but of the removal of the allergen pool and reduction in the mite population;

- patients are unlikely to benefit clinically from reductions in concentrations of allergens to which they are not allergic;
- there is little point in investing time, money and effort in reducing allergen concentrations in parts of the home where patients receive minimal exposure.

For example, controlling dust mite populations in a dining room that may be inhabited only for a total of ca. five hours per week is likely to be less cost-effective than doing so in a living room used for a total of ca. 30 hours per week or a bedroom used for ca. 60 hours per week.

One cannot expect to treat all houses for dust mites in the same way and get consistent, positive results. In many cases this will be because the degree of compliance with the recommended methods is uneven throughout the patient population. Compliance depends in part on economic capacity (Denson-Lino *et al.*, 1993; Joseph *et al.*, 2003), investment of time and money (Kniest *et al.*, 1992), whether the patients understand what they are being asked to do (K. Huss *et al.*, 1992) and perceive the methods are working (Ransom *et al.*, 1991).

Table 9.4 lists some relatively cheap, straightforward dust mite control methods which can be used in combination. The aim here is to provide a series of options for the design of flexible, integrated measures, ranging from those that are regarded to be of major benefit, though requiring most investment in resources from the patient, to those with less benefit, requiring fewer resources, but which are still capable of achieving sizeable reductions in dust mite populations.

10.   Conclusions and reflections

If one is pondering life or death, it is a terrible task to have to study mites.

W.N.P. Barbellion (B.F. Cummings), 1919
The Journal of a Disappointed Man.

In writing this book I have become more aware than ever of the gaps in our knowledge about the biology of house dust mites and how they affect human health, but do not propose to give a comprehensive listing of them here. I have never had much faith in the value of inventories with titles the likes of 'Future research directions', and have occasionally wondered idly about their purpose. Presumably they are aimed at researchers in the field. But these people are already only too familiar with the logistic, intellectual and experimental obstacles that confront them, and generally have a very good idea of what needs to be done next. Nor can I think of a duller cue to prod at scientific creativity than the optimistic prescription of some well-meaning author who, almost as a matter of course, feels obliged to provide suggestions for further investigation. What follows, then, is more reflection tinged with hope than recommendation tainted by duty.

10.1   Why are dust mites still a problem?

The research effort on dust mites and allergic diseases has declined somewhat in recent years, from its peak in the late 1980s and early 1990s. But the problem has not been solved or gone away. Part of the reason for the difficulty in controlling dust mites and their allergens is that it has taken a long time and many clinical trials for researchers to realise that measuring allergens and mites in reservoir dust is a poor indicator of clinically relevant allergen exposure, that some methods of

allergen avoidance are not very effective, and that reductions in allergen exposure that provide relief to some patients are too slight to make a difference to others. On the positive side, we now have methods for measuring personal allergen exposure, and a vastly improved knowledge of the behaviour and dynamics of allergens within homes. We have also witnessed the completion of some big, multifaceted clinical trials of allergen avoidance that resulted in improved clinical outcomes, and a dramatic fall in numbers of small, mono-faceted trials that tended to be less effective. There is still a fair way to go, but significant progress has been made, not least in recognising that because of regional differences in allergen exposure, intervention strategies that work in one part of the world may not be effective somewhere else. For secondary interventions, determination of region-specific fold-reductions in allergen exposure that are associated with improvement of symptoms may provide a more structured, targeted approach than hitherto. This concept is in line with the findings of Section 8.8 herein and of Peat *et al.* (1996) who found for six Australian towns and cities with markedly different levels of mean mite allergen exposure, the risk of mite-allergic asthma was roughly double with every doubling of allergen concentrations. It will also allow for setting of selection criteria for future intervention trials that are based on levels of allergen exposure typical for the region where the trial is being undertaken.

The finding detailed in Chapter 9, that clinically 'successful' trials were associated with average reductions in allergen concentrations of over 90%, is a new one and may help to provide a target for allergen avoidance interventions.

10.2 Patterns of mite species diversity and profiles of allergen exposure

This book has been about the biology and ecology of pyroglyphid, glycyphagoid and acaroid mites found in homes. Much of it has addressed the questions of Andrewartha and Birch (1954; see the quote at the beginning of chapter 4); namely, what factors determine the distribution and abundance of domestic mites and why are they abundant in some regions but not in others? The unifying theme is how distribution and abundance of mites influence allergen exposure and prevalence of atopic diseases. I have attempted to show how both temperature and the water content of the air are inextricably linked to the physiology of body water balance and thence to survival and population growth at a range of spatial scales, from the microhabitat to the whole planet. The metrics that provide the 'best' signal at different scales may change, but the key factors remain the same.

Analysis of life history parameters, their various strategies and trade-offs has provided some new insights into comparative population dynamics of some of the most important allergen-producing species. Yet there is very little known about the basic population biology of *Blomia tropicalis* (though it may well share characteristics of other glycyphagoids). *B. tropicalis* is probably the second most important species after *Dermatophagoides pteronyssinus* in terms of the number of people sensitised to its allergens. In order to resolve its population biology, the taxonomic status of *B. tropicalis* and *B. kulagini* first needs to be clarified (Chapter 1), ideally by using a combination of morphological and molecular approaches.

The compilation of comprehensive global records of mite and allergen distribution and abundance has not been undertaken previously. Only a preliminary analysis of these datasets has been presented herein. A more detailed treatment is beyond the scope of this book, and there is a lot of unpublished information that could improve and clarify the overall picture.

10.2.1 Resolving allergen exposure patterns

Species assemblages of dust mites vary regionally and, to a lesser extent, between homes. The repertoire (i.e. the diversity, as well as the amount) of allergens to which people are exposed will influence the pattern of sensitisation and the nature of the IgE antibody response. So is the immune response or disease expression different with exposure to 1000 *Dermatophagoides pteronyssinus* per gram versus the same total population density of *D. pteronyssinus*, *Euroglyphus maynei*, *Blomia tropicalis* and *Tyrophagus putrescentiae*? The specificity of IgE responses is a function of both local exposure to major allergens and genetic characteristics of the exposed population. Other environmental factors such as infections and parasite burdens are relevant too. Because sensitisation is usually determined by skin-prick testing using an extract of a mite species that will contain multiple allergens, or by IgE antibody assay for a single allergen like Der p 1, variation in IgE responses to individual allergens is overlooked. Component-resolved diagnosis (CRD) with a panel of purified natural and recombinant allergens has been used to identify variation in specificity of IgE binding within and between human populations (Pittner *et al.*, 2004). Distinctive IgE responses to mite allergens have been found between Europeans and Central Africans (Westritschnig *et al.*, 2003) and between Aboriginal and non-Aboriginal Australians (Hales *et al.*, 2007). In the latter paper, the IgE antibody binding activity of serum from Aboriginal people from an isolated community in the northern Kimberley region, who had quite high Der p 1 exposure, was directed not against group 1 or 2 allergens but predominantly against group 4 (amylase).

Although we have reasonably good information on diversity and co-occurrence of species at the regional scale, there is very little precise published information on differences in mean number of species between homes, or between homes between different regions, even though a great deal of such data has been collected. Currently, routine assays are available for very few mite allergens. Application of a component-resolved approach to allergen quantification in homes would give a much clearer picture of the variation in diversity of allergens to which people are routinely exposed.

Our perspective on mite abundance, diversity and allergen exposure has been from a viewpoint of urban homes, mainly in Westernised countries, in temperate latitudes at low altitude. From the few studies on rural–urban comparisons (e.g. Radon *et al.*, 2000), we know there are differences in homes, but what about non-Westernised societies? We have no information on dust mites from Tibet or Nepal,

Appendix 1a

A catalogue of the Family Pyroglyphidae

Family PYROGLYPHIDAE Cunliffe, 1958

Pyroglyphinae Cunliffe, 1958, *Proc. Ent. Soc. Wash.* 60, 85.

Type-genus: *Pyroglyphus* Cunliffe, 1958
Pyroglyphidae Cunliffe, 1958: Fain, 1965, *Rev. Zool. Bot. Afr.* 72, 259.

Subfamily PYROGLYPHINAE Cunliffe, 1968

Pyroglyphinae Cunliffe: Fain, 1967b, *Acarologia* 9, 871.

Genus *Asiopyroglyphus* Fain and Atyeo, 1990

Asiopyroglyphus Fain and Atyeo, 1990
Type-species: *Asiopyroglyphus thailandicus* Fain and Atyeo, 1990

Asiopyroglyphus thailandicus Fain and Atyeo, 1990

Comb./syn.: *Asiopyroglyphus thailandicus* Fain and Atyeo, 1990, *Acarologia* 31, 47, figs. 1–11.

Type depository: holotype, paratype: United States National Museum, Washington; other paratype: Institut Royal des Sciences naturelles de Belgique, Brussels.
Type locality and habitat data: from Buff-rumped Woodpecker, *Meiglyptes tristis* (Piciformes: Picidae), Khao Luang Nakornsithamaras, Thailand.
Habitat type: birds and their nests.
Distribution: Thailand.

Genus *Bontiella* Fain, 1965

Bontiella Fain, 1965, *Rev. Zool. Bot. Afr.* 72, 263.
Type-species: *Bontiella bouilloni* Fain, 1965

Bontiella bouilloni Fain, 1965

Comb./syn.: *Bontiella bouilloni* Fain, 1965, *Rev. Zool. Bot. Afr.* 72, 272, figs. 9–12, 22, 27.
Type depository: Musée d'Afrique Centrale, Tervuren, Belgium.
Type locality and habitat data: in nests of Bronze Mannikin, *Lonchura cucullatus* (Passeriformes: Estrildidae), Kinshasa, Zaire.
Habitat type: nests of birds and mammals.
Distribution: Zaire, Rwanda.

Genus *Campephilocoptes* Fain, Gaud and Pérez, 1982

Campephilocoptes Fain, Gaud and Pérez, 1982
Type-species: *Campephilocoptes atyeoi* Fain, Gaud and Pérez, 1982

Campephilocoptes atyeoi Fain, Gaud and Pérez, 1982

Comb./syn.: *Campephilocoptes atyeoi* Fain, Gaud and Pérez, 1982, *Acarologia* 23, 166, figs. 1–5.

Type depository: American Museum of Natural History, New York.
Type locality and habitat data: from Red-necked Woodpecker, *Phloeoceastes rubricollis* (Piciformes: Picidae), Suapure, Bolivar, Venezuela.
Habitat type: birds or their nests.
Distribution: Venezuela.

Campephilocoptes paraguayensis Fain, Gaud and Pérez, 1982

Comb./syn.: *Campephilocoptes paraguayensis* Fain, Gaud and Pérez, 1982, *Acarologia* 23, 167, figs. 6–9.

Type depository: American Museum of Natural History, New York.

Type locality and habitat data: from Cream-backed Woodpecker, *Pholeoceastes leucepogon* (Piciformes: Picidae), Gran Chaco, Paraguay.
Habitat type: birds or their nests.
Distribution: Venezuela.

Genus *Euroglyphus* Fain, 1965

Euroglyphus (Euroglyphus) Fain, 1965, *Rev. Zool. Bot. Afr.* 72, 276.

Type-species: *Mealia maynei* Cooreman, 1950
Euroglyphus Fain, 1965: Fain, 1988b in: Fain *et al.*, *Acariens et Allergies*, p. 24.

Euroglyphus maynei (Cooreman, 1950)

Comb./syn.: *Mealia maynei* Cooreman, 1950, *Bull. Ann. Soc. Ent. Belg.* 86, 164, figs. 1–4.

Dermatophagoides maynei: Hughes, 1954, *Proc. Zool. Soc. Lond.*, 124, 11.

Euroglyphus (Euroglyphus) maynei (Cooreman): Fain, 1965, *Rev. Zool. Bot. Afr.* 72, 263, Figs. 13–16, 25, 29.

Dermatophagoides scheremetewskyi Bogdanov, 1864, *Bull. Soc. Imp. Nat. Moscou*, 37, 343, pl. 7, fig. 2. Male. new synonym.

Type depository: Holotype: Institut Royal des Sciences naturelles de Belgique, Brussels; paratypes: C.E.R.E.A., Station Entomologique de l'Etat, Gembloux.
Type locality and habitat data: mouldy cotton-seed cake, Gembloux, Belgium.
Habitat type: house dust.
Distribution: cosmopolitan.
Remarks: Figures 2A and 2B of what Bogdanov (1865) tentatively assigned as the male of *Dermatophagoides scheremetewskyi* as 'Acarus de l'herpes farinosus (mâle du Dermatophagoides?)' are drawn to the same scale as those of the female (Bogdanov's Figs. 1A and 1B), and depict a male pyroglyphid mite considerably smaller than the male of *D. pteronyssinus*, but matching almost exactly the dimensions of the male of *E. maynei*. Other points of similarity with *E. maynei* are the broad genital area and stubby-tipped penis, the anal plate shaped like a broad inverted U, the pattern of ventral striae, especially its absence between legs I, legs III and IV almost subequal in length, the relatively well-spaced dorsal striations and

rectangular hysterosomal shield, the lack of any long dorsal hysterosomal setae, and the almost rectangular outline of the idiosoma. All of these characters fit the male of *E. maynei* far better than the male of *D. pteronyssinus*. Furthermore, *E. maynei* has, like *D. farinae* and *D. pteronyssinus*, been recorded in Moscow, the type locality of *D. scheremetewskyi*.

Genus *Gymnoglyphus* Fain, 1965

Euroglyphus (Gymnoglyphus) Fain, 1965, *Rev. Zool. Bot. Afr.* 72, 280.

Type-species: *Mealia longior* Trouessart, 1897
Gymnoglyphus Fain, 1965: Fain, 1988b in: Fain *et al.*, *Acariens et Allergies*, p. 25.

Gymnoglyphus longior (Trouessart, 1897)

Comb./syn.: *Mealia longior* Trouessart, in Berlese, 1897a. *Acari Myriapoda et Scorpiones in Italia Reperta*. I. Cryptostigmata. Patavii, p. 104.

Mealia longior: Trouessart, 1897 in: Berlese, 1897b. *Acari Myriapoda et Scorpiones in Italia Reperta*. Patavii, Fasc. 89, no. 10.

Mealia longior: Trouessart, 1897 in: Berlese, 1898. *Acari Myriapoda et Scorpiones in Italia Reperta*. Patavii, Fasc. 92, no. 4.

Pachylichus crassus Canestrini, 1894, *Prospetto Acarofauna Italiana – Parte VI*. Gli Epidermoptini. Stabilimento Prosperini, Padua [in part].

Dermatophagoides crassus (Canestrini, 1894): Thurman and Mulrennan, 1947, *J. Econ. Ent.*, 40, 591, fig. 1.

Dermatophagoides longior: Dubinin, 1953. *Fauna USSR. Analgesoidea*. Vol VI, no. 6.

Dermatophagoides dalarnensis Sellnick, 1958, *Statens Vaxtskanst. Medd.*, 11(71), 47, figs. 43–45.

Euroglyphus (Gymnoglyphus) longior (Trouessart, 1897): Fain, 1965, *Rev. Zool. Bot. Afr.* 72, 281, figs. 17–20.

Gymnoglyphus longior: Fain, 1988b in: Fain *et al.*, *Acariens et Allergies*, p. 26.

Type depository: Berlese Acaroteca, Florence.
Type locality and habitat data: decomposed small mammal remains, France.
Habitat type: house dust, birds' nests.
Distribution: Holarctic, Neotropics.

Gymnoglyphus osu (Fain and Johnston, 1973)

Comb./syn.: *Euroglyphus (Gymnoglyphus) osu* Fain and Johnston, 1973, *Bull. Ann. Soc. R. Belg. Ent.*, 109, 131, Figs. 1–2.

Gymnoglyphus osu: Fain, 1988b in: Fain *et al.*, *Acariens et Allergies*, p. 26.

Type depository: Holotype: U.S. National Museum; paratypes: Acarology Laboratory, Ohio State University, Columbus.
Type locality and habitat data: in barn dust and grain debris, Columbus, Ohio.
Habitat type: barn dust.
Distribution: USA.

Genus *Hughesiella* Fain, 1965

Pyroglyphus (Hughesiella) Fain, 1965. *Rev. Zool. Bot. Afr.* 72, 268.

Type species: *Dermatophagoides africanus* Hughes, 1954
Hughesiella Fain, 1965: Fain, 1988b in: Fain *et al.*, *Acariens et Allergies*, p. 24.

Hughesiella africana (Hughes, 1954)

Comb./syn.: *Dermatophagoides africanus* Hughes, 1954. *Proc. Zool. Soc. Lond.*, 124, 1, Figs. 1–17.

Pyroglyphus (Hughesiella) africanus: Fain, 1965. *Rev. Zool. Bot. Afr.* 72, 268, Figs. 5–8, 24, 28.

Pyroglyphus (Hughesiella) africanus: Fain and Rosa, 1982, *Rev. Brasil. Biol.*, 42, 320.

Hughesiella africana (Hughes, 1954): Fain, 1988b in: Fain *et al.*, *Acariens et Allergies*, p. 24.

Type depository: Natural History Museum, London.
Type locality and habitat data: fishmeal originating from Angola, stored in warehouses in England.
Habitat type: house dust, birds' nests.
Distribution: Africa, Middle East, Brazil.

Genus *Pyroglyphus* Cunliffe, 1958

Pyroglyphus Cunliffe, 1958
Type-species: *Pyroglyphus morlani* Cunliffe, 1958

Pyroglyphus morlani Cunliffe, 1958

Comb./syn.: *Pyroglyphus morlani* Cunliffe, 1958. *Proc. Ent. Soc. Wash.*, 60, 85, figs. 1–5.

Type depository: Holotype: U.S. National Museum, Washington.
Type locality and habitat data: in rodent nest, *Neotoma albigula*, nr. Santa Fe, New Mexico.
Habitat type: mammal nests.
Distribution: USA.

Genus *Weelawadjia* Fain and Lowry, 1974

Weelawadjia Fain and Lowry, 1974
Type-species: *Weelawadjia australis* Fain and Lowry, 1974

Weelawadjia australis Fain and Lowry, 1974

Comb./syn.: *Weelawadjia australis* Fain and Lowry, 1974. *Acarologia* 16, 334, Figs. 1–9.

Type depository: Holotype and allotype: Australian National Insect Collection, CSIRO Division of Entomology, Canberra.
Type locality and habitat data: Chocolate bat (*Chalinolobus morio*) guano and Welcome Swallow, *Hirundo neoxena* (Passeriformes: Hirundinidae) nest material, Weelawadji Cave, Eneabba, Western Australia.
Habitat type: bird and mammal associate.
Distribution: Australia.

Subfamily DERMATOPHAGOIDINAE Fain, 1963

Dermatophagoidinae Fain, 1963
Type-genus: *Dermatophagoides* Bogdanov, 1864
(cf. Fain, 1963, *Bull. Inst. r. Sci. nat. Belg.*, 39(32), 53.)

Guatemalichinae Fain, 1988b
Type-genus: *Guatemalichus* Fain and Wharton, 1970
(cf. Fain, 1988b, in: Fain *et al.*, *Acariens et Allergies*, p. 20; Gaud and Atyeo, 1996, *Mus. r. Afr. Centr. Ann. Sci. Zool.*, 277, 71.)

Onychalginae Fain, 1988b
Type-genus: *Onychalges* Gaud and Mouchet, 1959
(cf. Fain, 1988b, in: Fain *et al.*, *Acariens et Allergies*, p. 21; Gaud and Atyeo, 1996, *Mus. r. Afr. Centr. Ann. Sci. Zool.*, 277, 71.)

Genus *Dermatophagoides* Bogdanov, 1864

Dermatophagoides Bogdanov, 1864
Type-species: *Dermatophagoides scheremetewskyi* Bogdanov, 1864

Pachylichus Canestrini, 1894

Type-species: *Pachylichus crassus* Canestrini, 1894

Mealia Trouessart, 1897

Type-species: *Mealia pteronyssina* Trouessart, 1897 (synonymy by Baker and Wharton, 1952, *An Introduction to Acarology*, p. 374)

Visceroptes Sasa, 1948

Type-species: *Visceroptes satoi* Sasa, 1948

Paralgoides Gaud and Mouchet, 1959

Type-species: *Dermoglyphus (Paralges) pteronyssoides* Trouessart, 1886

Hullia Gaud, 1968 (synonymy by Fain, 1988b in: Fain et al., *Acariens et Allergies*, p. 30).

Type-species: *Hullia anisopoda* Gaud, 1968

Dermatophagoides alexfaini Cruz, 1988

Comb./syn.: *Dermatophagoides alexfaini* Cruz, 1988, *Poeyana*, No. 361, 3, figs. 1–4.

Type depository: Instituto de Zoologia, Academia de Ciencias de Cuba, Havana.

Type locality and habitat data: nest of Cave Swallow, *Petrochelidon fulva* (Passeriformes: Hirundinidae), Palacio Brunett, Trinidad, Sancti Spíritus, Cuba.

Habitat type: birds' nests.

Distribution: Cuba.

Dermatophagoides anisopoda (Gaud, 1968)

Comb./syn.: *Hullia anisopoda* Gaud, 1968, *Acarologia* 10, 299, fig. 1.

Dermatophagoides anisopoda: Fain, 1988b in: Fain et al., *Acariens et Allergies*, p. 39.

Type depository: J. Gaud personal collection.

Type locality and habitat data: on Red-faced Lovebird, *Agapornis pullaria* (Psittaciformes: Psittacide), Yaoundé, Cameroons.

Habitat type: on birds.

Distribution: Cameroons.

Dermatophagoides aureliani Fain, 1967b

Comb./syn.: *Dermatophagoides aureliani* Fain, 1967b, *Acarologia* 9, 873, figs. 1–5.

Type depository: holotype and allotype: Musée royal de l'Afrique Centrale, Tervuren, Belgium; paratypes: A. Fain, personal collection.

Type locality and habitat data: in nest of Gray-headed Sparrow, *Passer griseus* (Passeriformes: Ploceidae), Butare, Rwanda.

Habitat type: birds' nests.

Distribution: Rwanda.

Dermatophagoides evansi Fain, Hughes and Johnston, 1967

Comb./syn.: *Dermatophagoides (Dermatophagoides) evansi* Fain, Hughes and Johnston, in: Fain, 1967a, *Acarologia* 9, 205, figs. 26–38.

Dermatophagoides evansi Fain, Hughes and Johnston: Baker, Delfinado and Abbatiello, 1976, *J. NY Ent. Soc.*, 84, 53, figs. 15–25.

Type depository: holotype and paratypes: Natural History Museum, London; other paratypes: A. Fain, personal collection.

Type locality and habitat data: pillows made in Boston, England from feathers imported from Ghana.

Habitat type: feathers, birds' nests

Distribution: Ghana, USA.

Dermatophagoides farinae Hughes, 1961

Comb./syn.: *Dermatophagoides farinae* Hughes, 1961, *The Mites of Stored Food*, p. 148, figs. 205–207a, 208–210.

?*Visceroptes takeuchii* Sasa, 1947: Fain, 1988b in: Fain et al., *Acariens et Allergies*, p. 32.

Dermatophagoides culinae DeLeon, 1963, *Florida Ent.*, 46, 247 (cf. Fain, 1967, *Acarologia* 9, 194.)

Mealia farinae: Oshima, 1968, *Jpn. J. Sanit. Zool.*, 19, 170, figs. 8c, 9c, 10e, 11e, f, 12c, 18–22.

Type depository: Natural History Museum, London.

Type locality and habitat data: in poultry and pig-rearing meal, near Bristol, England.

Habitat type: house dust.

Distribution: cosmopolitan.

Dermatophagoides microceras Griffiths and Cunnington, 1971

Comb./syn.: *Dermatophagoides microceras* Griffiths and Cunnington, 1971, *J. Stored Prod. Res.*, 7, 1, figs. 1–30.

Dermatophagoides farinae Hughes, 1961, in part: Griffiths and Cunnington, 1971, *J. Stored Prod. Res.*, 7, 1, figs.1, 3, 5, 7, 9, 11, 13, 15, 17, 19–23, 25, 27, 28, 30.

Type depository: holotype: Natural History Museum, London; paratypes: United States National Museum, Washington and Acarology Laboratory, Ohio State University.
Type locality and habitat data: house dust, Greenwich, London, UK.
Habitat type: house dust.
Distribution: holarctic.

Dermatophagoides neotropicalis Fain and van Bronswijk, 1973

Comb./syn.: *Dermatophagoides neotropicalis* Fain and van Bronswijk, 1973, *Acarologia* 15, 181, figs. 1–5, 7–9, 11.
Type depository: holotype and paratypes: Rijksmuseum van Natuurlijke Historie, Leiden, the Netherlands; other paratypes: A. Fain, personal collection, J.E.M.H. van Bronswijk, personal collection.
Type locality and habitat data: dust from mattress, Paramaribo, Surinam.
Habitat type: house dust.
Distribution: Surinam.

Dermatophagoides pteronyssinus (Trouessart, 1897)

Comb./syn.: *Dermatophagoides scheremetewskyi* Bogdanov, 1864, *Bull. Soc. Imp. Nat. Moscou*, 37, 343, pl. 7, figs. 1, 2. Female.
Paralgoides pteronyssoides Trouessart, 1886, *Bull. Etude Sci. Angers*, 16, 119.
Mealia pteronyssina Trouessart, in Berlese, 1897a. *Acari Myriapoda et Scorpiones in Italia Reperta*. I. Cryptostigmata. Patavii, p. 104.
Mealia toxopei Oudemans, 1928, *Ent. Ber.*, 6, 293.
Visceroptes satoi Sasa, 1947, *Nisshin Igaku*, 34, 167.
Dermatophagoides scheremetewskyi Bogdanow, 1864: Baker *et al.*, 1956, *A Manual of Parasitic Mites of Medical or Economic Importance*, p. 146, fig. 51.
Dermatophagoides pteronyssinus: Fain, 1967a, *Acarologia* 9, 305, figs. 1–11, 13–28.
Mealia pteronyssina Oshima, 1968, *Jpn. J. Sanit. Zool.*, 19: 170, figs. 2–8a, 9a, 10a, 11a, b, 12a.
Type depository: lectotype: Berlese Acaroteca, Florence.
Type locality and habitat data: prepared mammalian hide, Paris.
Habitat type: house dust.
Distribution: cosmopolitan.

Dermatophagoides rwandae Fain, 1967a

Comb./syn.: *Dermatophagoides (Dermatophagoides) rwandae* Fain, 1967a, *Acarologia* 9, 211, figs. 39–46.
Type depository: holotype: Musée royal de l'Afrique Centrale, Tervuren, Belgium.
Type locality and habitat data: nest of Yellow-billed Oxpecker, *Buphagus africanus* (Passeriformes, Sturnidae), Butare, Rwanda.
Habitat type: birds' nests.
Distribution: Rwanda.

Dermatophagoides sclerovestibulatus Fain, 1975

Comb./syn.: *Dermatophagoides sclerovestibulatus* Fain, 1975, *Rev. Zool. Afr.*, 89, 254, figs. 1–6.
Type depository: holotype, allotype and paratypes: Musée royal de l'Afrique Centrale, Tervuren, Belgium; other paratypes: A. Fain, personal collection and South African Institute for Medical Research, Johannesburg.
Type locality and habitat data: nest of Red-billed Oxpecker, *Buphagus erythrorynchus* (Passeriformes: Sturnidae), nr. Satara, Kruger National Park, South Africa.
Habitat type: birds' nests.
Distribution: South Africa.

Dermatophagoides siboney Dusbabek, Cuervo and Cruz, 1982

Comb./syn.: *Dermatophagoides siboney* Dusbabek, Cuervo and Cruz, 1982, *Acarologia* 23, 55, figs. 1–4.
Type depository: holotype, allotype and paratypes: Institute of Zoology, Academy of Sciences of Cuba, Havana; other paratypes: Institute of Parasitology, Czech Academy of Sciences, Prague.
Type locality and habitat data: dust from mattress, Havana, Cuba.
Habitat type: house dust.
Distribution: Cuba, Panama, Puerto Rico, Algeria (OConnor, quoted by Miranda *et al.*, 2002).

Dermatophagoides simplex Fain and Rosa, 1982

Comb./syn.: *Dermatophagoides simplex* Fain and Rosa, 1982, *Rev. Brasil. Biol.*, 42, 317, figs. 1–6.
Type depository: holotype, allotype and paratypes: Department of Zoology, ESALQ, University of

São Paulo, Brazil; other paratypes: A. Fain, personal collection, A.E. Rosa, personal collection.
Type locality and habitat data: in nests of House Sparrow, *Passer domesticus* (Passeriformes: Ploceidae), Piracicaba, Brazil.
Habitat type: birds' nests.
Distribution: Brazil.

Genus *Fainoglyphus* Atyeo and Gaud, 1977

Fainoglyphus Atyeo and Gaud, 1977
Type-species: *Fainoglyphus magnasternus* Atyeo and Gaud, 1977

Fainoglyphus magnasternus Atyeo and Gaud, 1977

Comb./syn.: *Fainoglyphus magnasternus* Atyeo and Gaud, 1977, *Steenstrupia* 4, 122, figs. 1–3.
Type depository: holotype and paratypes: Zoological Museum, University of Copenhagen; other paratypes: Department of Entomology, University of Georgia.
Type locality and habitat data: from Red-faced Spinetail, *Certhiaxis erythrops* (Passeriformes: Furnariidae), Ecuador.
Habitat type: on birds and their nests.
Distribution: Ecuador.

Genus *Guatemalichus* Fain and Wharton, 1970

Guatemalichus Fain and Wharton, 1970
Type-species: *Guatemalichus bananae* Fain and Wharton, 1970

Guatemalichus bananae Fain and Wharton, 1970

Comb./syn.: *Guatemalichus bananae* Fain and Wharton, 1970, *Bull. Inst. Roy. Sci. Nat. Belg.*, 46, 2, figs. 1–2.
Type depository: holotype and paratypes: United States National Museum, Washington; other paratypes: A. Fain, personal collection; Acarology Laboratory, Ohio State University.
Type locality and habitat data: on a stem of bananas originating from Guatemala (U.S. Department of Agriculture: ?quarantine interception)

Habitat type: the type specimens are most probably 'accidental' in this habitat.
Distribution: Guatemala.

Guatemalichus tachornis Cruz, Cuervo and Dusbabek, 1984

Comb./syn.: *Guatemalichus tachornis* Cruz, Cuervo and Dusbabek, 1984, *Poeyana* No. 267, 1, figs. 1–7.
Guatemalichus tachornis: Cruz, 1988, *Poeyana* No. 361, 14.
Type depository: holotype and paratypes: Institute of Zoology, Academy of Sciences of Cuba, Havana. Paratypes: A. Fain, personal collection; F. Dusbabek collection, Institute of Parasitology, Czech Academy of Sciences, České Budějovice.
Type locality and habitat data: from nest of Antillean Palm Swift, *Tachornis phoenicobia iradii* (Apodiformes, Apodidae).
Habitat type: birds' nests.
Distribution: Cuba.

Genus *Hirstia* Hull, 1931

Hirstia Hull, 1931, *The Vasculum*, 17, 145.
Type-species: *Hirstia chelidonis* Hull, 1931

Hirstia chelidonis Hull, 1931

Comb./syn.: *Hirstia chelidonis* Hull, 1931, *The Vasculum*, 17, 145.
Dermatophagoides passericola Fain, 1964, *Rev. Zool. Bot. Afr.*, 69, 201.
Dermatophagoides passericola Fain, 1964: Fain, Oshima and Bronswijk, 1974, *Jpn. J. Sanit. Zool.*, 25, 197.
?*Dermatophagoides passericola* Fain, 1964: Fain, 1988b in: Fain *et al.*, *Acariens et allergies*, p. 39.
Type depository: types probably lost (cf. Fain *et al.*, 1974).
Type locality and habitat data: from nest of Common House Martin, *Delichon urbica* (Passeriformes: Hirundinidae), Belford, England.
Habitat type: birds' nests.
Distribution: England.
Remarks: Fain *et al.* (1974) compared *Dermatophagoides passericola* with non-type specimens of *H. chelidonis* Hull, 1931 in the Natural

History Museum, concluded they were conspecific and formally synonymised them. However, Fain (1990, pp. 30, 31) keys *Dermatophagoides passericola* as if it were a valid species and later (p. 43) queries the synonymy, pointing out it should be reconsidered if topotypic material becomes available.

Hirstia domicola Fain, Oshima and van Bronswijk, 1974

Comb./syn.: *Mealia passericola* Oshima, 1968, *Jpn. J. Sanit. Zool.*, 19, 178, figs. 13–17, not Fain, 1964.

Hirstia domicola Fain, Oshima and van Bronswijk, 1974, *Jpn. J. Sanit. Zool.*, 25, 198, figs. 1–9.

Hirstia chelidonis: Fain, Cunnington and Spieksma, 1969, not Hull, 1931 (misidentification).

Type depository: holotype, allotype and paratypes: National Science Museum, Tokyo; other paratypes: A. Fain, personal collection, S. Oshima, personal collection, J.E.M.H. van Bronswijk, personal collection.
Type locality and habitat data: in house dust, Tohuku District, Japan.
Habitat type: house dust.
Distribution: cosmopolitan.

Genus *Kivuicola* Fain, 1971

Sturnophagoides (Kivuicola) Fain, 1971

Kivuicola Fain, 1971: Fain, 1988b in: Fain *et al.*, *Acariens et Allergies*, p. 21.

Type-species: *Sturnophagoides (Kivuicola) kivuana* Fain, 1971

Kivuicola kivuana Fain, 1971

Comb./syn.: *Sturnophagoides (Kivuicola) kivuana* Fain, 1971, *Bull. Inst. Roy. Sci. Nat. Belg.*, 47 (8), 4, figs. 3–4.

Kivuicola kivuana Fain, 1971: Fain *et al.*, *Acariens et allergies*, p. 21.

Type depository: Musée royal de l'Afrique Centrale, Tervuren, Belgium.
Type locality and habitat data: dried skin of a Potto, *Perodicticus potto*, Mutwanga, Kivu, Zaire.
Habitat type: mammal pelt (probably an 'accidental').
Distribution: Zaire.

Genus *Malayoglyphus* Fain, Cunnington and Spieksma, 1969

Malayoglyphus Fain, Cunnington and Spieksma, 1969
Type-species: *Malayoglyphus intermedius* Fain, Cunnington and Spieksma, 1969

Malayoglyphus carmelitus Spieksma, 1973

Comb./syn.: *Malayoglyphus carmelitus* Spieksma, 1973, *Acarologia* 15, 174, figs. 1–11.

Type depository: holotype, allotype, paratypes: Rijksmuseum van Natuurlijke Historie, Leiden, the Netherlands; other paratypes: F.Th.M. Spieksma personal collection.
Type locality and habitat data: in house dust, Mt Carmel, Haifa, Israel.
Habitat type: house dust.
Distribution: Spain, Israel, India, Colombia.

Malayoglyphus intermedius Fain, Cunnington and Spieksma, 1969

Comb./syn.: *Malayoglyphus intermedius* Fain, Cunnington and Spieksma, 1969, *Acarologia* 11, 124, figs. 1–6.

Type depository: holotype, allotype, paratypes: Institut Royal des Sciences naturelles de Belgique, Brussels; other paratypes: Rijksmuseum van Natuurlijke Historie, Leiden, the Netherlands, A. Fain, personal collection, A.M. Cunnington, personal collection, F.Th.M. Spieksma, personal collection.
Type locality and habitat data: in house dust, Djatinegara, Java, Indonesia.
Habitat type: house dust.
Distribution: Greece, Indonesia, Singapore, Nigeria, South Africa, Tahiti, Malaysia, Hong Kong, Taiwan, Japan, Thailand, India, Cuba, Colombia, Surinam.

Genus *Onychalges* Gaud and Mouchet, 1959

Onychalges Gaud and Mouchet, 1959
Type-species: *Megninia longitarsus* Bonnet, 1924 (cf. Gaud and Mouchet, 1959, *Ann. Parasitol. Humaine et Compar.*, 34, 176.)
Neonychalges Gaud in Gaud and Atyeo, 1983, *J. Georgia Entomol. Soc.*, 18 (4), 518. (redundant

replacement name for *Onychalges*; cf. Fain, 1988b in: Fain *et al.*, *Acariens et allergies*, p. 21.)

Capitonoecius Fain and Gaud, 1984

Type-species: *Capitonoecius spinitarsis* Fain and Gaud, 1984, *Acarologia* 25, 51. (cf. Fain, in Fain *et al.*, 1988, p. 21.)

Onychalges asaphospathus Gaud, 1968

Comb./syn.: *Onychalges asaphospathus* Gaud, 1968, *Acarologia* 10, 302, figs. 3a, 4a.

Type depository: J. Gaud, personal collection.

Type locality and habitat data: on Brown Twin-spot, *Clytospiza monteiri* (Passeriformes: Estrildidae), Yaoundé, Cameroons.

Habitat type: on birds (other records include Black and White Mannekin, *Euschistospiza dybovskyi* and *Spermestes bicolor* [though the latter habitat is probably 'accidental']).

Distribution: Cameroons.

Onychalges longitarsus (Bonnet, 1924)

Comb./syn.: *Megninia longitarsus* Bonnet, 1924, *Bull. Soc. Zool. Fr.*, 49, 164, fig. 19.

Onychalges asaphospathus Gaud, 1968, *Acarologia* 10, 302, figs. 3a, 4a.

Type depository: Trouessart Collection, Muséum national d'Histoire naturelle, Paris.
Type locality and habitat data: on *Nigrita canicapillata* and *N. bicolor*, Congo, and *Pyrenestes ostrinus* (Bas Ogôué).
Habitat type: on birds.
Distribution: Congo.

Onychalges nidicola Fain and Rosa, 1982

Comb./syn.: *Onychalges nidicola* Fain and Rosa, 1982, *Rev. Brasil. Biol.*, 42, 319.

Type depository: holotype, allotype and paratypes: Department of Zoology, ESALQ, University of São Paulo, Brazil; other paratypes: A. Fain, personal collection, A.E. Rosa, personal collection.
Type locality and habitat data: in nests of House Sparrow, *Passer domesticus* (Passeriformes: Ploceidae), Piracicaba, Brazil.
Habitat type: birds' nests.
Distribution: Brazil.

Onychalges odonturus Gaud, 1968

Comb./syn.: *Onychalges odonturus* Gaud, 1968, *Acarologia* 10, 306, figs. 3c, 4c, 6b.

Type depository: J. Gaud, personal collection.
Type locality and habitat data: on Blue-bill, *Spermophaga haematina* (Passeriformes: Estrildidae), Yaoundé, Cameroons.
Habitat type: on birds.
Distribution: Cameroons.

Onychalges pachyspathus Gaud, 1968

Comb./syn.: *Onychalges pachyspathus* Gaud, 1968, *Acarologia* 10, 306, figs. 3d, 4e.

Type depository: J. Gaud, personal collection.
Type locality and habitat data: on Orange-cheeked Waxbill, *Estrilda melpoda* (Passeriformes: Estrildidae), Yaoundé, Cameroons.
Habitat type: on birds (other records include Black-headed Waxbill, *Estrilda atricapilla*, Black-crowned waxbill, *E. nonnula*, both from Cameroons, and Common Waxbill, *E. astrild*, from Transvaal, South Africa.
Distribution: Cameroons, South Africa.

Onychalges schizurus Gaud, 1968

Comb./syn.: *Onychalges schizurus* Gaud, 1968, *Acarologia* 10, 307, figs. 3e, 4d.

Type depository: J. Gaud, personal collection.
Type locality and habitat data: on African Waxbill, *Lagonostica rubricata* (Passeriformes: Estrildidae), Yaoundé, Cameroons.
Habitat type: on birds (other records include the same host species from Transvaal, South Africa.
Distribution: Cameroons, South Africa.

Onychalges spinitarsis (Fain and Gaud, 1984)

Comb./syn.: *Capitonoecius spinitarsis* Fain and Gaud, 1984, *Acarologia* 25, 51, figs. 7–8.

Onychalges spinitarsis: Fain, 1988b, in: Fain *et al.*, 1988, *Acariens et allergies*, p. 21.

Type depository: holotype: Musée royal de l'Afrique Centrale, Tervuren, Belgium.
Type locality and habitat data: on Speckled Tinkerbird, *Pogonoiulus scolopaceus* (Piciformes: Capitonidae), Lima, Zaire.

Habitat type: on birds.
Distribution: Zaire.

Genus *Paramealia* Gaud, 1968

Paramealia Gaud, 1968
Type-species: *Onychalges ovatus* Gaud and Mouchet, 1959 (cf. Gaud, 1968, *Acarologia* 10, 310.)

Paramealia ovata (Gaud and Mouchet, 1959)

Comb./syn.: *Onychalges ovatus* Gaud and Mouchet, 1959, *Ann. Parasitol. Humaine et Compar.*, 34, 176.
Paramealia ovata: Gaud, 1968, *Acarologia* 10, 310, fig. 9.
Type depository: Muséum national d'Histoire naturelle, Paris.
Type locality and habitat data: from Black-necked Weaver, *Ploceus nigricollis brachypterus* (Passeriformes: Ploceidae), Cameroons.
Habitat type: on birds.
Distribution: Cameroons.

Genus *Pottocola* Fain, 1971

Pottocola Fain, 1971
Type-species: *Pottocola scutata* Fain, 1971
Pottocola (*Capitonocoptes*) Fain and Gaud, 1984, *Acarologia* 25, 48.
Type-species: *Pottocola* (*Capitonocoptes*) *ventriscutata* Fain and Gaud, 1984
Pottocola: Fain, 1988b in: Fain *et al.*, *Acariens et allergies*, p. 20.

Subgenus *Pottocola* Fain, 1971

Pottocola (*Pottocola*): Fain and Gaud, 1984, *Acarologia* 25, 48.

Pottocola (*Pottocola*) *scutata* Fain, 1971

Comb./syn.: *Pottocola scutata* Fain, 1971, *Bull. Inst. roy. Sci. nat. Belg.*, 47 (8), 2, figs. 1–2.
Pottocola (*Pottocola*) *scutata* Fain, 1971: Fain and Gaud, 1984, *Acarologia* 25, 48.
Type depository: Musée royal de l'Afrique Centrale, Tervuren, Belgium.
Type locality and habitat data: dried skin of Potto, *Perodicticus potto*, Mutwanga, Kivu, Zaire.

Habitat type: mammal pelt (possibly an 'accidental').
Distribution: Zaire.

Subgenus *Capitonocoptes* Fain and Gaud, 1984

Pottocola (*Capitonocoptes*) Fain and Gaud, 1984, *Acarologia* 25, 48.
Type-species: *Pottocola* (*Capitonocoptes*) *ventriscutata* Fain and Gaud, 1984

Pottocola (*Capitonocoptes*) *longipilis* Fain and Gaud, 1984

Comb./syn.: *Pottocola* (*Capitonocoptes*) *longipilis* Fain and Gaud, 1984, *Acarologia* 25, 49, figs. 3–4.
Type depository: holotype: Musée royal de l'Afrique Centrale, Tervuren, Belgium.
Type locality and habitat data: on Speckled Tinkerbird, *Pogonoiulus scolopaceus* (Piciformes: Capitonidae), Ebeva, Togo.
Habitat type: on birds.
Distribution: Togo.

Pottocola (*Capitonocoptes*) *lybius* Fain and Gaud, 1984

Comb./syn.: *Pottocola* (*Capitonocoptes*) *lybius* Fain and Gaud, 1984, *Acarologia* 25, 51, figs. 5–6.
Type depository: holotype: Musée royal de l'Afrique Centrale, Tervuren, Belgium.
Type locality and habitat data: on Bearded Barbet, *Lybius dubius* (Piciformes: Capitonidae), Nanergou, Togo (other records include Viellot's Barbet, *Lybius vielloti*, from Ebeva, Togo and Kasengi, Zaire, and Black-collared Barbet, *Lybius torquatus cognicus*, and Red-faced Barbet, *Lybius rubrifacies*, from unknown central African localities.
Habitat type: on birds.
Distribution: Central Africa.

Pottocola (*Capitonocoptes*) *ventriscutata* Fain and Gaud, 1984

Comb./syn.: *Pottocola* (*Capitonocoptes*) *ventriscutata* Fain and Gaud, 1984, *Acarologia* 25, 48, figs. 1–2.
Type depository: holotype: Musée royal de l'Afrique Centrale, Tervuren, Belgium.
Type locality and habitat data: on *Lybius bidentatus* (Piciformes: Capitonidae), Rutshuru, Zaire.

Habitat type: on birds.
Distribution: Zaire.

Genus *Sturnophagoides* Fain, 1967a

Dermatophagoides (Sturnophagoides) Fain, 1967a
Type-species: *Dermatophagoides (Sturnophagoides) bakeri* Fain, 1967a

Sturnophagoides bakeri Fain, 1967

Comb./syn.: *Dermatophagoides (Sturnophagoides) bakeri* Fain, 1967, *Acarologia* 9, 215, figs. 47, 48.
Sturnophagoides bakeri Fain, 1967: Fain, 1967, *Acarologia* 9, 871.
Sturnophagoides bakeri Fain, 1967: Baker, Delfinado and Abbatiello, 1976, *J. NY Ent. Soc.*, 84, 53, figs. 15–25.
Type depository: United States National Museum, Washington.
Type locality and habitat data: from Common Starling, *Sturnus vulgaris* (Passeriformes: Sturnidae), Pulaski County, Virginia, USA.
Habitat type: birds and their nests.
Distribution: USA.
Remarks: the type series consists of two females. The male was first described by Baker *et al.* (1976).

Sturnophagoides brasiliensis Fain, 1967b

Comb./syn.: *Sturnophagoides brasiliensis* Fain, 1967b, *Acarologia* 9, 876, figs. 6–10.
Sturnophagoides halterophilus Fain and Feinberg, 1970, *Acarologia* 12, 164, figs. 1–7.
Type depository: holotype, allotype and paratypes: Institut Royal des Sciences naturelles de Belgique, Brussels; other paratypes: A. Fain, personal collection.
Type locality and habitat data: in house dust, Tejipio, nr. Recife, Brazil.
Habitat type: house dust.
Distribution: Spain, Brazil, Malaysia, Singapore, Indonesia.

Sturnophagoides petrochelidonis Cuervo and Dusbabek, 1987

Comb./syn.: *Sturnophagoides petrochelidonis* Cuervo and Dusbabek, 1987, *Poeyana* No. 335, 2, figs. 1–4.

Sturnophagoides (Sturnophagoides) petrochelidonis: Cruz, 1988, *Poeyana* No. 361, 12.
Type depository: holotype: Institute of Zoology, Academy of Sciences of Cuba, Havana; paratypes: Czech Academy of Sciences, České Budějovice.
Type locality and habitat data: nest of Cave Swallow, *Petrochelidon fulva* (Passeriformes: Hirundinidae), Cuba.
Habitat type: birds' nests.
Distribution: Cuba.

Subfamily PARALGOPSINAE Fain, 1988b

Paralgopsinae Fain, 1988b in: Fain *et al.*, *Acariens et allergies*, p. 21.

Genus *Paralgopsis* Gaud and Mouchet, 1959

Paralgopsis Gaud and Mouchet, 1959
Type-species: *Dermoglyphus (Paralges) paradoxus* Trouessart, 1899

Paralgopsis ctenodontus Gaud, 1968

Comb./syn.: *Paralgopsis ctenodontus* Gaud, 1968, *Acarologia* 10, 308, figs. 7, 8a.
Type depository: Muséum national d'Histoire naturelle, Paris.
Type locality and habitat data: on Scarlet Macaw, *Ara macao* (Psittaciformes: Psittacidae), Brazil.
Habitat type: on birds.
Distribution: Brazil.

Paralgopsis paradoxus (Trouessart, 1899)

Comb./syn.: *Dermoglyphus (Paralges) paradoxus* Trouessart, 1899, *Bull Etud. sci.* Angers 28, 15.
Paralgopsis paradoxus: Gaud, 1968, *Acarologia* 10, 308, fig. 8b.
Type depository: Trouessart Collection, Muséum national d'Histoire naturelle, Paris.
Type locality and habitat data: on White-eared Conure, *Pyrrhura leucotis* (Psittaciformes: Psittacidae), Colombia.
Habitat type: on birds.
Distribution: Colombia.

Appendix 1b

A catalogue of members of the genus *Blomia*

Blomia Oudemans, 1928

Blomia Oudemans, 1928

Type-species: *Glycyphagus tjibodas* Oudemans, 1910 (cf. Oudemans, 1928, *Ent. Ber.*, 7, 348.)

Blomia freemani Hughes, 1948

Comb./Syn.: *Blomia freemani* Hughes, 1948, *The Mites Associated with Stored Food Products*, p. 51, figs. 61–66.

Type depository: Slough Laboratories.
Type locality: Northern Ireland.
Habitat: stored wheat.
Distribution: U.K.

Blomia gracilipes (Banks, 1917)

Comb./Syn.: *Chortoglyphus gracilipes* Banks, 1917, *Ent. News* 28, 199, pl. 14, fig. 2.

Blomia gracilipes Banks, 1917: Fain, Hyland and Tadowski, 1977, *Ent. News* 88, 267, figs. 1–3.

Type depository: U.S. National Museum of Natural History, Washington.
Type locality: Tampa, Florida.
Habitat: in tobacco infested with the Cigarette Beetle.
Distribution: USA.

Blomia khaliovae Zachvatkin, 1949

Comb./Syn.: *Blomia khaliovae* Zachvatkin, 1949, *Entomologicheskoe Obozrenie*, Moscow, 30, 287.

Type depository: type specimens lost (cf. van Bronswijk, Cock and Oshima, 1973, p. 490).
Type locality: Baku, Azerbaijan.

Habitat: stored barley and millet.
Distribution: Azerbaijan.

Blomia kulagini Zachvatkin, 1936

Comb./Syn.: *Blomia kulagini* Zachvatkin, 1936, *Bull. Soc. Nat. Moscow Sect. Biol. N.S.* 45, 263.

Type depository: type specimens lost (cf. van Bronswijk, Cock and Oshima, 1973, p. 490).
Type locality: Moscow.
Habitat: stored wheat in granaries (type locality); crevices in school gymnasium floor.
Distribution: Russia, Japan.

Blomia thori Zachvatkin, 1936

Comb./Syn.: *Blomia thori* Zachvatkin, 1936, *Bull. Soc. Nat. Moscow Sect. Biol. N.S.* 45, 263.

Type depository: Entomological Institute, Czech Academy of Science, Prague.
Type locality: Smolensk.
Habitat: stored flax seed and other grains.
Distribution: Russia.

Blomia tjibodas Oudemans, 1910

Comb./Syn.: *Glycyphagus tjibodas* Oudemans, 1910, *Ent. Ber.* 3, 74.

Glycyphagus tjibodas: Oudemans, 1911, *Ent. Ber.* 4, 102.

Type depository: Rijksmuseum van Natuulijke Historie, Leiden.
Type locality: Tijbodas, Netherlands.
Habitat: house dust.
Distribution: Netherlands.

Blomia tropicalis Bronswijk, Cock and Oshima, 1973

Comb./Syn.: *Blomia tropicalis* Bronswijk, Cock and Oshima, 1973, *Acarologia* 15, 475, figs. 1–9.

Type depository: Rijksmuseum van Natuulijke Historie, Leiden.

Type locality: Bandung, Indonesia.

Habitat: house dust, dust in tobacco factory, stored rice.

Distribution: pantropics and subtropics.

Appendix 2

Distribution of species of domestic mites, based on published surveys

Altitude class: 1 = 0–200 m., 2 = 201–500 m., 3 = 501–1000 m., 4 = 1001– 2000 m., 5 = 2001–3000 m., 6 = 3001–5000 m.

Country & location	Alt. (m)	Alt. class	D. pt.	D. f	D. m	E. m	M. i	Hirstia spp.	A. s	Other A. spp.	T. p	Other T. spp.	A. o	S. m	L. d	G. d	B. t	C. a	G. f	C. e	Other C. spp.	Reference
ALGERIA																						
Ain-Roua	1100	4	1	1		1				1					1	1	1		1			Louadi & Robaux 1992
Ain-Touta	917	3							1													Louadi & Robaux 1992
Algiers	1	1	1	1		1																Abed-Benamara et al. 1983
Annaba	20	1	1	1		1			1	1	1											Louadi & Robaux 1992
Arris	1171	4												1								Louadi & Robaux 1992
Batna	1040	4	1	1																		Louadi & Robaux 1992
Bejaia	9	1	1	1		1			1	1	1				1		1					Louadi & Robaux 1992
Biskra	124	1	1			1																Louadi & Robaux 1992
Bou-Snib	886	3	1	1		1			1						1	1	1		1			Louadi & Robaux 1992
El-Kantara	513	3	1						1	1												Louadi & Robaux 1992
El-Oued	70	1	1																	1		Louadi & Robaux 1992
Jijei	6	1	1	1		1										1	1					Louadi & Robaux 1992
Oum-El-Bouaghi	950	3	1			1			1	1	1				1	1						Louadi & Robaux 1992
Serraidi	960	3	1	1		1			1	1									1			Louadi & Robaux 1992
Setif	1079	4	1	1		1			1						1	1	1					Louadi & Robaux 1992
Skikda	42	1	1	1		1			1	1	1					1			1			Louadi & Robaux 1992
Souk-Ahras	665	3	1			1			1						1	1		1				Louadi & Robaux 1992
Tebessa	885	3	1						1													Louadi & Robaux 1992
Touggourt	69	1	1																			Louadi & Robaux 1992
Zitouna	548	3	1						1	1	1				1	1	1		1			Louadi & Robaux 1992
ARGENTINA																						
San Miguel de Tucumán	431	2	1	1		1				1					1	1					1	Martínez Canzonieri et al. 1996
AUSTRALIA																						
New South Wales unspecified			1																			Trinca et al. 1969
Queensland, unspecified			1																			Trinca et al. 1969
Adelaide	72	1	1																			Morgan et al. 1974
Alice Springs	609	3	1	1																	1	Tovey et al. 2000a
Belmont	1	1	1	1																		Green et al. 1986
Brisbane	25	1	1									1										Domrow 1970
Broken Hill	304	2	1			1															1	Tovey et al. 2000a
Bunbury	1	1	1			1				1			1	1						1	1	Colloff et al. 1991
Busselton	1	1	1																		1	Green et al. 1986
Cairns	1	1	1																			Domrow 1970
Camden	291	2	1	1		1															1	Tovey et al. 2000a
Canberra	609	3	1			1																Colloff (unpublished)
Charleville	299	2	1																			Domrow 1970

continued...

Appendix 2 Continued

Country & location	Alt. (m)	Alt. class	D. pt.	D. f	D. m	E. m	M. i	Hirstia spp.	A. s	Other A. spp.	T. p	Other T. spp.	A. o	S. m	L. d	G. d	B. t	C. a	G. f	C. e	Other C. spp.	Reference
Edward River	3	1	1	1		1											1				1	Tovey et al. 2000a
Hobart	111	1	1																			Murton & Madden 1977
Hopevale	86	1	1	1		1											1				1	Tovey et al. 2000a
Lismore	134	1	1	1		1											1				1	Tovey et al. 2000a
Longreach	210	1	1																			Domrow 1970
Mackay	0	1	1																			Domrow 1970
Melbourne	115	1	1			1																Blythe 1976
Mt. Isa	439	2	1																			Domrow 1970
Narrabri	299	2	1																		1	Tovey et al. 2000a
Perth	33	1	1			1				1				1						1	1	Colloff et al. 1991
Rockhampton	1	1	1																			Domrow 1970
Sydney	1	1	1	1																		Green 1983
Sydney	1	1	1																			Tovey 1975
Townsville	1	1	1																			Domrow 1970
Wagga Wagga	148	1	1																			Green et al. 1986
Wagga Wagga	148	1	1												1						1	Tovey et al. 2000a
BARBADOS	214	2	1																			Fain 1966a
BARBADOS	214	2	1	1																		Pearson & Cunnington 1973
BELGIUM																						
Malines (Mechelen)	4	1	1																			Fain 1965; 1966a, b
Antwerp	4	1	1																			Fain 1965; 1966a, b
Brussels	77	1	1																			Fain 1965; 1966a, b
La Louviere	111	1	1																			Fain 1965; 1966a, b
Louvain	35	1	1																			Fain 1965; 1966a, b
Nethen	66	1	1	1		1										1			1			Gridelet & Lebrun 1973
Ostend	2	1	1																			Fain 1965; 1966a, b
BELARUS																						
Minsk	199	1	1	1																		Zheltikova et al. 1985
BRAZIL																						
Fernando do Noronha	1	1	1			1																Galvao & Guitton 1986
'Zona da Mata'	695	3	1	1		1											1	1				Oliveira & Daemon 2003
Aracaju	2	1	1														1	1				Rosa & Flechtman 1979
Belem	1	1	1														1					Rosa & Flechtman 1979
Belo Horizonte	876	3	1	1		1											1	1				Galvao & Guitton 1986
Belo Horizonte	876	3	1	1		1										1						Greco et al. 1974
Belo Horizonte	876	3	1	1		1											1	1				Moreira 1975
Belo Horizonte	876	3																1				Moreira 1978
Boa Vista do Rio Branco	75	1	1														1					Rosa & Flechtman 1979
Brasilia	1080	4	1																			Fain 1966b; 1967a
Campinas	680	3	1	1		1						1		1			1				1	Oliveira et al. 1999; 2003
Campo Grande	586	3	1	1													1					Galvao & Guitton 1986
Cuiaba	195	1	1														1					Galvao & Guitton 1986
Curitiba	914	3	1			1																Galvao & Guitton 1986
Curitiba	914	3	1														1				1	Rosário Filho et al. 1992
Florianopolis	27	1	1	1													1	1				Galvao & Guitton 1986
Fortaleza	1	1	1														1					Galvao & Guitton 1986

continued...

Appendix 2 Continued

Country & location	Alt. (m)	Alt. class	D. pt.	D. f	D. m	E. m	M. i	Hirstia spp.	A. s	Other A. spp.	T. p	Other T. spp.	A. o	S. m	L. d	G. d	B. t	C. a	G. f	C. e	Other C. spp.	Reference
Goiania	791	1	1	1													1					Galvao & Guitton 1986
Joao Pessoa	4	1	1														1					Galvao & Guitton 1986
Juiz de Fora	695	3	1	1		1					1				1		1	1			1	Ezequiel *et al.* 2001
Londrina	576	3	1	1		1					1						1	1				da Silva *et al.* 2005
Macapa	1	1	1														1					Galvao & Guitton 1986
Maceio	1	1	1	1													1					Galvao & Guitton 1986
Manaus	35	1	1	1													1					Galvao & Guitton 1986
Natal	1	1	1	1													1					Galvao & Guitton 1986
Porto Alegre	55	1	1								1						1				1	Bernd *et al.* 1994
Porto Alegre	55	1	1			1											1					Galvao & Guitton 1986
Porto Velho	5	1	1	1													1					Galvao & Guitton 1986
Recife	1	1	1																			Fain 1966b; 1967a
Recife	1	1	1														1					Galvao & Guitton 1986
Rio Branco	156	1	1														1					Galvao & Guitton 1986
Rio Claro	609	3	1	1		1											1	1				Rosa & Flechtman 1979
Rio de Janeiro	11	1	1	1		1											1	1				Galvao & Guitton 1986
Salvador	1	1	1	1				1			1						1		1		1	Baqueiro *et al.* 2006
Salvador	1	1	1																			Fain 1966b; 1967a
Salvador	1	1	1	1													1	1				Galvao & Guitton 1986
Salvador	1	1	1																		1	Serraville & Medeiros 1998
Sao Luis	7	1	1	1													1					Galvao & Guitton 1986
Sao Paulo	638	3	1																			Amaral 1968
Sao Paulo	638	3	1																			Fain 1966b; 1967a
Sao Paulo	638	3	1														1					Galvao & Guitton 1986
Teresina	48	1	1	1													1					Galvao & Guitton 1986
Vitoria	18	1	1	1													1					Galvao & Guitton 1986
BRUNEI DARUSSALAM			1	1													1	1	1			Woodcock & Cunnington 1980
Buau (Sukang)	89	1	1	1				1									1					Woodcock & Cunnington 1980
Bukit Masin	1	1	1					1														Woodcock & Cunnington 1980
Kampong Bebatek (Masin)	1	1	1	1				1									1					Woodcock & Cunnington 1980
Kampong Biang	128	1																				Woodcock & Cunnington 1980
Kampong Lopat	198	1	1					1									1					Woodcock & Cunnington 1980
Kuala Belait	3	1	1	1				1										1	1			Woodcock & Cunnington 1980
Labi (Kampong Teraja)	98	1	1	1				1									1	1	1			Woodcock & Cunnington 1980
Panchat			1	1													1					Woodcock & Cunnington 1980
BULGARIA																						
Silistra	7	1	1	1		1					1				1			1		1	1	Todorov 1978
BURMA																						
Rangoon	9	1		1																		Griffiths & Cunnington 1971
Rangoon	9	1	1	1			1	1									1				1	Htut *et al.* 1991
BURUNDI																						
Bujumbura	2542	5	1																			Fain 1988

continued...

Appendix 2 Continued

Country & location	Alt. (m)	Alt. class	D. pt.	D. f	D. m	E. m	M. i	Hirstia spp.	A. s	Other A. spp.	T. p	Other T. spp.	A. o	S. m	L. d	G. d	B. t	C. a	G. f	C. e	Other C. spp.	Reference
CANADA																						
Toronto	106	1		1																		Griffiths & Cunnington 1971
Vancouver	72	1	1	1																		Murray & Zuk 1981
CHILE																						
Ancud	1	1	1						1		1				1	1		1	1	1		Artigas & Casanueva 1983; C & A 1985
Antofagasta	276	2	1								1				1			1		1		Artigas & Casanueva 1983; C & A 1985
Arauco	5	1																1		1		Artigas & Casanueva 1983; C & A 1985
Arica	16	1		1														1				Artigas & Casanueva 1983; C & A 1985
Cauquenes	136	1									1				1							Artigas & Casanueva 1983; C & A 1985
Chiguayante	250	1	1								1				1			1	1			Artigas & Casanueva 1983; C & A 1985
Chillán	130	1	1	1							1				1	1		1		1		Artigas & Casanueva 1983; C & A 1985
Coelemu	34	1									1				1			1				Artigas & Casanueva 1983; C & A 1985
Concepción	151	1	1	1					1		1				1	1		1	1	1		Artigas & Casanueva 1983; C & A 1985
Constitucion	76	1	1																	1		Artigas & Casanueva 1983; C & A 1985
Coquimbo	16	1													1			1				Artigas & Casanueva 1983; C & A 1985
Coronel	84	1													1			1				Artigas & Casanueva 1983; C & A 1985
Curico	215	2									1				1					1		Artigas & Casanueva 1983; C & A 1985
Dichato	155	1	1	1							1				1	1		1		1		Artigas & Casanueva 1983; C & A 1985
Gorbea	120	1	1															1				Artigas & Casanueva 1983; C & A 1985
La Cruz	154	1													1					1		Artigas & Casanueva 1983; C & A 1985
La Serena	29	1	1	1												1			1	1		Artigas & Casanueva 1983; C & A 1985
Lanco	116	1							1						1							Artigas & Casanueva 1983; C & A 1985
Linares	165	1	1																			Artigas & Casanueva 1983; C & A 1985
Los Anjeles	137	1	1												1	1		1		1		Artigas & Casanueva 1983; C & A 1985
Nueva Imperial	46	1							1													Artigas & Casanueva 1983; C & A 1985
Ovalle	261	1							1											1		Artigas & Casanueva 1983; C & A 1985
Pitrufquén	81	1	1						1											1		Artigas & Casanueva 1983; C & A 1985
Puerto Montt	1	1													1			1				Artigas & Casanueva 1983; C & A 1985
Puerto Saavedra	7	1																		1		Artigas & Casanueva 1983; C & A 1985

continued...

Appendix 2 Continued

Country & location	Alt. (m)	Alt. class	D. pt.	D. f	D. m	E. m	M. i	Hirstia spp.	A. s	Other A. spp.	T. p	Other T. spp.	A. o	S. m	L. d	G. d	B. t	C. a	G. f	C. e	Other C. spp.	Reference
Puerto Varas	88	1													1			1				Artigas & Casanueva 1983; C & A 1985
Punta Arenas	35	1	1								1				1							Artigas & Casanueva 1983; C & A 1985
Purranque	235	2														1		1		1		Artigas & Casanueva 1983; C & A 1985
Quillota	141	1	1															1		1		Artigas & Casanueva 1983; C & A 1985
Quilpue	129	1													1					1		Artigas & Casanueva 1983; C & A 1985
Rancagua	518	3									1				1	1				1		Artigas & Casanueva 1983; C & A 1985
San Bernardo	545	3									1					1		1		1		Artigas & Casanueva 1983; C & A 1985
Santiago	522	3	1	1							1				1			1		1		Artigas & Casanueva 1983; C & A 1985
Talca	106	1													1							Artigas & Casanueva 1983; C & A 1985
Talcahuano	1	1	1								1				1			1	1	1		Artigas & Casanueva 1983; C & A 1985
Temuco	104	1	1								1				1	1		1	1	1		Artigas & Casanueva 1983; C & A 1985
Tomé	120	1		1							1				1			1	1	1		Artigas & Casanueva 1983; C & A 1985
Valdivia	5	1	1												1	1		1				Artigas & Casanueva 1983; C & A 1985
Valdivia	5	1	1	1			1				1				1	1		1	1	1		Franjola & Malonnek 1995
Valparaiso	164	1	1								1				1					1		Artigas & Casanueva 1983; C & A 1985
Vicuna	710	3	1	1							1									1		Artigas & Casanueva 1983; C & A 1985
Villa Alemana	155	2													1			1		1		Artigas & Casanueva 1983; C & A 1985
Vina del Mar	298	2		1														1		1		Artigas & Casanueva 1983; C & A 1985
Yaldad	4	1									1				1					1		Artigas & Casanueva 1983; C & A 1985
CHINA																						
Anhui Province			1																			Wen et al. 1988
Henan Province			1																			Wen et al. 1988
Hainan Is			1																			Wen et al. 1988
GuangZhou	1	1	1	1		1	1								1							Lai et al. 1982
Haikou	8	1	1														1					Zhu et al. 2007
Shanghai	8	1	1	1		1	1								1							Cai & Wen 1989
Shanghai	8	1	1	1		1	1	1		1		1	1		1							Wen et al. 1988; 1991
Shanghai	8	1	1	1		1	1								1							Wang & Wen 1997
Shenzhen	35	1	1																			Wen et al. 1988
COLOMBIA																						
8 climatic zones			1	1		1												1	1			Sanchez-Medina & S.-Gutierrez 1973
Apure	91	1										1						1				Charlet et al. 1977a, b
Bogotá	2620	5	1			1																Charlet et al. 1977c; 1978
Bogotá	2620	5	1	1		1																Charlet et al. 1979

continued...

Appendix 2 Continued

Country & location	Alt. (m)	Alt. class	D. pt.	D. f	D. m	E. m	M. i	Hirstia spp.	A. s	Other A. spp.	T. p	Other T. spp.	A. o	S. m	L. d	G. d	B. t	C. a	G. f	C. e	Other C. spp.	Reference
Buenaventura	1	1	1	1		1	1															Charlet *et al.* 1979
Cali	745	3	1	1		1	1															Charlet *et al.* 1979
Capa Rosa	1500	4															1					Charlet *et al.* 1977a, b
Cartagena	13	1	1														1					Fernandez-Caldas *et al.* 1993
Charta	2192	4	1									1					1					Charlet *et al.* 1977a, b
Charta	2192	5	1	1													1			1		Sanchez-Medina & Zarante 1996
Chinavita	1603	4															1					Charlet *et al.* 1977a, b
Chocontá	2689	5	1								1	1						1				Charlet *et al.* 1977a, b
Chocontá	2689	5	1	1													1	1				Sanchez-Medina & Zarante 1996
Concepcion	500	2																1				Charlet *et al.* 1977a, b
Cumaral	413	2	1	1			1										1					Charlet *et al.* 1977a, b
Dagua	1317	4	1	1		1	1															Charlet *et al.* 1979
Distracción	204	2	1																			Charlet *et al.* 1977a, b
El Cerrito	921	3															1					Charlet *et al.* 1977a, b
Espinal	319	2	1														1					Charlet *et al.* 1977a, b
Fuente de Oro	293	2								1												Charlet *et al.* 1977a, b
Fusagasugá	1983	4	1	1		1																Charlet *et al.* 1979
Garagoa	1928	4	1						1									1				Charlet *et al.* 1977a, b
Garagoa	1928	4	1	1													1	1		1		Sanchez-Medina & Zarante 1996
Girardot	276	2	1	1		1																Charlet *et al.* 1979
Granada	336	2															1					Charlet *et al.* 1977a, b
Granada	111	1	1														1					Charlet *et al.* 1977a, b
Guacarí	911	3	1			1											1	1				Charlet *et al.* 1977a, b
Guaitarilla	2701	5										1										Charlet *et al.* 1977a, b
Guamo	315	2	1														1					Charlet *et al.* 1977a, b
La Calera	2376	5	1															1				Charlet *et al.* 1977a, b
La Calera	2376	5	1	1														1		1		Sanchez-Medina & Zarante 1996
Montería	29	1		1							1						1					Charlet *et al.* 1977a, b
Morrocoy	60	1																1				Charlet *et al.* 1977a, b
Nayas	1850	4													1		1	1				Charlet *et al.* 1977a, b
Obenuco	2585	5	1	1																		Sanchez-Medina & Zarante 1996
Ocaña	1272	4	1	1													1					Charlet *et al.* 1977a, b
Ospina	2587	5																1		1		Charlet *et al.* 1977a, b
Pasca	2367	5	1								1											Charlet *et al.* 1977a, b
Piedecuesta	1178	4															1					Charlet *et al.* 1977a, b
Pitalito	1360	4	1															1				Charlet *et al.* 1977a, b
Pitalito	1360	4	1	1														1		1		Sanchez-Medina & Zarante 1996
Río Frio	750	3	1														1	1				Charlet *et al.* 1977a, b
Río Negro Sur	1088	4										1					1	1				Charlet *et al.* 1977a, b
Sabanilla	3000	5	1								1	1										Charlet *et al.* 1977a, b
Sahagún	72	1									1						1					Charlet *et al.* 1977a, b
Saladito	1673	4	1	1			1	1														Charlet *et al.* 1979
San Ciro	1370	4	1								1							1				Charlet *et al.* 1977a, b
San Juan de Arama	460	2	1																			Charlet *et al.* 1977a, b
Santiago de Tolú	1	1	1	1			1										1	1				Rodriguez Monterroza *et al.* 2006

continued...

Appendix 2 Continued

Country & location	Alt. (m)	Alt. class	D. pt.	D. f	D. m	E. m	M. i	Hirstia spp.	A. s	Other A. spp.	T. p	Other T. spp.	A. o	S. m	L. d	G. d	B. t	C. a	G. f	C. e	Other C. spp.	Reference
Sapuyes	3105	5									1							1				Charlet *et al.* 1977a, b
Silvia	2683	5	1	1																		Charlet *et al.* 1979
Sincelejo	220	4									1					1						Charlet *et al.* 1977a, b
Soatá	2076	5	1															1				Charlet *et al.* 1977a, b
Suratá	1887	4	1			1											1	1				Charlet *et al.* 1977a, b
Tenerife	6	1	1														1	1				Charlet *et al.* 1977a, b
Tenza	1630	4	1								1						1					Charlet *et al.* 1977a, b
Tenza	1630	4	1	1																1		Sanchez-Medina & Zarante 1996
Tibana	1991	4							1													Charlet *et al.* 1977a, b
Timaná	1121	4																1				Charlet *et al.* 1977a, b
Tipacoque	2174	5		1		1																Charlet *et al.* 1977a, b
Tuquerres	3150	6	1									1							1		1		Sanchez-Medina & Zarante 1996
Ubatoque	2500	5	1																			Charlet *et al.* 1977a, b
Viso Elias	1400	4	1														1	1				Charlet *et al.* 1977a, b
Viso Elias	1400	4	1	1																		Sanchez-Medina & Zarante 1996
Vista Hermosa	514	2															1	1				Charlet *et al.* 1977a, b
CONGO																						
Bokela	449	2	1																			Fain 1988
Bukavu	1559	4	1																			Fain 1988
Kimpako	598	3	1																			Fain 1967a
Kinshasa	178	2	1																			Fain, 1966a, b; 1967a
Kinsuka	117	2	1																			Fain 1967a
Lubumbashi	1209	4	1																			Fain 1988
Minkao			1																			Fain 1967a
COSTA RICA																						
San Jose	1149	4	1			1				1					1			1	1	1		Vargas & Mariena 1991
CROATIA																						
Pazin	278	2	1	1		1	1															Blythe 1976
Zadar	22	1	1	1		1				1					1		1				1	Macan *et al.* 2003
Zagreb	130	1	1	1		1			1	1					1						1	Macan *et al.* 2003
CUBA																						
Havana	5	1	1				1	1											1		1	Cuervo *et al.* 1983
Havana	5	1																				Dusbabek *et al.* 1982
Maniadero	3	1																				Dusbabek *et al.* 1982
Playa Larga	1	1																				Dusbabek *et al.* 1982
Santo Tomas	5	1																				Dusbabek *et al.* 1982
CZECH REPUBLIC																						
Unspecified localities			1	1											1							Samsinák *et al.* 1972; 1974
25 localities			1	1		1		1							1	1		1	1	1		Samsinák *et al.* 1978a, b
33 localities			1	1		1									1	1		1	1	1		Vobrazkova *et al.* 1986
Prague	245	2	1	1		1			1	1											1	Dusbabek 1975; 1979
Prague	245	2		1																		Vobrazkova *et al.* 1985
Prague & environs	245	2	1	1		1		1	1	1	1				1	1		1	1	1		Vobrazkova *et al.* 1979
DENMARK																						
Århus	1	1	1	1		1																Korsgaard 1979
Copenhagen	1	1	1	1		1																Haarløv & Alani 1970; Alani & Haarlov 1972

Appendix 2 Continued

Country & location	Alt. (m)	Alt. class	D. pt.	D. f	D. m	E. m	M. i	Hirstia spp.	A. s	Other A. spp.	T. p	Other T. spp.	A. o	S. m	L. d	G. d	B. t	C. a	G. f	C. e	Other C. spp.	Reference
Copenhagen	1	1	1	1																		Andersen 1984
Thisted	1	1		1												1						Hallas & Korsgaard 1983
ECUADOR																						
Cuenca	2450	5	1	1		1									1			1	1		1	Valdivieso et al. 2006
Guayaquil	46	1	1	1		1					1				1			1			1	Valdivieso et al. 2006
Quito	2763	5	1	1		1					1				1			1	1		1	Valdivieso et al. 2006
EGYPT																						
Giza Governorate	22	1	1																	1		Hafez et al. 1989
Alexandria	10	1	1	1							1					1					1	Sadaka et al. 2000
Alexandria	10	1	1	1							1		1	1		1			1		1	Rezk et al. 1996
Cairo	23	1		1																		Frankland & El Hefny 1971
Cairo	23	1		1																		Griffiths & Cunnington 1971
Cairo	23	1	1	1				1			1				1						1	Koraiem et al. 1999
Mansoura	13	1	1	1																		El-Shazly et al. 2006
Tanta	11	1	1	1							1						1					Gamal Eddin et al. 1982
EIRE																						
Various localities			1																			Spieksma 1967
Unspecified localities			1																			Fain 1988a
FAEROES																						
Torshavn	1	1	1																			Korsgaard, in Hallas et al 2004
FINLAND																						
Helsinki	24	1	1	1		1			1	1	1	1			1	1		1				Stenius & Cunnington 1972
Ilomantsi	147	1	1	1		1																Stenius & Cunnington 1972
Uukuniemi	84	1	1	1		1																Stenius & Cunnington 1972
FRANCE																						
Aspach	345	2	1			1														1		Arajou-Fontaine et al. 1973
Barr	199	1	1																			Arajou-Fontaine et al. 1973
Bénestroff	236	2	1													1						Arajou-Fontaine et al. 1973
Briançon	1572	4	1	1		1																Vervloet et al. 1982
Briançon	1572	4	1	1		1																Charpin et al. 1972; 1986
Carspach	366	2	1	1							1					1				1		Arajou-Fontaine et al. 1973
Clermont-Ferrand	363	2	1								1								1	1		Arajou-Fontaine et al. 1973
Dorlisheim	176	1	1																			Arajou-Fontaine et al. 1973
Eckbolsheim	145	1	1	1		1					1									1		Arajou-Fontaine et al. 1973
Eschau	144	1	1			1					1					1						Arajou-Fontaine et al. 1973
Fegersheim	146	1	1																			Arajou-Fontaine et al. 1973

continued...

Appendix 2 Continued

Country & location	Alt. (m)	Alt. class	D. pt.	D. f	D. m	E. m	M. i	Hirstia spp.	A. s	Other A. spp.	T. p	Other T. spp.	A. o	S. m	L. d	G. d	B. t	C. a	G. f	C. e	Other C. spp.	Reference
Grenoble	220	2	1	1		1																Lascaud 1978
Grenoble	220	2	1	1		1										1						Bruttmann 1975
Imling	259	2	1													1	1					Arajou-Fontaine et al. 1973
La Wantzenau	130	1	1																			Arajou-Fontaine et al. 1973
Lutzelhouse	293	2	1			1			1										1			Arajou-Fontaine et al. 1973
Marseille	54	1	1	1		1																Charpin et al. 1972; 1986
Marseille	54	1	1	1		1																Penaud et al. 1971; 1972
Marseille	54	1	1	1		1																Vervloet et al. 1982
Marseille	54	1		1																		Griffiths & Cunnington 1971
Molsheim	176	2	1	1		1										1		1		1		Arajou-Fontaine et al. 1973
Monswiller	202	2	1	1		1										1						Arajou-Fontaine et al. 1973
Montpellier & environs	30	1	1	1		1						1				1			1		1	Rousset 1971; Guy et al. 1972
Munster	229	2	1														1					Arajou-Fontaine et al. 1973
Nancy	200	2	1																			Percebois et al. 1972
Paris	35	1	1			1																Akoun et al. 1972
Robertsau	136	1	1	1		1											1		1			Arajou-Fontaine et al. 1973
Rothau	482	2																				Arajou-Fontaine et al. 1973
Saint-Dié	349	2	1																			Arajou-Fontaine et al. 1973
Sarrebourg	279	2	1	1		1					1							1	1			Arajou-Fontaine et al. 1973
Sarreguemines	246	2	1	1		1													1			Arajou-Fontaine et al. 1973
Strasbourg	140	1	1	1		1					1					1						Arajou-Fontaine et al. 1973
Strasbourg	140	1	1			1								1				1				Araujo-Fontaine et al. 1972
FRENCH POLYNESIA																						
Paea (Clipperton Is.)	177		1	1			1															Fain 1988
GEORGIA																						
Batum	45	1	1	1												1		1				Dubinina & Plietnev 1978
Sukhumi	12	1	1				1							1					1			Dubinina & Plietnev 1978
GERMANY																						
Heligoland	1	1	1	1		1																van Bronswijk & Jorde 1975
Berlin	35	1	1	1		1										1		1				Karg 1973
Hamburg	3	1	1	1		1								1	1						1	Keil 1983
Mönchengladbach	53	1	1	1		1																van der Lustgraaf & Jorde 1977
Munich	509	3														1						Dekker 1928
Rostock	3	1	1																			Bernhard et al. 1986

continued...

Appendix 2 Continued

Country & location	Alt. (m)	Alt. class	*D. pt.*	*D. f*	*D. m*	*E. m*	*M. i*	*Hirstia* spp.	*A. s*	Other *A.* spp.	*T. p*	Other *T.* spp.	*A. o*	*S. m*	*L. d*	*G. d*	*B. t*	*C. a*	*G. f*	*C. e*	Other *C.* spp.	Reference
GREECE																						
Attikí	142	1	1	1		1			1		1	1			1	1	1		1	1		Papaioannou-Souliotis 1991
Athens	154	1	1	1	1	1	1	1							1							Deliargyris *et al.* 1990
GREENLAND																						
Ilulissat (Jakobshavn)	13	1	1																			Korsgaard, in Hallas et al 2004
HONG KONG	18	1	1	1		1	1	1	1			1					1				1	Gabriel *et al.* 1982
HUNGARY																						
Budapest & environs	103	1	1	1		1		1														Halmai 1984
ICELAND																						
Reykjavik	15	1	1																			Hallas *et al.* 2004
INDIA																						
19 States			1	1													1					Dar *et al.* 1973
19 States			1	1					1		1				1	1	1				1	Dar & Gupta 1980
24 Parganas	1	1	1	1				1		1							1			1	1	Modak *et al.* 1991
Hooghly Dist., W. Bengal	1																1					van Bronswijk *et al.* 1973a
Aurangabad	565	3	1	1													1		1			Tilak & Jogdand 1989
Bangalore	914	3									1					1			1			Krishna Rao & ChannaBasavanna 1977
Bangalore	914	3	1	1		1											1					Krishna Rao *et al.* 1973
Bangalore	914	3	1	1		1	1	1			1				1		1	1	1		1	Krishna Rao *et al.* 1981
Bangalore	914	3															1		1			Maurya *et al.* 1983
Bangalore	914	3	1	1		1	1	1														Ranganath & ChannaBasavanna 1988
Barackpur	17	1	1																			Krishna Rao *et al.* 1981
Barddhaman	40	1	1	1				1									1			1		Modak *et al.* 1991
Bhilai	293	2															1					van Bronswijk *et al.* 1973a
Bhilai & environs	293	2	1	1	1																	Dixit & Mehta 1973
Calcutta	13	1		1																		Griffiths & Cunnington 1971
Calcutta	13	1	1		1																	Krishna Rao *et al.* 1981
Calcutta	13	1	1	1				1									1			1		Modak *et al.* 1991
Calcutta	13	1	1	1																		Saha *et al.* 1994
Calcutta	13	1	1	1																1		Tandon *et al.* 1988
Chandanbari	2868	5															1					Modak *et al.* 1992
Cherrapunji	1485	4													1			1				Krishna Rao *et al.* 1981
Chikmagalur	1037	4	1		1													1				Krishna Rao *et al.* 1981
Coimbatore	380	2	1														1					Krishna Rao *et al.* 1981
Darjiling	2038	5	1		1										1			1				Krishna Rao *et al.* 1981
Darjiling	2038	5													1			1				Maurya *et al.* 1983
Delhi	215	2															1	1				Maurya *et al.* 1983
Delhi	215	2	1	1								1									1	Nayar *et al.* 1974
Dibrugarh	94	1																1				Krishna Rao *et al.* 1981
Dibrugarh	94	1															1	1				Maurya *et al.* 1983
Durg	290	2	1	1	1																	Dixit & Mehta 1973
Gulbarga	455	2	1																			Krishna Rao *et al.* 1981
Jammu	900	3															1	1				Maurya *et al.* 1983
Jammu	900	3	1																			Modak *et al.* 1992

continued...

Appendix 2 Continued

Country & location	Alt. (m)	Alt. class	D. pt.	D. f	D. m	E. m	M. i	Hirstia spp.	A. s	Other A. spp.	T. p	Other T. spp.	A. o	S. m	L. d	G. d	B. t	C. a	G. f	C. e	Other C. spp.	Reference
Jamnagar	19	1		1																		Griffiths & Cunnington 1971
Kanpur	126	1															1					Maurya et al. 1983
Katra	753	3									1						1	1				Maurya et al. 1983
Kottayam	58	1	1			1											1					Krishna Rao et al. 1981
Lucknow	131	1	1			1					1											Krishna Rao et al. 1981
Lucknow	131	1	1	1																		Maurya & Jamil 1980
Lucknow	131	1									1		1		1	1	1	1				Maurya et al. 1983
Lucknow	131	1	1	1																		Nath et al. 1974
Madras	9	1	1	1	1	1	1										1					Kannan et al. 1996
Mudigere	915	3	1			1											1					Krishna Rao et al. 1981
Navsari	9	1									1						1					Krishna Rao et al. 1981
Pahalgam	2739	5	1	1																		Modak et al. 1992
Polambakkam	8	1	1																			Fain 1966b
Pune	570	3	1	1																		Maurya et al. 1983
Raipur	292	2	1	1	1																	Dixit & Mehta 1973
Sangli	545	3	1	1																		Griffiths & Cunnington 1971
Shillong	1526	4													1			1				Krishna Rao et al. 1981
Shimoga	569	3	1			1	1															Krishna Rao et al. 1981
Srinagar	1960	4	1																			Modak et al. 1992
Trichur	48	1	1														1					Krishna Rao et al. 1981
Trivandrum	1	1	1	1		1					1						1	1				Krishna Rao et al. 1981
Udhampur	756	4															1	1				Maurya et al. 1983
INDONESIA																						
Bandung	822	3															1					Bronswijk et al. 1973
Jakarta	1	1	1	1							1						1				1	Baratawidjaja et al. 1998
IRAN																						
Bushehr (Bushir; Bushire)	13	1	1	1						1	1						1			1		Sepasgosarian & Mumcuoglu 1979
Chalus (`Calûs, Châlûs)	20	1	1	1						1	1						1			1		Sepasgosarian & Mumcuoglu 1979
Isfahan (Esfahan)	1571	4	1							1	1						1			1		Sepasgosarian & Mumcuoglu 1979
Karaj (Kara`g, Karai)	1462	4	1							1	1						1			1		Sepasgosarian & Mumcuoglu 1979
Khorramabad	1298	4	1								1											Amoli & Cunnington 1977
Lahijan	25	1	1	1		1											1	1				Sepasgosarian & Mumcuoglu 1979
Ramsar (Sakht Sar)	1	1	1			1									1	1						Sepasgosarian & Mumcuoglu 1979
Sari	2071	5	1	1		1									1				1	1		Amoli & Cunnington 1977
Sharud	1345	4	1																			Amoli & Cunnington 1977
Shiraz (`Sîraâz)	1800	4	1																			Sepasgosarian & Mumcuoglu 1979
Tehran	1139	4	1						1	1	1	1										Amoli & Cunnington 1977
ISRAEL																						
13 localities			1	1		1													1			Mumcuoglu et al. 1999
Bat Yam	52	1	1	1		1			1						1		1					Feldman-Muhsam et al. 1985

continued...

Appendix 2 Continued

Country & location	Alt. (m)	Alt. class	D. pt.	D. f	D. m	E. m	M. i	Hirstia spp.	A. s	Other A. spp.	T. p	Other T. spp.	A. o	S. m	L. d	G. d	B. t	C. a	G. f	C. e	Other C. spp.	Reference
Be'er Sheva'	244	2	1	1		1		1							1							Feldman-Muhsam et al. 1985
Jerusalem	757	3	1	1		1		1							1		1					Feldman-Muhsam et al. 1985
Nes Ziyyona	35	1	1	1		1		1							1		1					Feldman-Muhsam et al. 1985
Nir Dawid	-111	0	1	1		1		1							1							Feldman-Muhsam et al. 1985
Tel Aviv	35	1	1	1																		Kivity et al. 1993
ITALY																						
Sardinia			1	1		1			1	1	1				1	1		1			1	Ottoboni et al. 1983
Castelfiorentino	60	1													1	1						Castagnoli et al. 1983
Figline Valdarno	157	1														1						Castagnoli et al. 1983
Firenze	91	1	1	1		1			1		1	1			1	1			1			Castagnoli et al. 1983
Firenze	91	1	1																			Fain 1966a, b
Fucecchio	18	1														1				–		Castagnoli et al. 1983
Gaeta	1	1	1																			Castagnoli et al. 1983
Livorno	5	1	1	1																	1	Goraccci et al. 1984
Martina Franca	404	2				1																Castagnoli et al. 1983
Milan	103	1	1	1		1																Blythe 1976
Naples	1	1	1	1		1																Blythe 1976
Naples	1	1	1	1		1																Noferi et al. 1974
Panzano	243	2	1	1		1																Blythe 1976
Pescia	64	1	1																			Castagnoli et al. 1983
Putignano	343	2				1																Castagnoli et al. 1983
Regello	512	3														1						Castagnoli et al. 1983
Rome	15	1	1	1		1										1				1		Bigliocchi & Maroli 1995
Salerno	1360	4	1			1									1	1						Castagnoli et al. 1983
San Piero a Sieve	192	1													1	1						Castagnoli et al. 1983
Sassari	256	2	1			1										1						Castagnoli et al. 1983
Turin	205	2	1	1		1																Blythe 1976
JAPAN																						
Tohoku District								1														Fain et al. 1974
Fukuoka	10	1	1	1				1							1			1				Oshima 1970; Miyamoto et al. 1970
Hiroshima	26	1	1	1				1								1	1	1				Miyamoto et al. 1970
Hiroshima	26	1	1	1												1		1				Takaoka & Okada 1984
Hiroshima	26	1									1											Horie et al. 1992
Kyushu	900	3	1																			Ishii et al. 1983
Nagoya	14	1	1	1		1		1								1		1				Oshima 1970; Miyamoto et al. 1970
Nagoya	14	1	1	1																		Suto et al. 1991a, b; Sakaki & Suto 1994
Nagoya	14	1	1	1		1		1	1		1				1	1		1		1	1	Suto et al. 1992a, b
Naha	2	1	1	1							1						1			1	1	Toma et al. 1993
Naha	2	1	1	1		1	1				1						1			1	1	Toma et al. 1998
Okayama	38	1	1	1				1			1					1				1	1	Hatsushika & Miyoshi 1992a, b
Okinawa	0	1	1	1		1	1				1						1			1	1	Toma et al. 1998
Osaka	0	1	1	1				1			1							1				Oshima 1970; Miyamoto et al. 1970
Sapporo	30	1	1	1		1		1			1				1	1		1				Oshima 1970; Miyamoto et al. 1970

continued...

Appendix 2 Continued

Country & location	Alt. (m)	Alt. class	D. pt.	D. f	D. m	E. m	M. i	Hirstia spp.	A. s	Other A. spp.	T. p	Other T. spp.	A. o	S. m	L. d	G. d	B. t	C. a	G. f	C. e	Other C. spp.	Reference
Sendai	54		1	1				1	1		1				1			1				Oshima 1970; Miyamoto et al. 1970
Tokushima	1	1	1	1		1		1			1				1	1		1				Oshima 1970; Miyamoto et al. 1970
Tokyo	43	1	1	1		1		1	1		1							1				Ishii et al. 1979
Tokyo	43	1															1					Bronswijk et al. 1973
Tokyo	43	1	1	1		1		1	1		1							1	1			Takaoka et al. 1977
Tokyo	43	1	1	1		1									1	1						Yoshikawa 1980
Tokyo	43	1	1	1		1		1	1						1	1		1			1	Yoshikawa et al. 1982
Tokyo	43	1	1	1																		Otani et al. 1984
Tokyo	43	1	1	1		1		1							1			1	1			Oshima 1970; Miyamoto et al. 1970
Tokyo	43	1	1	1		1		1			1				1	1		1		1	1	Miyamoto & Ouchi 1976
Tokyo	43	1						1														Fain et al. 1974
Tomakomai	3	1	1	1		1									1	1		1				Oshima 1970
Urawa	14		1	1		1	1	1			1							1			1	Takaoka & Okada 1984
Yokohama	1	1						1														Fain et al. 1974
Yokohama	1	1	1	1												1		1				Oshima 1964
Yokohama	1	1	1	1		1																Oshima 1967
Yokohama	1	1	1	1																		Oshima et al. 1972
KAZAKSTAN																						
Alma-Ata	862	3	1			1					1				1							Salykov & Ermekova 1987
Semipalatinsk	209	2	1	1		1											1	1				Yagofarov & Galikeev 1978
KENYA																						
Nairobi	1729	4	1	1		1																Blythe 1976
Nairobi	1729	4	1																			Gitoho & Rees 1971; Rees et al. 1974
KUWAIT	11	1	1																			Gamal-Eddin et al. 1985
LITHUANIA																						
Vilnius	184	1	1	1		1									1	1						Zheltikova et al. 1985
Vilnius	184	1	1	1		1			1	1	1				1	1		1	1			Dubinina et al. 1984
MALAYSIA																						
Cameron Highlands	1308	4	1	1	1	1	1	1			1							1	1		1	Ho & Nadchatram 1985
Cameron Highlands	1308	4																1				Mariana & Ho 1996
Gombak	100	1	1	1		1																Ho 1986
Jengka	65	1	1	1		1																Ho 1986
Jengka	65	1																1				Mariana & Ho 1996
Kangar Lenggor	29	1	1	1	1	1	1	1														Ho 1986
Kuala Lumpur	61	1	1	1							1							1			1	Rueda 1985
Kuala Lumpur	61	1	1	1	1	1	1	1														Ho 1986
Kuala Lumpur	61	1	1	1		1	1	1		1	1							1	1		1	Mariana & Ho 1996
Petaling Jaya	63	1	1																			Ho 1986
Petaling Jaya	63	1	1			1												1				Furumizo & Thomas 1977
Petaling Jaya	63	1	1								1			1				1				Thomas et al. 1976
Port Dickson	0	1	1	1	1	1	1	1														Ho 1986
Port Dickson	0	1																				Fain & Nadchatram 1985
Selangor	1	1	1	1		1	1	1		1	1							1	1		1	Mariana & Ho 1996
Templer Park	90		1	1		1	1															Ho 1986

continued...

Appendix 2 Continued

Country & location	Alt. (m)	Alt. class	D. pt.	D. f	D. m	E. m	M. i	Hirstia spp.	A. s	Other A. spp.	T. p	Other T. spp.	A. o	S. m	L. d	G. d	B. t	C. a	G. f	C. e	Other C. spp.	Reference
MARTINIQUE			1	1			1	1				1		1	1			1	1		1	Lafosse Marin *et al.* 2006
MEXICO																						
Acaponeta, Nay.	37	1	1	1																		González & Llorens 1974
Ahualco, Jal.	1336	4	1	1																		González & Llorens 1974
Amatlán de Cañas, Nay.	804	3		1																		González & Llorens 1974
Ameca, Jal.	1243	4	1	1																		González & Llorens 1974
Arandas, Jal.	2050	5		1																		González & Llorens 1974
Armeria, Col.	54	1	1	1																		González & Llorens 1974
Autlán, Jal.	925	3	1	1																		González & Llorens 1974
Barra de Navidad, Jal.	0	1	1	1																		González & Llorens 1974
Cacula, Jal.	1347	4	1	1																		González & Llorens 1974
Casimiro Castillo, Jal.	432	2	1	1																		González & Llorens 1974
Chapala, Jal.	602	3	1	1																		González & Llorens 1974
Cihuatlán, Jal.	91	1	1	1																		González & Llorens 1974
Ciudad Guzmán, Jal.	1571	4	1	1																		González & Llorens 1974
Colima, Col.	488	2	1	1																		González & Llorens 1974
Compostela, Nay.	875	3	1	1																		González & Llorens 1974
Cuernavaca, Mor.	1483	4	1																			Novoa Avilés 1975
Culiacán, Sin.	70	1	1																			Novoa Avilés 1975
El Salito, Jal.	1536	4	1	1																		González & Llorens 1974
Ensenada, B.C.	9	1	1																			Novoa Avilés 1975
Guadalajara, Jal.	1552	4	1	1																		González & Llorens 1974
Guadalajara, Jal.	1552	4	1																			Novoa Avilés 1975
Ixtlán del Río, Nay.	1087	4	1	1																		González & Llorens 1974
La Paz, B.C.	51	1	1	1																		Servin & Tejas 1991
Lagos, Jal.	1873	4		1																		González & Llorens 1974
León, Gto.	1791	4	1																			Novoa Avilés 1975
Magdalena, Jal.	1581	4	1	1																		González & Llorens 1974
Manzanillo, Col.	94	1	1	1																		González & Llorens 1974
Manzanillo, Col.	94	1	1																			Novoa Avilés 1975
Mascota, Jal.	1257	4	1	1																		González & Llorens 1974
Mazamitla, Jal.	2199	5		1																		González & Llorens 1974
Mérida, Yuc.	10	1	1																			Novoa Avilés 1975
Ocotlán	1550	4	1	1																		González & Llorens 1974
Pachuaca, Hgo.	2396	5	1																			Novoa Avilés 1975
Pihuamo, Jal.	778	3	1	1																		González & Llorens 1974
Puerto Vallarta, Jal.	0	1	1	1																		González & Llorens 1974
Querétaro, Qro.	1867	4	1																			Novoa Avilés 1975
Reynosa, Tam.	34	1	1																			Novoa Avilés 1975
Ruiz, Nay.	33	1	1	1																		González & Llorens 1974
San Martín Hidalgo, Jal.	1296	4	1	1																		González & Llorens 1974
Santa Maria del Oro, Nay.	1172	4		1																		González & Llorens 1974
Santiago Ixcuintla, Nay.	15	1	1	1																		González & Llorens 1974
Sayula, Jal.	1397	4		1																		González & Llorens 1974
Tala, Jal.	1351	4	1	1																		González & Llorens 1974

continued...

Appendix 2 Continued

Country & location	Alt. (m)	Alt. class	D. pt.	D. f	D. m	E. m	M. i	Hirstia spp.	A. s	Other A. spp.	T. p	Other T. spp.	A. o	S. m	L. d	G. d	B. t	C. a	G. f	C. e	Other C. spp.	Reference
Tamazula, Jal.	1055	4	1	1																		González & Llorens 1974
Tecomán, Col.	27	1	1	1																		González & Llorens 1974
Tecuala, Nay.	14	1	1	1																		González & Llorens 1974
Tepatitlán, Jal.	1846	4	1	1																		González & Llorens 1974
Tepic, Nay.	965	3	1	1																		González & Llorens 1974
Tepic, Nay.	965	3	1																			Novoa Avilés 1975
Tequila, Jal.	1225	4	1	1																		González & Llorens 1974
Tlaquepaque	1573	4	1	1																		González & Llorens 1974
Tuxpan	1235	4	1	1																		González & Llorens 1974
Tuxpan, Nay.	8	1	1																			González & Llorens 1974
Villa de Alvarez, Col.	495	2	1	1																		González & Llorens 1974
Zapopan	1564	4	1	1																		González & Llorens 1974
MOROCCO																						
Casablanca	17	1	1																			Fain 1988
MOZAMBIQUE																						
Maputo	63	1	1																			Ordman 1971
NAMIBIA																						
Oranjemund	30	1	1																			Ordman 1971
NETHERLANDS																						
Amsterdam	-2	1																				Bronswijk 1973
Amsterdam	-2	1	1	1																		de Boer 1990
Amsterdam	-2	1			1																	Cunnington et al. 1987
Delft	-2	1	1	1		1																Spieksma & S.-Boezeman 1967
Groesbeek	25	1	1			1																Lustgraaf 1978a, b
Leiden	-2	1	1																			Fain 1966a, b
Leiden	-2	1	1	1		1												1	1			Spieksma & S.-Boezeman 1967
Leiden	-2	1	1	1		1												1	1			Spieksma 1967; 1968
Leiden	-2	1	1																			Voorhorst et al. 1967
Leiden	-2	1	1	1		1																Voorhorst et al. 1969
Leiden	-2	1	1																			Spieksma et al. 1971; Spieksma 1973
Nijmegen	46	1	1			1																Bronswijk et al. 1971; Bronswijk 1973; 1974
Noordwijk	1	1		1																		Fain 1967a
Noordwijk	1	1	1	1		1																Spieksma & S.-Boezeman 1967
Oegstgeest	1	1	1	1		1																Spieksma & S.-Boezeman 1967
NEW ZEALAND																						
Auckland	26	1	1													1		1	1		1	Cornere 1971
Auckland	26	1	1																			Cornere 1972
Cambridge	140	1	1																			Cornere 1972
Christchurch	6	1	1																			Cornere 1972
Christchurch	6	1	1	1																		Ingham & Ingham 1976
Christchurch	6	1	1			1										1			1			Abbott et al. 1981
Dunedin	94	1	1																			Cornere 1972
Ellesmere	23	1	1																			Cornere 1972
Fairlie	335	2	1																			Cornere 1972
Foxton	15	1	1																			Cornere 1972
Gisborne	39	1	1																			Cornere 1972

continued...

Appendix 2 Continued

Country & location	Alt. (m)	Alt. class	D. pt.	D. f	D. m	E. m	M. i	Hirstia spp.	A. s	Other A. spp.	T. p	Other T. spp.	A. o	S. m	L. d	G. d	B. t	C. a	G. f	C. e	Other C. spp.	Reference
Glenfield	48	1	1																			Cornere 1972
Invercargill	9	1	1																			Cornere 1972
Invercargill	9	1	1			1																Blythe 1976
Lower Hutt	15	1	1																			Cornere 1972
Motueka	1	1	1																			Cornere 1972
Napier	1	1	1																			Cornere 1972
Nelson	24	1	1																			Cornere 1972
Palmerston North	98	1	1																			Cornere 1972
Paraparaumu	2	1	1																			Cornere 1972
Richmond	966	3	1																			Cornere 1972
Rotorua	417	2	1																			Cornere 1972
Shannon	17	1	1																			Cornere 1972
Taupo	391	2	1																			Cornere 1972
Tauranga	1	1	1																			Cornere 1972
Tawa Flat	123	1	1																			Cornere 1972
Te Awamutu	58	1	1																			Cornere 1972
Temuka	58		1																			Cornere 1972
Thames-Coromandel	50	1	1																			Cornere 1972
Timaru	1	1	1																			Cornere 1972
Waihi Beach	32	1	1																			Cornere 1972
Wairoa, Hawke's Bay	23	1	1																			Andrews et al. 1992
Wairoa, Hawke's Bay	23	1	1													1		1		1		Andrews et al. 1995
Waitara	222	2	1																			Cornere 1972
Wellington	21	1	1																			Cornere 1972
Wellington	21	1	1			1									1	1		1		1		Pike & Wickens 2008
NIGERIA																						
Lagos	35	1	1	1		1				1			1	1				1		1		Hunponu-Wusu & Somorin 1978
Lagos	35	1	1	1			1				1		1	1				1		1	1	Somorin et al. 1978
NORWAY																						
Unspecified locality			1																			Spieksma 1967
Nordland Fylke			1																			Mehl 1998
Troms Fylke			1																			Mehl 1998
Alta	0	1	1		1																	Mehl 1998
Baerum	133	1	1																			Mehl 1998
Bergen	7	1	1																			Mehl 1998
Drøbak	120	1	1		1																	Mehl 1998
Elverum	220	1	1																			Mehl 1998
Fana	40	1	1		1																	Mehl 1998
Flekkefjord	20	1	1																			Mehl 1998
Fredrikstad	15	1	1		1											1						Mehl 1998
Kristiansand	22	1	1																			Mehl 1998
Kvanne	1	1	1		1	1																Mehl 1998
Laksevåg	120	1	1																			Mehl 1998
Levanger	20	1	1		1	1									1	1						Mehl 1998
Moss	40	1	1		1										1	1						Mehl 1998
Namdalseid	158	1	1																			Mehl 1998
Namsos	95	1	1																			Mehl 1998
Namsskogan	235	2	1		1	1									1	1						Mehl 1998

continued...

Appendix 2 Continued

Country & location	Alt. (m)	Alt. class	D. pt.	D. f	D. m	E. m	M. i	Hirstia spp.	A. s	Other A. spp.	T. p	Other T. spp.	A. o	S. m	L. d	G. d	B. t	C. a	G. f	C. e	Other C. spp.	Reference
Oslo	13	1	1	1	1	1																Mehl 1998; Nafstad et al. 1998
Overhalla	122	1	1																			Mehl 1998
Ringerike	96	1	1	1											1	1						Mehl 1998
Sarpsborg	32	1	1	1											1	1						Mehl 1998
Sør-Varanger	3	1	1				1								1							Dotterud et al. 1995
Stavanger	9	1	1																			Mehl 1998
Stjørdal	13	1	1																			Mehl 1998
Trondheim	100	1	1												1	1		1				Mehl 1998
Valestrandfossen	79	1	1																			Mehl 1998
PALESTINE																						
Gaza	20	1	1			1			1	1	1							1		1		Mumcuoglu et al. 1994
PANAMA																						
La Chorrera	69	1	1							1							1	1			1	Miranda et al. 2002
PAPUA NEW GUINEA																						
Okapa District	1814	4	1	1																		Dowse et al. 1985
Asaro Valley	1540	4	1																			Turner et al. 1988
Baiyer River	1198	4	1																			Green et al. 1982
Goroka	1524	4	1			1																Anderson & Cunnington 1974
Lufa	1621	4	1	1		1																Anderson & Cunnington 1974
Waisa	1558	4	1																			Green et al. 1982
PERU																						
Huaraz & Ticapampa	3672	6	1							1					1	1		1	1			Caceres & Fain 1978; 1979
Lima	108	2	1			1				1					1	1		1	1			Caceres & Fain 1978; 1979
Lima	108	2	1			1				1							1	1			1	Croce et al. 2000
Lima	108	2	1							1					1	1						Villanueva et al. 2003
PHILIPPINES																						
Baybay	4	1	1															1				Corpuz-Raros et al. 1988
Dumaguette City	8	1	1															1				Bronswijk et al. 1973a
Los Banos	93	1	1															1				Corpuz-Raros et al. 1988
Manila	19	1	1															1				Corpuz-Raros et al. 1988
Manila	19	1	1															1			1	de las Llagas et al. 2005
POLAND																						
Bydgoszcz	65	1	1												1	1						Romanski et al. 1977
Bytom	269	2	1	1		1			1										1	1		Horak 1987
Gdansk & Gdynia	14		1	1							1	1								1	1	Racewicz 2001
Katowice	287	2	1	1		1																Solarz 1986
Katowice	287	2	1	1		1			1													Horak 1987
Katowice	287	2	1	1		1			1	1	1				1	1		1	1	1	1	Solarz 1998
Katowice	287	2	1	1		1			1	1								1				Solarz 2000
Sosnowiec	296	2	1	1		1																Solarz 1986; 1997
Warsaw	94	1	1	1		1												1	1			Samolinski et al. 1990
PORTUGAL																						
Coimbra	46	1	1																			Loureiro et al. 1990
Lisbon	16	1	1	1		1			1	1	1				1	1		1	1	1	1	Pinhão & Grácio 1978
Funchal	0	1	1																			Grácio & Quinta 2000
Lisbon	16	1	1																			Grácio & Quinta 2000

continued...

Appendix 2 Continued

Country & location	Alt. (m)	Alt. class	D. pt.	D. f	D. m	E. m	M. i	Hirstia spp.	A. s	Other A. spp.	T. p	Other T. spp.	A. o	S. m	L. d	G. d	B. t	C. a	G. f	C. e	Other C. spp.	Reference
Leira	78	1	1																			Grácio & Quinta 2000
Santarem	14	1	1																			Grácio & Quinta 2000
Setubal	0	1	1																			Grácio & Quinta 2000
PUERTO RICO																						
Arecibo	1	1		1													1					Montealegre *et al.* 1997
Canovanas	11	1		1													1					Montealegre *et al.* 1997
Guanica	81	1	1	1																		Montealegre *et al.* 1997
Guayama	43	1		1		1											1					Montealegre *et al.* 1997
Hormigueros	25	1		1		1											1					Montealegre *et al.* 1997
Lajas	46	1		1																		Montealegre *et al.* 1997
Lares	382	2	1	1													1					Montealegre *et al.* 1997
Loiza Valley	24	1	1	1													1					Montealegre *et al.* 1997
Mayagüez	23	1	1																		1	Santos 1978
Mayagüez	23	1	1	1		1									1			1	1			Fernandez-Caldas *et al.* 1995
Mayagüez	23	1	1	1													1					Montealegre *et al.* 1997
Ponce	13	1	1	1		1											1					Montealegre *et al.* 1997
San Juan	3	1	1	1													1					Montealegre *et al.* 1997
ROMANIA																						
Bucharest	71	1	1																			Popescu & Banescu 1975
RUSSIA																						
Chuvash			1	1		1		1	1				1		1	1	1	1	1			Dubinina & Plietniev 1978
Chuvash			1	1		1		1	1							1		1		1		Ivanova & Petrova 1984
Primorje Region			1	1				1	1		1	1	1		1			1	1			Tareev & Dubinina 1985
Gorky	135	1	1	1											1				1			Dubinina & Plietniev 1978
Krasnodar	36	1	1												1			1	1			Dubinina & Plietniev 1978
Moscow	150	1	1												1	1						Zheltikova *et al.* 1985
Moscow	150	1	1	1																		Zheltikova *et al.* 1986; 1994
Moscow	150	1	1	1		1		1	1						1	1			1			Zheltikova & Petrova 1990
Moscow	150	1	1	1											1	1						Petrova & Zheltikova 1990
Moscow	150	1	1	1												1			1			Dubinina & Plietniev 1978
Moscow	150	1	1	1		1		1			1				1	1		1	1			Petrova & Zheltikova 2000
Noril'sk	143	1	1																			Dubinina & Plietniev 1978
Rostov-on-Don	74	1	1	1		1										1		1				Dubinina & Plietniev 1978
St. Petersburg	5	1	1	1												1		1	1			Dubinina & Plietniev 1978
RWANDA																						
Kigali	1568	4	1																			Fain 1988
Butare	1734	4	1																			Fain 1988
SINGAPORE	1	1	1	1		1	1															Blythe 1976
SINGAPORE	1	1	1														1					Nadchatram *et al.* 1986

continued...

Appendix 2 Continued

Country & location	Alt. (m)	Alt. class	D. pt.	D. f	D. m	E. m	M. i	Hirstia spp.	A. s	Other A. spp.	T. p	Other T. spp.	A. o	S. m	L. d	G. d	B. t	C. a	G. f	C. e	Other C. spp.	Reference
SINGAPORE	1	1		1																		Griffiths & Cunnington 1971
SINGAPORE	1	1	1	1			1										1					Chew *et al.* 1999
SLOVAK REPUBLIC																						
Strbske Pleso	1204	4	1	1		1				1								1	1	1		Makovcová *et al.* 1982
SOUTH AFRICA																						
East coast						1	1															Fain 1988
Barberton	886	3	1																			Ordman 1971
Beaufort West	872	3	1																			Ordman 1971
Bredasdorp	52	1	1																			Ordman 1971
Burghersdorp	1388	4	1																			Ordman 1971
Cape Town	16	1	1																			Ordman 1971
Ceres	451	2	1																			Ordman 1971
Citrusdal	173	1	1																			Ordman 1971
Clanwilliam	75	1	1																			Ordman 1971
Cradock	959	3	1																			Ordman 1971
De Aar	1260	4	1																			Ordman 1971
Durban	4	1	1																			Ordman 1971
East London	100	1	1																			Ordman 1971
Empangeni	64	1	1																			Ordman 1971
Ermelo	1734	4	1																			Ordman 1971
Estcourt	1159	4	1																			Ordman 1971
George	253	2	1																			Ordman 1971
Graaf-Reinet	732	3	1																			Ordman 1971
Greytown	1099	4	1																			Ordman 1971
Humansdorp	8	1	1																			Ordman 1971
Jeffrey's Bay	7	1	1	1																		Ordman 1971
Johannesburg	1786	4	1																			Ordman 1971
Johannesburg	1786	4	1			1																Fain 1988
Kimberley	1218	4	1																			Ordman 1971
King William's Town	375	2	1																			Ordman 1971
Klerksdorp	1325	4	1																			Ordman 1971
Knysna	4	1	1																			Ordman 1971
Kroonstad	1348	4	1																			Ordman 1971
Ladysmith	1001	4	1																			Ordman 1971
Lambert's Bay	2	1	1																			Ordman 1971
Louis Trichardt	959	3	1	1																		Ordman 1971
Mafeking	1271	4	1																			Ordman 1971
Margate	15	1	1																			Ordman 1971
Messina	519	3	1																			Ordman 1971
Morgan's Bay	2	1	1																			Ordman 1971
Mossel Bay	61	1	1																			Ordman 1971
Nelspruit	665	3	1																			Ordman 1971
Oudtshoorn	332	2	1																			Ordman 1971
Piet Retief	1271	4	1																			Ordman 1971
Pietersburg	1294	4	1																			Ordman 1971
Port Alfred	61	1	1																			Ordman 1971
Port Elizabeth	55	1	1																			Ordman 1971
Pretoria	1379	4	1																			Ordman 1971
Queenstown	1077	4	1																			Ordman 1971

continued...

Appendix 2 Continued

Country & location	Alt. (m)	Alt. class	D. pt.	D. f	D. m	E. m	M. i	Hirstia spp.	A. s	Other A. spp.	T. p	Other T. spp.	A. o	S. m	L. d	G. d	B. t	C. a	G. f	C. e	Other C. spp.	Reference
Somerset East	766	3	1																			Ordman 1971
Springs	1688	4	1																			Ordman 1971
Stanger	43	1	1																			Ordman 1971
Steytlerville	405	3	1																			Ordman 1971
Uitenhage	100	1	1																			Ordman 1971
Umtata	679	3	1																			Ordman 1971
Volksrust	1558	4	1																			Ordman 1971
Vryburg	1188	4	1																			Ordman 1971
Vryheid	1194	4	1																			Ordman 1971
Welkom	1338	4	1																			Ordman 1971
Wilderness	4	1	1																			Ordman 1971
Worcester	248	2	1																			Ordman 1971
SOUTH KOREA																						
Chongju	40	1	1	1							1											Ree et al. 1997
Chonju	275	2	1	1							1											Ree et al. 1997
Chunchon	90	1	1	1							1											Ree et al. 1997
Inchon	75	1	1	1							1											Ree et al. 1997
Kwangju	164	1	1	1							1											Ree et al. 1997
Pusan	0	1	1	1							1											Ree et al. 1997
Seoul	34	1	1	1							1											Ree et al. 1997
Seoul	34	1													1	1						Kang & Chu 1975
Seoul	34	1	1																			Hong et al. 1987; 1988
Seoul	34	1	1	1				1														Paik et al. 1993
Taejon	104	1	1	1							1											Ree et al. 1997
Yonggwang	137		1	1																		Ree et al. 1997
SPAIN																						
Rioja Alta	620	3	1	1		1					1				1						1	Lobera et al. 2000
Rioja Media	380	2	1	1		1				1	1				1		1	1			1	Lobera et al. 2000
Rioja Baja	300	2	1	1		1					1							1			1	Lobera et al. 2000
Barcelona	1	1	1	1		1																Portus & Gomez 1976
Barcelona	1	1	1	1	1	1			1			1			1	1		1	1		1	Blasco & Portus 1975; Blasco et al. 1975
Barcelona	1	1	1	1	1																	Griffiths & Cunnington 1971
Barcelona	1	1																		1		Comamala de Florensa 1975
Cádiz	0	1	1			1			1							1		1				Del Rey & Garcia 1971; 1972a, b; 1973
Castelltersol	689	3	1	1		1																Portus & Gomez 1976
Huelva	49	1	1	1	1	1			1							1	1	1				Arias Irigoyen & García del Hoyo 2002
La Coruña	103	1	1	1		1				1	1				1	1		1	1		1	Boquete et al. 2006
La Laguna, Tenerife	583	3	1	1		1					1				1	1	1	1				Sanchez-Covisa et al. 1999
L'Ametlla de Merola	292	2	1	1		1			1	1		1			1	1			1		1	Gomez et al. 1981a, b
Lugo	444	2		1		1				1	1				1	1		1	1		1	Boquete et al. 2006
Madrid	588	3	1	1		1					1					1	1					Sastre et al. 2002
Orense	125	1		1		1				1	1				1	1		1	1		1	Boquete et al. 2006
Pontevedra	80	1		1		1				1	1				1	1		1	1		1	Boquete et al. 2006
Puigcerda	1140	4	1	1		1				1				1	1	1	1	1			1	Gomez et al. 1981a, b
Puigcerda	1140	4	1	1		1																Portus & Gomez 1976
Reus	93	1	1	1		1																Portus & Gomez 1976

continued...

Appendix 2 Continued

Country & location	Alt. (m)	Alt. class	D. pt.	D. f	D. m	E. m	M. i	Hirstia spp.	A. s	Other A. spp.	T. p	Other T. spp.	A. o	S. m	L. d	G. d	B. t	C. a	G. f	C. e	Other C. spp.	Reference
Reus	93	1	1	1		1			1		1	1			1				1		1	Gomez et al. 1981a, b
Salamanca	769	3	1	1		1				1	1	1			1	1		1				Lázaro & Igea 2000
San Cristóbal de la Laguna	547	3				1																	Sanchez-Covisa et al. 1999
Santa Cruz, Tenerife	1	1	1	1		1									1	1	1	1				Sanchez-Covisa et al. 1999
Santiago de Compostela	266	2	1	1		1					1	1			1	1						Agratorres et al. 1999
Torello	554	3	1	1																		Portus & Gomez 1976; 1980
Valladolid	703	3	1	1		1			1			1			1	1		1				Armentia et al. 1993
Zamora	616	3	1	1		1						1			1	1						Lázaro & Igea 2000
SURINAM																						
Paramaribo	2	1	1				1										1					Bronswijk 1972b
Paramaribo	2	1																				Bronswijk & Kok 1975
Paramaribo	2	1															1					Bronswijk et al. 1973a
Paramaribo	2	1						1														Fain et al. 1974
SWEDEN																						
Boden/Lulea	32	1	1			1									1							Wickman et al. 1993
Borlange/Falun	148	1	1			1									1							Wickman et al. 1993
Helsingborg	28	1	1	1	1	1			1	1					1	1						Warner et al. 1999
Linköping	43	1	1	1	1				1	1					1	1						Warner et al. 1999
Piteå	0	1	1			1									1							Wickman et al. 1993
Stockholm	16	1	1							1		1										Turos 1979
Stockholm	16	1	1	1	1																	Nordvall et al. 1988
Umeå	19	1	1			1									1							Wickman et al. 1993
Umeå	19								1	1						1						Warner et al. 1999
SWITZERLAND																						
Basel & environs	273			1																		Mumcuoglu 1976
Basel & environs	273			1																		Spieksma et al. 1971
Basel	273	2		1																		Spieksma 1973a
Basel	273	2		1																		Rufli 1970
Davos	1568	4	1	1		1																Spieksma et al. 1971
Davos	1568	4		1																		Spieksma 1973
Davos	1568	4		1																		Voorhorst et al. 1964
SYRIA																						
Dimashq (Damascus)	692	3		1																		Fain 1988
TADJIKISTAN																						
Dushanbe	810	3	1																			Takhirova et al. 1996
TAIWAN																						
Hualien	0		1	1					1		1						1			1		Wu 1999
Kaohsiung & Pingtung	4		1	1					1		1						1			1	1	Wu 1999
Nanao	14		1	1					1		1						1			1	1	Wu 1999
Taipei	6	1	1			1	1				1						1	1				Oshima 1970; Miyamoto et al. 1970
Taipei	6	1	1			1											1	1				Miyamoto
Taipei	6	1															1					Bronswijk et al. 1974
Taipei	6	1				1																Fain et al. 1974
Taipei	6	1	1	1		1	1		1		1						1					Chang & Hsieh 1989
Taipei	6	1	1	1		1			1		1						1			1	1	Wu 1999
Taichung	112	1	1	1													1					Sun & Lue 2000

continued...

Appendix 2 Continued

Country & location	Alt. (m)	Alt. class	D. pt.	D. f	D. m	E. m	M. i	Hirstia spp.	A. s	Other A. spp.	T. p	Other T. spp.	A. o	S. m	L. d	G. d	B. t	C. a	G. f	C. e	Other C. spp.	Reference
Taichung	112	1	1	1			1	1			1						1			1	1	Wu 1999
Taitung	0	1	1	1				1			1						1			1		Wu 1999
THAILAND																						
24 provinces			1	1		1																Malainual et al. 1995
Unspecified localities			1	1			1															Wongsathayong & Lakshana 1972
Ayutthaya			1																			Wongsathayong & Lakshana 1972
Chon Buri			1																			Wongsathayong & Lakshana 1972
Lop Buri			1																			Wongsathayong & Lakshana 1972
Nakhon Pathom			1																			Wongsathayong & Lakshana 1972
Nakhon Ratchasima			1																			Wongsathayong & Lakshana 1972
Ratchaburi			1																			Wongsathayong & Lakshana 1972
Samut Prakan			1																			Wongsathayong & Lakshana 1972
Samut Sakhon			1																			Wongsathayong & Lakshana 1972
Suphan Buri			1																			Wongsathayong & Lakshana 1972
Surat Thani			1																			Wongsathayong & Lakshana 1972
Thon Buri			1																			Wongsathayong & Lakshana 1972
Trang			1																			Wongsathayong & Lakshana 1972
Bangkok	2	1						1														Fain et al. 1974
Bangkok	2	1	1	1		1																Blythe 1976
Bangkok	2	1	1																			Wongsathayong & Lakshana 1972
TURKEY																						
Aegean Region			1	1																		Kalpaklioglu et al. 1997
Anatolia			1	1											1	1			1	1		Kalpaklioglu et al. 1997
Black Sea Region			1						1						1							Kalpaklioglu et al. 1997
Central Anatolia			1	1											1			1				Kalpaklioglu et al. 1997
Eastern Anatolia									1		1											Kalpaklioglu et al. 1997
Marmara Region			1	1												1						Kalpaklioglu et al. 1997
Mediterranean Region			1	1																1	1	Kalpaklioglu et al. 1997
Bursa	200	1	1	1							1					1						Gülegen et al. 2005
Istanbul	24	1	1	1		1																Blythe 1976
Istanbul	24	1		1																		Griffiths & Cunnington 1971
Kutahya	785	3	1						1		1				1	1					1	Akdemir & Gurdal 2005
Malatya	948	3	1									1			1			1			1	Atambay et al. 2006

continued...

Appendix 2 Continued

Country & location	Alt. (m)	Alt. class	D. pt.	D. f	D. m	E. m	M. i	Hirstia spp.	A. s	Other A. spp.	T. p	Other T. spp.	A. o	S. m	L. d	G. d	B. t	C. a	G. f	C. e	Other C. spp.	Reference
TRISTAN DA CUNHA	295	2	1			1												1				Mantle & Pepys 1974
TRISTAN DA CUNHA	295	2	1			1																Fain 1988
UNITED KINGDOM																						
Southeast England			1																			Hughes & Maunsell 1973
Aberdeen	36	1	1			1						1			1	1			1		1	Sesay & Dobson 1972
Birmingham	134	1	1			1										1			1			Blythe et al. 1974; Blythe 1976
Bristol	47	1	1			1															1	Carswell et al. 1982
Cardiff	16	1	1	1		1						1				1					1	Rao et al. 1975
Derby	67	1	1											1								Brown & Filer 1968
Edinburgh	32	1	1			1						1			1	1			1		1	Sesay & Dobson 1972
Glasgow	66	1	1			1						1			1	1			1		1	Sesay & Dobson 1972
Glasgow	66	1	1	1		1			1	1	1	1			1	1			1	1	1	Colloff 1987c; 1988a
Harpenden	87	1	1			1										1						Maunsell et al. 1968
Leeds	62	1	1																			Sarsfield 1974
Liverpool	5	1	1	1		1									1	1					1	Walshaw & Evans 1987
London	15	1	1			1										1						Cunnington & Gregory 1968
London	15	1	1		1																	Griffiths & Cunnington 1971
London	15	1	1			1																Tovey et al. 1981
London	15	1	1	1		1			1			1							1		1	Gabriel et al. 1982
London, S. Wales	15	1	1			1			1		1				1	1					1	Maunsell et al. 1968
Newbury & environs	75		1			1			1	1		1			1	1			1		1	Eaton et al. 1985a, b
Oxford & environs	68	1	1			1			1		1					1				1	1	Hart & Whitehead 1990
Truro & environs	38		1	1		1			1			1				1			1		1	Hewitt et al. 1973
URUGUAY																						
Montevideo	29	1	1																			Schuhl 1977
USA																						
California																						
Albion	59	1	1																			Furumizo 1973; 1975b
Bakersfield	121	1		1																		Furumizo 1973; 1975b
Briceland	182	1	1																			Furumizo 1973; 1975b
Carpinteria	23	1																				Furumizo 1973; 1975b
Corona	218	2	1																			Furumizo 1973; 1975b
Coronado	0	1		1																		Furumizo 1973; 1975b
Crannell	68	1	1																			Furumizo 1973; 1975b
Cummings	471	2		1																		Furumizo 1973; 1975b
Del Mar	40	2	1																			Furumizo 1973; 1975b
Descanso	1020	4		1																		Furumizo 1973; 1975b
Dixon	16	1	1																			Furumizo 1973; 1975b
El Portal	728	3		1																		Furumizo 1973; 1975b
Elmira	17	1	1	1																		Furumizo 1973; 1975b
Eureka	12	1	1																			Furumizo 1973; 1975b
Fortuna	53	1	1																			Furumizo 1973; 1975b
Fresno	90	1	1	1																		Furumizo 1973; 1975b
Goleta	19	1	1																			Furumizo 1973; 1975b

continued...

Appendix 2 Continued

Country & location	Alt. (m)	Alt. class	D. pt.	D. f	D. m	E. m	M. i	Hirstia spp.	A. s	Other A. spp.	T. p	Other T. spp.	A. o	S. m	L. d	G. d	B. t	C. a	G. f	C. e	Other C. spp.	Reference
Gonzales	42	1	1																			Furumizo 1973; 1975b
Grover Beach	33	1	1																			Furumizo 1973; 1975b
Hanford	64	1	1																			Furumizo 1973; 1975b
Indio	-3	1	1	1																		Lang & Mulla 1977c
Julian	1339	4	1																			Furumizo 1973; 1975b
Laguna Beach	0	1		1																		Furumizo 1973; 1975b
La Jolla	67	1	1																			Furumizo 1973; 1975b
Lake Arrowhead	1582	4	1	1																		Lang & Mulla 1977c
Long Beach	0	1	1																			Furumizo 1973; 1975b
Los Angeles	94	1	1	1																		Furumizo 1973; 1975b
Los Angeles	94	1	1	1																		Arlian et al. 1992
Madera	81	1		1																		Furumizo 1973; 1975b
Marysville	17	1	1																			Furumizo 1973; 1975b
Monterey	13	1	1																			Furumizo 1973; 1975b
Morro Bay	28	1	1																			Furumizo 1973; 1975b
National City	24	1			1																	Furumizo 1973; 1975b
Newport Beach	0	1			1																	Furumizo 1973; 1975b
Orange	59	1			1																	Furumizo 1973; 1975b
Orange County	13	1	1	1																		Mulla et al. 1975
Orange County	13	1	1	1																		Furumizo 1978
Oroville	57	1	1	1	1																	Furumizo 1973; 1975b
Palo Alto	30	1	1																			Furumizo 1973; 1975b
Pasadena	242	2	1																			Furumizo 1973; 1975b
Pescadero	59	1	1																			Furumizo 1973; 1975b
Pismo Beach	13	1	1																			Furumizo 1973; 1975b
Preston	120	1	1																			Furumizo 1973; 1975b
Redding	182	1	1																			Furumizo 1973; 1975b
Riverside	284	2		1																		Furumizo 1973; 1975b
Riverside	284	2		1																		Furumizo 1978
Riverside	284	2	1	1																		Lang & Mulla 1977a, b; 1978
Sacramento	4	1	1																			Furumizo 1973; 1975b
Salinas	15	1	1																			Furumizo 1973; 1975b
San Bernardino	315	2	1		1																	Furumizo 1973; 1975b
San Diego	25	1	1																			Furumizo 1973; 1975b
San Diego	25	1	1	1		1												1				Arlian et al. 1992
San Francisco	39	1	1	1																		Furumizo 1973; 1975b
San Jose	24	1	1	1																		Furumizo 1973; 1975b
Santa Barbara	28	1		1																		Furumizo 1973; 1975b
Santa Maria	88	1		1																		Furumizo 1973; 1975b
Santa Paula	96	1		1																		Furumizo 1973; 1975b
Scotia	118	1	1																			Furumizo 1973; 1975b
Sonoma	21	1	1																			Furumizo 1973; 1975b
Spring Valley	145	1		1																		Furumizo 1973; 1975b
Stockton	3	1	1																			Furumizo 1973; 1975b
Vacaville	48	1	1																			Furumizo 1973; 1975b
Ventura	0	1		1																		Furumizo 1973; 1975b
Yountville	26	1	1																			Furumizo 1973; 1975b
Colorado																						
Denver	1603	4	1	1																		Moyer et al. 1985
Padroni	1219	4		1																		Nelson & Fernandez-Caldas 1995

continued...

Appendix 2 Continued

Country & location	Alt. (m)	Alt. class	D. pt.	D. f	D. m	E. m	M. i	Hirstia spp.	A. s	Other A. spp.	T. p	Other T. spp.	A. o	S. m	L. d	G. d	B. t	C. a	G. f	C. e	Other C. spp.	Reference
Florida																						
Delray Beach	6	1	1	1		1											1					Arlian et al. 1992
Tampa	15	1	1	1		1					1						1	1				Fernandez-Caldas et al. 1990
Georgia																						
Atlanta	320	1	1	1		1																Smith et al. 1985
Hawaii																						
Hawaii Is.	5	1	1	1																		Sharp & Haramoto 1970
Honolulu	5	1	1	1		1																Massey et al. 1988
Honolulu & Kailua	5	1	1																			Nadchatram et al. 1981
Kauai Is.			1	1																		Sharp & Haramoto 1970
Manoa	91	1	1	1												1				1		Sharp & Haramoto 1970
Maui Is.			1	1																		Sharp & Haramoto 1970
Oahu Is.			1																			Nadchatram et al. 1981
Waikiki	2	1	1	1												1				1		Sharp & Haramoto 1970
Kansas																						
Eureka	330	2	1														1					Nelson & Fernandez-Caldas 1995
Kentucky																						
Lexington	303	2	1																			Fain 1966b; 1967a
Louisiana																						
New Orleans	3	1	1	1	1																	Griffiths & Cunnington 1971
New Orleans	3	1	1	1		1											1					Arlian et al. 1992
New Mexico																						
Artesia	1030	4		1																		Nelson & Fernandez-Caldas 1995
North Carolina																						
Greenville	17	1	1	1																		Arlian et al. 1992
North Dakota																						
Cooperstown	438	2	1																			Anonymous 1979
Ohio																						
Central Ohio			1	1																		Mitchell et al. 1969
Central Ohio																						Woodford 1980
Cincinnati	208	2	1	1																		Arlian et al. 1979a; 1982; 1983; 1992
Cincinnati	208	2	1	1																		Woodford et al. 1978
Columbus	244	2	1	1	1	1					1											Yoshikawa & Bennett 1979
Columbus	244	2						1														Fain et al. 1974
Columbus	244	2		1																		Fain 1967a
Columbus	244	2		1																		Larson et al. 1969
Columbus	244	2	1	1																		Mitchell et al. 1969
Dayton	229	2	1	1																		Arlian et al. 1978; 1979a; 1982; 1983
Dayton	229	2	1	1																		Woodford et al. 1978
Tennessee																						
Knocksville	271	2																				Shamiyeh et al. 1971; 1973
Memphis	77	1	1	1		1											1					Arlian et al. 1992
Oak Ridge	267	2	1	1																		Shamiyeh et al. 1971; 1973

continued...

Appendix 2 Continued

Country & location	Alt. (m)	Alt. class	D. pt.	D. f	D. m	E. m	M. i	Hirstia spp.	A. s	Other A. spp.	T. p	Other T. spp.	A. o	S. m	L. d	G. d	B. t	C. a	G. f	C. e	Other C. spp.	Reference
Texas																						
Bell County	210	2																				Outhouse & Castro 1990
Coryell County	280		1	1																		Outhouse & Castro 1990
Dallas	141	1	1																			Hall et al. 1971
Fort Worth	187	1		1																		Hall et al. 1971
Galveston	3	1	1	1		1											1					Arlian et al. 1992
Virginia																						
Alexandria	9	1	1																			Fain 1966b; 1967
Charlottesville	181	1	1	1																		Platts-Mills et al. 1987
Hampton Roads	3	1	1	1																		King et al. 1989
Williamsburg	26	1	1	1								1										Lassiter & Fashing 1990
Wisconsin																						
Milwaukee	193	2	1	1		1					1											Klein et al. 1986
UZBEKISTAN																						
Tashkent	460	3									1			1								Shcherbak & Nazrullayeva 1988
Tashkent	460	3	1											1								Nazrullayeva & Dubinina 1999
VENEZUELA																						
Caracas	263	2	1	1		1											1					Hurtado & Parini 1987
ZAMBIA																						
Ndola	1305	4	1	1																	1	Buchanan & Jones 1972; 1974
Total			824	449	35	260	41	74	61	40	151	41	9	9	154	139	192	163	81	92	89	2904
Minus repeat localities																						

Alt. = altitude; D. pt = *D. pteronyssinus*; D. f = *D. farinae*; D. m = *D. microceras*; E.m = *E. maynei*; M. i = *M. intermedius*; A. s = *Acarus siro*; Other A. spp. = other *Acarus* spp.; T. p = *Tyrophagus putrescentiae*; Other T. spp. = Other *Tyrophegus* spp.; A. o = *Aleuroglyphus ovatus*; S. m = *Suidasia medanensis*; L. d = *L. destructor*; G. d = *G. domesticus*; B. t = *Blomia tropicalis*; C. a = *C. arcuatus*; G. f = *G. fusca*; C. e = *C. eruditus*; Other C. spp. = Other Cheyletus spp.

Appendix 3

Abundance and frequency of occurrence of domestic mites in house dust, based on published surveys

A = asthmatic/atopic/clinical group; Po = population group; B = Beds; F = floors; s = swept sample

Country & location	Altitude	Volunteer group	Sample type	n (samples)	n (homes)	Yr. sampling began	Total mites mean no./g	% +ve samples	D. pteronyssinus mean no./g	% +ve samples	% of total mites	D. farinae mean no./g	% +ve samples	% of total mites	E. maynei mean no./g	% +ve samples	% of total mites	Blomia spp. mean no./g	% +ve samples	% of total mites	Reference
ALGERIA																					
Ain-Roua	1100	Po	B	6	6	1984	550		65		11.8	0		0.0	403		73.1	20		3.6	Louadi & Robaux 1992
Ain-Touta	917	Po	B	6	6	1984	10		0		0.0	0		0.0	0		0.0	0		0.0	Louadi & Robaux 1992
Algiers	1	Po	B	7	6	1984	4820	85.7		71.4	53.0		14.3	2.0		85.7	45.0	0	0.0	0.0	Abed-Benamara *et al.* 1983
Annaba	20	Po	B	6	6	1984	1260		1050		83.3	129		10.2	23		1.9	0		0.0	Louadi & Robaux 1992
Arris	1171	Po	B	6	6	1984	10		0		0.0	0		0.0	0		0.0	0		0.0	Louadi & Robaux 1992
Batna	1040	Po	B	6	6	1984	10		10		100	0		0.0	0		0.0	0		0.0	Louadi & Robaux 1992
Béjaia	9	Po	B	6	6	1984	6310		5982		94.8	19		0.3	105		1.7	2		0.03	Louadi & Robaux 1992
Biskra	124	Po	B	6	6	1984	10		8		83.3	0		0.0	2		16.7	0		0.0	Louadi & Robaux 1992
Bou-Snib	886	Po	B	6	6	1984	2290		1244		54.3	2		0.1	248		10.8	5		0.2	Louadi & Robaux 1992
El-Kantara	513	Po	B	6	6	1984	50		39		78.6	0		0.0	0		0.00	0		0.0	Louadi & Robaux 1992
El-Oued	70	Po	B	6	6	1984	10		10		100	0		0.0	0		0.00	0		0.0	Louadi & Robaux 1992
Jijel	6	Po	B	6	6	1984	6270		5518		88	401		6.4	27		0.4	3		0.05	Louadi & Robaux 1992
Oum-El-Bouaghi	950	Po	B	6	6	1984	50		39		77.4	0		0.0	2		3.2	0		0.0	Louadi & Robaux 1992
Seraidi	960	Po	B	6	6	1984	1540		1482		96.2	5		0.3	10		0.7	0		0.0	Louadi & Robaux 1992
Setif	1079	Po	B	6	6	1984	670		358		53.4	10		1.5	123		18.5	25		3.7	Louadi & Robaux 1992
Skikda	42	Po	B	6	6	1984	6630		6003		90.5	259		3.9	116		1.8	7		0.1	Louadi & Robaux 1992
Souk-Ahras	665	Po	B	6	6	1984	630		534		84.7	0		0.0	17		2.6	0		0.0	Louadi & Robaux 1992

continued…

Appendix 3 Continued

Country & location	Altitude	Volunteer group	Sample type	n (samples)	n (homes)	Yr. sampling began	Total mites mean no./g	% +ve samples	D. pteronyssinus mean no./g	% +ve samples	% of total mites	D. farinae mean no./g	% +ve samples	% of total mites	E. maynei mean no./g	% +ve samples	% of total mites	Blomia spp. mean no./g	% +ve samples	% of total mites	Reference
Tebessa	885	Po	B	6	6	1984	60		57		94.7	0		0.0	0		0.0	0		0.0	Louadi & Robaux 1992
Touggourt	69	Po	B	6	6	1984	10		10		100	0		0.0	0		0.0	0		0.0	Louadi & Robaux 1992
Zitouna	548	Po	B	6	6	1984	3940		3534		89.7	0		0.0	0		0.0	8		0.2	Louadi & Robaux 1992
AUSTRALIA																					
Cape York	3	A, Po	B	30	30	1995	375	100.0	286	100.0	76.3	1	3.0	0.2	17	27.0	4.5	51	53.0	13.7	Tovey et al. 2000a
Alice Springs	609	A, Po	B	30	30	1995	11	20.0	8	20.0	76.5	1	3.4	5.9	0	0.0	0.0	0	0.0	0.0	Tovey et al. 2000a
Belmont	1	A, Po	B	32	32	1982	441	94.0	33	60.0		284	100								Green et al. 1986
Brisbane	25	Po	F	48	4	1969	40	100.0													Domrow 1971
Broken Hill	304	A, Po	B	30	30		98	73.7	95	73.3	97.3	0	0.0	0.0	1	3.7	0.7	0	0.0	0.0	Tovey et al. 2000a
Bunbury	1	A	B	10	10	1986	585	100.0	399	100	48.4	0	0.0	0.0	399	100	48.4	0	0.0	0.0	Colloff et al. 1991
Bunbury	1	A	F	10	10	1986	992	100.0	909	100	96	0	0.0	0.0	909	100	96	0	0.0	0.0	Colloff et al. 1991
Busselton	1	A, Po	B	30	30		69	86.7	60	86.7	86.4	1	3.5	1.0	3	6.9	3.8	0	0.0	0.0	Tovey et al. 2000a
Busselton	1	A, Po	B	43	43	1983	275	100.0				0	0.0	0.0							Green et al. 1986; 1993
Camden	291	A, Po	B	30	30		389	96.7	368	96.7	94.6	1	3.9	0.2	1	3.9	0.2	0	0.0	0.0	Tovey et al. 2000a
Lismore	134	A, Po	B	30	30		476	100.0	436	100.0	91.7	1	7.0	0.3	9	7.0	1.8	2	7.0	0.4	Tovey et al. 2000a
Melbourne	115		B	8	8		6285		5810		92.4	0	0.0	0.0	1		0.2	0	0.0	0.0	Blythe 1976
Narrabri	33	A, Po	B	30	30		131	60.0	122	56.4	92.8	0	0.0	0.0	0	0.0	0.0	0	0.0	0.0	Tovey et al. 2000a
Perth	33	A	B	10	10	1986	480	100.0	347	100	64.4	0	0.0	0.0	347	100	64.4	0	0.0	0.0	Colloff et al. 1991
Perth	33	A	F	10	10	1986	263	100.0	201	100	95	0	0.0	0.0	201	100	95	0	0.0	0.0	Colloff et al. 1991
Sydney	1	Po	B	66	22	1973	546			78.0											Tovey et al. 1975
Sydney	1		B		1	1983	594		534		89.9	30		5.1							Green 1983
Sydney	1		B	7	7	1971	214	100.0													Mulvey 1972
Wagga Wagga	148	A, Po	B	30	30		103	76.7	70	59.8	68.4	0	0.0	0.0	0	0.0	0.0	0	0.0	0.0	Tovey et al. 2000a
Wagga Wagga	148	A, Po	B	67	67	1982	10	15.0	10	15.0		0	0.0	0.0	0	0.0	0.0	0	0.0	0.0	Green et al. 1986
Wagga Wagga	148	A, Po	F	20	20	1992	412	90.0													Green et al. 1993
BARBADOS	10	A	B	23		1970	4440					19	4.3	0.4							Pearson & Cunnington 1973

continued...

Appendix 3 Continued

Country & location	Altitude	Volunteer group	Sample type	n (samples)	n (homes)	Yr. sampling began	Total mites mean no./g	% +ve samples	D. pteronyssinus mean no./g	% +ve samples	% of total mites	D. farinae mean no./g	% +ve samples	% of total mites	E. maynei mean no./g	% +ve samples	% of total mites	Blomia spp. mean no./g	% +ve samples	% of total mites	Reference
BARBADOS	10	A	F	25		1970	777					4	4.0	0.5							Pearson & Cunnington 1973
BRAZIL																					
7 localities & Rio Claro	608		F	25				100.0		50	3.7					50	3.7		100	79.5	Rosa & Flechtman 1979
Belo Horizonte	876										9			39			9			36	Morera 1975
Brasilia	1080		Bs	122		1975	3652														Cardoso et al. 1979
Campinas	680		B	57	57	1996	2093			66.9			72.6						47.4		Oliveira et al. 1999; 2003
Juiz de Fora	695		B	160	160	1999		77.5		44.5	72		11.8	5		10.0	3.1		18.2	16.1	Ezequiel et al. 2001
Londrina	576	A	B	133	38	2001	583	50.0													da Silva et al. 2005
Recife	1	A, Po	B	40	40	1994	364														Sarinho et al. 1996
Salvador	1		B							39.9			2.5						71.8		Baqueiro et al. 2006
Salvador	1	Po	B	50	50	1998		98.0		70.0			8.0						30.0		Serravalle & Medeiros 1999
Sao Paulo	638						932														de Oliveira et al. 2003
Uberaba	747	A, Po	B, F	120	60	2000				15.60			12.3			7.9			4.4		Terra et al. 2004
'Zona da Mata'	695	Po	B	30	30	2000	647		224		46.5			0.2	16		3.3	203		42.1	Oliveira & Daemon 2003
BRUNEI DARUSSA-LAM	50	R	Bs	8	8	1978	250	100.0	40	87.5	17.4	6	50.0	2.0	0	0.0	0.0	9	75.0	7.0	Woodcock & Cunnington 1980
BULGARIA																					
Silistra	7	Po	B, F		173	1974		98.3	236	97.7	52.9	14	36.4	3.1	116	64.2	26				Todorov 1978
BURMA																					
Rangoon	9	A, Po	B	540	60	1986	72	18.0	23.0	5.9	3.2	1.4		2.0	0	0.0	0.0	49	94.1	69.0	Htut et al. 1991
CANADA																					
Vancouver	72	Po	Bs	72	2	1977		83.3		83.3	67.0			33.0							Murray & Zuk 1981
CHILE																					
53 localities			B	368	69	1980		68.0		68			18.9	3.8		68					Artigas & Casanueva 1983
Punta Arenas	35	Po	B	100	100	1989	10	29.1													Franjola & Malonnek 1999
Valdivia	5	Po	B	100	100	1989	6	70.0		15	4.5		3.0	1.0					23.0	19.2	Franjola & Malonnek 1995

continued...

Appendix 3 Continued

Country & location	Altitude	Volunteer group	Sample type	n (samples)	n (homes)	Yr. sampling began	Total mites mean no./g	% +ve samples	D. pteronyssinus mean no./g	% +ve samples	% of total mites	D. farinae mean no./g	% +ve samples	% of total mites	E. maynei mean no./g	% +ve samples	% of total mites	Blomia spp. mean no./g	% +ve samples	% of total mites	Reference
CHINA																					
Huaide & Kai-an		A	B	45	45	1989	25														Du *et al.* 1993
GuangZhou	1	A, Po	B	426		1979	1328														Lai *et al.* 1987
Haikou	8											9.3								71.8	Zhu *et al.* 2007
Shanghai	8	Po	B	1170	15	1984		100.0			67.5			4.6			10.3			0.0	Cai & Wen 1989
Shanghai	8	Po	B		25			66.7		85.7	49.2		1.7	8.0		1	5.5		0.0	0.0	Wen *et al.* 1991
Shanghai	8	A, Po	B	48		1995	62	68.7	53		85.7	1		1.7	1		1.0	0		0.0	Wang & Wen 1997
COLOMBIA																					
Cartagena	13	A	B, F	50	25	1991	418		105	90.0	35.7	72	8.0	17.2				147	96.0	40.1	Fernandez-Caldas *et al.* 1993
Charta	2191	Po	B, F	100	9	1992	288		158		54.9	102		35.4				12		4.2	Sanchez-Medina & Zarante 1996
Chocontá	2689	Po	B, F	100	9	1992	263		108		41.1	86		32.7				9		3.4	Sanchez-Medina & Zarante 1996
Garagoa	1928	Po	B, F	100	9	1992	237		116		49.0	74		31.2				14		5.9	Sanchez-Medina & Zarante 1996
La Calera	2376	Po	B, F	100	9	1992	229		50		21.8	169		73.8				0		0.0	Sanchez-Medina & Zarante 1996
Obenuco	2585	Po	B, F	100	9	1992	211		90		42.7	92		43.6				0		0.0	Sanchez-Medina & Zarante 1996
Pitalito	1360	Po	B, F	100	9	1992	673		357		53.0	213		31.6				0		0.0	Sanchez-Medina & Zarante 1996
Tenza	1630	Po	B, F	100	9	1992	100		60		60.0	33		28.0				0		0.0	Sanchez-Medina & Zarante 1996
Tuquerres	3150	Po	B, F	100	9	1992	119		72		60.5	0		0.0				16		13.5	Sanchez-Medina & Zarante 1996
Viso Elias	1400	Po	B, F	100	9	1992	242		148		61.2	81		33.5				0		0.0	Sanchez-Medina & Zarante 1996
Santiago de Tolú	1	A	B	28	28	2004	4866		1775		36.5	25		0.5				1366		28.1	Rodríguez Monterroza *et al.* 2006
COSTA RICA																					
Coast, centre		A	B	330		1995	2554														Soto-Quiros *et al.* 1998
Coast, centre		A	F	330		1995	989														Soto-Quiros *et al.* 1998
San Jose	4	Po	Bs		60	1985	13	100.0	2	68.3	15.2	0	0.0	0.0	0.2	11.7	0.01	4	95.0	27.9	Vargas & Mairena 1991

continued...

Appendix 3 Continued

Country & location	Altitude	Volunteer group	Sample type	n (samples)	n (homes)	Yr. sampling began	Total mites mean no./g	% +ve samples	*D. pteronyssinus* mean no./g	% +ve samples	% of total mites	*D. farinae* mean no./g	% +ve samples	% of total mites	*E. maynei* mean no./g	% +ve samples	% of total mites	*Blomia* spp. mean no./g	% +ve samples	% of total mites	Reference
CROATIA																					
Pazin	278	Po	B	18		1973	2582		1500		58.1	2		0.1	195		7.5				Blythe 1976
Zadar	22	Po	F	18		2001	120		83		69.2	3		2.5	6		5.0	9		7.5	Macan *et al.* 2003
Zagreb	130	Po	F	26		2001	94		53		56.4	15		16.0	12		12.8	0		0.0	Macan *et al.* 2003
CUBA																					
25 localities	5	A	B	54		1979	136	100.0	43	100	26.4	0	0.0	0.0	0	0.0	0.0	29	72.2	18.2	Cuervo *et al.* 1983
CZECH REPUBLIC																					
Unspecified localities			B, F	33		1976	12	69.7	9	66.7	72.0	1	12.1	3.0	1	18.2	4.5				Vobrazkova *et al.* 1986
25 localities			B	83		1972	64	90.4	7	12.1	10.3	25	36.0	38.4	136	100					Samsinak *et al.* 1972
33 localities		A, Po	B	338			29				31.3				17						Samsinak *et al.* 1978a
Prague	245		B	140	2	1972	30	95.7	30	95.7		26		88.4	30	95.7					Dusbabek 1975; 1979
Prague & environs	245	A	B	51	4		33	100.0	9	71.0	28.0	4	41.1	13.1	9	60.7	28.0				Vobrazkova *et al.* 1979
DENMARK																					
Århus	1	Po	B	75	75	1977	150	87.0	26		57.0	18		39.0	2		1.0				Korsgaard 1979
Århus	1		F	62	75	1977	30	55.0							30		55				Korsgaard 1979
Århus	1	A	B	25	25	1980	400														Korsgaard 1983b
Århus	1	A	F	25	25	1980	200														Korsgaard 1983b
Århus	1	A	F		34	1984	220														Harving *et al.* 1993
Århus	1	A	B		62	1984	75														Harving *et al.* 1993
Copenhagen	1	A	Bs	52	42	1969		67.3		38.5							1.9				Alani & Haarlov 1972
Copenhagen	1		B	25	5	1982		100.0		60.0	13.2		60.0	41.6							Andersen 1984
Copenhagen	1	Po	B	68	68	1999	10	58.8			22.1			51.5							Sidenius *et al.* 2002
Thisted	1		B	11	1	1980		100.0						72.0							Hallas & Korsgaard 1983
ECUADOR																					
Quito	2763	A, Po	B	20		2003	840	100.0	715	95.0		696	15.0		260	20.0		179	30.0		Valdivieso *et al.* 2006
Cuenca	2450	A, Po	B	23		2003	895	95.7	686	95.7		306	17.4		185	30.4		447	8.7		Valdivieso *et al.* 2006

continued...

Appendix 3 Continued

Country & location	Altitude	Volunteer group	Sample type	n (samples)	n (homes)	Yr. sampling began	Total mites mean no./g	% +ve samples	D. pteronyssinus mean no./g	% +ve samples	% of total mites	D. farinae mean no./g	% +ve samples	% of total mites	E. maynei mean no./g	% +ve samples	% of total mites	Blomia spp. mean no./g	% +ve samples	% of total mites	Reference
Guayaquil	46	A, Po	B	24		2003	4586	100.0	1120	100.0		306	64.0		1583	28.0		88	76.0		Valdivieso et al. 2006
EGYPT																					
Alexandria	10	A	B	30	30	1998	240		188		78.3	44		15.1	0		0.0	4		1.5	Sadaka et al. 2000
Alexandria	10	A	F	30		1998	126		87		68.9	19		18.3	0		0.0	4		3.4	Sadaka et al. 2000
Alexandria	10	A, Po	F	360	15	1991					0.7			0.6						0.8	Rezk et al. 1996
Tanta	11	Po	B, F	180	3	1980					14.6			24.4						6.8	Gamal-Eddin et al. 1982
FAROES																					
Torshavn	1	Po	B	8		1977				37.5											Korsgaard, in Hallas et al. 2004
FINLAND																					
Helsinki	24	A	Bs	37		1969	382	59.5	19	75	65.6	29	100								Stenius & Cunnington 1972
Uukuniem & Ilomantsi	84	Po	B	8		1971	29	100.0	7	32.4	22.9	3	50.0	8.5							Stenius & Cunnington 1972
FRANCE																					
Briançon & environs	1572	A	B	218		1979	30	18.8		17.0		7	24.0								Vervloet et al. 1982
Grenoble (0–300 m)	236	Po	B	16	4	1973	59	100.0													Lascaud 1978
Grenoble (300–1100 m)	690	Po	B	12	3	1973	10	91.7													Lascaud 1978
Grenoble (>1100 m)	1363	Po	B	12	3	1973	3	66.7													Lascaud 1978
Marseille & environs	54	A	B	77		1979	880	80.0	572	65.0		185	21.0								Vervloet et al. 1982
Montpellier & environs	30	Po	B	94		1970	20	100.0	18.0	92.9	58.1		7.5	4.6			4.2				Guy et al. 1972; Rousset 1971
Nancy & environs	200	Po	Bs	50		1968	180	100.0	54	100	62.1										Percebois et al. 1972
Strasbourg & environs	140	Po	B	33	33	1970	213	100.0	135	100.0	69.2	0.4	9.1	1.8	5.3	42.4	18.8				Araujo-Fontaine et al. 1973
GERMANY																					
Heligoland	1		B	48		1972	111														Bronswijk & Jorde 1975
Berlin	35		B	5	3	1972	103	100.0	45	60.0	20.0	10	20.0	1.4							Karg 1973
Hamburg	3			67	28	1981				86.5	72.8		32.8	12							Keil 1983
Hamburg	3		B	36	28	1981	163	88.9													Keil 1983
Hamburg	3		F	31	28	1981	209	96.8													Keil 1983

continued...

Appendix 3 Continued

Country & location	Altitude	Volunteer group	Sample type	n (samples)	n (homes)	Yr. sampling began	Total mites mean no./g	% +ve samples	*D. pteronyssinus* mean no./g	% +ve samples	% of total mites	*D. farinae* mean no./g	% +ve samples	% of total mites	*E. maynei* mean no./g	% +ve samples	% of total mites	*Blomia* spp. mean no./g	% +ve samples	% of total mites	Reference
GREECE																					
Attiki	142	Po	B	110		1989				75.5	48.7		51.8	16.8		2.7	0.26				Papaioannou-Souliotis 1991
Athens	154	A	B	25		1988					84.7			6.2			0.59				Delyargyris et al. 1990
GREENLAND																					
Ilulissat (Jakobshavn)	13	Po	B	7		1977				42.9											Korsgaard, in Hallas et al. 2004
HONG KONG	18	A	Bs	21	20	1976	1601	100.0	833	100	51.8	212	75.0	13.2	260	95.2	8.1	170	75.0	10.6	Gabriel et al. 1982
HUNGARY																					
Budapest & environs	103	A	B	318	256	1969	95		33		34.7	23	68.4	24.2							Halmai 1984
ICELAND																					
Reykjavik	15	A	B	20	10	2001	25	10.0	25	10.0	100										Hallas et al. 2004
INDIA																					
19 States		A	B, Fs	840	300	1972	116	78.0													Dar et al. 1973
19 States		A	B, Fs	170	50	1972	144	56.4													Dar & Gupta 1980
24 Parganas	1	A	Bs	10	10	1989	70														Modak et al. 1991
24 Parganas	1	A	Fs	10	10	1989	25														Modak et al. 1991
Aurangabad	565	A	Bs	205		1984	126														Tilak & Jogda 1989
Bangalore	914	Po	Bs	350		1978	874	100.0		100	42.5		20.7	1.2							Ranganath & Channa Basavanna 1988
Barddhaman	40	A	Bs	10	10	1989	100														Modak et al. 1991
Barddhaman	40	A	Fs	10	10	1989	39														Modak et al. 1991
Bhilai & environs	293	Po	B, Fs		5	1971	615	73.5			80.6			8.2							Dixit & Mehta 1973
Calcutta	13	A	Bs	10	10	1989	161														Modak et al. 1991
Calcutta	13	A	Fs	10	10	1989	24														Modak et al. 1991
Calcutta	13	A	Bs	82	82	1992	1014	100.0	419		41.3	194		19.1							Saha et al. 1994
Calcutta	13	A	Fs	82	82	1992	226	100.0	88.0		40.0	33		14.4							Saha et al. 1994
Chandanbari	2868	Po	B, Fs	5	5	1990	600	100.0	0		0.0	0		0.0				450		75.0	Modak et al. 1992

continued...

Appendix 3 Continued

Country & location	Altitude	Volunteer group	Sample type	n (samples)	n (homes)	Yr. sampling began	Total mites mean no./g	% +ve samples	D. pteronyssinus mean no./g	% +ve samples	% of total mites	D. farinae mean no./g	% +ve samples	% of total mites	E. maynei mean no./g	% +ve samples	% of total mites	Blomia spp. mean no./g	% +ve samples	% of total mites	Reference
Delhi	215	A	B	51		1970		52.9								52.9					Nayar et al. 1974
Delhi	215		Bs	88	15	1972	30	87.5							30	87.5					Dar & Gupta 1980
Dibrugarh	94	Po	B, F			1978														4.2	Maurya et al. 1983
Jammu	900	Po	B, Fs	5	5	1990	200	100.0	196		98.0	0		0.0							Modak et al. 1992
Jammu	900	Po	B, F			1978														6.0	Maurya et al. 1983
Katra	753	Po	B, F			1978														12.0	Maurya et al. 1983
Lucknow	131	Po	Bs	22		1976	1270														Maurya & Jamil 1980
Lucknow	131	Po	Fs	22		1976	634														Maurya & Jamil 1980
Lucknow	131	A	Bs	30		1972		63.3			36.7			10.0							Nath et al. 1974
Madras	9	Po	B	172	172	1992	492				97.1			18.0			7.6			1.7	Kannan et al. 1996
Pahalgam	2115	Po	B, Fs	5	5	1990	107	100.0	94		87.0	13		12.1							Modak et al. 1992
Srinagar	1960	Po	B, Fs	5	5	1990	800	100.0	600		75.0	0		0.0							Modak et al. 1992
Srinagar	1960	Po	B, F			1978														1.0	Maurya et al. 1983
Udampur	755	Po	B, F			1978														10.0	Maurya et al. 1983
INDONESIA																					
Jakarta	1	A	B, F	10	10	1995	255				25.4			39.4						14.1	Baratawidjaja et al. 1998
IRAN																					
Bushehr (Bushir; Bushire)	13	Po	B, F	15	15	1977	11		4	46.0	31	4	46.0	38.1	0	3.0	0.00				Sepasgosarian & Mumcuoglu 1979
Chalus (`Calûs, Châlûs)	20	Po	B, F	15	15	1977	3220	100	3073	100	95.5	6	20.0	0.2	0	2.0	0.00				Sepasgosarian & Mumcuoglu 1979
Isfahan (Esfahan)	1571	Po	B, F	15	15	1977	1		0.3	7.0	25	0	2.5	0.0	0	2.0	0.00				Sepasgosarian & Mumcuoglu 1979
Karaj (Kara`g, Karai)	1462	Po	B, F	15	15	1977	2		0.4	12.5	20	0	2.5	0.0	0	2.0	0.00				Sepasgosarian & Mumcuoglu 1979
Lahijan	25	Po	B, F	15	15	1977	8730	100	7645	100	87.6	17	6.0	0.2	640	40.0	7.30	1			Sepasgosarian & Mumcuoglu 1979
Lahijan & environs	25	A	B	40	40	1974	3322	90.0	1977	94.3	59.6	125	37.1	3.9	3	5.7	0.09	4	5.7	0.13	Amoli & Cunnington 1977

continued...

Appendix 3 Continued

Country & location	Altitude	Volunteer group	Sample type	n (samples)	n (homes)	Yr. sampling began	Total mites mean no./g	% +ve samples	D. pteronyssinus mean no./g	% +ve samples	% of total mites	D. farinae mean no./g	% +ve samples	% of total mites	E. maynei mean no./g	% +ve samples	% of total mites	Blomia spp. mean no./g	% +ve samples	% of total mites	Reference
Ramsar (Sakht Sar)	1	Po	B, F	15	15	1977	6785	100	4638	100	68.4	0	2.0	0.0	889	92.0	13.10	1			Sepasgosarian & Mumcuoglu 1979
Shiraz (`Sîraâz)	1800	Po	B, F	15	15	1977	2		0.5	12.5	33.3	0	2.5	0.0	0	2.0	0.00				Sepasgosarian & Mumcuoglu 1979
ISRAEL																					
13 localities		Po	B		65	1995	415	97.0	331	85.6	79.8	59	71.3	14.2	3	10.3	0.74	2	6.2	0.5	Mumcuoglu et al. 1999
Bat Yam	4	Po	B, F	59	2	1983	1209	100.0	1009	100	83.5	1	8.0	0.04	56	81	4.7	0	0.0	0.0	Feldman-Muhsam et al. 1985
Be'er Sheva'	244	Po	B, F	59	2	1983	73		58	80	68.2	19	53.0	22.4	18	57	5.8	0	0.0	0.0	Feldman-Muhsam et al. 1985
Jerusalem	757	Po	B, F	59	2	1983	82		62	75	87.1	1	16.0	2.0	0.2	4	0.2	0	0.0	0.0	Feldman-Muhsam et al. 1985
Nes Ziyyona	35	Po	B, F	59	2	1983	317		265	95	83.7	3	32.0	0.9	2	15	0.6	0	0.0	0.0	Feldman-Muhsam et al. 1985
Nir Dawid	−80	Po	B, F	59	2	1983	317		128	75	40.3	138	73.0	43.4	0.5	7	0.7	0	0.0	0.0	Feldman-Muhsam et al. 1985
Tel-Aviv	35	A	B	102	17	1989	1462		1350		92.4	95		6.5							Kivity et al. 1993
ITALY																					
Sardinia		A	B	55		1981	580	98.2	319	94.5	55.4	53	29.1	9.1	113	54.5	19.7				Ottoboni et al. 1983
Livorno	5	A, Po	B	37	37	1982	1238														Goraccci et al. 1984
Milan	103	Po	B	15		1973	191		3		14.7	160		83.8	0		0.0				Blythe 1976
Naples	1	Po	B	6		1973	440		20		45.5	180		40.9	40		9.1				Blythe 1976
Naples	1	A	B	35	37	1973	2232		187	88.6	12.2	1327	77.1	64.8	405	80.0	23.0				Noferi et al. 1974
Panzano	243	Po	B	18		1973	468		35		74.8	49		10.5	0.4		0.09				Blythe 1976
Rome	15	Po	B	17		1992	11	64.7		12.6	5.2		56.9	53.1		1.3	0.2				Bigliocchi & Maroli 1995
Rome	30	Po	B		10	1994	546	100.0	117		21.4	412		75.4	3		0.5				Bigliocchi et al. 1996
Turin	205	Po	B	12		1973	73		2		27.7	45		61.6	19		2.6				Blythe 1976
JAPAN																					
Fukuoka	10	Po	B, F			1968	724		327		45.2	1		0.1	0		0.0	0		0.0	Oshima 1970; Miyamoto et al. 1970
Hiroshima	26	Po	B, F			1968	435		228		52.4	8		1.8	0		0.0	30		6.9	Oshima 1970; Miyamoto et al. 1970

continued...

Appendix 3 Continued

Country & location	Altitude	Volunteer group	Sample type	n (samples)	n (homes)	Yr. sampling began	Total mites mean no./g	% +ve samples	D. pteronyssinus mean no./g	% +ve samples	% of total mites	D. farinae mean no./g	% +ve samples	% of total mites	E. maynei mean no./g	% +ve samples	% of total mites	Blomia spp. mean no./g	% +ve samples	% of total mites	Reference
Nagoya	14	Po	B, F			1968	670		100		14.9	311		46.4	1		0.1	0		0.0	Oshima 1970; Miyamoto et al. 1970
Nagoya	14	A, Po	F	176	22	1983	578	100	202	100	35.1	243	100	42.1	0.1	4.5	0.001	0	0.0	0.0	Suto et al. 1992a
Nagoya	14	A, Po	F	160	20	1983	703	100	309	100	43.7	271	100	38.3	21	15	2.9	0	0.0	0.0	Suto et al. 1992b
Nagoya	14	A, Po	F		33	1983			152			63										Suto et al. 1993
Nagoya	14	A, Po	F		28	1983			228			133										Suto et al. 1993
Naha	2	A, Po	F	78	6	1990	1148	100.0	1001		87.3	1		0.1	0		0.0	17		1.5	Toma et al. 1993
Naha	2	A, Po	F	65	5	1990	1604	100.0	1419		88.6	3		0.2	0		0.0	29		1.8	Toma et al. 1993
Osaka	0	Po	B, F			1968	2050		377		18.4	71		3.5	0		0.0	0		0.0	Oshima 1970; Miyamoto et al. 1970
Sapporo	31	Po	F			1968	827		393		47.5	1		0.1	7		0.9	0		0.0	Oshima 1970; Miyamoto et al. 1970
Sendai	54	Po	B, F			1968	1953		831		42.5	2		0.1	0		0.0	0		0.0	Oshima 1970; Miyamoto et al. 1970
Tokushima	1	Po	B, F			1968	549		192		35.0	9		1.6	9		1.6	0		0.0	Oshima 1970; Miyamoto et al. 1970
Tokyo	43	Po	B, F			1968	1546		194		12.6	3		0.2	2		0.1	0		0.0	Oshima 1970; Miyamoto et al. 1970
Tokyo	43	A, Po	B, F	32	32	1975	441	100.0	84	90.6	27.6	90	100	39.3	0.9	34.4	4.0	0	0.0	0.0	Takaoka et al. 1977; Ishii et al. 1979
Tsu	20	A	B, F	142	14	1989	109															Matsuoka et al. 1995
Urawa	14	Po	B, F	84	26	1981	315		82		26.0	78		24.8	7		2.2	0		0.0	Takaoka & Okada 1984
Wakayama	1	A, Po	B, F	24	24	1977	2840	100.0														Uchikoshi et al. 1982
Wakayama	1	A, Po	B, F	26	26	1977	998	100.0														Uchikoshi et al. 1982
KENYA																					
Nairobi	1729	Po	B	11		1973	7878		5240		66.5	500		6.3	1588		20.2				Blythe 1976
MALAYSIA																					
Cameron Highlands	1308	Po	B	130	48	1979			1440	100				42.7			40.5			68.0	Ho & Nadchatram 1985
Kuala Lumpur	61	Po	B	52	18	1984	235	51.0	11		12.8	0.8		1.5	0	0.0	0.0	34		66.2	Rueda 1985
Kuala Lumpur & Selangor	61	Po	B	120	120	1994	6990		281	100		21	66.8		0.7	2.2		341	100	46.1	Mariana & Ho 1996

continued...

Appendix 3 Continued

Country & location	Altitude	Volunteer group	Sample type	n (samples)	n (homes)	Yr. sampling began	Total mites mean no./g	% +ve samples	D. pteronyssinus mean no./g	% +ve samples	% of total mites	D. farinae mean no./g	% +ve samples	% of total mites	E. maynei mean no./g	% +ve samples	% of total mites	Blomia spp. mean no./g	% +ve samples	% of total mites	Reference
MARTINIQUE		Po	B	123		2000	383	100.0	171	100	44.7	25	6.5	1.6	0	0.0	0.0	94	24.5	95.9	Lafosse Marin et al. 2006
MEXICO																					
La Paz	51	Po	B		17	1985	550	100.0		100.0			78.6								Servín & Tejas 1991
NETHERLANDS																					
Leiden	−2		B		150	1965	18	100.0	16	100	87.6	0.2	2	1.2	2	53.0	11.2				Spieksma & S.-Boezeman 1967
Leiden	−2	Po	B, F	120		1965	5														Spieksma et al. 1971; Spieksma 1973
Leiden	−2	A	B	4		1962	81	100.0													Voorhorst et al. 1964
Nijmegen	46	A	B	36	3	1970	200	100.0	15	83											Bronswijk et al. 1971; Bronswijk 1973
Utrecht	1	A	B	60		1988	30														Bronswijk 1993
Utrecht	1	A	F	42		1988	9														Bronswijk 1993
NEW ZEALAND																					
Auckland	26	A	B	22		1970	20	100.0	12	100	59	0	0.0	0.0	0	0.0	0.0	0	0.0	0.0	Cornere 1971
Christchurch	6	A, Po	B	90		1979	833					0	0.0	0.0							Abbott et al. 1981
Invercargill	9		B	8		1973	854		710		83.1				10		1.2				Blythe 1976
Wairoa, Hawke's Bay	23	A, Po	B	42	42	1989	1095	100.0		100		0	0.0	0.0	0	0.0	0.0	0	0.0	0.0	Andrews et al. 1992; 1995
Wairoa, Hawke's Bay	23	A, Po	F	42	42	1989	630	100.0				0	0.0	0.0	0	0.0	0.0	0	0.0	0.0	Andrews et al. 1992; 1995
Wairoa, Hawke's Bay	23	A, Po	F	42	42	1989	1997	100.0				0	0.0	0.0	0	0.0	0.0	0	0.0	0.0	Andrews et al. 1992; 1995
Wairoa, Hawke's Bay	23	A, Po	F	42	42	1989	497	100.0				0	0.0	0.0	0	0.0	0.0	0	0.0	0.0	Andrews et al. 1992; 1995
Wellington	21	A, Po	B	79	79	1993	1922														Crane et al. 1995
Wellington	21	A, Po	F	79	79	1993	1137														Crane et al. 1995
Wellington	21	A, Po	F	72	72	1993	895														Crane et al. 1995
NIGERIA																					
Lagos	35	A, Po	B, F		16	1975	3063	100.0	1538	62.5	50.2		12.5			6.3		420	75.0	13.7	Hunponu-Wusu & Somorin 1978

continued...

Appendix 3   Continued

Country & location	Altitude	Volunteer group	Sample type	n (samples)	n (homes)	Yr. sampling began	Total mites mean no./g	% +ve samples	D. pteronyssinus mean no./g	% +ve samples	% of total mites	D. farinae mean no./g	% +ve samples	% of total mites	E. maynei mean no./g	% +ve samples	% of total mites	Blomia spp. mean no./g	% +ve samples	% of total mites	Reference
NORWAY																					
Fredrikstad	15				10	1994		80.0													Aas & Mehl 1996
Levanger	20				23	1994		82.6													Aas & Mehl 1996
Moss	40				10	1994		50.0													Aas & Mehl 1996
Namskogan	235				37	1994		81.1													Aas & Mehl 1996
Oslo	13	Po	B, F		540	1992				23.5			0.2		0.9						Mehl 1998
Ringerike	96				8	1994		75.0													Aas & Mehl 1996
Sarpsborg	32				10	1994		60.0													Aas & Mehl 1996
Sør–Varanger	3	A	B	19	19	1994	16	52.6													Dotterud et al. 1995
Trondheim	100				10	1994		70.0													Aas & Mehl 1996
PALESTINE																					
Gaza	20	A	B	40	40	1987	423	100.0	390	100.0	97.3	0	0.0	0.0	1	0.3	0.1	3	2.5	0.4	Mumcuoglu et al. 1994
PANAMA																					
La Chorrera	69	Po	Fs	70	20	1999	1230	100.0	17		1.38	0		0.0	0		0.0	464	100	37.7	Miranda et al. 2002
PAPUA NEW GUINEA																					
Okapa District	1814	A	Bs	32		1983	1380	100.0													Dowse et al. 1985
Okapa District	1814	A	Fs	30		1983	29	46.6													Dowse et al. 1985
Lufa & Goroka	1621	A	B	20	19	1972	213	100.0	131		100	73.2		20	0.42		76	20.0	22.4		Anderson & Cunnington 1974
PERU																					
Huaraz & Ticapampa	3672	Po	B	2	2	1973	15		2	100	10.5							1	50.0	2.6	Caceres & Fain 1978; 1979
Lima	108	Po	B	16	16	1973	141		23	87.5	16.5				17	50.0	11.9	11	37.5	8.0	Caceres & Fain 1978; 1979
Lima	108	Po	B, F	100		1987	20	92.0	3	15.9	17.4	0.0	0.0	0.0	0.05	1.0	0.2	12		59.4	Croce et al. 2000
Lima	108	Po	B	92	92	2002	9	56.5		48.1			34.6								Villanueva et al. 2003
PHILIPPINES																					
Manila	19	A	B		35	2002					77									13	de las Llagas et al. 2005

continued...

Appendix 3 Continued

Country & location	Altitude	Volunteer group	Sample type	n (samples)	n (homes)	Yr. sampling began	Total mites mean no./g	% +ve samples	D. pteronyssinus mean no./g	% +ve samples	% of total mites	D. farinae mean no./g	% +ve samples	% of total mites	E. maynei mean no./g	% +ve samples	% of total mites	Blomia spp. mean no./g	% +ve samples	% of total mites	Reference
Manila	19	A	F		35	2002						2								78	de las Llagas et al. 2005
Manila	19	A	F		35	2002						5								90	de las Llagas et al. 2005
POLAND																					
Gdansk & Gdynia	14	Po	B, F	134	17	1996	13	37.3	3	6.0	16.1	9	31.3	78.9	0	0.0	0.0	0	0.0	0.0	Racewicz 2001
Katowice & environs	287	Po	B, F	238	238	1981	74	51.3	33	34.0	45.1	30	34.9	40.2	2	2.9	2.6	0	0.0	0.0	Solarz 1998
Katowice & environs	287	A, Po	B	21	21	1985	8	61.9	2		30.4	5		62.7	0.1		1.6				Horak 1987
Sosnowiec	296	Po	B	3	2	1984	26	100.0													Solarz 1997
Warsaw	94	A	F	48	48	1989	14	72.9		47.9			10.4			18.5	18.2				Samolinsky et al. 1990
PUERTO RICO																					
Arecibo	1	Po	B	8		1994	245			37.0			0.0			0.0			62.5		Montealegre et al. 1997
Canovanas	11	Po	B	6		1994	67			33.0			0.0			0.0			33.0		Montealegre et al. 1997
Guanica	81	Po	B	4		1994	80			25.0			50.0			0.0			0.0		Montealegre et al. 1997
Guayama	81	Po	B	7		1994	160			43.0			0.0			29.0			43.3		Montealegre et al. 1997
Hormigueros	25	Po	B	5		1994	192			40.0			0.0			20.0			20.0		Montealegre et al. 1997
Lajas	46	Po	B	4		1994	110			50.0			0.0			0.0			0.0		Montealegre et al. 1997
Lares	382	Po	B	7		1994	366			71.4			25.0			0.0			43.9		Montealegre et al. 1997
Loiza	1	Po	B	2		1994	680			100.0			50.0			0.0			50.0		Montealegre et al. 1997
Mayagüez	397	Po	B	3		1994	267			67.0			67.0			0.0			33.0		Montealegre et al. 1997
Mayagüez	397	Po	F	31	31	1978	16	81.0														Santos 1978
Ponce	75	Po	B	4		1994	160			36.8			50.0			5.2			26.3		Montealegre et al. 1997
San Juan	5	Po	B	7		1994	217			57.0			14.2			0.0			29.0		Montealegre et al. 1997
RUSSIA																					
Moscow	150	A	B	30	15	1991	82	63.3	53	27.3		29	30.3								Zheltikova et al. 1994
SINGAPORE	1	Po	B	34		1973	8884		2180		24.5	2030		22.9	82		0.9				Blythe 1976
SINGAPORE	1	Po	B	50	50	1996	10 250	98.0	1150	80	5.8	500	20.0	0.5	0	0.0	0.0	7250	94	65.3	Chew et al. 1999a
SINGAPORE	1	Po	F	21	21	1996	8350	100.0	2200	86	32.5	700	67.0	6.6	0	0.0	0.0	3050	100	41.1	Chew et al. 1999a

continued...

Appendix 3 Continued

Country & location	Altitude	Volunteer group	Sample type	n (samples)	n (homes)	Yr. sampling began	Total mites mean no./g	% +ve samples	D. pteronyssinus mean no./g	% +ve samples	% of total mites	D. farinae mean no./g	% +ve samples	% of total mites	E. maynei mean no./g	% +ve samples	% of total mites	Blomia spp. mean no./g	% +ve samples	% of total mites	Reference
SINGAPORE	1	Po	F	50	50	1996	1550	94.0	900	52	18.4	100	20.0	1.0	0	0.0	0.0	1100	74	68.3	Chew *et al.* 1999a
SINGAPORE	1	Po	F	13	13	1996	1750	100.0	2150	100	15.3	1250	54.0	4.0	0	0.0	0.0	8250	100	68.6	Chew *et al.* 1999a
SLOVAK REPUBLIC																					
Strbske Pleso	1204	Po	B			1974					9.9	68	97.3	51.8			5.7				Makovcová *et al.* 1982
SOUTH AFRICA																					
Aliwal North	1331	Po	F	2		1969	0.2														Ordman 1971
Barberton	886	Po	F	6		1969	2.4														Ordman 1971
Barkly West	1094	Po	F	3		1969	0.2														Ordman 1971
Beaufort West	872	Po	F	4		1969	0.2														Ordman 1971
Bloemfontein	1397	Po	F	4		1969	0														Ordman 1971
Bredasdorp	52	Po	F	3		1969	7.2														Ordman 1971
Burghersdorp	1388	Po	F	1		1969	0.8														Ordman 1971
Calvinia	997	Po	F	3		1969	0.2														Ordman 1971
Cape Town	16	Po	F	10		1969	21.2														Ordman 1971
Ceres	451	Po	F	5		1969	0.6														Ordman 1971
Citrusdal	173	Po	F	4		1969	3.4														Ordman 1971
Clanwilliam	75	Po	F	1		1969	2.4														Ordman 1971
Cradock	872	Po	F	2		1969	0.2														Ordman 1971
De Aar	1260	Po	F	3		1969	0.2														Ordman 1971
Durban	4	Po	F	7		1969	28														Ordman 1971
East London	100	Po	F	5		1969	85.8														Ordman 1971
Empangeni	64	Po	F	2		1969	18.8														Ordman 1971
Ermelo	1734	Po	F	1		1969	10.6														Ordman 1971
Estcourt	1159	Po	F	4		1969	11.6														Ordman 1971
George	253	Po	F	1		1969	25.6														Ordman 1971
Graaf-Reinet	732	Po	F	5		1969	0.4														Ordman 1971
Greytown	1099	Po	F	4		1969	14.4														Ordman 1971
Humansdorp	8	Po	F	3		1969	7.4														Ordman 1971
Jeffrey's Bay	7	Po	F	3		1969	28.8														Ordman 1971
Johannesburg	1786	Po	F	9		1969	0.2														Ordman 1971
Kimberley	1218	Po	F	4		1969	0.4														Ordman 1971
King William's Town	375	Po	F	5		1969	2.2														Ordman 1971
Klerksdorp	1325	Po	F	3		1969	1.6														Ordman 1971
Knysna	4	Po	F	3		1969	5.8														Ordman 1971
Kroonstad	1348	Po	F	4		1969	0.4														Ordman 1971

continued...

Appendix 3 Continued

Country & location	Altitude	Volunteer group	Sample type	n (samples)	n (homes)	Yr. sampling began	Total mites mean no./g	% +ve samples	*D. pteronyssinus* mean no./g	% +ve samples	% of total mites	*D. farinae* mean no./g	% +ve samples	% of total mites	*E. maynei* mean no./g	% +ve samples	% of total mites	*Blomia* spp. mean no./g	% +ve samples	% of total mites	Reference
Ladysmith	1001	Po	F	1		1969	0.2														Ordman 1971
Laingsburg	668	Po	F	2		1969	0														Ordman 1971
Lambert's Bay	2	Po	F	4		1969	12.6														Ordman 1971
Louis Trichardt	959	Po	F	4		1969	26.6														Ordman 1971
Mafeking	1271	Po	F	3		1969	0.2														Ordman 1971
Margate	15	Po	F	4		1969	28.8														Ordman 1971
Messina	519	Po	F	3		1969	7.2														Ordman 1971
Morgans Bays Bay	2	Po	F	1		1969	4.2														Ordman 1971
Mossel Bay	61	Po	F	4		1969	39.8														Ordman 1971
Nelspruit	665	Po	F	5		1969	1.2														Ordman 1971
Oudtshoorn	332	Po	F	5		1969	0.6														Ordman 1971
Piet Retief	1271	Po	F	1		1969	1.2														Ordman 1971
Pietersburg	1294	Po	F	4		1969	0.6														Ordman 1971
Port Alfred	61	Po	F	4		1969	2.6														Ordman 1971
Port Elizabeth	55	Po	F	12		1969	44.8														Ordman 1971
Postmasburg	1324	Po	F	3		1969	0.2														Ordman 1971
Potchefstroom	1348	Po	F	3		1969	0														Ordman 1971
Pretoria	1379	Po	F	2		1969	1														Ordman 1971
Queenstown	1077	Po	F	6		1969	0.2														Ordman 1971
Richmond	1381	Po	F	3		1969	0														Ordman 1971
Somerset East	766	Po	F	2		1969	1.2														Ordman 1971
Springs	1688	Po	F	1		1969	1.8														Ordman 1971
Stanger	43	Po	F	4		1969	9.4									1					Ordman 1971
Steytlerville	405	Po	F	3		1969	1														Ordman 1971
Uitenhage	100	Po	F	4		1969	14.2														Ordman 1971
Umtata	679	Po	F	3		1969	1.8														Ordman 1971
Upington	799	Po	F	6		1969	0.2														Ordman 1971
Victoria West	1259	Po	F	4		1969	0														Ordman 1971
Volksrust	1558	Po	F	3		1969	5.6														Ordman 1971
Vryburg	1188	Po	F	3		1969	0.2														Ordman 1971
Vryheid	1194	Po	F	1		1969	40														Ordman 1971
Warmbaths	1174	Po	F	4		1969	0.2														Ordman 1971
Welkom	1338	Po	F	5		1969	0.2														Ordman 1971
Wilderness	4	Po	F	1		1969	4														Ordman 1971
Witbank	1470	Po	F	3		1969	0														Ordman 1971
Worcester	248	Po	F	4		1969	0.8														Ordman 1971

continued...

Appendix 3 Continued

Country & location	Altitude	Volunteer group	Sample type	n (samples)	n (homes)	Yr. sampling began	Total mites mean no./g	% +ve samples	*D. pteronysinus* mean no./g	% +ve samples	% of total mites	*D. farinae* mean no./g	% +ve samples	% of total mites	*E. maynei* mean no./g	% +ve samples	% of total mites	*Blomia* spp. mean no./g	% +ve samples	% of total mites	Reference
SOUTH KOREA																					
Chongju	40	Po	F		6	1993	9		0.03		0.4	8		95.3	0		0.0	0		0.0	Ree et al. 1997
Chonju	275	Po	F		5	1993	4		0.5		11.4	4		87.1	0		0.0	0		0.0	Ree et al. 1997
Chunchon	90	Po	F		9	1993	4		0.9		21.9	0.4		10.9	0		0.0	0		0.0	Ree et al. 1997
Inchon	75	Po	F		7	1993	27		0.2		0.9	26		96.5	0		0.0	0		0.0	Ree et al. 1997
Kwangju	164	Po	F		7	1993	26		3.6		14.0	16		63.6	0		0.0	0		0.0	Ree et al. 1997
Pusan (Inland)	100	Po	F		4	1993	13		0.8		6.5	10		79.6	0		0.0	0		0.0	Ree et al. 1997
Pusan (Yongdo)	0	Po	F		7	1993	14		9.1		64.9	2		14.9	0		0.0	0		0.0	Ree et al. 1997
Seoul	34	Po	B	74	7	1987	68	97.3	68	97.3	7.9	112	100	55.5	0	0	0.0	0	0.0	0.0	Paik et al. 1992
Seoul	34	A	B	74	76	1987	290														Hong 1991
Seoul	34	Po	F		9	1993	20		3.5		18.0	14		66.8	0		0.0	0		0.0	Ree et al. 1997
Taejon	104	Po	F		7	1993	8		0.2		3.1	6		83.9	0		0.0	0		0.0	Ree et al. 1997
Yonggwang	137	Po	F		4	1993	7		5.2		72.5	0.6		8.3	0		0.0	0		0.0	Ree et al. 1997
SPAIN																					
Rioja Alta	620	A	B	13	13	1998	519	92.3	311	92.3		96	7.7		463	38.5					Lobera et al. 2000
Rioja Media	380	A	B	61	61	1998	205	55.7	231	42.6		94	4.9		94	14.7		55	9.8		Lobera et al. 2000
Rioja Baja	300	A	B	26	26	1998	191	53.8	232	38.5		602	3.8		351	11.5					Lobera et al. 2000
Barcelona	1		B	53	53	1972	112	100.0	79	96.2	70.7	4	53.8	3.1	2	37.7	1.6				Blasco & Portus 1975
Barcelona	1	Po	B	93	6	1974	40														Portus & Blasco 1977
Barcelona	1	Po				1974	108		93		85.3	11		9.9							Portus & Gomez 1976
Castelltersol	689	Po		6		1974	145		16		18.3	73		81.1	1		0.7				Portus & Gomez 1976
La Coruña	103	A, Po	B	111		2002	1043	100.0	774	98.2		460	5.5		121	27.5					Boquete et al. 2006
La Laguna, Tenerife	583	A	B	55	65	1994		100.0	65		69.4										Sanchez-Covisa et al. 1999
L'Ametlla de Merola	292	Po	B	6	6	1976	93	100.0				7		7.7	0.1		0.1				Gomez et al. 1981a, b
Lugo	444	A, Po	B	84		2002	1072	100.0	705	98.8		103	4.8		147	41					Boquete et al. 2006
Madrid	588	A	B, F	36		2000	61	41.7	59	16.7	32.3	19	5.5	5.8	187	2.8	2.9			11.1	Sastre et al. 2002
Orense	125	A, Po	B	46		2002	489	98.6	237	93.3		133	6.7		132	56.2					Boquete et al. 2006

continued...

Appendix 3 Continued

Country & location	Altitude	Volunteer group	Sample type	n (samples)	n (homes)	Yr. sampling began	Total mites mean no./g	% +ve samples	D. pteronyssinus mean no./g	% +ve samples	% of total mites	D. farinae mean no./g	% +ve samples	% of total mites	E. maynei mean no./g	% +ve samples	% of total mites	Blomia spp. mean no./g	% +ve samples	% of total mites	Reference
Pontevedra	80	A, Po	B	91		2002	1639	100.0			100	442	2.2		82	12.8					Boquete et al. 2006
Puigcerda	1140			6		1974	26			5	29.6		2	12.3		9	58				Portus & Gomez 1976
Puigcerda	1140		B	6	6	1976	38			2	5.8		3	7.9		5	12.6	0.1		0.1	Gomez et al. 1981a, b
Reus	93			6		1974	396		362		98.5	5		1.3	1		0.2				Portus & Gomez 1976
Reus	93		B	5	5	1976	238		193		80.9	4		1.5	0.6		0.2				Gomez et al. 1981a, b
San Cristóbal de la Laguna	547	A	B	55	55	1994				98.2						21.8			16.4		Sanchez-Covisa et al. 1999
Santa Cruz, Tenerife	0	A	B	65	15	1998	2215	100.0								24.6			10.8		Sanchez-Covisa et al. 1999
Santiago de Compostella	266	Po	B	60	15	1998				70	73.3				88	15.5					Agratorres et al. 1999
Santiago de Compostela	266	Po	F	60	36	1998	60.9	44.1							12	5.2					Agratorres et al. 1999
Torello	554			3		1974	272		2		1.3	239		98.5							Portus & Gomez 1976
SURINAM																					
Paramaribo	2	Po	F	9	9	1969	51	100.0	33	100.0	65.3							8	88.9	18.5	Bronswijk 1972
SWEDEN																					
N.E. Sweden		A, Po	B		65	1990	14	35.0													Wickman et al. 1993
Helsingborg	28	A	B		19	1989	400	100.0			79.0				2	10.5					Warner et al. 1999
Linköping	43	A	B		20	1989	37	100.0													Warner et al. 1999
Stockholm	16	A, Po	B	201		1977		8.0			40.0										Turos 1979
Umeå	19	A	B		16	1989	4	87.5													Warner et al. 1999
SWITZERLAND																					
Arosa	1854	Po	B, F	10	10	1972	5	50.0													Mumcuoglu 1975
St Moritz	1853	Po	B, F	6	6	1972	7	66.7													Mumcuoglu 1975
Zermatt	1610	Po	B, F	4	4	1972	9	75.0													Mumcuoglu 1975
Davos	1561	Po	B, F	5	5	1972	6	40.0													Mumcuoglu 1975
Montana Vermala	1453	Po	B, F	5	5	1972	13	60.0													Mumcuoglu 1975

continued...

Appendix 3 Continued

Country & location	Altitude	Volunteer group	Sample type	n (samples)	n (homes)	Yr. sampling began	Total mites mean no./g	% +ve samples	D. pteronyssinus mean no./g	% +ve samples	% of total mites	D. farinae mean no./g	% +ve samples	% of total mites	E. maynei mean no./g	% +ve samples	% of total mites	Blomia spp. mean no./g	% +ve samples	% of total mites	Reference
Andermatt	1442	Po	B, F	5	5	1972	15	80.0													Mumcuoglu 1975
Adelboden	1345	Po	B, F	5	5	1972	209	100.0													Mumcuoglu 1975
Scuol	1253	Po	B, F	5	5	1972	23	100.0													Mumcuoglu 1975
Mont Soleil	1183	Po	B, F	6	6	1972	34	100.0													Mumcuoglu 1975
Airolo	1170	Po	B, F	5	5	1972	3	20.0													Mumcuoglu 1975
Beatenberg	1148	Po	B, F	5	5	1972	523	100.0													Mumcuoglu 1975
La Chaux-de-Fonds	990	Po	B, F	5	5	1972	85	100.0													Mumcuoglu 1975
Schiers	670	Po	B, F	5	5	1972	93	100.0													Mumcuoglu 1975
St. Gallen	664	Po	B, F	5	5	1972	373	100.0													Mumcuoglu 1975
Chur	663	Po	B, F	5	5	1972	70	100.0													Mumcuoglu 1975
Bern	572	Po	B, F	5	5	1972	314	100.0													Mumcuoglu 1975
Zurich	569	Po	B, F	5	5	1972	279	100.0													Mumcuoglu 1975
Lausanne	553	Po	B, F	5	5	1972	260	100.0													Mumcuoglu 1975
Sierre	552	Po	B, F	5	5	1972	143	100.0													Mumcuoglu 1975
Sion	549	Po	B, F	5	5	1972	43	100.0													Mumcuoglu 1975
Luzern	498	Po	B, F	5	5	1972	179	100.0													Mumcuoglu 1975
Glarus	480	Po	B, F	5	5	1972	42	100.0													Mumcuoglu 1975
Solothurn	470	Po	B, F	9	9	1972	340	100.0													Mumcuoglu 1975
Aarau	408	Po	B, F	5	5	1972	1192	100.0													Mumcuoglu 1975
Locarno	379	Po	B, F	5	5	1972	9	80.0													Mumcuoglu 1975
Basel	317	A, Po	B, F	10	10	1972	240	100.0													Mumcuoglu 1975
Baselland	300	A, Po	B, F	10	10	1972	508	100.0													Mumcuoglu 1975
Lugano	276	Po	B, F	5	5	1972	207	100.0													Mumcuoglu 1975
Bellinzona	237	Po	B, F	5	5	1972	25	100.0													Mumcuoglu 1975

continued...

Appendix 3 Continued

Country & location	Altitude	Volunteer group	Sample type	n (samples)	n (homes)	Yr. sampling began	Total mites mean no./g	% +ve samples	D. pteronyssinus mean no./g	% +ve samples	% of total mites	D. farinae mean no./g	% +ve samples	% of total mites	E. maynei mean no./g	% +ve samples	% of total mites	Blomia spp. mean no./g	% +ve samples	% of total mites	Reference
Basel & environs	273	A	B	190		1973	579	85.0	410	89.3	70.8	33	33.8	5.7	100	36.7	17.2				Mumcuoglu 1976
Basel & environs	273	Po	B, F		3	1966	3														Spieksma et al. 1971
Davos	1561	Po	B, F		4	1966	1.6														Spieksma et al. 1971
TAIWAN																					
Hualien	0	Po	B			1997				92.5	64.6		21.0	5.3			0.0		40.0	28.9	Wu 1999
Kaohsiung & Pingtung	4	Po	B			1997				100.0	49.2		34.0	30.9			0.0		37.0	18.9	Wu 1999
Nanao	14	Po	B			1997				87.5	49.8		5.0	2.2			0.0		60.0	41.9	Wu 1999
Taichung	112	Po	B			1997				66.0	74.7		23.5	19.6			0.6		19.0	3.9	Wu 1999
Taichung	112	A, Po	B	385	12	1998	119		91		77	16		13.0				3		1.8	Sun & Lue 2000
Taipei	6	Po	B			1997				81.0	82.2		21.0	7.0			1.3		34.0	7.3	Wu 1999
Taipei	6	Po	B			1968	735		13		1.8	0		0.0	2		0.3	547		74.4	Oshima 1970; Miyamoto et al. 1970
Taipei	6	A	B		61	1986	308		243		78.7	19		6.2	2		0.6	3		0.8	Chang & Hsieh 1989
Taipei	6	A	F		61	1986	393														Chang & Hsieh 1989
Taipei	6	A	F		61	1986	243														Chang & Hsieh 1989
Taipei	6	A	F		61	1986	326														Chang & Hsieh 1989
Taitung	0	Po	B			1997				86.5	44.6		18.0	29.0			0.0		43.0	25.4	Wu 1999
THAILAND																					
24 provinces		Po	B	630		1991	92	88.1	87		96.6	23		24.4	0.5		0.5				Malainual et al. 1995
15 provinces		Po	B	112		1970	729	84.8	70.5		40.8	53.6		46.5	0.0	0.0	0.0				Wongsatha-yong & Lakshana 1972
Central region			B	242		1991	19	93.8													Malainual et al. 1995
Northern region			B	209		1991	74	90.4													Malainual et al. 1995
Northeastern region			B	179		1991	61	77.6													Malainual et al. 1995
Bangkok	2		B	11		1973	2218		1180		53.2	510		23							Blythe 1976
Bangkok	2		B	76		1970	3486	80.3													Wongsatha-yong & Lakshana 1972
TURKEY																					
Aegean Region		A, Po	B, F	70		1995	106	15.7													Kalpaklioglu et al. 2004

continued...

Appendix 3 Continued

Country & location	Altitude	Volunteer group	Sample type	n (samples)	n (homes)	Yr. sampling began	Total mites mean no./g	% +ve samples	D. pteronyssinus mean no./g	% +ve samples	% of total mites	D. farinae mean no./g	% +ve samples	% of total mites	E. maynei mean no./g	% +ve samples	% of total mites	Blomia spp. mean no./g	% +ve samples	% of total mites	Reference
Black Sea Region		A, Po	B, F	113		1995	571	46.0													Kalpaklioglu et al. 2004
Central Anatolia		A, Po	B, F	244		1995	93	5.7													Kalpaklioglu et al. 2004
Eastern Anatolia		A, Po	B, F	170		1995	141	2.4													Kalpaklioglu et al. 2004
Southern Anatolia		A, Po	B, F	49		1995	0	0.0													Kalpaklioglu et al. 2004
Marmara Region		A, Po	B, F	220		1995	222	30.0													Kalpaklioglu et al. 2004
Mediterranean Region		A, Po	B, F	64		1995	858	48.4													Kalpaklioglu et al. 2004
Bursa	200	Po		32	32	2002				58.3			4.1								Gülegen et al. 2005
Istanbul	24		B	9		1973	989		470		47.5	400		40.4	0.9		0.09				Blythe 1976
Kutahya	785					2003		18.1			31										Akdemir & Gürdal 2005
Malatya	948			303		2000		23.1		100											Atambay et al. 2006
UNITED KINGDOM																					
Birmingham	134	A	Bs		10	1972	3290	100.0	1670	100	64	0.0	0.0	0.0	1290	76.0	36.0	0	0.0	0.0	Blythe 1976
Bristol	47	A	B	51	51	1979					63.2						13.2				Carswell et al. 1982
Cardiff	16	A, Po	Bs	50	50	1974	1464	100.0	1180	100	81	2	2	0.1	131	26.0	9.0	0	0.0	0.0	Rao et al. 1975
Glasgow	66	A, Po	B		60	1968	749	100.0	589	98.3					82	45.0			0	0.0	Sesay & Dobson 1972
Glasgow	66	A	F		9	1968	222								0	0.0			0	0.0	Sesay & Dobson 1972
Glasgow	66	A, Po	B	65	54	1982	819	100.0	591	100.0	72.2	1	3.0	0.1	188	35.8	22.9	0	0.0	0.0	Colloff 1987c
Glasgow	66	Po	F	27	23	1982	372	100.0	302	100.0	81.1	0	0.0	0.0	45	14.8	1.7	0	0.0	0.0	Colloff 1987c
Glasgow	66	Po	B	23	23	1985	520	100.0													Colloff 1992c
London	15	A	Bs	33	33	1976	1064	100.0	794	97	73.5	39	9.1	3.6	218	69.7	20.2	0	0.0	0.0	Gabriel et al. 1982
London	15	A	B, F	89		1979	300														Tovey et al. 1981
London, S. Wales	15	A	B	186	186	1966				82.3	67.0					40.3	15.0				Maunsell et al. 1968
London, S. Wales	15	A	B	22	22	1966			2568												Maunsell et al. 1968
London, S. Wales	15	A	F	22	22	1966			22												Maunsell et al. 1968
Oxford & environs	68	A	B	60	30	1988	724	100.0	530	89.7	75.1	0.0	0.0	0.0	161	24.1	20.3	0	0.0	0.0	Hart & Whitehead 1990

continued...

Appendix 3 Continued

Country & location	Altitude	Volunteer group	Sample type	n (samples)	n (homes)	Yr. sampling began	Total mites mean no./g	% +ve samples	D. pteronyssinus mean no./g	% +ve samples	% of total mites	D. farinae mean no./g	% +ve samples	% of total mites	E. maynei mean no./g	% +ve samples	% of total mites	Blomia spp. mean no./g	% +ve samples	% of total mites	Reference
USA																					
Arizona																					
Tucson	757	Po	B		82	1990		32.5													O'Rourke et al. 1996
Tucson	757	Po	F		82	1990		42.1													O'Rourke et al. 1996
California																					
Unspecified localities				106		1970		47.0		28.3			9.4								Furumizo & Mulla 1971
Orange Co. (coast)	10	A	B		42	1974		85.7		64.3			57.1								Mulla et al. 1975
Orange Co. (coast)	10	Po	B, F	510	14	1975	2890	90.9		73.3			46.7								Lang & Mulla 1977a
Los Angeles	94	A	B		13	1986		100.0		92.3			92.3			0.0			0.0		Arlian et al. 1992
Riverside	284	Po	B, F	306	9	1975	114	100.0		44.4			77.8								Lang & Mulla 1977a
San Diego	25	A	B		25	1986		100.0		100			84.0			8.0			44.0		Arlian et al. 1992
Colorado																					
Denver	1603	Po	B	80	20	1982	2	20.0													Moyer et al. 1985
Florida																					
Delray Beach	6	A	B		8	1986		100.0		75.0			100.0			12.5			25.0		Arlian et al. 1992
Tampa	15	A	B	20		1988	1513	85.0		70.0			40.0			5.0			30.0		Fernandez-Caldas et al. 1990
Tampa	15	A	F	40		1988	747	90.0		85.0			45.0			7.5			30.0		Fernandez-Caldas et al. 1990
Georgia																					
Atlanta	320	A	B	20	20	1983	100	100.0		100.0			85.0			5.0					Smith et al. 1985
Atlanta	320	A	F	40	20	1983	180	100.0		100.0											Smith et al. 1985
Hawaii																					
Honolulu & environs	50	Po	F	15	15	1986	444	93.3	364		81.9	50		11.2	2		0.4				Massey et al. 1988
Honolulu & Kailua	100	Po	B		21	1979	877	100.0	814		92.8										Nadcahtram et al. 1981
Louisiana																					
New Orleans	3	A	B		58	1986		100.0		98.3			81.0			31.0			5.2		Arlian et al. 1992

continued...

Appendix 3 Continued

Country & location	Altitude	Volunteer group	Sample type	n (samples)	n (homes)	Yr. sampling began	Total mites mean no./g	% +ve samples	D. pteronyssinus mean no./g	% +ve samples	% of total mites	D. farinae mean no./g	% +ve samples	% of total mites	E. maynei mean no./g	% +ve samples	% of total mites	Blomia spp. mean no./g	% +ve samples	% of total mites	Reference
North Carolina																					
Greenville	17	A	B		36	1986		100.0		94.4			100.0			0.0			0.0		Arlian et al. 1992
Ohio																					
Cincinnati	208	A	B		48	1986		100.0		81.3				95.8		0.0			0.0		Arlian et al. 1992
Columbus	244	Po	B		13	1977	153	92.3	42		27.5	104		68.2	0.1		0.1				Yoshikawa & Bennett 1979
Columbus & environs	244	A	B, F	112		1966		38.4				38.4									Larson et al. 1969
Columbus & environs	244	A	B	31		1967	50	70.0													Mitchell et al. 1969
Dayton	229	A	B	16	16	1976	43	56.3			6.3			81.3							Arlian et al. 1978
Dayton	229	A	F	31	16	1976	127	87.5													Arlian et al. 1978
Dayton	229	Po	F	198	87	1998	787	97.5													Arlian et al. 2002; Neal et al. 2002
Dayton & Cincinnati	229	A	B	513	19	1977	51														Arlian et al. 1979a; 1982
Dayton & Cincinnati	229	A	F	513	18	1977	269														Arlian et al. 1979a; 1982
Tennessee																					
Memphis	77	A	B		32	1986		100.0		77.4			93.5			12.9				3.2	Arlian et al. 1992
Oak Ridge, Knoxville	267	Po	F	180	15	1970	124	100.0													Shamiyeh et al. 1972; 1973
Texas																					
Bell Co. & Coryell Co.	210	Po	F	6	6	1989	40	83.3		50.0			50.0								
Galveston	3	A	B		32	1986		100.0		96.9			93.8			43.8				18.8	Arlian et al. 1992
Virginia																					
Williamsburg	26	Po	B	88	22	1982	1903	100.0	537	95.5	28.2	1326	100.0	69.7	0	0.0	0.0	0	0.0	0.0	Lassiter & Fashing 1990
Wisconsin																					
Milwaukee	193	Po	F	30	30	1983	113	93.0	16		14.2	94		83.2	1		0.5	0		0.0	Klein et al. 1986
UZBEKISTAN																					
Tashkent	460		B					62.1			28.5										Sherbak & Nazrullaeva 1988

continued...

Appendix 3 Continued

Country & location	Altitude	Volunteer group	Sample type	n (samples)	n (homes)	Yr. sampling began	Total mites mean no./g	% +ve samples	D. pteronyssinus mean no./g	% +ve samples	% of total mites	D. farinae mean no./g	% +ve samples	% of total mites	E. maynei mean no./g	% +ve samples	% of total mites	Blomia spp. mean no./g	% +ve samples	% of total mites	Reference
VENEZUELA																					
Caracas	263	Po	B	54		1984	15 600	100.0	12 854	100.0	82.4	3713		67	23.8	1217	22.0		7.8	2480	96.0	15.9	Hurtado & Parini 1987
Pueblo Llano & environs	2040	Po	Bs	147	147	1997	188	82.7															Rangel et al. 1998
ZAMBIA																					
Ndola	1305	A, Po	B	125	125	1971	124	81.3	112	90.4	71.2											Buchanan & Jones 1974
Total				23 611	9141																
Count				399	296		414	237	196	164	209	192	154	203	165	129	167	130	104	149	

Appendix 4

Concentrations of *Dermatophagoides* group 1 allergens in settled house dust, based on published surveys

A = asthmatic/atopic/clinical group; Po = population group; B = Beds; F = floors; s = swept sample

Country & location	Altitude	Patient group	Adults or children	Sample type	n (samples)	n (homes)	Yr. sampling began	Mean Der 1 λg/g	% +ve samples	Mean Der p 1 λg/g	% +ve samples	Mean Der f 1 λg/g	% +ve samples	Reference
ARGENTINA														
Santa Fe	18	A	A	B	52	52	1991	4.28						Neffen *et al.* 1996
AUSTRALIA														
Kimberley	1	A, Po	A, C	B, F	83	83	1995	6.04						Hales *et al.* 2007
La Trobe Valley		Po	A, C	B	133	80	1994	21.10		21.1				Garrett *et al.* 1998
La Trobe Valley		Po	A, C	F	213	80	1994	15.60		15.6				Garrett *et al.* 1998
Alice Springs	609	A	C	B	28	28	1995	0.84	28.0	0.84	28.0			Tovey *et al.* 2000a
Belmont	1	A	C	B	97	97	1994	36.50	100	36.5	100			Peat *et al.* 1996; Tovey *et al.* 2000a
Broken Hill	304	A	C	B	88	88	1994	0.79	97.0	0.79	97.0			Peat *et al.* 1996; Tovey *et al.* 2000a
Bunbury	1	A, Po	A	B	10	10	1986	3.80		3.8	100			Colloff *et al.* 1991
Bunbury	1	A, Po	A	F	10	10	1986	9.20		9.2	100			Colloff *et al.* 1991
Busselton	1	A, Po	C	B	104	104	1994	14.70	100	14.7	100			Tovey *et al.* 2000a
Campbelltown	104	A, Po	C	B	104	104	1990	14.90	100	14.9	100			Marks *et al.* 1995a; Tovey *et al.* 2000a
Edward River	3	A	A, C	B, F	35	35	1995	11.80	100.0	11.8	100			Veale *et al.* 1996; Tovey *et al.* 2000a
Hobart & environs	110	A	C	B	72	72	1995	2.03		2.03				Couper *et al.* 1998
Hobart & environs	110	A	C	F	72	72	1995	2.04		2.04				Couper *et al.* 1998
Hopevale	86	A	A, C	B, F	20	20	1995	14.90	100.0	14.9	100			Veale *et al.* 1996; Tovey *et al.* 2000a
Indulkana	384	A	A, C	B	29	29	1995	0.05		0.05	100			Veale *et al.* 1996
Lismore	134	A, Po	C	B	57	57	1991	56.20		56.2				Peat *et al.* 1993
Lismore	134	A, Po	C	F	114	57	1991	43.70		43.7				Peat *et al.* 1993
Lismore	134	A, Po	C	B	69	69	1994	47.80	100	47.8	100			Peat *et al.* 1996; Tovey *et al.* 2000a
Manly	1	A, Po	C	B	71	71	1994	22.50	100	22.5	100			Tovey *et al.* 2000a
Melbourne	115	Po	A	B	485	485	1996	20.30		20.3	96.7			Dharmage *et al.* 1999
Melbourne	115	Po	A	F	440	440	1996	17.20		17.2	90.3			Dharmage *et al.* 1999
Narrabri	33	A, Po	C	B	74	74	1991	7.20	100	7.2	100			Peat *et al.* 1993; Tovey *et al.* 2000a
Narrabri	33	A, Po	C	F	148	74	1991	3.35		3.35				Peat *et al.* 1993

continued...

Appendix 4 Continued

Country & location	Altitude	Patient group	Adults or children	Sample type	n (samples)	n (homes)	Yr. sampling began	Mean Der 1 λg/g	% +ve samples	Mean Der p 1 λg/g	% +ve samples	Mean Der f 1 λg/g	% +ve samples	Reference
Perth	33	A, Po	A	B	10	10	1986	4.20		4.2	100			Colloff et al. 1991
Perth	33	A, Po	A	F	10	10	1986	4.10		4.1	100			Colloff et al. 1991
Sydney	1	A, Po	C	B	68	68	1994	22.50		22.5				Peat et al. 1996
Sydney	1	A, Po	C	B	74	0	1988	39.00		39.00				Marks et al. 1995a, b
Sydney	1	A, Po	C	F	74	0	1988	17.60		17.6				Marks et al. 1995a, b
Utopia	490	A	A, C	B	26	26	1995	0.05		0.05	100			Veale et al. 1996
Wagga Wagga	148	A, Po	C	B	72	72	1994	1.37	100.0	1.37	100			Peat et al. 1996; Tovey et al. 2000a
BARBADOS	10	Po	A	B	16	17	1991	15.81	100	15.8	100	0.01	31.3	Barnes et al. 1997
BARBADOS	10	Po	A	F	17	17	1991	4.21	100	4.2	100	0.01	23.5	Barnes et al. 1997
BARBADOS	10	Po	A	F	16	17	1991	1.32	100	1.3	100	0.02	43.8	Barnes et al. 1997
BELGIUM														
Antwerp Centre	4	Po	A	B	96	96	2000	1.58	94.4	0.51	81.5	0.3	62.3	Zock et al. 2006
Antwerp South	4	Po	A	B	162	162	2000	2.43	90.6	0.96	80.2	0.2	66.7	Zock et al. 2006
BRAZIL														
Salvador	1	Po	A	B	376	101	2004	28.40		28.4				Baqueiro et al. 2006
Sao Paulo	638	A	C	B	20	20	1988	38.40		38.4		0.5		Arruda et al. 1991
Sao Paulo	638	A	C	F	40	20	1988	4.90		4.90				Arruda et al. 1991
Ribeirão Preto	589	A	A, C	B	24	24	2001	11.20		1.00		10.2		Tobias et al. 2003
Ribeirão Preto	589	A	A, C	F	24	24	2001	0.42		0.12		0.3		Tobias et al. 2003
Uberaba	747	A, Po	A, C	B	120	60	2000	25.65		0.95		24.7		Terra et al. 2004
Uberlândia	687	A, Po	A	B	64	64	1998	15.85		3.85		12.0		Sopelete et al. 2000
Uberlândia	687	A, Po	A	F	128	64	1998	0.45		0.1		0.35		Sopelete et al. 2000
CANADA														
Vancouver	72	A	A, C	B, F	57	57	1991	1.54		0.67		0.87		Chan-Yeung et al. 1995a, b
Winnipeg	227	A	A, C	B, F	63	63	1991	0.66		0.18		0.48		Chan-Yeung et al. 1995a, b
Wallaceburg	175	Po	C	B	59	59	1993	5.75		2.75		3.00		Lawton et al. 1998
CHILE														
Santiago	522	Po	C	B, F	36	36	2000	0.12		0.1	44.4	0.01	11.1	Wickens et al. 2004
Valdivia	5	Po	C	B, F	36	36	2000	9.30		9.28	97.2	0.01	11.1	Wickens et al. 2004
Valdivia	5	A	C	B	100	100	2003	18.90		18.3		0.6		Calvo et al. 2005
CHINA														
Shanghai	8	A, Po	A	B	48	48	1984	9.04		8.93		0.11		Wen et al. 1991; Wang & Wen 1997
COLOMBIA														
Cartagena	13	A	A	B	25	25	1991	1.40		1.1		0.3		Fernandez-Caldas et al. 1993
Cartagena	13	A	A	F	25	25	1991	0.50		0.4		0.1		Fernandez-Caldas et al. 1993
Cartagena	13	A	A, C	B	240	20	1994	0.61		0.61				Mercado et al. 1996

continued...

Appendix 4 Continued

Country & location	Altitude	Patient group	Adults or children	Sample type	n (samples)	n (homes)	Yr. sampling began	Mean Der 1 λg/g	% +ve samples	Mean Der p 1 λg/g	% +ve samples	Mean Der f 1 λg/g	% +ve samples	Reference
COSTA RICA														
Coast, centre		A	A, C	B	330	330	1995	17.40		16.95		0.45		Soto-Quiros *et al.* 1998
Coast, centre		A	A, C	F	330	330	1995	7.42		6.52		0.9		Soto-Quiros *et al.* 1998
CROATIA														
Zagreb	22	Po	A	F	71	71	2001	2.20		1.4		0.8		Macan *et al.* 2003
Zadar	130	Po	A	F	28	28	2001	4.50		4.5		0.0		Macan *et al.* 2003
ECUADOR														
Cuenca	2450	A, Po	A, C	B	23	23	2003	15.30		13.6		1.7		Valdivieso *et al.* 2006
Guayaquil	46	A, Po	A, C	B	24	24	2003	21.00		8.3		12.7		Valdivieso *et al.* 2006
Quito	2763	A, Po	A, C	B	20	20	2003	15.20		9.7		5.5		Valdivieso *et al.* 2006
DENMARK														
Fyn and Viborg		A, Po	C	B	47	47	1994	12.20						Halken *et al.* 2003
Copenhagen	1	Po	A	B	68	68	1999	3.77		0.06		2.41		Sidenius *et al.* 2002a
Copenhagen	1	Po	A	B	51	51	1989	0.47						Mosbech *et al.* 1991
ESTONIA														
Tartu	35	A	C	B	96	197	1993	0.37						Julge *et al.* 1998
Tartu	35	A	C	F	193	197	1993	0.10						Julge *et al.* 1998
Tartu	35	Po	A	B	182	182	2000	0.02	31.9	0.05	44.5	0.24	63.2	Zock *et al.* 2006
ETHIOPIA														
Jimma	1753	Po	A	B	467	467	1999	2.85		2.85				Scrivener *et al.* 2001
FINLAND														
Kuopio	83	Po	A	F	30	30	1996	0.01						Raunio *et al.* 1998
FRANCE														
Briançon	1571	A, Po	C	B	85	85	1988	0.36		0.24		0.18		Charpin *et al.* 1991
Grenoble	236	Po	A	B	174	174	2000	0.77	76.4	0.02	30.5	0.64	75.3	Zock *et al.* 2006
Marseille	54	A, Po	A	B	42	42	1988	20.80		20.80				Charpin *et al.* 1990
Marseille	54	A, Po	A	F	42	42	1988	4.35		4.35				Charpin *et al.* 1990
Martigues	39	A, Po	C	B	125	125	1988	15.80		5.00		5.50		Charpin *et al.* 1991
Paris	35	Po	A	B	166	166	2000	1.01	84.3	0.14	58.4	0.39	70.5	Zock *et al.* 2006
Strasbourg	140	A, Po	A	B	190	190	1988	39.42		24.00		20.7		Pauli *et al.* 1993
GERMANY														
Aue	438	A	C	B	68	68	1994	4.18						Hirsch *et al.* 1998
Aue	438	A	C	F	67	67	1994	0.90						Hirsch *et al.* 1998
Berlin	35	A, Po	C	B	183	183	1998	0.50		0.24	81.2	0.26		Wahn *et al.* 1997; Lau *et al.* 1989
Bitterfeld & environs	79	Po	C	B	454		1995	0.24		0.24				Gehring *et al.* 2005
Borken	38	Po	C	B	49	49	1997	4.60		3		1.6		Oppermann *et al.* 2001
Dresden/ Niederlausitz	109	A	C	B	95	95	1992	7.70						Hirsch *et al.* 1998
Dresden/ Niederlausitz	109	A	C	F	3	3	1992	0.25						Hirsch *et al.* 1998
Erfurt	193	Po	A	B	193	193	2000	0.57	73.6	0.05	40.4	0.26	62.7	Zock *et al.* 2006

continued...

Appendix 4 Continued

Country & location	Altitude	Patient group	Adults or children	Sample type	n (samples)	n (homes)	Yr. sampling began	Mean Der 1 λg/g	% +ve samples	Mean Der p 1 λg/g	% +ve samples	Mean Der f 1 λg/g	% +ve samples	Reference
Erfurt	193	Po	A	B	393	393	1995	0.25		0.08		0.16		Fahlbusch *et al.* 1999
Erfurt	193	Po	A	F	204	204	1995	0.09		0.02		0.06		Fahlbusch *et al.* 1999
Essen	106	Po	C	B	43	43	1997	3.40		1.4		2		Oppermann *et al.* 2001
Freiburg/Weil	346	A, Po	C	B	2483	1291	1990	1.48		1.48				Kuehr *et al.* 1994b
Halle	104	Po	A, C	B	68	68	1994	2.16						Hirsch *et al.* 1998
Halle	104	Po	A, C	F	59	59	1994	0.37						Hirsch *et al.* 1998
Halle	104	Po	C	B	41	41	1997	2.50		0.7		1.8		Oppermann *et al.* 2001
Hamburg	3	Po	A	B	168	168	2000	7.64	96.4	0.54	72.0	3.63	94.6	Zock *et al.* 2006
Hamburg	3	Po	A	B	377	377	1995	1.46		0.34		1.11		Fahlbusch *et al.* 1999
Hamburg	3	Po	A	F	201	201	1995	0.32		0.06		0.25		Fahlbusch *et al.* 1999
Magdeburg	43	Po	C	B	38	38	1997	0.80		0.2		0.6		Oppermann *et al.* 2001
Österburg	553	Po	C	B	37	37	1997	2.50		1.2		1.3		Oppermann *et al.* 2001
GHANA														
Kumasi	246	A	C	B	96	96	1999	14.70		13.3		1.4		Addo-Yobo *et al.* 2001
HONG KONG	18	Po	A	B	40	40	1996	8.83		8.83	100			Leung *et al.* 1998
HONG KONG	18	Po	A	F	80	40	1996	1.43		1.43	100			Leung *et al.* 1998
HONG KONG	18	Po	C	B, F	36	36	2000	0.98		0.16	44.4	0.6	63.9	Wickens *et al.* 2004
ICELAND														
Reykjavik	15	Po	A	B	184	184	2001	0.01	0.5	0	0.0	0.01	0.5	Hallas *et al.* 2004; Zock *et al.* 2006
INDIA														
Kerala	50	Po	C	B, F	36	36	2000	0.28		0.28	52.8	0.02	12.1	Wickens *et al.* 2004
Neyveli	38	Po	C	B, F	36	36	2000	4.25		4.04	88.9	0.03	19.4	Wickens *et al.* 2004
Pune	570	Po	C	B, F	33	33	2000	0.78		0.04	12.1	0.69	48.5	Wickens *et al.* 2004
ITALY														
Genoa	67	A	A	B, F	19	19	1990	3.69	100	0.7	100	2.93	100	Barber *et al.* 1992
Pavia	57	Po	A	B	44	44	1998	8.10		0.34		7.76		Moscato *et al.* 2000
Pavia	57	Po	A	F	44	44	1998	0.98		0.15		0.83		Moscato *et al.* 2000
Pavia	57	Po	A	B	72	72	2000	8.10	77.8	0.04	38.9	7.48	97.2	Zock *et al.* 2006
Turin	205	Po	A	B	75	75	2000	0.59	76	0.01	12.0	0.59	76	Zock *et al.* 2006
Verona	90	Po	A	B	110	110	2000	1.20	82.7	0.02	29.1	0.95	80	Zock *et al.* 2006
JAPAN														
Fukuoka	10	Po	C	B, F	35	35	2000	20.28		5.98	88.6	3.63	91.4	Wickens *et al.* 2004
Sagamihara	57	A	A	B	47	47	1998	4.64	100	2.15	100	1.05	100	Yasueda *et al.* 1989a
Tokyo	43	Po	A	B	96	8	1989	23.63						Miyazawa *et al.* 1996
Tokyo	43	Po	A	F	96	8	1989	13.90						Miyazawa *et al.* 1996
MALAYSIA														
Klang Valley	5	Po	C	B, F	35	35	2000	0.83		0.7	62.9	0.13	42.9	Wickens *et al.* 2004
MAURITIUS														
Port Louis	134	A	A	B, F	14	14	1989	17.30		17.3	100			Guerin *et al.* 1992
MOROCCO														
Casablanca	16	Po	A	B	20	20	1990	8.30				0.7		de Andrade *et al.* 1995

continued...

Appendix 4 Continued

Country & location	Altitude	Patient group	Adults or children	Sample type	n (samples)	n (homes)	Yr. sampling began	Mean Der 1 λg/g	% +ve samples	Mean Der p 1 λg/g	% +ve samples	Mean Der f 1 λg/g	% +ve samples	Reference
Marrakech	450	Po	A	B	20	20	1990	0.60				0.1		de Andrade *et al.* 1995
NETHERLANDS														
N. Holland, Gelderland	5	A, Po	C	B	512	512	1990	5.10		5.10				van Strien *et al.* 1994; Verhoeff *et al.* 1994
N. Holland, Gelderland	5	A, Po	C	F	1026	512	1990	2.29		2.29				van Strien *et al.* 1994; Verhoeff *et al.* 1994
Groningen	49	A	A	B	59	59	1995	5.20		5.2				van der Heide *et al.* 1997a
Groningen	49	A	A	B	49	49	1995	3.00		3.0				van der Heide 1997b
Hilversum	4	A	A	B	30	30	1996	7.35		7.35				Rijssenbeek-Nouwens *et al.* 2002
NEW ZEALAND														
Christchurch	6	A, Po	A, C	B	93	93	1994	5.90		5.9	100			Martin *et al.* 1996
Christchurch	6	A, Po	A, C	F	93	93	1994	14.70		14.7	100			Martin *et al.* 1996
Wellington	21	Po	C	B	467	467	1994	46.60		46.6				Wickens *et al.* 1997; Sawyer *et al.* 1998
Wellington	21	Po	C	F	938	938	1994	25.95		25.95				Wickens *et al.* 1997; Sawyer *et al.* 1998
NORWAY														
W. & E. Norway		A, Po	C	F	32	16	1993	0.06	46.9	0.05	46.9	c.0.01	46.9	Dybendal & Elsayed 1994
Sør-Varanger	3	A	C	B	19	19	1994	0.01		0.01	100			Dotterud *et al.* 1997
Sor-Varanger	3	A	C	F	19	19	1994	0.01		0.01	100			Dotterud *et al.* 1997
PALESTINE														
Ramallah	873	Po	C	B	112	112	2001	4.59		4.48		0.11		El Sharif *et al.* 2004
Ramallah	873	Po	C	F	112	112	2001	1.29		1.23		0.06		El Sharif *et al.* 2004
PORTUGAL														
Porto	90	A	A, C	B, F	59	59	1992	9.30		9.2		0.1		Plácido *et al.* 1996
RUSSIA														
Moscow	150	A	C	B	33	15	1991	11.00	100.0	7.1	93.9	3.9	93.9	Zheltikova *et al.* 1994
SAUDI ARABIA														
Qassim	600	A	C	F	120	120	1995	0.03	22.5	0.016	20.0	0.015	15.0	Al-Frayh *et al.* 1997
Abha	2261	A	C	F	40	40	1995	4.73	70.0	4.642	67.5	0.09	15.0	Al-Frayh *et al.* 1997
Jeddah	7	A	C	F	15	15	1995	0.01	6.7	0.003	6.7	0.01	6.7	Al-Frayh *et al.* 1997
Riyadh	624	A	C	F	203	203	1995	0.01	1.5	0.001	0.5	0.01	1.5	Al-Frayh *et al.* 1997
SINGAPORE	1	A, Po	A, C	B	130	103	1993	1.70	48	1.2	97.0	0.5	48	Zhang *et al.* 1997
SINGAPORE	1	A, Po	A, C	F	176	103	1993	1.45	52	1.1	76.0	0.35	52	Zhang *et al.* 1997
SINGAPORE	1	Po	C	B	51	51	2003	3.90		2.33		1.57		Lee *et al.* 2005
SOUTH AFRICA														
Cape Town	16	A	C	B	60	60	1993	13.46		13.46				Jooma *et al.* 1995
Cape Town	16	A	C	F	60	60	1993	8.46		8.46				Jooma *et al.* 1995
Cape Town	16	A	C	B	59	59	1993	31.56		31.56				Manjra *et al.* 1994
Cape Town	16	A	C	F	59	59	1992	20.38		20.38				Manjra *et al.* 1994

continued...

Appendix 4 Continued

Country & location	Altitude	Patient group	Adults or children	Sample type	n (samples)	n (homes)	Yr. sampling began	Mean Der 1 λg/g	% +ve samples	Mean Der p 1 λg/g	% +ve samples	Mean Der f 1 λg/g	% +ve samples	Reference
Johannesburg	1786	Po	A	B	56	30	1994	1.46	100	1.46	100			Cadman et al. 1998
Johannesburg	1786	Po	A	F	55	30	1994	0.90	100	0.90	100			Cadman et al. 1998
SOUTH KOREA														
Chungcheongbuk-do	39	A	A	B	42	42	2001	20.05		0.95		19.1		Lee 2003
Seoul	33	A, Po	A	B	191	191	1989	2.55	95.6	0.14	60.3	2.41	93.8	Hong & Lee 1992
Seoul	33	A, Po	A	F	50	50	1989	2.66				2.66		Hong & Lee 1992
SPAIN														
Alto Deva		A	A	B, F	40	40	1990	3.00		3.00				Echechipía et al. 1995
Menorca		Po	C	B	475	475	1997	3.12		3.12				Basagaña et al. 2002; Torrent et al. 2007
Menorca		Po	C	F	482	482	1997	9.06		9.06				Basagaña et al. 2002; Torrent et al. 2007
Rioja Alta	620	A	A	B	13	13	1998	8.5		8.5				Lobera et al. 2000
Rioja Media	380	A	A	B	61	61	1998	11.9		11.9				Lobera et al. 2000
Rioja Baja	300	A	A	B	26	26	1998	28.4		28.4				Lobera et al. 2000
Albacete	676	Po	A	B	177	177	2000	0.01	13	0.01	10.2	0.01	4.5	Zock et al. 2006
Barcelona	1	Po	A	B	179	179	2000	7.38	97.2	0.6	76.0	2.25	83.2	Zock et al. 2006
Barcelona	1	Po	C	B	363	363	1997	0.68		0.68				Basagaña et al. 2002; Torrent et al. 2007
Barcelona	1	Po	C	F	366	366	1997	0.77		0.77				Basagaña et al. 2002; Torrent et al. 2007
Bilbao	213	A	A	B, F	504	42	1992	7.15		6.69		0.46		Gatius et al. 1994
Galdakao	57	Po	A	B	189	189	2000	12.82	99.5	9.54	98.4	0.06	44.4	Zock et al. 2006
Huelva	49	Po	A	B	132	132	2000	2.99	89.4	2.78	89.4	0.01	11.4	Zock et al. 2006
La Coruña	103	A, Po	A	B	111	111	2002	17.83		16.7		1.13		Boquete et al. 2006
Las Palmas	113	A	A	B	10	10	1993	41.80		41.8	100			Cabrera et al. 1995
Las Palmas	113	A	A	B, F	18	18	1990	7.37		6.6		0.77		Álvarez et al. 1997
Lugo	444	A, Po	A	B	84	84	2002	13.80		13.5		0.3		Boquete et al. 2006
Madrid	588	A	A	B	10	10	1991	0.26	70.0	0.26	70.0			Chivato et al. 1993
Orense	125	A, Po	A	B	46	46	2002	9.70		8.1		1.6		Boquete et al. 2006
Oviedo	213	Po	A	B	196	196	2000	14.71	99.5	6.6	73.5	0.79	76	Zock et al. 2006
Pamplona	455	A	A	B, F	47	65	1990	0.73		0.7		0.03		Alvarez et al. 1997
Pamplona	455	A	A	B	27	27	1996	1.95	100	1.95	100			Alvarez et al. 2000
Pamplona	455	A	A	B, F	48	30	1991	0.79		0.78		0.006		Olaguibel et al. 1994
Pontevedra	76	A, Po	A	B	91	91	2002	18.50		18.5				Boquete et al. 2006
Seville	8	A, Po	A	B, F	17	17	1992	2.14		2.14				Medina Gallardo et al. 1994
Valencia	7	A	C	F	23	11	1993	2.25	100	1.72	100	0.53	95.7	Julia et al. 1995
Vitoria	511	A	A	B, F	40	40	1990	8.50		8.50				Echechipía et al. 1995
SWEDEN														
Goteborg	4	Po	A	B	191	191	2000	0.02	31.9	0.01	11.0	0.01	27.2	Zock et al. 2006
Helsingborg	28	A	C	B, F	45	45	1989	2.48	96		56.0		81	Munir et al. 1995; Warner et al. 1998

continued...

Appendix 4 Continued

Country & location	Altitude	Patient group	Adults or children	Sample type	n (samples)	n (homes)	Yr. sampling began	Mean Der 1 λg/g	% +ve samples	Mean Der p 1 λg/g	% +ve samples	Mean Der f 1 λg/g	% +ve samples	Reference
Linköping	43	A	C	B, F	40	40	1989	0.08	63		28.0		24	Munir *et al.* 1995; Warner *et al.* 1998
Linköping	43	A, Po	C	B, F	19	19	1994	0.17	100					Björkstén *et al.* 1996
Linköping	43	A, Po	C	B, F	19	19	1994	0.09	42.1					Björkstén *et al.* 1996
Stockholm	16	A, Po	C	B	160	160	1987	1.40	18.1	0.115	18.1	1.28	7.5	Wickman *et al.* 1991
Umeå	19	Po	A	B	189	189	2000	0.01	3.7	0.01	1.6	0.01	3.2	Zock *et al.* 2006
Umeå	19	A	C	B, F	41	41	1989	0.03	22		7.0		9	Munir *et al.* 1995; Warner *et al.* 1998
Uppsala	6	Po	A	B	183	183	2000	0.01	16.4	0.01	5.5	0.01	12.6	Zock *et al.* 2006
SWITZERLAND														
Basel	317	Po	A	B	130	130	2000	0.51	73.8	0.01	23.8	0.35	70	Zock *et al.* 2006
TAIWAN														
Taipei	6	A, Po		B, F	161	12	1992	2.10		2.1				Li *et al.* 1994
Tainan	10	A	C	B	125	40	2000	0.98						Chen *et al.* 2002
Tainan	10	A	C	F	235	40	2000	1.43						Chen *et al.* 2002
TANZANIA														
Ifakara	192	Po	C	B	128	128	1998	2.30						Sunyer *et al.* 2001
THAILAND														
Northern region				B	209	209	1991	7.40	97.6					Malainual *et al.* 1995
Northeastern region				B	179	179	1991	12.00	98.3					Malainual *et al.* 1995
Ban Thepha & environs	1	Po	C	B	50	50	2003	1.70		1.36		0.34		Lee *et al.* 2005
Chiang Mai	310	Po	C	B	36	36	1999	11.49	100	8.61	100	2.88	94.3	Trakultivakorn & K. 2004; Wickens *et al.* 2004
Chiang Mai	310	Po	C	F	36	36	1999	1.88	100	1.61	97.1	0.27	82.9	Trakultivakorn & K. 2004; Wickens *et al.* 2004
TOKELAU	2	Po	A	B	77	80	2000	0.16		0.04	27.3	0.12	68.8	Lane *et al.* 2005
TOKELAU	2	Po	A	F	80	80	2000	0.07		0.04	32.5	0.03	32.5	Lane *et al.* 2005
TURKEY														
Bursa	200	Po	A	F	60	60	2004	8.83		8.83				Sevki *et al.* 2006
UNITED KINGDOM														
Ashford	45	Po	C	B	619	618	1993	0.50		0.50	65.0			Atkinson *et al.* 1999
Ashford	45	Po	C	F	616	618	1993	1.40		1.40	89.0			Atkinson *et al.* 1999
Bristol	47	A	C	B	49	49	1994	0.48		0.48				Carswell *et al.* 1996
Ipswich	21	Po	A	B	104	104	2000	1.05	89.4	0.77	84.6	0.01	22.1	Zock *et al.* 2006
London	15	A	A	B	89	89	1979	2.50		2.50				Tovey *et al.* 1981
London	15	A	C	B	51	51	1988	4.02		4.02				Rieser *et al.* 1990
Manchester	72	A	A, C	B	200	40	1990	2.59		2.59				Kalra *et al.* 1992
Manchester	72	A	A, C	F	400	40	1990	1.81		1.81				Kalra *et al.* 1992
Manchester	72	Po	A	B	564	564	1995	1.19		1.19				Simpson *et al.* 2002

continued...

Appendix 4 Continued

Country & location	Altitude	Patient group	Adults or children	Sample type	n (samples)	n (homes)	Yr. sampling began	Mean Der 1 λg/g	% +ve samples	Mean Der p 1 λg/g	% +ve samples	Mean Der f 1 λg/g	% +ve samples	Reference
Manchester	72	Po	A	F	1128	564	1995	0.74		0.74				Simpson *et al.* 2002
Newport and environs	22	Po	C	F	240	120	1990	15.90		15.90				Arshad *et al.* 1992; Ridout *et al.* 1993
Norwich	8	Po	A	B	158	158	1996	2.00		2.0				Luczynska *et al.* 1998
Norwich	8	Po	A	F	316	316	1996	1.80		1.8				Luczynska *et al.* 1998
Norwich	8	Po	A	B	164	164	2000	1.54	90.9	1.3	88.4	0.01	22	Zock *et al.* 2006
Poole	0	A, Po	C	B	59	59	1979	18.40		18.4	96.6			Sporik *et al.* 1990
Poole	0	A, Po	C	F	59	59	1979	3.40		3.4	100			Sporik *et al.* 1990
Sheffield	95	A	A	B	30	30	1997	8.30		8.3				Htut *et al.* 2001
Southampton	0	A	C	F	44	52	1998	0.60		0.60				Price *et al.* 1990
Southampton	0	Po	A	B	217	58	1995	9.20		0.60				Doull *et al.* 1997
USA														
California		A	A, C	B	23	23	1989	5.19		2.47		2.93		Lintner & Brame 1993
Desert regions		A	A, C	B	21	21	1989	3.97		1.85		2.82		Lintner & Brame 1993
East Coast		A	A, C	B	179	179	1989	4.97		2.23		3.67		Lintner & Brame 1993
Front Range, CO		Po	A	B, F	190	38	1993	1.04	44.7					Ellingson *et al.* 1995
Great Plains		A	A, C	B	52	52	1989	4.16		2.91		1.66		Lintner & Brame 1993
Gulf Coast		A	A, C	B	177	177	1989	3.10		1.82		2.42		Lintner & Brame 1993
Pacific northwest		A	A, C	B	48	48	1989	3.42		2.96		1.35		Lintner & Brame 1993
S. New England		A	C	B	750	750	1996	0.87	79.5	0.32	60.3	0.55	79.5	van Strien *et al.* 2004
S. New England		A	C	F	750	750	1996	1.35	80.9	0.51	64.1	0.84	80.9	van Strien *et al.* 2004
Steppe		A	A, C	B	36	36	1989	3.18		1.67		2.63		Lintner & Brame 1993
Atlanta, GA	320	A, Po	C	B	20	20	1983	11.40						Smith *et al.* 1985
Atlanta, GA	320	A, Po	C	F	40	20	1983	6.85						Smith *et al.* 1985
Baltimore, MD	1	A, Po	A	B	42	42	1982	0.68		0.075	57.1	0.6	93.0	Lind *et al.* 1987
Baltimore, MD	1	A	C	F	120	120	2005	0.75						Simons *et al.* 2007
Charlottesville, VA	181	A, Po	A	B	68	6	1984	4.28	100					Platts-Mills *et al.* 1987
Charlottesville, VA	181	A, Po	A	F	71	6	1984	1.31	91.6					Platts-Mills *et al.* 1987
Cincinnati, OH	208	Po	C	F	774	774	2003	0.50				0.5	37.7	Cho *et al.* 2006
Dayton, OH	229	Po	A, C	F	198	87	1998	25.60						Neal *et al.* 2002
Denver, CO	1603	Po	C	B, F	86	86	1999	0.04		0.02		0.02		Gereda *et al.* 2001
Detroit, OH	173	A	C	F	91	91	1987	7.64		2.02	73.2	5.62	68.3	Peterson *et al.* 1999
Los Alamos, NM	2165	A, Po	C	B, F	111	111	1992	0.21		0.18		0.13		Sporik *et al.* 1995
Salem, MA	5	A	A	F	41	41	1995	36.10			22.0		92.7	Hannaway & Roundy 1997
San Diego, CA	25	A	C	B	31	31	1994	20.21		18.0		2.21		Christiansen *et al.* 1996
San Diego, CA	25	Po	C	F	23	23	1994	15.82		15.5		0.32		Christiansen *et al.* 1996
Seattle, WA	59	Po	A	B	38	19	1998	38.45						Vojta *et al.* 2001
St. Petersburg, FL	13	A, Po	C	B	30	30	1994	13.30		9.65		3.65		Nelson *et al.* 1996
St. Petersburg, FL	13	A, Po	C	F	30	30	1994	7.96		6.63		1.33		Nelson *et al.* 1996
Total					34345	28104								
Count					245	244		243	67	207	94	120	59	

References

Aalberse, R.C. (2000) Structural biology of allergens. *Journal of Allergy and Clinical Immunology*, 106, 228–238.

Aas, K. and Mehl, R. (1996) Påvisning og kvantitering av husstøvmidd. *Fagbladet Allergi i Praxis*, 1, 37–40.

Abbot, J., Cameron, J. and Taylor, B. (1981) House dust mite counts in different types of mattresses, sheepskins and carpets, and a comparison of brushing and vacuuming collection methods. *Clinical Allergy*, 11, 589–595.

Abe, T., Ohkido, M. and Yamamoto, K. (1978) Studies on skin surface barrier functions – skin surface lipids and transepidermal water loss in atopic skin during childhood. *Journal of Dermatology*, 5, 223–229.

Abed-Benamara, M., Fain, A. and Abed, L. (1983) Note preliminaire sur la faune acarologique des poussières de matelas d'Alger. *Acarologia*, 24, 79–83.

Adam, E., Hansen, K.K., Astudillo, O.F., Coulon, L., Bex, F., Duhant, X., Jaumotte, E., Hollenberg, M.D. and Jacquet, A. (2006) The house dust mite allergen Der p 1, unlike Der p 3, stimulates the expression of interleukin-8 in human airway epithelial cells via a proteinase-activated receptor-2-independent mechanism. *Journal of Biological Chemistry*, 281, 6910–6923.

Addo-Yobo, E.O.D., Custovic, A., Taggart, S.C.O., Craven, M., Bonnie, B. and Woodcock, A. (2001) Risk factors for asthma in urban Ghana. *Journal of Allergy and Clinical Immunology*, 108, 363–368.

Adinoff, A.D., Tellez, P., and Clark, R.A.F. (1988) Atopic dermatitis and aeroallergen contact sensitivity. *Journal of Allergy and Clinical Immunology*, 81, 736–742.

Agratorres, J.M., Pereira-Lorenzo, A., Fernandez-Fernandez, I. (1999) Population dynamics of house dust mites (Acari: Pyroglyphidae) in Santiago de Compostela (Galicia, Spain). *Acarologia*, 40, 59–63.

Ait-Khaled, N., Odhiambo, J., Pearce, N., Adjoh, K.S., Maesano, I.A., Benhabyles, B., Bouhayed, I.A., Bahati, E., Camara, L., Catteau, C., El Sony, A., Esamai, F.O., Hypolite, I.E., Melaku, K., Musa, O.A., Ng'ang'a, L., Onadeko, B.O., Saad, O., Jerray, M., Kayembe, J.M., Koffi, N.B., Khaldi, F., Kuaban, C., Voyi, K., M'Boussa, J., Sow, O., Tidjani, O. and Zar, H.J. (2007) Prevalence of symptoms of asthma, rhinitis, and eczema in 13- to 14-year old children in Africa: the International Study of Asthma and Allergies in Childhood Phase III. *Allergy*, 62, 247–258.

Akdemir, C. and Gürdal, H. (2005) [House dust mite in Kutahya, Turkey.] *Tükiye Parazitoloji Dergisi*, 29, 110–115. [In Turkish, English summary]

Aki, T., Fujikawa, A., Wada, T. Jyo, T., Shigeta, S., Murooka, Y., Oka, S. and Ono, K. (1994a) Cloning and expression of cDNA coding for a new allergen from the house dust mite, *Dermatophagoides farinae*: homology with human heat shock cognate proteins in the heat shock protein 70 family. *Journal of Biochemistry*, 115, 435–440.

Aki, T., Ono, K., Paik, S.-Y., Wada, T., Jyo, T., Shigeta, S., Murooka, Y. and Oka, S. (1994b) Cloning and characterization of cDNA coding for a new allergen from the house dust mite, *Dermatophagoides farinae*. *International Archives of Allergy and Immunology*, 103, 349–356.

Aki, T., Kodama, T., Fujikawa, A., Miura, K., Shigeta, S., Wada, T., Jyo, T., Murooka, Y., Oka, S. and Ono, K. (1995) Immunochemical characterization of recombinant and native tropomyosins as a new allergen from the house dust mite, *Dermatophagoides farinae*. *Journal of Allergy and Clinical Immunology*, 96, 74–83.

Akimov, I.A. (1973) On the morphological and physiological characteristics of the alimentary canal of the bulb mite *Rhizoglyphus echinopus* (Fumouze and Robin). In: Daniel, M. and Rosický, B. (eds), *Proceedings of the 3rd International Congress of Acarology, Prague, 1971*. Academia, Prague, pp. 703–706.

Akoun, G., Araujo-Fontaine, A., Basset, F., Borgard, and Cuzin, A. (1972) Miliaire pulmonaire et hypersensibilite aux acariens. *Revue de Tuberculose et Pneumonologie*, 36, 693–710.

Alani, M.D. and Haarlov, N. (1972) The house-dust mite: a possible source of allergen in the environment of patients with atopic dermatitis. *Journal of the National Medical Association*, 64, 302–305.

Alberti, G. (1995) Comparative spermatology of Chelicerata: review and perspective. In: Jamieson, B.G.M., Ausio, J. and Justine, J.-L. (eds), *Advances in Spermatozoal Phylogeny and Taxonomy. Mémoires de la Muséum nationale d'Histoire naturelle*, 166, 203–230.

Alberti, G. (1998) Fine structure of receptor organs in oribatid mites. In: Ebermann, E. (ed.) *Arthropod Biology: Contributions to Morphology, Ecology and Systematics*. Österreichische Akademie der Wissenschaften, Vienna, pp. 27–77.

Alberti, G. and Coons, L.B. (1999) Acari: Mites. In: Harrison, F.W. and Foelix, R.W. (eds), *Microscopic Anatomy of Invertebrates. Volume 8C: Chelicerate Arthropoda. Mites*. Wiley-Liss, New York, pp. 515–1215.

Alberti, G., Storch, V. and Renner, H. (1981) Über den feinstrukturellen aufbau der milbencuticula (Acari, Arachnida). *Zoologische Zahrbücher. Abteilung für Anatomie und Ontogenie der Tiere*, 105, 183–286.

Alexander, J.O'D. (1972) Mites and skin diseases. *Clinical Medicine*, 79, 14–19.

Alexander, J.O'D. (1984) *Arthropods and Human Skin*. Springer-Verlag, Berlin.

Al-Frayh, A.S., Hasnain, S.M., Gad-El-Rab, M.O., Schwartz, B., Al-Mobairek, K. and Al-Sedairy, S.T. (1997) House dust mite allergens in Saudi Arabia: regional variations and immune response. *Annals of Saudi Medicine*, 17, 156–160.

Allanach, D., Benisek, L., Bourn, W. and Rushforth, M.A. (1990) Dust mite repellent treatments. In: *Proceedings of the Eighth International Wool Textile Research Conference*, Christchurch, New Zealand, Vol. 4, pp. 643–652.

Allen, D.E. (1976) *The Naturalist in Britain. A Social History*. Allen and Lane, London.

Alm, J.S., Swartz, J., Lilja, G., Scheynius, A. and Pershagen, G. (1999) Atopy in children of families with an anthroposophic lifestyle. *Lancet*, 353, 1485–1488.

Almqvist, C., Marks, G., Li, Q., Daniel, C. and Tovey, E. (2007) The bell tolls for the relationship between house dust mite exposure and asthma in childhood.

European Respiratory Journal, 30 (Supplement 51), 684s. [abstract]

Altschul, S.F., Madden, T.L., Schäffer, A.A., Zhang, J., Zhang, Z., Miller, W. and Lipman, D.J. (1997) Gapped BLAST and PSI-BLAST: a new generation of protein database search programs. *Nucleic Acids Research*, 25, 3389–3402.

Álvarez, M.J., Olaguibel, J.M., Acero, S., Quirce, S., García, B.E., Carrillo, T., Cortés, C. and Tabar, A.I. (1997) Indoor allergens and dwelling characteristics in two cities in Spain. *Journal of Investigational Allergology and Clinical Immunology*, 7, 572–577.

Álvarez, M.J., Olaguibel, J.M., Acero, S., García, B.E., Tabar, A.I. and Urbiola, E. (2000) Effect of current exposure to Der p 1 on asthma symptoms, airway inflammation, and bronchial hyperresponsiveness in mite–allergic asthmatics. *Allergy*, 55, 185–190.

Amaral, V. (1968) Sôbre a ocorrência do ácaro *Dermatophagoides pteronyssinus* (Trouessart, 1897) no Brasil (Psoroptidae: Sarcoptiformes). *Revista de Medicina Veterinária*, 3, 296–300.

Amini, M. (1981) *Oriental Rugs: Care and Repair*. Macdonald, London.

Amoli, K. and Cunnington, A.M. (1977) House dust mites in Iran. *Clinical Allergy*, 7, 93–101.

Ancona, G. (1923) Asma epidemico da *Pediculoides ventricosus. Policlinico (Sezione Medica)* 30, 45–49.

Andersen, A. (1984) Abundance and spatial distribution of house-dust mites in their natural environment. *Entomologisk Meddelelser*, 52, 25–32.

Andersen, A. (1985) Microfungi in beds and their relation to house-dust mites. *Grana*, 24, 55–59.

Andersen, A. (1991) Nutritional value of yeast for *Dermatophagoides pteronyssinus* (Acari, Epidermoptidae) and the antigenic and allergenic composition of extracts during extended culturing. *Journal of Medical Entomology*, 28, 487–491.

Andersen, A. and Rosen, J. (1989) House dust mite *Dermatophagoides pteronyssinus* and its allergens: effects of washing. *Allergy*, 44, 396–400.

Anderson, H.R. (1974) The epidemiological and allergic features of asthma in the New Guinea Highlands. *Clinical Allergy*, 4, 171–183.

Anderson, H.R. and Cunnington, A.M. (1974) House dust mites in the highlands of Papua New Guinea. *Papua New Guinea Medical Journal*, 17, 304–308.

Anderson, H.R., Gupta, R., Strachan, D.P. and Limb, E.S. (2007) 50 years of asthma: UK trends from 1955 to 2004. *Thorax*, 62, 85–90.

Ando, T., Homma, R., Ino, Y., Ito, G., Miyahara, A., Yanagihara, T., Kimura, H., Ikeda, S., Yamakawa, H., Iwaki, M., Okumura, Y., Suko, M., Haida, M. and Okudaira H. (1993) Trypsin-like protease of mites: purification and characterisation of trypsin-like protease from mite faecal extract *Dermatophagoides farinae*. Relationship between trypsin-like protease and Der f III. *Clinical and Experimental Allergy*, 23, 777–784.

de Andrade, A.D., Bartal, M., Birnbaum, J., Lanteaume, A., Charpin, D. and Vervloet, D. (1995) House dust mite allergen content in two areas with large differences in relative humidity. *Annals of Allergy, Asthma and Immunology*, 74, 314–316.

Andrewartha, H.G. and Birch, L.C. (1954) *The Distribution and Abundance of Animals*. University of Chicago Press, Chicago.

Andrews, J.R.H., Crane, J., Phillips, L. and Robertson, B. (1992) House dust mites, asthma and skin prick tests in a group of Wairoa schoolchildren. *Entomological Society of New Zealand, 41st Annual Conference Proceedings, Central Institute of Technology, Heretaunga, Hutt Valley, N.Z., 17–20 May, 1992* [no pagination].

Andrews, J.R.H., Crane, J., Phillips, L. and Robertson, B. (1995) House dust mites, asthma and skin prick testing in a group of Wairoa schoolchildren. *New Zealand Entomologist*, 18, 71–75.

Andrews, M.L.A. (1976) *The Life That Lives on Man*. Faber and Faber, London.

Anonymous (1861) Editorial, Saturday October 19, 1861. *Lancet*, ii, 382–383.

Anonymous (1979) European house dust mite (*Dermatophagoides pteronyssinus*) – North Dakota – new state record. *Cooperative Plant Pest Report*, 4, 843.

Ansotegui, I.J., Härfast, B., Jeddi-Tehrani, M., Johansson, E., Johansson, S.G.O., van Hage-Hamsten, M. and Wigzell, H. (1991) Identification of a new major allergen of 39 kilodaltons of the storage mite *Lepidoglyphus destructor*. *Immunology Letters*, 27, 127.

Antens, C.J.M., Oldenwenig, M., Wolse, A., Gehring, U., Smit, H.A., Aalberse, R.C., Kerkhof, M., Gerritsen, J., de Jongste, J.C. and Brunekreef, B. (2006) Repeated measurements of mite and pet allergen levels in house dust over a time period of 8 years. *Clinical and Experimental Allergy*, 36, 1525–1531.

Antonicelli, L., Bilò, M.B., Pucci, S., Schou, C. and Bonifazi, F. (1991) Efficacy of an air-cleaning device equipped with a high efficiency particulate air filter in house dust mite respiratory allergy. *Allergy*, 46, 594–600.

Aoki, J.-I. (1973) Soil mites (oribatids) climbing trees. In: Daniel, M. and Rosicky, B. (eds), *Proceedings of the 3rd International Congress of Acarology*. Academia, Prague, pp. 59–65.

Apter, A., Bracker, A., Hodgson, M., Sidman, J. and Leung, W.-Y. (1994) Epidemiology of sick building syndrome. *Journal of Allergy and Clinical Immunology*, 94, 277–288.

Araujo-Fontaine, A., Wagner, M., Moreau, G. and Basset, A. (1972) Elevage des acariens de la poussière domestique et recherche des anticorps precipitants anti-poussière et anti-acariens par immunodiffusion en gelose. *Revue Française d'Allergologie*, 12, 231–238.

Araujo-Fontaine, A., Wagner, M. and Kremer, M. (1973) Contribution à l'étude des acariens de la poussière domestique en Alsace. Relations avec les conditions d'habitat. *Comptes Rendus des seances de la Société de Biologie*, 167, 371–378.

Araujo-Fontaine, A., Miltgen, F., Rombourg, H., Molet, B., Pauli, G. and Basset, A. (1974) Contribution à l'étude du role allergisant des acariens de la poussière. Étude immunologique des sérums des malades et des serums des lapins hyperimmunisés. *Revue Française d'Allergologie et d'Immunologie Clinique*, 14, 91–96.

Arias Irigoyen, J. and García del Hoyo, J.J. (2002) Skin sensitisation to mites and mite identification in the homes of an allergic population in the Huelva province. *Allergología y Immunología Clinica*, 17, 61–68.

Arlian, L.G. (1972) Equilibrium and non-equilibrium water exchange kinetics in an atracheate terrestrial arthropod, *Dermatophagoides farinae* Hughes. PhD thesis, The Ohio State University, Columbus.

Arlian, L.G. (1975a) Dehydration and survival of the European house dust mite *Dermatophagoides pteronyssinus*. *Journal of Medical Entomology*, 12, 437–442.

Arlian, L.G. (1975b) Water exchange and effect of water vapour activity on metabolic rate in the dust mite *Dermatophagoides*. *Journal of Insect Physiology*, 21, 1439–1442.

Arlian, L.G. (1977) Humidity as a factor regulating feeding and water balance of the house dust mites *Dermatophagoides farinae* and *D. pteronyssinus*. *Journal of Medical Entomology*, 14, 484–488.

Arlian, L.G. (1989) Biology and ecology of house dust mites, *Dermatophagoides* spp. and *Euroglyphus* spp. *Immunology and Allergy Clinics of North America*, 9, 339–356.

Arlian, L.G. (1991) House-dust mite allergens, a review. *Experimental and Applied Acarology*, 10, 167–186.

Arlian, L.G. (1992) Water balance and humidity requirements of house dust mites. *Experimental and Applied Acarology*, 16, 15–35.

Arlian, L.G. (2002) Arthropod allergens and human health. *Annual Review of Entomology*, 47, 395–433.

Arlian, L.G. and Dippold, J.S. (1996) Development and fecundity of *Dermatophagoides farinae* (Acari: Pyroglyphidae). *Journal of Medical Entomology*, 33, 257–260.

Arlian, L.G. and Veselica, M.M. (1979) Water balance in insects and mites. *Comparative Biochemistry and Physiology*, Series A, 64, 191–200.

Arlian, L.G. and Veselica, M.M. (1981a) Reevaluation of the humidity requirements of the house dust mite *Dermatophagoides farinae* (Acari: Pyroglyphidae). *Journal of Medical Entomology*, 18, 351–352.

Arlian, L.G. and Veselica, M.M. (1981b) Effect of temperature on the equilibrium body water mass in the mite *Dermatophagoides farinae*. *Physiological Zoology*, 54, 393–399.

Arlian, L.G. and Veselica, M.M. (1982) Relationship between transpiration rate and temperature in the mite *Dermatophagoides farinae*. *Physiological Zoology*, 55, 344–354.

Arlian, L.G. and Vyszenski-Moher, D.L. (1996) Responses of *Sarcoptes scabiei* (Acari: Sarcoptidae) to nitrogenous waste and phenolic compounds. *Journal of Medical Entomology*, 33, 236–243.

Arlian, L.G. and Wharton, G.W. (1974) Kinetics of active and passive components of water exchange between the air and a mite, *Dermatophagoides farinae*. *Journal of Insect Physiology*, 20, 1063–1077.

Arlian, L.G., Brandt, R.L. and Bernstein, R. (1978) Occurrence of house dust mites (Acari: Pyroglyphidae) during the heating season. *Journal of Medical Entomology*, 15, 35–42.

Arlian, L.G., Bernstein, I.L., Johnson, C.L. and Gallagher, J.S. (1979a) Ecology of house dust mites and dust allergy. In: Rodriguez, J.G. (ed.), *Recent Advances in Acarology*. Volume 2. Academic Press, New York, pp. 185–195.

Arlian, L.G., Bernstein, I.L., Johnson, C.L. and Gallagher, J.S. (1979b) A technique for separation of house dust mites (Acari: Pyroglyphidae) from culture media. *Journal of Medical Entomology*, 16, 128–132.

Arlian, L.G., Bernstein, I. L. and Gallagher, J.S. (1982) The prevalence of house dust mites, *Dermatophagoides* spp, and associated environmental conditions in homes in Ohio. *Journal of Allergy* and *Clinical Immunology*, 69, 527–532.

Arlian, L.G., Woodford, P.J., Bernstein, I.L. and Gallagher, J.S. (1983) Seasonal population structure of house dust mites, *Dermatophagoides* spp. (Acari: Pyroglyphidae). *Journal of Medical Entomology*, 20, 99–102.

Arlian, L.G., Geis, D.P., Vyszenski-Moher, D.L., Bernstein, I.L. and Gallagher, J.S. (1984a) Antigenic and allergenic properties of the storage mite *Tyrophagus putrescentiae*. *Journal of Allergy and Clinical Immunology*, 74, 166–172.

Arlian, L.G., Geis, D.P., Vyszenski-Moher, D.L., Bernstein, I.L. and Gallagher, J.S. (1984b) Cross antigenic and allergenic properties of the house dust mite *Dermatophagoides farinae* and the storage mite *Tyrophagus putrescentiae*. *Journal of Allergy and Clinical Immunology*, 74, 172.

Arlian, L.G., Vyszenski-Moher, D.L. and Gilmore, A.M. (1988) Cross-antigenicity between *Sarcoptes scabiei* and the house dust mite, *Dermatophagoides farinae* (Acari, Sarcoptidae and Pyroglyphidae). *Journal of Medical Entomology*, 25, 240–247.

Arlian, L.G., Rapp, C.M. and Ahmed, S.G. (1990) Development of *Dermatophagoides pteronyssinus* (Acari, Pyroglyphidae). *Journal of Medical Entomology*, 27, 1035–1040.

Arlian, L.G., Bernstein, D., Bernstein, I.L., Friedman, S., Grant, A., Lieberman, P., Lopez, M., Metzger, J., Platts-Mills, T.A.E., Schatz, M., Spector, S., Wassermann, S.I. and Zeiger, R.S. (1992) Prevalence of dust mites in the homes of people with asthma living in eight different geographic areas of the United States. *Journal of Allergy and Clinical Immunology*, 90, 292–300.

Arlian, L.G., Vyszenski-Moher, D.L. and Fernandez-Caldas, E. (1993) Allergenicity of the mite, *Blomia tropicalis. Journal of Allergy and Clinical Immunology*, 91, 1042–1050.

Arlian, L.G., Confer, P.D., Rapp, C.M., Vyszenski-Moher, D.L. and Chang, J.C.S. (1998a) Population dynamics of the house dust mites *Dermatophagoides farinae*, *D. pteronyssinus* and *Euroglyphus maynei* (Acari: Pyroglyphidae) at specific relative humidities. *Journal of Medical Entomology*, 35, 46–53.

Arlian, L.G., Neal, J.S. and Bacon, S.W. (1998b) Survival, fecundity and development of *Dermatophagoides farinae* (Acari: Pyroglyphidae) at fluctuating relative humidity. *Journal of Medical Entomology*, 35, 962–966.

Arlian, L.G., Morgan, M.S. and Houck, M.A. (1999a) Allergenicity of the mite *Hemisarcoptes cooremani*. *Annals of Allergy, Asthma and Immunology*, 83, 529–532.

Arlian, L.G., Neal, J.S. and Vyszenski-Moher, D.L. (1999b) Fluctuating hydrating and dehydrating relative humidities effects on the life-cycle of *Dermatophagoides farinae* (Acari: Pyroglyphidae). *Journal of Medical Entomology*, 36, 457–461.

Arlian, L.G., Neal, J.S., Morgan, M.S., Vyszenski-Moher, S.L., Rapp, C.M. and Alexander, A.K. (2001) Reducing relative humidity is a practical way to control dust mites and their allergens in homes in temperate climates. *Journal of Allergy and Clinical Immunology*, 107, 99–104.

Arlian, L.G., Morgan, M.S. and Neal, J.S. (2002) Dust mite allergens: ecology and distribution. *Current Allergy and Asthma Reports*, 2, 401–411.

Armentia, A., Perez-Santos, C., Fernandez, A., de la Fuente, R., Sánchez, P., Sanchis, E., Méndez, J., Tapias, J.A., Castrodeza, R. and Pascual, F. (1993) Estudio de prevalencia de los ácaros productores de alergia en la provincia de Valladolid. *Revista España Allergología y Immunología Clinica*, 8, 199–210.

Armitage, D.M. and George, C.L. (1986) The effect of three species of mites upon fungal growth on wheat. *Experimental and Applied Acarology*, 2, 111–124.

Armstrong, R.N. (1997) Structure, catalytic mechanism and evolution of the glutathione transferases. *Chemical Research in Toxicology*, 10, 2–18.

Arruda, L.K., Rizzo, M.C., Chapman, M.D., Fernandez-Caldas, E., Baggio, D., Platts-Mills, T.A.E. and Naspitz, C.K. (1991) Exposure and sensitisation to dust mite allergens among asthmatic children in Sao Paulo, Brazil. *Clinical and Experimental Allergy*, 21, 433–439.

Arruda, L.K. and Chapman, M.D. (1992) A review of recent immunochemical studies of *Blomia tropicalis* and *Euroglyphus maynei* allergens. *Experimental and Applied Acarology*, 16, 129–135.

Arruda, L.K., Fernandez-Caldas, E., Naspitz, C.K, Montealegre, F., Vailes, L.D. and Chapman, M.D. (1995) Identification of *Blomia tropicalis* allergen Blo t 5 by cDNA cloning. *International Archives of Allergy and Immunology*, 107, 456–457.

Arruda, L.K., Vailes, L.D., Platts-Mills, T.A.E., Fernandez-Caldas, E., Montealegre, F., Lin, K.L. Chua, K.Y., Rizzo, M.C., Naspitz, C.K. and Chapman MD (1997a) Sensitisation to *Blomia tropicalis* in patients with asthma and identification of allergen Blo t 5. *American Journal of Respiratory and Critical Care Medicine*, 155, 343–350.

Arruda, L.K., Vailes, L.D., Platts-Mills, T.A.E., Hayden, M.L. and Chapman, M.D. (1997b) Induction of IgE antibody responses by glutathione-S-transferase from the German cockroach (*Blatella germanica*). *Journal of Biological Chemistry*, 272, 20907–20912.

Arshad, S.H., Matthews, S., Gant, C. and Hide, D.W. (1992) Effect of food and house-dust mite allergen avoidance on development of allergic disorders in infancy. *The Lancet*, 339, 1493–1497.

Artamonov, M.I. (1969) *The Splendor of Scythian Art. Treasures From Scythian Tombs*. Frederick A. Praeger, New York.

Artigas, J.N. and Casanueva, M.E. (1983) Acaros del polvo de las habitaciones en Chile (Acari). *Gayana (Zoologia)*, 47, 3–106.

Asher, M.I., Montefort, S., Björkstén, B., Lai, C.K.W., Strachan, D.P., Weiland, S.K., Williams, H. and the ISAAC Phase Three Study Group (2006) Worldwide time trends in the prevalence of symptoms of asthma, allergic rhinoconjunctivitis, and eczema in childhood: ISAAC Phases One and Three repeat multicountry cross-sectional surveys. *Lancet*, 368, 733–743.

Aspaly, G., Stejskal, V., Pekár, S. and Hubert, J. (2007) Temperature-dependent population growth of three species of stored products mites (Acari: Acaridida). *Experimental and Applied Acarology*, 42, 37–46.

van Asperen, P. (1993) A review of domestic allergen avoidance methods. In: Tovey, E.R. and Mahmic, A. (eds), *Mites, Asthma and Domestic Design: proceedings of a conference held at the Powerhouse*

Museum, 15th March, 1993. University Printing Service, University of Sydney, pp. 23–25.

van Asselt, L. (1999) Interactions between domestic mites and fungi. *Indoor and Built Environment*, 8, 216–220.

Asturias, J.A., Arilla, M.C., Gómez-Bayón, N., Martínez, A., Martínez, J. and Palacios, R. (1998) Sequencing and high level expression in *Escherichia coli* of the tropomyosin allergen (Der p 10) from *Dermatophagoides pteronyssinus*. *Biochimica et Biophysica Acta*, 1397, 27–30.

Atambay, M., Aycan, O.M. and Dadal, N. (2006) [House dust mite fauna in Malatya.] *Tükiye Parazitoloji Dergisi*, 30, 205–208. [In Turkish, English summary]

Atkinson, W., Harris, J., Mills, P., Moffat, S., White, C., Lynch, O., Jones, M., Cullinan, P. and Newman Taylor, A.J. (1999) Domestic aeroallergen exposures in an English town. *European Respiratory Journal*, 13, 583–589.

Atyeo, W.T. and Gaud, J. (1977) A new dermatophagoidine mite from Ecuador (Acarina, Pyroglyphidae). *Steenstrupia*, 4, 121–124.

August, P.J. (1984) House dust mite causes atopic eczema. A preliminary study. *British Journal of Dermatology*, 111 (Supplement 26), 10–11.

Avarre, J.-C., Michelis, R., Tietz, A. and Lubzens, E. (2003) Relationship between vitellogenin and vitellin in a marine shrimp (*Penaeus semisulcatus*) and molecular characterisation of vitellogenin complementary cDNAs. *Biology of Reproduction*, 69, 355–364.

Babe, K.S., Arlian, L.G., Confer, P.D. and Kim, R. (1995) House dust mite (*Dermatophagoides farinae* and *Dermatophagoides pteronyssinus*) prevalence in the rooms and hallways of a tertiary care hospital. *Journal of Allergy and Clinical Immunology*, 95, 801–805.

Bahir, A., Goldberg, A., Mekori, Y.A., Confino-Cohen, C., Morag, H., Rosen, Y., Monakir, D., Rigler, S., Cohen, A.H., Horev, Z., Noviski, N. and Mandelberg, A. (1997) Continuous avoidance measures with or without acaricide in dust mite-allergic children. *Annals of Allergy, Asthma and Immunology*, 78, 506–512.

Baker, A.S. (1990) Two new species of *Lardoglyphus* Oudemans (Acari, Lardoglyphidae) found in the gut of human mummies. *Journal of Stored Products Research*, 26, 139–147.

Baker, E.W., Delfinado, M.D. and Abbatiello, M.J. (1976) Terrestrial mites of New York II. Mites in birds' nests (Acarina). *Journal of the New York Entomological Society*, 84, 48–66.

Baker, G.T. and Krantz, G.W. (1985) Structure of the male and female reproductive and digestive systems of *Rhizoglyphus robini* Claparède (Acari, Acaridae). *Acarologia*, 26, 55–65.

Baker, H. (1769) *The Microscope Made Easy, or the Nature, Uses, and Magnifying Powers of the Best Kinds of Microscopes Described, Calculated and Explained*. 5th Edition, J. Dodsley, London.

Baker, J.E. (1981) Resolution and partial characterisation of the digestive proteinases from larvae of the black carpet beetle *Attagenus megatoma*. *Insect Biochemistry*, 11, 583–591.

Baker, R.A. (1975) The structure and function of the alimentary canal in *Histiogaster carpio* (Kramer, 1881) Acari – Sarcoptiformes. *Acarologia*, 17, 126–137.

Baldo, B.A. and Uhlenbruck, G. (1977) Selective isolation of allergens I. Reaction of house dust mite extracts with tridacnin and concalavin A and examination of the allergenicity of the isolated components. *Clinical Allergy*, 7, 429–443.

Baldo, B.A., Quinn, E.H. and Turner, K.J. (1975) *In vitro* diagnosis of allergy. Radioallergosorbent test (RAST) studies with some allergens commonly encountered in Australia. *Medical Journal of Australia*, ii, 859–863.

Balendiran, G.K., Schnütgen, F., Scapin, G., Börchers, T., Xong, N., Lim, K., Godbout, R., Spener, F. and Sachettini, J.C. (2000) Crystal structure and thermodynamic analysis of human brain fatty acid binding protein. *Journal of Biological Chemistry*, 275, 27045–27054.

Baqueiro, T., Carvalho, F.M., Rios, C.F, dos Santos, N.M., Medical Student Group and Alcântra-Neves, N.M. (2006) Dust mite species and allergen concentrations in beds of individuals belonging to different urban socioeconomic groups in Brazil. *Journal of Asthma*, 43, 101–105.

Barabanova, V.V. and Zheltikova, T.M. (1985) [Digestive enzymes of acarids *Dermatophagoides pteronyssinus* Trouessart, 1897 and *D. farinae* Hughes, 1961 (Acariformes: Pyroglyphidae)]. Doklady Akademii Nauki CCCP, 283, 225–227. [In Russian]

Baratawidjaja, I.R., Baratawidjaja, P.P., Darwis, A., Yi, F.C., Chew, F.T., Lee, B.W. and Baratawidjaja,

K.G. (1998) Mites in Jakarta homes. *Allergy*, 53, 1226–1227.

Barber, D., Juan, F., Parmiani, S. and Pecora, S. (1992) Study on major mite allergens in Italian Homes. *Aerobiologia*, 8, 52–56.

Barber, D., Pernas, M., Chamorro, M.J., Careira, J., Arteaga, C., Sánchez-Monge, R., Polo, F. and Salcedo, G. (1996) Specific depletion of the house dust mite allergen Der p 1 by cereal flour prolamins. *Journal of Allergy and Clinical Immunology*, 97, 801–805.

Barber, E.W. (1994) *Women's Work: The First 20,000 Years. Women, Cloth, and Society in Early Times.* W.W. Norton, New York.

Barker, P.S. (1967) The effects of high humidity and different temperatures on the biology of *Tyrophagus putrescentiae* (Schrank) (Acarina: Tyroglyphidae). *Canadian Journal of Zoology*, 45, 91–96.

Barker, P.S. (1968) Bionomics of *Glycyphagus domesticus* (de Geer) (Acarina: Glycyphagidae), a pest of stored grain. *Canadian Journal of Zoology*, 46, 89–92.

Barker, P.S. (1983) Bionomics of *Lepidoglyphus destructor* (Schrank) (Acarina: Glycyphagidae), a pest of stored cereals. *Canadian Journal of Zoology*, 61, 355–358.

Barker, P.S. (1991) Bionomics of *Cheyletus eruditus* (Schrank) (Acarina: Cheyletidae), a predator of *Lepidoglyphus destructor* (Schrank) (Acarina: Glycyphagidae), at three constant temperatures. *Canadian Journal of Zoology*, 69, 2321–2325.

Barnes, K.C., Fernández-Caldas, E., Trudeau, W.L., Milne, D.E. and Brenner, R.J. (1997) Spatial and temporal distribution of house dust mite (Astigmata: Pyroglyphidae) allergens Der p 1 and Der f 1 in Barbadian homes. *Journal of Medical Entomology*, 34, 212–218.

Barnes, P., Grunstein, M., Leff, A.R. and Woolcock, A.J. (eds) (1997) *Asthma.* 2 Vols, Lippincott-Raven, Philadelphia.

Basagaña, X., Torrent, M., Atkinson, W., Puig, C., Barnes, M., Vall, O., Jones, M., Sunyer, J. and Cullinan, P. (2002) Domestic aeroallergen levels in Barcelona and Menorca. *Pediatric Allergy and Immunology*, 13, 412–417.

Beck, H.-I. and Bjerring, P. (1987) House dust mites and human dander. *Allergy*, 42, 471–472.

Beck, H.-I. and Korsgaard, J. (1989) Atopic dermatitis and house dust mites. *British Journal of Dermatology*, 120, 245–251.

Beggs, P.J. and Bambrick, H.J. (2005) Is the global rise of asthma an early impact of anthropogenic climate change? *Environmental Health Perspectives*, 113, 915–919.

Bell, D.M., Brink, E.W., Nitzkin, J.L., Hall, C.B., Wulff, H., Berkowitz, I.D., Feorino, P.M., Holman, R.C., Huntley, C.L., Meade, R.H., Anderson, L.J., Cheeseman, S.H., Fiumara, N.J., Gilfillan, R.F., Keim, D.E. and Modlin, J.F. (1981) Kawasaki syndrome, description of two outbreaks in the United States. *New England Journal of Medicine*, 304, 1568–1575.

Bennett, B.J. and Thomas, W.R. (1996) Cloning and sequencing of the group 6 allergen of *Dermatophagoides pteronyssinus*. *Clinical and Experimental Allergy*, 26, 1150–1154.

Bennett, I. (ed.) (1988) *Rugs and Carpets of the World.* Quarto Ltd., London.

Bereen, J.M. (1984) The functional response of *Cheyletus eruditus* Schrank to changes in the density of its prey *Acarus siro* L. In: Griffiths, D.A. and Bowman, C.E. (eds), *Acarology VI.* Volume 2. Ellis Horwood, Chichester, pp. 980–986.

Bereen, J.M. and Metwally, A.-S. M. (1984) Reproductive rates in *Cheyletus eruditus* (Schrank). In: Griffiths, D.A. and Bowman, C.E. (eds), *Acarology VI.* Volume 1. Ellis Horwood, Chichester, pp. 512–518.

Berlese, A. (1897a) *Acari, Myriapoda et Scorpiones Hucusque in Italia Reperta.* I. Cryptostigmata. Patavii, p. 104.

Berenbaum, M.R. (1995) *Bugs in the System: Insects and their Impact on Human Affairs.* Addison-Wesley, Reading, MA.

Berlese, A. (1897b) Ricerche sugli organi e sulla funzione della digestione negli acari. *Rivista Patologica Vegetale*, 5, 129–195.

Bernd, L.A., Baggio, D., Becker, A.B., Ambrózio, L.C. (1994) Identificação e estudo da atividade sensibilizante de ácaros domésticos em Porto Alegre (RS). *Revista Brasileira de Alergia e Imunopatologia*, 17, 23–33.

van den Bemt, L., van Knapen, L., de Vries, M.P., Jansen, M., Cloosterman, S. and van Schayck, C.P. (2004) Clinical effectiveness of a mite allergen–impermeable bed-covering system in asthmatic mite-sensitive patients. *Journal of Allergy and Clinical Immunology*, 114, 858–862.

Bernecker, C. (1970) The antigenicity of house dust and mites. *Acta Allergologica*, 25, 392–403.

Bernhard, K., Karg, W. and Steinbrink, H. (1986) Hausstaubmilben in bettstaub und am körper. *Agnewandte Parasitologie*, 27, 49–52.

Bernini, F. (1991) Fossil Acarida. Contribution of palaeontological data to acarid evolutionary history. In: Simonetta, A. and Conway Morris, S. (eds) *The Early Evolution of Metazoa and the Significance of Problematic Taxa*. Cambridge University Press, Cambridge, pp. 253–261.

Berrens, L. (1972) Heterogeneity of a crystalline protein from psoriatic scales. *Archiv für Dermatologische Forschung*, 2422, 233–238.

Bessot, J.C. and Pauli, G. (1986) Les allergènes de la poussière de maison. *Bulletin Européen de Physiopathologie Respiratoire*, 22, 1–8.

Biddulph, P., Crowther, D., Leung, B., Wilkinson, T., Hart, B.J., Oreszczyn, T., Pretlove, S., Ridley, I. and Ucci, M. (2007) Predicting the population dynamics of the house dust mite *Dermatophagoides pteronyssinus* (Acari: Pyroglyphidae) in response to a constant hygrothermal environment using a model of the mite life cycle. *Experimental and Applied Acarology*, 41, 61–86.

Bigliocchi, F. and Maroli, M. (1995) Distribution and abundance of house-dust mites (Acarina: Pyroglyphidae) in Rome, Italy. *Aerobiologia*, 11, 35–40.

Bigliocci, F., Frusteri, L., Carrieri, M.P. and Maroli, M. (1996) Distribution and density of house dust mites *Dermatophagoides* spp. (Acarina: Pyroglyphidae) in the mattresses of two areas of Rome, Italy. *Parassitologia*, 38, 543–546.

Biliotti, G., Romagnini, S. and Ricci, M. (1975) Mites and house dust allergy. IV. Antigen and allergen(s) of *Dermatophagoides pteronyssinus* extract. *Clinical Allergy*, 5, 69–77.

Binns, E.S. (1982) Phoresy as migration – some functional aspects of phoresy in mites. *Biological Reviews*, 57, 571–620.

Binotti, R.S., Muniz, J.R.O., Paschoal, I.A., do Prado, A.P. and Oliveira, C.H. (2001) House dust mites in Brazil – an annotated bibliography. *Memorias do Instituto Oswaldo Cruz*, 96, 911–916.

Birch, L.C. (1948) The intrinsic rate of natural increase of an insect population. *Journal of Animal Ecology*, 17, 15–26.

Bischoff, E. (1988) Methodes actuelles de quantifiation des acariens dans l'habitat. *Revue Française d'Allergologie et d'Immunologie Clinique*, 28, 115–122.

Bischoff, E. and van Bronswijk, J.E.M.H. (1986) Beiträge zur ökologie der hausstaubmilben I: über die erreichbarkeit von hausstaubmilben durch absaugen. *Allergologie*, 9, 375–378.

Bischoff, E. and Fischer, A. (1990) New methods for the assessment of mite numbers and results obtained from several textile objects. *Aerobiologia*, 6, 23–27.

Bischoff, E., Schirmacher, W. and Schober, G. (1985) Farbnachweis für allergenaltigen Hausstaub. *Allergologie*, 8, 97–99.

Bischoff, E., Krause-Michel, B. and Nolte, D. (1986a) Zur Bekampfung der Hausstaubmilben in Haushaltern von Patienten mit Milbenasthma. *Allergologie*, 9, 448–457.

Bischoff, E., Fischer, A. and Wetter, G. (1986b) Untersuchungen zur Okologie der Hausstaubmilben. *Allergologie*, 9, 45–54.

Bischoff, E., Fischer, A. and Liebenberg, B. (1992a) Assessment of mite numbers: new methods and results. *Experimental and Applied Acarology*, 16, 1–14.

Bischoff, E., Fischer, A., Kniest, F. and Liebenberg, B. (1992b) The numbers of living mites are much higher than hitherto assumed. *Allergy and Clinical Immunology News*, 4, 71–75.

Bischoff, E., Fischer, A. and Liebenberg, B. (1992c) Domestic mites, do they ingest acaricidal products based on soldified benzyl benzoate? A photographical study. *Journal of Allergy and Clinical Immunology*, 89, 256. [abstract]

Bisgaard, H. and Mygind, N. (1987) Nasal and aural allergy. In: Lessoff, M.H., Lee, T.H and Kemeny, D.M. (eds), *Allergy, an International Textbook*. John Wiley and Sons, Chichester, pp. 531–552.

Björkstén, B., Holt, B.J., Baron-Hay, M.J., Munir, A.K.M. and Holt, P. (1996) Low-level exposure to house dust mites stimulates T-cell responses during early childhood independent of atopy. *Clinical and Experimental Allergy*, 26, 775–779.

Björnsson, E., Norbäck, D., Janson, C., Widström, J., Palmgren, U., Ström, G. and Boman, G. (1995) Asthmatic symptoms and indoor levels of microorganisms and house dust mites. *Clinical and Experimental Allergy*, 25, 423–431.

Blackley, C.H. (1873) *Experimental Researches on the Causes and Nature of* Catarrhus Aestivus *(Hay Fever or Hay Asthma)*. Balliere, London.

Blainey, A.D., Topping, M.D., Ollier, S. and Davies, R.J. (1989) Allergic respiratory disease in grain workers: the role of storage mites. *Journal of Allergy and Clinical Immunology*, 84, 296–303.

Blasco, C. and Portus, M. (1975) Acarofauna. *Revista de la Real Academia de Farmacia Barcelona*, 11, 37–59.

Blasco, C., Gallego, J. and Portus, M. (1975) Estudio de la acarofauna del polvo domestico de Barcelona y poblaciones circundantes. *Allergologia et Immunopathologia*, 3, 403–418.

de Blay, F., Fourgaut, G. and N'Gom, S. (2002) Concepts architecturaux et réduction de la charge allergénique. *Revue Française d'Allergologie et d'Immunologie Clinique*, 42, 256–262.

de Blay, F., Fourgaut, G., Hedelin, G., Vervloet, D., Michel, F.-B., Godard, P., Charpin, D. and Pauli, G. (2003) Medical Indoor Environment Counselor (MIEC): role in compliance with advice on mite allergen avoidance and on mite allergen exposure. *Allergy*, 58, 27–33.

Blythe, M.E. (1976) Some aspects of the ecological study of house dust mites. *British Journal of Diseases of the Chest*, 70, 3–31.

Blythe, M.E., Williams, J.D. and Morrison-Smith, J. (1974) Distribution of pyroglyphid mites in Birmingham with particular reference to *Euroglyphus maynei*. *Clinical Allergy*, 4, 25–33.

Boczek, J. (1964) Artificial medium for rearing some stored products mites. *Acarologia*, 6, 392–398.

Boczek, J. (1974) Reproduction biology of *Tyrophagus putrescentiae* (Schr.) (Acarina: Acaridae). In: *Proceedings 1st International Conference of Stored Products Entomology, Savannah, GA, USA*. Sept. 7–11, 1974, pp. 154–159.

Boczek, J., Ignatowicz, S. and Davis, R. (1984) Some effects of mineral salts in the diet of the mould mite, *Tyrophagus putrescentiae*. *Journal of the Georgia Entomological Society*, 19, 235–248.

de Boer, R. (1990a) The control of house dust mite allergens in rugs. *Journal of Allergy and Clinical Immunology*, 86, 808–814.

de Boer, R. (1990b) Effect of heat treatment on the house-dust mites *Dermatophagoides pteronyssinus* and *D. farinae* (Acari, Pyroglyphidae) in a mattress-like polyurethane foam block. *Experimental and Applied Acarology*, 9, 131–136.

de Boer, R. (2003) The effect of sub-floor heating on house-dust-mite populations on floors and furniture. *Experimental and Applied Acarology*, 29, 315–330.

de Boer, R. and van der Geest, L.P.S. (1990) House dust mite (Pyroglyphidae) populations in mattresses, and their control by electric blankets. *Experimental and Applied Acarology*, 9, 113–122.

de Boer, R. and Kuller, K. (1995a) House dust mites (*Dermatophagoides pteronyssinus*) in mattresses: vertical distribution. *Proceedings of the Section Experimental and Applied Entomology of the Netherlands Entomological Society*, 5, 129–130.

de Boer, R. and Kuller, K. (1995b) Water balance of *Dermatophagoides pteronyssinus* in circumstances with only a few hours of moist air (above CEH) every day. In: Tovey, E, Fifoot, A. and Sieber, L. (eds), *Mites, Asthma and Domestic Design II*. University of Sydney, Sydney, pp. 22–25.

de Boer, R. and Kuller, K. (1995c) Winter survival of house dust mites (*Dermatophagoides* spp.) on the ground floor of Dutch houses. *Experimental and Applied Entomology*, 6, 47–51.

de Boer, R. and Kuller, K. (1997) Mattresses as a winter refuge for house dust mite populations. *Allergy*, 52, 299–305.

de Boer, R., Kuller, K. and Kahl, O. (1998) Water balance of *Dermatophagoides pteronyssinus* (Acari: Pyroglyphidae) maintained by brief daily spells of elevated air humidity. *Journal of Medical Entomology*, 35, 905–910.

de Boer, R., van der Hoeven, W.A.D. and Kuller, K. (1996) The control of house dust mites in rugs through wet cleaning. *Journal of Allergy and Clinical Immunology*, 97, 1214–1217.

Bogdanoff, A. (1864) Deux acariens, trouves par M. Scheremetewsky sur l'homme. *Bulletin de la Societe imperiale des naturalistes de Moscou*, 37, 341–345.

Boldbaatar, D., Sikasunge, C.S., Battsetseg, B., Xaun, X. and Fujisaki, K. (2006) Molecular cloning and functional characterisation of an aspartic protease from the hard tick *Haemaphysalis longicornis*. *Insect Biochemistry and Molecular Biology*, 36, 25–36.

Boner, A.L., Richelli, C., Vallone, G., Verga, A., Parotelli, R., Andri, L. and Piacentini, G.L. (1989) Skin and serum reactivity to some storage mites in children sensitive to *Dermatophagoides pteronyssinus*. *Annals of Allergy*, 63, 82–84.

Boquete, M., Iraola, V., Fernández-Caldas, E., Arenas Villaroel, L., Carballada, F.J., González de la Cuesta, C., López-Rico, M.R., Nuñez Orjales, R., Parra, A., Soto-Mera, M.T., Varela, S. and Vidal, C.

(2006) House dust mites species and allergen levels in Galicia, Spain: a cross-sectional, multi-center, comparative study. *Journal of Investigational Allergology and Clinical Immunology*, 16, 169–176.

Bottini, N., Ronchetti, M.P., Gloria-Bottini, F. and Fontana, L (1999) Malaria as a possible evolutionary cause of allergy. *Allergy*, 54, 188–189.

Boudreaux, H.B. and Dosse, G. (1963) The usefulness of new taxonomic characters in females of the genus *Tetranychus* Dufour (Acari: Tetranychidae). *Acarologia*, 5, 13–33.

Bousquet, J., Hale, R., Guerin, B. and Michel, F.-B. (1980) Enzymatic activities of house dust extracts. *Annals of Allergy*, 45, 316–321.

Bousquet, J., van Cauwenberge, P., Khaltaev, N. and ARIA Workshop Group (2001) Allergic rhinitis and its impact on asathma (ARIA). *Journal of Allergy and Clinical Immunology*, 108 (Supplement 5), S147–S333.

Bowler, S.D., Mitchell, C.A. and Miles, J. (1985) House dust mite control and asthma, a placebo control trial of cleaning and air filtration. *Annals of Allergy*, 55, 498–500.

Bowman, C.E. (1981) Hide protease in stored product mites (Astigmata: Acaridae). *Comparative Biochemistry and Physiology*, 71B, 803–805.

Bowman, C.E. (1984) Comparative enzymology of economically important astigmatid mites. In: Griffiths, D.A. and Bowman, C.E. (eds), *Acarology VI, Volume 2*, Ellis Horwood, Chichester, pp. 993–1001.

Bowman, C.E. and Childs, M. (1982) Polysaccharidases in astigmatid mites (Arthropoda: Acari). *Comparative Biochemistry and Physiology*, 72B, 551–557.

Bowman, C.E. and Lessiter, M.J. (1985) Amylase and esterase polymorphisms in economically important stored products mites. *Comparative Biochemistry and Physiology*, 81B, 353–360.

Bradford Hill, A. (1965) The environment and disease: association or causation? *Proceedings of the Royal Society of Medicine*, 58, 295–300.

Braun-Fahrländer, C., Riedler, J., Herz, U., Eder, W., Waser, M., Grize, L., Maisch, S., Carr, D., Gerlach, F., Bufe, D., Launer, R.P., Schierl, R., Renz, H., Nowak, D. and von Mutius, E. (2002) Environmental exposure to endotoxin and its relation to asthma in school-age children. *New England Journal of Medicine*, 347, 869–877.

Bremner, P.R., de Klerk, N.H., Ryan, G.F., James, A.L., Musk, M., Murray, C., Le Söuef, P.N., Young, S., Spargo, R. and Musk, A.W. (1998) Respiratory symptoms and lung function in Aborigines from tropical western Australia. *American Journal of Respiratory and Critical Care Medicine*, 158, 1724–1729.

Brenner, R.J. (1991a) Seasonality of peridomestic cockroaches (Blattoidea: Blattidae): mobility, winter reduction, and effect of traps and baits. *Journal of Economic Entomology*, 84, 1735–1745.

Brenner, R.J. (1991b) Insect pests, construction practices and humidity. In: *Proceedings Bugs, Mold and Rot. A Workshop on Residential Moisture Problems, Health Effects, Building Damage and Moisture Control*. National Institute of Building Science, Washington, pp. 19–26.

Brenner, R.J. (1993) Preparing for the 21st Century: research methods in developing management strategies for arthropods and allergens in the structural environment. In: Wildey, K.B. and Robinson, W.H. (eds), *Proceedings of the 1st International Conference on Insect Pests in the Urban Environment, Exeter, UK*; Wheatons Ltd, pp. 57–69.

Bridgman, H., Warner, R. and Dodson, J. (1995) *Urban Biophysical Environments*. Oxford University Press, Oxford.

Brody, A.R. and Wharton, G.W. (1970) *Dermatophagoides farinae*: ultrastructure of lateral opisthosomal dermal glands. *Transactions of the American Microscopical Society*, 89, 499–513.

Brody, A.R., McGrath, J.C. and Wharton, G.W. (1972) *Dermatophagoides farinae*, the digestive system. *Journal of the New York Entomological Society*, 80, 152–177.

Brody, A.R., McGrath, J.C. and Wharton, G.W. (1976) *Dermatophagoides farinae*: the supracoxal glands. *Journal of the New York Entomological Society*, 84, 34–47.

van Bronswijk, J.E.M.H. (1972a) Hausstaub-Okosystem und Hausstaub Allergen(e). *Acta Allergologica*, 27, 219–228.

van Bronswijk, J.E.M.H. (1972b) Parasitic mites of Surinam. X. Mites and fungi associated with house-floor dust. *Entomologische Berichten*, 32, 162–164.

van Bronswijk, J.E.M.H. (1972c) Food preference of pyroglyphid house-dust mites (Acari). *Netherlands Journal of Zoology*, 22, 335–340.

van Bronswijk, J.E.M.H. (1973) *Dermatophagoides pteronyssinus* (Trouessart) in mattress and floor dust in a temperate climate (Acari: Pyroglyphidae). *Journal of Medical Entomology*, 10, 63–70.

van Bronswijk, J.E.M.H. (1974) Colonization and its prevention on house floors and in mattresses with *Dermatophagoides pteronyssinus* (Acari, Sarcopti-formes) in a center for asthmatic children. *Entomologia Experimentalis et Applicata*, 17, 199–203.

van Bronswijk, J.E.M.H. (1979) House-dust as an ecosystem. In: Rodriguez, J.G. (ed.), *Recent Advances in Acarology*. Volume 2. Academic Press, New York, pp. 167–171.

van Bronswijk, J.E.M.H. (1981) *House Dust Biology*. Published by the author, Zoelmond.

van Bronswijk, J.E.M.H. (1984) Neues zur Okologie der Wohnungsmilben. *Allergologie*, 7, 438–445.

van Bronswijk, J.E.M.H. (1986) Guanine as a hygienic index for allergologically relevant mite infestations in mattress dust. *Experimental and Applied Acarology*, 2, 231–238.

van Bronswijk, J.E.M.H. (1993) Prevention and extermination strategies for house dust mites and their allergens in home textiles. In: Wildey, K.B. and Robinson, W.H. (eds), *Proceedings of the First International Conference of Insect Pests in the Urban Environment, Exeter, UK*; Wheatons Ltd. pp. 261–266.

van Bronswijk, J.E.M.H. and Berrens, L. (1979) The biological nature of the association between house-dust mites and house-dust allergens. In: Piffl, E. (ed.), *Proceedings IV International Congress of Acarology, 1974*. Akademai Kiado, Budapest, pp. 337–340.

van Bronswijk, J.E.M.H. and Jorde, W. (1975) Mites and allergenic activity in house dust in Heligoland. *Acta Allergologica*, 30, 209–215.

van Bronswijk, J.E.M.H. and Koekkok, H.H.M. (1972) Effect of low temperature on the survival of dust mites of the family Pyroglyphidae (Acari: Sarcoptiformes). *Netherlands Journal of Zoology*, 22, 207–211.

van Bronswijk, J.E.M.H., de Cock, A.W.A.M. and Oshima, S. (1973a) The genus *Blomia* Oudemans (Acari, Glycyphagidae) I. Description of *Blomia tropicalis* sp. n. from house dust in tropical and sub-tropical regions. *Acarologia*, 15, 477–489.

van Bronswijk, J.E.M.H., de Cock, A.W.A.M. and Oshima, S. (1973b) The genus *Blomia* Oudemans (Acari, Glycyphagidae) II. Comparison of its species. *Acarologia*, 15, 490–505.

van Bronswijk, J.E.M.H. and Kok, N.J.J. (1975) Parasitic mites of Surinam. XXV. Infestation of mattresses with Pyroglyphidae (Acari, Astigmata). *Entomolgische Berichten*, 35, 12–14.

van Bronswijk, J.E.M.H. and Sinha, R.N. (1971) Pyroglyphid mites (Acari) and house dust allergy, a review. *Journal of Allergy*, 47, 31–52.

van Bronswijk, J.E.M.H. and Sinha, R.N. (1973) Role of fungi in the survival of *Dermatophagoides* (Acarina: Pyroglyphidae) in the house-dust environment. *Environmental Entomology*, 2, 142–145.

van Bronswijk, J.E.M.H., Schoonen, J.M.C.P., Berlie, M.A.F. and Lukoschus, F.S. (1971) On the abundance of *Dermatophagoides pteronyssinus* (Trouessart, 1897) (Pyroglyphidae, Acarina) in house dust. *Researches in Population Ecology*, 13, 67–79.

van Bronswijk, J.E.M.H., de Saint Georges-Gridelet, D. and Lustgraaf, B. van de (1978) An evaluation of biological methods in house-dust allergen research. *Allergie und Immunologie*, 24, 18–28.

van Bronswijk, J.E.M.H., Rijckaert, G. and van der Lustgraaf, B. (1986) Indoor fungi, distribution and allergenicity. *Acta Botanica Neerlandica*, 35, 329–345.

van Bronswijk, J.E.M.H., Bischoff, E., Schirmacher, W. and Kneist, F.M. (1989) Evaluating mite (Acari) allergenicity of house dust by guanine quantification. *Journal of Medical Entomology*, 26, 55–59.

Brown, H.M. and Filer, J.L. (1968) Role of mites in allergy to house dust. *British Medical Journal*, iii, 646–647.

Brown, H.M. and Merrett, T.G. (1991) Effectiveness of an acaricide in management of house dust mite allergy. *Annals of Allergy*, 67, 25–31.

Brown, S.K. (1994) Optimisation of a screening procedure for house dust mite numbers in carpets and preliminary application to buildings. *Experimental and Applied Acarology*, 18, 423–434.

Brown, S.K. (1996) A lack of influence of permenthrin treatments of wool carpet on habitation by house dust mites. *Experimental and Applied Acarology*, 20, 355–357.

Brunekreef, B., van Strien, R., Pronk, A., Oldenwening, M., de Jongste, J.C., Wijga, A., Kerkhof, M. and Aalberse, R.C. (2005) La mano de DIOS…was the PIAMA intervention study intervened upon? *Allergy*, 60, 1083–1086.

Bruttmann, G. (1975) L'allergie à la poussière de maison valeur antigénique des acariens. *Cahiers Medicine*, 1(6), 407–416.

Buchanan, D.J. and Jones, I.G. (1972) Allergy to house dust mites in the tropics. *British Medical Journal*, ii, 764.

Buchanan, D.J. and Jones, I.G. (1974) Mites and house dust mite allergy in bronchial asthma in northern Zambia. *Postgraduate Medical Journal*, 50, 680–682.

Buckley, M.G., Variend, S. and Walls, A.F. (2001) Elevated serum concentrations of beta-tryptase, but not alpha-tryptase, in Sudden Infant Death Syndrome (SIDS). An investigation of anaphylactic mechanisms. *Clinical and Experimental Allergy*, 31, 1696–1704.

Buonomo, E., Jones, R., Huntingford, C. and Hannaford, J. (2005) On the robustness of changes in extreme precipitation over Europe from two high resolution climate change simulations. *Quarterly Journal of the Royal Meteorological Society*, 133, 65–81.

Burney, P.J.G., Luczynska, C., Chinn, S., Jarvis, D. for the The European Community Respiratory Health Survey (1994) The European Community Respiratory Health Survey. *European Respiratory Journal*, 7, 954–960.

Burr, M.L., St. Leger, A.S. and Neale, E. (1976) Anti-mite measures in mite sensitive adult asthma: a controlled trial. *The Lancet*, i, 333–335.

Burr, M.L., Dean, B.V., Merrett, T.G., Neale, E., St Leger, A.S. and Verrier-Jones, E.R. (1980a) Effects of anti-mite measures on children with mite-sensitive asthma: a controlled trial. *Thorax*, 35, 506–512.

Burr, M.L., Neale, E., Dean, B.V. and Verrier-Jones, E.R. (1980b) Effect of a change to mite-free bedding on children with mite-sensitive asthma: a controlled trial. *Thorax*, 35, 513–514.

Burr, M.L., Dean, B.V., Butland, B.K. and Neale, E. (1988) Prevention of mite infestation of bedding by means of an impregnated sheet. *Allergy*, 43, 299–302.

Butler, S., Nuttall, R.H. and Brown, O. (undated, but ca. 1986).*The Social History of the Microscope, Published to Accompany a Special Exhibition at the Whipple Museum of the History of Science*. The Whipple Museum of the History of Science, Cambridge.

Cabrera, P., Julià-Serdà, G., Rodriguez de Castro, F., Caminero, J., Barber, D. and Carillo, T. (1995) Reduction of house dust mite allergens after dehumidifier use. *Journal of Allergy and Clinical Immunology*, 95, 635–636.

Caceres, I. and Fain, A. (1978) Notes sur la faune acarologique des poussières de maisons du Perou. *Bulletin et Annales de la Societe royale d'entomolgie de Belgique*, 114, 301–303.

Caceres, I. and Fain, A. (1979) Notes sur la faune acarologique des poussières de maisons du Perou. In: Piffl, E. (ed.), *Proceedings IV International Congress of Acarology, 1974*. Akademai Kiado, Budapest, pp. 259–261.

Cadman, A., Prescott, R. and Potter, P.C. (1998) Year-round housedust mite levels on the highveld. *South African Medical Journal*, 88, 1580–1582.

Cai, L. and Wen, T. (1989) [Faunal survey and seasonal prevalence of house dust mites in the urban area of Shanghai]. *Acta Ecologica Sinica*, 9, 225–229. [In Chinese, English summary]

Cain, G., Elderfield, A.J., Green, R., Smillie, F.I., Chapman, M.D., Custovic, A. and Woodcock, A. (1998) The effects of dry heat on mite, cat and dog allergens. *Allergy*, 53, 1213–1215.

Cain, G., Elderfield, A.J., Green, R., Smillie, F.I., Chapman, M.D., Custovic, A. and Woodcock, A. (1998) The effects of dry heat on mite, cat and dog allergens. *Allergy*, 53, 1213–1215.

Call, R.S., Smith, T.F., Morris, E., Chapman, M.D. and Platts-Mills, T.A.E. (1992) Risk factors for asthma in inner city children. *Journal of Pediatrics*, 121, 862–866.

Callaini, G. and Mazzini, M. (1984) Fine structure of the egg shell of *Tyrophagus putrescentiae* (Schrank) (Acarina: Acaridae). *Acarologia*, 25, 359–364.

Calvo, M., Fernández-Caldas, E., Arellano, P., Marin, F., Carnes, J. and Hormaechea, A. (2005) Mite allergen exposure, sensitisation and clinical symptoms in Valdivia, Chile. *Journal of Investigational Allergology and Clinical Immunology*, 15, 189–196.

Cameron, M.M. (1997) Can house dust mite-triggered atopic dermatitis be alleviated using acaricides? *British Journal of Dermatology*, 137, 1–8.

Cameron, M.M. and Hill, N. (2002) Permethrin-impregnated mattress liners: a novel and effective intervention against house dust mites

(Acari: Pyroglyphidae). *Journal of Medical Entomology*, 39, 755–762.

Canestrini, G. and Kramer, P. (1899) Demodicidae and Sarcoptidae. In: Lohmann, H. (ed.), *Das Tierreich*. Part 7, Friedlander, Berlin.

Canestrini, G. (1894) Prospetto Acarofauna Italiana – Parte VI. Famiglia dei Psoroptidi (Psoroptidae) – Appendice ai Fitopidi italiana. Varia: *Psorergates, Nemisarcoptes, Histiogaster, Psoroptes ovis, Otodectes furonis. Gli Epidermoptini.* Stabilimento Prosperini, Padua, pp. 723–833.

Caraballo, L., Avjioglu, A., Marrugo, J., Puerta, L. and Marsh, D. (1996) Cloning and expression of complementary DNA coding for an allergen with common antibody-binding specificities with three allergens of the house dust mite *Blomia tropicalis. Journal of Allergy and Clinical Immunology*, 98, 573–579.

Caraballo, L., Puerta, L., Jiminéz, S., Mártinez, B., Mercardo, D., Avjiouglu, A. and Marsh, D. (1997) Cloning and IgE binding of a recombinant allergen from the mite *Blomia tropicalis*, homologous with fatty acid-binding proteins. *International Archives of Allergy and Immunology*, 112, 341–347.

Cardona, G., Guisantes, J., Eraso, E., Serna, L.A., Martínez, J. (2006) Enzymatic analysis of *Blomia tropicalis* and *Blomia kulagini* (Acari: Echymyopodidae) allergenic extracts obtained from different phases of culture growth. *Experimental and Applied Acarology*, 39, 281–288.

Cardoso, R.R. de A., Barbosa, C.A.A. and Nascimento, J.J. (1979) Dust mites. A 1-year study. *Allergy*, 34, 257–260.

Carey, J.R. (1982) Demography of the twospotted spider mite, *Tetranychus urticae* Koch. *Oecologia*, 52, 389–395.

Carey, J.R. (1993) *Applied Demography for Biologists.* Oxford University Press, New York.

Carey, J.R. (2001) Insect biodemography. *Annual Review of Entomology*, 46, 79–110.

Carey, J.R. and Krainacker, D.A. (1988) Demographic analyses of mite populations, extensions of stable theory. *Experimental and Applied Acarology*, 4, 191–210.

Carswell, F., Robinson, D.W., Oliver, J., Clark, J., Robinson, P. and Wadsworth, J. (1982) House dust mites in Bristol. *Clinical Allergy*, 12, 533–545.

Carswell, F., Birmingham, K., Oliver, J., Crewes, A. and Weeks, J. (1996) The respiratory effects of reduction of mite allergen in the bedrooms of asthmatic children – a double-blind controlled trial. *Clinical and Experimental Allergy*, 26, 386–396.

Carter, R.F., Haynes, M.E. and Morton, J. (1976) Rickettsia-like bodies and splenitis in Kawasaki disease. *The Lancet*, ii, December 4, 1254–1255.

Casanueva, M.E. and Artigas, J.N. (1985) Distribucion geografica y estacional de los acaros del polvo de habitacion en Chile (Arthropoda: Acari). *Gayana (Zoologia)*, 49, 3–75.

Castagnoli, M., Liguori, M. and Nannelli, R. (1983) Gli Acari della polvere delle case in Italia. In: *Atti XIII Congresso Nazionale Italiano di Entomologia, Sestriere – Torino, 1983*. pp. 577–582.

Castagnoli, M., Liguori, M. and Nannelli, R. (1989) A methodology for rearing *Euroglyphus (E.) maynei* (Coor.) (Acarina: Pyroglyphidae) in laboratory. *Redia*, 72, 127–131.

Caswell, H. (2001) *Matrix Population Models: Construction, Analysis and Interpretation*. 2nd Edition. Sinauer Associates, Sunderland, MA.

Caughey, G.H. and Nadel, J.A. (1997) Proteases. In: Barnes, P., Grunstein, M., Leff, A.R. and Woolcock, A.J. (eds), *Asthma*. Vol. 1. Lippincott-Raven, Philadelphia, pp. 609–625.

Chan, S.-L., Ong, S.-T., Ong, S.-Y., Chew, F.-T. and Mok, Y.-K. (2006) Nuclear magnetic resonance structure-based epitope mapping and modulation of dust mite group 13 allergen as a hypoallergen. *Journal of Immunology*, 176, 4852–4860.

Chang, J.H., Becker, A., Ferguson, A., Manfreda, J., Simons, E., Chan, H., Noertjojo, K. and Chang-Yeung, M. (1996) Effect of applications of benzyl benzoate on house dust mite allergen levels. *Annals of Allergy, Asthma and Immunology*, 77, 187–190.

Chang, Y.-C. and Hsieh, K.-H. (1989) The study of house dust mites in Taiwan. *Annals of Allergy*, 62, 101–106.

Chang-Yeung, M., Becker, A., Lam, J., Dimich-Ward, H., Ferguson, A., Warren, P., Simons, E., Broder, I. and Manfreda, J. (1995a) House dust mite allergen levels in two cities in Canada, effects of season, humidity, city and home characteristics. *Clinical and Experimental Allergy*, 25, 240–246.

Chang-Yeung, M., Manfreda, J., Dimich-Ward, H., Lam, J., Ferguson, A., Warren, P., Simons, E., Broder,

I., Chapman, M., Platts-Mills, T.A.E. and Becker, A. (1995b) Mite and cat allergen levels in homes and severity of asthma. *American Journal of Respiratory and Critical Care Medicine*, 152, 1805–1811.

Chapman, M.D. and Platts-Mills, T.A.E. (1978) Measurement of IgE, IgA and IgG antibodies to *Dermatophagoides pteronyssinus* by antigen-binding assay using a partially purified fraction of mite extract (F_4P_1). *Clinical and Experimental Immunology*, 34, 126–136.

Chapman, M.D. and Platts-Mills, T.A.E. (1980) Purification and characterization of the major allergen from *Dermatophagoides pteronyssinus* – Antigen P_1. *Journal of Immunology*, 125, 587–592.

Chapman, M.D., Rowntree, S., Mitchell, E.B., Di Prisco de Fuenmajor, M.C. and Platts-Mills, T.A.E. (1983) Quantitative assessments of IgG and IgE antibodies to inhalent allergens in patients with atopic dermatitis. *Journal of Allergy and Clinical Immunology*, 72, 27–33.

Charlet, L.D., Mulla, M.S. and Sánchez-Medina, M. (1977a) Domestic Acarina of Colombia: occurrence and distribution of Acari in housedust. *Acarologia*, 19, 302–317.

Charlet, L.D., Mulla, M.S. and Sánchez-Medina, M. (1977b) Domestic Acarina of Colombia: Acaridae, Glycyphagidae and Crytostigmata recovered from house dust. *International Journal of Acarology*, 3, 55–64.

Charlet, L.D., Mulla, M.S. and Sánchez-Medina, M. (1977c) Domestic Acari of Colombia: abundance of the European house dust mite, *Dermatophagoides pteronyssinus* (Acari, Pyroglyphidae) in homes in Bogota. *Journal of Medical Entomology*, 13, 709–712.

Charlet, L.D., Mulla, M.S. and Sánchez-Medina, M. (1978) Domestic Acari of Colombia: population trends of house dust mites (Acari, Pyroglyphidae) in homes in Bogota, Colombia. *International Journal of Acarology*, 4, 23–31.

Charlet, L.D., Mulla, M.S., Sánchez-Medina, M. and Reyes, M.A. (1979) Species composition and population trends of mites in various climatic zones in Colombia. *Journal of Asthma Research*, 16, 131–148.

Charpin, D., Kleisbauer, J.P., Lanteaume, A., Razzouk, H., Vervloet, D., Toumi, M., Faraj, F. and Charpin, J. (1988) Asthma and allergy to house-dust mites in populations living at high altitude. *Chest*, 93, 758–761.

Charpin, D., Birnbaum, J., Haddi, E., N'Guyen, A., Fondarai, J. and Vervloet, D. (1990) Evaluation de l'efficacite d'un acaricide: Acardust, dans le traitement de l'allergie aux acariens. *Revue Française d'Allergologie et d'Immunologie Clinique*, 30, 149–155.

Charpin, D., Birnbaum, J., Haddi, E., Genard, G., Lanteaume, A., Toumi, M., Faraj, F., van der Brempt, X. and Vervloet, D. (1991) Altitude and allergy to house dust mites. *American Review of Respiratory Disease*, 143, 983–986.

Charpin, J., Penaud, A., Nourrit, J., Autran, P. and Razzouk, H. (1971) Allergie aux poussières domestique et *Dermatophagoides*. *Revue Française d'Allergologie*, 11, 315–328.

Charpin, J., Penaud, A., Charpin, D., Faraj, F., Thibaudon, M., Vervloet, D. and Razzouk, H. (1986) *Euroglyphus maynei*, étude comparative des reactions cutanees a Marseilles et Briançon. *Revue Française d'Allergologie et d'Immunologie Clinique*, 26, 117–119.

Chen, C.-C. and Hsieh, K.-H. (1996) Effects of Microstop-treated anti-mite bedding on children with mite sensitive asthma. *Acta Paediatrica Sinica Taiwanica*, 37, 420–427.

Chen, C.M., Mielk, A., Fahlbusch, B., Bischof, W., Herbath, O., Borte, M., Wichmann, H.E. and Heinrich, J. (2007) Social factors, allergen, endotoxin and dust mass in mattresses. *Indoor Air*, 17, 384–393.

Chen, H.-L., Su, H.-J.J. and Lin, L.-L. (2002) Distribution variations of multi allergens at asthmatic children's homes. *Science of the Total Environment*, 289, 249–254.

Chen, J.-S. and Raikhel, A.S. (1996) Subunit cleavage of mosquito pro-vitellogenin by a subtilisin-like convertase. *Proceedings of the National Academy of Sciences*, 93, 6186–6190.

Chen, J.-S., Sappington, T.W. and Raikhel, A.S. (1997) Extensive sequence conservation among insect, nematode and vertebrate vitellogenins reveals ancient common ancestry. *Journal of Molecular Evolution*, 44, 440–451.

Chen, X. and Fu, C. (1992) Mites causing pulmonary, intestinal and urinary acariasis. In: Chen, X. and Ma, E. (eds), *Researches of Acarology in China*. Chongqing Publishing House, Chongqing, pp. 109–113.

Cheong, N., Yang, L., Lee, B.W. and Chua, K.Y. (2003) Cloning of a group 3 allergen from *Blomia tropicalis* mites. *Allergy*, 58, 352–356.

Chepstow-Lusty, A.J., Frogley, M.R., Bauer, B.S., Leng, M.J., Cundy, A.B., Boessenkool, K.P. and Gioda, A. (2007) Evaluating socio-economic change in the Andes using oribatid mite abundances as indicators of domestic animal densities. *Journal of Archaeological Science*, 34, 1178–1186.

Chew, F.T., Goh, D.Y.T. and Lee, B.W. (1996) Effects of an acaricide on mite allergen levels in the homes of asthmatic children. *Acta Paediatrica Japonica*, 38, 483–488.

Chew, F.T., Zhang, L., Ho, T.M. and Lee, B.W. (1999a) House dust mite fauna of tropical Singapore. *Clinical and Experimental Allergy*, 29, 201–206.

Chew, F.T., Lim, S.H., Goh, D.Y.T. and Lee, B.W. (1999b) Sensitisation to local dust mite fauna in Singapore. *Allergy*, 54, 1150–1159.

Chew, G.L., Higgins, K.M., Gold, D.R., Muilenberg, M.L. and Burge, H.A. (1999) Monthly measurements of indoor allergens and the influence of housing type in a northeastern US city. *Allergy*, 54, 1058–1066.

Childs, M. and Bowman, C.E. (1981) Lysozyme activity in six species of economically important astigmatid mites. *Comparative Biochemistry and Physiology*, 70B, 615–617.

Chino, H. (1985) Lipid transport: biochemistry of hemolymph lipophorin. In: Kerkut, G.A. and Gilbert, L.I. (eds), *Comprehensive Insect Physiology, Biochemistry and Pharmacology, Volume 10, Biochemistry*. Pergamon Press, Oxford, pp. 115–135.

Chivato, T., Barber, D., Laguna, R., Zubeldia, J., Martínez, D., Barrio, M. de and Rubio, M. (1993) Eficacia parasitológica y clínica del acaricida esbiol en aerosol. Estudia preliminar. *Revista España Immunologica Clínica*, 8, 65–71.

Chmielewski, W. (1987) Parametry rozwoju populacji *Glycyphagus destructor* (Schr.) (Acarida, Glycyphagidae) – gatunku roztoczy spotykanego w ulach pszczelich [Parameters of population development of *Glycyphagus destructor* (Schr.) (Acarida, Glycyphagidae) – mite species found in bee hives]. *Pszczelnicze Zeszyty Naukowe*, 31, 213–220.

Chmielewski, W. (1988) Potencjal biologiczny roztoczka domowego – *Glycyphagus domesticus* (De Geer) (Acarida: Glycyphagidae) – szkodnika pylku i pierzgi [Biological potential of house mite – *Glycyphagus domesticus* (De Geer) (Acarida: Glycyphagidae) – a pest of pollen and bee bread]. *Pszczelnicze Zeszyty Naukowe*, 32, 169–180.

Chmielewski, W. (1990) Porównanie wskazników wzrostu populacji *Glycyphagus destructor* (Schr.) i *Glycyphagus domesticus* (De Geer) (Glycyphagidae, Acarina) w podobnych warunkach otoczenia [Comparison of the indices of increase of populations of *Glycyphagus destructor* (Schr.) and *Glycyphagus domesticus* (De Geer) (Glycyphagidae, Acarina) under similar environmental conditions]. *Zeszyty Problemowe Postepów Naukowe Rolniczych*, 373, 93–100.

Chmielewski, W. (1995) Life history parameters of *Acarus siro* L. (Acari: Acaridae) bed buckwheat. *Fagopyrum*, 17, 73–75.

Cho, S.-H., Reponen, T., Bernstein, D.I., Olds, R., Levin, L., Liu, X., Wilson, K. and LeMasters, G. (2006) The effect of home characteristics on dust antigen concentrations and loads in homes. *Science of the Total Environment*, 371, 31–43.

Christeller, J.T. (1996) Degradation of wool by *Hoffmanophila pseudospretella* (Lepidoptera: Oeophoridae) larval midgut extracts under conditions simulating the midgut environment. *Archives of Insect Biochemistry and Physiology*, 32, 99–119.

Christiansen, S.C., Martin, S.B., Schleichter, N.C., Koziol, J.A., Hamilton, R.G. and Zuraw, B.L. (1996) Exposure and sensitisation to environmental allergen of predominantly Hispanic children with asthma in San Diego's inner city. *Journal of Allergy and Clinical Immunology*, 98, 288–294.

Chua, K.Y., Stewart, G.A., Thomas, W.R., Simpson, R.J., Dilworth, R.J., Plozza, T.M. and Turner, K.J. (1988) Sequence analysis of cDNA coding for a major house dust mite allergen, Der p I. Homology with cysteine proteases. *Journal of Experimental Medicine*, 167, 175–182.

Chua, K.Y., Doyle, C.R., Simpson, R.J., Turner, K.J., Stewart, G.A. and Thomas, W.R. (1990a) Isolation of cDNA coding for the major mite allergen Der p II by IgE plaque immunoassay. *International Archives of Allergy and Applied Immunology*, 91, 118–123.

Chua, K.Y., Dilworth, R.J. and Thomas, W.R. (1990b) Expression of *Dermatophagoides pteronyssinus* allergen Der p II in *Escherichia coli* and binding studies with human IgE. *International Archives of Allergy and Applied Immunology*, 91, 124–129.

Chua, K.Y., Kehal, P.K. and Thomas, W.R. (1993) Sequence polymorphisms of cDNA clones encoding the mite allergen Der p I. *International Archives of Allergy and Applied Immunology*, 101, 364–368.

Chua, K.Y., Huang, C.H., Shen, H.D. and Thomas, W.R. (1996) Analysis of sequence polymorphisms of a major mite allergen, Der p 2. *Clinical and Experimental Allergy*, 26, 829–837.

Clarke, C.C. (ed.) (1828) *Readings in Natural Philosophy; or, a Popular Display of the Wonders of Nature; Exclusively Selected from the Transactions of the Royal Society of London, From its Foundation to the present Time.* Horatio Phillips, London.

Cloosterman, S.G.M., Schermer, T.R.J., Bijl-Hofland, I.D., van der Heide, S., Brunekreef, B., van den Elshout, F.J.J., van Herwaarden, C.L.A. and van Schayck, C.P. (1999) Effects of house dust mite avoidance measures on Der p 1 concentrations and clinical condition of mild adult house dust mite-allergic asthmatic patients, using no inhaled steroids. *Clinical and Experimental Allergy*, 29, 1336–1346.

Coats, J.R. (ed.) (1982) *Insecticide Mode of Action.* Academic Press, New York.

Coca, A.F. and Cooke, R.A. (1922) On the classification of the phenomena of hypersensitivity. *Journal of Immunology*, 8, 163–182.

Cochran, D.G. (1995) Toxic effects of boric acid on the German cockroach. *Experientia*, 51, 561–563.

Codina, R., Lockey, R.F., Diwadkar, R., Mobly, L.L. and Godfrey, S. (2003) Disodium octaborate tetrahydrate (DOT) application and vacuum cleaning, a combined strategy to control house dust mites. *Allergy*, 58, 318–324.

Cole, L., Blum, M. and Roncadori, R. (1975) Antifungal properties of the insect alarm pheromone, citral, 2-heptanone and 4-methyl-3-heptanone. *Mycologia*, 67, 701–708.

Collier, C.G. (2006) The impact of urban areas on weather. *Quarterly Journal of the Royal Meterorological Society*, 132, 1–25.

Colloff, M.J. (1986) Use of liquid nitrogen in the control of house dust mite populations. *Clinical Allergy*, 16, 411–417.

Colloff, M.J. (1987a) Differences in development time, mortality and water loss between eggs from laboratory and wild populations of *Dermatophagoides pteronyssinus* (Trouessart, 1897) (Acari: Pyroglyphidae). *Experimental and Applied Acarology*, 3(3), 191–200.

Colloff, M.J. (1987b) Effects of temperature and relative humidity on development times and mortality of eggs from laboratory and wild populations of the European House Dust Mite *Dermatophagoides pteronyssinus* (Trouessart, 1897) (Acari: Pyroglyphidae). *Experimental and Applied Acarology*, 3(4), 279–289.

Colloff, M.J. (1987c) Mites from house dust in Glasgow. *Medical and Veterinary Entomology*, 1, 163–168.

Colloff, M.J. (1987d) Mite fauna from passenger trains in Glasgow. *Epidemiology and Infection*, 98, 127–130.

Colloff, M.J. (1988a) House dust mite ecology in my bed. In: De Weck, A.L. and Todt, A.L. (eds), *Mite Allergy, a Worldwide Problem.* UCB Institute of Allergy, Brussels, pp. 51–54.

Colloff, M.J. (1988b) Human semen as a dietary supplement for house dust mites. In: ChannaBasavanna, G.P. and Viraktamath, C. (eds), *Progress in Acarology*. Vol. 1. Oxford and IBH Publishing Co., New Delhi, pp. 141–146.

Colloff, M.J. (1989a) A new and rapid method for making permanent preparations of large numbers of house dust mites for light microscopy. *Experimental and Applied Acarology*, 7, 323–326.

Colloff, M.J. (1989b) House dust mites – Part I. A major worldwide problem. *Pesticide Outlook*, 1(1), 17–18.

Colloff, M.J. (1990) House dust mites – Part II. Chemical control. *Pesticide Outlook*, 1(2), 3–8.

Colloff, M.J. (1991a) A comparison between house dust mite populations in beds of patients with atopic dermatitis and healthy non-atopics. In: Dusbabek, F. and Bukva, V. (eds), *Modern Acarology*. Vol. 1. Academia, Prague and SPB Academic Publishing bv, The Hague, pp. 621–626.

Colloff, M.J. (1991b) Population studies on the house dust mite *Euroglyphus maynei* (Cooreman, 1950) (Pyroglyphidae). In: Schuster, R.H. and Murphy, P.W. (eds), *The Acari, Reproduction, Development and Life-History Strategies.* Chapman and Hall, London, pp. 497–506.

Colloff, M.J. (1991c) A review of the biology and allergenicity of the house dust mite *Euroglyphus maynei*. *Experimental and Applied Acarology*, 11, 177–198.

Colloff, M.J. (1991d) Practical and theoretical aspects of the ecology of house dust mites in relation to the study of mite-mediated allergy. *Review of Medical and Veterinary Entomology*, 79, 611–630.

Colloff, M.J. (1992a) Age structure and dynamics of house dust mite populations. *Experimental and Applied Acarology*, 16, 49–74.

Colloff, M.J. (1992b) House dust mite control with acarosan – an extreme test? *Clinical and Experimental Allergy*, 22, 657–658.

Colloff, M.J. (1992c) Exposure to house dust mites in homes of people with atopic dermatitis. *British Journal of Dermatology*, 127, 322–327.

Colloff, M.J. (1994a) Differences between the allergen repertoires of house dust mites and stored products mites. *Journal of Clinical Immunoassay*, 16, 213–221.

Colloff, M.J. (1994b) Dust mite control and mechanical ventilation: when the climate is right. *Clinical and Experimental Allergy*, 24, 94–96.

Colloff, M.J. (1995) Integrated strategies for dust mite control: a search for synergism. In: Tovey, E.R., Fifoot, A. and Sieber, L. (eds), *Mites, Asthma and Domestic Design II*. University of Sydney, Sydney, pp. 37–44.

Colloff, M.J. (1998) Distribution and abundance of dust mites within homes. *Allergy*, 53 (Supplement 48), 24–27.

Colloff, M.J. and Niedbala, W. (1996) Arboreal and terrestrial habitats of Phthiracaroid mites (Oribatida) in Tasmanian rainforests. In: Mitchell, R., Horn, D.J., Needham, G.R. and Welbourn, C. (eds), *Acarology IX*. Vol. 1. Ohio Biological Survey, Columbus, pp. 607–612.

Colloff, M.J. and Spieksma, F.Th.M. (1992) Pictorial keys for the identification of domestic mites. *Clinical and Experimental Allergy*, 22, 823–830.

Colloff, M.J. and Stewart, G.A. (1997) House dust mites. In: Barnes, P., Grunstein, M., Leff A.R. and Woolcock A.J. (eds), *Asthma* Vol. 2. Lippincott-Raven, Philadelphia, pp. 1089–1103.

Colloff, M.J., Lever, R.S. and McSharry, C. (1989) A controlled trial of house dust mite eradication using natamycin in homes of patients with atopic dermatitis: effect on clinical status and mite populations. *British Journal of Dermatology*, 121, 199–208.

Colloff, M.J., Stewart, G.A. and Thompson, P. (1991) House dust acarofauna and *Der p* I equivalent in Australia: the relative importance of *Dermatophagoides pteronyssinus* and *Euroglyphus maynei*. *Clinical and Experimental Allergy*, 21, 225–230.

Colloff, M.J., Howe, C.W., McSharry, C., Smith, H.V. (1992a) Characterization of IgE antibody binding profiles of sera from patients with atopic dermatitis to allergens of the domestic mites *Dermatophagoides pteronyssinus* and *Euroglyphus maynei* using enhanced chemiluminescent immunoblotting. *International Archives of Allergy and Immunology*, 97, 44–49.

Colloff, M.J., Ayres, J., Carswell, F., Howarth P., Merrett, T.G., Mitchell, E.B., Walshaw, M.J., Warner, J.A., Warner, J.O. and Woodcock, A.A. (1992b) The control of allergens of dust mites and domestic pets: a position paper. *Clinical and Experimental Allergy*, 22 (Supplement 2), 1–28.

Colloff, M.J.,Taylor, C. and Merrett, T.G. (1995) The use of domestic steam cleaning for the control of house dust mites. *Clinical and Experimental Allergy*, 25, 1061–1066.

Colloff, M.J., Merrett, T.G., Merrett, J., McSharry, C. and Boyd, G. (1997) Feather mites are potentially an important source of allergens for pigeon and budgerigar keepers. *Clinical and Experimental Allergy*, 27, 60–67.

Comamala de Florensa, C. (1970) Acaros que conviven con el hombre y su influencia en asma y alergias respiratorias. *Anales de Medicina, Sección Especialidades. Academia Ciencias y Medicina de Cataluña y Baleares*, 56, 361–370.

Combet, C., Blanchet, C., Geourjon, C. and Deléage, G. (2000) NPS@: Network Protein Sequence Analysis. *Trends in Biochemical Sciences*, 25, 147–150.

Cooper, P.J., Ayre, G., Martin, C., Rizzo, J.A., Ponte, E.V. and Cruz, A.A. (2008) Geohelminth infections: a review of the role of IgE and assessment of potential risks of anti-IgE treatment. *Allergy*, 63, 409–417.

Cooreman, J. (1950) Sur un acarien nouveau, préjudiciable aux matières alimentaires entreposées, *Mealia maynei* n. sp. *Bulletin et Annales de la Societe royale d'entomolgie de Belgique*, 86, 164–168.

Corente, C. and Knülle, W. (2003) Trophic determinants of hypopus induction in the stored-product mite *Lepidoglyphus destructor*. *Experimental and Applied Acarology*, 29, 89–107.

Cornere, B.M. (1971) The incidence of house dust mites in Auckland. *New Zealand Journal of Medical Laboratory Technology*, 25, 7–9.

Cornere, B.M. (1972) House dust mites, a national survey. *New Zealand Medical Journal*, 76, 270–271.

Corpuz-Raros, L.A., Sabio, G.C. and Velasco-Soriano, M. (1988) Mites associated with stored products, poultry houses and house dust in the Philippines. *Philippine Entomologist*, 7(3), 311–321.

Couper, D., Ponsonby, A.-L. and Dwyer, T. (1998) Determinants of dust mite allergen concentrations in infant bedrooms in Tasmania. *Clinical and Experimental Allergy*, 28, 715–723.

Coutts, A.M., Beringer, J. and Tapper, N.J. (2007) Impact of increasing urban density on local climate: spatial and temporal variations in the surface energy balance in Melbourne, Australia. *Journal of Applied Meteorology and Climatology*, 46, 477–493.

Cox, L.S., Linnemann, D.L., Nolte, H., Weldon, D., Finegold, I. and Nelson, H.S. (2006) Sublingual immunotherapy: a comprehensive review. *Journal of Allergy and Clinical Immunology*, 117, 1021–1035.

Craig, M.H., Sharp, B.L., Mabaso, M.L.H. and Kleinschmidt, I. (2007) Developing a spatial-statistical model and map of historical malaria prevalence in Botswana using a staged variable selection procedure. *International Journal of Health Geographics*, 6, 44. [http://www.ij–healthgeographics.com.content/6/1/44]

Crane, J., Grimmett, D., Wickens, K. and Kennedy, J. (1995) House dust mite allergen in Wellington homes – a preliminary report. In: Tovey, E.R., Fifoot, A. and Sieber, L. (eds), *Mites, Asthma and Domestic Design II*. University of Sydney, Sydney, pp. 107–110.

Crisafulli, D., Almqvist, C., Marks, G. and Tovey, E. (2007) Seasonal trends in house dust mite allergen in children's beds over a 7-year period. *Allergy*, 62, 1394–1400.

Croce, M., Costa-Manso, E., Baggio, D. and Croce, J. (2000) House dust mites in the city of Lima, Peru. *Investigative Allergology and Clinical Immunology*, 10, 286–288.

Crowson, R.A. (1970) *Classification and Biology*. Heinemann Educational Books Ltd, London.

Crowther, D., Wilkinson, T., Biddulph, P., Oreszczyn, T., Pretlove, S. and Ridley, I. (2006) A simple model for predicting the effect of hygrothermal conditions on populations of house dust mite *Dermatophagoides pteronyssinus* (Acari: Pyroglyphidae). *Experimental and Applied Acarology*, 39, 127–148.

de la Cruz, J. (1988) Acaros nidàícolas de Cuba. II. Familia Pyroglyphidae. *Poeyana*, 361, 1–24.

de la Cruz, J., Cuervo, N. and Dusbabek, F. (1984) Nueva especie de ácaro (Acarina: Pyroglyphidae) de los nidos del Vencejo de Palma, *Tachornis phoenicopia iradii* (Aves: Apodidae) de Cuba. *Poeyana*, 267, 1–12.

Cuervo, N., Dusbabek, F., de la Cruz, J. and Abreu, R. (1983) Los acaros (Acarina, Pyroglyphidae, Cheyletidae, Saproglyphidae y Glycyphagidae) de los polvos domesticos en Cuba. *Revista Cubano Medicina Tropical*, 35, 83–103.

Cunliffe, F. (1958) *Pyroglyphus morlani*, a new genus and species of mite forming a new family, Pyroglyphidae in the Acaridiae. *Proceedings of the Entomological Society of Washington*, 60, 85–86.

Cunningham, M.J. (1996) Controlling dust mites psychrometrically – a review for building scientists and engineers. *Indoor Air*, 6, 249–258.

Cunningham, M.J. (1998) Direct measurement of temperature and humidity in dust mite microhabitats. *Clinical and Experimental Allergy*, 28, 1104–1112.

Cunningham, M.J. (1999) Development and performance of a small relative humidity sensor for indoor microclimate measurements. *Building and Environment*, 34, 349–352.

Cunningham, M.J. (2000) A proposed experimental programme towards control of dust mites by microclimate modification. In: Siebers, R., Cunningham, M., Fitzharris, P. and Crane, J. (eds), *Mites, Asthma and Domestic Design III*. Wellington School of Medicine, New Zealand, pp. 69–78.

Cunningham, M.J., Roos, C., Go, L. and Spolek, G. (2004) Predicting psychrometric conditions in biocontaminant microenvironments with a microclimating heat and moisture transfer model – description and field comparison. *Indoor Air*, 14, 235–242.

Cunnington, A.M. (1965) Physical limits for the complete development of the grain mite, *Acarus siro* (Acarina, Acaridae) in relation to its world distribution. *Journal of Applied Ecology*, 2, 295–306.

Cunnington, A.M. (1969) Physical limits for the complete development of the copra mite, *Tyrophagus putrescentiae* (Shrank) (Acarina,

Acaridae) in relation to its world distribution. In: Evans, G.O. (ed.), *Proceedings of the Second International Congress of Acarology*, Akademai Kiado, Budapest, pp. 241–248.

Cunnington, A.M. (1971) House dust mites and respiratory allergy: a note on the identity of *Dermatophagoides farinae*. *Clinical Allergy*, 1, 447–449.

Cunnington, A.M. (1984) Resistance of the grain mite *Acarus siro* L. (Acarina: Acaridae) to unfavourable physical conditions beyond the limits of its development. *Agriculture, Ecosystems and Environment*, 11, 319–339.

Cunnington, A.M. (1985) Factors affecting oviposition and fecundity in the grain mite *Acarus siro* L. (Acarina: Acaridae), especially temperature and relative humidity. *Experimental and Applied Acarology*, 1, 327–344.

Cunnington, A.M. and Gregory, P.H. (1968) Mites in bedroom air. *Nature*, 217, 1271–1272.

Cunnington, A.M., Lind, P. and Spieksma, F.Th.M. (1987) Taxonomic and immunochemical identification of two house dust mites *Dermatophagoides farinae* and *Dermatophagoides microceras*. *Journal of Allergy and Clinical Immunology*, 79, 410–411.

Custovic, A., Taggart, S.C.O. and Woodcock, A. (1994) House dust mite and cat antigen in different indoor environments. *Clinical and Experimental Allergy*, 24, 1164–1168.

Custovic, A., Taggart, S.C.O. Niven, R.M.L. and Woodcock, A. (1995) Evaluating exposure to mite allergens. *Journal of Allergy and Clinical Immunology*, 96, 134–135.

Custovic, A., Taggart, S.C.O., Kennaugh, J.H. and Woodcock, A. (1995b) Portable dehumidifiers in the control of house dust mites and mite allergens. *Clinical and Experimental Allergy*, 25, 312–316.

Custovic, A., Green, R., Smith, A., Chapman, M.D. and Woodcock, A. (1996) New mattresses: how fast do they become a significant source of exposure to house dust mite allergens? *Clinical and Experimental Allergy*, 26, 1243–1245.

Custovic, A., Woodcock, H., Craven, M., Hassall, R., Hadley, E., Simpson, A. and Woodcock, A. (1999) Dust mite allergens are carried on not only large particles. *Pediatric Allergy and Immunology*, 10, 258–260.

Custovic, A., Simpson, B.M., Murray, C.S., Lowe, L. and Woodcock, A. (2002) The National Asthma Campaign Manchester Asthma and Allergy Study. *Pediatric Allergy and Immunology*, 13, 32–37.

Cutcher, J. (1973) The critical equilibrium activity of nonfeeding *Tyrophagus putrescentiae* (Acarina: Acaridae). *Annals of the Entomological Society of America*, 66, 609–611.

Cuthbert, O.D. (1990) Storage mite allergy. *Clinical Reviews in Allergy*, 8, 69–86.

Cuthbert, O.D., Jeffrey, I.G., McNeill, H.B., Wood, J. and Topping, M.D. (1984) Barn allergy among Scottish farmers. *Clinical Allergy*, 14, 197–206.

Czajkowska, B. and Kropcynska, D. (1991) The influence of different host plants on the reproductive potential of *Tyrophagus putrescentiae* (Schrank) and *Tyrophagus nieswanderi* Johnston and Bruce (Acaridae). In: Schuster, R.H. and Murphy, P.W. (eds), *The Acari, Reproduction, Development and Life–History Strategies*, Chapman and Hall, London, pp. 313–317.

Czernecki, N. and Kraus, H. (1978) Milbendermatitis durch *Tyrophagus dimidiatus*. *Zeitschrift für Hautkrankheiten*, 53, 414–416.

Dandeu, J.-P., Le Mao, J., Lux, M., Rabillon, J. and David, B. (1982) Antigens and allergens in *Dermatophagoides farinae mite* II. Purification of Ag 11, a major allergen in *Dermatophagoides farinae*. *Immunology*, 46, 679–687.

Danielsen, C., Hansen, L.S., Nachman, G. and Herling, C. (2004) The influence of temperature and relative humidity on the development of *Lepidoglyphus destructor* (Acari: Glycyphagidae) and its production of allergens: a laboratory experiment. *Experimental and Applied Acarology*, 32, 151–170.

Dar, N. and Gupta, V. (1979) Studies on the house dust mites of India and their role in causation of bronchial asthma and allergic rhinitis. Part I. The mites. *Oriental Insects*, 13, 261–298.

Dar, N., Menon, M.P.S., Shivpuri, D.N. and Gupta, V.K. (1973) The mite fauna of Indian house dusts. *Aspects of Allergy and Applied Immunology*, 6, 51–60.

Dasgupta, A. and Cunliffe, A.C. (1970) Common antigenic determinants in extracts of house dust and *Dermatophagoides* species. *Clinical and Experimental Immunology*, 6, 891–898.

Davis, G., Luyt, D., Prescott, R. and Potter, P.C. (1994) House dust mites in Soweto. *Current Allergy and Clinical Immunology*, 7, 16–17.

Davis, R. and Brown, S.W. (1969) Some population parameters for the grain mite, *Acarus siro. Annals of the Entomological Society of America*, 62, 1161–1166.

Dean, F. (1993) Chemical applications on wool textiles to control insects and mites. In: Tovey, E.R. and Mahmic, A. (eds), *Mites, Asthma and Domestic Design: proceedings of a conference held at the Powerhouse Museum, 15th March, 1993.* University Printing Service, University of Sydney, pp. 85–89.

Dekker, H. (1928) Asthma und Milben. *München Meditziner Wochenschrift*, 75, 515–516 [reprinted in English, (1971) *Journal of Allergy and Clinical Immunology*, 48, 251–252].

Del Rey Calero, J. and García de Lomas, J. (1971) Los acaros sensibilizates del polvo domestico. *Medicina Tropical*, 47, 312–323.

Del Rey Calero, J. and García de Lomas, J. (1972a) Alergia respiratoria provocada por los acaros sensibilizantes del polvo domestico. *Revista Clínica Española*, 126, 215–222.

Del Rey Calero, J. and García de Lomas, J. (1972b) Neumoalergenos extrinsicos con estudio preferencial del polvo de casa y sus ácaros. *Medicamenta*, 59, 499.

Del Rey Calero, J. del and García de Lomas, J. (1973) Fauna Acarina alergógena del polvo doméstico. *Anales de la Real Academia de Farmacia*, 35, 103–148.

Deliargyris, N., Galatas, J. and Kontoy-Fili, K. (1990) House dust mites in Athens, Greece. *Clinical and Experimental Allergy*, 20, Supplement 1, 120. [abstract]

Denson-Lino, J.M., Willies-Jacobo, L.J., Rosas, A., O'Connor, R.D. and Wilson, N.W. (1993) Effect of economic status on the use of house dust mite avoidance measures in asthmatic children. *Annals of Allergy*, 71, 130–132.

Derewenda, U., Li, J., Derewenda, Z., Dauter, Z., Mueller, G.A., Rule, G.S. and Benjamin, D.C. (2002) The crystal structure of a major dust mite allergen Der p 2, and its biological implications. *Journal of Molecular Biology*, 318, 189–197.

Dharmage, S., Bailey, M., Raven, J., Cheng, A., Rolland, J., Thien, F., Forbes, A., Abramson, M. and Walters, E.H. (1999) Residential characteristics influence Der p 1 levels in homes in Melbourne, Australia. *Clinical and Experimental Allergy*, 29, 461–469.

Dharmage, S., Walters, E.H., Thien, F., Bailey, M., Raven, J., Wharton, C., Rolland, J., Light, L.,

Freezer, N. and Abramson, M. (2006) Encasement of bedding does not improve asthma in atopic adult asthmatics. *International Archives of Allergy and Immunology*, 139, 132–138.

Dietemann, A., Bessot, J.-C., Hoyet, C., Ott, M., Verot, A. and Pauli, G. (1993) A double-blind, placebo controlled trial of solidified benzyl benzoate applied in dwellings of asthmatic patients sensitive to mites: Clinical efficacy and effect on mite allergens. *Journal of Allergy and Clinical Immunology*, 91, 738–746.

Dilworth, R.J., Chua, K.Y. and Thomas, W.R. (1991) Sequence analysis of cDNA coding for a major house dust mite allergen, *Der f I. Clinical and Experimental Allergy*, 21, 25–32.

Dimarcq, J.L., Bulet, P., Hetru, C. and Hoffmann, J. (1998) Cysteine-rich antimicrobial peptides in invertebrates. *Biopolymers*, 47, 465–477.

Dixit, I.P. (1973) Dust-mite urticaria. *The Practitioner*, 210, 664.

Dixit, I.P. and Mehta, R.S. (1973) Prevalence of *Dermatophagoides* sp. Bogdanov, 1864 in India and its role in causation of bronchial asthma. *Journal of the Association of Physicians of India*, 21, 31–37.

Dixon, M. and Webb, C.E. (1979) *Enzymes.* 3rd Edition. Longman, London.

Dobson, R.M. (1979) Some effects of microclimate on the longevity and development of *Dermatophagoides pteronyssinus* (Trouessart). *Acarologia*, 21, 482–486.

Dodin, A. and Rak, H. (1993) Influence of low temperature (−30°C) on the different stages of the human allergy mite *Dermatophagoides pteronyssinus* (Acari: Epidermoptidae). *Journal of Medical Entomology*, 30, 810–811.

Domrow, R. (1970) Seasonal variation in numbers of the house-dust mite in Brisbane. *Medical Journal of Australia*, ii, 1248–1250.

Domrow, R. (1992) Acari Astigmata (excluding feather mites) parasitic on Australian Vertebrates: an annotated checklist, keys and bibliography. *Invertebrate Taxonomy*, 6, 1459–1606.

Dornelas de Andrade, A., Birnbaum, J., Lanteaume, A., Izard, J.L., Corget, P., Artillau, M.F., Vervloet, D. and Charpin, D. (1995) Housing and house dust mites. *Allergy*, 50, 142–146.

Dornelas de Andrade, A., Charpin, D., Birnbaum, J., Lanteaume, A., Chapman, M. and Vervloet, D.

(1995) Indoor allergen levels in day nurseries. *Journal of Allergy and Clinical Immunology*, 95, 1158–1163.

Dorward, A.J., Colloff, M.J., MacKay, N.S., McSharry, C. and Thompson, N.C. (1988) Effect of house dust mite avoidance measures in adult atopic asthma. *Thorax*, 43, 98–102.

Dotterud, L.K., Korsgaard, J. and Falk, E.S. (1995) House-dust mite content in mattresses in relation to residential characteristics and symptoms in atopic and nonatopic children in northern Norway. *Allergy*, 50, 788–793.

Dotterud, L.K., Van, T.D., Kvammen, B., Dybendal, T., Elsayed, S. and Falk, E.S. (1997) Allergen content in dust from homes and schools in northern Norway in relation to sensitisation and allergy symptoms in schoolchildren. *Clinical and Experimental Allergy*, 27, 252–261.

Dougall, A., Holt, D.C., Fischer, K., Currie, B.J., Kemp, D.J. and Walton, S.F. (2005) Identification and characterization of *Sarcoptes scabiei* and *Dermatophagoides pteronyssinus* glutathione s-transferases: implication as a potential major allergen in crusted scabies. *American Journal of Tropical Medicine and Hygiene*, 73, 977–984.

Douglas, A.E. and Hart, B.J. (1989) The significance of the fungus *Aspergillus penicilloides* to the house dust mite *Dermatophagoides pteronyssinus*. *Symbiosis*, 7, 105–116.

Douglas, A.E. and Smith, D.C. (1989) Are symbioses mutualistic? *Trends in Ecology and Evolution*, 4, 350–352.

Doull, I.J.M., Bright, J., Yongeswaran, P. and Howarth, P.H. (1997) House-dust mite allergen Der p 1: amount or concentration? *Allergy*, 52, 220–223.

Dowse, G.K., Turner, K.J., Stewart, G.A., Alpers, M.P. and Woolcock, A.J. (1985) The association between *Dermatophagoides* mites and the increasing prevalence of asthma in village communities within the Papua New Guinea Highlands. *Journal of Allergy and Clinical Immunology*, 75, 75–83.

Du, E.-C., Li, Z.-M., Sui, C.-S., Wang, W. and Zhang, Q.-X. (1993) Relationship between asthma and allergenic antigens in rural houses. *Biomedical and Environmental Sciences*, 6, 27–30.

Dubinin, V.B. (1953) [*Arachnida, Vol. 6, No. 6: Analgesoidea Mites, Part 2: Families Epidermoptidae and Freyanidae.*] *Fauna of the USSR*, New Series, No. 55. Academy of Sciences of the USSR, Moscow and Leningrad. [In Russian]

Dubinin, V.B., Guselnikova, M.I. and Raznatovsky, I.M. (1956) [Discovery of skin mites (*Dermatophagoides scheremetewskyi* Bogdanov, 1864) in some human skin diseases. *Bulletin de la Societe des Sciences naturelles du Moscou, Section Biologique*, 61, 43–50. [In Russian]

Dubinina, H.V. and Pletnev, B.D. (1978) [Acarofauna of dust of human dwellings.] *Parazitologicheskii Sbornik*, 28, 37–46. [In Russian]

Dubinina, H.V., Vaitsekuaskaite, R.L., Baiorinaite, A. and Razgauskas, A.F. (1984) [The fauna of house dust mites in the homes of bronchial asthma patients in the Lithuanian SSR.] *Meditsinskaia Parazitologiia i Parazitarnye Bolezni*, 5, 63–67. [In Russian, English summary]

Dusbabek, F. (1975) Population structure and dynamics of the house dust mite *Dermatophagoides farinae* (Acarina, Pyroglyphidae) in Czechoslovakia. *Folia Parasitologia (Praha)*, 22, 219–231.

Dusbabek, F. (1979) Dynamics and structure of mixed populations of *Dermatophagoides farinae* and *D. pteronyssinus*. In: Rodriguez, J.G. (ed.), *Recent Advances in Acarology*, Volume 2. Academic Press, New York, pp. 173–177.

Dusbabek, F., Cuervo, N. and de la Cruz, J. (1982) *Dermatophagoides siboney* sp. n. (Acarina: Pyroglyphidae), a new house dust mite from Cuba. *Acarologia*, 29, 285–295.

Dutau, G. (2002) House dust mites: new food allergens. *Revue Française d'Allergologie et d'Immunologie Clinique*, 42, 171–177.

Dutau, G. and Rochiccioli, P. (1979) Traitement 'ecologique' de l'allergie aux acariens par le Paragerm AK. *Revue de Medecine de Toulouse*, 15, 457–460.

Dybendal, T., Hetland, T., Vik, H., Apold, J. and Elsayed, S. (1989a) Dust from carpeted and smooth floors I. Comparative measurements of antigenic and allergenic proteins in dust vacuumed from carpeted and non-carpeted classrooms in Norwegian schools. *Clinical and Experimental Allergy*, 19, 217–224.

Dybendal, T., Vik, H. and Elsayed, S. (1989b) Dust from carpeted and smooth floors II. Antigenic and allergenic content of dust vacuumed from carpeted and smooth floors in schools under routine cleaning schedules. *Allergy*, 44, 401–411.

Dybendal, T. and Elsayed, S. (1994) Dust from carpeted and smooth floors VI. Allergens in homes compared with those in schools in Norway. *Allergy*, 49, 210–216.

Eaton, K.K., Downing, F.S., Griffiths, D.A., Hockland, S. and Lynch, S. (1985a) House dust mites (*D. pteronyssinus*) in pets' beds and their relation to dust allergy. *Clinical Allergy*, 15, 151–154.

Eaton, K.K., Downing, F.S., Griffiths, D.A., Lynch, S., Hockland, S. and McNulty, D.W. (1985b) Storage mites: culturing, sampling and their role in housedust allergy in rural areas in the United Kingdom. *Annals of Allergy*, 55, 62–67.

Echechipía, S., Ventas, P., Audícana, M., Urutia, I., Gastaminza, G., Polo, F. and Fernández de Corres, L. (1995) Quantitation of major allergens in dust samples from urban populations collected in different seasons in two climatic areas of the Basque region (Spain). *Allergy*, 50, 478–482.

Edgar, R.C. (2004) MUSCLE: multiple sequence alignment with high accuracy and high throughput. *Nucleic Acids Research*, 32, 1792–1797.

Edney, E.B. (1977) *Water Balance in Land Arthropods.* Springer-Verlag, Berlin.

Edston, E. and van Hage-Hamsten, M. (2003) Death in anaphylaxis in a man with house dust mite allergy. *International Journal of Legal Medicine*, 117, 299–301.

Edwards, T.B., Trudeau, W.L., Fernández-Caldas, E., Lee, D.K., Seleznick, M.J. and Lockey, R.F. (1992) Proteinases in extracts of the storage mite *Aleuroglyphus ovatus. Journal of Allergy and Clinical Immunology*, 90, 129–131.

Ehnert, B., Lau-Schadendorf, S., Weber, A., Buettner, P., Schou, C. and Wahn, U. (1992) Reducing domestic exposure to dust mite allergen reduces bronchial hyperreactivity in sensitive children with asthma. *Journal of Allergy and Clinical Immunology*, 90, 135–138.

Eldefrawi, M.E. (1985) Nicotine. In: Kerkut, G.A. and Gilbert, L.I. (eds), *Comprehensive Insect Physiology, Biochemistry and Pharmacology, Volume 12, Insect Control.* Pergamon Press, Oxford, pp. 263–272.

Elias, P.M. (1981) Lipids and the epidermal permeability barrier. *Archives of Dermatological Research*, 280, 95–117.

Elixmann, J., Bischoff, E., Jorde, W. and Linskens, H.F. (1988) Eimalige Acarosan – Application zur Sanierung von Wohentextilien in Haushalten von Patienten mit Milbenallergie. *Allergologie*, 11, 274–279.

Elixmann, J.H., Bischoff, E., Jorde, W. and Liskens, H.F. (1991) Changements [sic] during 2 years in populations of different mite species in house dust before and after a single acaricidal treatment. *Acarologia*, 32, 385–398.

Ellingsen, I. (1974) Comparison of active and quiescent protonymphs of the American house-dust mite. PhD thesis, The Ohio State University, Columbus.

Ellingsen, I.J. (1975) Permeability to water in different adaptive phases of the same instar in the American house-dust mite. *Acarologia*, 17, 734–744.

Ellingsen, I.J. (1978) Oxygen consumption in active and quiescent protonymphs of the American house dust mite. *Journal of Insect Physiology*, 24, 13–16.

Ellingson, A.R., LeDoux, R.A., Vedanthan, P.K. and Weber, R.W. (1995) The prevalence of *Dermatophagoides* mite allergen in Colorado homes utilizing central evaporative coolers. *Journal of Allergy and Clinical Immunology*, 96, 473–479.

Ellul-Micallef, R. (1997) History of asthma. In: Barnes, P., Grunstein, M., Leff A.R. and Woolcock A.J. (eds), *Asthma* Vol. 1. Lippincott-Raven, Philadelphia, pp. 9–25.

Ellwood, P., Asher, M.I., Björkstén, B., Burr, M., Pearce, N., Robertson, C.F. and the ISAAC Phase One Study Group (2001) Diet and asthma, allergic rhinoconjunctivitis and atopic eczema symptom prevalence: an ecological analysis of the International Study of Asthma and Allergies in Childhood (ISAAC) data. *European Respiratory Journal*, 17, 436–443.

El Sharif, N., Douwes, J., Hoet, P.M.H., Doekes, G. and Nemery, B. (2004) Concentrations of domestic mite and pet allergens and endotoxin in Palestine. *Allergy*, 59, 623–631.

El-Shazly, A.M., El-Beshbishi, S.N., Azab, M.S., El-Nahas, H.A., Soliman, M.E., Fouad, M.A.H. and Monib, M.E.-S.M.M. (2006) Present situation of house dust mites in Dakahlia Governorate, Egypt. *Journal of the Egyptian Society of Parasitology*, 36, 113–126.

Emekçi, M. and Toros, S. (1989) *Acarus siro* L. (Acarina, Acaridae) 'nun degisik sicaklik ve nem ortamlarindaki gelismesi üzerinde arastirmalar. *Türkiye Entomolji Dergisi*, 13, 217–228.

Emenius, G., Korsgaard, J. and Wickman, M. (2000) Window pane condensation and high indoor vapour contribution – markers of an unhealthy indoor climate? *Clinical and Experimental Allergy*, 30, 418–425.

Epstein, H.M., Duke, K.M. and Wharton, G.W. (1979) Laser-generated X-ray microradiography applied to entomology. *Transactions of the American Microscopical Society*, 98, 427–436.

Epton, M.J., Dilworth, R.J., Smith, W., Hart, B.J. and Thomas, W. (1999) High molecular weight allergens of the house dust mite: an apolipophorin-like cDNA has sequence identity with the major M-177 allergen and the IgE-binding peptide fragments Mag 1 and Mag 3. *International Archives of Allergy and Immunology*, 120, 185–191.

Epton, M.J., Dilworth, R.J., Smith, W. and Thomas, W.R. (2001a) Sensitisation to the lipid-binding apolipophorin allergen Der p 14 and the peptide Mag-1. *International Archives of Allergy and Immunology*, 124, 57–60.

Epton, M.J., Malainual, N., Smith, W. and Thomas, W.R. (2001b) Vitellogenin-apolipophorin like allergen Der p 14 is a major specificity in house dust mite sensitisation. *Journal of Allergy and Clinical Immunology*, 107(2), Supplement 1, S107. [abstract]

Eraky, S.A. (1985a) Some biological spects of *Tyrophagus putrescentiae* (Schrank) (Acari: Acaridae). In: Kropcynska, D., Boczek, J. and Tomczyk, A. (eds), *The Acari: Physiological and Ecological Aspects of Acari–Host Relationships*. Oficyna Dabor, Warsaw, pp. 198–204.

Eraky, S.A. (1985b) Effect of temperature on the development of the copra mite *Tyrophagus putrescentiae* (Schrank) (Acari: Acaridae). In: Kropcynska, D., Boczek, J. and Tomczyk, A. (eds), *The Acari: Physiological and Ecological Aspects of Acari–Host Relationships*. Oficyna Dabor, Warsaw, pp. 206–210.

Erban, T. and Hubert, J. (2008) Digestive function of lysozyme in synanthropic acaridid mites enables utilization of bacteria as a food source. *Experimental and Applied Acarology*, 44, 199–212.

Erben, A.M., Rodribuez, J.L., McCullough, J. and Ownby, J.R. (1993) Anaphylaxis after ingestion of beignets contaminated with *Dermatophagoides farinae*. *Journal of Allergy and Clinical Immunology*, 92, 846–849.

Eriksson, T.L., Johansson, E., Whitley, P., Schmidt, M., Elsayed, S., van Hage-Hamsten, M. (1998) Cloning and characterisation of a group II allergen from the dust mite *Tyrophagus putrescentiae*. *European Journal of Biochemistry*, 251, 443–447.

Eriksson, T.L., Whitley, P., Johansson, E., van Hage-Hamsten, M. and Gafvelin, G. (1999) Identification and characterization of two allergens from the dust mite Acarus siro, homologous with fatty-acid binding proteins. *International Archives of Allergy and Immunology*, 119, 275–281.

Eriksson, T.L., Rasool, O., Huecas, S., Whitley, P., Crameri, R., Appenzeller, U., Gafvelin, G. and van Hage-Hamsten, M. (2001) Cloning of three new allergens from the dust mite *Lepidoglyphus destructor* using phage surface display technology. *European Journal of Biochemistry*, 268, 287–294.

European Community Respiratory Health Survey (1996) Variations in the prevalence of respiratory symptoms, self–reported asthma attacks, and the use of asthma medication in the European Community Respiratory Health Survey (ECRHS). *European Respiratory Journal*, 9, 687–695.

European Community Respiratory Health Survey II Steering Committee (2002) The European Community Respiratory Health Survey II. *European Respiratory Journal*, 20, 1071–1079.

Evans, G.O. (1992) *Principles of Acarology*. CAB International, Wallingford.

Evans, G.O., Sheals, J.G. and Macfarlane, D. (1961) *The Terrestrial Acari of the British Isles. An Introduction to their Morphology, Biology and Classification*. Vol. 1. British Museum (Natural History), London.

Ezequiel, O. da S., Gazeta, G.S., Amorim, M. and Serra-Freire, N.M. (2001) Evaluation of the acarofauna of the domicilary ecosystem in Juiz de Fora, State of Minas Gerais, Brazil. *Memorias do Instituto Oswsaldo Cruz*, 96, 911–916.

Fahlbusch, B., Heinrich, J., Gross, I., Jäger, L., Richter, K. and Wichmann, H.-E. (1999) Allergens in house-dust samples in Germany: results of an East–West German comparison. *Allergy*, 54, 1215–1222.

Fain, A. (1963) Les acariens producteurs de gale chez les lemuiens et les singes avec une étude des Psoroptidae (Sarcoptiformes). *Bulletin de l'Institute Royal des Sciences Naturelles de Belgique*, 39(32), 1–125.

Fain, A. (1965) Les acariens nidicoles et detriticoles de la famille Pyroglyphidae Cunliffe (Sarcoptiformes). *Revue Zoologique et de Botanique Africaines*, 52, 257–286.

Fain, A. (1966a) Nouvelle description de *Dermatophagoides pteronyssinus* (Trouessart, 1897). Importance de cet acarien en pathologie humaine (Psoroptidae, Sarcoptiformes). *Acarologia*, 8, 302–327.

Fain, A. (1966b) Allergies respiratoires produites par un Acarien (*Dermatophagoides pteronyssinus*) vivant dans les poussières des habitations. *Bulletin de l'Academie Royale Medicale du Belgique*, 6, 479–499.

Fain, A. (1967a) Le genre *Dermatophagoides* Bogdanov, 1864. Son importance dans les allergies respiratoires et cutanees chez l'homme (Psoroptidae, Sarcoptiformes). *Acarologia*, 9, 179–225.

Fain, A. (1967b) Deux nouvelles especes de Dermatophagoidinae, rattachement de cette sous-famille aux Pyroglyphidae. *Acarologia*, 9, 870–881.

Fain, A. (1971) Deux nouvelles especes du Dermatophagoidinae de la region du Kivu (Republique Democratique du Congo) (Acarina, Sarcoptiformes). *Bulletin de l'Institute Royal des Sciences Naturelles de Belgique*, 47(8), 1–5.

Fain, A. (1975) *Dermatophagoides sclerovestibulatus*, une nouvelle espèce provenant du nid de *Buphagus erythrorhynchus* d'Afrique du Sud. *Revue Zoologique africaine*, 89, 252–256.

Fain, A. (1976) Le genre *Austroglycyphagus* Fain et Lowry, 1974 (Acarina, Astigmata, Glycyphagidae), description d'especes nouvelles. *Acarologia*, 17, 709–729.

Fain, A. (1977) The prelarva in the Pyroglyphidae (Acarina: Astigmata). *International Journal of Acarology*, 3, 115–116.

Fain, A. (1978) *Austroglycyphagus asthmaticus* spec. nov. (Acari, Glycyphagidae) vivant dans le guano de chiroptères à Bujumbura (Burundi). *Revue Zoologique africaine*, 92, 953–958.

Fain, A. (1979) Specificity, adaptation and parallel host–parasite evolution in acarines. In: Rodriguez, J.G. (ed.), *Recent Advances in Acarology,* Volume 2. Academic Press, New York, pp. 321–328.

Fain, A. (1980) A method for remounting old preparations of acarines without raising or displacing the cover slip. *International Journal of Acarology*, 6, 169–170.

Fain, A. (1986) Observations sur les Glycyphagidae (Acari, Astigmata) avec description de deux espèces nouvelles. *Bulletin et Annales de la Societe royale d'entomolgie de Belgique*, 122, 155–169.

Fain, A. (1988a) Nouvelles localities pour des acariens de la famille Pyroglyphidae. *Bulletin et Annales de la Societe royale d'entomolgie de Belgique*, 124, 267–268.

Fain, A. (1988b) Morphologie, systématique et distribution géographique des acariens responsables des allergies respiratoires ches l'homme. In: Fain, A., Guerin, B. and Hart, B.J. *Acariens et Allergies.* Allerbio, Varennes-en-Argonne, pp. 13–55.

Fain, A. (1990) Morphology, systematics and geographical distribution of mites responsible for allergic diseases in man. In: Fain, A., Guerin, B. and Hart, B.J., *Mites and Allergic Disease.* Allerbio, Varennes-en-Argonne, pp. 13–134.

Fain, A. and Atyeo, W.T. (1990) A new pyroglyphid mite (Acari : Pyroglyphidae) from a woodpecker (Picidae) in Thailand. *Acarologia*, 31, 43–50.

Fain, A. and van Bronswijk, J.E.M.H. (1973) On a new species of *Dermatophagoides* (*D. neotropicalis*) from house dust, producing both normal and heteromorphic males (Sarcoptiformes, Pyroglyphidae). *Acarologia*, 15, 181–187.

Fain, A. and Feinberg, J.G. (1970) Un nouvel acarien provenant des poussières d'une maison à Singapour (Sarcoptiformes: Pyroglyphidae) *Acarologia*, 12, 164–167.

Fain, A. and Galloway, T.D. (1993) Mites (Acari) from nests of sea birds in New Zealand II. Mesostigmata and Astigmata. *Bulletin de l'Institute Royal des Sciences Naturelles de Belgique, Entomologie*, 63, 95–111.

Fain, A. and Gaud, J. (1984) Sur un nouveau groupe d'acariens, dans la famille Pyroglyphidae (Astigmates), inféodé aux pics afrotropicaux des familles Capitonidae et Picidae. *Acarologia*, 25, 47–53.

Fain, A. and Hart, B.J. (1986) A new, simple technique for extraction of mites, using the difference in density between ethanol and saturated NaCl (preliminary note). *Acarologia*, 27, 255–256.

Fain, A. and Hérin, A. (1978) La prélarve chez les Astigmates (Acari). *Acarologia*, 20, 566–571.

Fain, A. and Johnston, D. (1973) *Euroglyphus* (*Gymnoglyphus*) *osu* new species from barn floor in USA (Acarina: Pyroglyphidae: Sarcoptiformes). *Bulletin et Annales de la Societe royale d'entomolgie de Belgique*, 109, 131–134.

Fain, A. and Lowry, W. J. (1974) A new pyroglyphid mite from Australia (Acarina: Sarcoptiformes: Pyroglyphidae). *Acarologia*, 16, 331–339.

Fain, A. and Nadchatram, M. (1980) New house dust mites from Malaysia I. Two new species of *Austroglycyphagus* Fain and Lowry, 1974 (Astigmata, Glycyphagidae). *International Journal of Acarology*, 6, 1–8.

Fain, A. and Rosa, A.E. (1982) Pyroglyphid mites from nests of sparrows *Passer domesticus* L., 1758 in Brasil. *Revista Brasileira Biologia*, 42, 317–320.

Fain, A. and Wharton, G.W. (1970) Un nouveau Dermatophagoidinae du Guatemala (Pyroglyphidae: Sarcoptiformes). *Bulletin de la Institute royale des Sciences naturelles de Belgique*, 46(27), 1–4.

Fain, A., Cunnington, A.M. and Spieksma, F.Th.M. (1969) *Malayoglyphus intermedius* n.g., n.sp. a new mite from house dust in Singapore and Djakarta. *Acarologia*, 11, 121–126.

Fain, A., Oshima, S. and van Bronswijk, J.E.M.H. (1974) *Hirstia domicola* sp. n. from house dust in Japan and Surinam (Acarina, Sarcoptiformes, Pyroglyphidae). *Japanese Journal of Sanitary Zoology*, 25, 197–203.

Fain, A., Hyland, K. and Tadkowski, T. (1977) *Blomia gracilipes* (= *Chortoglyphus gracilipes*): redescription and status (Acarina: Glycyphagidae). *Entomological News*, 88, 267–269.

Fain, A., Feldman-Muhsam, B. and Mumcuoglu, K.Y. (1980) *Cheyletus tenuipilis* (n. sp.) (Acari, Cheyletidae) nouvel acarien des poussières de maisons en Europe Occidentale et en Israel. *Bulletin et Annales de la Societe royale d'entomolgie de Belgique*, 116, 35–44.

Fain, A., Gaud, J. and Perez, T.M. (1982) A new genus and two new species of Pyroglyphinae (Acari, Astigmata, Pyroglyphidae) from South American Birds. *Acarologia*, 23, 165–170.

Fain, A., Guerin, B. and Hart, B.J. (1988) *Acariens et Allergies*. Allerbio, Varennes-en-Argonne.

Fain, A., Guerin, B. and Hart, B.J. (1990) *Mites and Allergic Disease*. Allerbio, Varennes-en-Argonne.

Falk, E.S. and Bolle, R. (1980a) IgE antibodies to house dust mite in patients with scabies. *British Journal of Dermatology*, 102, 283–288.

Falk, E.S. and Bolle, R. (1980b) *In vitro* demonstration of specific immunological hypersensitivity to scabies mite. *British Journal of Dermatology*, 103, 367–373.

Fan, Q.-H. and Zhang, Z.-Q. (2007) *Tyrophagus* (Acari: Astigmata: Acaridae). *Fauna of New Zealand*, 56, 1–291.

Fashing, N. (1989) Fine structure of the Claparède organs and genital papillae of *Naidacarus arboricola* (Astigmata: Acaridae), an inhabitant of water-filled treeholes. In: ChannaBasavanna, G.P. and Viraktamath, C. (eds), *Progress in Acarology. Vol. 1*. Oxford and IBH Publishing Co., New Delhi, pp. 219–228.

Fashing, N.A. (1994) Life-history patterns of Astigmatid inhabitants of water-filled treeholes. In: Houck, M.A. (ed.), *Mites: Ecological and Evolutionary Analyses of Life-History Patterns*. Chapman and Hall, New York, pp. 160–185.

Fejt, R. and Ždárková, E. (2001) Bionomics of *Acarus siro* L. (Acarina: Acaridae) on oilseeds. *Plant Protection Science*, 37, 110–113.

Feldman-Muhsam, B., Mumcuoglu, K.Y. and Osterovitch, T. (1985) A survey of house dust mites (Acari, Pyroglyphidae and Cheyletidae) in Israel. *Journal of Medical Entomology*, 22, 663–669.

Fell, P., Mitchell, B. and Brostoff, J. (1992) Wet vacuum-cleaning and housedust-mite allergen. *Lancet*, 340, 788–789.

Fernández-Caldas, E. and Lockey, R.F. (1995) *Blomia tropicalis*, a mite whose time has come. *Allergy*, 1161–1164.

Fernández-Caldas, E., Fox, R.W., Bucholz, G.A., Trudeau, W.L., Ledford, D.K. and Lockey, R.F. (1990) House dust mite allergy in Florida. Mite survey in households of mite-sensitive individuals in Tampa, Florida. *Allergy Proceedings*, 11, 263–267.

Fernández-Caldas, E., Puerta, L., Mercado, D., Lockey, R.F. and Caraballo, L.R. (1993) Mite fauna, *Der p* I, *Der f* I and *Blomia tropicalis* allergen in a tropical environment. *Clinical and Experimental Allergy*, 23, 292–297.

Fernández-Caldas, E., Sepulveda, A. and Montealegre, F. (1995) Identification of the domestic mite fauna of Puerto Rico. *Allergy*, 50 (Supplement 26), 72. [abstract]

Fernández-Caldas, E., Puerta, L., Caraballo, L., Mercado, D. and Lockey, R.F. (1996) Sequential determinations of *Dermatophagoides* spp. allergens in a tropical city. *Journal of Investigational Allergology and Clinical Immunology*, 6, 98–102.

Ferrándiz, R. (1997) Allergic Characterization of the Domestic Mite *Dermatophagoides siboney*. Doctoral thesis, University of Linköping, Sweden.

Ferrándiz, R., Casas, R., Dreborg, S., Einarsson, R., and Ferrándiz, B. (1995a) Crossreactivity between *Dermatophagoides siboney* and other house dust mite allergens in sensitised asthmatic patients. *Clinical and Experimental Allergy*, 25, 929–934.

Ferrándiz, R., Casas, R., Dreborg, S., Einarsson, R., Bonachea, I. and Chapman, M.D. (1995b) Characterization of allergenic components from house dust mite *Dermatophagoides siboney*, Purification of *Der s* 1 and *Der s* 2 allergens. *Clinical and Experimental Allergy*, 25, 922–928.

Ferrándiz, R., Casas, R. and Dreborg, S. (1996) Sensitisation to *Dermatophagoides siboney*, *Blomia tropicalis*, and other domestic mites in asthmatic patients. *Allergy*, 51, 501–505.

Ferrándiz, R., Casas, R. and Dreborg, S. (1997) Purification and IgE binding capacity of Der s 3, a major allergen from *Dermatophagoides siboney*. *Clinical and Experimental Allergy*, 27, 700–704.

Ferrándiz, R., Casas, R. and Dreborg, S. (1998) Crossreactivity between *Dermatophagoides siboney* and other domestic mites. II. Analysis of individual crossreacting allergens by Western blotting inhibition. *International Archives of Allergy and Immunology*, 116, 206–214.

Fischer, K., Holt, D.C., Harumal, P., Currie, B.J., Walton, S.F. and Kemp, D.J. (2003) Generation and characterisation of cDNA clones from *Sarcoptes scabiei* var *hominis* for an expressed sequence tag library: identification of homologues of house dust mite allergens. *American Journal of Tropical Medicine and Hygiene*, 68, 61–64.

Fisher, R.C. (1938) Studies on the biology of the death-watch beetle, *Xestobium rufovillosum* De G. II. The habits of the adult with special reference to the factors affecting oviposition. *Annals of Applied Biology*, 25, 155–180.

Flannery, E.M., Holdaway, M.D., Herbison, G.P., Jones, D.T., Hewitt, C.J. and Sears, M.R. (1994) Sheepskins and bedding in childhood, and the risk of development of bronchial asthma. *Australian and New Zealand Medical Journal*, 24, 687–692.

Flechtmann, C.H.W., Da Costa, C.P. and Maielli, J.A. (1998) *A Residência para o Alérgico: Construção e Adaptação*. Editoria UNIMEP (Universidad Metodista de Piracicaba), Piracicaba, Brazil.

Fletcher, A.M., Pickering, C.A.C., Custovic, A., Simpson, J., Kennaugh, J. and Woodcock, A. (1996) Reduction in humidity as a method of controlling mites and mite allergens: the use of mechanical ventilation in British domestic dwellings. *Clinical and Experimental Allergy* 26, 1051–1056.

Flores, I., Mora, C., Rivera, E., Donnelly, R. and Montealegre, F. (2003) Cloning and molecular characterization of a cDNA from *Blomia tropicalis* homologues to dust mite group 3 allergens (trypsin-like proteases). *International Archives of Allergy and Immunology*, 130, 12–16.

Flower, D.R. (1996) The lipocalin protein family: structure and function. *Biochemical Journal*, 318, 1–14.

Forbes-Smith, M.R., Morris, S.C. and Scott, K.J. (1989) An inexpensive and accurate humidity generating system. *Laboratory Practice*, 38(9), 99–103.

Ford, B.J. (1981a) Leeuwenhoek's specimens discovered after 307 years. *Nature*, 392, 407.

Ford, B.J. (1981b) Found – van Leeuwenhoek's original specimens. *New Scientist*, 30 July, 307.

Ford, B.J. (1985) *Single Lens. The Story of the Simple Microscope*. Harper and Row, New York.

Ford, P.J.R. (1989) *Oriental Carpet Design. A Guide to Traditional Motifs, Patterns and Symbols*. Revised edition, Thames and Hudson, London.

Fortey, R. (1997) *Life: An Unauthorised Biography*. Harper Collins, London.

Franjola, R. and Malonnek, M. (1995) Acaros del polvo de habitaciones de la ciudad de Valdivia, Chile. *Boletín Chileno de Parasitologia*, 50, 16–20.

Franjola, T. and Rosinelli, M. (1999) Acaros del polvo de habitaciones en la ciudad de Punta Arenas, Chile. *Boletín de Parasitologia*, 54, 82–88.

Frankland, A.W. and Easty, D. (1971) Vernal kerato-conjunctivitis: an atopic disease. *Transactions of the Opthalmological Society, UK*, 91, 479–482.

Frankland, A.W. and El-Hefney, A. (1971) House dust and mites as causes of inhalant allergic problems in the United Arab Republic. *Clinical Allergy*, 1, 257–260.

Fraser, R.D.B., MacRae, T.P. and Rogers, G.E. (1972) *Keratins, Their Composition, Structure and Biosynthesis*. Charles C. Thomas, Springfield, Illinois.

Frazier, M.R., Huey, R.B. and Berrigan, D. (2006) Thermodynamics constrains the evolution of insect population growth rates: 'warmer is better'. *American Naturalist*, 168, 512–520.

Frenken, J.H. (1962) *Dermanyssus gallinae* (*D. avium*). Strophulus or 'Insect' bites, diet or D.D.T. *Dermatologica*, 125, 322–331.

Friedland, N., Liou, H.-L., Lobel, P. and Stock, A.M. (2003) Structure of a cholesterol-binding protein deficient in Niemann–Pick type C2 disease. *Proceedings of the National Academy of Sciences*, 100, 2512–2517.

Fromme, H., Mattulat, A., Lahrz, T. and Rüden, H. (2005) Occurrence of organotin compounds in house dust in Berlin (Germany). *Chemosphere*, 58, 1377–1383.

Fuchs, E. and Hanukoglu, I. (1996) Epidermal α-keratins: structural diversity and changes during tissue differentiation. In: Bereiter-Hahn, J., Matoltsy, A.G. and Richards, K.S. (eds), *Biology of the Integument. Vol. 2 Vertebrates*. Springer-Verlag, Berlin, pp. 644–665.

Fujikawa, A., Ishimaru, N., Seto, A., Yamada, H., Aki, T., Shigeta, S., Wada, T., Jyo, T., Murooka, Y., Oka, S. and Ono, K. (1996) Cloning and characterization of a new allergen, Mag 3, from the house dust mite, *Dermatophagoides farinae*: cross-reactivity with high-molecular-weight allergen. *Molecular Immunology*, 33, 311–319.

Fujikawa, A., Uchida, K., Yanagidani, A., Kawamoto, S., Aki, T., Shigeta, S., Wada, T., Suzuki, O., Jyo, T. and Ono, K. (1998) Altered allergenicity of M-177, a 177 kDa allergen from house dust mite, *Dermatophagoides farinae*, in stored extracts. *Clinical and Experimental Allergy*, 28, 1549–1558.

Fumimoto, T., Kato, H., Ichiose, E. and Sasaguri, Y. (1982) Immune complex and mite antigen in Kawasaki disease. *The Lancet*, October 30th, 980–981.

Furmonaviciene, R. and Shakib, F. (2001) The molecular basis of allergenicity: comparative analysis of the three dimensional structures of diverse allergens reveals a common structural motif. *Journal of Clinical Pathology: Molecular Pathology*, 54, 155–159.

Fürstenburg, M.H.F. (1861) *Die Krätzmilben der Menschen und Thiere*. Leipzig.

Furumizo, R.T. (1973) The Biology and Ecology of the House-dust Mite *Dermatophagoides farinae* Hughes, 1961 (Acarina, Pyroglyphidae). PhD thesis, University of California, Riverside.

Furumizo, R.T. (1975a) Laboratory observations on the life history and biology of the American House Dust Mite, *Dermatophagoides farinae* (Acarina, Pyroglyphidae). *California Vector News*, 22, 49–59.

Furumizo, R.T. (1975b) Geographical distribution of house dust mites (Acarina, Pyroglyphidae) in California. *California Vector Views*, 22, 89–95.

Furumizo, R.T. (1975c) Collection and isolation of mites from house-dust samples. *California Vector Views*, 22, 19–27.

Furumizo, R.T. (1978) Seasonal abundance of *Dermatophagoides farinae* Hughes 1961 (Acarina, Pyroglyphidae) in house dust in Southern California. *California Vector Views*, 25, 13–19.

Furumizo, R.T. and Mulla, M.S. (1971) Distribution of *Dermatophagoides* mites in house dust samples. *Proceedings of the North Central Branch of the Entomological Society of America*, 26, 67–68.

Furumizo, R.T. and Wharton, G.W. (1975) A case of postimaginal moult in the American house dust mite *Dermatophagoides farinae* Hughes, 1961 (Acari: Pyroglyphidae). *Acarologia*, 17, 730–733.

Furumizo, R.T. and Thomas, V. (1977) Mites of house dust. *South East Asian Journal of Tropical Medicine and Public Health*, 8, 411–412.

Furusho, K., Ohba, T., Soeda, T., Kimoto, K., Okabe, T. and Hirota, T. (1981) Possible role for mite antigen in Kawasaki disease. *The Lancet*, July 25, 194–195.

Gabriel, M., Ng, H.-K., Allan, W.H.L., Hill, L.E. and Nunn, A.J. (1977) A study of prolonged hyposensitisation with *D. pteronyssinus* extract in allergic rhinitis. *Clinical Allergy*, 7, 325–336.

Gabriel, M., Cunnington, A.M., Allan, W.G.L., Pickering, C.A.C. and Wraith, D.G. (1982) Mite allergy in Hong Kong. *Clinical Allergy*, 12, 157–171.

Gaddie, J., Skinner, C. and Palmer, K.N.V. (1976) Hyposensitisation with house dust mite vaccine in bronchial asthma. *British Medical Journal*, ii, 561–562.

Gaede, K. and Knülle, W. (1987) Water vapour uptake from the atmosphere and critical equilibrium humidity of a feather mite. *Experimental and Applied Acarology*, 3, 45–52.

Gafvelin, G., Johansson, E., Lundin, A., Smith, A.M., Chapman, M.D., Benjamin, B.C., Derewenda, U. and van Hage-Hamsten, M. (2001) Cross-reactivity studies of a new group 2 allergen from the house dust mite *Glycyphagus domesticus*, Gly d 2 and group 2 allergens from *Dermatophagoides pteronyssinus, Lepidoglyphus destructor* and *Tyrophagus putrescentiae* with recombinant allergens. *Journal of Allergy and Clinical Immunology*, 107, 511–518.

Galväo, A.B. and Guitton, N. (1986a) *Dermatophagoides deani* sp. n., nova espécie de acaro piroglifído encontrada no Brasil, em poeira domicilar. *Memorias Instituto Oswaldo Cruz*, 81, 241–244.

Galvão, A.B. and Guitton, N. (1986b) Acaros em poeira domicilar das capitais Brasilieras e Ilha Fernando Noronha. *Memorias Instituto Oswaldo Cruz*, 81, 417–430.

Gamal-Eddin, F.M., Tayel, S.E., Abou-Senna, F.M. and Shehata, K.K. (1982) Present status and ecology of house dust mites in Egypt as approaches to environmental control of mites and preparation of specific diagnostic antigen before resort to any desensitising vaccine. *Journal of the Egyptian Society of Parasitology*, 12, 253–282.

Gamal-Eddin, F.M., Abou-Senna, F.M., Tayel, S.E., Aboul-Atta, A.M., Seif, A.M. and Gaafar, S.M. (1983a) Longevity of adult mites in the laboratory as determinal factor in indicating the peaks of environmental pollution with house dust mite allergens. *Journal of the Egyptian Society of Parasitology*, 13, 31–41.

Gamal-Eddin, F.M., Abou-Sinna, F.M., Tayel, S.E., Aboul-Atta, A.M., Seif, A.M. and Gaafar, S.M. (1983b) Duration of the developmental stages of house dust mites *Dermatophagoides farinae* and *D. pteronyssinus* under controlled temperatures and relative humidities to pave the way in front of the workers in the field of house–dust mite bronchial asthma. 1. Pre-imaginal Period. *Journal of the Egyptian Society of Parasitology*, 13, 319–334.

Gamal-Eddin, F.M., Shehata, K.K., Tayel, S.E., Abou-Sinna, F.M., Aboul-Atta, A.M., Seif, A.I., Imam, M.H. and Hafez, A.H. (1983c) Duration of the developmental stages of house dust mites *Dermatophagoides farinae* and *D. pteronyssinus* under controlled conditions, to pave the way in front of the workers in the field of house-dust mite asthmatic bronchitis. 2. Oviposition period, fecundity and oval duration. *Journal of the Egyptian Society of Parasitology*, 13, 557–581.

Gamal-Eddin, F.M., Shehata, K.K., Abou-Sinna, F.M., Tayel, S.E., Aboul-Atta, A.M., El-Ahmedaawy, B.-E.A., Fayed, M.A., Hafez, A.H. and Imam, M.H. (1983d) Duration of the developmental stages of house dust mites *Dermatophagoides farinae* and *D. pteronyssinus* under controlled conditions to pave the way in front of the workers in the field of house-dust mite asthmatic bronchitis. 3. Larval duration. *Journal of the Egyptian Society of Parasitology*, 13, 583–595.

Gamal-Eddin, F.M., Aboul-Atta, A.M., Hamad, M.G.M., Eraki, A.-S.M. and Tubeileh, F.M.S. (1985) House dust mites in Kuwait, preliminary pilot survey to shed some light on diagnosis and treatment of mite allergy. *Journal of the Egyptian Society of Parasitology*, 15, 313–322.

Gamble, C. (1999) *The Palaeolithic Societies of Europe*. Cambridge University Press, Cambridge.

Gao, Y.F., Wang, D.Y., Ong, T.C., Tay, S.L., Yap, K.H. and Chew, F.T. (2007) Identification and characterisation of a novel allergen from *Blomia tropicalis*: Blo t 21. *Journal of Allergy and Clinical Immunology*, 120, 105–112.

Gao, Z., Tseng, C., Pei, Z. and Blaser, M.J. (2007) Molecular analysis of human forearm superficial skin bacterial biota. *Proceedings of the National Academy of Sciences*, 104, 2927–2932.

Garcia-Robaina, J.C., Eraso, E., De la Torre, F., Guisantes, J., Martinez, A., Palacios, R. and Martinez, J. (1998) Extracts from various mite species contain cross-reactive and noncross-reactive IgE epitopes. A RAST inhibition study. *Journal of Investigational Allergology and Clinical Immunology*, 8, 285–289.

Garrett, M.H., Hooper, B.M. and Hooper, M.A. (1998) Indoor environmental factors associated with house-dust-mite allergen (Der p 1) levels in south-eastern Australian houses. *Allergy*, 53, 1060–1065.

Gatius, C., Gamboa, P., Jáuregui, I., Antépara, I., Urrutia, I., González, G., Zaballa, J. and Delgado, A. (1994) Annual distribution of *Dermatophagoides* (Der) major and minor allergens in houses from perennial asthmatics. *Allergy and Clinical Immunology News*, Supplement 2, 464. [abstract]

Gaud, J. (1969) Acariens de la sous-famille des Dermatophagoidinae (Psoroptidae) récoltés dans les plumages d'oiseaux. *Acarologia*, 10, 292–312.

Gaud, J. and Atyeo, W.T. (1996) Feather mites of the World (Acarina: Astigmata): the supraspecific taxa. *Musee Royal de l'Afrique Centrale Annales Sciences Zoologiques*, 277: Part I. Text, 1–193; Part II. Illustrations, 1–436.

Gaud, J. and Mouchet, J. (1959) Acariens plumicoles des oiseaux du Cameroun II. Analgesidae. *Annales de Parasitologie humaine et comparée*, 34, 149–208.

Gause, G.F., Smaragova, N.P. and Witt, A.A. (1936) Further studies on the interaction between a predator and its prey. *Journal of Animal Ecology*, 5, 1–18.

Gehring, U., Bischof, W., Schlenvoigt, G., Richter, K., Fahlbusch, B., Wichmann, H.E. and Heinrich, J. (2004) Exposure to house dust endotoxin and allergic sensitisation in adults. *Allergy*, 59, 946–952.

Gehring, U., Brunekreef, B., Fahlbusch, B., Wichmann, H.-E. and Heinrich, J. (2005) Are house dust mite allergen levels influenced by cold winter weather? *Allergy*, 60, 1079–1082.

Geiger, R. (1957) *The Climate Near The Ground*. (Translation of the second German edition of *Das Klima der Bodennahen Luftschicht*). Harvard University Press, Cambridge.

Gelber, L.E., Seltzer, L.H., Bouzoukis, J.K., Pollart, S.M., Chapman, M.D. and Platts-Mills, T.A.E. (1993) Sensitisation and exposure to indoor allergens as risk factors for asthma among patients presenting to hospital. *American Review of Respiratory Disease*, 147, 573–578.

Geller-Bernstein, C., Pibourdin, J.M., Dornelas, A. and Fondarai, J. (1995). Efficacy of the acaricide Acardust for the prevention of asthma and rhinitis due to dust mite allergy, in children. *Allergie et Immunologie*, 27, 147–154.

Gentry, T.J., Wickham, G.S., Schadt, C.W., He, Z. and Zhou, J. (2006) Microarray applications in microbial ecology research. *Microbial Ecology*, 52, 159–175.

Gereda, J.E., Klinnert, M.D., Price, M.R., Leung, D.Y.M. and Liu, A.H. (2001) Metropolitan home living conditions associated with indoor endotoxin levels. *Journal of Allergy and Clinical Immunology*, 107, 790–796.

Gershwin, M.E. and Albertson, T.A. (2001) *Bronchial Asthma: Principles of Diagnosis and Treatment*. Humana Press, Totowa, NJ.

Ghazala, L., Schmid, F., Helbling, A., Pichler, W.J. and Pichler, C.E. (2004) Efficacy of house dust mite- and allergen-impermeable encasings in patients with house dust mite allergy. *Allergologie*, 27, 26–34.

Gillies, D.R.N., Littlewood, J.M. and Sarsfield, J.K. (1987) Controlled trial of house dust mite avoidance in children with mild to moderate asthma. *Clinical Allergy*, 17, 105–111.

Gislason, D. and Gislason, T. (1999) IgE-mediated allergy to *Lepidoglyphus destructor* in an urban population – an epidemiological study. *Allergy*, 54, 878–883.

Gitoho, F. and Rees, H.P. (1971) High altitude and house dust mites. *British Medical Journal*, ii, 475.

Glass, E.V. and Needham, G.R. (2004) Eliminating *Dermatophagoides farinae* spp. (Acari: Pyroglyphidae) and their allergens through high temperature treatment of textiles. *Journal of Medical Entomology*, 41, 529–532.

Glass, E.V., Yoder, J.A. and Needham, G.R. (1998) Clustering reduces water loss by adult American house dust mites *Dermatophagoides farinae*. *Experimental and Applied Acarology*, 22, 31–37.

Glob, P.V. (1974) *The Mound People*. Faber and Faber, London.

Gomes, C., Freihaut, J. and Bahnfleth, W. (2007) Resuspension of allergen-containing particles under mechanical and aerodynamic disturbances from human walking. *Atmospheric Environment*, 41, 5257–5270.

Gomez, M.S., Portus, M. and Gallego, J. (1981a) Factores que influyen en la composicion de la fauna de acaros del polvo domestico. IV. Altitud. *Allergologia et Immunopathologia*, 9, 123–130.

Gomez, M.S., Portus, M. and Gallego, J. (1981b) Relacion existente entre la fisiografia de una localidad y la variacion estacional de su Acarofauna pulvicola. *Circular Farmaceuica*, 29, 243–253.

González Rodríguez, J. and Llorens Delgado, M. (1974) Identificacíon de los ácaros causantes de la sensibilización al polvo casero. Perimer informe en México. *Revista Médicale del Instituto Mexicano del Seguro Social*, 13, 49–57.

Goodman, N. and Hughes, J.F. (2002) Long-range destruction of Der p 1 using experimental and commercially available ionisers. *Clinical and Experimental Allergy*, 32, 1613–1619.

Goodman, N. and Hughes, J.F. (2004) The effect of corona discharge on dust mite and cat allergens. *Journal of Electrostatics*, 60, 69–91.

Goraccci, E., Lazzeri, S., Magnisi, M.T., Sondini, M.L. and Duranti, A. (1984) Gli allergoacari della polvere dei materassi della città di Livorno. *Folia Allergologia y Immunologia Clinica*, 31, 415–421.

Gore, R.B., Hadi, E.A., Craven, M., Smillie, F.I., O'Meara, T.J., Tovey, E.R., Woodcock, A. and Custovic, A. (2002) Personal exposure to house dust mite allergen in bed: nasal air sampling and reservoir allergen levels. *Clinical and Experimental Allergy*, 32, 856–859.

Gøtzsche, P. and Johansen, H.K. (2008) House dust control measures for asthma (review). *Cochrane Database of Systematic Reviews* 2008, Issue 2, Article No. CD001187.

Gøtzsche, P., Hammarquist, C. and Burr, M.L. (1998) House dust control measures in the management of asthma: meta-analysis. *British Medical Journal*, 317, 1105–1110.

Gøtzsche, P., Johansen, H.K., Schmidt, L.M. and Burr, M.L. (2004) House dust control measures for asthma (review). *Cochrane Database of Systematic Reviews* 2004, Issue 4, Article No. CD001187. (Also published in: *The Cochrane Library*, 2007, Issue 4).

Gough, L., Schulz, O., Sewell, H.F. and Shakib, F. (1999) The cysteine protease activity of the major dust mite allergen Der p 1 selectively enhances the immunoglobulin E antibody response. *Journal of Experimental Medicine*, 190, 1897–1901.

Gough, L., Sewell, H.F. and Shakib, F. (2001) The proteolytic activity of the major dust mite allergen Der p 1 enhances the IgE antibody response to a bystander allergen. *Clinical and Experimental Allergy*, 31, 1594–1598.

Gould, S.J. (1991) Chapter 5. In: *Bully for Brontosaurus*. Penguin Books, London, pp. 79–83.

Grácio, A.J. dos S. and Quinta, M.J. (2000) Contribution to the knowledge of the house dust mites in Portugal (Arachnida: Acari). *Acta Parasitologica Portuguesa*, 7, 39–42.

Graham, J.A.H., Pavlicek, P.K., Sercombe, J.K., Xavier, M.L. and Tovey, E.R. (2000) The nasal air sampler: a device for sampling inhaled aeroallergens. *Annals of Allergy, Asthma and Immunology*, 84, 599–604.

Grandjean, F. (1937) *Otodectes cynotis* (Hering) et les prétendues trachées des Acaridiae. *Bulletin de la Societè Zoologique de France*, 62, 280–290.

Greco, D.B., Moreira, N.S., Filogonio, C.J.B. and Greco, J.B. (1974) Demonstracão da presenca de acaros em po domicilar de Belo Horizonte e outras Cidades de Minas Gerais. *14th Congresso Brasiliero do Alergia y Immunopathologia, Recife*, p. 1.

Green, W.F. (1983) The house dust mite, *Dermatophagoides farinae*, in Australia. *Medical Journal of Australia*, ii, 259–260.

Green, W.F. and Woolcock, A.J. (1978) *Tyrophagus putrescentiae*, an allergenically important mite. *Clinical Allergy*, 8, 135–144.

Green, W.F., Woolcock, A.J. and Dowse, G. (1982) House dust mites in the highlands of Papua New Guinea. *Papua New Guinea Medical Journal*, 25, 219–222.

Green, W.F., Woolcock, A.J., Stuckey, M., Segewick, C. and Leeder, S.R. (1986) House dust mites and skin tests at different Australian localities. *Australian and New Zealand Journal of Medicine*, 16, 639–643.

Green, W.F., Nicholas, N.R., Salome, C.M. and Woolcock, A.J. (1989) Reduction of house dust mites and mite allergens, effects of spraying carpets and blankets with Allersearch agent. *Clinical and Experimental Allergy*, 19, 203–207.

Green, W.F., Marks, G.B., Tovey, E.R., Toelle, B.G. and Woolcock, A.J. (1992) House dust mites and mite allergens in public places. *Journal of Allergy and Clinical Immunology*, 89, 1196–1197.

Green, W.F., Toelle, B. and Woolcock, A.J. (1993) House dust mite increase in Wagga Wagga houses. *Australian and New Zealand Journal of Medicine*, 23, 409.

Grice, K.A. (1980) Transepidermal water loss. In: Jarrett, A. (ed.), *The Physiology and Pathophysiology of the Skin*. Volume 6. Academic Press, London, pp. 2115–2146.

Gridelet, D. and Lebrun, P. (1973) Contribution a l'étude écologique des acariens des poussières de maisons. *Acarologia*, 15, 461–467.

Gridelet-de-Saint Georges, D. (1975) Techniques d'extraction applicables a l'étude écologique des acariens des poussières de maison. Comparison qualitative et quantitative de divers types de poussières. *Acarologia*, 17, 693–708. [cf. also D. de Saint Georges-Gridelet below.]

Griendel von Ach, J.F. (1687) *Micrographia Nova: Oder Neu-Curieuse Beschriedener Verschiedener kleiner Körper, welche vermittelst eines Absonderlichen von dem Authore neuerfundenen Vergrösser-Glases Verwunderlich gross vorgestellet werden*. Johann Ziegers, Nürnberg.

Griffin, D.H. (1994) *Fungal Physiology*. 2nd edition. Wiley-Liss, New York.

Griffin, P., Ford, A.W., Alterman, L., Thompson, J., Parkinson, C., Blainey, A.D., Davies, R.J. and Topping, M.D. (1989) Allergenic and antigenic relationship between three species of storage mite and the house dust mite, *Dermatophagoides pteronyssinus*. *Journal of Allergy and Clinical Immunology*, 84, 108–117.

Griffiths, D.A. (1960) Some field habitats of mites of stored food products. *Annals of Applied Biology*, 48(1), 134–144.

Griffiths, D.A. (1964) A revision of the genus *Acarus* L., 1758 (Acaridae, Acarina). *Bulletin of the British Museum (Natural History) Zoology*, 11(6), 415–463.

Griffiths, D.A. (1970) A further systematic study of the genus *Acarus* L., 1758 (Acaridae, Acarina), with a key to species. *Bulletin of the British Museum (Natural History) Zoology*, 19(2), 85–118.

Griffiths, D.A. (1984) The morpho-species and its relationship to the biological species in the genus *Tyrophagus* (Acaridae, Acarina). In: Griffiths, D.A. and Bowman, C.E. (eds), *Acarology VI. Volume 1*. Ellis Horwood, Chichester, pp. 199–212.

Griffiths, D.A. and Boczek, J. (1977) Spermatophores of some acaroid mites. *International Journal of Morphology and Embryology*, 6, 231–238.

Griffiths, D.A. and Cunnington, A.M. (1971) *Dermatophagoides microceras* sp. n., a description and comparison with its sibling species, *D. farinae* Hughes, 1961. *Journal of Stored Products Research*, 7, 1–14.

Gross, I., Heinrich, J., Fahlbusch, B., Jäger, L., Bischof, W. and Wichmann, H.E. (2000) Indoor determinants of Der p 1 and Der f 1 in house dust are different. *Clinical and Experimental Allergy*, 30, 376–382.

Gu, P.-L., Gunawardine, Y.I., Chow, B.-C., He, J.-G. and Chan, S.-M. (2002) Characterisation of a novel cellular retinoic acid/retinol binding protein from shrimp: expression of the recombinant protein for immunohistochemical detection and binding assay. *Gene*, 288, 77–84.

Guerin, B., Levy, D.A., LeMao, J., Leynardier, F., Baligadoo, G., Fain, A. and Dry, J. (1992) The house dust mite *Dermatophagoides pteronyssinus* is the most important allergen on the island of Mauritius. *Clinical and Experimental Allergy*, 22, 533–539.

Guerra, R.M.S.N.C., Gazêta, G.S., Amorim, M., Duarte, A.N. and Serra-Freire, N.M. (2003) Ecological analysis of Acari recovered from coprolites from archaeological site of Northeast Brazil. *Memorias do Instituto Oswaldo Cruz*, 98 (Supplement 1), 181–190.

Gülegen, E., Girisgin, O., Kütükoglu, F., Girisgin, A.O. and Coskun, S.Z. (2005) [Mite species found in dust in houses in Bursa.] *Tükiye Parazitoloji Dergisi*, 29, 185–187. [In Turkish, English summary]

Gupta, R., Sheikh, A., Strachan, D.P. and Anderson, H.R. (2007) Time trends in allergic disorders in the UK. *Thorax*, 62, 91–96.

Gupta, S.K. and Roy, R.K.D. (1975) Occurrence of house dust mites in West Bengal (India). *Newsletter of the Zoological Survey of India*, 1, 48–50.

Gutgesell, C., Heise, S., Seubert, S., Seubert, A., Domhof, S., Brunner, E. and Neumann, C. (2001) Double-blind placebo-controlled house dust mite control measures in adult patients with atopic dermatitis. *British Journal of Dermatology*, 145, 70–74.

Guy, Y., Rioux, J.-A., Guin, J.-J. and Rousset, G. (1972) La faune acarologique de la poussière de maison. Résultats d'une enquête en Languedoc-Roussillon. *Bulletin de la Societé du Pathologie exotique*, 65, 472–481.

Haarløv, N. and Alani, M. (1970) House-dust mites (*Dermatophagoides pteronyssinus* (Trt.), *D. farinae* Hughes, *Euroglyphus maynei* (Cooreman) Fain in Denmark. *Entomologia Scandinavica*, 1, 301–306.

Hadley, N.F. (1994) *Water Relations of Terrestrial Arthropods*. Academic Press, San Diego.

Hafez, S.M. and Salem, M.S. (1988) Aleuroglyphid mites inhabiting house dust (Acari: Acaridae). *Annals of Agricultural Science, Cairo*, 33, 1403–1409.

Hafez, S.M., Seoudi, M.M. and Salem, M.S. (1989) Occurrence of five mite species in house-dust at Giza Governorate, Egypt. *Annals of Agricultural Science, Cairo*, 34, 427–434.

Hafiz, H.A. (1935) The embryonic development of *Cheyletus eruditus* (a mite). *Proceedings of the Royal Society* (Series B), 117, 174–201.

Hagan, L.L., Goetz, D.W., Revercomb, C.H. and Garriott, J. (1998) Sudden infant death syndrome: a search for allergen hypersensitivity. *Annals of Allergy, Asthma and Immunology*, 80, 227–231.

van Hage-Hamsten, M. and Johansson, S.G.O. (1989) Clinical significance and allergenic cross-reactivity of *Euroglyphus maynei* and other nonpyroglyphid and pyroglyphid mites. *Journal of Allergy and Clinical Immunology*, 83, 581–589.

van Hage-Hamsten, M., Johansson, S.G.O., Höglund, S., Tüll, P., Wirén, A. and Zetterstrom, O. (1985) Storage mite allergy is common in a farming community. *Clinical Allergy*, 15, 555–564.

van Hage-Hamsten, M., Johansson, S.G.O., Johansson, E. and Wiren, A. (1987) Lack of allergenic cross-reactivity between storage mites and *Dermatophagoides pteronyssinus*. *Clinical Allergy*, 17, 23–31.

van Hage-Hamsten, M., Machado, L., Barros M.T. and Johansson, S.G.O. (1990a) Comparison of clinical significance and allergenic cross-reactivity of storage mites *Blomia kulagini* and *Lepidoglyphus destructor* in Sweden and Brazil. *Allergy*, 45, 409–417.

van Hage-Hamsten, M., Machado, L., Barros, M. and Johansson, S.G.O. (1990b) Immune response to *Blomia kulagini* and *Dermatophagoides pteronyssinus* in Sweden and Brazil. *International Archives of Allergy and Applied Immunology*, 91, 186–191.

van Hage-Hamsten, M., Scheynius, A., Härfast, B., Wirén, A. and Johansson, S.G.O. (1992a) Localisation of allergens in the domestic mite *Lepidoglyphus destructor*. *Clinical and Experimental Allergy*, 22, 251–256.

van Hage-Hamsten, M., Bergman, T. and Johansson, E. (1992) N-terminal aminoacid sequence of principal allergen of storage mite *Lepidoglyphus destructor*. *Lancet*, 340, 614.

van Hage-Hamsten, M., Olsson, S., Emilson, A., Härfast, B., Svensson, A. and Scheynius, A. (1995) Localisation of major allergens in the dust mites *Lepidoglyphus destructor* with confocal scanning microscopy. *Clinical and Experimental Allergy*, 25, 536–542.

Haida, M., Okudaira, H., Ogita, T., Ito, K., Miyamoto, T., Nakajima, T. and Hongo, O. (1985) Allergens of the house dust mite *Dermatophagoides farinae* – immunochemical studies of four allergic fractions. *Journal of Allergy and Clinical Immunology*, 75, 686–692.

Haldane, J.B.S. (1927) *Possible Worlds and Other Essays* . Chatto and Windus, London.

Hales, B.J., Hazell, L.A., Smith, W. and Thomas, W.R. (2002) Genetic variation of Der p 2 allergens: effects on T cell responses and immunoglobulin E

binding. *Clinical and Experimental Allergy*, 32, 1461–1467.

Hales, B.J., Laing, I.A., Pearce, L.J., Hazell, L.A., Mills, K.L., Chua, K.Y., Thornton, R.B., Richmond, P., Musk, A.W., James, A.L., LeSouëf, P.N. and Thomas, W.R. (2007) Distinctive immunoglobulin E anti-house dust allergen-binding specificities in a tropical Australian Aboriginal community. *Clinical and Experimental Allergy*, 37, 1357–1363.

Halken, S., Høst, A., Niklassen, U., Hansen, L.G., Nielsen, F., Pedersen, S., Østerballe, O., Veggerby, C. and Poulsen, L.K. (2003) Effect of mattress and pillow encasings on children with asthma and house dust mite allergy. *Journal of Allergy and Clinical Immunology*, 111, 169–176.

Hall, C.C., McMahon, B. and Sams, J.T. (1971) Collecting and rearing *Dermatophagoides farinae* Hughes, a mite from house dust. *Annals of Allergy*, 29, 81–85.

Hallas, T.E. (1998) Pitfalls in evaluating mite exposure from house-dust samples. *Respiratory Medicine*, 92, 1099–1101.

Hallas, T.E. and Gudmundsson, B. (1987) Storage mites in hay in Iceland. *European Journal of Respiratory Diseases*, 71 (Supplement 154), 60–64.

Hallas, T.E. and Korsgaard, J. (1983) Annual fluctuations of mites and fungi in Danish house dust, an example. *Allergologia et Immunopathologia*, 11, 195–200.

Hallas, T.E., Gislason, D., Björnsdottir, U.S., Jörundsdottir, K.B., Janson, C., Luczynska, C.M., Gislason, T. (2004) Sensitisation to house dust mites in Reykjavik, Iceland, in the absence of domestic exposure to mites. *Allergy*, 59, 515–519.

de Halleux, S., Stura, E., VanderElst, L., Carlier, V., Jacquemin, M. and Saint-Remy, J.-M. (2006) Three-dimensional structure and IgE-binding properties of mature fully active Der p 1, a clinically relevant major allergen. *Journal of Allergy and Clinical Immunology*, 117, 571–576.

Halmai, Z. (1980) Morphological variant of a *Dermatophagoides* species occurring in house dust (Acari, Pyroglyphidae). *Parasitologia Hungarica*, 13, 103–106.

Halmai, Z. (1984) Changes in the composition of house-dust mite fauna in Hungary. *Parasitologia Hungarica*, 17, 59–70.

Halmai, Z. (1989) Postimaginal moults in house-dust mites *Dermatophagoides pteronyysinus* (Acari,

Pyroglyphidae) associated with various defects of development. *Parasitologia Hungarica*, 22, 137–142.

Halmai, Z. (1994) The phenomenon 'cannibalism' in *Dermatophagoides farinae* (Acari) populations. *Parasitologia Hungarica*, 27, 69–72.

Halmai, Z. and Alexander, F.A.R. (1971) Studies on the house-dust allergen (preliminary report). *Allergie und Immunologie*, 17, 69–77.

Halmai, Z. and Szocska, M. (1983) On the correlation between the allergic tests with house dust extracts and the mite content of the house dust samples. *Parasitologia Hungarica*, 16, 111–117.

Halmerbauer, G., Gartner, C., Schierl, M., Arshad, M., Dean, T., Kolloer, D.Y., Karmaus, W., Kuehr, J., Forster, J., Urbanek, R. and Frischer, T. (2003) Study on the Prevention of Allergy in Children in Europe (SPACE): Allergic sensitisation at 1 year of age in a controlled trial of allergen avoidance from birth. *Pediatric Allergy and Immunology*, 14, 10–17.

Hamashima, Y., Kishi, K. and Tasaka, K. (1973) Ricketsia-like bodies in infantile acute febrile mucocutaneous lymph-node syndrome. *The Lancet*, July 7, 42.

Hamashima, Y., Tasaka, K., Hoshino, T., Nagata, N., Furukawa, F., Kao, T. and Tanaka, H. (1982) Mite-associated particles in Kawasaki disease. *The Lancet*, July 31, 266.

Hamilton, G. (1998) Let them eat dirt. *New Scientist*, 18 July, 26–31.

Hannaway, P.J. and Roundy, C. (1997) Distribution of *Dermatophagoides* spp., *D. farinae* and *D. pteronyssinus* antigen in homes of patients with asthma in eastern Massachusetts. *Allergy and Asthma Proceedings*, 18, 177–180.

Hanski, I.A. and Gilpin, M.E. (eds) (1997) *Metapopulation Biology: Ecology, Genetics and Evolution*. Academic Press, New York.

Härfast, B., Hage-Hamsten, M. van, Ansotegui, I.J., Johansson, E., Jeddi-Tehrani, M. and Johansson, S.G.O. (1992) Monoclonal antibodies to *Lepidoglyphus destructor*, delineation of crossreactivity between storage mites and house dust mites. *Clinical and Experimental Allergy*, 22, 1032–1038.

Härfast, B., Johansson, E., Johansson, S.G.O. and van Hage-Hamsten, M. (1996) ELISA method for detection of mite allergens in barn dust: comparison with mite counts. *Allergy*, 51, 257–261.

Harris, J. (ed.) (1993) *5000 Years of Textiles*. British Museum Press, London.

Harrop, J., Chinn, S., Verlato, G., Olivieri, M., Norback, D., Wjst, M., Janson, C., Zock, J.-P., Leynart, B., Gislason, D., Ponzio, M., Villani, S., Carosso, A., Svannes, C., Heinrich, J. and Jarvis, D. (2007) Eczema, atopy and allergen exposure in adults: a population-based study. *Clinical and Experimental Allergy*, 37, 526–535.

Hart, B.J. (1990) Ecology and biology of allergenic mites. In: Fain, A., Guerin, B. and Hart, B.J., *Mites and Allergic Diseases*. Allerbio, Varennes-en-Argonne, pp. 137–152.

Hart, B.J. (1995) The biology of allergenic domestic mites: an update. *Clinical Reviews in Allergy and Immunology*, 13, 115–133.

Hart, B.J. and Fain, A. (1987) A new technique for the isolation of mites exploiting the differences in density between ethanol and saturated NaCl, Qualitative and quantitative studies. *Acarologia*, 28, 251–254.

Hart, B.J. and Fain, A. (1988) Morphological and biological studies of medically important house-dust mites. *Acarologia*, 29, 285–295.

Hart, B.J. and Whitehead, L. (1990) Ecology of house dust mites in Oxfordshire. *Clinical and Experimental Allergy*, 20, 203–209.

Hart, B.J., Guérin, B. and Nolard, N. (1992) *In vitro* evaluation of acaricidal and fungicidal activity of the house dust mite acaricide, Allerbiocid. *Clinical and Experimental Allergy*, 22, 923–928.

Hart, B.J., Crowther, D., Wilkinson, T., Biddulph, P., Ucci, M., Pretlove, S., Ridley, I. and Oreszczyn, T. (2007) Reproduction and development of laboratory and wild house dust mites (Acari: Pyroglyphidae) and their relationship to the natural dust ecosystem. *Journal of Medical Entomology*, 44, 568–574.

Harumal, P., Morgan, M., Walton, S.F., Holt, D.C., Rode, J., Arlian, L.G., Currie, B.J. and Kemp, D.J. (2003) Identification of a homologue of a house dust mite allergen in a cDNA library from *Sarcoptes scabiei* var. *hominis* and evaluation of its vaccine potential in a rabbit/*S. scabiei* var. *canis* model. *American Journal of Tropical Medicine and Hygiene*, 68, 54–60.

Harving, H., Korsgaard, J., Dahl, R., Beck, H.-I. and Bjerring, P. (1990) House dust mites and atopic dermatitis. A case-control study on the significance of house dust mites as etiologic allergens in atopic dermatitis. *Annals of Allergy*, 65, 25–31.

Harving, H., Korsgaard, J. and Dahl, R. (1993) House-dust mites and associated environmental conditions in Danish homes. *Allergy*, 48, 106–109.

Harving, H., Korsgaard, J. and Dahl, R. (1994) House-dust mite exposure reduction in specially designed, mechanically ventilated 'healthy' homes. *Allergy*, 49, 713–718.

Hatsushika, R. and Miyoshi, K. (1992a) [Mite fauna found in house-dust from the residences of suspected acarine dermatitis patients.] *Japanese Journal of Sanitary Zoology*, 43(2), 125–127. [In Japanese, English summary]

Hatsushika, R. and Miyoshi, K. (1992b) [Studies on house-dust mites. The itching dermatitis in Okayama Prefecture, Japan.] *Kawasaki Medical Journal*, 18, 1–9. [In Japanese, English summary]

Hauptmann, A. (1657) *Uhralter Wolkensteinischer Warmer Badt- und Wassersschatz*. Leipzig.

Hay, D.B. (1995) An 'in situ' coring technique for estimating the population size of house dust mites in their natural habitat. *Acarologia*, 36, 341–345.

Hay, D.B., Hart, B.J., Pearce, R.B., Kozakiewicz, Z. and Douglas, A.E. (1992) How relevant are house dust mite–fungal interactions in laboratory culture to the natural dust system? *Experimental and Applied Acarology*, 16, 37–48.

Hay, D.B., Hart, B.J. and Douglas, A.E. (1993) Effects of the fungus *Aspergillus penicilloides* on the house dust mite *Dermatophagoides pteronyssinus*: an experimental re-evaluation. *Medical and Veterinary Entomology*, 7, 271–274.

Hayden, M.L., Rose, G., Diduch, K.B., Domson, P., Chapman, M.B., Heymann, P.W. and Platts-Mills, T.A.E. (1992). Benzyl benzoate moist powder: investigation of acaricidal activity in cultures and reduction of dust mite allergens in carpets. *Journal of Allergy and Clinical Immunology*, 89, 536–545.

Hegarty, J.M., Rouhbakhsh, S., Warner, J.A. and Warner, J.O. (1995) A comparison of the effect of conventional and filter vacuum cleaners on airborne house dust mite allergen. *Respiratory Medicine*, 89, 279–284.

Hegarty, S.A. (1988) Effect of indoor environment and clinical status of allergy on house dust mite populations. Unpublished BSc Honours Dissertation, Department of Zoology, University of Glasgow.

van der Heide, S., Kauffman, H.F., Dubois, A.E.J. and de Monchy, J.G.R. (1997a) Allergen-avoidance measures in homes of house-dust-mite-allergic asthmatic patients: effects of acaricides and mattress encasings. *Allergy*, 52, 921–927.

van der Heide, S., Kauffman, H.F., Dubois, A.E.J. and de Monchy, J.G.R. (1997b) Allergen reduction measures in houses of allergic asthmatic patients: effects of air-cleaners and allergen-impermeable mattress covers. *European Respiratory Journal*, 10, 1217–1223.

van der Heide, S., De Monchy, J., De Vries, K., Dubois, A. and Kauffman, H. (1997c) Seasonal differences in airway hyper-responsiveness in asthmatic patients: relationship with allergen exposure and sensitisation to house dust mites. *Clinical and Experimental Allergy*, 27, 627–633.

Heinrich, J., Popescu, M.A., Wjst, M., Goldstein, I.F. and Wichmann, H.-E. (1998) Atopy in children and parental social class. *American Journal of Public Health*, 88, 1319–1324.

Heinze, K. (1952) Polyvinylalkohol-Lactophenol-Gemisch als Einbettungsmittel für Blattläuse. *Naturwissenshaften*, 39, 285–286.

Heniger, J. (ed.) (1976) *The Collected Works of Antoni Van Leeuwenhoek. Volume 9*. Swets and Zeitlinger, Amsterdam, pp. 269–317.

Herbert, C.A., King, C.M., Ring, P.C., Holgate, S.T., Stewart, G.A., Thompson, P.J. and Robinson, C. (1995) Augmentation of permeability in the bronchial epithelium by the house dust mite allergen Der p 1. *American Journal of Respiratory Cellular and Molecular Biology*, 12, 369–378.

Hewitt, C.R.A., Brown, A.P., Hart, B.J. and Pritchard, D.I. (1995) A major house dust mite allergen disrupts the immunoglobulin E network by selectively cleaving CD23: innate protection by antiproteases. *Journal of Experimental Medicine*, 182, 1537–1544.

Hewitt, M., Barrow, G.I., Miller, D.C., Turk, F. and Turk, S. (1973) Mites in the personal environment and their role in skin disorders. *British Journal of Dermatology*, 89, 401–409.

Heymann, P.W., Chapman, M.D., Platts-Mills, T.A.E. (1986) Antigen *Der f* I from the dust mite *Dermatophagoides farinae*, structural comparison with *Der p* I from *Dermatophagoides pteronyssinus* and epitope specificity of murine IgG and human IgE antibodies. *Journal of Immunology*, 137, 2841–2847.

Heymann, P.W., Chapman, M.D., Aalberse, R.C., Fox, J.W. and Platts-Mills, T.A.E. (1989) Antigenic and structural analysis of group II allergens (*Der f* II and *Der p* II) from house dust mites (*Dermatophagoides* spp.). *Journal of Allergy and Clinical Immunology*, 83, 1055–1067.

Heyraud, J.D., Marotel, C., Natali, F., L'Her, P., Vaylet, F. and Allard, P. (1989) Etude de la sensibilisation cutanée au *Lepidoglyphus destructor* et au *Tyrophagus putrescentiae* chez 200 asthmatiques. *Revue Française d' Allergologie et d'Immunologie Clinique*, 29, 181–183.

Hickey, D., Benkel, B.F., Boer, P.H., Genest, Y., Abukashawa, S. and Ben-David, G. (1987) Enzyme-coding genes as molecular clocks: the molecular evolution of animal α-amylases. *Journal of Molecular Evolution*, 26, 252–256.

Hide, D.W., Matthews, S., Tariq, S. and Arshad, S.H. (1996) Allergen avoidance in infancy and allergy at 4 years of age. *Allergy*, 51, 89–93.

Hirsch, Th., Range, U., Walther, K.U., Hederer, B., Lässig, S., Frey, G. and Leupold, W. (1998) Prevalence and determinants of house dust mite allergen in East German homes. *Clinical and Experimental Allergy*, 28, 956–964.

Hirsch, U. (1991) The fabric of deities and kings. *Hali*, 58, 104–111.

van der Hoeven, W.A.D., de Boer, R. and Bruin, J. (1992) The colonisation of new houses by house dust mites (Acari: Pyroglyphidae). *Experimental and Applied Acarology*, 16, 75–84.

van der Hoeven, W.A.D., Bruin, J. and de Boer, R. (1995) How fast do dust mites colonize new houses? In: Kropczynska, D., Boczek, J. and Tomczyk, A. (eds), *The Acari. Physiological and Ecological Aspects of Acari – Host Relationships* . Oficyna DABOR, Warsaw, pp. 211–216.

Ho, T.M. (1986) Pyroglyphid mites found in house dust in Peninsular Malaysia. *Tropical Biomedicine*, 3, 89–93.

Ho, T.M. and Nadchatram, M. (1984) Life-cycle and longevity of *Dermatophagoides pteronyssinus* (Trouessart) (Acarina, Astigmata, Pyroglyphidae) in a tropical laboratory. *Tropical Biomedicine*, 1, 159–162.

Ho, T.M. and Nadchatram, M. (1985) Distribution of *Dermatophagoides pteronyssinus* (Astigmata: Pyroglyphidae) in Cameron Highlands, Malaysia. *Tropical Biomedicine*, 2, 54–58.

Hodgson, R.K. (1976a) Sex ratio and sex determination in the American house dust mite, *Dermatophagoides farinae. Annals of the Entomological Society of America*, 69, 1085–1086.

Hodgson, R.K. (1976b) *Tyrophagus putrescentiae* (Acarina, Tyroglyphidae) in cultures of *Dermatophagoides farinae* (Acarina, Pyroglyphidae) controlled by low humidity. *Journal of Medical Entomology*, 13, 223–224.

Hoeppli, R. (1959) *Parasites and Parasitic Infections in Early Medicine and Science.* University of Malaya Press, Singapore.

Hoffman, D.R. (1994) Allergens in hymenoptera venom. XXVI: the complete amino acid sequences of two vespid venom phospholipases. *International Archives of Allergy and Immunology*, 104, 184–190.

Holdridge, L.R. (1947) Determination of world plant formations from simple climatic data. *Science*, 105, 367–368.

Holm, L., van Hage-Hamsten, M., Öhman, S. and Scheynius, A. (1999) Sensitisation to allergens of house-dust mite in adults with atopic dermatitis in a cold temperate region. *Allergy*, 54, 708–715.

Holm, L., Öhman, S., Bengtsson, Å., van Hage-Hamsten, M. and Scheynius, A. (2001) Effectiveness of occlusive bedding in the treatment of atopic dermatitis – a placebo-controlled trial of 12 months' duration. *Allergy*, 56, 152–158.

Holt, D.C., Fischer, K., Allen, G.E., Wilson, D., Slade, R., Currie, B.J., Walton, S.F. and Kemp, D.J. (2003) Mechanisms for a novel immune evasion strategy in the scabies mite *Sarcoptes scabiei*: a multigene family of inactivated serine proteases. *Journal of Investigative Dermatology*, 121, 1419–1424.

Holt, D.C., Fischer, K., Pizzutto, S.J., Currie, B.J., Walton, S.F. and Kemp, D.J. (2004) A multigene family of inactivated cysteine proteases in *Sarcoptes scabiei. Journal of Investigative Dermatology*, 123(1), 240–241.

Holt, P.G., McMenamin, C. and Nelson, D. (1990) Primary sensitisation to inhalant allergens during infancy. *Pediatric Allergy and Immunology*, 1, 3–13.

Holt, P.G., Macaubas, C., Stumbles, P.A. and Sly, P.D. (1999) The role of allergy in the development of asthma. *Nature*, 402 (Supplement), B12– B17.

Holz, O., Mücke, M., Paasch, K., Böhme, S., Timm, P., Richter, K., Magnussen, H. and Jörres, R.A. (2002) Repeated ozone exposures enhace bronchial allergen responses in subjects with rhinitis or asthma. *Clinical and Experimental Allergy*, 32, 681–689.

Hong, C.S. (1991) [Sensitisation of house dust mites in the allergic patients and mite ecology in their house dusts.] *Journal of the Korean Society of Allergology*, 11, 457–475. [In Korean, English summary]

Hong, C.S., Park, H.S. and Oh, S.H. (1987) *Dermatophagoides farinae*, an important allergenic substance in buckwheat husk pillows. *Yonsei Medical Journal*, 28, 274–281.

Hong, C.S. and Lee, M.K. (1992) [Measurement of group I allergen of house dust mites in dusts of Seoul and monthly variations of *Der f* I.] *Journal of the Korean Society of Allergology*, 12, 482–492. [In Korean, English summary]

Hooke, R. (1665) *Micrographia, or Some Physiological Descriptions of Minute Bodies Made by Magnifying Glasses With Observations and Inquiries Thereon.* The Royal Society, London, pp. 213–215.

Hoole, S. (1798, 1807) *The Select Works of Antony van Leeuwenhoek, Containing his Microscopical Discoveries in Many of the Works of Nature.* 2 Vols, Henry Fry, London.

Hooper, R., Calvert, J., Thompson, R.L., Deetlefs, M.E. and Burney, P. (2008) Urban/rural differences in diet and atopy in South Africa. *Allergy*, 63, 425–431.

Hora, A.M. (1934) On the biology of the mite, *Glycyphagus domesticus* De Geer (Tyroglyphidae, Acarina). *Annals of Applied Biology*, 21, 483–494.

Horák, B. (1987) Composition of bacteria, fungi and mites in samples of house dust from Silesia (Poland). *Allergologia et Immunopathologia*, 15, 161–166.

Horák, B., Dutkiewic, J. and Solarz, K. (1996) Microflora and acarofauna of bed dust from homes in Upper Silesia, Poland. *Annals of Allergy, Asthma and Immunology*, 76, 41–50.

Horie, M., Ushiyama, M., Tanaka, T., Adachi, M., Aoki, S. and Oharu, S. (1992) [Investigations of the species of mite in the dust of detached houses in Hiroshima area.] *Japanese Journal of Pediatric Allergy and Clinical Immunology*, 6, 27–32. [In Japanese, English summary]

Howard, R.W., Kuwahara Y., Suzuki, H., Suzuki, T. (1988) Pheromone study on acarid mites. XII. Characterization of hydrocarbons and external gland morphology of the opisthonotal glands of six species of mites (Acari: Astigmata). *Applied Entomology and Zoology*, 23, 58–66.

Howarth, P.H., Lunn, A. and Tomkins, S. (1992) Bedding barrier intervention in house dust mite respiratory allergy. *Clinical and Experimental Allergy*, 22, 140. [abstract]

Howe, R.W. (1953) The rapid determination of the intrinsic rate of increase of an insect population. *Annals of Applied Biology*, 40, 134–151.

Howe, R.W. (1971) A parameter for expressing the suitability of an environment for insect development. *Journal of Stored Products Research*, 7, 63–65.

Htut, T., Nyunt, D.W.W. and Mitchell, S. (1991) A study of the mite fauna in the homes of asthmatics with positive skin tests for *Dermatophagoides pteronyssinus* in Burma. *International Journal of Environmental Health Research*, 1, 117–122.

Htut, T., Higenbottam, T.W., Gill, G.W., Darwin, S.R., Anderson, P.B. and Syed, N. (2001) Eradication of house dust mite from homes of atopic asthmatic subjects: a double-blind trial. *Journal of Allergy and Clinical Immunology*, 107, 55–60.

Huang, C.H., Liew, L.M., Mah, K.W., Kuo, I.C., Lee, B.W. and Chua, K.Y. (2006) Characterisation of glutathione s-transferase from dust mite, Der p 8 and its immunoglobulin E cross-reactivity with cockroach glutathione s-transferase. *Clinical and Experimental Allergy*, 36, 369–376.

Huang, Y., Tan, J.M.W.L., Kini, R.M. and Ho, S.H. (1997) Toxic and antifeedant action of nutmeg oil against *Tribolium castaneum* (Herbst) and *Sitophilus zeamais* Motsch. *Journal of Stored Products Research*, 33, 289–298.

Hughes, A.M. (1948) *Mites Associated with Stored Food Products.* Her Majesty's Stationery Office, London.

Hughes, A.M. (1954) On a new species of *Dermatophagoides* belonging to the family Psoroptidae Canestrini, 1892 (Acarina). *Proceedings of the Zoological Society of London*, 124, 1–12.

Hughes, A.M. (1956) The mite genus *Lardoglyphus* Oudemans, 1927 (= *Hoshikadania* Sasa and Asanuma, 1951). *Zoologische Mededelingen*, 34(20), 271–285.

Hughes, A.M. (1961) *The Mites of Stored Food.* Technical Bulletin of the Ministry of Agriculture, London, no. 9, Her Majesty's Stationery Office, London.

Hughes, A.M. (1976) *The Mites of Stored Food and Houses.* Her Majesty's Stationery Office, London.

Hughes, A.M. and Hughes, T.E. (1939) The internal anatomy and post-embryonic development *of Glycyphagus domesticus* de Geer. *Proceedings of the Zoological Society of London*, Series B, 108, 715–733.

Hughes, A.M. and Maunsell, K. (1973) A study of a population of house dust mites in its natural environment. *Clinical Allergy*, 3, 127–131.

Hughes, T.E. (1950) The physiology of the alimentary canal of *Tyroglyphus farinae*. *Quarterly Journal of Microscopical Science*, 91, 45–61.

Hughes, T.E. (1959) *The Acarina, or the Mites*. Athlone Press, London.

Hunponu–Wusu, O.O. and Somorin, A.O. (1978) Epidemiological aspects of house dust mite allergy in Nigeria. *Journal of Dermatology*, 5, 27–32.

Hunter, C.A., Grant, C., Flannagan, B. and Bravery, A.F. (1988) Moulds in buildings: the air spora of domestic dwellings. *International Biodeterioration*, 24, 81–101.

Huntley, J.F., Machell, J., Nisbet, A.J., van den Broek, A., Chua, K.Y., Cheong, N., Hales, B.J. and Thomas, W.R. (2004) Identification of tropomyosin, paramyosin and apolipophorin/vitellogenin as three major allergens of the sheep scab mite, *Psoroptes ovis*. *Parasite Immunology*, 26, 335–342.

Hurtado, I. and Parini, M. (1987) House dust mites in Caracas, Venezuela. *Annals of Allergy*, 59, 128–130.

Huss, K., Squire, E.N., Carpenter, G.B., Smith, L.J., Huss, R.W., Salata, K., Salerno, M., Agostinelli, D. and Hershey, J. (1992) Effective education of adults with asthma who are allergic to dust mites. *Journal of Allergy and Clinical Immunology*, 89, 836–843.

Huss, R.W., Huss, K., Squire, E.N., Carpenter, G.B., Smith, L.J., Salata, K. and Hershey, J. (1994) Mite allergen control with acaricide fails. *Journal of Allergy and Clinical Immunology*, 94, 27–32.

Hutchinson, G.E. (1958) Concluding remarks. *Cold Spring Harbour Symposia in Quantitative Biology*, 22, 415–427.

Hyndman, S.J., Vickers, L.M., Htut, T., Maunder, J.W., Peock, A. and Higenbottam, T.W. (2000) A randomized trial of dehumidification in the control of house dust mite. *Clinical and Experimental Allergy*, 30, 1172–1180.

Ichikawa, S., Hatanaka, H., Yuuki, T., Iwamoto, N., Kojima, S., Nishiyama, C., Ogura, K., Okamura, Y.

and Inagi, F. (1998) Solution structure of Der f 2, the major mite allergen for atopic diseases. *Journal of Biological Chemistry*, 273, 356–360.

Ichikawa, S., Takai, T., Inoue, T., Yuuki, T., Okumura, Y., Ogura, K., Inagaki, F. and Hatanaka, H. (2005) NMR study of the major mite allergen Der f 2: its refined tertiary structure, epitopes for monoclonal antibodies and characteristics shared by ML protein group members. *Journal of Biochemistry*, 137, 255–263.

Illi, S., von Mutius, E., Lau, S., Niggermann, B., Grüber, C. and Wahn, U. (2006) Perennial allergen sensitisation early in life and chronic asthma in children: a birth cohort study. *Lancet*, 368, 763–770.

Ingham, P.E. and Ingham, D.M. (1976) House dust mites and infant-use sheepskins. *Medical Journal of Australia*, i, 302–304.

Ingram, C.G., Symington, I.S., Jeffrey, I.G. and Cuthbert, O.D. (1979) Bronchial provocation studies in farmers allergic to storage mites. *Lancet*, December 22–29, 1330–1332.

International Commission on Zoological Nomenclature (1999) *International Code of Zoological Nomenclature*. Fourth Edition. International Trust for Zoological Nomenclature, London.

International Study of Asthma and Allergies in Childhood (ISAAC) Steering Committee (Writing Group: Asher, M.I., Anderson, H.R., Stewart, A.W. and Crane, J.) (1998) Worldwide variations in the prevalence of asthma symptoms: the International Study of Asthma and Allergies in Childhood (ISAAC). *European Respiratory Journal*, 12, 315–335.

International Union of Biochemistry and Molecular Biology Nomenclature Committee (1992) *Enzyme Nomenclature. Recommendations (1992) of the Nomenclature Committee of the International Union of Biochemistry and Molecular Biology*. Academic Press, San Diego.

International Workshop Report (1988) Dust mite allergens and asthma: a worldwide problem. *Bulletin of the World Health Organization*, 66, 769–780.

Ishii, A., Noda, K., Nagai, Y., Ohsawa, T., Kato, I., Yokota, M., Nakamura, S., Matuhasi, T. and Sasa, M. (1973) Biological and chemical properties of the house dust mite extract, *Dermatophagoides farinae*. *Japanese Journal of Experimental Medicine*, 43, 495–501.

Ishii, A., Takaoka, M., Ichinoe, M., Kabasawa, Y. and Ouchi, T. (1979) Mite fauna and fungal flora in house dust from homes of asthmatic children. *Allergy*, 34, 379–387.

Ishii, A., Yatani, T., Kato, H. and Fujimoto, T. (1983) Mite fauna, house dust and Kawasaki disease. *The Lancet*, ii, 102–103.

Ivanova, L.N. and Petrova, Y.I. (1984) [The ecology and fauna of house-dust mites.] *Meditsinskaya Parazitologiya Parazitarnye Bolezni*, no. 5, 78–82. [In Russian, English summary]

Jacobson, M., Adler, V.E., Kishaba, A.M. and Priesner, E. (1976) 2-Phenylethanol, a presumed sexual stimulant produced by male cabbage looper moth, *Trichoplusia ni*. *Experientia*, 32, 964–966.

James, D.L. and Mulla, M.S. (1978) A chemically-defined diet for the American House Dust Mite, *Dermatophagoides farinae*. *Annals of the Entomological Society of America*, 71, 785–787.

Janecek, S. (1997) a-Amylase family: molecular biology and evolution. *Progress in Biophysical and Molecular Biology*, 67, 67–97.

Janko, M., Gould, D.C., Vance, L., Stengel, C.C. and Flack, J. (1995) Dust mite allergens in the office environment. *American Industrial Hygiene Association Journal*, 56, 1133–1140.

Janson, C., Antó, J., Burney, P., Chinn, S., de Marco, R., Heinrich, J., Jarvis, D., Kuenzli, N., Leynaert, B., Luczynska, C., Neukirch, F., Svannes, C., Sunyer, J. and Wjst, M. on behalf of the European Community Respiratory Health Survey II (2001) The European Community Respiratory Health Survey: what are the main results so far? *European Respiratory Journal*, 18, 598–611.

Järvinen, M. (1978) Purification and some characteristics of the human epidermal SH-protease inhibitor. *Journal of Investigative Dermatology*, 72, 114–118.

Jarvis, D., Luczynska, C., Chinn, S., Potts, J., Sunyer, J., Janson, C., Svanes, C., Künzli, N., Leynaert, B., Heinrich, J., Kerkoff, M., Ackerman-Liebrich, U., Antó, J.M., Cerveri, I., de Marco, R., Gislason, T., Neukirch, F., Vermire, P., Wjst, M. and Burney, P. (2005) Change in prevalence of IgE sensitisation and mean total IgE with age and cohort. *Journal of Allergy and Clinical Immunology*, 116, 675–682.

Jeffrey, C. (1989) *Biological Nomenclature*. Edward Arnold, London.

Jeong, K.Y., Lee, I.Y., Ree, H.I., Hong, C.S. and Yong, T.S. (2002) Localisation of Der f 2 in the gut and faecal pellets of *Dermatophagoides farinae*. *Allergy*, 57, 729–731.

Jeong, K.Y., Kim, W.K., Lee, J.S., Lee, J., Lee, I.Y., Kim, K.E., Park, J.W., Hong, C.S., Ree, H.I. and Yong, T.S. (2005a) Immunoglobulin E reactivity of recombinant allergen Tyr p 13 from *Tyrophagus putrescentiae*. Homologous to fatty acid binding protein. *Clinical and Diagnostic Laboratory Immunology*, 12, 581–585.

Jeong, K.Y., Lee, H., Lee, J.S., Lee, J., Lee, I.Y., Ree, H.I., Hong, C.S., Park, J.W. and Yong, T.S. (2005b) Immunoglobulin E binding reactivity of recombinant allergen homologous to alpha tubulin from *Tyrophagus putrescentiae*. *Clinical and Diagnostic Laboratory Immunology*, 12, 1451–1454.

Jeong, K.Y., Hong, C.S. and Yong, T.S. (2006) Allergenic tropomyosins and their cross-reactivities. *Protein and Peptide Letters*, 13, 835–845.

Jin, K., Higaki, Y., Tagaki, Y., Higuchi, K., Yada, Y., Kawashima, M. and Imokawa, G. (1994). Analysis of beta-glucocerebrosidase and ceramidase activities in atopic and aged dry skin. *Acta Dermatologia et Venereologia (Stockholm)*, 74, 337–340.

Johannessen, B.R., Skov, L.K., Kastrup, J.S., Kristensen, O., Bolwig, C., Larsen, J.N., Spangfort, M., Lund, K. and Gajhede, M. (2005) Structure of the house dust mite allergen Der f 2: implications for function and molecular basis of IgE cross-reactivity. *FEBS Letters*, 579, 1208–1212.

Johansson, E., van Hage-Hamsten, M. and Johansson, S.G.O. (1988) Demonstration of allergen components in the storage mite *Lepidoglyphus destructor* by an immunoblotting technique. *International Archives of Allergy and Applied Immunology*, 85, 8–13.

Johansson, E., Borga, A., Johansson, S.G.O. and Hage-Hamsten, M. van. (1991) Immunoblot multi-allergen inhibition studies of allergenic cross-reactivity of the house dust mites *Lepidoglyphus destructor* and *Dermatophagoides pteronyssinus*. *Clinical and Experimental Allergy*, 21, 511–518.

Johansson, E., Johansson, S.G.O. and van Hage-Hamsten, M. (1994) Allergenic characterization of *Acarus siro* and *Tyrophagus putrescentiae* and their crossreactivity with *Lepidoglyphus destructor* and *Dermatophagoides pteronyssinus*. *Clinical and Experimental Allergy*, 24, 743–751.

Johansson, S.G.O., Hourihane, J.O'B., Bousquet, J., Bruijnzeel-Koomen, C., Dreborg, S., Haatela, T., Kowalski, M.L., Mygind, N., Ring, J., van Cauwenhabe, P., van Hage-Hamsten, M. and Wüthrich, B. (2001) A revised nomenclature for allergy: an EAACI position statement for the EAACI nomenclature task force. *Allergy*, 56, 813–824.

Johns, T.C., Gregory, J.M., Ingram, W.J., Johnson, C.E., Jones, A., Lowe, J.A., Mitchell, J.F.B., Roberts, D.L., Sexton, D.M.H., Stevenson, D.S., Tett, S.F.B. and Woodage, M.J. (2003) Anthropogenic climate change from 1860 to 2100 simulated with the HadCM3 model under updated emissions scenarios. *Climate Dynamics*, 20, 583–612.

Johnsen, C.R., Mosbech, H. and Heinig, J.H. (1997) Can ionisers in bedrooms help asthmatics? *Indoor and Built Environment*, 6, 174–178.

Johnson, C.G. (1937) Thermograph records in rooms of some London dwelling-houses throughout the year 1935–36 and their comparison with temperatures recorded in outdoor meteorological stations. *Journal of Hygiene*, 38, 222–232.

Johnston, D.E. and Bruce, W.A. (1965) *Tyrophagus nieswanderi*, a new acarid mite of agricultural importance. *Research Bulletin Ohio Agricultural Research and Development Center*, no. 977, 1–15.

Jones, B.M. (1950) Acarine growth: A new edysial mechanism. *Nature*, 166, 908–909.

Jones, B.M. (1954) On the role of the integument in acarine development and its bearing on pupa-formation. *Quarterly Journal of Microscopical Science*, 95, 169–181.

Jones, H. (1990) *Population Geography*. Paul Chapman Publishing Ltd, London.

de Jong, N.W., Groenwoud, C.G.M., van Ree, R., van Leeuwen, A., Vermeulen, A.M., van Toorenberg, A.W., de Groot, H. and van Wijk, R.G. (2004) Immunoblot and radioallergosorbent test inhibition studies of allergenic cross-reactivity of the predatory mite *Amblyseius cucumeris* with the house dust mite *Dermatophagoides pteronyssinus*. *Annals of Allergy, Asthma and Immunology*, 93, 281–287.

Jooma, O.F., Weinberg, E.G., Berman, D., Manjra, A.I. and Potter, P.C. (1995). Acccumulation of house-dust mite (Der-p-1) levels on mattress covers. *South African Medical Journal*, 85, 1002–1005.

Jordan, S.C., Platts-Mills, T.A.E., Mason, W., Takahashi, M., Rawle, F. and Wilkins, S. (1983) Lack of evidence for mite-antigen-mediated pathogenesis in Kawasaki disease. *The Lancet*, April 23, 931.

Joseph, K.E., Adams, C.D., Cottrell, L., Hogan, M.B. and Wilson, N.W. (2003) Providing dust mite-proof covers improves adherence to dust mite control measures in children with mite allergy and asthma. *Annals of Allergy, Asthma and Immunology*, 90, 550–553.

Julge, K., Munir, A.K.M., Vasar, M. and Björkstén, B. (1998) Indoor allergen levels and other environmental risk factors for sensitisation in Estonian homes. *Allergy*, 53, 388–393.

Juliá, J.C., Martorell, A., Ventas, P., Cerdá, J.C., Torró, I., Carreira, J., Guinot, E., Sanz, J. and Álvarez, V. (1995) *Lepidoglyphus destructor* acarus in the urban house environment. *Journal of Investigational Allergology and Clinical Immunology*, 5, 318–321.

Jurenka, R.A. and Roelofs, W.L. (1993) Biosynthesis and endocrine regulation of fatty acid derived sex pheromones in moths. In: Stanley-Samuelsen, D.W. and Nelson, D. (eds), *Insect Lipids: Chemistry, Biochemistry and Biology*. University of Nebraska Press, Lincoln, Nebraska.

Kabasawa, Y. and Ishii, A. (1979) Studies on the physicochemical properties of the house dust mite (*Dermatophagoides pteronyssinus*) allergens involved in reaginic reaction. *Japanese Journal of Experimental Medicine*, 49, 51–57.

Kaiser, L., Gafvelin, G., Johansson, E., van Hage-Hamsten, M. and Rasool, O. (2003) Lep 2 polymorphisms in wild and cultured *Lepidoglyphus destructor* mites. *European Journal of Biochemistry*, 270, 646–653.

Kalma, J.D. (1968) A comparison of methods for computing daily mean air temperature and humidity. *Weather*, 23, 248–252.

Kalpaklioglu, A.F., Emekçi, M., Ferizli, A.G. and Misirligil, Z. (1997) House dust mite fauna in Turkey. *Journal of Investigative Allergology and Clinical Immunology*, 7, 578–582.

Kalpaklioglu, A.F., Ferizli, A.G., Misirhgil, Z., Demirel, Y.S. and Gürbüz, L. (1996) The effectiveness of benzyl benzoate and different chemicals as acaricides. *Allergy*, 51, 164–170.

Kalpaklioglu, A.F., Emeki, M., Ferizli, A. and Misirligil, Z. (2004) A survey of acarofauna in Turkey: comparison of seven different geographic regions. *Allergy and Asthma Proceedings*, 25, 185–190.

Kalra, S., Owen, S.J., Hepworth, J. and Woodcock, A. (1990) Airborne house dust mite antigen after vacuum cleaning. *The Lancet*, 336, 449.

Kalra, S., Owen, P., Hepworth, J., Pickering, C.A.C., Pandit, A. and Woodcock, A. (1991) *Der p* I concentrations six months after application of the acaricide Acarosan. *Thorax*, 46, 749. [abstract]

Kalra, S., Crank, P., Hepworth, J., Pickering, C.A.C. and Woodcock, A.A. (1992) Absence of seasonal variation in concentrations of the house dust mite allergen Der p 1 in South Manchester homes. *Thorax*, 47, 928–931.

Kalra, S., Crank, P., Hepworth, J., Pickering C.A.C. and Woodcock, A.A. (1993) Concentrations of the domestic house dust mite allergen der-p-i after treatment with solidified benzyl benzoate (acarosan) or liquid-nitrogen. *Thorax*, 48, 10–13.

Kang, S.Y. and Chu, J.K. (1975) [Studies on the dust mite (*Tarsonemus*) in Korea.] *Journal of the Korean Medical Association*, 18, 1081–1089. [In Korean, English summary]

Kannan, I., Rajendran, P., Selvarani, P. and Thyagarajan, S.P. (1996) A survey of house dust mites in Madras (India) in relation to mite density, seasonal variation and niche. *Biomedicine*, 16, 58–62.

Kanungo, K. (1969) Acarine moulting – the migration of hemocytes through the epidermis of *Caloglyphus berlesei*. *Annals of the Entomological Society of America*, 62, 155–157.

Kanungo, K. and Naegle, J.A. (1964) The hemocytes of the acarid mite *Caloglyphus berlesei* (Mich. 1903). *Journal of Insect Physiology*, 10, 651–655.

Karg, W. (1973) Hausstaubmilben in der Deutschen Demokratische Republik. *Allergie und Immunologie*, 19, 81–85.

Karlsson, A.-S., Hendrén, M., Almqvist, C., Larsson, K. and Renström, A. (2002) Evaluation of Petri dish sampling for assessment of cat allergen in airborne dust. *Allergy*, 57, 164–168.

Karn, R.C. and Malacinski, G.M. (1978) The comparative biochemistry, physiology and genetics of animal alpha-amylases. *Advances in Comparative Physiology and Biochemistry*, 7, 1–103.

Karouna-Renier, N.K., Yang, W.-J. and Ranga Rao, K. (2003) Cloning and characterization of a 70 kDa heat shock cognate gene (HSC70) from two species of *Chironomus*. *Insect Molecular Biology*, 12, 19–26.

Kato, H., Fujimoto, T., Inoue, O., Kondo, M., Koga, Y., Yamamoto, S., Shingu, M., Tominaga, K. and Sasaguri, Y. (1983). Variant strain of *Proprionibacterium acnes*: a clue to the aetiology of Kawasaki disease. *Lancet*, ii, 1383–1387.

Kato, Y., Katsuno, T., Aoki, M., Kato, M., Tanaka, K., Fujii, T., Hirose, Y., Tabei, A., Inoue, K., Jing, S., Ono, T. and Ishii, A. (1991) [Effect of intensive vacuum cleaning in reducing house dust mite antigen in bedrooms of asthmatic children.] *Nippon Koshu Eisei Zasshi* [*Japanese Journal of Public Health*], 38, 801–807. [In Japanese, English summary]

Kawamoto, S., Mizuguchi, Y., Morimoto, K., Aki, T., Shigeta, S., Yasueda, H., Wada, T., Suzuki, O., Jyo, T. and Ono, K. (1999) Cloning and expression of Der f 6, a serine protease allergen from the house dust mite, *Dermatophagoides farinae*. *Biochimica Biophysica Acta*, 1454, 201–207.

Kawamoto, S., Suzuki, T., Aki, T., Katsutani, T., Tsuboi, S., Shigeta, S. and Ono, K. (2002a) Der f 16: a novel gelsolin-related molecule identified as an allergen from the house dust mite, *Dermatophagoides farinae*. *FEBS Letters*, 516, 234–238.

Kawamoto, S., Aki, T., Yamasita, M., Tategaki, A., Fijimura, T., Tsuboi, A., Katsutani, T., Suzuki, O., Shigeta, S., Murooka, Y. and Ono, K. (2002b) Towards elucidating the full spectrum of mite allergens – state of the art. *Journal of Bioscience and Bioengineering*, 94, 285–298.

Keber, M.M., Gradisar, H. and Jerala, R. (2005) MD-2 and Der p 2 – a tale of two cousins or distant relatives? *Journal of Endotoxin Research*, 11, 186–192.

Keil, H. (1983) Ökofaunistische Untersuchungen der Hausstaubmilben in Hamburg. *Entomologische Mitteilungen aus dem Zoologischen Museum Hamburg*, 7, 343–386.

Kent, N.A., Hill, M.R., Keen, J.N., Holland, P.W.H. and Hart, B.J. (1992) Molecular characterisation of Group I Allergen *Eur m* I from house dust mite *Euroglyphus maynei*. *International Archives of Allergy and Immunology*, 99, 150–152.

Keren, G. and Wolman, M. (1984) Can Pseudomonas infection in experimental animals mimic Kawasaki's disease? *Journal of Infection*, 9, 22–29.

Kerkut, G.A. and Gilbert, L.I. (eds) (1985) *Comprehensive Insect Physiology, Biochemistry and Pharmacology* (Vols. 1–13). Pergamon Press, Oxford.

Kern, R. A. (1921) Dust sensitization in bronchial asthma. *Medical Clinics of North America*, 5, 751–761.

Kersten, W., Stollewerk, D. and Musken, H. (1988) Klinische studie zur wirksamkeit der akariziden substanz Acarosan bei hausstaubmilbenallergiken. *Allergologie*, 11, 371–90.

Kim, Y.K., Lee, M.H., Jee, Y.K., Hong, S.C., Bae, J.M., Chang, Y.S., Jung, J.W., Lee, B.J., Son, J.W., Cho, S.H., Min, K.U. and Kim, Y.Y. (1999) Spider mite allergy in apple-cultivating farmers: European red mite (*Panonychus ulmi*) and the two–spotted spider mite (*Tetranychus urticae*) may be important allergens in the development of work-related asthma and rhinitis symptoms. *Journal of Allergy and Clinical Immunology*, 104, 1285–1292.

King, C.M., Simpson, R.J., Moritz, R.L., Reid, G.E., Thompson, P.J. and Stewart, G.A. (1996) The isolation and characterisation of a novel collagenolytic serine protease allergen (Der p 9) from the dust mite *Dermatophagoides pteronyssinus*. *Journal of Allergy and Clinical Immunology*, 98, 739–747.

King, M.J., Sonenshine, D.E. and Betts, L.S. (1989) House dust mites in naval ships, military barracks and homes in the Hampton Roads area of Virginia. *Military Medicine*, 154, 467–473.

King, T.P., Hoffman, D., Løwenstein, H., Marsh, D.G., Platts-Mills, T.A.E. and Thomas, W. (1994) Allergen nomenclature. *International Archives of Allergy and Immunology*, 105, 221–233.

Kinnaird, C.H. (1974) Thermal death point of *Dermatophagoides pteronyssinus* (Trouessart, 1897) (Astigmata, Pyroglyphidae), the house dust mite. *Acarologia*, 16, 340–342.

Kirchoff, C., Osterhoff, C. and Young, L. (1996) Molecular cloning and characterization of HE1, a major secretory protein of the human epididymis. *Biology of Reproduction*, 54, 847–856.

Kirschke, H. and Barrett, A.J. (1987) Chemistry of lysosomal proteases. In: H. Glaumann, H. and Ballard, F.J. (ed.), *Lysosomes: their Role in Protein Breakdown*. Academic Press, London, pp. 193–238.

Kitch, B.T., Chew, G., Burge, H.A., Muilenberg, M.L., Weiss, S.T., Platts-Mills, T.A.E., O'Connor, G. and Gold, D.R. (2000) Socioeconomic predictors of high allergen levels in homes in the Greater Boston area. *Environmental Health Perspectives*, 108, 301–307.

Kivity, S., Solomon, A., Soferman, R., Schwarz, Y., Mumcuoglu, K.Y. and Topilsky, M. (1993) Mite asthma in childhood, a study of the relationship between exposure to house dust mites and disease activity. *Journal of Allergy and Clinical Immunology*, 91, 844–849.

Klag, J. (1971) The fine structure of the cuticle of *Acarus siro* L. (Acarina). *Acta Biologica Cracoviensia, Series Zoologia*, 14, 307–316.

Klein, B.S., Rodgers, L.A., Patrican, L.A., Burgdorfer, W., Schell, W.L., Kochel, R.L., Marchette, N.J., McPherson, J.T., Nelson, D.B., Yolken, R.H., Wortmann, D. and Davis, J.P. (1986) Kawasaki syndrome, a controlled study of an outbreak in Wisconsin. *American Journal of Epidemiology*, 124, 306–316.

Kliks, M.Mc.K. (1988) Palaeoparasitological analysis of faecal material from Amerindian (or New World) mummies, evaluation of saprophytic arthropod remains. *Palaeopathology Newsletter*, no. 64, 7–11.

Kniest, F.M., Young, E., van Praag, M.C.G., Vos, H., Kort, H.S.M., de Maat-Bleeker, F. and van Bronswijk, J.E.M.H. (1991) Clinical evaluation of a double-blind dust-mite avoidance trial with mite-allergic rhinitic patients. *Clinical and Experimental Allergy*, 21, 39–47.

Knülle, W. (1962) Die abhängigkeit der luftfeuchtereaktionen der mehlmilbe (*Acarus siro* L.) von wassergehalt des körpers. *Zeitschrift für vergleichende Physiologie*, 45, 233–246.

Knülle, W. (1965) Die sorbtion und transpiration des wasserdampfes bei der mehlmilbe (*Acarus siro* L.). *Zeitschrift für vergleichende Physiologie*, 49, 586–604.

Knülle, W. (1984) Water vapour uptake in mites and insects: an ecophysiological and evolutionary perspective. In: Griffiths, D.E. and Bowman, C.E. (eds), *Acarology VI*. Vol. 1. Ellis Horwood, Chichester, pp. 71–82.

Knülle, W. (1987) Genetic variability and ecological adaptability of hypopus formation in as stored product mite. *Experimental and Applied Acarology*, 3, 21–32.

Knülle, W. (1991) Genetic and environmental determinants of hypopus duration in the stored-product mite *Lepidoglyphus destructor*. *Experimental and Applied Acarology*, 10, 231–258.

Knülle, W. (1995) Expression of a dispersal trait in a guild of mites colonising transient habitats. *Evolutionary Ecology*, 9, 341–353.

Knülle, W. (2003) Interaction between genetic and inductive factors controlling the expression of dispersal and dormancy morphs in dimorphic astigmatid mites. *Evolution*, 57, 828–838.

Knülle, W. and Wharton, G.W. (1964) Equilibrium humidities in arthropods and their ecological significance. *Acarologia* 6, special issue: *Proceedings of the 1st International Congress of Acarology*, Fort Collins, USA, 1963, pp. 299–306.

Kobayashi, K., Matsumoto, K. and Wada, Y. (1979) Studies on lipids of adult, nymph and larva of the grain mites. *Japanese Journal of Sanitary Zoology*, 30, 237–241.

Koekkoek, H.H.M. and van Bronswijk, J.E.M.H. (1972) Temperature requirements of a house-dust mite *Dermatophagoides pteronyssinus* compared with the climate in different habitats of houses. *Entomologia Experimentalis et Applicata*, 15, 438–442.

Kohomoto, S., Kodera, Y., Takahashi, K., Nishimura, H., Matsushima, A. and Inada, Y. (1991) Activation of the kallikrein-kinin system in human plasma by a serine protease from mites. *Journal of Clinical Biochemistry and Nutrition*, 10, 15–20.

Konishi, E. and Uehara, K. (1995) Distribution of *Dermatophagoides* mite (Acari: Pyroglyphidae) antigens in homes of allergic patients in Japan. *Experimental and Applied Acarology*, 19, 275–286.

Konishi, E. and Uehara, K. (1999) Contamination of public facilities with *Dermatophagoides* mites (Acari: Pyroglyphidae) in Japan. *Experimental and Applied Acarology*, 23, 41–45.

Köppen, W. (1931) *The Climates of the Earth.* DeGruyter, Berlin.

Koraiem, M.K. and Fahmy, I.A. (1999) Studies on the house dust mites in Great Cairo, Egypt. *Journal of the Egyptian Society of Parasitology*, 29, 131–138.

Koren, L.G.H. (1993) Long term efficacy of acaricides against house dust mites (*Dermatophagoides pteronyssinus*) in a semi-natural test system. In: Wildey, K.B. and Robinson, W.H. (eds), *Proceedings of the First International Conference on Insect Pests in the Urban Environment.* Exeter, UK; Wheatons Ltd, pp. 367–371.

Koren, L.G.H. and Eckhardt, M. (1995) Suitability of furnishings-in-use for mite population development. In: Tovey, E., Fifoot, A. and Sieber, L. (eds), *Mites, Asthma and Domestic Design II.* University of Sydney, Sydney, pp. 48–51.

Korsgaard, J. (1979) Husstovmider (Pyroglyphidae, Acari) i danskeboliger. *Ugeskrift for Laeger*, 141, 888–892.

Korsgaard, J. (1982) Preventitive measures in house-dust allergy. *American Review of Respiratory Diseases*, 125, 80–84.

Korsgaard, J. (1983a) House dust mites and absolute indoor humidity. *Allergy*, 38, 85–92.

Korsgaard, J. (1983b) Mite asthma and residency. A case-control study on the impact of exposure to house dust mites in dwellings. *Annual Review of Respiratory Diseases*, 128, 231–235.

Korsgaard, J. (1983c) Preventive measures in mite asthma. A controlled trial. *Allergy*, 38, 93–102.

Korsgaard, J. (1991) Mechanical ventilation and house dust mites, a controlled investigation. In: van Moerbeke, D. (ed.), *Dust Mite Allergens and Asthma. Report of the 2nd International Workshop, Minster Lovell, Oxfordshire, England, September 19–21, 1990.* UCB Institute of Allergy, Brussels, pp. 87–89.

Korsgaard, J. (1998) Epidemiology of house dust mites. *Allergy*, 53 (Supplement 48), 36–40.

Korsgaard, J. and Iversen, M. (1991) Epidemiology of house dust mite allergy. *Allergy*, 46, Supplement 11, 14–18.

Kort, H.S.M. (1990) Mites, dust lice, fungi and their interrelationships on damp walls and room partitions. In: Bruin, J. (ed.), *Proceedings of the Section Experimental and Applied Entomology of the Netherlands Entomological Society,* 11, 63–68.

Kort, H.S.M., Schober, G., Koren, L.G.H. and Scharinga, J. (1997) Mould-devouring mites differ in guanine excretion from dust-eating Acari, a possible error source in mite allergen exposure studies. *Clinical and Experimental Allergy*, 27, 921–925.

Kramer, K.J. and Muthukrishnan, S. (1997) Insect chitinases: molecular biology and potential use as biopesticides. *Insect Biochemistry and Molecular Biology*, 27, 887–900.

Kramer, K.J., Corpuz, L., Choi, H. and Muthukrishnan, S. (1993) Sequence of a cDNA and expression of a gene encoding epidermal and gut chitinases of *Manduca sexta. Insect Biochemistry and Molecular Biology*, 23, 691–701.

Krantz, G.W. (1978) *A Manual of Acarology.* Oregon State University Bookstore, Corvallis, Oregon.

Krebs, C.J. (1972) *Ecology. The Experimental Analysis of the Distribution and Abundance of Animals* (Harper International Edition). Harper and Row, New York.

Krebs, C.J. (1989) *Ecological Methodology*. Harper and Row, New York.

Krebs, C.J. (2001) *Ecology. The Experimental Analysis of the Distribution and Abundance*. 5th edition. Addison Wesley Longman, San Francisco.

Krishna Rao, N.S. and ChannaBasavanna, G.P. (1977) Some unrecorded mites from house dust samples in Bangalore, Karnataka, India. *Acarology Newsletter*, no. 5, 5–7.

Krishna Rao, N.S., Khuddus, C.A. and ChannaBasavanna, G.P. (1973) Pyroglyphid mites in man and his surroundings. *Current Science*, 42, 33.

Krishna Rao, N.S., Ranganath, H.R. and ChannaBasavanna, G.P. (1981) Housedust mites from India. *Indian Journal of Acarology*, 5, 85–94.

Kroidl, R., Maasch, H.J. and Wahl, R. (1992) Respiratory allergies (bronchial asthma and rhinitis) due to sensitisation of type I allergy to red spider mite (Koch). *Clinical and Experimental Allergy*, 22, 958–962.

Kroidl, R.F., Göbel, D., Balzer, D., Trendelenburg, F. and Schwichtenberg, U. (1998) Clinical effects of benzyl benzoate in the prevention of house-dust-mite allergy. Results of a prospective, double-blind, multicenter study. *Allergy*, 53, 435–440.

Kronqvist, M., Johansson, E., Magnusson, C.G., Olsson, S., Eriksson, E.L., Gafvelin, G. and Van Hage-Hamsten, M. (2000) Skin prick tests and serological analysis with recombinant group 2 allergens of the house dust mites *L. destructor* and *T. putrescentiae*. *Clinical and Experimental Allergy*, 30, 670–676.

Kronqvist, M., Johansson, E., Kolmodin-Hedman, B., Oman, H., Svartengren, M. and van Hage-Hamsten, M. (2005) IgE sensitisation to predatory mites and respiratory symptoms in Swedish greenhouse workers. *Allergy*, 60, 521–526.

Kuehr, J., Frischer, T., Meinert, R., Barth, R., Forster, J., Schraub, S., Urbanek, R. and Karmaus, W. (1994a) Mite allergen exposure is a risk for the incidence of specific sensitisation. *Journal of Allergy and Clinical Immunology*, 94, 44–52.

Kuehr, J., Frischer, T., Karmaus, W., Meinert, R., Barth, R., Schraub, S., Daschner, A., Urbanek, R. and Forster, J. (1994b) Natural variation in mite antigen density in house dust and relationship to residential factors. *Clinical and Experimental Allergy*, 24, 229–237.

Kuehr, J., Frischer, T., Meinert, R., Barth, R., Schraub, S., Urbanek, R., Karmaus, W. and Forster, J. (1995) Sensitisation to mite allergens is a risk factor for early and late onset of asthma and for persistence of asthmatic signs in children. *Journal of Allergy and Clinical Immunology*, 95, 655–662.

Kuo, I.C., Yi, F.C., Cheong, N., Shek, L.P., Chew, F.T., Lee, B.W. and Chua, K.Y. (1999) Sensitisation to *Blomia tropicalis* and *Dermatophagoides pteronyssinus* – a comparative study between Singapore and Taiwan. *Asian Pacific Journal of Allergy and Immunology*, 17, 179–188.

Kuo, I.C., Yi, F.C., Cheong, N., Shek, L.P., Chew, F.T., Lee, B.W. and Chua, K.Y. (2003) An extensive study of human IgE cross-reactivity of Blo t 5 and Der p 5. *Journal of Allergy and Clinical Immunology*, 111, 603–609.

Kuo, J.S. and Nesbitt, H.H.J. (1970) The internal anatomy and histology of adult *Caloglyphus mycophagus* (Mégnin) (Acarina: Acaridae). *Canadian Journal of Zoology*, 48, 505–518.

Kusakawa, S. and Heiner, D. (1976) Elevated levels of immunoglobulin E in the acute febrile mucocutaneous lymph node syndrome. *Pediatric Research*, 10, 108–111.

Kuwahara, Y. (1991) Pheromone studies on astigmatid mites – alarm, aggregation and sex. In: Dusbabek, F. and Bukva, V. (eds), *Modern Acarology*. Vol. 1. Academia, Prague and SPB Academic Publishing bv, The Hague, pp. 43–52.

Kuwahara, Y. (1995) [Sex pheromone study of *Caloglyphus* sp. (Astigmata: Acaridae).] *Reports of the Chemical Materials Research Development Foundation*, 10, 45–52. [In Japanese]

Kuwahara, Y. (1999) [Chemical ecology in astigmatid mites.] In: Hikada, T., Matsumoto, Y. and Honda, K. *Environmental Entomology: Behaviour, Physiology and Chemical Ecology*. University of Tokyo Press, Tokyo, 380–393. [In Japanese]

Kuwahara, Y. (2004) Chemical ecology of astigmatid mites. In: Cardé, R.T. and Millar, J.G. (eds), *Advances in Insect Chemical Ecology*. Cambridge University Press, Cambridge, pp. 76–109.

Kuwahara, Y. and Sakuma, L. (1982) Pheromone study on acarid mites VIII. Primary (Z)-2-alkenyl formate responsible for the alarm pheromone activity against the mold mite, *Tyrophagus putrescentiae* (Schrank) (Acarina: Acaridae). *Applied Entomology and Zoology*, 17, 263–268.

Kuwahara, Y., Ishii, S. and Fukami, H. (1975) Neryl formate: alarm pheromone of the cheese mite, *Tyrophagus putrescentiae* (Schrank) (Acarina: Acaridae). *Experientia*, 31, 1115–1116.

Kuwahara, Y., Fukami, H., Ishii, S., Matsumoto, K. and Wada, Y. (1979) Pheromone study on acarid mites II. Presence of the alarm pheromone in the mould mite, *Tyrophagus putrescentiae* (Schrank) (Acarina: Acaridae) and the site of its production. *Japanese Journal of Sanitary Zoology*, 30, 309–314.

Kuwahara, Y., Matsumoto, K. and Wada, Y. (1980) Pheromone study on acarid mites IV. Citral, composition and function as an alarm pheromone and its secretory gland in four species of acarid mites. *Japanese Journal of Sanitary Zoology*, 31, 73–80.

Kuwahara, Y., My-Yen, L.T., Tominaga, Y., Matsumoto, K. and Wada, Y. (1982) 1,3,5,7-Tetramethyldecyl formate, lardolure: aggregation pheromone of the acarid mite, *Lardoglyphus konoi* (Sasa et Asanuma) (Acarina: Acaridae). *Agricultural and Biological Chemistry*, 46, 2283–2291.

Kuwahara, Y., Leal, W.S., Kurosa, K., Sato, M., Matsuyama, S. and Suzuki, T. (1992) Chemical ecology of astigmatid mites XXXIII. Identification of (Z,Z)-6,9-heptadecadiene in the secretion of *Carpoglyphus lactis* (Acarina, Carpoglyphidae) and its distribution among astigmatid mites. *Journal of the Acarological Society of Japan*, 1, 95–104.

Kuwahara, Y., Sato, M., Koshii, T. and Suzuki, T. (1992) Chemical ecology of astigmatid mites XXXII. 2-Hydroxy-6-methyl-benzaldehyde, the sex pheromone of the brown-legged grain mite *Aleuroglyphus ovatus* (Troupeau) (Acarina: Acaridae). *Applied Entomology and Zoology*, 27, 253–260.

Lafosse Marin, S., Iraola, V., Merle, S. and Fernández-Caldas, E. (2006) Étude de la faune acarologique des matelas de l'île de la Martinique. *Revue Française d'Allergologie et d'Immunologie Clinique*, 46, 62–67.

Lai, N.Q. (1982) [A report of the distribution of dust mites in civil houses in Guang Zhou and some other cities.] *Journal of the Guang Zhou Medical College*, 3 (Supplement), 20–25. [In Chinese]

Lai, N.Q. and Wen, T.-H. (1982) [A report of the distribution of dust mites in civil houses in Guang Zhou and some other cities.] *Journal of the Guang Zhou Medical College*, 3 (Supplement), 20–25. [In Chinese]

Lake, F.R., Ward, L.D., Simpson, R.J., Thompson, P.J. and Stewart, G.A. (1991) House dust mite derived amylase: allergenicity and physicochemical characterisation. *Journal of Allergy and Clinical Immunology*, 87, 1035–1042.

Lambert, H.P., Fisher-Hoch, S.P. and Grover, S.A. (1985) Kawasaki disease and *Coxiella burneti*. *The Lancet*, October 12, 844.

Lane, J., Siebers, R., Pene, G., Howden-Chapman, P. and Crane, J. (2005) Tokelau: a unique low allergen environment at sea level. *Clinical and Experimental Allergy*, 35, 479–482.

Lang, J.D. and Mulla, M.S. (1977a) Distribution and abundance of house dust mites, *Dermatophagoides* spp. in different climatic zones of Southern California. *Environmental Entomology*, 6, 213–216.

Lang, J.D. and Mulla, M.S. (1977b) Abundance of house dust mites, *Dermatophagoides* spp. influenced by environmental conditions in homes in Southern California. *Environmental Entomology*, 6, 643–648.

Lang, J.D. and Mulla, M.S. (1978) Seasonal dynamics of house dust mites, *Dermatophagoides* spp., in homes in Southern California. *Environmental Entomology*, 7, 281–286.

Lang, J.D., Charlet, L.D. and Mulla, M.S. (1976) Bibliography (1864 to 1974) of house-dust mites *Dermatophagoides* spp. (Acarina: Pyroglyphidae). *Science of Biology Journal*, March–April, 1976, 62–83.

Larson, D.G. (1969) The critical equilibrium activity of adult females of the house-dust mite *Dermatophagoides farinae* Hughes. PhD thesis, The Ohio State University, Columbus.

Larson, D.G., Mitchell, W.F. and Wharton, G.W. (1969) Preliminary study on *Dermatophagoides farinae* Hughes, 1961 (Acari) and house dust allergy. *Journal of Medical Entomology*, 6, 295–299.

Lascaud, D. (1978) Étude écologique des Acariens Pyroglyphides de la poussière de maison dans la region Grenobloise. *Annales de Parasitologie (Paris)*, 53, 675–695.

Lassiter, M.T. and Fashing, N.J. (1990) House dust mites in Williamsburg, Virginia. *Virginia Medicine*, April, 1990, 152–157.

Lau, S., Falkenhorst, G., Weber, A., Werthman, I., Lind, P., Buettner-Goetz, P. and Wahn, U. (1989) High mite-allergen exposure increases the risk of sensitisation in atopic children and young adults. *Journal of Allergy and Clinical Immunology*, 84, 718–725.

Lau, S., Weber, A.-K., Werthmann, I. and Wahn, U. (1990) Saisonale schwankungen von hausstaub-milbenallergenen bedeutung für atopische kinder. *Monatsschrift Kinderheilkunde*, 138, 58–61.

Lau, S., Luck, W., Kulig, M. and Wahn, U. (1997) Indoor allergen exposure and environmental tobacco smoke exposure during the first five years of life. *Journal of Allergy and Clinical Immunology*, 99, S85.

Lau, S., Illi, S., Sommerfeld, C., Niggemann, B., Bergmann, R., von Mutius, E. and Wahn, U. (2000) Early exposure to house-dust mite and cat allergens and development of childhood asthma: a cohort study. *Lancet*, 356, 1392–1397.

Lau, S., Wahn, J., Schulz, G., Sommerfeld, C. and Wahn, U. (2002) Placebo-controlled study of the mite allergen-reducing effect of tannic acid plus benzyl benzoate on carpets in homes of children with house dust mite sensitization and asthma. *Pediatric Allergy and Immnuology*, 13, 31–36.

Lau-Schadendorf, S., Rusche, A.F., Weber, A.-K., Buettner-Goetz, P. and Wahn, U. (1991) Short-term effect of solidified benzyl benzoate on mite-allergen concentrations in house dust. *Journal of Allergy and Clinical Immunology*, 87, 41–47.

Lawton, M.D., Dales, R.E. and White, J. (1998) The influence of house characteristics in a Canadian community on microbiological contamination. *Indoor Air*, 8, 2–11.

Lázaro, M. and Igea, J.M. (2000) Ácaros en viviendas de Salamanca y Zamora. *Allergología y Immunología Clínica*, 15, 215–219.

Leal, W.S. and Mochizuki, F. (1990) Chemoreception in astigmatid mites. *Naturwissenschaften*, 77, 593–594.

Leal, W.S. and Kuwahara, Y. (1991) Cuticle wax chemistry of Astigmatid mites. In: Dusbabek, F. and Bukva, V. (eds), *Modern Acarology*. Vol. 1. Academia, Prague and SPB Academic Publishing bv, The Hague, pp. 419–423.

Leal, W.S., Nakano, Y., Kuwahara, Y., Nakao, H. and Suzuki, T. (1988) Pheromone study of acarid mites XVII. Identification of 2-hydroxy-6-methyl-benzaldehyde as the alarm pheromone of the acarid mite *Tyrophagus perniciosus* (Acarina: Acaridae), and its distribution among related mites. *Applied Entomology and Zoology*, 23, 422–427.

Leal, W.S., Kuwahara, Y. and Suzuki, T. (1989) β-Acaridial, the sex pheromone of the acarid mite *Caloglyphus polyphyllae*. *Naturwissenschaften*, 76, 332–333.

Leal, W.S., Kuwahara, Y. and Suzuki, T. (1990) Hexyl 2-formyl-3-hydroxybenzoate, a fungitoxic cuticular constituent of the bulb mite *Rhizoglyphus robini*. *Agricultural and Biological Chemistry*, 54, 2593–2597.

Lebrun, P. and de Saint Georges-Gridelet, D. (1984) Process for combating and/or preventing allergic diseases employing natamycin. *United .States Patent*, April 10, 1984, No. 4,448,091.

Leclerq-Foucart, J., de Saint Georges-Gridelet, D., Geubelle, F. and Lebrun, P. (1985) Controle de l'acarien des poussières (*Dermatophagoides pteronyssinus*) par utilisation d'un fongicide. Observations experimentales. Essai clinique chez l'enfant allergie au *Dermatophagoides*. *Revue Medicine Liege*, 40, 91–99.

Lee, A., Sangsupawanich, P., Ma, S., Tan, T.-N., Shek, L.P., Goh, D.L.M., Ho, B., van Bever, H. and Lee, B.W. (2005) Endotoxin levels in rural Thai and urban Singaporean homes. *International Archives of Allergy and Immunology*, 141, 396–400.

Lee, A.J., Isaac, R.E. and Coates, D. (1999) The construction of a cDNA expression library for the sheep scab mite *Psoroptes ovis*. *Veterinary Parasitology*, 83(3–4), 241–252.

Lee, A.J., Machell, J., van den Broek, A.H.M., Nisbet, A.J., Miller, H.R.P., Isaac, R.E. and Huntley, J.F. (2002) Identification of an antigen from the sheep scab mite, *Psoroptes ovis*, homologous with house dust mite group 1 allergens. *Parasite Immunology*, 24, 413–422.

Lee, C.S., Tsai, L.C., Chao, P.L., Lin, C.Y., Hung, M.W., Chien, A.I., Chiang, Y.T. and Han, S.H. (2004) Protein sequence analysis of a novel 103 kDa *Dermatophagoides pteronyssinus* mite allergen and prevalence of serum immunoglobulin E reactivity to rDer p 11 in allergic adult patients. *Clinical and Experimental Allergy*, 34, 354–362.

Lee, I.S. (2003) Effect of bedding control on amount of house dust mite allergens, asthma symptoms and peak expiratory flow rate. *Yonsei Medical Journal*, 44, 313–322.

Lee, M.H., Cho, S.H., Park, H.S., Bahn, J.W., Lee, B.J., Son, J.W., Kim, Y.K., Koh, Y.Y., Min, K.U. and Kim, Y.Y. (2000) Citrus red mite (*Panonychus citrilis*) is a common sensitising allergen among children living around citrus orchards. *Annals of Allergy, Asthma and Immunology*, 85, 200–204.

van Leeuwen, J. and Aalberse, R.C. (1991) To sieve or not to sieve. In: Todt, A. (ed.), *Dust Mite Allergens and Asthma. Report of the Second International Workshop*. UCB Institute for Allergy, Brussels, pp. 70–74.

Lefkovitch, L.P. (1965) The study of population growth in organisms grouped by stages. *Biometrics*, 21, 1–18.

Leiros, H.-K.S., Brandsdal, B.O., Andersen, O.A., Os, V., Leiros, I., Helland, R., Otlewski, J., Willassen, N.P. and Smalas, A.O. (2004) Trypsin specificity as elucidated by LIE calculations, X-ray structures and association constant measurements. *Protein Science*, 13, 1056–1070.

Lell, B., Borrmann, S., Yazdanbakhsh, M. and Kremsner, P.G. (2001) Atopy and malaria. *Wiener Klinische Wochenschrift*, 113, 927–929.

Le Mao, J., Dandeu, J.-P., Rabillon, J., Lux, M. and David, B. (1981) Antigens and allergens in *Dermatophagoides farinae* mite. I. Immunochemical and physicochemical study of two allergenic fractions from a partially purified *Dermatophagoides farinae* mite extract. *Immunology*, 44, 239–247.

Le Mao, J., Dandeu, J.-P., Rabillon, J., Lux, M. and David, B. (1983) Comparison of antigenic and allergenic composition of two partially purified extracts of *Dermatophagoides farinae* and *D. pteronyssinus* mite cultures. *Journal of Allergy and Clinical Immunology*, 71, 588–596.

Le Mao, J., Mayer, C.E., Peltre, G., Desvaux, F.X., David, B., Weyer, A. and Senechal, H. (1998) Mapping of *Dermatophagoides farinae* mite allergens by two-dimensional immunoblotting. *Journal of Allergy and Clinical Immunology*, 102, 631–636.

Leung, D.Y.M. (1995) Atopic dermatitis: the skin as a window into the pathogenesis of chronic allergic diseases. *Journal of Allergy and Clinical Immunology*, 96, 302–319.

Leung, R., Lam, C.W.K., Chan, A., Lee, M., Chan, H.S., Pang, S.W. and Lai, C.K.W. (1998) Indoor environment of residential homes in Hong Kong – relevance to asthma and allergic disease. *Clinical and Experimental Allergy*, 28, 585–590.

Leupen, M. J. and Varekamp, H. (1966) Some constructional and physical considerations concerning the micro-climatological conditions affecting growth of the house-dust mite (*Dermatophagoides*). In: *Interasma 5. Congress Proceedings. May, 1966, Utrecht*. Pressa Trajectina, Utrecht, pp. 2.44–2.55.

Levinson, H.Z. (1984) Contemplations on the origin of insect species inhabiting the storage environment. *Proceedings of the XVII International Congress of Entomology, Hamburg, 1984*. Paul Parey Scientific Publishers, New York, p. 626.

Levinson, H.Z. and Levinson, A.R. (1994) Origin of grain storage and insect species consuming desiccated food. *Anzeiger für Schädlingskunde, Pflanzenschutz, Umweltshulz*, 67, 47–59.

Levinson, H.Z., Levinson, A.R. and Muller, K. (1991) The adaptive function of ammonia and guanine in the biocenotic association between Ascomycetes and flour mites. *Naturwissenschaften*, 78, 174–176.

Levinson, H.Z., Levinson, A.R. and Offenberger, M. (1992) Effect of dietary antagonists and corresponding nutrients on growth and reproduction of the flour mite (*Acarus siro* L.). *Experientia*, 48, 721–729.

Li, C.-S., Hsu, C.-W., Chua, K.-Y., Hsieh, K.-H. and Lin, R.-H. (1996) Environmental distribution of house dust mite allergen (Der p 5). *Journal of Allergy and Clinical Immunology*, 97, 857–859.

Li, C.-S., Wan, G.-H., Hsieh, K.-H., Chua, K.-Y. and Lin, R.-H. (1994) Seasonal variation of house dust mite allergen (Der p 1) in a subtropical climate. *Journal of Allergy and Clinical Immunology*, 94, 131–134.

Li, C.P., Cui, Y.B., Wang, J., Yang, Q.G. and Tian, Y. (2003) Acaroid mites, intestinal and urinary acariasis. *World Journal of Gastroenterology*, 9, 874–877.

Liaw, S.-H., Chen, H.-Z., Liu, G.-G. and Chua, K.-Y. (2001) Acid-induced polymerization of the Group 5 mite allergen from *Dermatophagoides pteronyssinus*. *Biochemical and Biophysical Research Communications*, 285, 308–312.

Lin, K.L., Hsieh, K.H., Thomas, W.R., Chiang, B.L. and Chua, K.Y. (1994) Characterization of *Der p* V allergen, cDNA analysis, and IgE-mediated reactivity to the recombinant protein. *Journal of Allergy and Clinical Immunology*, 94, 989–996.

Linacre, E. (1977) A simple formula for estimating evaporation rates in various climates, using temperature data alone. *Agricultural Meteorology*, 18, 409–424.

Linacre, E. (1992) *Climate Data and Resources: A Reference and Guide*. Routledge, London.

Linacre, E. and Geerts, B. (1997) *Climates and Weather Explained*. Routledge, London.

Lincoln, R., Boxhall, G. and Clark, P. (1998) *A Dictionary of Ecology, Evolution and Systematics (Second Edition)*. University Press, Cambridge.

Lind, P. (1985) Purification and partial characterization of two major allergens from the house dust mite *Dermatophagoides pteronyssinus*. *Journal of Allergy and Clinical Immunology*, 76, 753–761.

Lind, P. (1986a) Enzyme-linked immunosorbent assay for determination of major excrement allergens of house dust mite species *D. pteronyssinus*, *D. farinae* and *D. microceras*. *Allergy*, 41, 442–451.

Lind, P. (1986b) Demonstration of close physico-chemical similarity and partial immunochemical identity between the major allergen, Dp42, of the house dust mite, *D. pteronyssinus* and corresponding antigens of *D. farinae* (Df6) and *D. microceras* (Dm6). *International Archives of Allergy and Applied Immunology*, 79, 60–65.

Lind, P. and Løwenstein, H. (1983) Identification of allergens in *Dermatophagoides pteronyssinus* mite body extract by crossed radioimmunoelectrophoresis with two different antibody pools. *Scandinavian Journal of Immunology*, 17, 263–273.

Lind, P., Norman, P.S., Newton, M., Løwenstein, H. and Schwartz, B. (1987) The prevalence of indorr allergens in the Baltimore area: house dust-mite and animal-dander antigens measured by immunochemical techniques. *Journal of Allergy and Clinical Immunology*, 80, 541–547.

Lindquist, E.E. (1972) A new species of *Tarsonemus* from stored grain (Acarina, Tarsonemidae). *Canadian Entomologist*, 104, 1699–1708.

Lintner, T.J. and Brame, K.A. (1993) The effects of season, climate and air-conditioning on the prevalence of *Dermatophagoides* mite allergens in household dust. *Journal of Allergy and Clinical Immunology*, 91, 862–867.

List, R.J. (1949) *Smithsonian Meterological Tables*. Smithsonian Miscellaneous Collections, Volume 114. Smithsonian Institution Press, Washington.

Liu, T., Jin, D.-C., Guo, J.-J. and Li, L. (2006) [Development and growth of *Tyrophagus putrescentiae* (Schrank) (Acarina: Acaridae) bred under different temperatures with different nutrients. *Acta Entomologica Sinica*, 49, 714–718. [In Chinese, English summary]

Ljunggren, E.L., Bergström, K., Morrison, D.A. and Mattson, J.G. (2006) Characterisation of an atypical antigen from *Sarcoptes scabiei* containing an MADF domain. *Parasitology*, 132, 117–126.

de las Llagas, L. A., Mistica, M.S. and Bertuso, A.G. (2005) House dust mites collected from houses of persons with respiratory allergies. *Philippine Entomologist*, 19, 193–200.

Loan, R., Sievers, R., Fitzharris, P. and Crane, J. (1998) House dust mite allergen and cat allergen variability within carpeted living room floors in domestic dwellings. *Indoor Air*, 13, 232–236.

Lobera, T., Blasco, A., Del Pozo, M.D., Etxenagusia, M., Iraola, V.M. and Fernández-Caldas, E. (2000) Sensibilización a ácaros en la Rioja. *Zubía Monográfico*, no. 12, 39–48.

Lombardero, M., Heymann, P.W., Platts, M.T., Fox, J.W. and Chapman, M.D. (1990) Conformational stability of B cell epitopes on group I and group II *Dermatophagoides* spp. allergens. Effect of thermal and chemical denaturation on the binding of murine IgG and human IgE antibodies. *Journal of Immunology*, 144, 1353–1360.

Lotka, A.J. (1922) The stability of the normal age distribution. *Proceedings of the National Academy of Sciences*, 8, 339–345.

Louadi, K and Robaux, P. (1992) Étude des populations d'acariens pulvicoles dans l'est Algérien selon les gradients climatiques propres a cette région. *Acarologia*, 33, 177–191.

Lucas, F.S., Broennimann, O., Febbraro, I. and Heeb, P. (2003) High diversity among feather-degrading bacteria from a dry meadow soil. *Microbial Ecology*, 45, 282–290.

De Lucca, S., Sporik, R., O'Meara, T.J. and Tovey, E.R. (1999) Mite allergen (Der p 1) is not only carried on mite feces. *Journal of Allergy and Clinical Immunology*, 103, 174–175.

Luczynska, C. (1994) Risk factors for indoor allergen exposure. In: Todt, A. (ed.), *Environmental Measures in the Prevention of Allergy*. UCB Institute of Allergy, Braine-l'Alleud, pp. 35–42.

Luczynska, C.M. (1997) Sampling and assay of indoor allergens. *Journal of Aerosol Science*, 28, 393–399.

Luczynska, C.M., Arruda, L.K., Platts-Mills, T.A.E., Miller, J.D., Lopez, M. and Chapman, M.D. (1989) A two-site monoclonal antibody ELISA for the quantitation of the major *Dermatophagoides* spp. allergens *Der p* I and *Der f* I. *Journal of Immunological Methods*, 118, 227–235.

Luczynska, C., Li, Y., Chapman, M.D. and Platts-Mills, T.A.E. (1990a) Airborne concentrations and particle distribution of allergen derived from domestic cats (*Felis domesticus*), measurements using a cascade impactor, liquid impinger and a two site monoclonal antibody assay for *Fel d* I. *American Review of Respiratory Diseases*, 141, 361–367.

Luczynska, C.M., Griffin, P., Davies, R.J. and Topping, M.D. (1990b) Prevalence of specific IgE to storage mites (*A. siro*, *L. destructor* and *T. longior*) in an urban population and crossreactivity with the house dust mites (*D. pteronyssinus*). *Clinical and Experimental Allergy*, 20, 403–406.

Luczynska, C., Sterne, J., Bond, J., Azima, H. and Burney, P. (1998) Indoor factors associated with concentrations of house dust mite allergen Der p 1, in a random sample of houses in Norwich, UK. *Clinical and Experimental Allergy*, 28, 1201–1209.

Luczynska, C., Tredwell, E., Smeeton N. and Burney, P. (2003) A randomized controlled trial of mite allergen-impermeable bed covers in adult mite-sensitized asthmatics. *Clinical and Experimental Allergy* 33, 1648–1653.

van der Lustgraaf, B. (1977) Xerophilic fungi in mattress dust. *Mycosen*, 20, 101–106.

van der Lustgraaf, B. (1978a) Ecological relationships between xerophilic fungi and house-dust mites (Acarida, Pyroglyphidae). *Oecologia*, 33, 351–359.

van der Lustgraaf, B. (1978b) Seasonal abundance of xerophilic fungi and house-dust mites (Acarida, Pyroglyphidae) in mattress dust. *Oecologia*, 36, 81–91.

van der Lustgraaf, B. and Jorde, W. (1977) Pyroglyphid mites, xerophilic fungi and allergenic activity in dust from hospital mattresses. *Acta Allergologia*, 32, 406–412.

Lynch, C.A. (1989) Two new species of the genus *Tyrophagus* (Acari: Acaridae). *Journal of Zoology, London*, 219, 545–567.

van Lynden-van Nes, A.M.T., Kort, H.S.M., Koren, L.G.H., Pernot, C.E.E. and van Bronswijk, J.E.M.H. (1996) Limiting factors for the growth and development of domestic mites. In: van Bronswijk, J.E.M.H. and Pauli, G. (eds), *An Update on Long-lasting Mite Avoidance. Dwelling Construction, Humidity Management, Cleaning. A Symposium held at the 11996 Annual Congress of the European Respiratory Society in Stockholm, Sweden.* GuT, Gemeinschaft umweltfreunlicher Teppichboden b.v., Aachen, pp. 13–20.

Macan, J., Kanceljak, B., Plavek, D. and Milkovic-Kraus, S. (2003) Differences in mite fauna between the continental and Mediterranean climates of Croatia: microscopy and Dustscreen™ test findings. *Allergy*, 58, 780–783.

Macdonald, C., Sternberg, A. and Hunter, P.R. (2007) A systematic review and meta-analysis of interventions used to reduce exposure to house dust and their effect on the development and severity of asthma. *Environmental Health Perspectives*, 115, 1691–1695.

Maele, B. van der (1983) A new strategy in the control of house dust mite allergy. *Pharmatherapeutica*, 3, 441–444.

Mahakittikun, V., Komoltri, C., Nochot, H., Angus, A.C. and Chew, F.T. (2003) Laboratory assessment of the efficiency of encasing materials against house dust mites and their allergens. *Allergy*, 58, 981–985.

Mahakittikun, V., Boitano, J.J., Tovey, E., Bunnag, C., Ninsanit, P., Matsumoto, T. and André, C. (2006) Mite penetration of different types of material claimed as mite proof by the Siriraj chamber method. *Journal of Allergy and Clinical Immunology*, 118, 1164–1168.

Makovcová, S., Samsinak, K. and Vobrázková, E. (1982) Biologický aspekt roztocov z domáceho prachu a ich vyznam v alergii na domaci prach. I. Cast. *Fysiatrickya Reumatologicky Vestnik*, 60, 69–74.

Malainual, N., Vichyanond, P. and Phan-Urai, P. (1995) House dust fauna in Thailand. *Clinical and Experimental Allergy*, 25, 554–560.

Manjra, A., Berman, D., Toerien, A., Weinberg, E.G. and Potter, P.C. (1994) The effects of a single treatment of an acaricide, Acarosan, and a detergent, Metsan, on Der p 1 allergen levels in the carpets and mattresses of asthmatic children. *South African Medical Journal*, 84, 278–280.

Mansfield, S.G., Cammer, S., Alexander, S.C., Muehleisen, D.P., Gray, R.S., Tropsha, A. and

Bollenbacher, W.E. (1998) Molecular cloning and characterisation of an invertebrate cellular retinoic acid binding protein. *Proceedings of the National Academy of Sciences*, 95, 6285–6830.

Mantle, J. and Pepys, J. (1974) Asthma amongst Tristan da Cunha Islanders. *Clinical Allergy*, 4, 161–170.

Marchler-Bauer, A. and Bryant, S.H. (2004) CD-Search: protein domain annotations on the fly. *Nucleic Acids Research*, 32, 327–331.

Marcucci, F., Sensi, L.G. and Bizzarri, G. (1985) Specific IgE to food and inhalent allergens in intestinal washings of children affected by atopic eczema. *Clinical Allergy*, 15, 345–354.

Mariana, A. and Ho, T.M. (1996) Distribution of *Blomia tropicalis* in Malaysia. *Tropical Biomedicine*, 13, 85–88.

Mariana, A., Ho, T.M. and Heah, S.K. (1996) Life-cycle, longevity and fecundity of *Blomia tropicalis* in a tropical laboratory. *Southeast Asian Journal of Tropical Medicine and Public Health*, 27, 392–395.

Markarian, J. (2006) Steady growth predicted for biocides. *Plastics, Additives and Compounding*, 8(1), 30–33.

Marks, G.B. (1998) House dust mite exposure as a risk factor for asthma: benefits of avoidance. *Allergy*, 53, 108–114.

Marks, G.B., Tovey, E.R., Green, W., Shearer, M., Salome, C.M. and Woolcock, A.J. (1994) House dust mite allergen avoidance: a randomized controlled trial of surface chemical treatment and encasement of bedding. *Clinical and Experimental Allergy*, 24, 1078–1083.

Marks, G., Tovey, E.R., Peat, J.K., Salome, C.M. and Woolcock, A.J. (1995a) Variability and repeatability of house dust mite allergen measurement: implications for study design and interpretation. *Clinical and Experimental Allergy*, 25, 1190–1197.

Marks, G., Tovey, E.R., Toelle, B.G., Wachinger, S., Peat, J.K. and Woolcock, A.J. (1995b) Mite allergen (Der p 1) concentration in houses and its relation to the presence and severity of asthma in a population of Sydney schoolchildren. *Journal of Allergy and Clinical Immunology*, 96, 441–448.

Marples, M.J. (1965) *The Ecology of the Human Skin.* Charles C. Thomas, Springfield, IL.

Marsh, D.G., Goodfriend, L., King, T.P., Løwenstein, H. and Platts-Mills, T.A.E. (1987) Allergen nomenclature. *Journal of Allergy and Clinical Immunology*, 80, 639–645.

Martin, I., Henwood, J., Wilson, F., Koning, M., Pike, A. and Town, I. (1996) House dust mite and cat allergen levels and housing characteristics in Christchurch, New Zealand. In: *New Zealand Society for Parasitology Conference Proceedings, 25th Jubilee Conference, Lakeside Motor Inn, Taupo, New Zealand, 28–30 August, 1996*, p. 15. [abstract]

Martínez, J., Eraso, E., Palacios, R. and Guisantes, J.A. (1999) Enzymatic analysis of house dust mite extracts from *Dermatophagoides pteronyssinus* and *Dermatophagoides farinae* (Acari: Pyroglyphidae) during different phases of culture growth. *Journal of Medical Entomology*, 36, 370–375.

Martínez Canzonieri, C.F., Baggio, D. and Lizarralde de Grosso, M. (1996) Nuevos hallazgos de fauna acarológia ne polvo de habitacion ne la provincial de Tucumán. *Acta Zoológia Lilloana*, 43, 289–291.

Mason, K., Riley, G., Siebers, R., Crane, J. and Fitzharris, P. (1999) Hot tumble drying and mite survival in duvets. *Journal of Allergy and Clinical Immunology*, 104, 499–500.

Massey, D.G. (1981) House dust mites in Hawaii. *Annals of Allergy*, 46, 197–200.

Massey, D.G., Fournier-Massey, G. and James, R.H. (1993). Minimising acarians and house dust in the tropics. *Annals of Allergy*, 71, 439–444.

Massey, D.G., Furumizo, R.T., Fournier-Massey, G., Kwock, D., Harris, S.T. (1988) House dust mites in university dormitories. *Annals of Allergy*, 61, 229–232.

Massey, J.E. and Massey, D.G. (1984) Effect of vacuum cleaning on house dust mites. *Hawaii Medical Journal*, 43, 404–406.

Mathaba, L.T., Pope, C.H., Lenzo, J., Hartofillis, M., Peake, H., Moritz, R.L., Simpson, R.J., Bubert, A., Thompson, P.J. and Stewart, G.A. (2002) Isolation and characterisation of a 13.8-kDa bacteriolytic enzyme from house dust mite extracts: homology with prokaryotic proteins suggests that the enzyme could be bacterially-derived. *FEMS Immunology and Medical Microbiology*, 33, 77–88.

Matheson, M.C., Dharmage, S.C., Forbes, A.B., Raven, J.M., Woods, R.K., Thien, F.C.K., Guest, D.I., Rolland, J.M., Walters, E.H. and Abramson, M.J. (2003) Residential characteristics predict changes in Der p 1, Fel d 1 and ergosterol, but not fungi over time. *Clinical and Experimental Allergy*, 33, 1281–1288.

Matsumoto, K. (1964) Studies on the environmental factors for the breeding of grain mites. V. Comparison of the breeding rate of *Carpoglyphus lactis* and *Tyrophagus dimidiatus*. *Japanese Journal of Sanitary Zoology*, 15, 17–24.

Matsumoto, K. (1965a) [Studies on the environmental factors for the breeding of grain mites. VI. Digestive enzymes of the grain mites, *Carpoglyphus lactis, Aleuroglyphus ovatus* and *Tyrophagus dimidiatus*.] *Japanese Journal of Sanitary Zoology*, 16, 86–89. [In Japanese, English summary]

Matsumoto, K. (1965b) [Studies on the environmental factors for the breeding of grain mites. VII. Relationships between reproduction of mites and kind of carbohydrates in the diet.] *Japanese Journal of Sanitary Zoology*, 16, 118–122. [In Japanese, English summary]

Matsumoto, K. (1966) [Studies on the environmental factors for the breeding of grain mites. VIII. The effect of relative humidity on the age composition of the population of *Lardoglyphus konoi*.] *Japanese Journal of Sanitary Zoology*, 16, 86–89. [In Japanese, English summary]

Matsumoto, K. (1975) [Studies on the environmental requirement for breeding the dust mites, *Dermatophagoides farinae* Hughes, 1961. Part 3. Effect of the lipids in the diet on the population growth of the mites.] *Japanese Journal of Sanitary Zoology*, 26, 121–127. [In Japanese, English summary]

Matsumoto, K. (1978) Studies on the environmental factors for the breeding of grain mites. XII. Observations on the mode of breeding and of hypopus appearance of *Lardoglyphus konoi* (Sasa and Asanuma, 1951) in various kinds of diet. *Japanese Journal of Sanitary Zoology*, 29, 287–294.

Matsumoto, K., Okamoto, M. and Wada, Y. (1986) [Effect of relative humidity on life cycle of the house dust mites, *Dermatophagoides farinae* and *D. pteronyssinus*.] *Japanese Journal of Sanitary Zoology*, 37, 79–90. [In Japanese, English summary]

Matsumoto, T., Mike, T. and Ono, T. (1996) Mite-related allergy and oral tolerance. *Allergy*, 51, 276–277.

Matsuoka, H., Maeda, N., Atsuta, Y., Ando, K. and Chinzei, Y. (1995) Seasonal fluctuations of *Dermatophagoides* mite population in house dust. *Japanese Journal of Medical Science and Biology*, 48, 103–115.

Mattsson, J.G., Bergström, K. and Näslund, K. (1999) Molecular cloning and expression of a major antigen from *Sarcopes scabiei*. In: *17th International Conference of the World Association for the Advancement of Veterinary Parasitology, Denmark*, 15–19 August, 1999.

Mattsson, J.G., Ljunggren, E.L. and Bergström, K. (2001) Paramyosin from the parasitic mite *Sarcoptes scabiei*: cDNA cloning and heterologous expression. *Parasitology*, 122, 555–562.

Maunsell, K. (1951) Quantitative aspects of allergy to house dust. *Proceedings of the 1st International Congress of Allergy, Zürich.* p. 306.

Maunsell, K., Wraith, D.G. and Cunnington, A.M. (1968) Mites and house-dust allergy in bronchial asthma. *The Lancet*, ii, 1267–1270.

Maunsell, K., Wraith, D.G. and Hughes, A.M. (1971) Hyposensitisation in mite asthma. *Lancet*, 8 May 1971, 967–968.

Mauro, K., Akaike, T., Ono, T., Okamoto, T. and Maeda, H. (1997) Generation of anaphylotoxins through propteolytic processing of C3 and C5 by house dust mite protease. *Journal of Allergy and Clinical Immunology*, 100, 253–260.

Maurya, K.R. and Jamil, Z. (1980) Factors affecting the distribution of house-dust mites under domestic conditions in Lucknow. *Indian Journal of Medical Research*, 72, 284–292.

Maurya, K.R., Jamil, Z. and Dev, B. (1983) Prevalence of astigmatid mites (Acarina) in the domestic environment of north-eastern and northern India. *Biological Memoirs*, 8, 207–215.

Mayr, E. (1982) *The Growth of Biological Thought.* Belknap Press of Harvard University Press, Cambridge, USA.

McCall, C., Hunter, S., Stedman, K., Weber, E., Hillier, A., Bozic, C., Rivoire, B. and Olivry, T. (2001) Characterization and cloning of a major high molecular weight house dust mite allergen (Der f 15) for dogs. *Veterinary Immunology and Immunopathology*, 78, 231–247.

McCarney, R., Warner, J., Iliffe, S., van Hasselen, R., Griffin, M. and Fisher, P. (2007) The Hawthorn Effect: a randomised controlled trial. *BMC Medical Research Methodology*, 7, Article No. 7 (www.biomedcentral.com/1471-2288/7/30).

McDonald, E., Cook, D., Newman, T., Griffith, L., Cox, G. and Guyatt, G. (2002) Effect of air filtration systems on asthma. A systematic review of randomised controlled trials. *Chest*, 122, 1535–1542.

McDonald, L. and Tovey, E. (1992) The role of water temperature and laundry procedures in reducing house dust mite populations and allergen content of bedding. *Journal of Allergy and Clinical Immunology*, 90, 599–608.

McDonald, L. and Tovey, E. (1993) The effectiveness of benzyl benzoate and some essential plant oils as laundry additives for killing house dust mites. *Journal of Allergy and Clinical Immunology*, 92, 771–772.

McDowell, J.A. (1993) A Survey of World Textiles. The Ancient World. 1. Introduction. 2. The Mediterranean. In: Harris, J. (ed.), *5000 Years of Textiles*. British Museum Press, London, pp. 54–65.

McGregor, P. (1995) Population dynamics for house dust mite researchers. In: Tovey, E., Fifoot, A. and Sieber, L. (eds), *Mites, Asthma and Domestic Design II*. University of Sydney, Sydney, pp. 22–25.

McGregor, P. and Peterson, P.G. (2000) Population growth in house dust mite cultures: an inversely density-dependent effect. In: Siebers, R., Cunningham, M., Fitzharris, P. and Crane, J. (eds), *Mites, Asthma and Domestic Design III*. Wellington School of Medicine, New Zealand, pp. 27–31.

McMurry, J. (2000) *Organic Chemistry*. 5th ed. Brooks/Cole, Pacific Grove, CA.

Medina Gallardo, J.F., Castillo Gomez, J., Capote Gil, F., Ayerbe Garcia, R., Sanchez Armenggol, M.A. and Munoz Biedma, M.L. (1994) Utilidad de los dispositivos deshumidifacores en la reducción de la concentración de los ácaros. *Archivos de Bronchopneumonologica*, 30, 287–290.

Mégnin, P. (1876) Note sur la faculté qu'ont certains Acariens avec ou sans bouche de vivre sans nourriture pendant des phases entières de leur existence et même pendant toute leur vie. *Journal d'Anatomie et Physiologie*, 12, 603–606.

Mehl, R. (1998) Occurrence of mites in Norway and the rest of Scandinavia. *Allergy*, 53 (Supplement 48), 28–35.

Mekie, E.C. (1926) Parasitic infection of the urinary tract. *Edinburgh Medical Journal*, 33, 708–719.

Melnick, B.C., Hollman, J., Hoffman, U., Yuh, M.-S. and Plewig, G. (1990) Lipid composition of outer stratum corneum and nails in atopic and control subjects. *Archives of Dermatological Research*, 282, 549–551.

Meltzer, E.O. (2007) Allergic rhinitis: the impact of discordant perspectives of patient and physician on treatment decisions. *Clinical Therapeutics*, 29, 1428–1440.

Meno, K., Thorsted, P.B., Ipsen, H., Kristensen, O., Larsen, J., Spangfort, M.D., Gajhede, M. and Lund, K. (2005) The crystal structure of recombinant proDer p 1, a major house dust mite proteolytic allergen. *Journal of Immunology*, 175, 3835–3845.

Mercado, D., Puerta, L. and Caraballo, L. (1996) Niveles de alergenos de acaros en el polvo de habitacion en Cartagena. *Biomedica*, 16, 307–314.

Merzendorfer, H. and Zimoch, L. (2003) Chitin metabolism in insects: structure, function and regulation of chitin synthases and chitinases. *Journal of Experimental Biology*, 206, 4393–4412.

Mészáros, M. and Morton, D.B. (1996a) Identification of a developmentally regulated gene, esr16, in the tracheal epithelium of *Manduca sexta*, with homology to a protein from human epididymis. *Insect Biochemistry and Molecular Biology*, 26, 711.

Mészáros, M. and Morton, D.B. (1996b) Comparison of the expression patterns of five developmentally regulated genes in *Manduca sexta* and their regulation by 20-hydroxyecdysone *in vitro*. *Journal of Experimental Biology*, 199, 1555–1561.

Michael, A.D. (1884) The Hypopus Question, or the life-history of certain Acarina. *Journal of the Linnean Society, Zoology*, 17, 371–394.

Michael, A.D. (1895) On the form and proportions of the brain in the Oribatidae and in other Acarina. *Journal of the Royal Microscopical Society*, 1895, 274–282.

Michael, A.D. (1901, 1903) *British Tyroglyphidae*. Volumes I and II, Ray Society, London.

Michel, F.B., Guin, J.J., Seignalet, C., Rambier, A., Marty, J.C., Caula, F. and Laveil, G. (1977) Allergie a *Panonychus ulmi* (Koch). *Revue Française d'Allergologie et d'Immunologie Clinique*, 17, 93–97.

Mihrshahi, S., Peat, J.K., Webb, K., Tovey, E.R., Marks, G.B., Mellis, C.M. and Leeder, S.R. (2001) The Childhood Asthma Prevention Study (CAPS): design and research protocol of a randomized trial for the primary prevention of asthma. *Controlled Clinical Trials*, 22, 333–354.

Mihrshahi, S., Marks, G.B., Vanlaar, C., Tovey, E. and Peat, J. (2002) Predictors of high house dust mite allergen concentrations in residential homes in Sydney. *Allergy*, 57, 137–142.

Mihrshahi, S., Marks, G.B., Criss, S., Tovey, E.R., Vanlaar, C.H. and Peat, J.K. for the CAPS Team (2003) Effectiveness of an intervention to reduce house dust mite allergen levels in children's beds. *Allergy*, 58, 784–789.

Mills, J. (1991) Carpets in paintings. The 'Bellini', 'Keyhole' or 'Re-entrant' rugs. *Hali*, no. 58, 86–103.

Mills, K.L., Hart, B.J., Lynch, N.R., Thomas, W.R. and Smith, W. (1999) Molecular characterisation of the group 4 house dust mite allergen from *Dermatophagoides pteronyssinus* and its amylase homologue from *Euroglyphus maynei*. *International Archives of Allergy and Immunology*, 120, 100–107.

Miner, C.S. and Dalton, N.N. (1953) Physical properties of glycerol and its solutions. *American Chemical Monograph Series*, no. 117, p. 269.

Miranda, R.J., Quintero, D.A. and Almanza, A. (2002) House dust mites from urban and rural houses on the lowland Pacific slopes of Panama. *Systematic and Applied Acarology*, 7, 23–30.

Mitchell, E.A. and Elliott, R.B. (1980) Controlled trial of an electrostatic precipitator in childhood asthma. *The Lancet*, ii, 559.

Mitchell, E.B., Crow, J., Chapman, M.D., Jouhal, S.S., Pope, F.M., Platts-Mills, T.A.E. (1982) Basophils in allergen-induced patch test sites in atopic dermatitis. *Lancet*, i, 127–130.

Mitchell, E.B., Crow, J., Rowntree, S., Webster, A.D. and Platts-Mills, T.A.E. (1984) Cutaneous basophil hypersensitivity to inhalant allergens in atopic dermatitis patients, elicitation of delayed responses containing basophils following local transfer of immune serum but not IgE antibody. *Journal of Investigative Dermatology*, 83, 290–295.

Mitchell, E.B., Wilkins, S., Deighton, J. and Platts-Mills, T.A.E. (1985) Reduction of house dust mite allergen levels in the home, use of the acaricide, pirimiphos methyl. *Clinical Allergy*, 15, 235–240.

Mitchell, W.F. (1969) House dust mites in Hawaii. *Annals of Allergy*, 46, 197–200.

Mitchell, W.F., Wharton, G.W., Larson, D.G. and Modic, R. (1969) House dust, mites and insects. *Annals of Allergy*, 27, 93–99.

Miyamoto, J. and Ouchi, T. (1976) [Ecological studies of house dust mites. Seasonal changes in mite populations in house dust in Japan.] *Japanese Journal of Sanitary Zoology*, 27, 251–259. [In Japanese]

Miyamoto, J., Ishii, A. and Sasa, M. (1975) A successful method for mass culture of the house dust mite *Dermatophagoides pteronyssinus* (Trouessart, 1897). *Japanese Journal of Experimental Medicine*, 45, 133–138.

Miyamoto, T., Oshima, S., Ishizaki, T. and Sato, S. (1968) Allergenic identity between the common floor mite (*Dermatophagoides farinae* Hughes, 1961) and house dust as a causative agent in bronchial asthma. *Journal of Allergy*, 42, 14–28.

Miyamoto, T., Oshima, S., Mizuno, K., Sasa, M. and Ishizaki, T. (1969) Cross-antigenicity among six species of dust mites and house dust antigens. *Journal of Allergy*, 44, 228.

Miyamoto, T., Oshima, S., Domae, A., Takahashi, K., Izeki, M., Tanaka, T. and Ishizaki, T. (1970) Allergenic potency of different house dusts in relation to contained mites. *Annals of Allergy*, 28, 405–412.

Miyazawa, H., Sakaguchi, M., Inouye, S., Ikeda, K., Honbo, Y., Yasueda, H. and Shida, T. (1996) Seasonal changes in mite allergen (Der I and der II) concentrations in Japanese homes. *Annals of Allergy, Asthma and Immunology*, 76, 170–174.

Mocellin, J. and Foggin, P. (2008) Health status and geographic mobility among semi-nomadic pastoralists in Mongolia. *Health and Place*, 14, 228–242.

Modak, A., Saha, G.K., Tandon, N. and Gupta, S.K. (1991) Dust mite fauna in houses of bronchial asthma patients – a comparative study of three zones of West Bengal. *Entomon*, 16, 115–120.

Modak, A., Saha, G.K. and Choudri, D.K. (1992) High altitude house dust mite fauna of Kashmir Valley, India. *Annals of Entomology*, 10, 43–46.

Moerbeke, D. van (ed.) (1991) *Dust Mite Allergens and Asthma. Report of the 2nd International Workshop, Minster Lovell, Oxfordshire, England, September 19–21, 1990.* UCB Institute of Allergy, Brussels.

Molina, C., Aiache, J.M., Tourreau, A. and Jeanneret, A. (1975) Enquête épidémiologique et immunologique chez les fromagers. *Revue Française d'Allergologie et d'Immunologie Clinique*, 15, 89–91.

Molina, C., Tourreau, A., Aiache, J.M., Brun, J., Jeanneret, A. and Roche, G. (1977) Manifestations allergiques chez les fromagers. *Revue Française d'Allergologie et d'Immunologie Clinique*, 17, 235–245.

Mollet, J.A. (1996) Dispersal of American house dust mites (Acari: Pyroglyphidae) in a residence. *Journal of Medical Entomology*, 33, 844–847.

Mollet, J. and Robinson, W. (1995) Use of marked mites to study the dispersal of the American house dust mite (*Dermatophagoides farinae* Hughes). In: Tovey, E., Fifoot, A. and Sieber, L. (eds), *Mites, Asthma and Domestic Design II*. University of Sydney, Sydney, pp. 19–21.

Montealegre, F., Sepulveda, A., Bayona, M., Quiñones, C. and Fernández-Caldas, E. (1997) Identification of the domestic mite fauna of Puerto Rico. *Puerto Rico Health Sciences Journal*, 16, 109–116.

Moore, J.C., Walter, D.E. and Hunt, H.W. (1988) Athropod regulation of micro- and mesobiota in below-ground detrital foodwebs. *Annual Review of Entomology*, 33, 419–439.

Mora, C., Flores, I., Montealegre, F. and Diaz, A. (2003) Cloning and expression of Blo t 1, a novel allergen from the dust mite *Blomia tropicalis*, homologous to cysteine proteinases. *Clinical and Experimental Allergy*, 33, 28–34.

Moreira, N.S. (1975) Acarinos Pyroglyphidae e outros Sarcoptiformes emamostras de po domicilar em Belo Horizonte, Minas Gerais. MS thesis, Federal University of Minas Gerais.

Moreira, N.S. (1978) Redescrição de *Chortoglyphus arcuatus* (Troupeau, 1878) (Sarcoptiformes, Chortoglyphidae). *Revista Brasileira Biologia*, 38(2), 245–249.

Morgan, F.D., Hopkins, D., Morgan, B.J., Johnston, M. and Ford, R.M. (1974) The house dust mite in Adelaide, South Australia. *Medical Journal of Australia*, ii, 224–225.

Morgan, W.J., Crain, E.F., Gruchalla, R.S., O'Connor, G., Kattan, M., Evans, R., Stout, J., Malindzak, G., Smartt, E., Plaut, M., Walter, M., Vaughan, P. and Mitchell, H. (2004) Results of a home-based environmental intervention among urban children with asthma. *New England Journal of Medicine*, 351, 1068–1080.

Mori, K. and Kuwahara, S. (1986a) Synthesis of both the enantiomers of lardolure, the aggregation pheromone of the acarid mite, *Lardoglyphus konoi*. *Tetrahedron*, 42, 5539–5544.

Mori, K. and Kuwahara, S. (1986b) Stereochemistry of lardolure, the aggregation pheromone of the acarid mite, *Lardoglyphus konoi*. *Tetrahedron*, 42, 5545–5550.

Mori, K. and Sajiki, J. (1997) Phospholipid metabolism in European House Dust Mite (Acari: Pyroglyphidae). *Journal of Medical Entomology*, 34, 538–543.

Mori, N. and Kuwahara, Y. (2000) Comparative studies of the ability of males to discriminate between sexes in *Caloglyphus* spp. *Journal of Chemical Ecology*, 26, 1299–1309.

Mori, N., Kuwahara, Y., Kurosa, K., Nishida, R. and Fukushima, T. (1995) Chemical ecology of astigmatid mites. XLI. Undecane: the sex pheromone of the acarid mite *Caloglyphus rodriguezi* Samsinak (Acarina: Acaridae). *Applied Entomology and Zoology*, 30, 415–423.

Morita, Y., Miyamoto, T., Horiuchi, Y., Oshima, S., Katsuhata, A. and Kawal, S. (1975) Further studies in allergenic identity between house dust and the house dust mite, *Dermatophagoides farinae*, Hughes, 1961. *Annals of Allergy*, 35, 361–366.

Morrison Smith, J., Disney, M.E., Williams, J.D. and Goels, Z.A. (1969) Clinical significance of skin reactions to mite extracts in children with asthma. *British Medical Journal*, I (659), 723–726.

Mosbech, H., Korsgaard, J. and Lind, P. (1988) Control of house dust mites by electrical heating blankets. *Journal of Allergy and Clinical Immunology*, 81, 706–710.

Mosbech, H., Jensen, A., Heinig, J.H. and Schou, C. (1991) House dust mite allergens on different types of mattresses. *Clinical and Experimental Allergy*, 21, 351–355.

Moscato, G., Perfetti, L., Galdi, E. and Minoia, C. (2000) Levels of house-dust-mite allergen in homes of nonallergic people in Pavia, Italy. *Allergy*, 55, 873–878.

Moser, J.C. and Roton, L.M. (1970) Tagging mites with aerosol paint. *Annals of the Entomological Society of America*, 63, 1784.

Mösges, R. and Klimek, L. (2007) Today's allergic rhinitis patients are different: new factors that may play a role. *Allergy*, 62, 969–975.

Mothes-Wagner, U. (1984) Fine structure of the cuticle and structural changes occurring during moulting in the mite *Tetranychus urticae*. II. Moulting process (Chelicerata, Acarina). *Zoomorphology*, 104, 105–110.

Mound, L.A. (1983) 'For a taxonomist you seem to know a lot about biology!' *Antenna*, 7(1), 3–5.

Moyer, D.B., Nelson, H.S. and Arlian, L.G. (1985) House dust mites in Colorado. *Annals of Allergy*, 55, 680–682.

Mueller, G.A., Benjamin, D.C. and Rule, G.S. (1998) Tertiary structure of the major house dust mite allergen Der p 2: sequential and structural homologies. *Biochemistry*, 37, 12707–12714.

Mueller, G.A., Smith, A.M., Williams, D.C., Hakkart, G.A.J., Aalberse, R.C., Chapman, M.D., Rule, G.S. and Benjamin, D.C. (1997) Expression and secondary structure determination by NMR methods of the major house dust mite allergen Der p 2. *Journal of Biological Chemistry*, 272, 26893–26898.

Mueller, G.A., Smith, A.M., Chapman, M.D., Rule, G.S. and Benjamin, D.C. (2001) Hydrogen exchange nuclear magnetic resonance spectroscopy mapping of antibody epitopes on the house dust mite allergen Der p 2. *Journal of Biological Chemistry*, 276, 9359–9365.

Mulcahy, R., Jean-Pastor, M.J. and Maunder, J.W. (1988) Efficacité d'Acardust sur les acariens de la poussière de maison dans des conditions semi-naturelles. *Comptes Rendus de Thérapeutique et de Pharmacologie Clinique*, 6, 3–9.

Mulla, M.S. and Sanchez Medina, M. (eds) (1980) *Domestic Acari of Colombia. Bionomics, Ecology and Distribution of Allergenic Mites, their Role in Allergic Diseases.* Editoria Guadalupe, Bogota.

Mulla, M.S., Harkrider, J.R., Galant, S.P. and Amin, L. (1975) Some house dust control measures and abundance of *Dermatophagoides* mites in Southern California. *Journal of Medical Entomology*, 12, 5–9.

Mulligan, M. (2000) Downscaled climate change scenario for Colombia and their hydrological consequences. *Advances in Environmental Monitoring and Modelling*, 1, 3–35.

Mullins, D.E. and Cochran, D.G. (1974) Nitrogen metabolism in the American cockroach: an examination of whole body and fat body regulation of cations in response to nitrogen balance. *Journal of Experimental Biology*, 61, 557–570.

Mulvey, P.M. (1972) Cot death survey. Anaphylaxis and the house dust mite. *Medical Journal of Australia*, ii, 1290–1244.

Mumcuoglu, K.Y. (1975) Zur biologie der hausstaub-milbe *Dermatophagoides pteronyssinus* II. Milbenhäufigkeit in den verschiedenen Regionen der Schweiz und ihre Abhängigkeit von Klima.

Schweizerische Meditziner Wochenschrift, 105, 1013–1020.

Mumcuoglu, K.Y. (1976) House dust mites in Switzerland I. Distribution and Taxonomy. *Journal of Medical Entomology*, 13, 361–373.

Mumcuoglu, K.Y. (1977a) House dust mites in Switzerland, II. Culture and control. *International Journal of Acarology*, 3, 19–25.

Mumcuoglu, K.Y. (1977b) House dust mites in Switzerland, III. Allergenic properties of the mites. *Acta Allergologica*, 32, 339–349.

Mumcuoglu, K.Y. (1977c) Immunologische untersuchungen von hausstaub und hausstaubmilben. I. Schultz-Dale test. *Allergie und Immunologie*, 23, 107–110.

Mumcuoglu, K.Y. (1988) Biologie und ökologie der hausstaubmilben. *Allergologie*, 11, 223–228.

Mumcuoglu, K.Y. and Rufli, Th. (1979) Immunological investigations of house-dust and house-dust mites II. Localisation of the antigen in then body of the house-dust mite *Dermatophagoides pteronyssinus* by means of the indirect immunofluorescence method. In: Rodriguez, J.G. (ed.), *Recent Advances in Acarology*. Volume 2. Academic Press, New York, pp. 205–210.

Mumcuoglu, K.Y. and Rufli, Th. (1981) Pyroglyphidae/Hausstaubmilben. *Schweizerische Rundschau für Medizin Praxis*, 70, 1039–1049.

Mumcuoglu, K.Y., Zavaro, A., Samra, Z. and Lazarowitz, Z. (1988) House dust mites and vernal keratoconjunctivitis. *Opthalmologica*, 196, 175–181.

Mumcuoglu, K.Y., Abed, Y., Armenios, B., Shaheen, S., Jacobs, J., Bar-Sela, S. and Richter, E. (1994) Asthma in Gaza refugee camp children and its relationship with house dust mites. *Annals of Allergy*, 72, 163–166.

Mumcuoglu, K.Y., Gat, Z., Horowitz, T., Miller, J., Bar-Tana, R., Ben-Zvi, A. and Naparstek, Y. (1999) Abundance of house dust mites in relation to climate in contrasting agricultural settlements in Israel. *Medical and Veterinary Entomology*, 13, 252–258.

Munir, A.K.M., Einarsson, R. and Dreborg, S.K.G. (1993) Vacuum cleaning decreases the levels of mite allergens in house dust. *Pediatric Allergy and Immunology*, 4, 136–143.

Munir, A.K.M., Björkstén, B., Einarsson, R., Ekstrand-Tobin, A., Möller, C., Warner, A. and

Kjellmann, N.-I.M. (1995) Mite allergens in relation to home conditions and sensitisation of asthmatic children from three climatic regions. *Allergy*, 50, 55–64.

Murray, A.B. and Ferguson, A.C. (1983) Dust free bedrooms in the treatment of asthmatic children with house dust or house dust mite allergy, a controlled trial. *Pediatrics*, 71, 418–422.

Murray, A.B. and Zuk, P. (1979) The seasonal variation in a population of house dust mites in a North American city. *Journal of Allergy and Clinical Immunology*, 64, 266–269.

Murray, A.B., Ferguson, A.C. and Morrison, B.J. (1985) Sensitization to house dust mites in different climatic areas. *Journal of Allergy and Clinical Immunology*, 76, 108–112.

Murray, A.B., Laxer, R.M., Petty, R.E. and Speert, D.P. (1984) Role of house dust mites in Kawasaki disease. *Canadian Medical Association Journal*, 131, 720.

Murton, J.J. and Madden, J.L. (1977) Observations on the biology, behaviour and ecology of the house-dust mite, *Dermatophagoides pteronyssinus* (Trouessart) (Acarina: Pyroglyphidae) in Tasmania. *Journal of the Australian Entomological Society*, 16, 281–287.

Musken, H., Wahl, R., Franz, J.T., Masuch, G. and Bergman, K.C. (1996) Sensibilizierungen gegen die raubmilben *Cheyletus eruditus* und vor-ratsmilben bei patienten mit hausstaubmilben sensibilisierung. *Allergologie*, 19, 29–34.

Musken, H., Franz, J.T., Wahl, R., Paap, A., Cromwell, O., Masuch, G. and Bergmann, K.C. (2003) Sensitisation to different mite species in German farmers. *Journal of Investigational Allergology and Clinical Immunology*, 13, 26–35.

Muzinich, S.A., Lake, F., Thompson, P.J. and Stewart, G.A. (1997) Characterization of a mite allergen with glucoamylase activity. In: van Moerbeke, D. (ed.), *Report of the Third International Workshop on Indoor Allergens and Asthma, Cuenca, Spain.* UCB Institute for Allergy, Brussels, p. 50.

My-Yen, L.T., Matsumoto, K., Wada, Y. and Kuwahara, Y. (1980a) Pheromone study on acarid mites V. Presence of citral as a minor component of the alarm pheromone in the mold mite, *Tyrophagus putrescentiae* (Schrank, 1781) (Acarina: Acaridae). *Applied Entomology and Zoology*, 15, 474–477.

My-Yen, L.T., Wada, Y., Matsumoto, K. and Kuwahara, Y. (1980b) Pheromone study on acarid mites VI. Demonstration and isolation of an aggregation pheromone in *Lardoglyphus konoi* Sasa et Asanuma. *Japanese Journal of Sanitary Zoology*, 31, 249–254.

Nadchatram, M., Fournier, G., Gordon, B.L. and Massey, D.G. (1981) House dust mites in Hawaii. *Annals of Allergy*, 46, 197–200.

Nadchatram, M., Chan, K.L. and Yeow, M.N. (1986) The mite fauna of house dust in Singapore with studies on the life history of *Austroglycyphagus malaysiensis*. *Bulletin of the Faculty of Science, National University of Singapore*, 6, 27–28.

Nafstad, P., Øie, L., Mehl, R., Gaarder, P.I., Lødrup-Carlsen, K.C., Botten, G., Magnus, P. and Jaakola, J.J.K. (1998) Residential dampness problems and symptoms and signs of bronchial obstruction in young Norwegian children. *American Journal of Respiratory and Critical Care Medicine*, 157, 410–414.

Naik, M.T., Chang, C.-F., Kuo, I.-C., Kung, C.C.-H., Yi, F.-C., Chua, K.-Y. and Huang, T.-H. (2008) Roles of structure and structural dynamics in the antibody recognition of the allergen proteins: an NMR study on *Blomia tropicalis* major allergen. *Structure*, 16, 125–136.

Nakagawa, T., Kudo, K., Okudaira, H., Miyamoto, T. and Horiuchi, Y. (1977) Characterisation of the allergic components of the house dust mite, *Dermatophagoides farinae*. *International Archives of Allergy and Applied Immunology*, 55, 47–53.

Nakamura, K. and Yoshida, T. (1977) [Fluctuations of serum IgE levels in house dust-sensitive children who received hyposensitisation therapy with house dust extract.] *Japanese Journal of Allergology*, 26, 573–580. [In Japanese, English summary]

Nalepa, A. (1884) Die anatomie der Tyroglyphen I. *Sitzungsberichten der Akademie der Wissenschaften in Wien, Mathematische – Naturwissenschaften Klasse*, 90, 197–228.

Nalepa, A. (1885) Die anatomie der Tyroglyphen II. *Sitzungsberichten der Akademie der Wissenschaften in Wien, Mathematische – Naturwissenschaften Klasse*, 92, 116–127.

Nandy, A., Graefe, L., Buchhop, S., Kniest, F.M., Kahlert, H., Cromwell, O. and Fiebig, H. (2003) Novel alleles of *Dermatophagoides farinae* house dust mite allergen Der f 2. In: Marone, G. (ed.), *Clinical Immunology and Allergy in Medicine.* JGC Editions, Naples, pp. 735–740.

Nannelli, R., Liguori, M. and Castagnoli, M. (1983) Osservazioni preliminari sulla biologia di *Euroglyphus (E.) maynei* (Coor.) (Acarina, Pyroglyphidae) e sua distribuzione in Italia. *Redia*, 66, 3–10.

Naspitz, C.K., Szefler, S.J., Tinkelman, D.G. and Warner, J.O. (eds) (2001) *Textbook of Pediatric Asthma. An International Perspective.* Informa Healthcare, London.

Nath, P., Gupta, R. & Jameel, Z. (1974) Mite fauna in the house dust of bronchial asthmatic patients. *Indian Journal of Medical Research*, 62, 1140–1145.

Natuhara, Y. (1989) New wet sieving method for isolating house dust mites. *Japanese Journal of Sanitary Zoology*, 40, 333–336.

Natuhara, Y., Horiuchi, Y., Kidera, K., Shino, K., Ozaki, H., Funamoto, J., Ueno, S., Yoshimura, A., Sugahara, T., Fujitani, H., Gen, T., Saraie, M., Nakajima, S. and Isshiki, G. (1991) Preliminary test of encasement of bedding with the high-density fabric for reducing house-dust mites (Acari: Pyroglyphidae). *Japanese Journal of Sanitary Zoology*, 42, 305–309.

Nayar, E., Lal, M. and Gupta, D.A. (1974) The prevalence of mite *Dermatophagoides pteronyssinus* and its association with house dust allergy. *Indian Journal of Medical Research*, 62, 11–13.

Nazrullaveva, M.F. and Dubinina, Y.V. (1999) [The house-dust mites *Glycyphagus cadaverum* are a powerful source of allergen in Uzbekistan.] *Meditsinskaya Parazitologiya i Parazitarnye Bolezni*, no. 2, 35–40. [In Russian, English summary]

Neal, J.S., Arlian, L.G. and Morgan, M.J. (2002) Relationship among house-dust mites, Der 1, Fel d 1 and Can f 1 on clothing and automobile seats with respect to densities in houses. *Annals of Allergy, Asthma and Immunology*, 88, 410–415.

Needham, G., Begg, C. and Buchanan, S. (2006) Ultraviolet C exposure is fatal to American house dust mite eggs. *Journal of Allergy and Clinical Immunology*, 117(2), Supplement 1, S28. [abstract]

Neffen, H.E., Fernández-Caldas, E., Predolini, N., Trudeau, W.L., Sánchez-Guerra, M.E. and Lockey, R.F. (1996) Mite sensitivity and exposure in the city of Santa Fe, Argentina. *Journal of Investigational Allergology and Clinical Immunology*, 6, 278–282.

Nelson, H.S. and Fernández-Caldas, E. (1995) Prevalence of house dust mites in the Rocky Mountain States. *Annals of Allergy, Asthma and Immunology*, 75, 337–339.

Nelson, H.S., Hirsch, R., Ohman, J.L., Platts-Mills, T.A.E., Reed, C.E. and Solomon W.R. (1988) Recommendations for the use of residential air-cleaning devices in the treatment of allergic respiratory diseases. *Journal of Allergy and Clinical Immunology*, 82, 661–669.

Nelson, R.P., DiNicolo, R., Fernández-Caldas, E., Seleznick, M.J., Lockey, R.F. and Good, R.A. (1996) Allergen-specific IgE levels and mite allergen exposure in children with acute asthma first seen in an emergency department and in nonasthmatic control subjects. *Journal of Allergy and Clinical Immunology*, 98, 258–263.

Neville, A.C. (1975) *Biology of the Arthropod Cuticle.* Springer Verlag, Berlin.

Ninomiya, T. and Kawasaki, T. (1988) Acarid attractant. *Japan Kokai Tokkyo Koho*, Syo 63-230605, Sept. 27, 1988. [Japanese Patent]

Nisbet, A.J., MacKellar, A., Wright, H.W., Brennan, G.P., Chua, K.Y., Cheong, N., Thomas, J.E. and Huntley, J.F. (2006a) Molecular characterisation, expression and localisation of tropomyosin and paramyosin immunodominant allergens from sheep scab mites (*Psoroptes ovis*). *Parasitology*, 133, 515–523.

Nisbet, A.J., Huntley, J.F., MacKellar, A., Sparks, N. and McDevitt, R. (2006b) A house dust mite allergen homologue from poultry red mite *Dermanyssus gallinae* (De Geer). *Parasite Immunology*, 28, 401–405.

Nishiyama, C., Yuuki, T., Takai, T., Okumura, Y. and Okudaira, H. (1993) Determination of the three disulphide bonds in a major house dust mite allergen, *Der f* II. *International Archives of Allergy and Immunology*, 101, 159–166.

Nishiyama, C., Yasuhara, T., Yuuki, T. and Okumura, Y. (1995) Cloning and expression in *Escherichia coli* of cDNA encoding house dust mite allergen Der f 3, serine protease from *Dermatophagoides farinae*. *FEBS Letters*, 377, 62–66.

Niven, R.M., Fletcher, A.M., Pickering, A.C., Custovic, A., Sivour, J.B., Preece, A.R., Oldham, L.A. and Francis, H.C. (1999) Attempting to control mite allergens with mechanical ventilation and dehumidification in British houses. *Journal of Allergy and Clinical Immunology*, 103, 756–762.

Noferi, A., Serino, W., Sacerdoti, G. and Blythe, M.E. (1974) Survey on concentration and type of house dust mites (HDM) in dusts of specifically sensitised people in Naples. Preliminary data. *Folia Allergologia Immunologia Clinica*, 21, 466–469.

Nogrady, S.G. and Furnass, S.B. (1983) Ionisers in the management of bronchial asthma. *Thorax*, 38, 919–922.

Nordvall, S.L., Eriksson, M., Rylander, E. and Schwartz, B. (1988) Sensitisation of children in the Stockholm area to house dust mite. *Acta Paediatrica Scandinavica*, 77, 716–720.

Norman, J. (1990) *The Complete Book of Spices*. Dorling Kindersley, London.

Norton, R.A. (1985) Aspects of the biology and systematics of soil arachnids, particularly saprophagous and mycophagous mites. *Quaestiones Entomologicae*, 21, 523–541.

Norton, R.A. (1998) Morphological evidence for the evolutionary origin of the Astigmata (Acari: Acariformes). *Experimental and Applied Acarology*, 22, 559–594.

Norton, R.A., Bonamo, P., Grierson, J.D. and Shear, W. (1988) Oribatid mite fossils from a terrestrial Devonian deposit near Gilboa, New York. *Journal of Palaeontology*, 62, 259–269.

Norton, R.A., Kethley, J.B., Johnston, D.E. and OConnor, B.M. (1993) Phylogenetic perspectives on genetic systems and reproductive modes of mites. In: Wrensch, D. and Ebbert, M. (eds), *Evolution and Diversity of Sex Ratio in Insects and Mites*. Chapman and Hall, New York, pp. 8–99.

Notredame, C., Higgins, D. and Heringa, J. (2000) T-Coffee: A novel method for multiple sequence alignments. *Journal of Molecular Biology*, 302, 205–217.

Nourrit, J., Penaud, A., Timon-David, P., Autran, P. and Nicoli, R.M. (1975) Méthode simplifiée pour la recherche des acariens allergisant des poussières de maison sur des échantillons réduits. *Bulletin de la Société de Pathologie Exotique*, no. 3, 318–320.

Noverr, M.C. and Huffnagle, G.B. (2005) The 'microflora hypothesis' of allergic diseases. *Clinical and Experimental Allergy*, 35, 1511–1520.

Novoa Avilés, D. (1975) Acaros del polvo en la Republica Mexicana. *Alergia*, 22, 59–77.

Nowak, R.M. and Paradiso, J.L. (1983) *Walker's Mammals of the World, Volume 2*. 4th edition, Johns Hopkins University Press, Baltimore.

Nuttall, T.J., Hill, P.B., Bensignor, E. and Willemse, T. (2006) House dust and forage mite allergens and their role in human and canine atopic dermatitis. *Veterinary Dermatology*, 17, 223–235.

Nylandsted, J., Gyrd-Hansen, M., Danielwicz, A., Fehrenbacher, N., Lademann, U., Høyer-Hansen, M., Weber, E., Multhoff, G., Rohde, M. and Jäätelä, M. (2004) Heat shock protein 70 promotes cell survival by inhibiting lysosomal membrane permeabilisation. *Journal of Experimental Medicine*, 200, 425–435.

Ober, C. (1997) Genetics of atopy. In: Barnes, P., Grunstein, M., Leff A.R. and Woolcock, A. J. (eds), *Asthma*, Volume 1. Lippincott-Raven, Philadelphia, pp. 129–144.

OConnor, B.M. (1979) Evolutionary origins of astigmatid mites inhabiting stored products. In: Rodriguez, J.G. (ed.), *Recent Advances in Acarology*. Volume 1. Academic Press, New York, pp. 273–278.

OConnor, B.M. (1981) A systematic revision of the family-group taxa of the non-psoroptid Astigmata (Acari: Acariformes). PhD thesis, Cornell University.

OConnor, B.M. (1982a) Evolutionary ecology of astigmatic mites. *Annual Review of Entomology*, 27, 385–409.

OConnor, B.M. (1982b) Acari: Astigmata. In: Parker, S.P. (ed.), *Synopsis and Classification of Living Organisms*. Vol. 1. McGraw-Hill, New York, pp. 146–169.

OConnor, B.M. (1984) Acarine–fungal relationships: the evolution of symbiotic associations. In: Wheeler, Q. and Blackwell, M. (eds), *Fungus–Insect Relationships. Perspectives in Ecology and Evolution*. Columbia University Press, New York, pp. 354–381.

OConnor, B.M. (1994) Life-history modifications in astigmatid mites. In: Houck, M. (ed.), *Mites. Ecological and Evolutionary Analysis of Life-History Patterns*. Chapman and Hall, New York, pp. 136–159.

O'Connor, G.T. (2005) Allergen avoidance in asthma: what do we do now? *Journal of Allergy and Clinical Immunology*, 116, 26–30.

Office for National Statistics (2004) *The National Statistics Socio-economic Classification User Manual*. Version No. 1.2. Her Majesty's Stationery Office, London.

Oh, H., Ishii, A., Tongu, Y. and Itano, K. (1986) Microorganisms associated with the house dust mite, *Dermatophagoides*. *Japanese Journal of Sanitary Zoology*, 37, 229–235.

Ohashi, Y. and Kida, H. (2004) Local circulations developed in the vicinity of both coastal and inland urban areas. Part II. Effects of urban and mountain areas on moisture transport. *Journal of Applied Meteorology*, 43, 119–133.

Ohga, K., Yamanaka, R., Kinumaki, H., Awa, S. and Kobayashi, N. (1983) Kawasaki disease and rug shampoo. *The Lancet*, April 23, 930.

Okamoto, M., Matsumoto, K., Wada, Y. and Nakano, H. (1978) [Studies on antifungal effect of mite alarm pheromone Citral 1. Evaluation of antifungal effects of citral.] *Japanese Journal of Sanitary Zoology*, 29, 255–260. [In Japanese, English summary]

Okamura, K., Kiuchi, S., Tamba, M., Kashima, T., Hiramoto, S., Baba, T., Dacheux, F., Dacheux, J.L., Sugita, Y. and Jin, Y.Z. (1999) A porcine homolog of the major secretory protein of human epididymis, HE1, specifically binds cholesterol. *Biochimica Biphysica Acta*, 1438, 377–387.

Oke, R. (1982) The energetic basis of the urban heat island. *Quarterly Journal of the Royal Meteorological Society*, 108, 1–24.

Okuhira, H. (1994) [Molecular biology of mite allergens.] *Arerugi*, 43, 435–440. [In Japanese]

Olaguibel, J.M., Quirce, S., García Figueroa, B.E., Barber, D., Rico, P. and Tabar, A.I. (1994) Grado de exposición alergénica a *Dermatophagoides* spp. y eficacia de las medidas de desalergenización en el área de Pamplona. *Revista España Immunologica Clínica*, 9, 83–90.

Oliveira, A.J. and Daemon, E. (2003) Qualitative and quantitative assessment of mites (Acari) in domicilary dust in rural dwellings in the 'Zona da Mata' region, Minas Gerais, Brazil. *Revista Brasileira de Zoologia*, 20, 675–679.

Oliveira, C.H., Binotti, R.S., Muniz, J.R.O., Pinho Jr., A.J., Prado, A.P. and Lazzarini, S. (1999). Fauna acarina da poeira de colchões na cidade de Campinas-SP. *Revista Brasileira de Alergia e Imunopatologia* 22, 188–197.

Oliveira, C.H., Binotti, R.S., Muniz, J.R.O., dos Santos, J.C., do Prato, A.P. and de Pino, A.J. (2003) Comparison of house dust mites found on different mattress surfaces. *Annals of Allergy, Asthma and Immunology*, 91, 559–562.

Olsson, S., Harfast, B., Johansson, S.G.O. and Hage-Hamsten, M. van. (1994) Detection of at least one high-molecular-mass, IgE-binding component of the dust mite *Lepidoglyphus destructor*. *Allergy*, 49, 620–625.

O'Meara, T. and Tovey, E.R. (2000) Monitoring personal allergen exposure. *Clinical Reviews in Allergy and Immunology*, 18, 341–395.

O'Neill, G., Donovan, G.R. and Baldo, B.A. (1994a) Cloning and characterization of a major allergen of the house dust mite, *Dermatophagoides pteronyssinus*, homologous with glutathione S-transferase. *Biochimica et Biophysica Acta*, 1219, 521–528.

O'Neill, G., Donovan, G.R. and Baldo, B.A. (1994b) Sequence analyses of the native form of a major house dust mite allergen and its nascent polypeptide. *Molecular Immunology*, 31, 1447–1448.

O'Neill, S.E., Heinrich, T.K., Hales, B.J., Hazell, L.A., Holt, D.C., Fischer, K. and Thomas, W.R. (2006) The chitinase allergens Der p 15 and Der p 18 from *Dermatophagoides pteronyssinus*. *Clinical and Experimental Allergy*, 36, 831–839.

Oosting, A.-J., de Bruin-Weller, M.S., Terreehorst, I., Tempels-Pavlica, Z., Aalberse, R.C., de Monchy, J.G.R., van Wijk, R.G. and Bruijnzeel-Koomen, C.A.F.M. (2002) Effect of mattress encasings on atopic dermatitis outcome measures in a double-blind, placebo-controlled study: the Dutch Mite Avoidance Study. *Journal of Allergy and Clinical Immunology*, 110, 500–506.

Opie, J. (1992) *Tribal Rugs. Nomadic and Village Weavings from the Near East and Central Asia.* Laurence King Publishing, London.

Oppermann, H., Doering, C., Sobottka, A., Krämer, U. and Thriene, B. (2001) Belastungssituation ost- und westdeutscher haushalte mit hausstaubmilben und schimmelpilzen. *Geshundheitswesen*, 63, 85–89.

Ordman, D. (1971) The incidence of 'climate asthma' in South Africa, its relation to the distribution of mites. *South African Medical Journal*, 45, 739–743.

O'Rourke, M.K., Moore, C.L. and Arlian, L.G. (1996) Prevalence of house dust mites from homes in the Sonoran Desert, Arizona. In: Muilenberg, M. and Burge, H. (eds), *Aerobiology*. CRC Press, Boca Raton, pp. 67–70.

Ortego, F., Sánchez-Ramos, I., Ruiz, M. and Castañera, P. (2000) Characterisation of proteases from a stored product mite, *Tyrophagus putrescentiae*.

Archives of Insect Biochemistry and Physiology, 43, 116–124.

Oshima, S. (1964) [Observations of floor mites collected in Yokohama. I. On the mites found in several schools in summer.] *Japanese Journal of Sanitary Zoology*, 15, 233–244. [In Japanese, English summary]

Oshima, S. (1967) Studies on the genus *Dermatophagoides* (Psoroptidae, Acarina) as floor-mites, with special reference to the medical importance. *Japanese Journal of Sanitary Zoology*, 18, 213–215.

Oshima, S. (1968) [Redescription of three species of *Mealia* from house dust in Japan.] *Japanese Journal of Sanitary Zoology*, 19, 165–192. [In Japanese, English summary]

Oshima, S. (1970) Studies on the mite fauna of the house-dust of Japan and Taiwan with special reference to house-dust allergy. *Japanese Journal of Sanitary Zoology*, 21, 1–17.

Oshima, S. (1979) New acaroid mites in house dust. *Annotationes Zoologicae Japonensis*, 52, 240–245.

Oshima, S. and Sugita, K. (1965) [Notes on the life history of *Dermatophagoides farinae* Hughes, 1961 (Acarina, Epidermoptidae).] *Bulletin of the Yokohama Institute of Public Health*, 4, 66–69. [In Japanese, English summary]

Oshima, S., Nakamura, Y., Sugita, K., Yoneyama, E., Kitazume, M. and Yoshinaga, Y. (1972) [Population density of the mites of floor affected by the moisture content of the Tatami mat, with particular reference for practical prevention.] *Annual Report of the Yokohama Institute of Health*, no. 11, 62–69. [In Japanese, English summary]

Østergaard, P.A. (1979) Effekten af en støvsanering af soverum hos asthmabørn med husstøvmiddeallergi. *Ugeskrift for Laeger*, 141, 2928–2931.

Otani, T., Kinugawa, N., Iikura, Y. and Hishi, F. (1984) [Distribution of house-dust mites in the houses of asthmatic children in Japan.] *Aerugi, Japanese Journal of Allergology*, 33, 454–462. [In Japanese, English summary]

Ott, C.M., Bruce, R.J. and Pierson, R.L. (2004) Microbial characterization of free floating condensate aboard the Mir space station. *Microbial Ecology*, 47, 133–136.

Ottoboni, F. and Piu, G. (1990) Gli Acari allergenici. *Guida al Loro Riconoscimento*. UTET Periodici Scientifici S.r.l., Milan.

Ottoboni, F., Falagiani, P. and Centanni, S. (1983) Gli Acari allergenici. *Bolletin Istituto Sieroterico Milano*, 63, 389–419.

Ottoboni, F., Falagiani, P., Lorenzini, E., Piu, G., Carluccio, A. and Centanni, S. (1983) Gli Acari domestici in Sardegna. *Bolletin Istituto Sieroterico Milano*, 62, 363–369.

Ouborg, N.J. and Vriezen, W.H. (2007) An ecologist's guide to ecogenomics. *Journal of Ecology*, 95, 8–16.

Ouchi, T., Miyamoto, J. and Ishii, A. (1976) [Mite fauna of house dust collected from newly-built apartment houses.] *Japanese Journal of Sanitary Zoology*, 27, 427–429. [In Japanese, English summary]

Oudemans, A.C. (1910) Acarologische Aanteekeningen XXXII. *Entomologische Berichten*, 3, 67–74.

Oudemans, A.C. (1911) Acarologische Aanteekeningen XXXVII. *Entomologische Berichten*, 3, 165–175.

Oudemans, A.C. (1926) *Kritisch Historisch Overzicht der Acarologie. Eerste Gedeelte, 850 vC-1758.* E.J. Brill, Leiden.

Oudemans, A.C. (1928) Acarologische Aanteekeningen XCIII. *Entomologische Berichten*, 7, 346–348.

Outhouse, S.L. and Castro, S.D. (1990) Occurrence of house dust mites in Central Texas. *Texas Journal of Science*, 42, 104–106.

Owen, S., Morganstern, M., Hepworth, J. and Woodcock, A. (1990) Control of house dust mite antigen in bedding. *The Lancet*, 335, 396–397.

Paik, Y., Takaoka, M., Matsuoka, H. and Ishii, A. (1992) Mite fauna and mite antigen in house dust from houses in Seoul, Korea. *Japanese Journal of Sanitary Zoology*, 43(1), 29–35.

Papaioannou-Souliotis, P. (1991) House dust mites in Attiki. *Annals of the Institute of Phytopathology of Benaki*, 16, 105–114.

Park, G.-M., Lee, S.-M., Lee, I.-Y., Ree, H.-I., Kim, K.-S., Hong, C.-S. and Yong, T.-S. (2000) Localisation of a major allergen, Der p 2, in the gut and faecal pellets of *Dermatophagoides pteronyssinus*. *Clinical and Experimental Allergy*, 30, 1293–1297.

Parkinson, C.L. (1992) Culturing free-living astigmatid mites. In: Cooper, J.E., Pearce-Kelly, P. and Williams, D.L. (eds), *Arachnida, Proceedings of a Symposium on Spiders and their Allies*. Published by the editors, London, pp. 62–70.

Parkinson, C.L., Jamieson, N., Eborall, J. and Armitage, D.M. (1991) Comparison of the fecundity of three species of grain store mites on fungal diets. *Experimental and Applied Acarology*, 12, 297–302.

Patriarca, P.A., Rodgers, M.F., Morens, D.M., Schonberger, L.B., Kaminski, R.M., Burns, J.C. and Glode, M.P. (1982) Kawasaki disease and rug shampoo. *The Lancet*, April 23, 930.

Paul, T.C. and Sinha, R.N. (1972) Low-temperature survival of *Dermatophagoides farinae*. *Environmental Entomology*, 1, 547–549.

Pauli, G., Dietemann, A., Hoyet, C., Eckert, F. and Bessot, J.C. (1989) Effets d'un nouvel acaricide sur les symptoms des asthmatiques allergiques aux acariens et influence de l'environment domestique. *Revue Française d'Allergologie et d'Immunologie Clinique*, 29, 175–180.

Pauli, G., Quoix, E., Hedelin, G., Bessot, J.C., Ott, M. and Dietemann, A. (1993) Mite allergen content of *Dermatophagoides*-allergic asthmatics/rhinitics and matched controls. *Clinical and Experimental Allergy*, 23, 606–611.

Pearce, N., Pekkanen, J. and Beasley, R. (1999) How much asthma is really attributable to atopy? *Thorax*, 54, 268–272.

Pearce, N., Douwes, J. and Beasley, R. (2000) Is allergen exposure the major primary cause of asthma? *Thorax*, 55, 424–431.

Pearce, N., Sunyer, J., Cheng, S., Chinn, S., Björkstén, B., Burr, M., Keil, U., Anderson, H.R. and Burney, P. on behalf of the ISAAC Steering Committee and the European Community Respiratory Health Survey (2000) Comparison of asthma prevalence in the ISAAC and the ECRHS. *European Respiratory Journal*, 16, 420–426.

Pearce, N., Aït-Khaled, N., Beasley, R., Mallol, J., Keil, U., Mitchell, E., Robertson, C. and the ISAAC Phase Three Study Group (2007) Worldwide trends in the prevalence of asthma symptoms: phase III of the International Study of Asthma and Allergies in Childhood (ISAAC). *Thorax*, 62, 758–766.

Pearson, D. (1995) Modification of the bed microenvironment: Allergen-barrier bed covers. In: Tovey, E., Fifoot, A. and Sieber, L. (eds), *Mites, Asthma and Domestic Design II*. University of Sydney, Sydney, pp. 52–56.

Pearson, R.S.B. and Cunnington, A.M. (1973) The importance of mites in house dust sensitivity in Barbadian asthmatics. *Clinical Allergy*, 3, 299–306.

Peat, J.K. (1995) Epidemiological evidence that house dust mites cause asthma. In: Tovey, E., Fifoot, A. and Sieber, L. (eds), *Mites, Asthma and Domestic Design II*. University of Sydney, Sydney, pp. 1–8.

Peat, J.K., Britton, W.J., Salome, C.M. and Woolcock, A.J. (1987) Bronchial responsiveness in two populations of Australian schoolchildren III. Effect of exposure to environmental allergens. *Clinical Allergy*, 17, 297–300.

Peat, J.K., Tovey, E.R., Mellis, C.M., Leeder, S.R. and Woolcock, A.J. (1993) Importance of house dust mite and *Alternaria* allergens in childhood asthma, an epidemiological study in two climatic regions of Australia. *Clinical and Experimental Allergy*, 23, 812–820.

Peat, J.K., Toelle, B.G., Gray, E.J., Haby, M.M., Belousova, E., Mellis, C.M. and Woolcock, A.J. (1995) Prevalence and severity of childhood asthma and allergic sensitisation in seven climatic regions of New South Wales. *Medical Journal of Australia*, 163, 22–26.

Peat, J.K., Tovey, E.R., Toelle, B.G., Haby, M.M., Gray, E.J., Mahmic, A. and Woolcock, A.J. (1996) House dust mite allergens – a major risk factor for childhood asthma in Australia. *American Journal of Respiratory and Critical Care Medicine*, 153, 141–146.

Pedersen, L.C., Yee, V.C., Bishop, P.D., Trong, I.L., Teller, D.C. and Stenkamp, R.E. (1994) Transglutaminase factor XIII uses proteinase-like catalytic triad to crosslink macromolecules. *Protein Science*, 3, 1131–1135.

Pekár, S. and Žd'árková, E. (2004) A model of the biological control of *Acarus siro* by *Cheyletus eruditus* (Acari: Acaridae, Cheyletidae) on grain. *Journal of Pesticide Science*, 77, 1–10.

Pelikan, Z. (2007) The role of nasal allergy in chronic secretory otitis media. *Annals of Allergy, Asthma and Immunology*, 99, 401–407.

Penaud, A., Nourrit, J., Timon-David, P., Autran, P., and Charpin, J. (1971) Acariens pyroglyphides dans les poussières domestique. *Marseille Médical*, no. 5, 363–368.

Penaud, A., Nourrit, J., Autran, P., Timon-David, P. and Nicoli, R.-M. (1972) Données actuelles sur les Acariens Pyroglyphides des poussières de maison. *Annales de Parasitologie*, 47, 631–662.

Penaud, A., Nourrit, J., Timon-David, P., Jacquet-Francillon, M. and Charpin, J. (1976a) Methods of destroying house dust pyroglyphid mites. *Clinical Allergy*, 5, 109–114.

Penaud, A., Nourrit, J., Timon-David, P., Jacquet-Francillon, M., Autran, P. and Charpin, J. (1976b)

Effets sur les acariens pyroglyphides des poussière domestiques du Paragerm AK appliqué au domicile des malades allergiques. *Revue Française d'Allergologie et d'Immunologie Clinique*, 26, 125–129.

Penaud, A., Nourrit, J., Timon-David, P. and Charpin, J. (1977) Results of a controlled trial of the acaricide Paragerm on *Dermatophagoides* spp. in dwelling houses. *Clinical Allergy*, 7, 49–53.

Penders, J., Stobberingh, E.E., van den Brandt, P.A. and Thijs, C. (2007) The role of the intestinal microbiota in the development of atopic disorders. *Allergy*, 62, 1223–1236.

Penry, D.L. and Jumars, P.A. (1987) Modelling animal guts as chemical reactors. *The American Naturalist*, 129, 69–96.

Pepys, J., Chan, M. and Hargreave, F.E. (1968) Mites and house-dust allergy. *Lancet*, June 15, 1270.

Percebois, G., Gallois, A.-M. and Denvis, P. (1972) Mise en évidence de *Dermatophagoides pteronyssinus* dans des poussières de matelas en Lorraine. *Annales Médicales de Nancy*, 11, 79–82.

Pérez Lozano, A. (1979) Environmental control in asthmatic homes. The role of *Cheyletus* mites. Preliminary report. *Allergologia et Immunopathologia*, 7, 303–306.

Perozich, J., Nicholas, H., Wang, B.-C., Lindahl, R. and Hempel, J. (1999) Relationships within the aldehyde dehydrogenase extended family. *Protein Science*, 8, 137–146.

Perron, R. (1954) Untersuchungen über Bau, Entwicklung und Physiologie der Milbe *Histiostoma laboratorium*. *Acta Zoologica (Stuttgart)*, 35, 71–176.

Peters, W. (1992) *Peritrophic Membranes*. Springer-Verlag, Berlin.

Peterson, E.L., Ownby, D.R., Kallenbach, L. and Johnson, C.C. (1999) Evaluation of air and dust sampling schemes for Fel d 1, Der f 1, and Der p 1 allergens in homes in the Detroit area. *Journal of Allergy and Clinical Immunology*, 104, 348–355.

Peterson, R.D.A., Wicklund, P.E. and Good, R.A. (1964) Endotoxin activity of a house dust extract. *Journal of Allergy*, 35, 134–142.

Petrova, A.D. and Zheltikova, T.M. (1990) [The seasonal dynamics of allergenic mite numbers (Acariformes, Pyroglyphidae) in house dust of three apartments in Moscow.] *Biologicheskie Nauki* no. 10, 37–45. [In Russian, English summary]

Petrova, A.D. and Zheltikova, T.M. (2000) [Long-term dynamics and structure of house dust mite community in Moscow (Acariformes, Astigmata).] *Zoologicheskii Zhurnal*, 79, 1402–1408. [In Russian, English summary]

Phiriyangkul, P. and Utarabhand, P. (2006) Molecular characterisation of a CDNA encoding vitellogenin in the banana shrimp, *Penaeus* (*Litopenaeus*) *merguiensis* and sites of vitellogenin mRNA expression. *Molecular Reproduction and Development*, 73, 410–423.

Pianka, E.R. (1970). On *r*- and *K*-selection. *American Naturalist*, 104, 592–597.

Pigatto, P.D., Bigardi, A., Riboldi, A., Legori, A., Altomare, G.F. and Ottoboni, F. (1991) Occupational atopic contact dermatitis from *Rhizoglyphus*. *Contact Dermatitis*, 25, 193–194.

Pike, A.J. and Wickens, K. (2008) The house dust mite and storage mite fauna of New Zealand dwellings. *New Zealand Entomologist*, 31, 17–22.

Pike, A.J., Cunningham, M.J. and Lester, P.J. (2005) Development of *Dermatophagoides pteronyssinus* (Acari: Pyroglyphidae) at constant and simultaneously fluctuating temperature and humidity conditions. *Journal of Medical Entomology*, 42, 266–269.

Pillai, A., Ueno, S., Zhang, H. and Kato, Y. (2003) Induction of ASABF (*Ascaris suum* antibacterial factor)–type antimicrobial peptides by bacterial injection: novel members of ASABF in the nematode *Ascaris suum*. *Biochemical Journal*, 371, 663–668.

Pinhão, R.C. and Grácio, A.J.S. (1978) House dust mites in Lisbon – a short notice. *Anais do Instituto Higiene e Medicine tropicale (Lisboa)*, 5, 361.

Pitten, F.A., Kalveram, C.M., Krüger, U., Müller, G. and Kramer, A. (2000) Reduktion der Besieddlung neuwertiger Matratzen mit Bakterien, Schimmelpilzen und Hausstaubmilben durch Matratzenganzbezüge. *Der Hautarzt*, 51, 655–660.

Pittner, G., Vrtala, S., Thomas, W.R., Weghofer, M., Kundi, M., Horak, F., Kraft, D. and Valenta, R. (2004) Component-resolved diagnosis of house-dust mite allergy with purified natural and recombinant allergens. *Clinical and Experimental Allergy*, 34, 597–603.

Plácido, J.L., Cuesta, C., Delgado, L., Moreira da Silva, J.P., Miranda, M., Ventas, P. and Vaz, M. (1996) Indoor mite allergens in patients with respiratory allergy living in Porto, Portugal. *Allergy*, 51, 633–639.

Platts-Mills, T.A.E. (1987) The biological role of allergy. In: Lessoff, M.H., Lee, T.H. and Kemeney, D.M. (eds), *Allergy: An International Textbook*. Wiley, New York, pp. 1–35.

Platts-Mills, T.A.E. and Chapman, M.D. (1987) Dust mites, immunology, allergic disease, and environmental control. *Journal of Allergy and Clinical Immunology*, 80, 755–775; and erratum (1988), *Journal of Allergy and Clinical Immunology*, 82, 841.

Platts-Mills, T.A.E., Mitchell, E.B., Nock, P., Tovey, E.R., Moszoro, H. and Wilkins, S.R. (1982) Reduction of bronchial hyperreactivity during prolonged allergen avoidance. *Lancet*, 25 September, 675–678.

Platts-Mills, T.A.E., Mitchell, E.B., Rowntree S., Chapman, M.D. and Wilkins, S.R. (1983) The role of dust mite allergens in atopic-dermatitis. *Clinical and Experimental Dermatology*, 8, 233–247.

Platts-Mills, T.A.E, Heymann, P.W., Chapman, M.D., Hayden, M.L. and Wilkins, S.R. (1986a) Cross-reacting and species-specific determinants on a major allergen from *Dermatophagoides pteronyssinus* and *D. farinae*, development of a radioimmunoassay for antigen P_1 equivalent in house dust and dust mite extracts. *Journal of Allergy and Clinical Immunology*, 78, 398–407.

Platts-Mills, T.A.E., Heyman, P.W., Longbottom, J. and Wilkins, S.R. (1986b) Airborne allergens associated with asthma: particle size measured with a cascade impactor. *Journal of Allergy and Clinical Immunology*, 77, 850–857.

Platts-Mills, T.A.E., Hayden, M.L., Chapman, M.D. and Wilkins, S.P. (1987) Seasonal variation in dust mite and grass pollen allergens in dust from the houses of patients with asthma. *Journal of Allergy and Clinical Immunology*, 79, 781–791.

Platts-Mills, T.A.E, Chapman, M.D., Heyman, P.W. and Luczynska, C.M. (1989a) Measurements of airborne allergen using immunoassays. *Immunology and Allergy Clinics of North America*, 9, 269–283.

Platts-Mills, T.A.E., de Weck, A., Aalberse, R.C., Bessot, J.C., Bjorksten, B., Bischoff, E., Bousquet, J., van Bronswijk, J.E.M.H., ChannaBasavanna, G.P., Chapman, M.D., Colloff, M.J., Goldstein, R.A., Guerin, B., Hart, B.J., Hong, C.-S., Ito, K., Jorde, W.,

Korsgaard, J., Le Mao, J., Miyamoto, T., Lind, P., Lowenstein, H., Mitchell, E.B., Murray, A.B., Nolte, D., Norman, P.S., Pauli, G., Ranganath, H.R., Reed, C., Reiser, J., Stewart, G.A., Turner, K.J., Vervloet, D. and Wen, T. (1989b) Dust mite allergens and asthma – a worldwide problem. *Journal of Allergy and Clinical Immunology*, 83, 416–427.

Platts-Mills, T.A.E, Chapman, M.D., Mitchell, E.B., Heymann, P.W. and Dewell, B. (1991) Role of inhalent allergens in atopic eczema. In: Ruzicka, T., Ring, J. and Przybilla, B. (eds), *Handbook of Atopic Eczema*. Springer-Verlag, Berlin, pp. 192–203.

Platts-Mills, T.A.E., Thomas, W.R., Aalberse, R.C., Vervloet, D. and Chapman, M.D. (1992) Dust mite allergens and asthma: report of a second international workshop. *Journal of Allergy and Clinical Immunology*, 89, 1046–1060.

Platts-Mills, T.A.E., Sporik, R., Wheatley, L.M. and Heyman, P.W. (1995) Is there a dose-response relationship between exposure to indoor allergens and symptoms of asthma? *Journal of Allergy and Clinical Immunology*, 96, 435–440.

Platts-Mills, T.A.E, Woodfolk, J.A., Chapman, M.D., Heymann, P.W. (1996) Changing concepts of allergic disease – the attempt to keep up with real changes in lifestyles. *Journal of Allergy and Clinical Immunology*, 98, Supplement S, S297–S306.

Platts-Mills, T.A.E., Vervloet, D., Thomas, W.R., Aalberse, R.C. and Chapman, M.D. (1997) Indoor allergens and asthma: report of the Third International Workshop. *Journal of Allergy and Clinical Immunology*, 100(6) (Supplement 1), S1–S24.

Platts-Mills, T.A.E., Vaughan, J.W., Carter, M.C. and Woodfolk, J.A. (2000) The role of intervention in established allergy: avoidance of indoor allergens in the treatment of chronic allegic disease. *Journal of Allergy and Clinical Immunology*, 106, 787–804.

Popescu, I.G. and Banescu, O. (1975) Presence of *Dermatophagoides pteronyssinus* in the houses of asthmatics sensitised to house dust. A one-year study. *Revue Roumaine de Medecine Interne*, 13, 293–295.

Popplewell, E.J., Innes, V.A., Lloyd-Hughes, S., Jenkins, E.L., Khdir, K., Bryant, T.N., Warner, J.O. and Warner, J.A. (2000) The effect of high-efficiency and standard vacuum-cleaners on mite, cat and dog allergen levels and clinical progress. *Pediatric Allergy and Immunology*, 11, 142–148.

Portus, M. and Blasco, C. (1977) Factores que influyen en la composicion cuali y cuantitativa de la fauna de acaros del polvo domestico III. Epoca del ano. *Allergologia et Immunopathologia*, 5, 645–652.

Portus, M. and Gomez, M.S. (1976) Variaciones cuali y cuantitivos dela fauna de acaros del polvo domestico en diversas localidades catalanas. *Circular Farmacéutica*, no. 253, 547–551.

Portus, M., Gallego, J. and Blasco, C. (1977) Factores que influyen en la composicion de la fauna de acaros del polvo domestico. I. Temperatura y humedad de la vivienda. *Allergologia et Immunopathologia*, 5, 27–34.

Posse, J.M. (1946) *Contribution al estudio de la alergia respiratoria in Vizcaya*. Archivos del Hospital, Santo Hospital Civil del Gen. Franco, Bilbao.

Potter, P.C., Davis, G., Manjra, A. and Luyt, D. (1996). House dust mite allergy in Southern Africa – historical perspective and current status. *Clinical and Experimental Allergy*, 26, 132–137.

Poulos, L.M., O'Meara, T.J., Sporik, R. and Tovey, E.R. (1999) Detection of inhaled Der p 1. *Clinical and Experimental Allergy*, 29, 1232–1238.

Prassé, J. (1967) Zur anatomie und histologie der Acaridae mit besonders Berücksichtung von *Caloglyphus berlesei* (Michael, 1903) und *C. michaeli* (Oudemans, 1924). I. Das Darmsystem. *Wissenschaftliche Zeitschrift der Universität, Halle*, 16, 789–812.

Price, J.A., Pollock, I., Little, S.A., Longbottom, J.L. and Warner, J.O. (1990) Measurement of airborne mite antigen in homes of asthmatic children. *Lancet*, 336, 895–897.

Price, J.B. and Holditch, D.M. (1980) Changes in osmotic pressure and sodium concentration of the haemolymph of woodlice with progressive dehydration. *Comparative Biochemistry and Physiology, Series A*, 66, 297–305.

Pruett, J.H. (1999) Identification and purification of a 16 kDa allergen from *Psoroptes ovis* (Acarina: Psoroptidae). *Journal of Medical Entomology*, 36, 544–550.

Puerta, L., Fernández-Caldas, E., Caraballo, L.R. and Lockey, R.F. (1991) Sensitisation to *Blomia tropicalis* and *Lepidoglyphus destructor* in *Dermatophagoides* spp.-allergic individuals. *Journal of Allergy and Clinical Immunology*, 88, 943.

Puerta, L., Férnandez-Caldas, E., Lockey, R.F. and Caraballo, L.R. (1993) Sensitisation to *Chortoglyphus arcuatus* and *Aleuroglyphus ovatus* in *Dermatophagoides* spp. allergic individuals. *Clinical and Experimental Allergy*, 23, 117–123.

Puerta, L., Caraballo, L.R., Fernández-Caldas, E., Avijoglu, A., Marsh, D.G., Lockey, R.F. and Dao, M.L. (1996a) Nucleotide sequence analysis of a complementary DNA coding for a *Blomia tropicalis* allergen. *Journal of Allergy and Clinical Immunology*, 98, 932–937.

Puerta, L., Fernández-Caldas, E., Mercado, D., Lockey, R.F. and Caraballo, L.R. (1996b) Sequential determinations of *Blomia tropicalis* allergens in mattress and floor dust in a tropical city. *Journal of Allergy and Clinical Immunology*, 97, 689–691.

Puerta, L., Lagares, A., Mercado, D., Fernández-Caldas, E. and Caraballo, L. (2005) Allergenic composition of the mite *Suidasia medanensis* and cross-reactivity with *Blomia tropicalis*. *Allergy*, 60, 41–47.

Quek, S.C., Chong, A., Connet, G.J. and Lee, B.W. (1994) Effects of an acaricide on asthmatic children with house dust mite allergy. *Acta Paediatrica Japonica*, 36, 669–672.

Raab, W.P. (1971) *Natamycin (Pimaricin): Its Properties and Possibilities in Medicine*. Georg Thieme, Stuttgart.

Racewicz, M. (2001) House dust mites (Acari: Pyroglyphidae) in the cities of Gdansk and Gdynia (Northern Poland). *Annals of Agricultural and Environmental Medicine*, 8, 33–38.

Radon, K., Schottky, A., Garz, S., Koops, F., Szadkowski, D., Radon, K., Nowak, D. and Luczynska, C. (2000) Distribution of dust-mite allergens (Lep d 2, Der p 1, Der f 1, Der 2) in pig-farming environments and sensitisation of the respective farmers. *Allergy*, 55, 219–225.

Radovsky, F.R. (1970) Mites associated with coprolites and mummified human remains in Nevada. *Contributions of the University of California Archaeological Research Facility*, no. 10, 186–190.

Radwan, J. (1991a) Sperm competition in the mite *Caloglyphus berlesei*. *Behavioural Ecology and Sociobiology*, 29, 291–296.

Radwan, J. (1991b) Sperm competition. *Nature*, 352, 671–672.

Radwan, J. (1993) The adaptive significance of male polymorphism in the acarid mite *Caloglyphus berlesei. Behavioural Ecology and Sociobiology*, 33, 201–208.

Radwan, J. (1994) Male morph determination in two species of acarid mites. *Heredity*, 74, 669–673.

Radwan, J. (1996) Intraspecific variation in sperm competition success in the bulb mite: a role for sperm size. *Proceedings of the Royal Society Series B*, 263, 855–859.

Ramos, J.D., Cheong, N., Lee, B.W. and Chua, K.Y. (2001) cDNA cloning and expression of Blot 11, the *Blomia tropicalis* allergen homologous to paramyosin. *International Archives of Allergy and Applied Immunology*, 126, 286–293.

Ranganath, H.R. and ChannaBasavanna, G.P. (1988) House dust mites from Bangalore, India – A quantitative analysis. In: de Weck, A.L. and Todt, A. (eds), *Mite Allergy a Worldwide Problem*. UCB Institute of Allergy, Brussels, pp. 21–22.

Rangel, A., Salmen, S., Muñoz, J., García, F. and Hernández, M. (1998) *Dermatophagoides* sp. and IgE anti-*D.pteronyssinus* and *D. farinae* detection in a Venezuelan community at more than 2000 m above the sea level. *Clinical and Experimental Allergy*, 28, 1100–1103.

Ransom, J.H., Leonard, J. and Wasserstein, R.L. (1991) Acarex test correlates with monoclonal antibody test for dust mites. *Journal of Allergy and Clinical Immunology*, 87, 886–888.

Rao, N.S.K., Ranganath, H.R. and ChannaBasavanna, G.P. (1981) Housedust mites from India. *Indian Journal of Acarology*, 5, 85–94.

Rao, V.R.M., Dean, B.V., Seaton, A. and Williams, D.A. (1975) A comparison of mite populations in mattress dust from hospital and from private houses in Cardiff, Wales. *Clinical Allergy*, 5, 209–215.

Raunio, P., Pasanen, A.L., Reiman, M. and Virtanen, T. (1998) Cat, dog and house-dust-mite allergen levels of house dust in Finnish apartments. *Allergy*, 53, 195–199.

Rawle, F.C., Mitchell, E.B. and Platts-Mills, T.A.E. (1984) T cell responses to the major allergen from the house dust mite *Dermatophagoides pteronyssinus*, Antigen P_1 – comparison of patients with asthma, atopic dermatitis and perennial rhinitis. *Journal of Immunology*, 133, 195–201.

Rawlings, N.D. and Barrett, A.J. (1994) Families of serine peptidases. In: Barrett, A.J. (ed.), *Proteolytic Enzymes: Serine and Cysteine Peptidases. Methods in Enzymology*, 244, 19–61.

Rebmann, H., Weber, A.K., Focke, I., Rusche, A., Lau, S., Ehnert, B. and Wahn, U. (1996) Does benzyl benzoate prevent colonization of new mattresses by mites? A prospective study. *Allergy*, 51, 876–882.

Recer, G.M. (2004) A review of the effects of impermeable bedding encasements on dust mite-allergen exposure and bronchial hyperresponsiveness in dust-mite-sensitised patients. *Clinical and Experimental Allergy*, 34, 268–275.

Ree, H.-I., Lee, I.-Y., Kim, T.-E., Jeon, S.-H. and Hong, C.-S. (1997a) Mass culture of house dust mites, *Dermatophagoides farinae* and *D. pteronyssinus* (Acari: Pyroglyphidae). *Medical Entomology and Zoology*, 48, 109–116.

Ree, H.-I., Jeon, S.-H., Lee, I.-Y., Hong, C.-S. and Lee, D.-K. (1997b) Fauna and geographical distribution of house dust mites in Korea. *Korean Journal of Parasitology*, 35, 9–17.

Reed, C.E. and Kito, H. (2004) The role of protease activation of inflammation in allergic respiratory diseases. *Journal of Allergy and Clinical Immunology*, 114, 997–1008.

Rees, J.A., Carter, J., Sibley, P. and Merrett, T.G. (1992) Localisation of the major house dust mite allergen *Der p I* in the body of *Dermatophagoides pteronyssinus* by Immunostain®. *Clinical and Experimental Allergy*, 22, 640–641.

Rees, P.H., Gitoho, F., Mitchell, H.S. and Rees, C. (1974) Some aspects of the aetiology of asthma in Nairobi with special reference to patients and the house dust mite. *East African Medical Journal*, 51, 729–733.

Reese, G., Ayuso, R. and Lehrer, S.B. (1999) Tropomyosin: an invertebrate pan–allergen. *International Archives of Allergy and Immunology*, 119, 347–258.

Reiser, J., Ingram, D., Mitchell, E.B. and Warner, J.O. (1990) House dust mite allergen levels and an anti-mite mattress spray (natamycin) in the treatment of childhood asthma. *Clinical and Experimental Allergy*, 20, 561–567.

Reisman, R.E., Mauriello, P.M., Davis, G.B., Georgitis, J.M. and DeMasi, J.M. (1990) A double-blind study of the effectiveness of a high-efficiency particulate air (HEPA) filter in the treatment of patients with perennial allergic rhinitis and asthma. *Journal of Allergy and Clinical Immunology*, 85, 1050–1057.

Reitamo, S., Visa, K., Kahonen, K., Kayhko, K., Stubbs, S. and Salo, O.P. (1986) Eczematous reactions in atopic patients caused by epicutaneous testing with inhalant allergens. *British Journal of Dermatology*, 114, 303–309.

Reka, S.A., Suto, C. and Yamaguchi, M. (1992) Evidence of aggregation pheromone in the feces of house dust mite, *Dermatophagoides farinae*. *Japanese Journal of Sanitary Zoology*, 43, 339–341.

Renwick, J.A.A., Pitman, G.B. and Vité, J.P. (1976) 2-Phenylethanol isolated from bark beetles. *Naturwissenschaften*, 63, 198.

Reunala, T., Björkstén, F., Förström, L. and Kanerva, L. (1983) IgE-mediated occupational allergy to a spider mite. *Clinical Allergy*, 13, 383–388.

Revsbech, P. and Dueholm, M. (1990) Storage mite allergy among bakers. *Allergy*, 45, 204–208.

Rezk, H.A., El-Hamid, M.M.A. and El-Latif, M.A.A. (1996) House dust mites in Alexandria region, Egypt. *Alexandria Journal of Agricultural Research*, 41, 209–216.

Rhode, C.J. and Oemick, D.A. (1967) Anatomy of the digestive and reproductive systems in an acarid mite (Sarcoptiformes). *Acarologia*, 9, 608–616.

Ricci, G., Patrizi, A., Speccia, F., Menna, L., Bottau, P., D'Angelo, V. and Masi, M. (2000) Effect of house dust mite avoidance measures in children with atopic dermatitis. *British Journal of Dermatology*, 143, 379–384.

Richards, A.G. and Richards, P.A. (1977) The peritrophic membrane of insects. *Annual Review of Entomology*, 22, 219–240.

Ridout, S., Twiselton, R., Matthews, S., Stevens, M., Matthews, L., Arshad, S.H. and Hide, D.W. (1993) Acarosan and the Acarex test in the control of house dust mite allergens in the home. *British Journal of Clinical Practice*, 47, 141–144.

Riechel-Dolmatoff, G. (1975) *The Shaman and the Jaguar. A Study of Narcotic drugs Among the Indians of Colombia.* Temple University Press, Philadelphia.

Rigamonti, I.E., Caserio, F., Lozzia, G.C. and Ottoboni, F. (1993) Enzimi degli acari derrate. In: *5° Simposio Sulla Difresa Antiparassitaria Nelle Industrie Alimentari e Sulla Protezione Degli Alimenti.* Chiriotti Editori SpA, Pinerolo, pp. 423–428.

Rijkaert, G., van Bronswijk, J.E.M.H. and Liskens, H.F. (1981) House-dust community (fungi, mites) in different climatic regions. *Œcologia (Berlin)*, 48, 83–185.

Rijssenbeek-Nouwens, L.H.M., Oosting, A.J., de Bruin-Weller, M.S., Bregman, I., de Monchy, J.G.R. and Postma, D.S. (2002) Clinical evaluation of the effect of anti-allergic mattress covers in patients with moderate to severe asthma and house dust mite allergy: a randomised double blind placebo controlled study. *Thorax*, 57, 784–790.

Rivard, I. (1961a) Influence of temperature and humidity on mortality and rate of development of immature stages of the mite *Tyrophagus putrescentiae* (Schrank) (Acarina, Acaridae) reared on mould cultures. *Canadian Journal of Zoology*, 39, 419–426.

Rivard, I. (1961b) Influence of temperature and humidity on longevity, fecundity and rate of increase of the mite *Tyrophagus putrescentiae* (Schrank) (Acarina, Acaridae) reared on mould cultures. *Canadian Journal of Zoology*, 39, 869–876.

Roberts, D.L.L. (1984) House dust mite avoidance and atopic dermatitis. *British Journal of Dermatology*, 110, 735–736.

Robertson, P.L. (1944) A technique for biological studies of cheese mites. *Bulletin of Entomological Research*, 35, 251–255.

Robertson, P.L. (1952) Cheese mite infestation. An important storage problem. *Journal of the Society of Dairy Technology*, 5(2), 86–95.

Robertson, P.L. (1959) A revision of the genus *Tyrophagus*, with a discussion on its taxonomic position in the Acarina. *Australian Journal of Zoology*, 7, 146–181.

Robertson, P.L. (1961) A morphological study of variation in *Tyrophagus* (Acarina) with particular reference to populations infesting cheese. *Bulletin of Entomological Research*, 52, 501–529.

Robinson, C., Kalsheker, N.A., Srinivasan, N., King, C.M., Garrod, D.R., Thompson, P.J. and Stewart, G.A. (1997) On the potential significance of the enzymatic activity of mite allergens to immunogenicity. Clues to structure and function revealed by molecular characterisation. *Clinical and Experimental Allergy*, 27, 10–21.

Rodd, E.H. (ed.) (1951) *Chemistry of Carbon Compounds. Volume 1 Part A. General Introduction and Aliphatic Compounds.* Elsevier, Amsterdam.

Rodriguez, J.G. (1972) Inhibition of acarid mite development by fatty acids. In: Rodriguez, J.G. (ed.),

Insect and Mite Nutrition. North Holland Publishing Co., Amsterdam, pp. 637–650.

Rodriguez, J.G. (1979) Allelochemical effects of some flavouring components on the acarid, *Tyrophagus putrescentiae.* In: Rodriguez, J.G. (ed.), *Recent Advances in Acarology.* Volume. 1. Academic Press, New York, pp. 251–261.

Rodriguez, J.G. and Blake, D.F. (1979) Culturing *Dermatophagoides farinae* in a meridic diet. In: Rodriguez, J.G. (ed.), *Recent Advances in Acarology. Volume 2.* Academic Press, New York, pp. 211–216.

Rodriguez, J.G. and Lasheen, A.M. (1971) Axenic culture of *Tyrophagus putrescentiae* in a chemically defined diet and determination of essential amino acids. *Journal of Insect Physiology*, 17, 979–985.

Rodriguez, J.G. and Rodriguez, L.D. (1987) Nutritional ecology of stored-product and house dust mites. In: Slansky, F. and Rodriguez, J.G. (eds), *Nutritional Ecology of Insects, Mites, Spiders and Related Invertebrates.* John Wiley, New York, pp. 345–368.

Rodriguez, J.G., Potts, M.F. and Rodriguez, L.D. (1979) Effects of tobacco smoke condensates on protein synthesis in Acarid mites. In: Piffl, E. (ed.), *Proceedings of the 4th International Congress of Acarology.* Akadémiai Kiadó, Budapest, pp. 587–591.

Rodriguez, J.G., Potts, M.F. and Patterson, C.G. (1984) Mycotoxin-producing fungi: effects on stored products mites. In: Griffiths, D.E. and Bowman, C.E. (eds), *Acarology VI.* Vol. 1. Ellis Horwood, Chichester, pp. 343–350.

Rodríguez Monterroza, S., Rivera Herrera, I., Martinez, A.C., Navarro, J.M., Arroyo, L., Mendoza Meza, D.L. and Tuiran, P.B. (2006) Asma alérgica, niveles de IgE total y exposición a los acaros del polvo casero en el municipo de Santiago de Tolú. *Duazary (Revista de la Facultad de Ciencias de Salud, Universidad de Magdalena, Colombia)*, 3, 10–17.

Rohde, C.J. and Oemick, D.A. (1967) Anatomy of the digestive and reproductive systems in an acarid mite (Sarcoptiformes). *Acarologia*, 9, 608–616.

Romanski, B., Dziedziczko, A., Pawlik, K., Wilewska-Klubo, T., Wojtanowski, I. and Zbikowska, M. (1977) Alergia na kurz domowy u chorych na dychawice oskrzelowa. I. Czestosc wystepowania i problem charakteru alergenu kurzu domowego. *Polskie Archiwum Medycyny Wewnetrznej*, 57, 21– 26.

Rook, G.A.W. and Stanford, J.L. (1998) Give us this day our daily germs. *Immunology Today*, 19, 113–116.

Rosa, A.E. and Flechtmann, C.H.W. (1979) Mites in house dust from Brazil. *International Journal of Acarology*, 5, 195–198.

Rosário Filho, N.A., Baggio, D., Suzuki, M.M., Thomaz, P.C.P., Sugisawa, S. and Hanggi, V. (1992) Ácaros na poeira domiciliar em Curitiba. *Revista Brasileira de Alergia e Imunopatologia*, 15, A25. [abstract]

Rosenberg, N.J. (1974) *Microclimate: The Biological Environment.* John Wiley & Sons, New York.

Rousset, G. (1971) Contribution a l'étude systematique et ecologie des acariens impliques dans l'allergie respiratoire a la poussière de maison. Thesis, Faculty of Medicine, University of Montpellier.

Rudolph, D. and Knülle, W. (1974) Site and mechanism of water uptake from the atmosphere in ixodid ticks. *Nature*, 249, 84–85.

Rueda, L.M. (1985) House dust mites (Acari) in Malaysia. *Philippine Entomologist*, 6, 428–434.

Rufli, Th. (1970) Die bedeutung der milbenfauna als asthmaallergen. *Dermatologia*, 140, Supplement 2, 46–64.

Russell, D.W., Fernández-Caldas, E., Swanson, M.C., Seleznick, M.J., Trudeau, W.L., Lockey, R.F. (1991) Caffeine, a naturally occurring acaricide. *Journal of Allergy and Clinical Immunology*, 87, 107–110.

Saarne, T., Kaiser, L., Rsool, O., Huecas, S. and van Hage-Hamsten, M. (2003) Cloning and characterisation of two IgE-binding proteins, homologous to tropomyosin and alpha tubulin, from the mite *Lepidoglyphus destructor. International Archives of Allergy and Immunology*, 130, 258–265.

Sadaka, H.A.H., Allam, S.R., Rezek, H.A., El-Nzar, S.Y. and Shola, A.Y. (2000) Isolation of dust from houses of Egyptian allergic patients and induction of experimental sensitivity by *Dermatophagoides pteronyssinus. Journal of the Egyptian Society of Parasitology*, 30, 263–276.

Saha, G.K., Modak, A., Tandon, N. and Choudry, D.K. (1994) Prevalence of house dust mites (*Dermatophagoides* spp.) in homes of asthmatic patients in Calcutta. *Annals of Entomology*, 12, 13–17.

de Saint Georges-Gridelet, D. (1981a) Bioécologie et stratégie de controle de l'acarien des poussières domestiques *Dermatophagoides pteronyssinus.* Unpublished PhD thesis, University of Louvain, Belgium.

de Saint Georges-Gridelet, D. (1981b) Mise au point d'une stratégie de contrôle de l'acarien des poussières domestiques *Dermatophagoides pteronyssinus* par utilisation d'un fongicide. *Acta Oecologia/Oecologia Applicata*, 2, 117–126.

de Saint Georges-Gridelet, D. (1984) Effects of dietary lipids on the population growth of *Dermatophagoides pteronyssinus.* In: Griffiths, D.E. and Bowman, C.E. (eds), *Acarology VI. Vol. 1.* Ellis Horwood, Chichester, pp. 351–357.

de Saint Georges-Gridelet, D. (1987a) Vitamin requirements of the European House Dust Mite, *Dermatophagoides pteronyssinus* (Acari, Pyroglyphidae), in relation to its fungal association. *Journal of Medical Entomology*, 24, 408–411.

de Saint Georges-Gridelet, D. (1987b) Physical and nutritional requirements of house-dust mite *Dermatophagoides pteronyssinus* and its fungal association. *Acarologia*, 28, 345–353.

de Saint Georges-Gridelet, D. (1988) Optimal efficacy of a fungicide preparation, natamycin, in the control of the house-dust mite, *Dermatophagoides pteronyssinus. Experimental and Applied Acarology*, 4, 63–72.

de Saint Georges-Gridelet, D., Kneist, F.M., Schober, G., Penaud, A. and van Bronswijk, J.E.M.H. (1988) Lutte chimique contre les acariens de la poussière de maison. *Revue Française d'Allergologie et d'Immunologie Clinique*, 28, 131–138.

Sakagami, Y., Taki, K., Matsuhisha, T. and Marumo, S. (1992) Identification of 2-deoxyecdysone from the mite, *Tyrophagus putrescentiae. Experientia*, 48, 793–795.

Sakaguchi, M., Inouye, S., Yasueda, H. and Shida, T. (1992) Concentrations of airborne mite allergens (Der I and Der II) during sleep. *Allergy*, 47, 55–57.

Sakaguchi, M., Inouye, S., Sasaki, R., Hashimoto, M., Kobayashi, C. and Yasueda, H. (1996) Measurement of airborne mite allergen exposure in individual subjects. *Journal of Allergy and Clinical Immunology*, 97, 1040–1044.

Sakaki, I. and Suto, C. (1991) The termination of prolonged quiescence (diapause) in house dust mite, *Dermatophagoides farinae* at room temperature. *Japanese Journal of Sanitary Zoology*, 42, 93–97.

Sakaki, I. and Suto, C. (1994) [Multiple regression analysis of effect of housing conditions on the prevalence of domestic mites in wooden houses in Nagoya, Japan]. *Japanese Journal of Sanitary Zoology*, 45, 341–351. [In Japanese, English summary]

Sakaki, I., Ito, H. and Suto, C. (1991) [An improved method with saturated sodium chloride for isolating house dust mites]. *Japanese Journal of Sanitary Zoology*, 42, 43–46. [In Japanese, English summary]

Saleh, S.M., Abdel-Hamid, M.M. and Rezk, H.A. (1991) Biology of the European house dust mite, *Dermatophagoides pteronyssinus* (Trouessart). *Acarologia*, 32, 57–60.

Salykov, A.D. and Ermekova, R.K. (1987) [House dust mites in Alma-Ata and the Alma-Ata Region.] *Izvestiya Akademii Nauki Kazakhskoi SSR. Seriya Biologicheskaya*, no. 6, 38–42. [In Russian, English summary]

Samolinski, B., Zawisza, E., Gozdzik-Zolnierkiewicz, T. and Samolinska, U. (1990) Mite allergy in Warsaw, species and their influence on laboratory and clinical picture of atopic patients. *Clinical and Experimental Allergy*, 21, Supplement 2, 116. [abstract]

Samsinák, K. (1962) Beitrage zur kenntnis der gattung *Tyrophagus* Oudemans. *Acta Societatis Entomologicae Ceshosloveniae*, 59(3), 266–280.

Samsinák, K. and Vobrazkova, E. (1985) Mites from the City Pavement. *Zentralblatt für Bakteriologie, Parasitenkunde, Infektionskrankheiten und Hygiene*, B181, 132–138.

Samsinák, K., Dusbabek, F. and Vobrázková, E. (1972) Note on the house dust mites in Czechoslovakia. *Folia Parasitologica (Praha)*, 19, 383–384.

Samsinák, K., Vobrázková, E. and Spicak, V. (1978a) Investigations on the fauna of beds in flats, children's sanatoria and old-age homes. *Folia Parasitologica (Praha)*, 25, 157–163.

Samsinák, K., Palicka, P., Vobrazkova, E., Malis, L. and Zitek, K. (1978b) Findings of *Sarcoptes scabiei* (Linnaeus, 1758) in the beds of patients with scabies (Sarcoptiformes, Sarcoptidae). *International Journal of Acarology*, 4, 33–37.

Samsinák, K., Vobrazkova, E. and Dubinina, H.V. (1982) Contribution to the taxonomic status of *Dermatophagoides sheremetewskyi* Bogdanoff, 1864. *Folia Parasitologica (Praha)*, 29, 375–376.

Sánchez, F. and Alvarez, S. (2004) Modelling micro-climate in urban environments and assessing its influence on the performance of surrounding buildings. *Energy and Buildings*, 36, 403–413.

Sánchez, F. and Fernández-Caldas, E. (1994) Ácaros en Colombia y su relación con las alergias respiratorias. *Revista Brasileira Medecina*, 51, 199–208.

Sánchez-Borges, M., Capriles-Hulet, A., Fernandez-Caldas, E., Suarez-Chacon, R., Carabello, F., Castillo, S. and Sotillo, E. (1997) Mite contaminated foods as a cause of anaphylaxis. *Journal of Allergy and Clinical Immunology*, 99, 738–743.

Sánchez-Borges, M., Capriles-Hulet, A., Suarez-Chacon, R. and Fernandez-Caldas, E. (2001) Oral anaphylaxis from mite ingestion. *Allergy and Clinical Immunology International*, 13, 33–35.

Sánchez-Borges, M., Suarez-Chacon, R., Capriles-Hulet, A. and Carabello-Fonseca, F. (2005) An update on oral anaphylaxis from mite ingestion. *Annals of Allergy, Asthma and Immunology*, 94, 216–220.

Sánchez-Covisa, A., Rodriguez-Rodriguez, J.A., de la Torre, F., Garcia-Robaina, J.C. (1999) Faune d'acariens de la poussière domestique dans l'Ile Tenerife. *Acarologia*, 40, 55–58.

Sánchez-Medina, M. and Sánchez-Gutiérrez, G. (1973) Acaros en polvo de habitaciones a diferentes Alturas y climas de Colombia. *Alergia*, 20, 171–188.

Sánchez-Medina, M., Trudeau, W.L., Fernandez-Caldas, E. and Lockey, R.F. (1993). Cutaneous sensitivity to inhalant allergens in Colombian Indian tribes. *Journal of Allergy and Clinical Immunology*, 91(1), Part 2, 279. [abstract]

Sánchez-Medina, M. and Zarante, I. (1996) Dust mites at high altitude in a tropical climate. *Journal of Allergy and Clinical Immunology*, 97, 1167–1168.

Sánchez-Monge, R., García-Casado, G., Barber, D. and Salcedo, G. (1996) Interaction of allergens from house-dust mite and from cereal flours: *Dermatophagoides pteronyssinus* α-amylase (Der p 4) and wheat and rye α-amylase inhibitors. *Allergy*, 51, 176–180.

Sánchez-Ramos, I. and Castañera, P. (2001) Acaricidal activity of natural monoterpenes on *Tyrophagus putrescentiae* (Schrank), a mite of stored food. *Journal of Stored Products Research*, 37, 93–101.

Sánchez-Ramos, I. and Castañera, P. (2005) Effect of temperature on reproductive parameters and longevity of *Tyrophagus putrescentiae* (Acari: Acaridae). *Experimental and Applied Acarology*, 36, 93–105.

Sanda, T., Yasue, T., Oohashi, M. and Yasue, A. (1992) Effectiveness of housedust-mite allergen avoidance through clean room therapy in patients with atopic dermatitis. *Journal of Allergy and Clinical Immunology*, 89, 653–657.

Santos, M.A. (1978) Allergy-inducing mites in the bedroom dust of houses on Puerto-Rico. *Boletin de la Asociacion Medica de Puerto Rico*, 70, 419–422.

Sarinho, E., Fernández-Caldas, E. and Solé, D. (1996) Acaros da poira domicilar em residências de crianças asthmáticas e controles da cidade de Recife – Pernambuco. *Revista Brasileira Alergia Imunopatologia*, 19, 228–230.

Sarsfield, J.K. (1974) Role of house dust mites in childhood asthma. *Archives of Diseases of Childhood*, 49, 711–715.

Sarsfield, J.K., Gowland, G., Toy, R., Norman, A.L.E. (1974) Mite sensitive asthma of childhood. Trial of avoidance measures. *Archives of Diseases of Childhood*, 49, 716–721.

Sasa, M. (1950) Mites of the genus *Dermatophagoides* Bogdanov, 1864 found from three cases of human acariasis. *Japanese Journal of Experimental Medicine*, 20, 519–525.

Sasa, M. (1951) Further notes on mites of the genus *Dermatophagoides* Bogdanov from human acariasis (Acarina, Epidermoptidae). *Japanese Journal of Experimental Medicine*, 21, 199–203.

Sasa, M., Miyamoto, J., Shinohara, S., Suzuki, H. and Katsuhata, A. (1970) Studies on mass culture and isolation of *Dermatophagoides farinae* and some other mites associated with house dust and stored food. *Japanese Journal of Experimental Medicine*, 40, 367–382.

Sastre, J., Iraola, V., Figueredo, E., Tornero, P. and Fernandez-Caldas, E. (2002) Mites in Madrid. *Allergy*, 57, 58–59.

Sato, M., Kuwahara, Y., Matsuyama, S. and Suzuki, T. (1993) Chemical ecology of astigmatid mites XXXVII. Fatty acid as food attractant of astigmatid mites, its scope and limitation. *Applied Entomology and Zoology*, 28, 565–569.

Sato, M., Kuwahara, Y., Matsuyama, S., Suzuki, T., Okamoto, M. and Matsumoto, K. (1993) Male and female sex pheromones produced by *Acarus immobilis* Griffiths (Acaridae: Acarina). Chemical ecology of astigmatid mites XXXIV. *Naturwissenschaften*, 80, 34–36.

Scala, G. (1995) House-dust mite ingestion can induce allergic intestinal syndrome. *Allergy*, 50, 517–519.

van Schayck, O.C.P., Maas, T., Kaper, J., Knotterus, A.J.A. and Sheikh, A. (2007) Is there any role for allergen avoidance in the primary prevention of childhood asthma? *Journal of Allergy and Clinical Immunology*, 119, 1323–1328.

Scherbak, V.P. and Nazrullayeva, M.F. (1998) [Mites in domestic dust in the town of Tashkent.] *Biologicheskii Zhurnal Uzbekistana*, no. 3, 52–54. [In Russian]

Scherr, M.S. and Peck, L.W. (1977) Effects of high-efficiency air filtration system on nighttime asthma attacks. *West Virginia Medical Journal*, 73, 144–148.

Schmidt, L.M. and Gøtzsche, P.C. (2005) Of mites and men: reference bias in narrative review articles. A systematic review. *The Journal of Family Practice*, 54, 334–338.

Schmidt, M., Olsson, S.,Vanderploeg, I. and Hage-Hamsten, M. van (1995) cDNA analysis of the mite allergen Lep d 1 identifies two different isoallergens and variants. *FEBS Letters*, 370, 11–14.

Schmidt-Nielsen, K. (1975) *Animal Physiology*. Cambridge University Press, Cambridge.

Schober, G., Kneist, F.M., Kort, H.S.M., de Saint Georges-Gridelet, D.M.O.G. and van Bronswijk, J.E.M.H. (1992) Comparative efficacy of house dust mite extermination products. *Clinical and Experimental Allergy*, 22, 618–626.

Schram-Bijkerk, D., Doekes, G., Boeve, M., Douwes, J., Riedler, J., Üblagger, E., von Mutius, E., Budde, J., Pershagen, G., van Hage, M., Wickman, M., Braun-Fahrländer, C., Waser, M. and Brunekreef, B. (2006) Nonlinear relations between house dust mite allergen levels and mite sensitisation in farm and nonfarm children. *Allergy*, 61, 640–647.

Schuhl, J.F. (1977) Asma y alergia a acaros en Montevideo. *Allergologia et Immunopathologia*, 5, 117–122.

Schwartz, B., Lind, P. and Løwenstein, H. (1987) Level of indoor allergens in dust from homes of allergic and non-allergic individuals. *International Archives of Allergy and Applied Immunology*, 82, 447–449.

Scrivener, S.,Yemanerberham, H., Zebenigus, M., Tilahun, D., Girma, S., Ali, S., McElroy, P., Custovic, A., Woodcock, A., Pritchard, D., Venn, A. and Britton, J. (2001) Independent effects of intestinal parasite infection and domestic allergen exposure on risk of wheeze in Ethiopia: a nested case-control study. *Lancet*, 358, 1493–1499.

Seethaler, H., Knülle, W. and Devine, T.L. (1979) Water vapour intake and body water (3HOH) clearance in the housemite *Glycyphagus domesticus*. *Acarologia*, 21, 442–450.

Sehra, S., Tuana, F.M.B., Holbreich, M., Mousdicas, N., Tepper, R.S., Chang, C.H., Travers, J.B. and Kaplan, M.H. (2008) Scratching the surface: towards understanding the pathogenesis of atopic dermatitis. *Critical Reviews in Immunology*, 28, 15–43.

Selnes, A., Bolle, R., Holt, J. and Lund, E. (1999) Atopic diseases in Nami and Norse schoolchildren living in northern Norway. *Pediatric Allergy and Immunology*, 10, 216–220.

Sepasgosarian, H. and Mumcuoglu, K.Y. (1979) Faunistische und Ökologische studien der hausstaubmilben in Iran. *International Journal of Acarology*, 4, 131–137.

Sercombe, J.K., Liu–Brennan, D., Garcia, M.L. and Tovey, E.R. (2005) Evaluation of home allergen sampling devices. *Allergy*, 60, 515–520.

Serravalle, K., Medeiros Jr, M. (1999) Ácaros da poeira domiciliar na cidade de Salvador, BA. *Revista Brasileira de Alergia e Imunopatologia*, 22, 19–24.

Servín, V.R. and Tejas, A. (1991) Presencia de *Dermatophagoides* en Baja California Sur, Mexico. *Southwestern Entomologist*, 16, 156–161.

Sesay, H.R. and Dobson, R.M. (1972) Studies on the mite fauna of house dust in Scotland with special reference to that of bedding. *Acarologia*, 14, 384–392.

Sette, L., Comis, A., Marcucci, F., Sesi, L., Piacentini, G.L. and Boner, A.L. (1994) Benzyl benzoate foam: effects on mite allergens in mattress, serum and nasal secretory IgE to *Dermatophagoides pteronyssinus*, and bronchial hyperreactivity in children with allergic asthma. *Pediatric Pulmonology*, 18, 218–227.

Sevki, C., Levent, A., Ender, G. and Firdevs, M. (2006) Reduction of house-dust mite allergen concentrations in carpets by aluminium potassium sulphate dodecahydrate (alum). *Allergy and Asthma Proceedings*, 27, 350–353.

Sexton, D.J. and Haynes, B. (1975) Bird-mite infestation in a university hospital. *Lancet*, February 22, 445.

Shakib, F., Schulz, O. and Sewell, H. (1998) A mite subversive: cleavage of CD23 and CD25 by Der p 1 enhances allergenicity. *Immunology Today*, 19, 313–316.

Shamiyeh, N.B., Bennett, S.E., Hornsby, R.P. and Woodiel, N.L. (1971) Isolation of mites from house dust. *Journal of Economic Entomology*, 64, 53–55.

Shamiyeh, N.B., Hornsby, R.P., Bennett, S.E. and Woodiel, N.L. (1973) Distribution of house dust mites in the environs of a Tennessee river valley. *Journal of Economic Entomology*, 66, 998–999.

Shapiro, G.G., Wighton, T.G., Chinn, T., Zuckerman, J., Eliassen, H., Picciano, J.F. and Platts-Mills, T.A.E. (1995) House dust mite avoidance for children with asthma in homes of low-income families. *Journal of Allergy and Clinical Immunology*, 103, 1069–1074.

Sharma, S., Lackie, P.M. and Holgate, S.T. (2003) Uneasy breather: the implications of dust mite allergens. *Clinical and Experimental Allergy*, 33, 163–165.

Sharp, J.L. and Haramoto, F.H. (1970) *Dermatophagoides pteronyssinus* (Trouessart) and other Acarina in house dust in Hawaii. *Proceedings of the Hawaiian Entomological Society*, 20, 583–589.

Shatrov, A.B. (1997) The ultrastructure of haemocytes in trombiculid mites. (Acariformes: Trombiculidae). *Experimental and Applied Acarology*, 21, 49–64.

Shcherbak, V.P. and Nazrullaeva, M.F. (1988) [Mites in domestic dust in the town of Tashkent.] *Biologicheskii Zhurnal Uzbekistana*, no. 3, 52–54. [In Russian, English summary]

Sheals, J.G. (1956) Notes on a collection of soil Acari. *Entomologist's Monthly Magazine*, 92, 99.

Sheikh, A., Hurwitz, B. and Shehata, Y. (2007) House dust mite avoidance measures for perennial allergic rhinitis. *Cochrane Database of Systematic Reviews*, 2007, Issue 1, Article No. CD001563.

Shen, H.-D., Chua, K.-Y., Lin, K.-L., Hsieh, K.-H. and Thomas, W.R. (1993) Molecular cloning of a house dust mite allergen with common antibody binding specificities with multiple components in mite extracts. *Clinical and Experimental Allergy*, 23, 934–940.

Shen, H.D., Chua, K.Y., Lin, W.L., Hsieh, K.H. and Thomas, W.R. (1995a) Characterization of the house dust mite allergen Der p 7 by monoclonal antibodies. *Clinical and Experimental Allergy*, 25, 416–422.

Shen, H.D., Chua, K.Y., Lin, W.L., Hsieh, K.H. and Thomas, W.R. (1995b) Molecular cloning and immunological characterisation of the house dust mite allergen Der f 7. *Clinical and Experimental Allergy*, 25, 1000–1006.

Shen, H.D., Chua, K.Y., Lin, W.L., Chen, H.L., Hsieh, K.H. and Thomas, W.R. (1996) IgE and monoclonal antibody binding by the mite allergen Der p 7. *Clinical and Experimental Allergy*, 26, 308–315.

Shibashaki, M. and Takita, H. (1994) Effect of electric-heating carpet on house-dust mites. *Annals of Allergy*, 72, 541–545.

Shibasaki, M., Ikeda, H., Isoyama, S., Imoto, N., Takeda, K., Noguchi, E. and Takita, H. (1996) Treatment of whole houses with liquid nitrogen for control of dust mites. *Journal of Medical Entomology*, 33, 906–910.

Shimomura, H., Ishii, A., Takaoka, M. and Kano, R. (1982) [Collection of each stage of the house dust mite from a culture by sieving and observation on the mode of breeding.] *Japanese Journal of Sanitary Zoology*, 33, 1–7. [In Japanese, English summary]

Shulman, S.T. and Rowley, A.H. (1986) Does Kawasaki disease have a retroviral aetiology? *The Lancet*, September 6th, 545–546.

Sibley, C.J. and Ahlquist, J.O. (1990) *Phylogeny and Classification of Birds: A Study in Molecular Evolution*. Yale University Press, New Haven, Connecticut.

Sidenius, K.E., Hallas, T.E., Poulsen, L.K. and Mosbech, H. (2002a) House dust mites and their allergens in Danish mattresses – results from a population based study. *Annals of Agricultural and Environmental Medicine*, 9, 33–39.

Sidenius, K.E., Hallas, T.E., Poulsen, L.K. and Mosbech, H. (2002b) A controlled intervention study concerning the effect of intended temperature rise on house dust mite load. *Annals of Agricultural and Environmental Medicine*, 9, 163–168.

Siebers, R., Weinstein, P., Fitzharris, P. and Crane, J. (1999) House-dust mite and cat allergens in the Antarctic. *Lancet*, 353, 1942.

Silacci, P., Mazzolai, L., Gauci, C., Stergiopulos, N., Yin, H.L. and Hayoz, D. (2004) Gelsolin superfamily proteins: key regulators of cellular functions. *Cellular and Molecular Life Sciences*, 61, 2614–2623.

Silton, R.P., Fernández-Caldas, E., Trudeau, W.L., Swanson, M. and Lockey, R. (1991), Prevalence of specific IgE to the storage mite, *Aleuroglyphus ovatus. Journal of Allergy and Clinical Immunology*, 88, 595–603.

da Silva, R.R., Binotti, R.S., da Silva, C.M., de Oliveira, C.H., Condino-Neto, A. and de Capitani, E.M. (2005) Mites in dust samples from mattress surfaces from single beds or cribs in the south Brazilian city of Londrina. *Pediatric Allergy and Immunology*, 16, 132–136.

Simons, E., Curtin-Brosnan, J., Buckley, T., Breysse, P. and Eggleston, P.A. (2007) Indoor environmental differences between inner city and suburban homes of children with asthma. *Journal of Urban Health–Bulletin of the New York Academy of Medicine*, 84, 577–590.

Simpson, A., Woodcock, A. and Custovic, A. (2001) Housing characteristics and mite allergen levels: to humidity and beyond. *Clinical and Experimental Allergy*, 31, 803–805.

Simpson, A., Hassall, R., Custovic, A. and Woolcock, A. (1998) Variability of house-dust-mite allergen levels within carpets. *Allergy*, 53, 602–607.

Simpson, A., Simpson, B., Custovic, A., Gain, G., Craven, M. and Woodcock, A. (2002) Household characteristics and mite allergen levels in Manchester, UK. *Clinical and Experimental Allergy*, 32, 1413–1419.

Simpson, E.H. (1949) Measurement of diversity. *Nature*, 163, 688.

Simpson, R.J., Nice, E.C. and Stewart, G.A. (1989) Structural studies on the allergen *Der p* 1 from the house dust mite *Dermatophagoides pteronyssinus*: similarity with cysteine proteinases. *Protein Sequence Data Analysis*, 2, 17–21.

Singh, P. (1977) *Artificial Diets for Insects, Mites and Spiders*. Plenum Press, New York.

Sinha, R.N. (1964) Ecological relationships of stored-products mites and seed-borne fungi. *Acarologia*, 6, 372–389.

Skerrow, D. (1986) Epidermal α-keratin: structure and chemical composition. In: Bereiter–Hahn, J., Matoltsky, A.G. and Richards, K.S. (eds), *Biology of the Integument. Vol. 2 Vertebrates*. Springer Verlag, Berlin, pp. 621–643.

Slansky, F. and Rodriguez, J.G. (1987) Nutritional ecology of insects, mites, spiders and related invertebrates: an overview. In: Slansky, F. and Rodriguez, J.G. (eds), *Nutritional Ecology of Insects, Mites, Spiders and Related Invertebrates*. John Wiley, New York, pp. 1–23.

Slansky, F. and Scriber, M. (1985) Food consumption and utilization. In: Kerkut, G.A. and Gilbert, L.I. (eds), *Comprehensive Insect Physiology, Biochemistry and Pharmacology, Volume 4, Regulation: Digestion, Nutrition, Excretion*. Pergamon Press, Oxford, pp. 88–163.

Sly, R.M., Josephs, S.H. and Eby, D.M. (1985) Dissemination of dust by central and portable vacuum cleaners. *Annals of Allergy*, 54, 209–212.

Smith, A.M., Benjamin, D.C., Hozic, N., Derewenda, U., Smith, W.A., Thomas, W.R., Gafveling, G., van Hage-Hamsten, M. and Chapman, M.D. (2001) The molecular basis of antigenic cross-reactivity between the group 2 mite allergens. *Journal of Allergy and Clinical Immunology*, 107, 977–984.

Smith, K.G.V (1986) *A Manual of Forensic Entomology*. British Museum (Natural History), London.

Smith, R.F. and Allen, W.W. (1954) Insect control and the balance of nature. *Scientific American*, 190(6), 38–42.

Smith, R.F., Apple, J.L. and Bottrell, D.G. (1976) The origins of integrated pest management concepts for agricultural crops. In: Apple, J.L. and Smith, R.F. (eds), *Integrated Pest Management*. Plenum Press, New York, pp. 1–16.

Smith, T.F., Kelly, L.B., Heyman, P.W., Wilkins, S.R. and Platts-Mills, T.A.E. (1985) Natural exposure and serum antibodies to house dust mite of mite – allergic children with asthma in Atlanta. *Journal Allergy and Clinical Immunology*, 76, 782–788.

Smith, W., Mills, K., Hazell, L., Hart, B.J. and Thomas, W.R. (1999) Molecular analysis of group 1 and 2 allergens from the house dust mite, *Euroglyphus maynei. International Archives of Allergy and Immunology*, 118, 15–22.

Smith, W.A. and Thomas, W.R. (1996) Sequence polymorphisms of the Der p 3 house dust mite allergen. *Clinical and Experimental Allergy*, 26, 571–579.

Smith, W.A., Chua, K.Y., Kuo, M.C., Rogers, B.L. and Thomas, W.R. (1994) Cloning and sequencing of the *Dermatophagoides pteronyssinus* group III allergen, Der p III. *Clinical and Experimental Allergy*, 24, 220–228.

Smith, W.A., Hales, B.J., Jarnicki, A.G. and Thomas, W.R. (2001) Allergens of wild house dust mites: environmental Der p 1 and Der p 2 sequence polymorphisms. *Journal of Allergy and Clinical Immunology*, 107, 985–992.

Smrž, J. and Catská, V. (1987) Food selection of the field population of *Tyrophagus putrescentiae* (Schrank) (Acari, Acarida). *Zeitschrift für angewandte Entomologie*, 104, 329–335.

Šbotník, J., Alberti, G., Weyda, F. and Hubert, J. (2008) Ultrastructure of the digestive tract of *Acarus siro* (Acari: Acaridida). *Journal of Morphology*, 269, 54–71.

Solarz, K. (1986) Alergogenna akarofauna pylu domowego wybranych miast Górnego Slaska. *Wiadomósci Parazytologiczne*, 32, 431–433.

Solarz, K. (1997) Seasonal dynamics of house dust mite populations in bed/mattress dust from two dwellings in Sosnowiec (Upper Silesia, Poland): an attempt to assess exposure. *Annals of Agricultural and Environmental Medicine*, 4, 253–261.

Solarz, K. (1998) The allergenic acarofauna of house dust from dwellings, hospitals, libraries and institutes in Upper Silesia (Poland). *Annals of Agricultural and Environmental Medicine*, 5, 73–85.

Solarz, K. (2000) Annual fluctuations in the number of the developmental stages of *Dermatophagoides* spp. (Astigmata: Pyroglyphidae) in the Upper Silesia Region, Poland. *International Journal of Acarology*, 26, 371–377.

Solomon, M.E. (1966) Moisture gains, losses and equilibria of flour mites, *Acarus siro* L. in comparison with larger arthropods. *Entomologia Experimentata et Applicatis*, 9, 25–41.

Solomon, M.E. (1945) Tyroglyphid mites in stored products. Methods for the study of population density. *Annals of Applied Biology*, 32, 71–75.

Solomon, M.E. (1946a) Tyroglyphid mites in stored products. Ecological studies. *Annals of Applied Biology*, 33, 82–97.

Solomon, M.E. (1946b) Tyroglyphid mites in stored products. Nature and amount of damage to wheat. *Annals of Applied Biology*, 33, 280–289.

Solomon, M.E. (1951) Control of humidity with potassium hydroxide, sulphuric acid, or other solutions. *Bulletin of Entomological Research*, 42, 543–554.

Solomon, M.E. (1961) Mites in houses, shops and other occupied buildings. *The Sanitarian*, March, 1961.

Solomon, M.E. (1962) Ecology of the flour mite *Acarus siro* (= *Tyroglyphus farinae* De G.). *Annals of Applied Biology*, 50, 178–184.

Solomon, M.E. (1969) *Population Dynamics*. Edward Arnold, London.

Solomon, M.E. and Cunnington, A.M. (1964) Rearing acaroid mites. *Acarologia*, 6, 399–403.

Somorin, A.O., Hunponu-Wusu, O.O., Mumcuoglu, K.Y. and Heiner, D.C. (1978) Mite allergy in Nigerians, studies on house dust mites in houses of allergic patients in Lagos. *Irish Journal of Medical Science*, 147, 26–30.

Sompolinsky, D., Samra, Z., Zavaro, A. and Barishak, Y. (1984) Allergen-specific immunoglobulin E in tears and serum of vernal conjunctivitis patients. *International Archives of Allergy and Applied Immunology*, 75, 317–321.

Sonenshine, D.E., Taylor, D. and Carson, K.A. (1986) Chemically mediated behaviour in Acari: adaptations for finding hosts and mates. *Journal of Chemical Ecology*, 12, 1091–1108.

Sopolete, M.C., Silva, D.A.O., Arruda, L.K., Chapman, M.D. and Taketomi, E.A. (2000) *Dermatophagoides farinae* (Der f 1) and *Dermatophagoides pteronyssinus* (Der p 1) allergen exposure among subjects living in Uberlandia, Brazil. *International Archives of Allergy and Immunology*, 122, 257–263.

Soto-Quiros, M.E., Ståhl, A., Calderon, O., Sánchez, C., Hanson, L.Å. and Belin, L. (1998) Guanine, mite, and cockroach allergens in Costa Rican homes. *Allergy*, 53, 499–505.

Spencer, T.S., Linamen, C.E., Akers, W.A. and Jones, H.E. (1975) Temperature dependence of water content of stratum corneum. *British Journal of Dermatology*, 93, 159–168.

Spieksma, F.Th.M. (1967) The house dust mite *Dermatophagoides pteronyssinus* (Trouessrt, 1897) producer of the house-dust allergen (Acari, Psoroptidae). Doctoral thesis, State University of Leiden.

Spieksma, F.Th.M. (1968) Die huistofmijt *Dermatophagoides pteronyssinus* (Trouessart, 1897) de producent van het huisstofallergen. *Entomologische Berichten*, 28, 27–38.

Spieksma, F.Th.M. (1969) Cultivation of house dust mites. *Journal of Allergy*, 43, 152–154.

Spieksma, F.Th.M. (1973) *Malayoglyphus carmelitus* n.sp. a new mite from dust from a house on Mount Carmel (Pyroglyphidae, Sarcoptiformes). *Acarologia*, 15, 171–180.

Spieksma, F.Th.M. (1973a) Ecological distribution of house-dust mites in Europe. In: Daniel, M. and Rosický B. (eds), *Proceedings of the 3rd International Congress of Acarology, Prague, 1971.* Academia, Prague, pp. 551–556.

Spieksma, F.Th.M. (1976) Cultures of house dust mites on animal skin scales. *Allergologia et Immunopathologia*, 4, 419–428.

Spieksma, F.Th.M. (1991) Domestic mites, their role in respiratory allergy. *Clinical and Experimental Allergy*, 21, 655–660.

Spieksma, F.Th.M. (1992) Introduction. *Experimental and Applied Acarology*, 16(1–2), ix–xiii.

Spieksma, F.Th.M. and Dieges, P.H. (2004) The history of the finding of the house dust mite. *Journal of Allergy and Clinical Immunology*, 113, 573–576.

Spieksma, F.Th.M. and Spieksma-Boezeman, M.I.A. (1967) The mite fauna of house dust with particular reference to the house dust mite *Dermatophagoides pteronyssinus* (Trouessart, 1897) (Psoroptidae, Sarcoptiformes). *Acarologia*, 9, 226–241.

Spieksma, F.Th.M. and Voorhorst, R. (1969) Comparison of skin reactions to extracts of house dust, mites and human skin scales. *Acta Allergologica*, 24, 124–146.

Spieksma, F.Th.M., Zuidema, P. and Leupen, M.J. (1971) High altitude and house dust mites. *British Medical Journal*, i, 82–84.

Sporik, R. and Platts-Mills, T.A.E. (1992) Epidemiology of dust-mite-related disease. In *House Dust Mites* [Special Issue] (ed. B. Hart). *Experimental and Applied Acarology*, 16, 141–151.

Sporik, R., Holgate, S.T., Platts-Mills, T.A.E. and Cogswell, J.J. (1990) Exposure to house-dust mite allergen (*Der p* I) and the development of asthma in childhood. A prospective study. *New England Journal of Medicine*, 323, 502–507.

Sporik, R., Platts-Mills, T.A.E. and Cogswell, J.J. (1993) Exposure to house-dust mite allergen of children admitted to hospital with asthma. *Clinical and Experimental Allergy*, 23, 740–746.

Sporik, R., Ingram, J.M., Price, W., Sussman, J.H., Honsinger, R.W. and Platts-Mills, T.A.E. (1995) Association of asthma with serum IgE and skin-test reactivity to allergens among children living at high altitude: tickling the dragon's breath. *American Journal of Respiratory and Critical Care Medicine*, 151, 1388–1392.

Sporik, R., Hill, D.J., Thompson, P.J., Stewart, G.A., Carlin, J.B., Nolan, T.M., Kemp, A.S. and Hosking, C.S. (1998) The Melbourne house dust mite study: long-term efficacy of house dust mite reduction strategies? *Journal of Allergy and Clinical Immunology*, 101, 451–456.

Stanaland, B.E., Fernández-Caldas, E., Jacinto, C.M., Trudeau, W.L. and Lockey, R.F. (1996) Positive nasal challenge responses to *Blomia tropicalis*. *Journal of Allergy and Clinical Immunology*, 97, 1045–1049.

Stearns, S.C. (1992) *The Evolution of Life Histories.* Oxford University Press, New York.

Stefaniak, O. and Seniczak, S. (1976) The microflora of the alimentary canal of *Achiptera coleoptrata* (Acarina, Oribatei). *Pedobiologia*, 16, 185–194.

Stepien, Z.A. and Rodriguez, J.G. (1972) Food utilisation by acarid mites. In: Rodriguez, J.G. (ed.), *Insect and Mite Nutrition.* North Holland, Amsterdam, pp. 127–151.

Stenius, B. and Cunnington, A.M. (1972) House dust mites and respiratory allergy: a qualitative survey of species occurring in Finnish house dust. *Scandinavian Journal of Respiratory Diseases*, 53, 338–348.

Stewart, A.W., Mitchell, E.A., Pearce, N., Strachan, D.W. and Weiland, S.K. on behalf of the ISAAC Steering Committee (2001) The relationship of per capita gross national product to the prevalence of symptoms of asthma and other atopic diseases in children (ISAAC). *International Journal of Epidemiology*, 30, 173–179.

Stewart, G.A. (1982) Isolation and characterization of the allergen Dpt 12 from *Dermatophagoides pteronyssinus* by chromatofocusing. *International Archives of Allergy and Applied Immunology*, 69, 224–230.

Stewart, G.A. (1994) Molecular biology of allergens. In: Busse, W.W. and Holgate, S.T. (eds), *Asthma and Rhinitis.* Oxford, Blackwell Scientific Publications, pp. 898–932.

Stewart, G.A. and Fisher, W.F. (1986) Cross-reactivity between the house dust mite *Dermatophagoides pteronyssinus* and the mange mites *Psoroptes cuniculi* and *Psoroptes ovis*. I. Demonstration of antibodies to the house dust mite allergen Dpt 12 in sera from *P. cuniculi*-infested rabbits. *Journal of Allergy and Clinical Immunology*, 78, 293–299.

Stewart, G.A. and McWilliam, A.S. (2001) Endogenous function and biological significance of aeroallergens: an update. *Current Opinion in Allergy and Clinical Immunology*, 1, 95–103.

Stewart, G.A. and Robinson, C. (2003) The immunobiology of allergic peptidases. *Clinical and Experimental Allergy*, 33, 3–6.

Stewart, G.A. and Thompson, P.J. (1996) The biochemistry of common aeroallergens. *Clinical and Experimental Allergy*, 26, 1020–1044.

Stewart, G.A. and Turner, K.J. (1980) Physicochemical and immunochemical characterisation of the allergens from the mite *Dermatophagoides pteronyssinus*. *Australian Journal of Experimental Biology and Medical Sciences*, 58, 259–274.

Stewart, G.A. and Turner, K.J. (1980b) Physicochemical and immunochemical characterisation of the high molecular weight allergens from *Dermatophagoides pteronyssinus* with particular reference to the tridacnin, Con-A and S107 reactive components. *Australian Journal of Experimental Biology and Medical Sciences*, 58, 275–288.

Stewart, G.A., Butcher, A. and Turner, K.J. (1983) Characterization of a very high density lipoprotein allergen, Dpt 4, from the house dust mite *Dermatophagoides pteronyssinus*. *International Archives of Allergy and Applied Immunology*, 71, 340–345.

Stewart, G.A., Butcher, A., Lees, K. and Ackland, J. (1986) Immunochemical and enzymatic analyses of extracts of the house dust mite *Dermatophagoides pteronyssinus*. *Journal of Allergy and Clinical Immunology*, 77, 14–24.

Stewart, G.A., Simpson, R.J., Moritz, R.L., Thomas, W.R. and Turner, K.J. (1988) Physiochemical characterization of allergens *Der p* I (Dpt 12), Dpt 22 and Dpt 36 from the house dust mite. In: de Weck, A.L. and Todt, A. (eds), *Mite Allergy, a World Wide Problem*. UCB Institute for Allergy, Brussels, pp. 35–38.

Stewart, G.A., Thompson, P.J. and Simpson, R.J. (1989) Protease antigens from house dust mite. *Lancet*, ii, 154–155, and correction, *Lancet*, ii, 462.

Stewart, G.A., Bird, C.H. and Thompson, P.J. (1992) Do the group II dust mite allergens correspond to lysozyme? *Journal of Allergy and Clinical Immunology*, 90, 141–142.

Stewart, G.A., Lake, F.R. and Thompson, P.J. (1991) Faecally derived hydrolytic enzymes from *Dermatophagoides pteronyssinus*: physicochemical characterisation of potential allergens. *International Archives of Allergy and Applied Immunology*, 95, 248–256.

Stewart, G.A., Bird, C.H., Krska, K.D., Colloff, M.J. and Thompson, P.J. (1992) A comparative study of allergenic and potentially allergenic enzymes from *Dermatophagoides pteronyssinus*, *D. farinae* and *Euroglyphus maynei*. *Experimental and Applied Acarology*, 16, 165–180.

Stewart, G.A., Ward, L.D., Simpson, R.J. and Thompson, P.J. (1992) The group III allergen from the house dust mite *Dermatophagoides pteronyssinus* is a trypsin-like enzyme. *Immunology*, 75, 29–35.

Stewart, G.A., Kollinger, M.R., King, C.M. and Thompson, P.J. (1994) A comparative study of three serine proteases from *Dermatophagoides pteronyssinus* and *D. farinae*. *Allergy*, 49, 553–560.

Stewart, G.A., van Hage-Hamsten, M., Krska, K., Thompson, P.J. and Olsson, S. (1998) An enzymatic analysis of the storage mite *Lepidoglyphus destructor*. *Comparative Biochemistry and Physiology*, 119B, 341–347.

Storm van Leeuwen, W. (1924) Bronchial asthma in relation to climate. *Proceedings of the Royal Society of Medicine*, 17, 19–26.

Storm van Leeuwen, W. (1927) Asthma and Tuberculosis in relation to 'climate allergens'. *British Medical Journal*, August 27, 344–347.

Storm van Leeuwen, W., Bein, Z. and Varekamp, H. (1924) Experimentelle allergische krankheiten (Asthma bronchiale, Rhinitis vasomotoria). *Zeitschrift für Immunitätsforschung und Experimentelle Therapie*, 40, 552.

Storm van Leeuwen, W., Einthoven, J.W. and Kremer, W. (1927) The allergen-proof chamber in the treatment of bronchial asthma and other respiratory diseases. *Lancet*, June 1927, 1287–1289.

Strachan, D.P. (1998) House dust mite allergen avoidance in asthma. Benefits unproved but not yet excluded. *British Medical Journal*, 317, 1096–1097.

Strachan, D.P. (1989) Hay fever, hygiene and household size. *British Medical Journal*, 299, 1259–1260.

Strachan, D.P. (2000) Family size, infection and atopy: the first decade of the 'hygiene hypothesis'. *Thorax*, 55 (Supplement 1), S2–S10.

Stratil, H.U., Stratil, H.H. and Knülle, W. (1980) Untersuchungen über die spezifische Vermehrunsrate von populationen der im lagergetride lebenden milbe *Glycyphagus destructor* (Schranck, 1781) bei verschiedenen temperature- und luftfeuchtebedingungen. *Zeitschrift für Angewandte Entomologie*, 90, 209–220.

van Strien, R.T., Verhoeff, A.P., Brunekreef, B. and Wijnen, J.H. van (1994) Mite antigen in house dust: relationship with different housing characteristics in the Netherlands. *Clinical and Experimental Allergy*, 24, 843–853.

van Strien, R.T., Verhoeff, A.P., Wijnen, J.H. van, Doeks, G., Meer, G.E.A. de and Brunekreef, B. (1995) *Der p* I concentrations in mattress surface and floor dust collected from infants' bedrooms. *Clinical and Experimental Allergy*, 25, 1184–1189.

van Strien, R.T., Koopman, L.P., Kerkof, M., Spithoven, J., de Jongste, J.C., Gerritsen, J., Neijens, H.J., Aalberse, R.C., Smit, H.A. and Brunekreef, B. (2002) Mite and pet allergen levels in homes of children born to allergic and nonallergic parents: the PIAMA study. *Environmental Health Perspectives*, 110, A693–698.

van Strien, R.T., Gehring, U., Belanger, K., Triche, E., Gent, J., Bracken, M.B. and Leaderer, B.P. (2004) The influence of air conditioning, humidity, temperature and other household characteristics on mite allergen concentrations in the northeastern United States. *Allergy*, 59, 645–652.

Summers, C.B. and Felton, G.W. (1996) Peritrophic envelope as a functional antioxidant. *Archives of Insect Biochemistry and Physiology*, 32, 131–142.

Sun, G., Stacey, M.A., Schmidt, M., Mori, L. and Mattoli, S. (2001) Interaction of mite allergens Der p 3 and Der p 9 with protease-activated receptor-2 expressed by lung epithelial cells. *Journal of Immunology*, 167, 1014–1021.

Sun, H.-L. and Lue, K.-H. (2000) Household distribution of house dust mite in Central Taiwan. *Journal of Microbiology, Immunology and Infection*, 33, 233–236.

Sun, S., Lindström, I., Boman, H.G., Faye, I. and Schmidt, O. (1990) Hemolin: an insect-immune protein belonging to the immunoglobulin superfamily. *Science*, 250, 1729–1732.

Sundell, J., Wickmann, M., Pershagen, G. and Nordvall, S.L. (1995) Ventilation in homes infested by house-dust mites. *Allergy*, 50, 106–112.

Sunyer, J., Menendez, C., Ventura, P.J., Aponte, J.J., Schellenberg, D., Kahigwa, E., Acosta, C., Antó, J.M. and Alonso, P.L. (2001) Prenatal risk factors of wheezing at the age of four years in Tanzania. *Thorax*, 56, 290–295.

Sunyer, J., Jarvis, D., Pekkanen, J., Chinn, S., Janson, C., Leynart, B., Luczynska, C., Garcia-Esteban, R., Burney, P., Antó, J.M. for the European Community Respiratory Health Survey Study Group (2004) Geographic variation in the effect of atopy on asthma in the European Community Respiratory Health Study. *Journal of Allergy and Clinical Immunology*, 114, 1033–1039.

Sutherst, R.W. and Maywald, G.F. (1985) A computerised system for matching climates in ecology. *Agriculture, Ecosystems and Environment*, 13, 281–299.

Suto, C. and Sakaki, I. (1990) Studies on the factors influencing the induction, persistence and termination of prolonged quiescence (diapause) in house dust mite, *Dermatophagoides farinae*. *Japanese Journal of Sanitary Zoology*, 41, 375–381.

Suto, C., Sakaki, I., Ito, H. and Mitibata, M. (1991a) [Comparative studies on the population dynamics of house dust mites, *Dermatophagoides farinae* and *D. pteronyssinus* (Acari: Pyroglyphidae) in homes in Nagoya.] *Japanese Journal of Sanitary Zoology*, 42, 217–228. [In Japanese, English summary]

Suto, C., Sakaki, I., Ito, H. and Mitibata, M. (1991b) [Distribution and abundance of house dust mites, *Dermatophagoides* spp. (Acari: Pyroglyphidae), influenced by environmental conditions in wooden and apartment houses in Nagoya.] *Japanese Journal of Sanitary Zoology*, 42, 217–228. [In Japanese, English summary]

Suto, C., Sakaki, I., Ito, H. and Mitibata, M. (1992a) [Studies on the ecology of house–dust mites in wooden houses in Nagoya, with special reference to the influence of room ratios on the prevalence of mites and allergy.] *Japanese Journal of Sanitary Zoology*, 43, 217–228. [In Japanese, English summary]

Suto, C., Sakaki, I., Ito, H. and Mitibata, M. (1992b) [Influence of floor levels on the prevalence of house-dust mites in apartments in Nagoya.] *Japanese Journal of Sanitary Zoology*, 43(4), 307–318. [In Japanese, English summary]

Suto, C., Sakaki, I., Ito, H. and Mitibata, M. (1993) [Influence of types of rooms and flooring on the prevalence of house dust mites *Dermatophagoides farinae* and *D. pteronyssinus* (Acari: Pyroglyphidae) in houses in Nagoya.] *Japanese Journal of Sanitary Zoology*, 43, 307–318. [In Japanese, English summary]

Svendsen, U.G., Olsen, O.T. and Nuchel Petersen, L. (1990) House dust mites and allergy. Position papers, Danish Society for Allergology 1990. *Allergy*, 46, Supplement 11, 1–46.

Svoboda, J.A. and Thompson, M.J. (1985) Steroids. In: Kerkut, G.A. and Gilbert, L.I. (eds), *Comprehensive Insect Physiology, Biochemistry and Pharmacology, Volume 10, Biochemistry*. Pergamon Press, Oxford, pp. 137–175.

Swanson, M.C., Agarwal, M.K. and Reed, C.E. (1985) An immunochemical approach to indoor aeroallergen quantitation with a new volumetric air sampler: studies with mite, roach, cat, mouse and guinea pig. *Journal of Allergy and Clinical Immunology*, 76, 724–729.

Swanson, M.C., Campbell, A.R., Klauck, M.J. and Reed, C.E. (1989) Correlations between levels of mite and cat allergens in settled and airborne dust. *Journal of Allergy and Clinical Immunology*, 83, 776–783.

Syme, S.L. (1986) Social determinants of health and disease. In: Last, J.M. (ed.), *Maxcy-Rosenau Preventive Medicine and Public Health* (12th edition), Appleton-Century-Crofts, Norwalk, CT, pp. 953–970.

Szent-Györgyi, A. (1971) Biology and pathology of water. *Perspectives in Biology and Medicine*, 14, 239–249.

Tafforeau, M., Charpin, J., Vervloet, D., Charpin, D. and Jean-Pastor, M.J. (1988) Efficacite d'Acardust sur les acariens de la poussière de maison au domicile de patients (etude preliminaire). *Comptes Rendus de Therapeutique et de Pharmacologie Clinique*, 6, 3–6.

Takahashi, K., Aoki, T., Kohmoto, S., Nishimura, H., Kodera, Y., Matsushima, A. and Inada, Y. (1990) Activation of kallikrein-kinin system in human plasma with purified serine protease from *Dermatophagoides farinae*. *International Archives of Allergy and Applied Immunology*, 91, 80–85.

Takaoka, M. and Okada, S. (1984) [Ecological studies on mites in Saitama Prefecture Japan. Seasonal occurrence of the pyroglyphid mites in house dusts in 1981 to 1982.] *Japanese Journal Sanitary Zoology*, 35, 129–137. [In Japanese, English summary]

Takaoka, M., Ishii, A., Kabasawa, Y. and Ouchi, T. (1977) [Mite fauna in house dust of asthmatic children and their immunologic reactions to the mite extract.] *Japanese Journal of Sanitary Zoology*, 28, 355–361. [In Japanese, English summary]

Takatori, K., Saito, A., Yasueda, H. and Akiyama, K. (2001) The effect of house design and environment on fungal movement in homes of bronchial asthma patients. *Mycopathologia*, 152, 41–49.

Takhirova, G.S., Umarova, A.A. and Odinayev, F.I. (1996) [*Dermatophagoides pteronyssinus* in the Republic of Tadjikistan.] *Meditsinskaya Parazitologiya i Parazitarnye Bolezni*, 4, 16–18. [In Russian, English summary]

Tan, B.B., Weald, D., Strickland, I. and Friedmann, P.S. (1996) Double-blind controlled trial of effect of housedust-mite allergen avoidance on atopic dermatitis. *The Lancet*, 347, 15–18.

Tanaka, H., Sekimoto, K. and Naoe, S. (1976) Kawasaki disease. Relationship with infantile periarteritis nodosa. *Archives of Pathology and Laboratory Medicine*, 100, 81–86.

Tanaka, Y., Tanaka, M., Anan, S. and Yoshida, H. (1989) Immunohistochemical studies on dust mite antigen in positive reaction site of patch test. *Acta Dermatologia et Venereologia (Stockholm)*, Supplement 144, 93–96.

Tandon, N., Chatterjeee, H., Gupta, S.K. and Hati, A.K. (1988) Some observations on house dust mites in relation to naso-bronchial asthma in Calcutta. In: ChannaBasavanna, G.P. and Viraktamath, C.A. (eds), *Progress in Acarology*. Vol. 1. Oxford and IBH Publishing Co., New Delhi, pp. 163–168.

Tareev, V.N. and Dubinina, H.V. (1985) [On the fauna of dust inhabiting mites from Primorje.] *Parazytologia*, 19, 27–30. [In Russian, English summary]

Tategaki, A., Kawamoto, S., Aki, T., Jyo, T., Suzuki, O., Shigeta, S. and Ono, K. (2000). Newly described mite allergens. *Allergy and Clinical Immunology International*, Supplement 1, 74–76.

Taylor, R.N. (1975) Contributions to the biology and ecology of house dust mites. Unpublished PhD thesis, Department of Zoology, University of Glasgow, UK.

Tellam, R.L., Wijffels, G. and Willasden, P. (1999) Peritrophic matrix proteins. *Insect Biochemistry and Molecular Biology*, 29, 87–101.

Temeyer, K.B., Soileau, L.C. and Pruett, J.H. (2002) Cloning and sequence analysis of a cDNA encoding Pso II, a mite group II allergen of the sheep scab mite (Acari: Psoroptidae). *Journal of Medical Entomology*, 39, 384–391.

Teo, A.S.M., Ramos, J.D.A., Lee, B.W., Cheong, N. and Chua, K.W. (2006) Expression of the *Blomia tropicalis* paramyosin Blo t 11 and its immunodominant peptide in insect cells. *Biotechnology and Applied Biochemistry*, 45, 13–21.

Terra, S.A., Silva, D.A.O., Sopelete, M.C., Mendes, J., Sung, S.J. and Taketomi, E.A. (2004) Mite allergen levels and acarologic analysis in house dust samples in Uberaba, Brazil. *Journal of Investigational Allergology and Clinical Immunology*, 14, 232–237.

Terra, W.R. (1990) Evolution of digestive systems in insects. *Annual Review of Entomology*, 35, 181–200.

Terra, W.R. (2001) The origin and functions of the insect peritrophic membrane and peritrophic gel. *Archives of Insect Biochemistry and Physiology*, 47, 47–61.

Terra, W.R. and Ferreira, C. (1994) Insect digestive enzymes: properties, compartmentalisation and function. *Comparative Biochemistry and Physiology*, 109B, 1–62.

Terreehorst, I., Hak, E., Oosting, A.J., Tempels-Pavlica, Z., de Monchy, J.G.R., Bruijnzeel-Koomen, C.A.F.M., Aalberse, R.C. and van Wijk, R.G. (2003) Evaluation of impermeable covers for bedding in patients with allergic rhinitis. *New England Journal of Medicine*, 349, 237–246.

Thiam, D.G.Y., Tim, C.F., Hoon, L.S., Lei, Z. and Lee, B.W. (1999) An evaluation of mattress encasings and high efficiency particulate filters on asthma control in the tropics. *Asian Pacific Journal of Allergy and Immunology*, 17, 169–174.

Thind, B.B. (2000) Determination of low levels of mite and insect contaminants in food and feedstuffs by a modified flotation method. *Journal of the Association of Official Analytical Chemists International*, 83, 113–119.

Thind, B.B. and Dunn, J.A. (2003) Application of a rapid, sensitive flotation method for the determination of mites in disparate types of dust and debris samples. *Acarologia*, 43, 113–120.

Thind, B.B. and Griffiths, D.A. (1979) Flotation technique for quantitative determination of mite populations in powdered and compacted foodstuffs. *Journal of the Association of Official Analytical Chemists*, 62, 278–282.

Thind, B.B. and Wallace, D.J. (1984) Modified flotation technique for quantitative determination of mite populations in feedstuffs. *Journal of the Association of Official Analytical Chemists*, 67, 866–868.

Thomas, B., Heap, P. and Carswell, F. (1991) Ultrastructural localization of the allergen Der p I in the gut of the house dust mite *Dermatophagoides pteronyssinus*. *International Archives of Allergy and Applied Immunology*, 94, 365–367.

Thomas, V., Hay, T.B. and Leng, Y.P. (1976) Mass rearing of mites collected from house dust samples. *Medical Journal of Malaysia*, 30, 331–333.

Thomas, W.R. and Chua, K.Y. (1995) The major mite allergen Der p 2 – a secretion of the male mite reproductive tract? *Clinical and Experimental Allergy*, 25, 667–669.

Thomas, W.R., Stewart, G.A., Simpson, R.J., Chua, K.Y., Plozza, T.M., Dilworth, R.J., Nisbet, A. and Turner, K.J. (1988) Cloning and expression of DNA coding for the major house dust mite allergen Der p 1 in *Escherichia coli*. *International Archives of Allergy and Applied Immunology*, 85, 127–129.

Thomas, W.R., Chua, K.Y. and Smith, W.-A. (1992) Molecular polymorphisms of house dust mite allergens. *Experimental and Applied Acarology*, 16, 153–164.

Thomas, W.R., Smith, W.A., Hales, B.J., Mills, K.L. and O'Brien, R.M. (2002) Characterization and immunology of house mite allergens. *International Archives of Allergy and Applied Immunology*, 129, 1–18.

Thompson, P.J. and Stewart, G.A. (1989) House-dust mite reduction strategies in the treatment of asthma. *Medical Journal of Australia*, 151, 408, 411.

Thompson, P.J., Gillon, R.L., Bird, C., Krska, K. and Stewart, G.A. (1991) The effect of a combined acaricide/cleaning agent on house dust mite allergen load in carpet and mattress. *Australian and New Zealand Journal of Medicine*, 21, 660. [abstract]

Thompson, S.J. and Carswell, F. (1988) The major allergen of the house dust mite, *Dermatophagoides pteronyssinus*, is synthesised and secreted into its alimentary canal. *International Archives of Allergy and Applied Immunology*, 85, 312–315.

Tilak, S.T. and Jogdand, S.B. (1989) House dust mites. *Annals of Allergy*, 63, 392–397.

Tissot van Patot, P.N. (1929) Het Uitschakelen van de Exogene Oorzaken van het Klimaatasthma als Therapeutische Maatregel. Thesis, Leiden.

Tobias, K.R.C., Ferriani, V.P.L., Chapman, M.D. and Arruda, L.K. (2004). Exposure to indoor allergens in homes of patients with asthma and/or rhinitis in southeast Brazil: effect of mattress and pillow covers on mite allergen levels. *International Archives of Allergy and Immunology*, 133, 365–370.

Todorov, D.A. (1979) [Mites of the family Pyroglyphidae (Sarcoptiformes) in house dust and methods for their isolation.] *Acta Zoologica Bulgarica*, 13, 64–71. [In Bulgarian, English summary]

Toma, T., Miyagi, I., Kishimoto, M., Nagama, T. and Tamanaha, I. (1993) [Fauna and seasonal appearances of mites in house dusts collected from the residences including asthmatic children in Okinawa Prefecture, Ryukyu Archipelago.] *Japanese Journal of Sanitary Zoology*, 44(3), 223–235. [In Japanese, English summary]

Toma, T., Miyagi, I., Takeda, F., Kishimoto, M. and Ahagon, A. (1998) Mite fauna and abundance in dust collected from bedding and bedrooms in Okinawa, Japan. *Medical Entomology and Zoology*, 49, 309–319.

Tongu, Y., Ishii, A. and Oh, H. (1986) Ultrastructure of house-dust mites, *Dermatophagoides farinae* and *D. pteronyssinus*. *Japanese Journal of Sanitary Zoology*, 37, 237–244.

Topham, C.M., Srinivasen, N., Thorpe, C.J., Overington, J.P. and Kalsheker, N.A. (1994) Comparative modelling of major house dust mite allergen *Der p* I. Structure validation using an extended environmental amino acid propensity table. *Protein Engineering*, 7, 869–894.

Topp, R., Wimmer, K., Fahlbusch, B., Bischof, W., Richter, K., Wichman, H.E. and Heinrich, J. for the INGA study group (2003) Repeated measures of allergens and endotoxin in settled house dust over a time period of 6 years. *Clinical and Experimental Allergy*, 33, 1659–1666.

Torrent, M., Sunyer, J., Garcia, R., Harris, J., Iturriaga, M.V., Puig, C., Vall, O., Antó, J.M., Newman Taylor, A.J. and Cullinan, P. (2007) Early-life allergen exposure and atopy, asthma and wheeze up to 6 years of age. *American Journal of Respiratory and Critical Care Medicine*, 176, 446–453.

Tovey, E.R. (1992) Allergen exposure and control. *Experimental and Applied Acarology*, 16, 181–202.

Tovey, E.R. (1995) Mite allergen measurement. In: Tovey, E., Fifoot, A. and Sieber, L. (eds), *Mites, Asthma and Domestic Design II*. University of Sydney, Sydney, pp. 14–18.

Tovey, E.R. (1997) Environmental control. In: P. Barnes, M. Grunstein, A.R. Leff and A.J. Woolcock (eds), *Asthma* Vol. 2. Lippincott-Raven, Philadelphia, pp. 1883–1904.

Tovey, E.R. and Baldo, B.A. (1990) Localization of antigens and allergens in thin sections of the house dust mite, *Dermatophagoides pteronyssinus* (Acari, Pyroglyphidae). *Journal of Medical Entomology*, 27, 368–376.

Tovey, E.R. and Marks, G. (1999) Methods and effectiveness of environmental control. *Journal of Allergy and Clinical Immunology*, 103, 179–191.

Tovey, E.R. and Woolcock, A.J. (1994) Direct exposure of carpets to sunlight can kill all mites. *Journal of Allergy and Clinical Immunology*, 93, 1072–1074.

Tovey, E.R., Giunan, J.E. and Vandenberg, R. (1975) Mite populations in Sydney household bedding with particular reference to nursery sheepskins. *Medical Journal of Australia*, ii, 770–772.

Tovey, E.R., Chapman, M.D. and Platts-Mills, T.A.E. (1981) Mite faeces are a major source of house dust mite allergens. *Nature*, 289, 592–593.

Tovey, E.R., Chapman, M.D., Wells, C.W. and Platts-Mills, T.A.E. (1981) The distribution of dust mite allergen in the homes of patients with asthma. *American Review of Respiratory Diseases*, 124, 630–635.

Tovey, E.R., Johnson, M.C., Roche, A.L, Cobon, G.S. and Baldo, B.A. (1989) Cloning and sequencing of a cDNA expressing a recombinant house dust mite protein that binds human IgE and corresponds to an important low molecular weight allergen. *Journal of Experimental Medicine*, 170, 1457–1462, and erratum, *Journal of Experimental Medicine*, (1990) 171, 1387.

Tovey, E.R., Marks, G.B., Matthews, M., Green, W.F. and Woolcock, A. (1992) Changes in mite allergen *Der p* I in house dust following spraying with a tannic acid/acaricide solution. *Clinical and Experimental Allergy*, 22, 67–74.

Tovey, E.R., Marks, G., Shearer, M. and Woolcock, A. (1993) Allergens and occlusive bedding covers. *Lancet*, 342, 126.

Tovey, E.R., Mahmic, A. and McDonald, L.G. (1995) Clothing – an important source of mite allergen exposure. *Journal of Allergy and Clinical Immunology*, 96, 999–1001.

Tovey, E.R., McDonald, L., Peat, J. and Marks, G. (2000a) Domestic mite species and Der p 1 allergen levels in nine locations in Australia. *Allergy Clinical Immunology International*, 12, 226–231.

Tovey, E.R., Taylor, D.J.M., Graham, A.H., O'Meara, T.J., Lovborg, U., Jones, A. and Sporik, R. (2000b) New immunodiagnostic system. *Aerobiologia*, 16, 113–118.

Tovey, E.R., Taylor, D.J., Mitakakis, T.Z. and De Lucca, S.D. (2001) Effectiveness of laundry washing agents and conditions in the removal of cat and dust mite allergen from bedding dust. *Journal of Allergy and Clinical Immunology*, 108, 369–374.

Tovey, E.R., Mitakakis, T.Z., Sercombe, J.K., Vanlaar, C.H. and Marks, G.B. (2003) Four methods of sampling for dust mite allergen: differences in 'dust'. *Allergy*, 58, 790–794.

Tovey, E.R., Almqvist, C., Li, Q., Crisafulli, D. and Marks, G.B. (2008) Nonlinear relationship of mite allergen exposure to mite sensitisation in a birth cohort. *Journal of Allergy and Clinical Immunology*, 122, 114–118.

Trakultivakorn, M. and Krudtong, S. (2004) House dust mite allergen levels in Chiang Mai homes. *Asian Pacific Journal of Allergy and Immunology*, 22, 1–6.

Trang, R.B., Tsai, L.-C., Hwang, H.-M., Hwang, B., Wu, K.-G., Hung, M.-W. (1990) The prevalence of allergic disease and IgE antibodies to house dust mite in school-children in Taiwan. *Clinical and Experimental Allergy*, 20, 33–38.

Traver, J.R. (1951) Unusual scalp dermatitis in humans caused by the mite *Dermatophagoides* (Acarina: Epidermoptidae). *Proceedings of the Entomological Society of Washington*, 53, 1–25.

Travers, S.E., Smith, M.D., Bai, J., Hulbert, S.H., Leach, J.E., Schnable, P.S., Knapp, A.K., Milliken, G.A., Fay, P.A., Saleh, A. and Garrett, K.A. (2007) Ecological genomics: making the leap from model systems in the lab to native populations in the field. *Frontiers in Ecology and Environment*, 5, 19–24.

Treat, A.E. (1975) *Mites of Moths and Butterflies.* Comstock Publishing Associates, Ithaca.

Trinca, J.C., Stringer, K.C., Drummond, F.H. and Bristow, V.G. (1969) Allergenicity of the house-dust mite *Dermatophagoides pteronyssinus*. *Medical Journal of Australia*, ii, 1177–1179.

Trouessart, E.L. (1887) Diagnose d'espèces nouvelles de Sarcoptides plumicoles (Analgesinae). *Bulletin de la Société d'études scientifiques d'Angers*, 16, 85–156.

Trouessart, E.L. (1901) Sur deux espèces, formant un genre nouveau, de Sarcoptides detriticoles parasites des fourrures. *Bulletin de la Société Zoologique de France*, 26, 82–84.

Trudinger, M., Chua, K.Y. and Thomas, W.R. (1991) cDNA encoding the major mite allergen Der f II. *Clinical and Experimental Allergy*, 21, 33–37.

Tsai, L.C., Chao, P.L., Shen, H.D., Tang, R.B., Chang, T.C., Chang, Z.N., Hung, M.W., Lee, B.L. and Chua, K.Y. (1998). Isolation and characterisation of a novel 98-kd *Dermatophagoides farinae* mite allergen. *Journal of Allergy and Clinical Immunology*, 102, 295–303.

Tsai, L.C., Peng, H.J., Lee, C.S., Chao, P.L., Tang, R.B., Tsai, J.J., Shen, H.D., Hung, M.W. and Han, S.H. (2005) Molecular cloning and characterization of full-length cDNAs encoding a novel high-molecular-weight *Dermatophagoides pteronyssinus* mite allergen, Der p 11. *Allergy*, 60, 927–937.

Tsang, W.-S., Quackenbush, L.S., Chow, B.K.C., Tiu, S.H.K., He, J.-G. and Chan, S.-M. (2003) Organisation of the shrimp vitellogenin gene: evidence of multiple genes and tissue specific expression by the ovary and hepatopancreas. *Gene*, 303, 99–109.

Tsundona, T., Mori, H. and Shimada, K. (1992) [Studies on supercooling and cold-hardiness in the house dust mite *Dermatophagoides pteronyssinus* (Astigmata, Pyroglyphidae).] Japanese Journal of Applied Entomology, 36, 1–4. [In Japanese, English summary]

Türk, E. and Türk, F. (1957) Systematik und Ökologie der Tyroglyphiden Mitteleuropas. In: Stammer, H.-J. (ed.), *Beiträge Zur Systematik und Ökologie der Mitteleuropäischer Acarina. Band I Tyroglyphidae und Tarsonemini.* Geest and Portig K.-G., Leipzig, pp. 1–231.

Turner, K.J., Baldo, B.A. and Hilton, J.M.N. (1975) IgE antibodies to *Dermatophagoides pteronyssinus* (house dust mite), *Aspergillus fumigatus* and β-lactoglobulin in sudden infant death syndrome. *British Medical Journal*, I, 357–360.

Turner, K.J., Stewart, G.A., Woolcock, A.J., Green, W. and Alpers, M.P. (1988) Relationship between mite densities and the prevalence of asthma, comparative studies in two populations in the Eastern Highlands of Papua New Guinea. *Clinical Allergy*, 18, 331–340.

Turos, M. (1979) Mites in house dust in the Stockholm area. *Allergy*, 34, 11–18.

Turunen, S. (1979) Digestion and absorption of lipids in insects. *Comparative Biochemistry and Physiology*, 63A, 455–460.

Ucci, M., Pretlove, S.E.C., Biddulph, P., Oreszczyn, T., Wilkinson, T., Crowther, D., Scadding, G., Hart, B. and Mumovic, D. (2007) The psychrometric control of house dust mites: a pilot study. *Building Service Engineering Research and Technology*, 28, 347–356.

Uchikoshi, S., Kimura, H., Nomura, K., Chien, C., Iida, M. and Miyake, H. (1982) A study of the ecology of the house dust mite in dwelling houses. *Tokai Journal of Experimental Clinical Medicine*, 7, 233–243.

Uehara, K., Toyoda, Y. and Konishi, E. (2000) Contamination of passenger trains with *Dermatophagoides* (Acari: Pyroglyphidae) mite antigen in Japan. *Experimental and Applied Acarology*, 24, 727–734.

Upton, M.S. (1993) Aqueous gum-chloral slide mounting media: an historical review. *Bulletin of Entomological Research*, 83, 267–274.

Ushijima, H., Park, R., Yoshino, K., Ohta, K., Fujii, R. and Kobayashi, N. (1983) Microorganisms in house-dust mites in Kawasaki disease. *Acta Paediatrica Japonensis*, 25, 127–129.

Valdivieso, R., Iraola, V., Estupiñán, M. and Fernández-Caldas, E. (2006) Sensitization and exposure to house dust and storage mites in high-altitude areas of Ecuador. *Annals of Allergy Asthma and Immunology*, 97, 532–538.

Valera, J., Ventas, P., Carreira, J., Barbas, J.A., Giminez-Gallego, G. and Polo, F. (1994) Primary structure of *Lep d I*, the main *Lepidoglyphus destructor* allergen. *European Journal of Biochemistry*, 225, 93–98.

Valerio, C.R., Murray, P., Arlian, L.G. and Slater, J.E. (2005) Bacterial 16S ribosomal DNA in house dust mite cultures. *Journal of Allergy and Clinical Immunology*, 116, 1296–1300.

Vanhalen, M., Vanhalen-Fastré, R. and Geeraerts, J. (1980) Occurrence in mushrooms (Homobasidiomycetes) of cis- and trans-octa-1,5-dien-3-ol, attractants to the cheese mite *Tyrophagus putrescentiae* (Schrank) (Acarina, Acaridae). *Experientia*, 36, 406–407.

Vanlaar, C.H., Peat, J.K., Marks, G.B., Rimmer, J. and Tovey, E.R. (2000) Domestic control of house dust mite allergen in children's beds. *Journal of Allergy and Clinical Immunology*, 105, 1130–1133.

Varekamp, H. (1925) *Der Exogene Oorzaken van Asthma Bronchiale*. Doctoral thesis, University of Leiden.

Varekamp, H. and Voorhorst, R. (1960) Interrelation of vasomotor rhinitis and asthma, occurring in patients with house-dust atopy. II. Influence of houses lived in. *Acta Allergologica*, 15, 248–255.

Varekamp, H., Spieksma, F.Th.M., Leupen, M.J. and Lyklema, A.W. (1966) House dust mites in their relation to dampness in houses and the allergen content of house dust. In: *Interasthma V. Congress Proceedings*, Pressa Trajectina, Utrecht, 2.56–2.68.

Vargas, V.M. and Mairena, A.H. (1991) House dust mites from the metropolitan area of San José, Costa Rica. *International Journal of Acarology*, 17, 141–144.

Vargas, V.M. and Smiley, R.L. (1994) A new species of *Hughesiella* (Acari: Astigmata: Pyroglyphidae) from Costa Rica. *International Journal of Acarology*, 20, 123–131.

Vaughan, J.W., Woodfolk, J.A. and Platts-Mills, T.A.E. (1999) Assessment of vacuum cleaners and vacuum cleaner bags recommended for allergic subjects. *Journal of Allergy and Clinical Immunology*, 104, 1079–1083.

Veale, A.J., Peat, J.K., Tovey, E.R., Salome, C.M., Thompson, J.E. and Woolcock, A.J. (1996) Asthma and atopy in four rural Australian Aboriginal communities. *Medical Journal of Australia*, 165, 192–196.

Ventas, P., Carreira, J. and Polo, F. (1991) Identification of IgE-binding proteins from *Lepidoglyphus destructor* and production of monoclonal antibodies to a major allergen. *Immunology Letters*, 29, 229–234.

Verhoeff, A.P. (1995) House dust mite, allergens and fungi in homes in relation to residential characteristics. In: Tovey, E., Fifoot, A. and Sieber, L. (eds), *Mites, Asthma and Domestic Design II*. University of Sydney, Sydney, pp. 85–88.

Verhoeff, A.P., Strien, R.T. van, Wijnen, J.H. van and Brunekreef, B. (1994) House dust mite allergen (*Der p* I) and respiratory symptoms in children: a case-control study. *Clinical and Experimental Allergy*, 24, 1061–1069.

Verhoeff, A.P., Strien, R.T. van, Wijnen, J.H. van and Brunekreef, B. (1995) Damp housing and childhood respiratory symptoms: the role of sensitisation to dust mites and moulds. *American Journal of Epidemiology*, 141, 103–110.

Verlato, G., Calabrese, R. and de Marco, R. (2002) Correlation between asthma and climate in the European Community Respiratory Health Survey. *Archives of Environmental Health*, 57, 48–52.

Verrall, B., Muir, D.C.F., Wilson, W.M., Milner, R., Johnston, M. and Dolovich, J. (1988) Laminar-flow air cleaner bed attachment – a controlled trial. *Annals of Allergy*, 61, 117–122.

Vervloet, D., Bongrand, P., Arnaud, A., Boutin, C. and Charpin, J. (1979) Donnees objectives cliniques et immunologiques observees au cours d'une cure d'altitude a Briançon chez des enfants asthmatiques allergeniques a la poussière de maison et a *Dermatophagoides pteronyssinus*. *Revue Française Maladies Respiratoires*, 7, 19–27.

Vervloet, D., Penaud, A., Razzouk, H., Senff, M., Arnaud, A., Boutin, C. and Charpin, J. (1982) Altitude and house dust mites. *Journal of Allergy and Clinical Immunology,* 69, 290–296.

Vicentini, L., Peroni, D., Miraglia del Giudice, M., Mazzi, P., Bodini, A. and Piacentini, G.L. (2002) Comparison of vacuum cleaners. *Allergy*, 57, 555.

Vichayond, P., Uthaisangsook, S., Ruangruk, S. and Malainual, N. (1999) Complete mattress encasing is not superior to partial encasing in the reduction of mite allergen. *Allergy*, 54, 736–741.

Viinanen, A., Munhbayarlah, S., Zevgee, T., Narantsetseg, L., Naidansuren, Ts., Koskenvuo, M., Helenius, H. and Terho, E.O. (2005) Prevalence of asthma, allergic rhinoconjunctivitis and allergic sensitisation in Mongolia. *Allergy*, 60, 1370–1377.

Viinanen, A., Munhbayarlah, S., Zevgee, T., Narantsetseg, L., Naidansuren, Ts., Koskenvuo, M., Helenius, H. and Terho, E.O. (2007) The protective effect of rural living against atopy in Mongolia. *Allergy*, 62, 272–280.

Vijayambika, V. and John, P.A. (1977) Digestive system in Aleuroglyphus ovatus (Acarina: Tyroglyphidae). *Indian Journal of Acarology*, 2, 24–29.

Villanueva, E., Wong, P.A., Yengle, M.A., Yoshida, I., Ysmodes, Y., Vílchez, F., Yoshioka, D., Yamunaqué, P. and Yana, E. (2003) Prevalencia de ácaros del polvo en habitaciones de la Communidad '7 de Octubre' de El Agustino, Lima. Octobre 2002. *Revista Peruana de Epidemiología*, 11, 1–6.

Villaveces, J.W., Rosengren, H. and Evans, J. (1977) Use of laminar flow portable filter in asthmatic children. *Annals of Allergy*, 38, 400–404.

Vitzthum, H. Graf. (1940) *Bronn's Klassen und ordnung des Tierreichs, Acarina.* Becker & Erler, Leipzig.

Vobrázková, E., Samsinák, K. and Spicák, V. (1979) Allergogenous mites in private recreation houses. *Folia Parasitologia (Praha)*, 26, 343–349.

Vobrazkova, E., Samsinak, K. and Spicak, V. (1985) The possibility of sensitization to inhalatory allergens in nursery schools. *Zoologischer Anzeiger*, 215, 195–200.

Vobrázková, E., Kasiaková, A. and Samsinák, K. (1986) Analysis of dust samples from the clinical environment of children with eczemas. *Agnewandte Parasitologie*, 27, 53–55.

Vojta, P.J., Randels, S.P., Stout, J., Muilenberg, M., Burge, H.A., Lynn, H., Mitchell, H., O'Connor, G.T. and Zeldin, D.C. (2001) Effects of physical interventions on house dust mite allergen levels in carpet, bed, and upholstery dust in low-income, urban homes. *Environmental Health Perspectives*, 109, 815–819.

Voorhorst, R. (1970) Quantitative aspects of the problem of house-dust atopy and house-dust mites. *Acta Allergologica*, 25(4), 237–254.

Voorhorst, R., Spieksma-Boezeman, M.I.A. and Spieksma, F.Th.M. (1964) Is a mite the producer of the house dust allergen? *Allergie und Asthma*, 10, 329–334.

Voorhorst, R., Spieksma, F.Th.M., Varekamp, H., Leupen, M.J. and Lykema, A.W. (1967) The house-dust mite (*Dermatophagoides pteronyssinus*) and the allergens it produces. Identity with the house-dust allergen. *Journal of Allergy*, 39, 325–339.

Voorhorst, R., Spieksma, F.Th.M. and Varekamp, N. (1969) *House Dust Atopy and the House Dust Mite,* Dermatophagoides pteronyssinus *(Trouessart, 1897).* Stafleu's Scientific Publishing Co., Leiden.

de Vries, M.P., van den Bemt, L., Aretz, K., Thoonen, B.P.A., Muris, J.W.M., Kester, A.D.M., Cloosterman, S. and van Schayck, C.P.O. (2007) House dust mite allergen avoidance and self-management in allergic patients with asthma: randomised controlled trial. *British Journal of General Practice*, 57, 184–190.

Vyszenski-Moher, D.L. and Arlian, L.G. (2003) Effects of wet cleaning with disodium octaborate tetrahyhdrate on dust mites (Acari: Pyroglyphidae) in carpet. *Journal of Medical Entomology*, 40, 508–511.

Vyszenski-Moher, D.L., Arlian, L.G., Bernstein, I.L. and Gallagher, J.S. (1986) Prevalence of house dust mites in nursing homes in southwest Ohio. *Journal of Allergy and Clinical Immunology*, 77, 745–748.

Wahn, U., Lau, S., Bergmann, R., Kulig, M., Forster, J., Bergmann, K., Bauer, C.-P. and Guggenmoos-Holzmann, I. (1997) Indoor allergen exposure is a risk factor for sensitisation during the first three years of life. *Journal of Allergy and Clinical Immunology*, 99, 763–769.

Waki, S. and Matsumoto, K. (1973a) [Studies on the environmental requirements for the breeding of the dust mite, *Dermatophagoides farinae* Hughes, 1961 Part 1. Observations on the mode of breeding under various temperature and humidity conditions.] *Japanese Journal of Sanitary Zoology*, 23, 159–163. [In Japanese, English summary]

Waki, S. and Matsumoto, K. (1973b) [Studies on the environmental requirements for the breeding of the dust mite, *Dermatophagoides farinae* Hughes, 1961 Part 2. Observations on the mode of breeding in various kinds of food.] *Japanese Journal of Sanitary Zoology*, 24, 117–121. [In Japanese, English summary]

Wallace, D.R.J. (1960) Observations on hypopus development in the Acarina. *Journal of Insect Physiology*, 5, 216–229.

Walshaw, M.J. and Evans, C.C. (1986) Allergen avoidance in house dust mite sensitive adult asthma. *Quarterly Journal of Medicine*, 58, 199–215.

Walshaw, M.J. and Evans, C.C. (1987) The effect of seasonal and domestic factors on the distribution of *Euroglyphus maynei* in the homes of *Dermatophagoides pteronyssinus* allergic patients. *Clinical Allergy*, 17, 7–14.

Walter, D.E. and Proctor, H.C. (1999) *Mites: Ecology, Evolution and Behaviour*. University of New South Wales Press, Sydney and CABI Publishing, Wallingford.

Walter, F., Fletcher, D.J.C., Chautems, D., Cherix, D., Keller, L., Francke, W., Fortelius, W., Rosengren, R. and Vargo, E.L. (1993) Identification of the sex pheromone of an ant, *Formica lugubris* (Hymenoptera, Formicidae). *Naturwissenschaften*, 80, 30–34.

Walzl, M.G. (1978) Die kopulationsorgane von *Dermatophagoides pteronyssinus* und *D. farinae*. Eine rasterelektronen-mikroskopische untersuchungen. *Zoologischer Anzeiger*, 201, 44–48.

Walzl, M.G. (1988) A scanning electron microscopic study of the embryonic development of the house dust mite, *Dermatophagoides farinae* (Pyroglyphidae, Actinotrichida) from blastula to the hatching of the larva. *Institute of Physics Conference Series*, 3(93), Chapter 6, pp. 159–160.

Walzl, M.G. (1991a) Comparison of the chitinous parts of the reproductive organs of house dust mites by means of scanning electron microscopy. In: Schuster, R.H. and Murphy, P.W. (eds), *The Acari, Reproduction, Development and Life-History Strategies*, Chapman and Hall, London, pp. 355–362.

Walzl, M.G. (1991b) Microwave treatment of mites (Acari, Arthropoda) for extruding hidden cuticular parts of the body for scanning electron microscopy. *Micron Microscopica Acta*, 22, 9–15.

Walzl, M.G. (1992) Ultrastructure of the reproductive system of the house dust mites *Dermatophagoides farinae* and *Dermatophagoides pteronyssinus* (Acari, Pyroglyphidae) with remarks on spermatogenesis and oogenesis. *Experimental and Applied Acarology*, 16, 85–116.

Walzl, M.G. (2001) Ultrasonication, a tool for microdissection of astigmatid mites investigated by SEM. In: Halliday, R.B., Walter, D.E., Proctor, H.C., Norton, R.A. and Colloff, M.J. (eds), *Acarology: Proceedings of the 10th International Congress*. CSIRO Publishing, Melbourne, pp. 226–229.

Wan, H., Winton, H.L., Soeller, C., Tovey, E.R., Gruenert, D.C., Thompson, P.J., Stewart, G.A., Taylor, G.W., Garrod, D.R., Mark, B., Cannell, M.B. and Robinson, C. (1999) Der p 1 facilitates transepithelial allergens delivery by disruption of tight junctions. *Journal of Clinical Investigation*, 104, 123–133.

Wan, H., Winton, H.L., Soeller, C., Taylor, G.W., Gruenert, D.C., Thompson, P.J., Cannell, M.B., Stewart, G.A., Garrod, D.R. and Robinson, C. (2001) The transmembrane protein occluding of epithelial tight junctions is a functional target for serine peptidases from faecal pellets of *Dermatophagoides pteronyssinus*. *Clinical and Experimental Allergy*, 31, 279–294.

Wang, H. and Wen, T. (1997) Mite allergens Der p 1 and Der f 1 in bedding dust determined by monoclonal antibodies in Shanghai. *Systematic and Applied Acarology*, 2, 135–140.

Wang, P., Li, G. and Granados, R.R. (2004) Identification of two new peritrophic membrane proteins from larval *Trichoplusia ni*: structural characteristics and their functions in the protease rich insect gut. *Insect Biochemistry and Molecular Biology*, 34, 215–227.

Ward, C.W. (1975) Properties and specificity of the major metal-chelator-sensitive proteinase in the keratinolytic larvae of the webbing clothes moth. *Biochimica Biophysica Acta*, 384, 215–227.

Ward, N. (2006) New directions and interactions in metagenomics research. *FEMS Microbiology and Ecology*, 55, 331–338.

Warner, A., Boström, S., Munir, A.K.M., Möller, C., Schou, C. and Kjellmann, N.-I.M. (1998) Environmental assessment of *Dermatophagoides* mite-allergen levels in Sweden should include Der m 1. *Allergy*, 53, 698–704.

Warner, A., Boström, S., Möller, C., Kjellman, N.-I.M. (1999) Mite fauna in the home and sensitivity to house-dust and storage mites. *Allergy*, 54, 681–690.

Warner, J.A., Marchant, J.L. and Warner, J.O. (1993a) Allergen avoidance in the homes of atopic asthmatic children: the effect of Allersearch DMS. *Clinical and Experimental Allergy*, 23, 279–286.

Warner, J.A., Marchant, J.L. and Warner, J.O. (1993b) Double-blind trial of ionizers in children with asthma sensitive to the house dust mite. *Thorax*, 48, 330–333.

Warner, J.A., Frederick, J.M., Bryant, T.N., Weich, C., Raw, G.J., Hunter, C., Stephen, F.R., McIntyre, D.A. and Warner, J.O. (2000) Mechanical ventilation and high-efficiency vacuum cleaning: A combined strategy of mite and mite allergen reduction in the control of mite-sensitive asthma. *Journal of Allergy and Clinical Immunology*, 105, 75–82.

Warner, J.O. (1978) Mites and asthma in children. *British Journal of Diseases of the Chest*, 72, 79–87.

Wassenaar, D.P.J. (1988a) Effectiveness of vacuum cleaning and wet cleaning in reducing house-dust mites, fungi and mite allergen in a cotton carpet, a case study. *Experimental and Applied Acarology*, 4, 53–62.

Wassenaar, D.P.J. (1988b) Reducing house dust mites by vacuuming. *Experimental and Applied Acarology*, 4, 167–171.

Wasylik, A. (1959) Mite fauna (Tyroglyphoidea) in the nests of the common sparrow (*Passer domesticus* L.). *Ekologia Polska*, 5, 187–190.

Wasylik, A. (1963) A method of continuous analysis of the Acarina of birds' nests. *Ekologia Polska*, 9, 219–224.

Weast, R.C. and Astle, M.J. (1980) *CRC Handbook of Chemistry and Physics*. 61st edition. CRC Press, Boca Raton.

Weber, E., Hunter, S., Stedman, K., Dreitz, S., Olivry, T., Hillier, A. and McCall, C. (2003) Identification, characterization, and cloning of a complementary DNA encoding a 60-kd house dust mite allergen (Der f 18) from human beings and dogs. *Journal of Allergy and Clinical Immunology*, 112, 79–86.

Weghofer, M., Thomas, W.R., Pittner, G., Horak, F., Valenta, R. and Vrtala, S. (2005) Comparison of purified *Dermatophagoides pteronyssinus* allergens and extract by two-dimensional immunoblotting and quantitative immunoglobulin E inhibitions. *Clinical and Experimental Allergy*, 35, 1384–1391.

Weightman, G. and Humphries, S. (1984). *The Making of Modern London 1815–1914*. Sidgwick and Jackson, London.

Weiland, S.K., Hüsing, A., Strachan, D.P., Rzehak, P., Pearce, N. and the ISAAC Phase One Study Group (2004) Climate and the prevalence of symptoms of asthma, allergic rhinitis, and atopic eczema in children. *Occupational and Environmental Medicine*, 61, 609–615.

Weinberg, E.G. (1999) Urbanisation and childhood asthma: an African perspective. *Journal of Allergy and Clinical Immunology*, 105, 224–231.

Weinmayr, G., Weiland, S.K., Björkstén, B., Brunekreef, B., Büchele, G., Cookson, W.O.C., Garcia-Marcos, L., Gotua, C., van Hage, M., von Mutius, E., Riikjärv, M.-A., Rzehak, P., Stein, R., Strachan, D.P., Tsanakas, J., Wickens, K., Wong, G. and the ISAAC Phase Two Study Group (2007) Atopic sensitisation and the international variation of asthma symptom prevalence in children. *American Journal of Respiratory and Critical Care Medicine*, 176, 565–574.

Weiss, K.B., Gergen, P.J. and Hodgson, T.A. (1992) An economic valuation of asthma in the United States. *New England Journal of Medicine*, 326, 862–866.

Wen, D.C., Shyur, S.D., Ho, C.M., Chiang, Y.C., Huang, L.H., Lin, M.T., Yang, H.C. and Liang, P.H. (2005) Systemic anaphylaxis after the ingestion of pancake contaminated with the storage mite *Blomia freemani*. *Annals of Allergy, Asthma and Immunology*, 95, 612–614.

Wen, T. and Wang, H. (1988) Studies on mite population and mite allergens in beds in China. In: de Weck, A.L. and Todt, A. (eds), *Mite Allergy: A Worldwide Problem*. UCB Institute of Allergy, Brussels, pp. 83–86.

Wen, T., Hong, S., Shen, S., Cal, L., Xu, W., Wang, H., Liao, M., Zhuang, Y. and Liu, J. (1988) Preliminary survey of mite allergy in China. In: de Weck, A.L. and Todt, A. (eds), *Mite Allergy: A Worldwide Problem*. UCB Institute of Allergy, Brussels, pp. 23–24.

Wen, T.-H., Cal, L., Xu, W.-H., Wang, H., Xiang, L., Hong, S.-S., Zhung, J.-M. and Zhuang, Y.-J. (1991) Mite prevalence and mite sensitivity in China. *Journal of the Korean Society of Allergology*, 11, 201–214.

Wenzel, S.E. (2006) Asthma; defining of the persistent adult phenotype. *Lancet*, 368, 804–813.

Westritschnig, K., Sibanda, E., Thomas, W., Auer, H., Aspöck, H., Pittner, G., Vrtala, S., Spitzauer, S., Kraft, D. and Valenta, R. (2003) Analysis of the sensitisation profile towards allergens in central Africa. *Clinical and Experimental Allergy*, 33, 22–27.

Wharton, G.W. (1963) Equilibrium humidity. In: *Advances in Acarology. Volume 1*, Cornell University Press, pp. 201–208.

Wharton, G.W. (1973) Spatial relations of house dust mites. In: Daniel, M. and Rosicky, B. (eds), *Proceedings of the 3rd International Congress of Acarology*. Academia, Prague, pp. 557–559.

Wharton, G.W. (1976) House dust mites. *Journal of Medical Entomology*, 12, 577–621.

Wharton, G.W. (1978) Uptake of water vapour by mites and mechanisms utilised by the Acaridei. In: Schmidt-Nielsen, K., Bolis, L. and Maddrell, S.H.P. (eds), *Comparative Physiology: Water, Ions and Fluid Mechanics*. Cambridge University Press, Cambridge, pp. 79–95.

Wharton, G.W. (1985) Water balance of insects. In: Kerkut, G.A. and Gilbert, L.I. (eds), *Comprehensive Insect Physiology, Biochemistry and Pharmacology, Volume 4, Regulation: Digestion, Nutrition, Excretion*. Pergamon Press, Oxford, pp. 565–601.

Wharton, G.W. and Arlian, L.G. (1972b) Predatory behaviour of the mite *Cheyletus aversor*. *Animal Behaviour*, 20, 719–723.

Wharton, G.W. and Arlian, L.G. (1972a) Utilization of water by terrestrial mites and insects. In: Rodriguez, J.G. (ed.), *Insect and Mite Nutrition*. North Holland Publishing Co., Amsterdam, pp. 153–165.

Wharton, G.W. and Brody, A.R. (1972) The peritrophic membrane of the mite *Dermatophagoides farinae* (Acariformes). *Journal of Parasitology*, 58, 801–804.

Wharton, G.W. and Fuller, H.S. (1952) A manual of the Chiggers. *Memoirs of the Entomological Society of Washington*, No. 4, 1–175.

Wharton, G.W. and Furumizo, R.T. (1977) Supracoxal gland secretions as a source of fresh water for Acaridei. *Acarologia*, 19, 112–116.

Wharton, G.W. and Richards, A.G. (1978) Water vapour exchange kinetics in insects and acarines. *Annual Review of Entomology*, 23, 309–328.

Wharton, G.W., Duke, K. and Epstein, H.M. (1979) Water and the physiology of house dust mites. In: Rodriguez, J.G. (ed.), *Recent Advances in Acarology. Volume 1*. Academic Press, New York, pp. 325–335.

Wheeler, Q.D. (2007) Invertebrate systematics or spineless taxonomy? *Zootaxa*, 1668, 11–18.

Wickens, K., Siebers, R., Ellis, I., Lewis, S., Sawyer, G., Tohill, S., Stone, L., Kent, R., Kennedy, J., Slater, T., Crothall, A., Trethowen, H., Pearce, N., Fitzharris, P. and Crane, J. (1997) Determinants of house dust mite allergen in homes in Wellington, New Zealand. *Clinical and Experimental Allergy*, 27, 1077–1085.

Wickens, K., Mason, K., Fitzharris, P., Siebers, R., Hearfield, M., Cunningham, M. and Crane, J. (2001) The importance of housing characteristics in determining Der p 1 levels in carpets in New Zealand homes. *Clinical and Experimental Allergy*, 31, 827–835.

Wickens, K., de Bruyne, J., Calvo, M., Choon-Kook, S., Jayaraj, G., Lai, C.K.W., Lane, J., Maheshwari, R., Mallol, J., Nishima, S., Purdie, G., Siebers, R., Sukumaran, T., Trakultivakorn, M. and Crane, J. (2004) The determinants of dust mite allergen and its relationship to the prevalence of symptoms of asthma in the Asia-Pacific region. *Pediatric Allergy and Immunology*, 15, 55–61.

Wickman, M. (1993) Residential characteristics and allergic sensitization in children, especially to mites. Doctoral thesis, Karolinska Institute, Stockholm.

Wickman, M. (1995) Residential characteristics and allergen sensitization in children, especially to mites: A Swedish perspective. In: Tovey, E., Fifoot, A. and Sieber, L. (eds), *Mites, Asthma and Domestic Design II*. University of Sydney, Sydney, pp. 81–84.

Wickman, M., Nordvall, S.L., Pershagen, G., Sundell, J. and Schwartz, B. (1991) House dust mite sensitisation in children and residential character-istics in a temperate region. *Journal of Allergy and Clinical Immunology*, 88, 89–95.

Wickman, M., Nordvall, S.L., Pershagen, G., Korsgaard, J. and Johansen, N. (1993) Sensitisation to domestic mites in a cold temperate region. *American Review of Respiratory Disease*, 148, 58–62.

Wickman, M., Emenius, G., Egmar, A.-C., Axelsson, G. and Pershagen, G. (1994a) Reduced mite allergen levels in dwellings with mechanical exhaust and supply ventilation. *Clinical and Experimental Allergy*, 24, 109–114.

Wickman, M., Nordvall, S.L., Pershagen, G., Korsgaard, J., Johansen, N. and Sundell, J. (1994b) Mite allergens during 18 months of intervention. *Allergy*, 49, 114–119.

Wickman, M., Paues, S. and Emenius, G. (1997) Reduction of the mite-allergen reservoir within mattresses by vacuum-cleaning. A comparison of three vacuum-cleaning systems. *Allergy*, 52, 1123–1127.

Wigglesworth, V.B. (1972) *The Principles of Insect Physiology*. 7th edition, Chapman and Hall, London.

Williams, S.G., Brown, C.M., Falter, K.H., Alverson, C.J., Gotway-Crawford, C., Homa, D., Jones, D.S., Adams, E.K. and Redd, S.C. (2006) Does a multifaceted environmental intervention alter the impact of asthma on inner-city children? *Journal of the National Medical Association*, 98, 249–260.

Wilson, E.O. (1994) *Naturalist*. Allen Lane, The Penguin Press, London.

Winston, P.W. and Bates, D.H. (1960) Saturated solutions for the control of relative humidity. *Ecology*, 41, 232–237.

Wold, A.E. (1998) The hygiene hypothesis revised: is the rising frequency of allergy due to changes in intestinal flora? *Allergy*, 53 (Supplement 46), 20–25.

Wolden, B., Smestad Paulsen, B., Wold, J.K. and Grimmer, O. (1982) Characterisation of the carbohydrate moiety in a purified allergen preparation from the mite *Dermatophagoides farinae* and its importance for allergenic activity as tested by the RAST-inhibition method. *International Archives of Allergy and Applied Immunology*, 68, 144–151.

Wongsathuaythong, S. and Lakshana, P. (1972) House-dust mite survey in Bangkok and other provinces of Thailand. *Journal of the Medical Association of Thailand*, 55, 272–286.

Wood, R.A., Eggleston, P.A., Mudd, K.E. and Adkinson, N.F. (1989) Indoor allergen levels as a risk factor for allergic sensitisation. *Journal of Allergy and Clinical Immunology*, 83, 199. [abstract]

Woodcock, A.A. and Cunnington, A.M. (1980) The allergenic importance of house dust and storage mites in asthmatics in Brunei, S.E. Asia. *Clinical Allergy*, 10, 609–615.

Woodcock, A., Forster, L., Matthews, E., Martin, J., Letley, L., Vickers, M., Britton, J., Strachan, D., Howarth, P., Altmann, D., Frost, C. and Custovic, A. (2003) Control of exposure to mite allergen and allergen-impermeable bed covers for adults with asthma. *New England Journal of Medicine*, 349, 225–236.

Woodcock, A., Lowe, L.A., Murray, C.S., Simpson, B.M., Pipis, S.D., Kissen, P., Simpson, A. and Custovic, A. (2004) Early life environmental control. Effect on symptoms, sensitisation, and lung function at age 3 years. *American Journal of Respiratory and Critical Care Medicine*, 170, 433–439.

Woodfolk, J.A., Hayden, M.L., Miller, J.D., Rose, G., Chapman, M. and Platts-Mills, T.A.E. (1994) Chemical treatment of carpets to reduce allergen: a detailed study of the effects of tannic acid on indoor allergens. *Journal of Allergy and Clinical Immunology*, 94, 19–26.

Woodfolk, J.A., Hayden, M.L., Couture, N. and Platts-Mills, T.A.E. (1995) Chemical treatment of carpets to reduce allergen: comparison of the effects of tannic acid and other treatments on proteins derived from dust mites and cats. *Journal of Allergy and Clinical Immunology*, 96, 325–333.

Woodford, P.J. (1980) The house dust mite, *Dermatophagoides farinae* as a causative agent of delusive dermatitis. *Annals of Allergy*, 45, 248–250.

Woodford, P.J., Arlian, L.G., Bernstein, I.L., Johnson, C.L. and Gallagher, J.S. (1979) Population dynamics of *Dermatophagoides* spp. in southwest Ohio homes. In: Rodriguez, J.G. (ed.), *Recent Advances in Acarology. Volume 2.* Academic Press, New York, pp. 185–195.

Woodring, J.P. (1963) The nutrition and biology of saprophytic sarcoptiformes. In: *Advances in Acarology*, Volume 1. Cornell University Press, pp. 89–111.

Woodring, J.P. (1968) An automatic collecting device for tyroglyphid (Acaridae) mites. *Annals of the Entomological Society of America*, 61, 1030–1031.

Woodring, J.P. (1969) Environmental regulation of andropolymorphism in tyroglyphids. In: Evans, G.O. (ed.), *Proceedings of the 2nd International Congress of Acarology, Sutton Bonnington, UK, 1967.* Akademai Kiado, Budapest, pp. 433–440.

Woodring, J.P. and Cook, E.F. (1962) The internal anatomy, reproductive physiology, and moulting process of *Ceratozetes cisalpinus* (Acarina: Oribatei). *Annals of the Entomological Society of America*, 55, 164–181.

Woodring, J.P. and Cutcher, J.J. (1967) Vital dye marking of tyroglyphid (Acaridae) mites. *Annals of the Entomological Society of America*, 61, 1031–1032.

Woodroffe, G.E. (1953) An ecological study of the insects and mites in the nests of certain birds in Britain. *Bulletin of Entomological Research*, 44, 739–772.

Woodroffe, G.E. (1954) An additional note on the fauna of birds' nests in Britain. *Bulletin of Entomological Research*, 45, 135–136.

Woolcock, A.J. (1972) Asthma in the highlands of Papua New Guinea. *Australian and New Zealand Journal of Medicine*, 2, 310 [abstract].

Woolcock, A.J., Marks, G., Matthews, M. and Tovey, E. (1991) Prevalence of dust mite allergens in different parts of the world. In: Todt, A. (ed.), *Dust Mite Allergens and Asthma. Report of the Second International Workshop.* UCB Institute for Allergy, Brussels, pp. 75–76.

Woolcock, A.J., Green, W. and Alpers, M.P. (1981) Asthma in a rural highland area of Papua New Guinea. *American Review of Respiratory Disease*, 123, 565–567.

Woolcock, A.J., Peat, J.K. and Trevillion, L.M. (1995) Is the increase in asthma prevalence linked to increase in allergen load? *Allergy*, 50, 935–940.

Wooley, T.A. (1988) *Acarology. Mites and Human Welfare.* John Wiley, New York.

Wortmann, F. (1977) Oral hyposensitisation of children with pollinosis or house-dust asthma. *Allergologia et Immunopathologia*, 5, 15–26.

Wraith, D.G., Cunnington, A.M. and Seymour, W.M. (1979) The role and allergenic importance of storage mites in house dust and other environments. *Clinical Allergy*, 9, 545–561.

Wright, G. (1963) The improvement of some asthmatics with oiling of bedding. *Medical Journal of Australia*, ii, 12–14.

Wu, H.-H. (1999) The most common house dust mites of houses in Taiwan. *Chinese Journal of Entomology*, Special Publication No. 12, 179–191.

Wyatt, I.J. and White, P.F. (1977) Simple estimation of intrinsic increase rates for aphids and tetranychid mites. *Journal of Applied Ecology*, 14, 757–766.

Yagofarov, F.F. and Galikeev, Kh.L. (1987) [Some data on the ecology of *Dermatophagoides* mites in the town of Semipalatinsk.] *Parazitologiya*, 21(2), 148–151. [In Russian]

Yamada, M., Kawakami, M., Hidano, A., Okamoto, M. and Matsumoto, K. (1988) [Evaluation of mites in house dust collected from dwellings of the patients with prurigo or insect bites.] *Journal of the Tokyo Women's Medical College*, 58, 1127–1131. [In Japanese, English summary]

Yamashita, H., Haida, M., Suko, M., Okudaira, H. and Miyamoto, T. (1989) Allergens of the house dust mite *Dermatophagoides farinae*. II. Immunological characterisation of four allergic molecules. *International Archives of Allergy and Applied Immunology*, 88, 173–175.

Yasueda, H., Mita, H., Yui, Y. and Shida, T. (1986) Isolation and characterisation of two allergens from *Dermatophagoides farinae*. *International Archives of Allergy and Applied Immunology*, 81, 214–223.

Yasueda, H., Mita, H., Yui, Y. and Shida, T. (1989a) Measurement of allergens associated with dust mite allergy I. Development of sensitive radioimmunoassays for the two groups of *Dermatophagoides* mite allergens, *Der* I and *Der* II. *International Archives of Allergy and Applied Immunology*, 90, 182–189.

Yasueda, H., Mita, H., Yui, Y. and Shida, T. (1989b) Comparative analysis of the physicochemical and immunochemical properties of the two major allergens from *Dermatophagoides pteronyssinus* and the corresponding allergens from *Dermatophagoides farinae*. *International Archives of Allergy and Applied Immunology*, 88, 402–407.

Yasueda, H., Mita, H., Akiyama, K., Shida, T., Ando, T., Sugiyama, S. and Yamakawa, H. (1993) Allergens from *Dermatophagoides* mites with chymotryptic activity. *Clinical and Experimental Allergy*, 23, 384–390.

Yasuhara, T., Takai, T., Yuuki, T., Okudira, H. and Okumura, Y. (2001) Cloning and expression of cDNA encoding the complete prepro-form of an isoform of Der f 1, the major group 1 allergen from house dust mite *Dermatophagoides farinae*. *Bioscience, Biotechnology and Biochemistry*, 65, 563–569.

Yi, F.C., Cheong, N., Shek, P.C., Wang, D.Y., Chua, K.Y. and Lee, B.W. (2002) Identification of shared and unique immunoglobulin E epitopes of highly conserved tropomyosins in *Blomia tropicalis* and *Dermatophagoides pteronyssinus*. *Clinical and Experimental Allergy*, 32, 1203–1210.

Yi, F.C., Chua, K.Y., Cheong, N., Shek, L.P. and Lee, B.W. (2004a) Immunoglobulin E reactivity of native Blo t 5, a major allergen of *Blomia tropicalis*. *Clinical and Experimental Allergy*, 34, 1762–1767.

Yi, F.C., Lee, B.W., Cheong, N. and Chua, K.Y. (2004) Quantification of Blo t 5 in mite and dust extracts by two-site ELISA. *Allergy*, 60, 108–112.

Yoshikawa, M. (1979) The critical equilibrium activity (minimum humidity for survival) and water loss kinetics of *Dermatophagoides microceras*. *Japanese Journal of Sanitary Zoology*, 30, 271–276.

Yoshikawa, M. (1980) Epidemic of dermatitis due to a cheyletid mite, *Chelacaropsis* sp., in tatami rooms (I). *Annual Report of the Tokyo Metropolitan Research Laboratory of Public Health*, 31, 253–260.

Yoshikawa, M. (1985) Skin lesions of papular urticaria induced experimentally by *Cheyletus malaccensis* and *Chelacaropsis* sp. (Acari: Cheyletidae). *Journal of Medical Entomology*, 22, 115–117.

Yoshikawa, M. (1987) Feeding of *Cheyletus malaccensis* (Acari: Cheyletidae) on human body fluids. *Journal of Medical Entomology*, 24, 46–53.

Yoshikawa, M. and Bennett, P.H. (1979) House dust mites in Columbus, Ohio. *Ohio Journal of Science*, 79, 280–282.

Yoshikawa, M., Hanaoka, K. and Yamada, Y. (1982) Seasonal changes of mite fauna and population in four concrete apartment houses. *Annual Report of the Tokyo Metropolitan Research Laboratory of Public Health*, 33, 299–306.

Yoshikawa, M., Hanaoka, K., Yamada, Y. and Mikata, T. (1983) Experimental proof of itching papules caused by *Cheyletus malaccensis* Oudemans. *Annual Report of the Tokyo Metropolitan Research Laboratory of Public Health*, 34, 264–276.

Yoshikawa, M., Wauke, T. and Morozumi, S. (1988) Effects of a dehumidifier on mites and fungi. *Annual Report of the Tokyo Metropolitan Research Laboratory of Public Health*, 39, 237–244.

Yu, C.-J., Lin, Y.-F., Chiang, B.-L. and Chow, L.-P. (2003) Proteomics and immunological analysis of a novel shrimp allergen, Pen m 2. *Journal of Immunology*, 170, 445–453.

Yuuki, T., Okumura, Y., Ando, T., Yamakawa, H., Suko, M., Haida, M. and Okudaira, H. (1991) Cloning and expression of cDNA coding for the major house dust mite allergen *Der f* II in *Escherichia coli*. *Agricultural and Biological Chemistry*, 55, 1233–1238.

Yuuki, T., Okumura, Y. and Okudaira, H. (1997) Genomic organisation and polymorphisms of the major house dust mite allergen Der f 2. *International Archives of Allergy and Immunology*, 112, 44–48.

Zbikowska, M. (1977) Alergia na kurz domowy u chorych na dychawice oskrzelowa. I. Czestosc wystepowania i problem charakteru alergenu kurzu domowego. *Polskie Archiwum Medycyny-Wewnetrznej*, 57, 21–26.

Zachvatkin, A.A. (1941) *Fauna of the USSR, Arachnoidea. Vol VI, No. 1 Tyroglyphoidea (Acari)*. Academy of Science, Moscow [Translated by A. Ratcliffe and A.M. Hughes, American Institute of Biological Science, Washington, 1959].

Zhang, H. and Kato, Y. (2003) Common structural properties specifically found in CSaβ-type antimicrobial peptides in nematodes and molluscs: evidence for the same evolutionary origin? *Developmental and Comparative Immunology*, 27, 499–503.

Zhang, L., Chew, F.T., Soh, S.Y., Yi, F.C., Law, S.Y., Goh, D.Y.T. and Lee, B.W. (1997) Prevalence and distribution of indoor allergens in Singapore. *Clinical and Experimental Allergy*, 27, 876–885.

Zheltikova, T.M. and Petrova, A.D. (1990) [The fauna, number and spatial distribution of mites (Acari) in house dust in Moscow.] *Biologicheskie Nauki*, no. 1, 42–52. [In Russian]

Zheltikova, T.M., Petrova-Nikitina, A.D., Kanchurin, A.Kh., Berzhets, V.M. and Muzyleva, I.L. (1985) [Mites in house dust and human allergosis]. *Biologicheskie Nauki*, no. 2, 12–30. [In Russian, English summary]

Zheltikova, T.M., Berzhets, V.M., Petrova-Nikitina, A.D. and Kanchurin, A.Kh. (1986) [The detection of house dust mite antigens in Moscow.] *Zhurnal Mikrobiologii, Epidemiologii i Immunobiologii*, no. 6, 80–83. [In Russian]

Zheltikova, T.M., Petrova-Nikitina, A.D., Kanchurin, A.Kh., Berzhets, V.M. and Muzyleva, I.L. (1987) [Allergenic house dust mites (Acariformes, Pyroglyphidae).] *Biologicheskie Nauki*, no. 2, 4–21. [In Russian, English summary]

Zheltikova, T.M., Ovsyannikova, I.G., Gervazieva, V.B., Platts-Mills, T.A.E., Chapman, M.D., Petrova-Nikitina, A.D. and Stepanova, G.N. (1994) Comparative detection of mite allergens in house dust of homes in Moscow by enzyme-linked immunosorbent assay and acarologic analysis. *Allergy*, 49, 816–819.

Zhu, H., Cui, Y.-B. and Rao, L.-Y. (2007) [Acarine fauna and correlation between mite density and allergen concentration in house dust samples from asthma patients sin the city of Haikou.] *Chinese Journal of Vector Biology and Control*, 18, 381–383. [In Chinese, English summary]

Zinsstag, J., Ould Taleb, M. and Craig, P.S. (2006) Health of nomadic pastoralists: new approaches towards equity effectiveness. *Tropical Medicine and International Health*, 11, 565–568.

Zock, J.-P., Brunekreef, B., Hazebroek-Kamschreur, A.A.J.M. and Roosjen, C.W. (1994) House dust mite allergen in bedroom floor dust and respiratory health of children with asthmatic symptoms. *European Respiratory Journal*, 7, 1254–1259.

Zock, J.-P., Heinrich, J., Jarvis, D., Verlato, G., Norbäck, D., Plana, E., Sunyer, J., Chinn, S., Olivieri, M., Soon, A., Villani, S., Ponzio, M., Dahlman-Hoglund, A., Svannes, C. and Luczynska, C. for the Indoor Working Group of the European Community Respiratory Health Survey (2006) Distribution and determinants of house dust mite allergens in Europe: the European Community Respiratory Health Survey II. *Journal of Allergy and Clinical Immunology*, 118, 682–690.

Zock, J.-P., Plana, E., Jarvis, D., Anto, J.M., Kromhout, H., Kennedys, S.M., Kuenzli, N., Villani, S., Olivieri, M., Toren, K., Radon, K., Sunyer, J., Dahlman-Hoglund, A., Norbaeck, D. and Kogevinas, M. (2007) The use of household cleaning sprays and adult asthma – an international longitudinal study. *American Journal of Respiratory and Critical Care Medicine*, 176, 735–741.

Zwemer, R.J. and Karibo, J. (1973) Use of laminar control device as adjunct to standard environmental control measures in symptomatic asthmatic children. *Annals of Allergy*, 31, 284–290.

Index